Diagnostic, Prognostic and Predictive Biomarkers in Prostate Cancer

Special Issue Editor
Carsten Stephan

MDPI • Basel • Beijing • Wuhan • Barcelona • Belgrade

MDPI

Special Issue Editor
Carsten Stephan
Universitätsmedizin Berlin
Germany

Editorial Office
MDPI AG
St. Alban-Anlage 66
Basel, Switzerland

This edition is a reprint of the Special Issue published online in the open access journal *International Journal of Molecular Sciences* (ISSN 1422-0067) from 2016–2017 (available at: http://www.mdpi.com/journal/ijms/special_issues/prostate_cancer_2016).

For citation purposes, cite each article independently as indicated on the article page online and as indicated below:

Author 1; Author 2. Article title. *Journal Name* **Year**, *Article number*, page range.

First Edition 2017

ISBN 978-3-03842-632-5 (Pbk)
ISBN 978-3-03842-633-2 (PDF)

Table of Contents

About the Special Issue Editor

Carsten Stephan, Prof. Dr. Med., studied medicine at the Humboldt University Berlin from 1991–1998 and has worked as a physician since 1999 at the Department of Urology in the Charité Hospital. As a participant of the Biomedical Science Exchange Programme, he isolated PSA isoforms at The John Hopkins Hospital in Baltimore (USA) from 1997–1998. In 1998, he finished his Doctoral thesis with summa cum laude. He won several prizes including the Charité Research Prize, and the "Science around thirty" Award from the German Urological Association. Furthermore, with the MSD-grant "Urology 2000", he worked as a postdoc at the Mt Sinai Hospital, Toronto (Canada) in 2001. In 2006, he became a certified specialist of urology and qualified as a university lecturer (Privat-Dozent) with his postdoctoral thesis: Improved Diagnosis of Prostate Cancer by Using New Tumor Markers and Applying ANNs. Since 2009, he has been business manager of the "Berliner Forschungsinstitut für Urologie" of the "Stiftung Urologische Forschung Berlin". Professor Stephan continued his studies from 2009–2011 at the Berlin School of Economics and Law/IMB Institute of Management Berlin to become a master of business administration (MBA). Since 2014, he has been a full professor "außerplanmäßiger Professor" at the Charité Department of Urology, Berlin. In 2017, he became a deputy chairman of the "Stiftung Urologische Forschung Berlin". Prof. Stephan is author and coauthor of more than 230 scientific publications including book chapters and one book on prostate cancer diagnostics. He is married and has two daughters.

Preface to "Diagnostic, Prognostic and Predictive Biomarkers in Prostate Cancer"

In recent years, many new developments in prostate cancer (PCa) diagnostics have been published. Numerous new biomarkers in serum, plasma or urine have been described. However, it is important to know not only the diagnostics, but also the risk stratification and prognostic factors for the progress of this disease.

This Special Issue contains 26 papers consisting of 22 original research studies and four reviews. An editorial summarizes all these contributions. The studies cover basic research data using PCa cell lines but the vast majority of articles include serum markers with a focus on PCa diagnostic, prognostic or evaluation of advanced or metastatic disease stages. Reviews further summarize biomarkers for PCa diagnostic and active surveillance or have special topics such as nanoparticles as theranostic vehicles or PCa stem cells. Other groups have published results by using immunohistochemistry in prostate, lymph node or bone tissue.

Further specialized topics include seminal plasma biomarkers, improved PET contrast in bone metastatic lesions and the preanalytical impact on CTC-mRNA (AR-V7) analysis. Two clinical studies on biomarkers for prostate brachytherapy complete the large number of publications within this Special Issue of the International Journal of Molecular Sciences on diagnostic, prognostic and predictive biomarkers in PCa.

Carsten Stephan
Special Issue Editor

International Journal of
Molecular Sciences

MDPI

Editorial

Advances in Biomarkers for PCa Diagnostics and Prognostics—A Way towards Personalized Medicine

Carsten Stephan [1,2,*] and Klaus Jung [1,2]

1 Department of Urology, Charité University Hospital, 10117 Berlin, Germany; klaus.jung@charite.de
2 Berlin Institute for Urologic Research, 10115 Berlin, Germany
* Correspondence: carsten.stephan@charite.de; Tel.: +49-30-450-515-052; Fax: +49-30-450-515-904

Received: 10 October 2017; Accepted: 18 October 2017; Published: 20 October 2017

1. Introduction

Prostate cancer (PCa) is, with an estimated number of 161,360 cases and 26,730 deaths in 2017, the most common malignancy in the USA [1]. Worldwide, the mortality rates tend to be higher in less developed regions, including parts of South America, the Caribbean, and Africa [2]. The overall PCa incidence in all countries with available data for 2008 was counted with almost 900,000 cases while the mortality rate was estimated with approximately 258,000 deaths from PCa [2]. Thus, worldwide almost 30% of men with PCa will die of this malignancy.

In recent years, many new developments in PCa diagnostics were reviewed. Numerous new protein- and nucleic acid-based biomarkers in whole blood and its fractions serum or plasma as well in urine and its different fractions were described based on novel technologies [3–7]. To reduce overdiagnosis and overtreatment there is a clear focus to preferentially detect clinical significant PCa because indolent PCa often do not need treatment or treatment can be delayed. Here a multimodal risk score was developed in a study on urine samples with two independent prospective multicenter cohorts [8]. The multimodal approach including the *HOXC6* and *DLX1* mRNA levels reached excellent area under the Receiver-Operating-Characteristic (ROC) curve (AUCs) of 0.90 and 0.86 in the training and validation cohort [8]. Beside the diagnostics, also the risk stratification [9] and prognostic factors [10] are important for the progress of the disease.

The treatment options with several newly available substrates for advanced PCa, including castration-resistant prostate cancer (CRCP), have been changed dramatically within recent years, but no biomarker has been validated so far [11]. The androgen-receptor splice variant 7 (AR-V7) protein expression and its localization in circulating tumor cells (CTCs) should also be highlighted [12], but there is a need for further prospective studies to validate this biomarker. As an example, the measurement of epidermal growth factor receptor (EGFR) on CTCs might be promising as a prognostic marker in metastatic PCa patients [13].

2. Overview

This Special Issue contains 26 papers consisting of 22 original research studies and four reviews. Five studies present their basic research data using PCa cell lines [14–17] or summarize data on cancer/testis antigens (CTAs) as potential PCa biomarkers [18]. Serum markers are the subject in the vast majority of publications (*n* = 8) with focus on PCa diagnostic and prognostic [19–23] or evaluation of advanced/metastatic disease stages [24–26]. Two of the four reviews further summarize biomarkers for PCa diagnostic [27] or active surveillance [28] while the other two reviews have special topics like nanoparticles as theranostic vehicles [29] or PCa stem cells [30]. Four working groups published results by using immunohistochemistry in prostate, lymph node or bone tissue [31–34]. This Special Issue also covered studies on seminal plasma biomarkers [35], improved PET contrast between osteolytic and osteoblastic bone metastatic lesions with fasting [36] and the impact of the blood collection tube and storage time on CTC-mRNA (AR-V7) analysis [37]. Two clinical studies on different biomarkers for prostate brachytherapy [38,39] complete the large number of publications.

3. Basic Research

Metformin, known as an anti-diabetes drug, has been shown to have anti-neoplastic effects in several tumors including PCa. Metformin targets Hedgehog (Hh) signaling, which is an important target for radiosensitization. Gonnissen et al. [14] evaluated the combination of metformin and the Hh inhibitor GANT61 with or without ionizing radiation in the three PCa cell lines PC3, DU145, and 22Rv1. Although this drug combination reduced the cell growth and enhanced the radiosensitization effect compared to single agent in cell lines, both in vitro effects could not be confirmed in vivo [14]. This observation shows the limitation of in vitro testing and the importance of careful translation of successful in vitro experiments under in vivo setting conditions.

Dayal et al. [15] showed that mutations in RNase L may promote PCa by increasing expression of androgen receptor (AR)-responsive genes and cell motility and they identified novel roles of RNase L as a PCa susceptibility gene. For instance, the activity of the two matrix metalloproteinases (MMP)-2 and -9 is significantly increased in cells where RNase L levels are ablated [15]. The imbalance between MMP-9 activity and its inhibitory counterparts in prostate cancerous tissue already implicated a rationale of using synthetic inhibitors of MMPs as potential therapeutic tools [40] that was also confirmed in a prostate cancer standard rat model [41].

In advanced PCa, small ubiquitin-like modifier (SUMO)-specific cysteine protease 1 (SENP1) is up-regulated. In their study, Zhang et al. [16] developed a lentiviral vector, to silence SENP1 in prostate cancer cells with high metastatic characteristics (PC3M). The researcher further created an adenovirus vector to over-express SENP1 in prostate cancer cells with low metastatic potential (LNCaP). The authors could show that silencing of SENP1 promoted cellular apoptosis and they concluded finally that SENP1 is a potential target for treatment of advanced PCa [16]. Further studies including first clinical trials are necessary.

Bascetta et al. [17] isolated PCa cells resistant to docetaxel (DCTR) clones from different PCa cell lines and performed through next-generation sequencing of the released miRNAs. They identified several miRNAs, which were differentially released in the growth medium [17]. The authors proposed that the utilization of clones resistant to a given drug as in vitro model to identify the differentially released miRNAs, could be tested in perspective as predictive biomarkers of drug resistance in tumor patients under therapy [17]. This is an interesting theory. However, measurements of these miRNAs in plasma/serum must follow to find evidence of this approach.

As a basic step for numerous experiments, Australian researchers determined the ideal blood storage conditions to preserve CTC-specific mRNA biomarkers [37]. Luk et al. [37] tested the preservation of tumor cells and CTC-mRNA over the time in ethylene-diamine-tetra-acetic acid (EDTA) and acid citrate dextrose solution B (Citrate) blood collection tubes in comparison to special cell-free DNA, RNA and Cyto-Chex blood collection tubes (Streck, Omaha, NE, USA). Tumor mRNA biomarkers were readily detectable after 48 h storage in conventional EDTA and Citrate tubes, but not in the three specially developed preservative-containing collection tubes. Notably, AR-V7 expression was detected in PCa patient blood samples after 48 h storage in EDTA tubes at room temperature, leaving a more feasible timeframe compared to previous recommendations [37].

The cancer/testis antigens (CTAs) are a group of proteins that are typically restricted to the testis in the normal adult but are aberrantly expressed in several other types of cancers including PCa. Using prostate-associated gene 4 (PAGE4) as an example of a disordered CTA, Kulkami and Uversky [18] highlighted how intrinsically disordered proteins (IDP) conformational dynamics may regulate phenotypic heterogeneity in PCa cells, and how it may be exploited both as a potential biomarker as well as a promising therapeutic target in PCa. The authors favor the theranostic potential of CTAs that is latent in their lack of structure and casts them as next generation or "smart" biomarker candidates [18]. This idea should be supported but the markers should be disease specific.

4. Serum Biomarkers

Hagiwara and colleagues [19] evaluated the performance of Wisteria floribunda agglutinin (WFA)-reactive glycan-carrying PSA-glycosylation isomer (PSA-Gi) in serum by using an automated immunoassay system in 244 patients with PCa and 184 with biopsy-proven benign prostate hyperplasia (BPH). The area under the ROC curve (AUC) for PSA-Gi was almost 0.8, which was higher than for PSA (0.64) or the PSA-Gi/PSA ratio (AUC 0.73). The correlation between PSA-Gi and the Gleason grade is a further very positive aspect that implies that PSA-Gi is a promising marker not only for detecting PCa but also for assessing its aggressiveness [19]. A tissue-based nomogram was also developed as a predictive tool for determining the PSA-free survival probability [19].

Another research group from Japan worked on the PCa-associated α2,3-linked sialyl N-glycan-carrying PSA (S2,3PSA) [20]. The authors estimated PSA and S2,3PSA in each of 50 age and PSA matched biopsy-proven patients with PCa and BPH (from a larger cohort of >550 biopsied men) by a newly developed automated micro-total immunoassay system with a detection limit of 0.05 ng/mL [20]. The authors can be congratulated for the development of a robust and reproducible immunoassay with a coefficient of variation within 15% that is superior to their earlier magnetic bead-based S2,3PSA assay [42]. The AUC for the S2,3PSA to PSA ratio (%S2,3PSA) was 0.834 and much superior to PSA alone (AUC: 0.506) [20]. However, a comparison of %S2,3PSA with the FDA-approved and currently best available PSA-based serum marker prostate health index (PHI; formula: -2proPSA/free PSA $\times$ $\sqrt{}$PSA) has not been performed. It would be very interesting if %S2,3PSA could further improve PHI in future prospective multicenter studies.

An independent research group also published data with their 2015 patented PSA glycoform assay [43], based on the determination of the α2,3-sialic acid percentage of serum PSA (%α2,3-SA). The current study compared PHI with %α2,3-SA in a cohort of 79 patients that included 50 PCa and 29 with BPH [22]. The %α2,3-SA could distinguish high-risk PCa patients from the rest of patients better than PHI (AUC 0.971 vs. 0.840), although PHI correlated better with the Gleason score than the %α2,3-SA [22]. The combination of both markers increased the AUC up to 0.985 resulting in 100% sensitivity and 94.7% specificity to differentiate high-risk PCa from the other low and intermediate-risk PCa and BPH patients [22]. The editor is eagerly awaiting multicenter data on the new %α2,3-SA assay. It would be further interesting to compare both α2,3-sialic acid assays from Ishikawa et al. [20] and Ferrer-Batalle et al. [22] in view of its introduction as soon as possible in clinical practice.

Friedersdorff et al. [21] investigated the relationship between PHI, Gleason Score and prostate tumor volume in almost 200 prostatectomy specimen. With an AUC of 0.79, PHI was the most accurate predictor of a tumor volume > 0.5 cm^3 [21]. Most important, PHI correlated significantly with the tumor volume ($r = 0.588$), which is significantly better ($p = 0.008$) than the correlation of the Gleason score with tumor volume ($r = 0.385$). This shows that the gold standard of Gleason Score has been surpassed regarding its value on tumor size. Further, our own data regarding PCa prognosis also show the value of PHI with an improved prediction of biochemical recurrence (BCR) [44]. PHI obviously outperforms other diagnostic PCa biomarkers (urinary PCA3, TMPRSS2:ERG) with its good association to tumor aggressiveness, tumor volume and prognosis [45–47].

PHI was again the topic of a study by Schlack and colleagues [25]. In 25 patients with metastatic castration resistant prostate cancer (mCRPC), following initiation of Abiraterone-therapy, the PSA-subforms were analyzed before and at 8–12 weeks under therapy as prognosticators of progression-free-survival (PFS) and overall survival (OS). Comparing patients with a PFS < vs. ≥ 12, the relative-median-change of PSA, free PSA and -2proPSA differed significantly [25]. Decreasing free PSA and -2proPSA values indicated an OS of 32 months compared with 21 months in men with rising values [25]. Univariate and multivariate Cox regression analyses could not prove all these tests as suitable predictive PFS and OS markers [25]. However, the authors stressed the limitation of their study due to the small sample size in a single center.

In the last diagnostic study with serum, high throughput sequencing of small RNAs extracted from blood from 28 untreated PCa patients and 12 healthy controls was used to identify microRNAs as PCa biomarker [23]. Four microRNAs (miR-127-3p, miR-204-5p, miR-329-3p and miR-487b-3p) were upregulated, three miRNAs (miR-32-5p, miR-20a-5p and miR-454-3p) were downregulated. ROC curves exhibited a better correlation with PCa than for PSA [23]. It should be emphasized that the selection of four miRNA as normalization standard is a very positive additional part of this study. However, the AUCs for the single miRNAs are between 0.75 and 0.95, but the number of patients is quite low and comparing not only with PSA but also at least %free PSA or better PHI would improve this comparison. In addition, some of the suggested miRNAs (miR-20a-5p, miR-32-5p, and miR-454-3p) are hemolysis-affected [48]. The use of such hemolysis-affected circulating miRNAs has been recently questioned without a special control of hemolysis [49].

Geng et al. [24] checked first whether genetic variants in the Wnt pathway influence clinical outcomes for advanced PCa patients receiving androgen deprivation therapy (ADT). In 465 PCa patients, two common single nucleotide polymorphisms (SNPs), the adenomatous polyposis coli (APC) rs2707765 and rs497844, were significantly associated with both PCa progression and all-cause mortality [24]. This may implement a preclinical rationale for using APC as a prognostic marker in advanced PCa by identifying patients who would not benefit from ADT [24].

In 96 patients with mCRPC under a median time of 10 months with Abiraterone treatment a serum pre-treatment neutrophil-to lymphocyte-ratio (NLR) < 5 was associated with better survival outcomes [26]. Contrary, the authors found that after eight weeks of Abiraterone therapy, a change of initially elevated NLR of >5 to <5 was associated with worse survival. Therefore, a deeper understanding of the underlying immune mechanisms in this setting is highly warranted [26].

5. Reviews

Filella and Foj [27] extensively reviewed all important biomarkers for early detection of PCa, also regarding the important point of overdetection and false positive results. This excellent compilation provides an overview of all PSA subforms and the timeline of PCa biomarkers since 1970 [27]. Beside the emerging role of PHI, also the use of the urine based markers PCA3 (FDA approved in 2012 for men older than 50 who have at least one previous negative biopsy) and TMPRSS2:ERG fusion gene is provided. Furthermore, aberrant microRNA and exosomal biomarkers are reviewed [27].

Ferro and colleagues [28] provided a comprehensive overview of biomarkers as predictors of clinical significant PCa and for PCa patients under active surveillance. The main topics were further epigenetic signatures with DNA methylation, histone modifications, and noncoding RNA, which all could potentially provide new tools for PCa prognosis [28].

Individualized targeted theranostic nanomedicine has emerged by using nanoparticles as vehicles carrying both diagnostic and therapeutic molecular entities. Nanomedicine can increase sensitivity and specificity on diagnosis and might be used for improved survival or prolonged survival after therapy [29]. A large review by Dr. Elgqvist [29] presents and discusses important and promising different kinds of nanoparticles, as well as imaging and therapy options, suitable for theranostic applications. Beside breast cancer, PCa is presented in detail regarding diagnosis, staging, recurrence, metastases, and treatment options that are available today, followed by possible ways to move forward applying theranostics. This very comprehensive review encompasses 53 pages and almost 600 references [29].

The hot topic of the last of the four reviews by Zhang et al. [30] included cancer stem cells biomarkers including a few novel markers discovered recently. Those biomarkers might play an important role to detect resistance to traditional cancer therapies [30]. Further studies of cancer stem cells (including specific isolation and targeting on those cells) might be helpful for the discovery of novel treatments for prostate cancer, especially castration resistant disease [30].

6. Immunohistochemistry and Other Methods

In the first accepted paper of this Special issue by Chen et al. [31] the authors used tissue microarray and immunohistochemistry to estimate the phosphorylated levels of Akt (p-Akt) in 53 radical prostatectomy (RP) specimen. Within a Cox proportional hazard model a high p-Akt image score better predicted (hazard ratio (HR): 3.12) the risk of BCR than a high Gleason score (HR: 1.18) or a high PSA (HR: 0.62, $p = 0.57$) [31]. It should be noted that the initial 76 RP specimen (23 had to be excluded) were collected over a relatively long time from 1999 to 2011 [31]. A limitation of this second study from Taiwan was a much higher mean PSA (almost 30 ng/mL) in the group of high p-Akt Scores ≥ 8 as compared with the group with a p-Akt Score ≤ 6, where the mean PSA was only 12 ng/mL. Therefore, the conclusion that p-Akt activation can potentially determine BCR in pT2 PCa patients after RP should be taken with caution. Regarding BCR after RP, a very recent publication of our own working group by Zhao et al. [50] comprehensively reviews biomarkers with a special focus on miRNAs and its combinations to improve PCa prognosis. There is obviously a theranostic utility and a diagnostic, prognostic and future therapeutic potential of miRNAs in prostate cancer [51].

Campos et al. [32] investigated in a pilot study the expression of the epithelial cell adhesion molecule (EpCAM) in PCa lymph nodes and matched normal lymph nodes, in PCa bone metastases, and in normal bone by immunohistochemistry. EpCAM was expressed in 100% of lymph node metastases ($n = 21$), in 0% of normal lymph nodes (0 out of 21), in 95% of bone metastases (19 out of 20), and in 0% of normal bone (0 out of 14) [32]. Based on these results, EpCAM may be a feasible imaging target in PCa lymph node and bone metastases [32]. If prospective trials can confirm these promising results, EpCAM may help to improve pretherapeutical staging and to detect possible micro metastasis.

Another survey by Genitsch et al. [33] also tested the neuroendocrine differentiation (NED) by Chromogranin A expression in lymph node metastases, as well as primary tumors, from 119 consecutive PCa patients. The mean percentage of NED cells increased significantly ($p < 0.001$) from normal prostate glands (0.4%), to primary prostate cancer (1.0%) and nodal metastases (2.6%). However, in primary tumors and nodal metastases, tumor areas with higher Gleason patterns tended to display a higher NED, but no significance was reached [33].

A third study in lymph node metastases evaluated the homeobox protein Hox-B13 (HOXB13), which has been suggested as a new marker for the detection of prostatic origin [52]. Kristiansen et al. [34] semi-quantitatively compared the diagnostic value of different immunohistochemically markers such as PSA, Prostatic acid phosphatase (PSAP), prostate specific membrane antigen (PSMA), homeobox gene *NKX3.1*, protein, androgen receptor (AR), HOXB13, and the ETS-related gene (ERG) in 64 lymph node metastasis. The detection rate of prostate origin of metastasis for single markers was 100% for NKX3.1, 98.1% for AR, 84.3% for PSMA, 80.8% for PSA, 66% for PSAP, 60.4% for HOXB13, 59.6% for prostein, and 50.0% for ERG [34]. Thus, HOXB13 alone lacks sensitivity for the detection of prostatic origin, so the combination of PSA and NKX3.1 should be preferred.

Based on their previous work on PCa detection with naturally occurring fragments of the larger seminal proteins semenogelin 1 and 2 in seminal plasma by using CE-MS/MS [53], the group of Neuhaus et al. [35] identified proteases putatively involved in PCa specific protein cleavage, and further examined gene expression and tissue protein levels. They found different MMP3 and MMP7 activity in PCa compared with BPH due to fine regulation by their tissue inhibitor TIMP1 [35]. These data support the old idea of non-invasive seminal plasma biomarkers as additional tool for PCa detection and risk stratification. However, tests in seminal plasma seem to be of little acceptance in clinical practice.

A third Japanese research group investigated the influence of fasting on fluciclovine-PET using a triple-tracer autoradiography in a rat breast cancer model of mixed osteolytic/osteoblastic bone metastases in which the animals fasted overnight [36]. Their in vivo and in vitro results suggest that fasting before 18F-fluciclovine-PET improves the contrast between osteolytic and osteoblastic bone metastatic lesions and background, which might facilitate a clearer visualization of lesions in fluciclovine-PET imaging [36].

7. Brachytherapy

The last two studies of this Special issue present data on patients receiving prostate brachytherapy [38,39]. In the sixth study from Germany by PD Ecke et al. [38] PSA, PSA density and other clinical data were evaluated before and after brachytherapy with external beam radiation in

Int. J. Mol. Sci. **2017**, *18*, 2193

79 high-risk PCa patients treated between 2009 and 2016. PSA density and PSA at time of diagnosis ($p = 0.009$ and 0.033), and PSA on date of first follow-up after one year ($p = 0.025$) were significant predictors for local recurrence during follow-up [38]. The authors further concluded that their specific radiation therapy for high-risk PCa resulted in high biochemical control rates with minimal side-effects [38].

In the fourth contribution from Japan, Tsumura et al. [39] measured CTCs before and during brachytherapy in 30 high-risk and 29 low-risk PCa patients. While no preoperative sample showed CTCs (0%), they were detected in intraoperative samples in 7 of the 59 patients [39]. The authors could not find any association of intraoperative CTC increases with clinicopathological data at all [39]. This may reflect the fact that all patients undergoing brachytherapy have a certain risk of intraoperative haematogenous spillage of prostate cancer cells, irrespective of use of neoadjuvant hormonal therapy, type of brachytherapy, PSA and other clinical data. This lowers a possible impact of CTCs in the diagnosis or prognosis of PCa. Further, there is also the possibility that normal epithelial cells are transiently released into the blood stream due to brachytherapy procedures and therefore additional tests would need to confirm if cells released in this manner are indeed real CTCs.

8. Summary

In summary, this Special Issue of the International Journal of Molecular Sciences on diagnostic, prognostic and predictive biomarkers in PCa is an excellent compilation of 26 publications accepted between July 2016 and July 2017 with authors from 15 different countries.

As recently reviewed by Ali et al. [54], serum, urine, tissue and imaging biomarkers have been widely evaluated to improve the identification of clinically significant PCa. Importantly, changes in MRI technology such as multiparametric MRI (mpMRI) have realized a quantum change, and this facility is now becoming more widely incorporated into diagnostic and disease risk-stratification protocols [54]. An example on combining biomarker and mpMRI results has been published very recently (October 2017) by Hendriks et al. [55]. The researchers determined the association between a urinary biomarker-based risk score (SelectMDx) and the mpMRI based prostate imaging reporting and data system (PI-RADS) score regarding prostate biopsy results [55]. With an AUC of 0.83 for SelectMDx (compared to 0.66 for PSA and 0.65 for PCA3) it was further not surprising that there was a positive association between the SelectMDx score and the PI-RADS score with significant differences between PI-RADS 3 and 4 ($p < 0.01$) and between PI-RADS 4 and 5 ($p < 0.01$) in the SelectMDx score [55]. Analogous to the older concept of PSA density (PSA divided by prostate volume) the new term of PHI density (PHI/prostate volume) was described in 2017 as another marker combination by Tosoian et al. [56].

Beside these new developments in urinary markers [5,8,55] and the PHI density [56], the serum immunoassays with the PCa-associated aberrant glycosylation of PSA (S2,3PSA) published here should be highlighted [20,22].

To conclude, the identification of clinically significant PCa by biomarkers and image modalities (mpMRI) is a step towards personalized diagnosis, which will be the future. Also for advanced PCa, results of this Special issue [24] might influence personalized PCa management.

Conflicts of Interest: The authors declare no conflict of interest.

References

1. Siegel, R.L.; Miller, K.D.; Jemal, A. Cancer statistics, 2017. *CA Cancer J. Clin.* **2017**, *67*, 7–30. [CrossRef] [PubMed]
2. Center, M.M.; Jemal, A.; Lortet-Tieulent, J.; Ward, E.; Ferlay, J.; Brawley, O.; Bray, F. International variation in prostate cancer incidence and mortality rates. *Eur. Urol.* **2012**, *61*, 1079–1092. [CrossRef] [PubMed]
3. Stephan, C.; Jung, K.; Ralla, B. Current biomarkers for diagnosing of prostate cancer. *Future Oncol.* **2015**, *11*, 2743–2755. [CrossRef] [PubMed]

4. Hendriks, R.J.; van Oort, I.M.; Schalken, J.A. Blood-based and urinary prostate cancer biomarkers: A review and comparison of novel biomarkers for detection and treatment decisions. *Prostate Cancer Prostatic Dis.* **2017**, *20*, 12–19. [CrossRef] [PubMed]
5. Wu, D.; Ni, J.; Beretov, J.; Cozzi, P.; Willcox, M.; Wasinger, V.; Walsh, B.; Graham, P.; Li, Y. Urinary biomarkers in prostate cancer detection and monitoring progression. *Crit. Rev. Oncol. Hematol.* **2017**, *118*, 15–26. [CrossRef] [PubMed]
6. Smits, M.; Mehra, N.; Sedelaar, M.; Gerritsen, W.; Schalken, J.A. Molecular biomarkers to guide precision medicine in localized prostate cancer. *Expert Rev. Mol. Diagn.* **2017**, *17*, 791–804. [CrossRef] [PubMed]
7. Tanase, C.P.; Codrici, E.; Popescu, I.D.; Mihai, S.; Enciu, A.M.; Necula, L.G.; Preda, A.; Ismail, G.; Albulescu, R. Prostate cancer proteomics: Current trends and future perspectives for biomarker discovery. *Oncotarget* **2017**, *8*, 18497–18512. [CrossRef] [PubMed]
8. Van Neste, L.; Hendriks, R.J.; Dijkstra, S.; Trooskens, G.; Cornel, E.B.; Jannink, S.A.; de Jong, H.; Hessels, D.; Smit, F.P.; Melchers, W.J.; et al. Detection of high-grade prostate cancer using a urinary molecular biomarker-based risk score. *Eur. Urol.* **2016**, *70*, 740–748. [CrossRef] [PubMed]
9. Reiter, R.E. Risk stratification of prostate cancer 2016. *Scand. J. Clin. Lab. Investig. Suppl.* **2016**, *245*, S54–S59. [CrossRef] [PubMed]
10. Pugliese, D.; Palermo, G.; Totaro, A.; Bassi, P.F.; Pinto, F. Clinical, pathological and molecular prognostic factors in prostate cancer decision-making process. *Urol. J.* **2016**, *83*, 14–20. [CrossRef] [PubMed]
11. Seisen, T.; Roupret, M.; Gomez, F.; Malouf, G.G.; Shariat, S.F.; Peyronnet, B.; Spano, J.P.; Cancel-Tassin, G.; Cussenot, O. A comprehensive review of genomic landscape, biomarkers and treatment sequencing in castration-resistant prostate cancer. *Cancer Treat. Rev.* **2016**, *48*, 25–33. [CrossRef] [PubMed]
12. Scher, H.I.; Lu, D.; Schreiber, N.A.; Louw, J.; Graf, R.P.; Vargas, H.A.; Johnson, A.; Jendrisak, A.; Bambury, R.; Danila, D.; et al. Association of AR-V7 on circulating tumor cells as a treatment-specific biomarker with outcomes and survival in castration-resistant prostate cancer. *JAMA Oncol.* **2016**, *2*, 1441–1449. [CrossRef] [PubMed]
13. Josefsson, A.; Linder, A.; Flondell, S.D.; Canesin, G.; Stiehm, A.; Anand, A.; Bjartell, A.; Damber, J.E.; Welen, K. Circulating tumor cells as a marker for progression-free survival in metastatic castration-naive prostate cancer. *Prostate* **2017**, *77*, 849–858. [CrossRef] [PubMed]
14. Gonnissen, A.; Isebaert, S.; McKee, C.M.; Muschel, R.J.; Haustermans, K. The effect of Metformin and GANT61 combinations on the radiosensitivity of prostate cancer cells. *Int. J. Mol. Sci.* **2017**, *18*, 399. [CrossRef] [PubMed]
15. Dayal, S.; Zhou, J.; Manivannan, P.; Siddiqui, M.A.; Ahmad, O.F.; Clark, M.; Awadia, S.; Garcia-Mata, R.; Shemshedini, L.; Malathi, K. RNase L suppresses androgen receptor signaling, cell migration and matrix metalloproteinase activity in prostate cancer cells. *Int. J. Mol. Sci.* **2017**, *18*, 529. [CrossRef] [PubMed]
16. Zhang, X.; Wang, H.; Wang, H.; Xiao, F.; Seth, P.; Xu, W.; Jia, Q.; Wu, C.; Yang, Y.; Wang, L. SUMO-specific cysteine protease 1 promotes epithelial mesenchymal transition of prostate cancer cells via regulating SMAD4 DeSUMOylation. *Int. J. Mol. Sci.* **2017**, *18*, 808. [CrossRef] [PubMed]
17. Bascetta, L.; Oliviero, A.; D'Aurizio, R.; Evangelista, M.; Mercatanti, A.; Pellegrini, M.; Marrocolo, F.; Bracarda, S.; Rizzo, M. The prostate cancer cells resistant to docetaxel as in vitro model for discovering micrornas predictive of the onset of docetaxel resistance. *Int. J. Mol. Sci.* **2017**, *18*, 1512. [CrossRef] [PubMed]
18. Kulkarni, P.; Uversky, V.N. Cancer/Testis Antigens: "Smart" biomarkers for diagnosis and prognosis of prostate and other cancers. *Int. J. Mol. Sci.* **2017**, *18*, 740. [CrossRef] [PubMed]
19. Hagiwara, K.; Tobisawa, Y.; Kaya, T.; Kaneko, T.; Hatakeyama, S.; Mori, K.; Hashimoto, Y.; Koie, T.; Suda, Y.; Ohyama, C.; et al. Wisteria Floribunda Agglutinin and its Reactive-Glycan-Carrying Prostate-Specific Antigen as a novel diagnostic and prognostic marker of prostate cancer. *Int. J. Mol. Sci.* **2017**, *18*, 261. [CrossRef] [PubMed]
20. Ishikawa, T.; Yoneyama, T.; Tobisawa, Y.; Hatakeyama, S.; Kurosawa, T.; Nakamura, K.; Narita, S.; Mitsuzuka, K.; Duivenvoorden, W.; Pinthus, J.H.; et al. An automated micro-total immunoassay system for measuring cancer-associated alpha2,3-Linked sialyl N-glycan-carrying prostate-specific antigen may improve the accuracy of prostate cancer diagnosis. *Int. J. Mol. Sci.* **2017**, *18*, 470. [CrossRef] [PubMed]
21. Friedersdorff, F.; Gross, B.; Maxeiner, A.; Jung, K.; Miller, K.; Stephan, C.; Busch, J.; Kilic, E. Does the prostate health index depend on tumor volume?—A study on 196 patients after radical prostatectomy. *Int. J. Mol. Sci.* **2017**, *18*, 488. [CrossRef] [PubMed]

22. Ferrer-Batalle, M.; Llop, E.; Ramirez, M.; Aleixandre, R.N.; Saez, M.; Comet, J.; de Llorens, R.; Peracaula, R. Comparative study of blood-based biomarkers, alpha2,3-sialic acid PSA and PHI, for high-risk prostate cancer detection. *Int. J. Mol. Sci.* **2017**, *18*, 845. [CrossRef] [PubMed]

23. Daniel, R.; Wu, Q.; Williams, V.; Clark, G.; Guruli, G.; Zehner, Z. A panel of MicroRNAs as diagnostic biomarkers for the identification of prostate cancer. *Int. J. Mol. Sci.* **2017**, *18*, 1281. [CrossRef] [PubMed]

24. Geng, J.H.; Lin, V.C.; Yu, C.C.; Huang, C.Y.; Yin, H.L.; Chang, T.Y.; Lu, T.L.; Huang, S.P.; Bao, B.Y. Inherited variants in Wnt pathway genes influence outcomes of prostate cancer patients receiving androgen deprivation therapy. *Int. J. Mol. Sci.* **2016**, *17*, 1970. [CrossRef] [PubMed]

25. Schlack, K.; Krabbe, L.M.; Fobker, M.; Schrader, A.J.; Semjonow, A.; Boegemann, M. Early prediction of therapy response to abiraterone acetate using PSA subforms in patients with castration resistant prostate cancer. *Int. J. Mol. Sci.* **2016**, *17*, 1520. [CrossRef] [PubMed]

26. Boegemann, M.; Schlack, K.; Thomes, S.; Steinestel, J.; Rahbar, K.; Semjonow, A.; Schrader, A.J.; Aringer, M.; Krabbe, L.M. The role of the neutrophil to lymphocyte ratio for survival outcomes in patients with metastatic castration-resistant prostate cancer treated with Abiraterone. *Int. J. Mol. Sci.* **2017**, *18*, 380. [CrossRef] [PubMed]

27. Filella, X.; Foj, L. Prostate Cancer Detection and Prognosis: From prostate specific antigen (PSA) to exosomal biomarkers. *Int. J. Mol. Sci.* **2016**, *17*, 1784. [CrossRef] [PubMed]

28. Ferro, M.; Ungaro, P.; Cimmino, A.; Lucarelli, G.; Busetto, G.M.; Cantiello, F.; Damiano, R.; Terracciano, D. Epigenetic signature: A new player as predictor of clinically significant prostate cancer (PCa) in patients on active surveillance (AS). *Int. J. Mol. Sci.* **2017**, *18*, 1146. [CrossRef] [PubMed]

29. Elgqvist, J. Nanoparticles As Theranostic vehicles in experimental and clinical applications-focus on prostate and breast cancer. *Int. J. Mol. Sci.* **2017**, *18*, 1102. [CrossRef] [PubMed]

30. Zhang, K.; Zhou, S.; Wang, L.; Wang, J.; Zou, Q.; Zhao, W.; Fu, Q.; Fang, X. Current stem cell biomarkers and their functional mechanisms in prostate cancer. *Int. J. Mol. Sci.* **2016**, *17*, 1163. [CrossRef] [PubMed]

31. Chen, W.Y.; Hua, K.T.; Lee, W.J.; Lin, Y.W.; Liu, Y.N.; Chen, C.L.; Wen, Y.C.; Chien, M.H. Akt activation correlates with snail expression and potentially determines the recurrence of prostate cancer in patients at stage T2 after a radical prostatectomy. *Int. J. Mol. Sci.* **2016**, *17*, 1194. [CrossRef] [PubMed]

32. Campos, A.K.; Hoving, H.D.; Rosati, S.; van Leenders, G.J.; de Jong, I.J. EpCAM expression in lymph node and bone metastases of prostate carcinoma: A pilot study. *Int. J. Mol. Sci.* **2016**, *17*, 1650. [CrossRef] [PubMed]

33. Genitsch, V.; Zlobec, I.; Seiler, R.; Thalmann, G.N.; Fleischmann, A. Neuroendocrine differentiation in metastatic conventional prostate cancer is significantly increased in lymph node metastases compared to the primary tumors. *Int. J. Mol. Sci.* **2017**, *18*, 1640. [CrossRef] [PubMed]

34. Kristiansen, I.; Stephan, C.; Jung, K.; Dietel, M.; Rieger, A.; Tolkach, Y.; Kristiansen, G. Sensitivity of HOXB13 as a diagnostic immunohistochemical marker of prostatic origin in prostate cancer metastases: Comparison to PSA, prostein, androgen receptor, ERG, NKX3.1, PSAP, and PSMA. *Int. J. Mol. Sci.* **2017**, *18*, 1151. [CrossRef] [PubMed]

35. Neuhaus, J.; Schiffer, E.; Mannello, F.; Horn, L.C.; Ganzer, R.; Stolzenburg, J.U. Protease expression levels in prostate cancer tissue can explain prostate cancer-associated seminal biomarkers-an explorative concept study. *Int. J. Mol. Sci.* **2017**, *18*, 976. [CrossRef] [PubMed]

36. Oka, S.; Kanagawa, M.; Doi, Y.; Schuster, D.M.; Goodman, M.M.; Yoshimura, H. Fasting enhances the contrast of bone metastatic lesions in 18F-Fluciclovine-PET: Preclinical study using a rat model of mixed osteolytic/osteoblastic bone metastases. *Int. J. Mol. Sci.* **2017**, *18*, 934. [CrossRef] [PubMed]

37. Luk, A.W.S.; Ma, Y.; Ding, P.N.; Young, F.P.; Chua, W.; Balakrishnar, B.; Dransfield, D.T.; Souza, P.; Becker, T.M. CTC-MRNA (AR-V7) analysis from blood samples-impact of blood collection tube and storage time. *Int. J. Mol. Sci.* **2017**, *18*, 1047. [CrossRef] [PubMed]

38. Ecke, T.H.; Huang-Tiel, H.J.; Golka, K.; Selinski, S.; Geis, B.C.; Koswig, S.; Bathe, K.; Hallmann, S.; Gerullis, H. Prostate Specific Antigen (PSA) as predicting marker for clinical outcome and evaluation of early toxicity rate after high-dose rate brachytherapy (HDR-BT) in combination with additional external beam radiation therapy (EBRT) for high risk prostate cancer. *Int. J. Mol. Sci.* **2016**, *17*, 1879. [CrossRef] [PubMed]

39. Tsumura, H.; Satoh, T.; Ishiyama, H.; Tabata, K.I.; Takenaka, K.; Sekiguchi, A.; Nakamura, M.; Kitano, M.; Hayakawa, K.; Iwamura, M. Perioperative search for circulating tumor cells in patients undergoing prostate brachytherapy for clinically nonmetastatic prostate cancer. *Int. J. Mol. Sci.* **2017**, *18*, 128. [CrossRef] [PubMed]

40. Lichtinghagen, R.; Musholt, P.B.; Stephan, C.; Lein, M.; Kristiansen, G.; Hauptmann, S.; Rudolph, B.; Schnorr, D.; Loening, S.A.; Jung, K. MRNA expression profile of matrix metalloproteinases and their tissue inhibitors in malignant and non-malignant prostatic tissue. *Anticancer Res.* **2003**, *23*, 2617–2624. [PubMed]

41. Lein, M.; Jung, K.; Ortel, B.; Stephan, C.; Rothaug, W.; Johannsen, M.; Deger, S.; Schnorr, D.; Loening, S.; Krell, H.-W. The new synthetic matrix metalloproteinase inhibitor (Roche 28–2653) reduces tumor growth and prolongs survival in a prostate cancer standard rat model. *Oncogene* **2002**, *21*, 2089–2096. [CrossRef] [PubMed]

42. Yoneyama, T.; Ohyama, C.; Hatakeyama, S.; Narita, S.; Habuchi, T.; Koie, T.; Mori, K.; Hidari, K.I.; Yamaguchi, M.; Suzuki, T.; et al. Measurement of aberrant glycosylation of prostate specific antigen can improve specificity in early detection of prostate cancer. *Biochem. Biophys. Res. Commun.* **2014**, *448*, 390–396. [CrossRef] [PubMed]

43. Llop, E.; Ferrer-Batalle, M.; Barrabes, S.; Guerrero, P.E.; Ramirez, M.; Saldova, R.; Rudd, P.M.; Aleixandre, R.N.; Comet, J.; de Llorens, R.; et al. Improvement of prostate cancer diagnosis by detecting psa glycosylation-specific changes. *Theranostics* **2016**, *6*, 1190–1204. [CrossRef] [PubMed]

44. Maxeiner, A.; Kilic, E.; Matalon, J.; Friedersdorff, F.; Miller, K.; Jung, K.; Stephan, C.; Busch, J. The Prostate Health Index PHI predicts oncological outcome and biochemical recurrence after radical prostatectomy—Analysis in 437 patients. *Oncotarget* **2017**, *8*, 79279–79288. [CrossRef]

45. Dani, H.; Loeb, S. The role of prostate cancer biomarkers in undiagnosed men. *Curr. Opin. Urol.* **2017**, *27*, 210–216. [CrossRef] [PubMed]

46. Loeb, S. Prostate Cancer: Predicting prostate biopsy results—PCA3 versus Phi. *Nat. Rev. Urol.* **2015**, *12*, 130–131. [CrossRef] [PubMed]

47. Stephan, C.; Jung, K.; Semjonow, A.; Schulze-Forster, K.; Cammann, H.; Hu, X.; Meyer, H.A.; Bogemann, M.; Miller, K.; Friedersdorff, F. Comparative assessment of urinary prostate cancer antigen 3 and TMPRSS2:ERG gene fusion with the serum [−2]proprostate-specific antigen-based prostate health index for detection of prostate cancer. *Clin. Chem.* **2013**, *59*, 280–288. [CrossRef] [PubMed]

48. MacLellan, S.A.; MacAulay, C.; Lam, S.; Garnis, C. Pre-profiling factors influencing serum MicroRNA levels. *BMC Clin. Pathol.* **2014**, *14*, 27. [CrossRef] [PubMed]

49. Fendler, A.; Stephan, C.; Yousef, G.M.; Kristiansen, G.; Jung, K. The translational potential of MicroRNAs as biofluid markers of urological tumours. *Nat. Rev. Urol.* **2016**, *13*, 734–752. [CrossRef] [PubMed]

50. Zhao, Z.; Stephan, C.; Weickmann, S.; Jung, M.; Kristiansen, G.; Jung, K. Tissue-based MicroRNAs as predictors of biochemical recurrence after radical prostatectomy: What can we learn from past studies? *Int. J. Mol. Sci.* **2017**, *18*, 2023. [CrossRef] [PubMed]

51. Matin, F.; Jeet, V.; Clements, J.A.; Yousef, G.M.; Batra, J. MicroRNA theranostics in prostate cancer precision medicine. *Clin. Chem.* **2016**, *62*, 1318–1333. [CrossRef] [PubMed]

52. Edwards, S.; Campbell, C.; Flohr, P.; Shipley, J.; Giddings, I.; Te-Poele, R.; Dodson, A.; Foster, C.; Clark, J.; Jhavar, S.; et al. Expression analysis onto microarrays of randomly selected CDNA clones highlights HOXB13 as a marker of human prostate cancer. *Br. J. Cancer* **2005**, *92*, 376–381. [CrossRef] [PubMed]

53. Neuhaus, J.; Schiffer, E.; von Wilcke, P.; Bauer, H.W.; Leung, H.; Siwy, J.; Ulrici, W.; Paasch, U.; Horn, L.C.; Stolzenburg, J.U. Seminal plasma as a source of prostate cancer peptide biomarker candidates for detection of indolent and advanced disease. *PLoS ONE* **2013**, *8*, e67514. [CrossRef] [PubMed]

54. Ali, A.; Hoyle, A.; Baena, E.; Clarke, N.W. Identification and evaluation of clinically significant prostate cancer: A step towards personalized diagnosis. *Curr. Opin. Urol.* **2017**, *27*, 217–224. [CrossRef] [PubMed]

55. Hendriks, R.J.; van der Leest, M.M.G.; Dijkstra, S.; Barentsz, J.O.; Van, C.W.; Hulsbergen-van de Kaa, C.A.; Schalken, J.A.; Mulders, P.F.A.; van Oort, I.M. A urinary biomarker-based risk score correlates with multiparametric MRI for prostate cancer detection. *Prostate* **2017**, *77*, 1401–1407. [CrossRef] [PubMed]

56. Tosoian, J.J.; Druskin, S.C.; Andreas, D.; Mullane, P.; Chappidi, M.; Joo, S.; Ghabili, K.; Mamawala, M.; Agostino, J.; Carter, H.B.; et al. Prostate health index density improves detection of clinically significant prostate cancer. *BJU Int.* **2017**. [CrossRef] [PubMed]

International Journal of
Molecular Sciences

MDPI

Article

The Effect of Metformin and GANT61 Combinations on the Radiosensitivity of Prostate Cancer Cells

Annelies Gonnissen [1,2], Sofie Isebaert [1,2,*], Chad M. McKee [3], Ruth J. Muschel [3] and Karin Haustermans [1,2]

1   Laboratory of Experimental Radiotherapy, Department of Oncology, KU Leuven—University of Leuven, 3000 Leuven, Belgium; annelies.gonnissen@kuleuven.be (A.G.); karin.haustermans@uzleuven.be (K.H.)
2   Department of Radiation Oncology, University Hospitals Leuven, 3000 Leuven, Belgium
3   Department of Oncology, CRUK/MRC Oxford Institute for Radiation Oncology, University of Oxford, Oxford OX3 7DQ, UK; chad.mckee@oncology.ox.ac.uk (C.M.M.); ruth.muschel@oncology.ox.ac.uk (R.J.M.)
*   Correspondence: sofie.isebaert@uzleuven.be; Tel.: +32-16-345-018

Academic Editor: Carsten Stephan
Received: 5 December 2016; Accepted: 7 February 2017; Published: 13 February 2017

Abstract: The anti-diabetes drug metformin has been shown to have anti-neoplastic effects in several tumor models through its effects on energy metabolism and protein synthesis. Recent studies show that metformin also targets Hedgehog (Hh) signaling, a developmental pathway re-activated in several tumor types, including prostate cancer (PCa). Furthermore, we and others have shown that Hh signaling is an important target for radiosensitization. Here, we evaluated the combination of metformin and the Hh inhibitor GANT61 (GLI-ANTagonist 61) with or without ionizing radiation in three PCa cell lines (PC3, DU145, 22Rv1). The effect on proliferation, radiosensitivity, apoptosis, cell cycle distribution, reactive oxygen species production, DNA repair, gene and protein expression was investigated. Furthermore, this treatment combination was also assessed in vivo. Metformin was shown to interact with Hh signaling by inhibiting the effector protein glioma-associated oncogene homolog 1 (GLI1) in PCa cells both in vitro and in vivo. The combination of metformin and GANT61 significantly inhibited PCa cell growth in vitro and enhanced the radiation response of 22Rv1 cells compared to either single agent. Nevertheless, neither the growth inhibitory effect nor the radiosensitization effect of the combination treatment observed in vitro was seen in vivo. Although the interaction between metformin and Hh signaling seems to be promising from a therapeutic point of view in vitro, more research is needed when implementing this combination strategy in vivo.

Keywords: prostate cancer; metformin; hedgehog pathway; radiosensitization; xenograft mouse model

1. Introduction

The biguanide metformin, commonly used for the treatment of patients with type 2 diabetes, has been associated with a decreased incidence and mortality in several tumor types, including prostate cancer (PCa) [1–3]. Metformin has also been shown to have anti-neoplastic effects [4], potentially due to its effects on energy metabolism and protein synthesis by activation of adenosine monophosphate (AMP)-activated protein kinase (AMPK) signaling and inhibition of the mammalian target of rapamycin (mTOR) [5–7]. Additionally, several preclinical studies have shown that metformin can increase the effect of radiotherapy [8–12]. The mechanism of these radiosensitizing effects are not completely understood but include reoxygenation of hypoxic regions [9], impairment of DNA damage response [11] and induction of reactive oxygen species (ROS) [12]. The radiosensitizing effect of metformin has been shown to be highly context-dependent and more research is needed to identify the appropriate treatment conditions and cell populations to induce radiosensitization by metformin.

Some recent studies have identified the Hedgehog (Hh) signaling pathway as a novel target of metformin [13,14]. This pathway is a potential candidate for anti-cancer treatment, since several tumor types are known to have deregulated Hh signaling, including PCa [15–18]. Of note, metformin has been shown to inhibit Hh signaling in breast [13] and pancreatic [14] cancer cells.

We have previously shown that inhibition of Hh signaling at the level of the transcription factors using the small molecule inhibitor GANT61 (GLI-ANTagonist 61) is highly effective in PCa cells and increased radiosensitivity both in vitro and in vivo [19].

In this study, the link between AMPK and Hh signaling was explored in several PCa cell lines. Additionally, we investigated whether simultaneous targeting of both pathways with metformin and GANT61, respectively, could enhance PCa cytotoxicity and/or radiosensitivity both in vitro and in vivo.

2. Results

2.1. Metformin Decreases Cell Growth and Survival of Prostate Cancer (PCa) Cells

Metformin significantly decreased PCa cell growth in a time- and dose-dependent manner (Figure 1A). A high dose of metformin (5 mM) was able to induce a significant decrease in cell survival in all cell lines (Figure 1B). The PC3 cells were the most sensitive with a half maximal inhibitory concentration (IC_{50}) value around 5 mM, whereas the IC_{50} in 22Rv1 and DU145 was between 10–20 mM metformin (Figure S1).

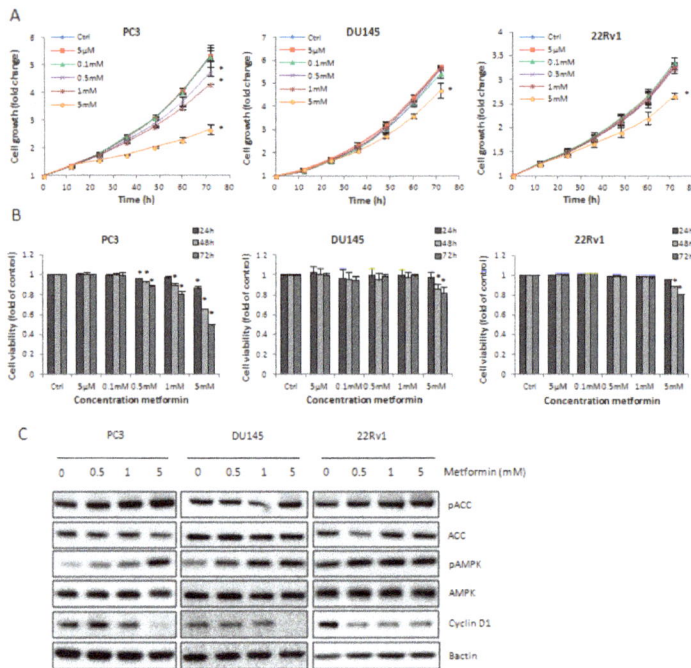

Figure 1. Effect of metformin on prostate cancer (PCa) cells. (**A**) Cell growth and (**B**) cell viability after metformin treatment. Means ± standard error of means (SEM) of three independent experiments. * $p < 0.05$ vs. control; (**C**) Protein expression of the downstream signaling molecules after 72-h metformin treatment. ACC, Acetyl-CoA carboxylase; AMPK, adenosine monophosphate (AMP)-activated protein kinase; pACC, phospho-Acetyl-CoA carboxylase; pAMPK, phospho-Adenosine monophosphate (AMP)-activated protein kinase.

In line with this, cyclin D1 protein expression was drastically decreased after treatment with 5 mM metformin, especially in the rapidly proliferating PC3 and DU145 cell lines (Figure 1C). Additionally, metformin activated its downstream signaling components AMPK and Acetyl-CoA carboxylase (ACC) in a dose-dependent manner in all PCa cell lines (Figure 1C).

2.2. Metformin Increases Radiosensitivity of PCa Cells Independent of Adenosine Monophosphate (AMP)-Activated Protein Kinase (AMPK) Activation

Metformin (5 mM) increased radiosensitivity of DU145 and 22Rv1 cells with a dose-enhancement factor (DEF) of 1.6 ± 0.15 ($p < 0.05$) and 1.36 ± 0.08 ($p < 0.05$) respectively. In contrast, the radiosensitivity of PC3 cells was not affected by metformin (Figure 2A). To evaluate the role of AMPK in the metformin-induced radiosensitization effect in the DU145 and 22Rv1 cells, AMPK was silenced by means of silencing RNA (siRNA). Downregulation of (phospho)AMPK did not affect the intrinsic radiosensitivity of either cell line nor did it change the metformin-induced radiosensitization (Figure 2B).

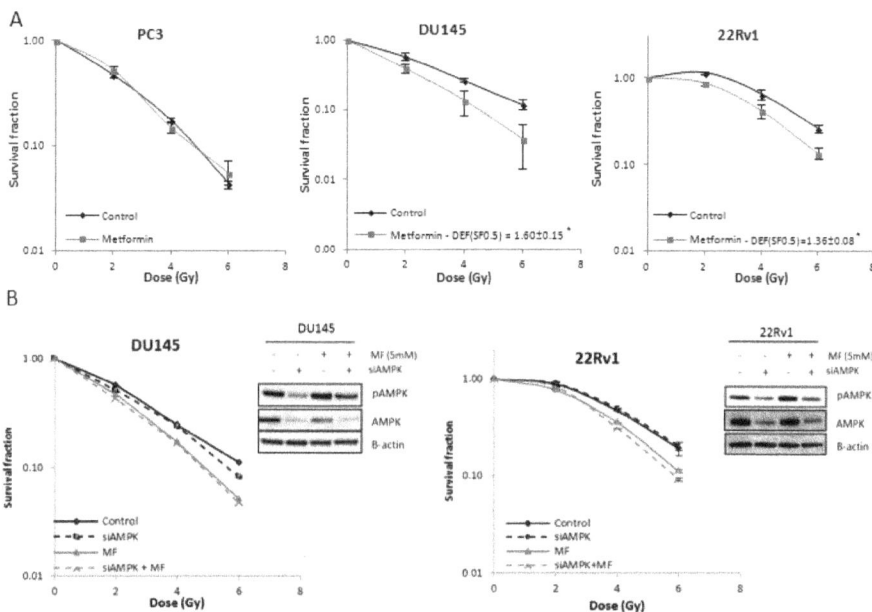

Figure 2. Effect of metformin (MF) on radiosensitivity of PCa cells. (**A**) Clonogenic survival after 72-h treatment with metformin (5 mM) prior to/during ionizing radiation (IR); (**B**) Clonogenic survival of DU145 and 22Rv1 cells transfected with AMPK silencing RNA (siRNA) and 72-h treatment with metformin (5 mM) prior to/during IR. Knockdown was verified with western blotting. Means ± SEM of three independent experiments. * $p < 0.05$ vs. control. DEF: dose-enhancement factor.

2.3. Metformin Regulates Hedgehog Signaling in an AMPK-Dependent Manner

Next, we investigated if there was a link between metformin and Hh signaling in PCa cells. Indeed, metformin (5 mM) significantly decreased glioma-associated oncogene homolog 1 (*GLI1*) and patched 1 (*PTCH1*) gene and protein expression in all cell lines (Figure 3A,B). Although metformin only significantly decreased glioma-associated oncogene homolog 2 (*GLI2*) gene expression in the DU145 cells, we did observe decreased GLI2 protein expression after metformin treatment in all PCa cell lines (Figure 3A,B). In addition, AMPK activation was shown to be inversely correlated with GLI1 protein expression. Silencing AMPK expression with siRNA increased GLI1 expression, whereas activation of AMPK by metformin decreased GLI1 expression in 22Rv1 cells (Figure 3C). This indicates that metformin regulates Hh signaling, probably through AMPK signaling.

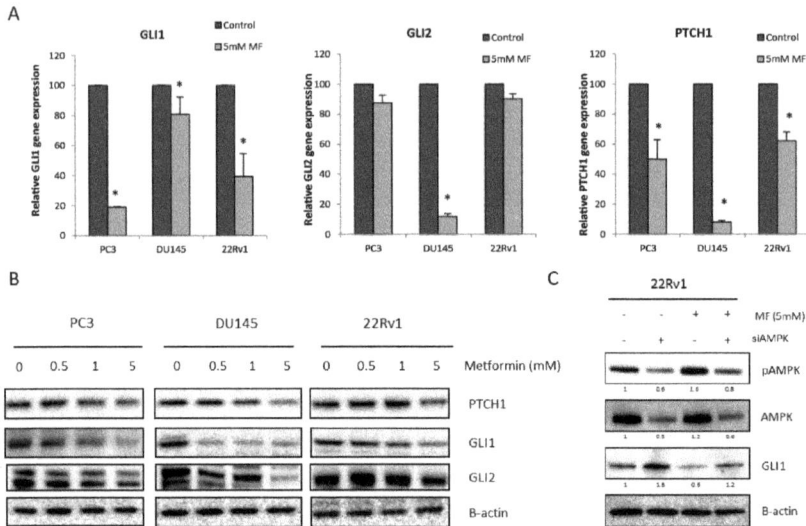

Figure 3. Link between metformin and Hedgehog signaling. (**A**) *GLI1*, *GLI2* and *PTCH1* gene expression after 72-h metformin treatment. Means ± SEM of two independent experiments. * $p < 0.05$ vs. control; (**B**) PTCH1, GLI1 and GLI2 protein expression after 72-h metformin treatment; (**C**) (p)AMPK protein and GLI1 expression in 22Rv1 cells transfected with AMPK siRNA and treated with metformin (5 mM) 72-h prior to protein lysis. GLI1, glioma-associated oncogene homolog 1; GLI2, glioma-associated oncogene homolog 2; PTCH1, patched 1.

2.4. Combination of Metformin and GANT61 (GLI-ANTagonist 61) Synergistically Decreases PCa Cell Growth

The link between AMPK and GLI1 led to the question as to whether the combination of metformin with Hh inhibitors could enhance the cytotoxic effect of the individual drugs. We have previously shown that the GLI1/2 inhibitor GANT61 significantly decreased cell survival of PC3 and 22Rv1 cells [19]. Indeed, combining metformin and GANT61 significantly decreased cell growth of all PCa cell lines, resulting in an almost complete blockage of cell growth in PC3 and 22Rv1 cells (Figure 4A). Additionally, we confirmed decreased *GLI1* gene expression in all cells treated with the drug combination (Figure S2). Cell cycle analyses revealed that the drug combination in the PC3 cells led to a G2/M-arrest after only 24 h, which persisted until 72 h of treatment (Figure 4B). This corresponds to the dramatic decrease in cell growth already observed after 24 h of treatment. The drug combination also significantly increased the sub-G1 population which peaked at 48 h (Figure 4C). In the DU145 cells, no significant cell cycle effects were observed after 24–72 h of either treatment (Figure 4B), whereas the combination treatment did significantly increase apoptosis after 72 h compared to either single agent (Figure 4C). In the 22Rv1 cells, GANT61 induced a G1-arrest after only 24 h. Metformin alone did not have a significant effect on cell cycle, however, the combination of both drugs resulted in a much more pronounced G1-arrest after 72 h of treatment compared to GANT61 alone (Figure 4B). Apoptosis was also significantly induced by both individual drugs and even more by the drug combination (Figure 4C). Moreover, we observed that both drugs and their combination induced DNA damage in all cell lines after 72 h of treatment assessed by means of γH2AX staining (Figure S3).

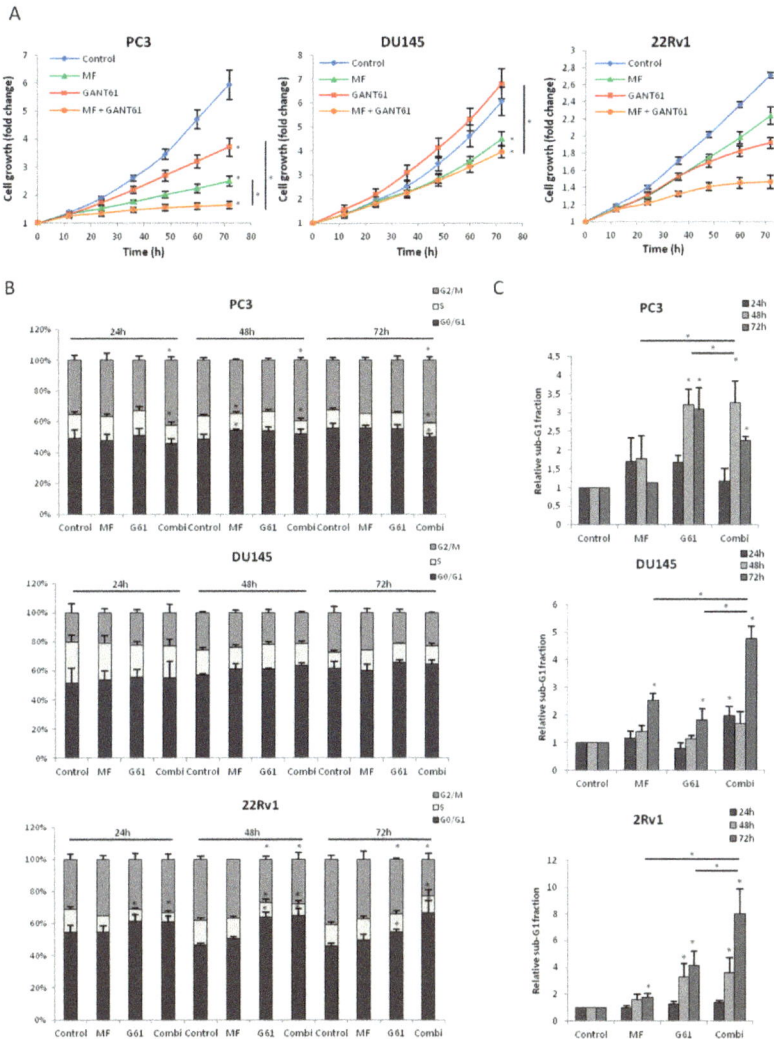

Figure 4. Combination of metformin and GANT61 (GLI-ANTagonist 61) in PCa cells (**A**) Cell growth; (**B**) cell cycle distribution and (**C**) sub-G1 population after treatment with metformin (5 mM), GANT61 (10 μM) or combination; (**B,C**) Cells were fixed at 24, 48 and 72 h of treatment. Means ± SEM of three independent experiments. * $p < 0.05$ vs. control.G61, GANT61.

2.5. Combining Metformin and GANT61 Enhances the Effect of Ionizing Radiation In Vitro

Short-term survival analyses illustrated that the combination of metformin and GANT61 with ionizing radiation (IR) significantly decreased cell survival in all cell lines (Figure 5A). In the PC3 and DU145 cells, the response to IR in the combination treatment is most likely due to metformin since addition of GANT61 did not significantly decrease cell survival any further. In contrast, in the 22Rv1 cells, a significant decrease in cell survival in the combination group compared to metformin alone was observed. To assess the effect on the intrinsic radiosensitivity, clonogenic survival assays were performed. Consistent with previous results, the combination of metformin and GANT61 significantly enhanced radiosensitivity of 22Rv1 cells compared to either single agent (Figure 6B). In the PC3 cells, we did not observe an effect of the combination treatment or either single agent, whereas only metformin had a radiosensitizing effect in the DU145 cells (Figure 5B).

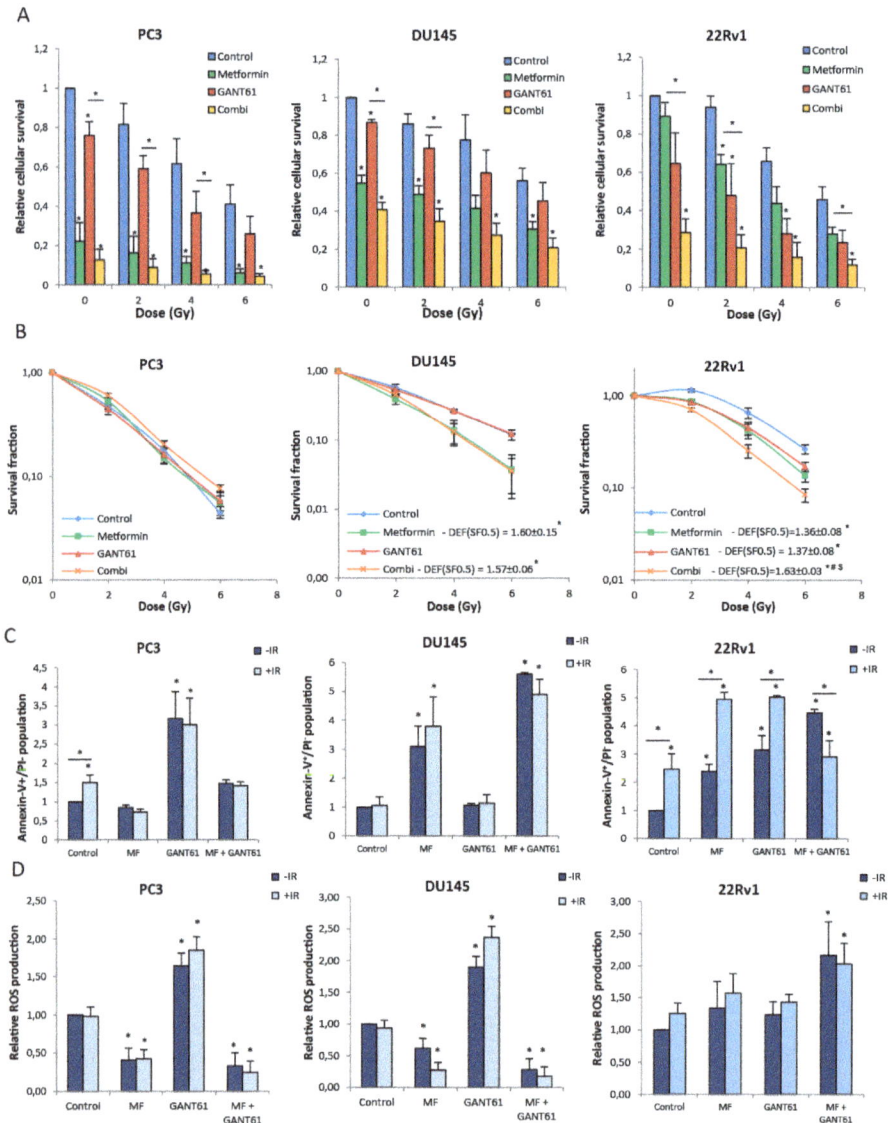

Figure 5. Effect of the combination of metformin and GANT61 on radiosensitivity of PCa cells. (**A**) Relative short-term cell survival 7 days after treatment with increasing doses of ionizing radiation after 72-h pretreatment with metformin, GANT61 or combination; (**B**) Clonogenic survival after 72-h treatment with metformin, GANT61 or combination prior to/during IR; (**C**) AnnexinV$^+$/propidium iodide (PI)$^-$ cells and (**D**) Reactive oxygen species (ROS) production at 24-h post-IR after 72-h pretreatment with metformin (5 mM), GANT61 (10 μM) or combination. Means ± SEM of three independent experiments. * $p < 0.05$ vs. control.

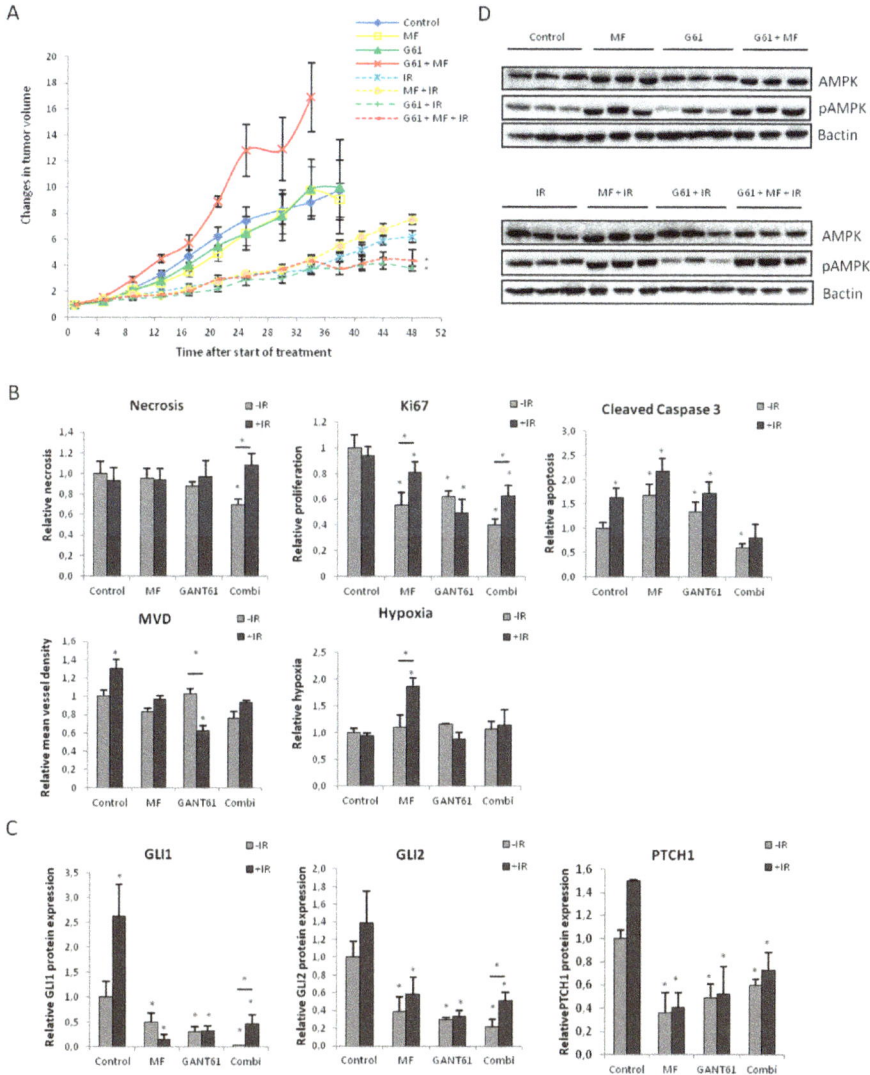

Figure 6. Effect of the combination of metformin and GANT61 on radiosensitivity in a 22Rv1 xenograft model. (**A**) Relative tumor growth of 22Rv1 xenograft mice treated with metformin (250 mg/kg), GANT61 (50 mg/kg) or combination. At treatment day 5, tumors were irradiated with a single dose of 6Gy (six mice/group); (**B**) Immunohistochemical analyses of necrosis, Ki67, cleaved caspase 3, mean vessel density (MVD), hypoxia and (**C**) Hedgehog (Hh) target proteins GLI1, GLI2 and PTCH1 (six mice/group, $n \leq 12$ tumors). * $p < 0.05$ vs. control; (**D**) Protein expression of pAMPK and AMPK in 22Rv1 xenograft tumors ($n = 3$ tumors/treatment group).

The combination of IR with either single agent increased apoptosis in the 22Rv1 cells. However, in the drug combination treatment, addition of IR did not further increase apoptosis; on the contrary, apoptosis was decreased in this condition indicating that other mechanisms might be important here (Figure 5C). In the PC3 and DU145 cells, apoptosis was not (further) increased upon addition of IR

(Figure 5C). Furthermore, we observed that ROS were increased in the 22Rv1 cells when using the drug combination; an effect which could attribute to the increased radiosensitivity (Figure 5D).

2.6. Combination of Metformin and GANT61 Stimulates Tumor Growth In Vivo

Since the drug combination enhanced the radiosensitivity of 22Rv1 cells in vitro, we used a 22Rv1 xenograft model to assess the effect of the drug combination in vivo. Surprisingly, although we observed a robust decrease of cell proliferation in vitro, the combination of GANT61 and metformin appeared to be pro-proliferative in vivo. Metformin and GANT61 alone did not have any effect on tumor growth (Figure 6A). To investigate why the combination of both drugs resulted in a paradoxical effect, we looked deeper into the molecular changes that occurred in these tumors. Tumors treated with the drug combination were less necrotic and had fewer apoptotic cells compared to control tumors or those treated with either single agent (Figure 6B). In addition, these tumors contained fewer blood vessels compared to controls or the GANT61-treated tumors. No changes in hypoxia were observed. Although the drug combination increased tumor growth, this was not reflected by Ki67 expression, which was significantly lower in these tumors compared to controls and either single agent (Figure 6B). This implies that the growth stimulatory effect of the drug combination is more likely due to pro-survival effects rather than effects on proliferation.

As for the combination with IR, metformin did not increase radiosensitivity of 22Rv1 tumors, while GANT61 did result in enhanced radiosensitivity. The combination of metformin and GANT61 with IR also significantly decreased tumor growth, however this was most likely attributed to the effect of GANT61 (Figure 6A). As previously shown [19], the GANT61-induced radiosensitization is attributed to decreased proliferation and increased apoptosis. In the tumors treated with the drug combination, the radiosensitizing effect can also, at least partially, be ascribed to effects on proliferation. In contrast, the amount of apoptosis is lower in these tumors compared to controls or tumors treated with a single agent, but we observed more necrosis in the tumors treated with the combined modality, which could also attribute to the decreased tumor growth (Figure 6B).

Finally, we verified whether both drugs reached their target in the tumors to verify their specificity in vivo. Immunohistochemical analyses of GLI1, GLI2 and PTCH1 demonstrated that GANT61 was able to specifically inhibit Hh signaling in the tumor (Figure 6C). We also provide evidence that metformin targets AMPK signaling as phospho-AMPK (pAMPK) was significantly increased in all tumor of metformin-treated mice (Figure 6D). Additionally, we showed that the link between metformin and Hh signaling also exists in vivo. Both metformin also significantly decreased protein expression of GLI1, GLI2 and PTCH1 (Figure 6C).

3. Discussion

Contradicting findings have been reported in PCa patients regarding an association between metformin use in diabetic PCa patients and biochemical recurrence (BCR) and metastasis. Several studies have shown that metformin had no effect on BCR-free survival of PCa patients after radical prostatectomy [20–23]. In contrast, a retrospective study by He et al. [24] did find a beneficial effect of metformin on overall survival (OS) of PCa patients treated with radical prostatectomy [24]. In a study by Spratt et al. [25] the authors specifically looked at PCa patients treated with external-beam radiotherapy (EBRT) and found a significant positive correlation between metformin use and BCR-free survival and OS [25]. These studies demonstrate that metformin might be an important treatment option for PCa patients, especially in the combination setting with EBRT.

In this study, we first evaluated the effect of metformin on cell growth and radiosensitivity of PCa cells. In line with a previous study by Sahra et al. [4], we found that metformin (5 mM) decreased cell growth of all cell lines, which coincided with reduced protein levels of cyclin D1. The fact that we could only observe a significant inhibition of cell growth at high doses of metformin is in accordance with multiple studies using different tumor cell types [4,8,26–28]. Several preclinical studies have already investigated the effect of metformin on the radiosensitivity of different tumor types and multiple mechanisms have been associated with the enhanced radiation sensitivity after metformin treatment [8–12]. Here, a radiosensitizing effect of metformin was observed in the DU145 and 22Rv1 cells, but not in the PC3 cells. This lack of radiosensitization in the PC3 cells could be a result of the higher sensitivity of PC3 cells to metformin leading to the absence of any additional effect upon IR. Data also suggest that metformin may enhance radiation response specifically in certain genetic backgrounds (p53, liver kinase B1 (LKB1)), but more research is needed to gain better insight in the specific interactions between metformin and important oncogenic and tumor suppressor genes [28].

We show that metformin also interacts with the Hh signaling pathway, specifically by decreasing gene and protein expression of the Hh target genes *GLI1* and *PTCH1*. Similar interactions were described recently in breast [13] and pancreatic [14] cancer cells. Our data demonstrate that this interaction is mediated by the activation of AMPK signaling, which is in line with the data of Nakamura et al. [14]. Accordingly, two recent studies have shown the AMPK phosphorylates GLI1, which results in the proteasomal degradation of GLI1 [29,30]. On top of that, we demonstrate that metformin also decreased Hh signaling in vivo. Interestingly, we observed a differential inhibition of Hh signaling in the three PCa cell lines at the gene level. In the PC3 and 22Rv1 cells, metformin significantly inhibited *GLI1* and *PTCH1* gene expression, whereas in the DU145 cells, metformin strongly inhibited *GLI2* gene expression in addition to inhibiting *GLI1* and *PTCH1*. Nevertheless, we did not observe a different phenotype in these cell lines in response to metformin treatment.

In a previous study by our group, we demonstrated that Hh inhibition increased the intrinsic radiosensitivity of 22Rv1 cells and not of PC3 and DU145 cells, which appeared to be dependent on functional p53 signaling [19]. Furthermore, it seems unlikely that the effect of metformin on radiosensitivity is through its effects on Hh signaling, as we have shown that metformin regulates Hh signaling in an AMPK-dependent manner, whereas it increases radiosensitivity of PCa cells independent of AMPK activation. Therefore, Hh inhibition and metformin both increase radiosensitivity of PCa cells but in a different manner.

This interesting interaction between metformin and Hh signaling suggests that the combination of metformin with Hh inhibition has the potential to further enhance the cytotoxic and/or radiosensitizing effect of either single agent. To our best knowledge, this combination has never been tested before. Our results show that this combination treatment is highly effective against PCa proliferation in vitro. This was associated with an increase of apoptosis and the induction of DNA damage in all cell lines. Additionally, we observed G2/M-arrest after just 24 h in the PC3 cells, these being the most sensitive to this drug combination. In the 22Rv1 cells, the cells treated with the drug combination experienced G1-arrest, which was at least partially ascribed to GANT61. In contrast with the high cytotoxicity of the drug combination observed in vitro, combining metformin and GANT61 appeared to be pro-proliferative in a 22Rv1 xenograft mouse model. Strangely, the amount of proliferation in these tumors was significantly decreased compared to control or either single agent, suggesting that the increased tumor growth was more likely attributed to decreased apoptosis and necrosis.

Our data also suggest that the combination of metformin and GANT61 only increases radiosensitivity of cells that are sensitized by both individual drugs, as was the case with the 22Rv1 cells. The PC3 cells were not radiosensitized by either drug or the combination, whereas the DU145 cells were only sensitized by metformin and not by GANT61 due to its p53 mutation [19], resulting in no extra benefit upon drug combination. In the 22Rv1 cells we observed a radiosensitizing effect of both metformin and GANT61 which was further enhanced by the drug combination. This was not the case in the in vivo situation, where we only observed a radiosensitizing effect of GANT61 and no additive effect on radiosensitivity by metformin. In line with our previously published data, the GANT61-induced radiosensitization was associated with decreased proliferation and increased cell death [19]. In tumors treated with the drug combination and IR, the amount of necrosis was significantly increased. One potential reason for the lack of radiosensitization of metformin in vivo might be the dosage of metformin used. In vitro, high-doses of metformin (5 mM) are needed to induce cytotoxic and radiosensitizing effects. These high levels of metformin are typically not achieved in an in vivo situation. We used 250 mg/kg intraperitoneally, which is also considered quite high compared to the typical metformin dosing in diabetic patients which lies around 30 mg/kg orally. However, considering the interspecies differences between mice and humans, the Food and Drug Administration (FDA) applies a standard scaling factor of 12.3 [31] which makes our in vivo dosage used definitely acceptable.

Several Smoothened (SMO) inhibitors are currently in the early stages of clinical investigation in PCa patients. Despite the high efficacy of SMO inhibition in patients with basal cell carcinoma and medulloblastoma, which are characterized by the presence of Hh mutations, the efficacy of these inhibitors in ligand-dependent Hh-activated cancer types such as PCa has not yet been established. Early phase clinical trials demonstrated little or no responsiveness to Hh inhibition in these tumor types so far. Multiple clinical trials are currently ongoing to investigate the potential of Hh inhibition as monotherapy or in combination with hormonal therapy in PCa [32]. The use of Hh inhibitors more downstream the signaling cascade has not yet been investigated in patients. Based on our results, the combination of GANT61 with radiotherapy might also be a promising strategy to test in the clinic [19].

Although the combination of metformin and GANT61 therapy appeared very effective in an in vitro situation, the interaction of metformin and GANT61 in the in vivo setting resulted in a paradoxical (pro-survival) effect. This suggests that the tumor microenvironment could play an important role in the anti-tumor activity of drug combinations in vivo. We have previously shown that Hh inhibition indeed also targets the stromal compartment of the tumor and that this results in an increased efficacy of radiotherapy and Hh inhibition in two PCa xenograft models [19]. This was in line with other reports indicating that the tumor-associated stroma is also influenced by Hh inhibition and this might have an impact on the tumor cells [33,34]. Additionally, metformin has also been shown to influence the tumor microenvironment. A study by Martin et al. [35] has found that metformin induced vascular endothelial growth factor A (VEGF-A) expression in a BRAF-driven melanoma tumor model. This resulted in increased angiogenesis and accelerated tumor growth [35]. Although we also observed an enhanced tumor growth in the combination group, we did not observe any changes in microvessel density in our study.

This observation highlights the importance of testing drug interactions in an in vivo setting. The effect of drug interactions is often overlooked in preclinical studies, resulting in the failure of many novel medications in clinical trials. Therefore, more research into the interaction between Hh inhibition and metformin should be performed in multiple tumor models to guarantee the safety of Hh inhibition in diabetic patients using metformin.

4. Materials and Methods

4.1. Cell Culture and Drug Exposure

The androgen-unresponsive PCa cell lines PC3 and DU145 were obtained from the American Type Culture Collection (ATCC; Manassas, VA, USA). The PC3 cells were grown in minimal essential medium (MEM, Life Technologies, Carlsbad, CA, USA) supplemented with 10% fetal bovine serum (FBS; Life Technologies). The DU145 cells were cultured in MEM (Life Technologies) supplemented with 10% FBS, 1% sodium pyruvate (Life Technologies) and 1% non-essential amino acids (Life Technologies). The androgen-responsive 22Rv1 cells (European Collection of Cell Cultures, ECACC, Salisbury, UK) were cultured in RPMI 1640 medium without phenol red (Sigma-Aldrich, St. Louis, MA, USA), supplemented with 10% FBS, 1% L-glutamine and 1% 4-(2-hydroxyethyl)-1-piperazineethanesulfonic acid (HEPES) buffer (Life Technologies). All cells were maintained at 37 °C in a humidified incubator with 5% CO_2/95% O_2 atmosphere.

Stock solutions of metformin (Sigma-Aldrich) were prepared in sterile water or saline for in vitro and in vivo experiments, respectively. For in vitro experiments, stock solutions of GANT61 were prepared in dimethyl sulfoxide (DMSO; Adipogen, San Diego, CA, USA). For the in vivo experiment,

GANT61 (Tocris, Bristol, UK) was dissolved in 100% EtOH and was further dissolved in saline (9:1 saline:EtOH). Control conditions were treated with the corresponding drug solvent.

4.2. Cell Growth and Survival

Cells were seeded in a 96-well plate at a density of 2.5–45 $\times$ 10^4 cells per well and treated for 72 h with different concentrations of the inhibitors. Cell growth was assessed using the Incucyte Zoom system (Essen BioScience, Ann Arbor, MI, USA). Short-term survival assays were performed by pretreating the cells with GANT61 (10 μM) and metformin (5 mM) for 72 h followed by IR (2, 4, or 6 Gy). After 24 h, fresh medium was added and cell survival was assessed 7 days thereafter by means of sulforhodamine B (SRB) assay [36].

4.3. Quantitative Real-Time Polymerase Chain Reaction (qPCR)

RNA isolation and quantitative Polymerase Chain Reaction (qPCR) reactions were performed as previously described [19]. Primer sequences for glyceraldehyde 3-phosphate dehydrogenase (GAPDH), PTCH1, GLI1 and GLI2 are enlisted in Supplemental Table S1. Gene expression was calculated as expression per 100,000 copies of the household gene *GAPDH*.

4.4. Immunoblot Analysis

Immunoblotting was performed as done before [19]. Primary antibodies against ACC (#3662, 1:1000), phospho-acetyl-CoA carboxylase (pACC) (#3661, 1:500), AMPK (#2532, 1:1000), pAMPK (#2535, 1:500), GLI1 (#2534, 1:500) from Cell Signaling Technologies (Beverly, MA, USA), cyclin D1 (CCND1) (sc-8396, 1:200) and PTCH1 (sc-6149, 1:200) from Santa Cruz (Dallas, TX, USA) and GLI2 (600-401-845, 1:1000) from Rockland Immunochemicals (Limerick, PA, USA) were used. B-actin (Cell Signaling Technologies, #4967, 1:1000) was used as loading control. An enhanced chemiluminescence detection system (Perkin Elmer, Waltham, MA, USA) was used to visualize immune-reactive proteins using Fujifilm LAS-3000 mini camera (Fujifilm, Germany). Protein expression was quantified using ImageJ 1.50.

4.5. Flow Cytometry

Apoptotic cell populations (AnnexinV$^+$/PI$^-$),DNA damage and cell cycle distribution were measured as previously described [19]. Reactive oxygen species (ROS) were detected using 2′,7′-dichlorodihydrofluorescein diacetate [37]. FACSVerse Flow cytometer (BD Biosciences, Franklin Lakes, NJ, USA) was used for flow cytometric analysis.

4.6. Colony Formation

Clonogenic formation assays were performed as previously described [19]. After 72-h drug treatment, cells were seeded at low density and irradiated with 2, 4 and 6 Gy using a Baltograph (Balteau NDT, Hermalle-sous-argenteau, Belgium) or mock irradiated. Fresh medium was added 16 h post-irradiation. After 11–21 days post-irradiation, cells were fixed with 2.5% glutaraldehyde in phosphate buffered saline (PBS) and stained with 0.4% crystal violet. The colonies containing $\geq$50 cells were counted with ColCount (Oxford Optronix, Oxford, UK). Survival fractions normalized for drug-induced toxicity. Dose-enhancement factor (DEF) was calculated as the ratio of the dose needed for the control cells to the dose needed for the treated cells to reach a survival fraction of 0.5 (DEF0.5).

4.7. Animal Experiments

Animal experiments were approved by the ethics committee of KU Leuven (P131/2014). Male NMRI Nu/Nu mice (Janvier, France) were inoculated in both flanks with 2 $\times$ 10^6 22Rv1 cells in 100:100 μL medium/Matrigel (VWR, Radnor, PA, USA). Once tumors reached a volume of 150 mm^3, mice were treated by intraperitoneal injection with solvent (9:1 saline/EtOH), metformin (250 mg/kg, every day), GANT61 (50 mg/kg every other day) or the combination of both drugs for 7 weeks. At day 5 of drug treatment, tumors were irradiated with a single dose of 6 Gy. During the entire treatment period, tumor growth was followed by 2-weekly caliper measurements and tumor volumes were calculated (V = (length × width × height) × π/6). In addition, the body weight of the mice was monitored to assess potential treatment toxicity. Mice were euthanized at the end of drug treatment or when tumors reached the maximum ethically permitted volume of 2×10^3 mm^3. Thirty min before euthanasia, pimonidazole was intraperitoneally injected. Afterwards, tumors were excised and half of the tumor was fixed in formalin and embedded in paraffin for immunohistochemical analysis and the other half was snap-frozen for protein analysis.

4.8. Immunohistochemistry

Immunohistochemistry for Ki67, cleaved caspase3, pimonidazole, cluster of differentiation 31 (CD31), GLI1, GLI2 and PTCH1 was performed as previously described [19]. Protein expression was quantified using ImageJ.

4.9. Statistical Analysis

One-way ANOVA with Tukey's multiple comparison test or a two-tailed student's *t*-test were used for the in vitro experiments. For the in vivo experiment, a Kolmogorov–Smirnov method was used to test for normality. Thereafter, either a two-tailed student's *t*-test was used when the data were normally distributed with equal variance or nonparametric analysis using the Mann–Whitney rank-sum test in other conditions. All statistical tests were performed using the software package Statistica 12 (StatSoft Inc., Tulsa, OK, USA). A *p*-value of <0.05 was considered statistically significant.

Supplementary Materials: Supplementary materials can be found at www.mdpi.com/1422-0067/18/2/399/s1.

Acknowledgments: Annelies Gonnissen and Sofie Isebaert were sponsored by a grant from the National Cancer Plan Action 29 Belgium (KPC_29_023). Karin Haustermans is a clinical research fellow of the Research Foundation Flanders.

Author Contributions: Annelies Gonnissen, Sofie Isebaert, Chad M. McKee, Ruth J. Muschel and Karin Haustermans conceived and designed the experiments; Annelies Gonnissen performed the experiments; Annelies Gonnissen and Sofie Isebaert analyzed the data; Karin Haustermans contributed reagents/materials/ analysis tools; Annelies Gonnissen, Sofie Isebaert, Chad M. McKee, Ruth J. Muschel and Karin Haustermans wrote the paper.

Conflicts of Interest: The authors declare no conflict of interest.

Abbreviations

ACC	Acetyl-CoA carboxylase
AMP	Adenosine monophosphate
AMPK	AMP-activated protein kinase
CCND1	Cyclin D1
CD31	Cluster of differentiation 31
qPCR	Quantitative Polymerase Chain Reaction
DEF	Dose-enhancement factor
DMSO	Dimethyl sulfoxide
EBRT	External-beam radiotherapy
ECACC	European Collection of Cell Cultures
FBS	Fetal bovine serum
GANT61	GLI-ANTagonist 61
GAPDH	Glyceraldehyde 3-phosphate dehydrogenase
GLI	Glioma-associated oncogene homolog
HEPES	4-(2-hydroxyethyl)-1-piperazineethanesulfonic acid
Hh	Hedgehog

IC$_{50}$	Half maximal inhibitory concentration
IR	Ionizing radiation
LKB1	Liver kinase B1
MEM	Minimal essential medium
MF	Metformin
mTOR	Mammalian target of rapamycin
MVD	Microvessel density
OS	Overall survival
pACC	Phospho Acetyl-CoA carboxylase
pAMPK	Phospho AMP-activated protein kinase
PBS	Phosphate buffered saline
PI	Propidium iodide
PTCH1	Patched 1
PCa	Prostate cancer
ROS	Reactive oxygen species
SEM	Standard error of means
siRNA	Silencing RNA
SMO	Smoothened
VEGF-A	Vascular endothelial growth factor A

References

1. Noto, H.; Goto, A.; Tsujimoto, T.; Noda, M. Cancer risk in diabetic patients treated with metformin: A systematic review and meta-analysis. *PLoS ONE* **2012**, *7*, e33411. [CrossRef] [PubMed]
2. Pollak, M.N. Investigating metformin for cancer prevention and treatment: The end of the beginning. *Cancer Discov.* **2012**, *2*, 778–790. [CrossRef] [PubMed]
3. Del, B.S.; Vazquez-Martin, A.; Cufi, S.; Oliveras-Ferraros, C.; Bosch-Barrera, J.; Joven, J.; Martin-Castillo, B.; Menendez, J.A. Metformin: Multi-faceted protection against cancer. *Oncotarget* **2011**, *2*, 896–917.
4. Ben, S.I.; Laurent, K.; Loubat, A.; Giorgetti-Peraldi, S.; Colosetti, P.; Auberger, P.; Tanti, J.F.; Le Marchand-Brustel, Y.; Bost, F. The antidiabetic drug metformin exerts an antitumoral effect in vitro and in vivo through a decrease of cyclin D1 level. *Oncogene* **2008**, *27*, 3576–3586.
5. Kourelis, T.V.; Siegel, R.D. Metformin and cancer: New applications for an old drug. *Med. Oncol.* **2012**, *29*, 1314–1327. [CrossRef] [PubMed]
6. Dowling, R.J.; Zakikhani, M.; Fantus, I.G.; Pollak, M.; Sonenberg, N. Metformin inhibits mammalian target of rapamycin-dependent translation initiation in breast cancer cells. *Cancer Res.* **2007**, *67*, 10804–10812. [CrossRef] [PubMed]
7. Zakikhani, M.; Dowling, R.; Fantus, I.G.; Sonenberg, N.; Pollak, M. Metformin is an AMP kinase-dependent growth inhibitor for breast cancer cells. *Cancer Res.* **2006**, *66*, 10269–10273. [CrossRef] [PubMed]
8. Song, C.W.; Lee, H.; Dings, R.P.; Williams, B.; Powers, J.; Santos, T.D.; Choi, B.H.; Park, H.J. Metformin kills and radiosensizes cancer cells and preferentially kills cancer stem cells. *Sci. Rep.* **2012**, *2*, 362. [CrossRef] [PubMed]
9. Zannella, V.E.; Dal, P.A.; Muaddi, H.; McKee, T.D.; Stapleton, S.; Sykes, J.; Glicksman, R.; Chaib, S.; Zamiara, P.; Milosevic, M.; et al. Reprogramming metabolism with metformin improves tumor oxygenation and radiotherapy response. *Clin. Cancer Res.* **2013**, *19*, 6741–6750. [CrossRef] [PubMed]
10. Storozhuk, Y.; Hopmans, S.N.; Sanli, T.; Barron, C.; Tsiani, E.; Cutz, J.C.; Pond, G.; Wright, J.; Singh, G.; Tsakiridis, T. Metformin inhibits growth and enhances radiation response of non-small cell lung cancer (NSCLC) through ATM and AMPK. *Br. J. Cancer* **2013**, *108*, 2021–2032. [CrossRef] [PubMed]
11. Zhang, T.; Zhang, L.; Zhang, T.; Fan, J.; Wu, K.; Guan, Z.; Wang, X.; Li, L.; Hsieh, J.T.; He, D.; et al. Metformin sensitizes prostate cancer cells to radiation through EGFR/p-DNA-PKCS in vitro and in vivo. *Radiat. Res.* **2014**, *181*, 641–649. [CrossRef] [PubMed]
12. Kim, E.H.; Kim, M.S.; Cho, C.K.; Jung, W.G.; Jeong, Y.K.; Jeong, J.H. Low and high linear energy transfer radiation sensitization of HCC cells by metformin. *J. Radiat. Res.* **2014**, *55*, 432–442. [CrossRef] [PubMed]

13. Fan, C.; Wang, Y.; Liu, Z.; Sun, Y.; Wang, X.; Wei, G.; Wei, J. Metformin exerts anticancer effects through the inhibition of the Sonic hedgehog signaling pathway in breast cancer. *Int. J. Mol. Med.* **2015**, *36*, 204–214. [CrossRef] [PubMed]

14. Nakamura, M.; Ogo, A.; Yamura, M.; Yamaguchi, Y.; Nakashima, H. Metformin suppresses sonic Hedgehog expression in pancreatic cancer cells. *Anticancer Res.* **2014**, *34*, 1765–1769. [PubMed]

15. Karhadkar, S.S.; Bova, G.S.; Abdallah, N.; Dhara, S.; Gardner, D.; Maitra, A.; Isaacs, J.T.; Berman, D.M.; Beachy, P.A. Hedgehog signalling in prostate regeneration, neoplasia and metastasis. *Nature* **2004**, *431*, 707–712. [CrossRef] [PubMed]

16. Gonnissen, A.; Isebaert, S.; Haustermans, K. Hedgehog signaling in prostate cancer and its therapeutic implication. *Int. J. Mol. Sci.* **2013**, *14*, 13979–14007. [CrossRef] [PubMed]

17. Teglund, S.; Toftgard, R. Hedgehog beyond medulloblastoma and basal cell carcinoma. *Biochim. Biophys. Acta* **2010**, *1805*, 181–208. [CrossRef] [PubMed]

18. Onishi, H.; Katano, M. Hedgehog signaling pathway as a therapeutic target in various types of cancer. *Cancer Sci.* **2011**, *102*, 1756–1760. [CrossRef] [PubMed]

19. Gonnissen, A.; Isebaert, S.; McKee, C.M.; Dok, R.; Haustermans, K.; Muschel, R.J. The hedgehog inhibitor GANT61 sensitizes prostate cancer cells to ionizing radiation both in vitro and in vivo. *Oncotarget* **2016**, *7*, 84286–84298. [CrossRef] [PubMed]

20. Kaushik, D.; Karnes, R.J.; Eisenberg, M.S.; Rangel, L.J.; Carlson, R.E.; Bergstralh, E.J. Effect of metformin on prostate cancer outcomes after radical prostatectomy. *Urol. Oncol.* **2014**, *32*, 43–47. [CrossRef] [PubMed]

21. Patel, T.; Hruby, G.; Badani, K.; Abate-Shen, C.; McKiernan, J.M. Clinical outcomes after radical prostatectomy in diabetic patients treated with metformin. *Urology* **2010**, *76*, 1240–1244. [CrossRef] [PubMed]

22. Rieken, M.; Kluth, L.A.; Xylinas, E.; Fajkovic, H.; Becker, A.; Karakiewicz, P.I.; Herman, M.; Lotan, Y.; Seitz, C.; Schramek, P.; et al. Association of diabetes mellitus and metformin use with biochemical recurrence in patients treated with radical prostatectomy for prostate cancer. *World J. Urol.* **2014**, *32*, 999–1005. [CrossRef] [PubMed]

23. Allott, E.H.; Abern, M.R.; Gerber, L.; Keto, C.J.; Aronson, W.J.; Terris, M.K.; Kane, C.J.; Amling, C.L.; Cooperberg, M.R.; Moorman, P.G.; et al. Metformin does not affect risk of biochemical recurrence following radical prostatectomy: Results from the SEARCH database. *Prostate Cancer Prostatic Dis.* **2013**, *16*, 391–397. [CrossRef] [PubMed]

24. He, X.X.; Tu, S.M.; Lee, M.H.; Yeung, S.C. Thiazolidinediones and metformin associated with improved survival of diabetic prostate cancer patients. *Ann. Oncol.* **2011**, *22*, 2640–2645. [CrossRef] [PubMed]

25. Spratt, D.E.; Zhang, C.; Zumsteg, Z.S.; Pei, X.; Zhang, Z.; Zelefsky, M.J. Metformin and prostate cancer: Reduced development of castration-resistant disease and prostate cancer mortality. *Eur. Urol.* **2013**, *63*, 709–716. [CrossRef] [PubMed]

26. Liu, J.; Li, M.; Song, B.; Jia, C.; Zhang, L.; Bai, X.; Hu, W. Metformin inhibits renal cell carcinoma in vitro and in vivo xenograft. *Urol. Oncol.* **2013**, *31*, 264–270. [CrossRef] [PubMed]

27. Zhang, T.; Guo, P.; Zhang, Y.; Xiong, H.; Yu, X.; Xu, S.; Wang, X.; He, D.; Jin, X. The antidiabetic drug metformin inhibits the proliferation of bladder cancer cells in vitro and in vivo. *Int. J. Mol. Sci.* **2013**, *14*, 24603–24618. [CrossRef] [PubMed]

28. Koritzinsky, M. Metformin: A Novel Biological Modifier of Tumor Response to Radiation Therapy. *Int. J. Radiat. Oncol. Biol. Phys.* **2015**, *93*, 454–464. [CrossRef] [PubMed]

29. Li, Y.H.; Luo, J.; Mosley, Y.Y.; Hedrick, V.E.; Paul, L.N.; Chang, J.; Zhang, G.; Wang, Y.K.; Banko, M.R.; Brunet, A.; et al. AMP-Activated Protein Kinase Directly Phosphorylates and Destabilizes Hedgehog Pathway Transcription Factor GLI1 in Medulloblastoma. *Cell Rep.* **2015**, *12*, 599–609. [CrossRef] [PubMed]

30. Di, M.L.; Basile, A.; Coni, S.; Manni, S.; Sdruscia, G.; D'Amico, D.; Antonucci, L.; Infante, P.; De, S.E.; Cucchi, D.; et al. The energy sensor AMPK regulates Hedgehog signaling in human cells through a unique Gli1 metabolic checkpoint. *Oncotarget* **2016**, *7*, 9538–9549.

31. Sharma, V.; McNeill, J.H. To scale or not to scale: The principles of dose extrapolation. *Br. J. Pharmacol.* **2009**, *157*, 907–921. [CrossRef] [PubMed]

32. Suzman, D.L.; Antonarakis, E.S. Clinical Implications of Hedgehog Pathway Signaling in Prostate Cancer. *Cancers* **2015**, *7*, 1983–1993. [CrossRef] [PubMed]

33. Zeng, J.; Aziz, K.; Chettiar, S.T.; Aftab, B.T.; Armour, M.; Gajula, R.; Gandhi, N.; Salih, T.; Herman, J.M.; Wong, J.; et al. Hedgehog Pathway Inhibition Radiosensitizes Non-Small Cell Lung Cancers. *Int. J. Radiat. Oncol. Biol. Phys.* **2013**, *86*, 143–149. [CrossRef] [PubMed]

34. Karlou, M.; Lu, J.F.; Wu, G.; Maity, S.; Tzelepi, V.; Navone, N.M.; Hoang, A.; Logothetis, C.J.; Efstathiou, E. Hedgehog signaling inhibition by the small molecule smoothened inhibitor GDC-0449 in the bone forming prostate cancer xenograft MDA PCa 118b. *Prostate* **2012**, *72*, 1638–1647. [CrossRef] [PubMed]

35. Martin, M.J.; Hayward, R.; Viros, A.; Marais, R. Metformin accelerates the growth of BRAF V600E-driven melanoma by upregulating VEGF-A. *Cancer Discov.* **2012**, *2*, 344–355. [CrossRef] [PubMed]

36. De, S.H.; Kimpe, M.; Isebaert, S.; Nuyts, S. A systematic assessment of radiation dose enhancement by 5-Aza-2′-deoxycytidine and histone deacetylase inhibitors in head-and-neck squamous cell carcinoma. *Int. J. Radiat. Oncol. Biol. Phys.* **2009**, *73*, 904–912.

37. Isebaert, S.F.; Swinnen, J.V.; McBride, W.H.; Begg, A.C.; Haustermans, K.M. 5-aminoimidazole-4-carboxamide riboside enhances effect of ionizing radiation in PC3 prostate cancer cells. *Int. J. Radiat. Oncol. Biol. Phys.* **2011**, *81*, 1515–1523. [CrossRef] [PubMed]

International Journal of
Molecular Sciences

MDPI

Article

RNase L Suppresses Androgen Receptor Signaling, Cell Migration and Matrix Metalloproteinase Activity in Prostate Cancer Cells

Shubham Dayal, Jun Zhou, Praveen Manivannan, Mohammad Adnan Siddiqui, Omaima Farid Ahmad, Matthew Clark, Sahezeel Awadia, Rafael Garcia-Mata, Lirim Shemshedini and Krishnamurthy Malathi *

Department of Biological Sciences, 2801 W. Bancroft St., University of Toledo, Toledo, OH 43606, USA;
Shubham.Dayal@rockets.utoledo.edu (S.D.); Jun.zhou@rockets.utoledo.edu (J.Z.);
Praveen.Manivannan@rockets.utoledo.edu (P.M.); Mohammad.Siddiqui@rockets.utoledo.edu (M.A.S.);
Omaima.Ahmad@rockets.utoledo.edu (O.F.A.); clark.matt11@gmail.com (M.C.);
Sahezeel.Awadia@rockets.utoledo.edu (S.A.); Rafael.GarciaMata@utoledo.edu (R.G.-M.);
Lirim.Shemshedini@utoledo.edu (L.S.)
* Correspondence: Malathi.Krishnamurthy@utoledo.edu; Tel.: +1-419-530-2135; Fax: +1-419-530-7737

Academic Editor: Carsten Stephan
Received: 22 January 2017; Accepted: 20 February 2017; Published: 1 March 2017

Abstract: The interferon antiviral pathways and prostate cancer genetics converge on a regulated endoribonuclease, RNase L. Positional cloning and linkage studies mapped Hereditary Prostate Cancer 1 (*HPC1*) to *RNASEL*. To date, there is no correlation of viral infections with prostate cancer, suggesting that RNase L may play additional roles in tumor suppression. Here, we demonstrate a role of RNase L as a suppressor of androgen receptor (AR) signaling, cell migration and matrix metalloproteinase activity. Using RNase L mutants, we show that its nucleolytic activity is dispensable for both AR signaling and migration. The most prevalent HPC1-associated mutations in RNase L, R462Q and E265X, enhance AR signaling and cell migration. RNase L negatively regulates cell migration and attachment on various extracellular matrices. We demonstrate that RNase L knockdown cells promote increased cell surface expression of integrin β1 which activates Focal Adhesion Kinase-Sarcoma (FAK-Src) pathway and Ras-related C3 botulinum toxin substrate 1-guanosine triphosphatase (Rac1-GTPase) activity to increase cell migration. Activity of matrix metalloproteinase (MMP)-2 and -9 is significantly increased in cells where RNase L levels are ablated. We show that mutations in RNase L found in HPC patients may promote prostate cancer by increasing expression of AR-responsive genes and cell motility and identify novel roles of RNase L as a prostate cancer susceptibility gene.

Keywords: RNase L; androgen receptor; filamin A; prostate cancer

1. Introduction

RNase L is an interferon-regulated endoribonuclease that provides cellular defense against virus infections by targeting diverse RNA substrates [1,2]. Other non-canonical roles of RNase L in regulating barrier function, cellular differentiation, senescence, development of diabetes, lipid storage and demyelination of axons indicate broader roles than the established antiviral functions [3–7]. Genetic association of Hereditary Prostate Cancer 1 (*HPC1*) to *RNASEL* expands the role of RNase L to include tumor suppression [8]. However, beyond the established antiviral effect via nucleolytic function, little is known about the antitumor function of RNase L. Four germline mutations in *HPC1/RNASEL* have been identified in hereditary prostate cancer cases: M1I (start codon substitution), E265X (stop codon at 265), 471ΔAAAG (deletion causing a frameshift and stop codon) and R462Q (missense mutation at 462) [8–11]. The variant RNase L R462Q, which is defective in inducing apoptosis and has a three-fold decrease in enzymatic activity, was reported in 43% of early onset cases of hereditary prostate cancer [12]. However, in some studies, no clear correlation of prostate cancer with RNase L R462Q mutation has been observed indicating heterogeneous disease with more complex etiology involving multiple genes and factors [13–15]. Previous studies show that prostate cancer cells depleted of RNase L were resistant to apoptosis by the combined treatment of anti-cancer drugs, TNF-related apoptosis-inducing ligand (TRAIL) and Camptothecin, suggesting that mutations in RNase L may render tumor cells refractory to cell death by conventional therapies [16].

RNase L is expressed in all cell types as a latent enzyme. It is activated by a unique and specific oligonucleotide ligand, 2–5A, that is produced from cellular adenosine 5'-triphosphate (ATP) by oligoadenylate synthetase (OAS) and double-strand RNA (dsRNA) during interferon exposure or viral infections [2,17]. In the absence of 2–5A, RNase L exists as an inactive monomer. Binding to the activator, 2–5A, induces conformational change and dimerization to produce an active endoribonuclease which cleaves diverse RNA substrates. The cleaved RNA products amplify interferon production [18], activate inflammasome [19] and promote a switch from autophagy to apoptosis [20]. Recent reports show that RNase L negatively regulates cell migration and downregulates messenger RNAs (mRNAs) for cell adhesion [21,22]. While these established roles of RNase L may contribute to tumor development, they do not provide understanding of how mutations in RNase L predispose to prostate cancer.

RNase L interacts with several cellular proteins like Filamin A, IQ (isoleucineglutamine) motif containing GTPase activating protein 1 (IQGAP1), ligand of numb protein X (LNX), androgen receptor (AR), extracellular matrix (ECM) and cytoskeletal proteins that may provide alternative mechanisms by which it mediates biological functions [3,23–26]. Recently, we have shown a nuclease-independent role of RNase L in regulating actin dynamics by interacting with an actin-binding protein, Filamin A, to regulate virus entry [3]. RNase L was also reported to interact with AR in breast cancer cells [25]. Filamin A interacts with AR, and a cleaved fragment of Filamin A colocalizes with AR in the nucleus to repress AR-responsive gene expression suggesting important roles for these interactions in regulating androgen signaling [27–29]. Several studies demonstrate the importance of microtubules and actin cytoskeleton in shuttling of AR from cytoplasm to the nucleus in cell lines and in clinical samples of prostate cancers [30–32]. Considering the requirement of AR to promote prostate cancer and the association of RNase L with genetic predisposition to HPC, we explored the mechanisms that underlie tumor suppression. In this study, we demonstrate the role of RNase L, which did not rely on enzyme activity, as a suppressor of AR signaling, cell migration and matrix metalloproteinase activity. The most prevalent HPC1-associated mutations in RNase L, R462Q and E265X, enhanced AR signaling and cell migration and our studies identify a novel role of RNase L as a prostate cancer susceptibility gene.

2. Results

2.1. RNase L Negatively Regulates Androgen Signaling

Mutations in RNase L correlate with HPC and RNase L interacts with AR and Filamin A (FLNA) [3,25]. To determine the role of RNase L in HPC, we first examined the effect of androgen, R1881, on the interaction of RNase L with AR and FLNA. Androgen-responsive LNCaP cells were transfected with Flag-RNase L and treated with R1881 (1 nM), and the interaction with AR and FLNA was analyzed by coimmunoprecipitation. In untreated cells, Flag-RNase L interacts with AR and FLNA (Figure 1A). Following treatment with R1881 for 1 h, AR dissociates from Flag-RNase L and there was reduced FLNA associated with Flag-RNase L which decreased further at 24 h. In the absence of ligand, AR remains in the cytoplasm and translocates to the nucleus on binding to androgens to regulate transcription of androgen-responsive genes [33,34]. To determine the impact of RNase L on AR subcellular localization, RNase L was depleted in LNCaP cells using short hairpin RNA (shRNA) and stimulated with R1881 (1 nM) for 24 h and analyzed by confocal microscopy. Increased nuclear AR staining was observed only after R1881 treatment (Figure 1B, top) as quantified by measuring fluorescence intensity from three or more fields from three independent experiments (Figure 1B, bottom). Since RNase L interacts with FLNA in addition to AR, we knocked-down expression of FLNA or both RNase L and FLNA in LNCaP cells (Figure 1E) and stimulated with R1881 for 24 h. Cells lacking FLNA expression showed increased nuclear AR staining which was further increased when both RNase L and FLNA were depleted (Figure 1B). To test if the effect of RNase L on AR nuclear accumulation impacts AR-responsive gene expression, mRNA levels of AR target genes *PSA*, *ETV1* and *sGCα1* were determined in response to R1881 (Figure 1C). The effect of RNase L on AR-transcriptional activity was determined in cells expressing prostate specific antigen (PSA)-luciferase reporter construct, which has copies of AR-response elements fused to luciferase reporter, after stimulating with R1881 (Figure 1D). Increased expression of AR-responsive genes and AR-transcriptional activity correlated with increase in AR nuclear localization in cells lacking RNase L or FLNA and the effect was potentiated in cells lacking both RNase L and FLNA. These results suggest that RNase L negatively regulates androgen signaling and these effects may be mediated, in part, by interaction with FLNA.

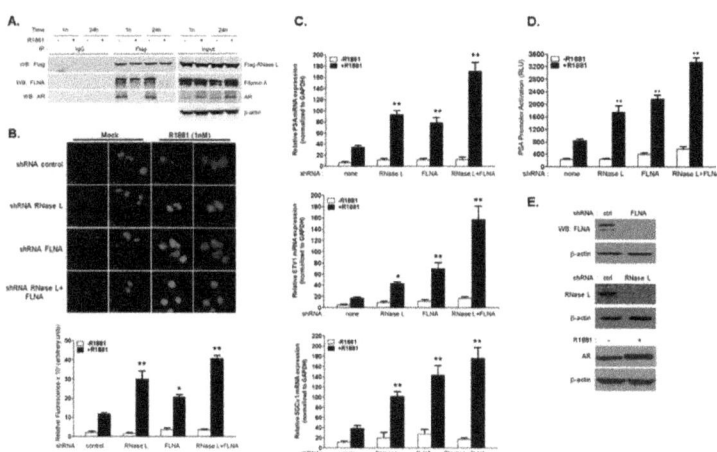

Figure 1. RNase L negatively regulates androgen signaling. (**A**) RNase L interacts with androgen receptor (AR) and Filamin A (FLNA) and dissociates on androgen treatment. LNCaP cells expressing Flag-RNase L were treated with vehicle or R1881 (1 nM) for 1 h or 24 h and immunoprecipitated with Flag-M2 agarose beads or isotype control immunoglobulin G (IgG) beads. The samples were separated on sodium dodecyl sulfate-polyacrylamide gel electrophoresis (SDS-PAGE), and the presence of AR or FLNA was determined using specific antibodies by Western blot analysis (WB). (**B**) Localization of endogenous AR in LNCaP cells expressing short hairpin RNA (shRNA) to knockdown RNase L, Filamin A or both compared to control shRNA. Cells grown in 2% charcoal stripped serum media were treated with vehicle (mock) or R1881 (1 nM) for 24 h, stained with AR antibody followed by Alexa 647-conjugated secondary antibodies and analyzed under confocal microscope. Images are representative of experiments performed in triplicate. AR localization in the nucleus (stained with diamidino-2-phenylindole (DAPI) in blue) was quantitated by measuring fluorescence intensity using Image J software. More than 10 cells (from at least three fields) were analyzed from three independent experiments. RNase L inhibits AR-mediated gene expression. LNCaP cells expressing shRNAs (as in B) were treated or mock-treated and (**C**) quantitative reverse transcriptase-polymerase chain reaction (RT-PCR) of messenger RNA (mRNA) levels of *PSA*, *ETV1* and *sGCα1* was determined and normalized to glyceraldehyde-3-phosphate dehydrogenase (*GAPDH*) mRNA levels. (**D**) Prostate specific antigen (PSA)-luciferase promoter activity was determined 18 h after R1881 treatment and normalized to β-galactosidase levels. Data shown are mean values ± standard deviation (SD) of experiments performed in triplicate from three independent experiments. Student's *t*-test was used to determine *p*-values. * $p < 0.01$, ** $p < 0.001$ and compared to cells expressing control shRNA and treated with R1881. (**E**) Cell lysates were analyzed on immunoblots for knockdown of RNase L or Filamin A and increase in AR on R1881 treatment normalized to β-actin levels. Scale bar 10 μm; Magnification ×63.

RNase L has an N-terminal ankyrin repeat domain, a pseudokinase domain in the middle and a C-terminal ribonuclease domain. RNase L binds to its activator, 2–5A, via the ankyrin repeat and pseudokinase domains allowing dimerization which is required for nuclease activity. To identify the region of RNase L that is required for suppression of androgen signaling, we over-expressed full-length Flag-RNase L (FL 1–741), N-terminal 1–335 amino acid residues (lacking nuclease domain (ΔC (1–335)) or C-terminal 386–741 amino acid residues (lacking ankyrin repeats (ΔN (386–741)) or vector alone in LNCaP cells (Figure 2B) and stimulated with R1881 for 24 h. Compared to cells expressing endogenous levels of RNase L (Figure 2, labeled as none), over-expression of RNase L suppressed AR nuclear localization two-fold (Figure 2A) which correlated with decrease in expression of AR-responsive genes (Figure 2C) and AR-transcriptional activity (Figure 2D). Interestingly, the N-terminal domain of RNase L which lacks nuclease activity suppressed androgen signaling, whereas expression of C-terminal domain alone increased AR nuclear localization, mRNA levels of AR target genes as well as AR-transcriptional activity (Figure 2A–D). Taken together, our results show that RNase L suppresses androgen signaling in LNCaP cells and the N-terminal ankyrin repeat domain is required for this effect.

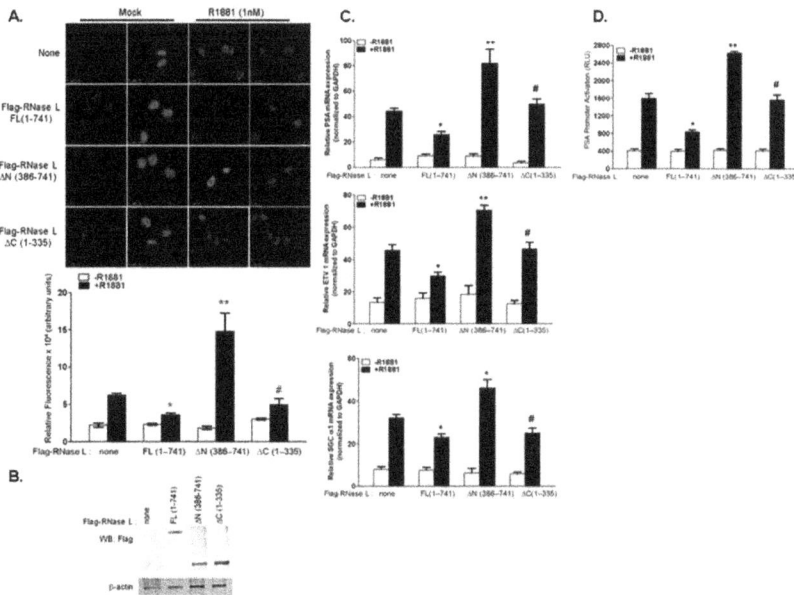

Figure 2. Over-expression of RNase L suppresses AR signaling. (**A**) LNCaP cells expressing full-length Flag-RNase L (FL 1–741), Flag-RNase L lacking N-terminal 385 amino acid residues (ΔN 386–741), Flag-RNase L lacking C-terminal 336–741 amino acid residues (ΔC 1–335) or vector alone (none) were grown in 2% charcoal stripped serum containing media and treated with R1881 or vehicle for 24 h, analyzed under confocal microscope for AR nuclear localization and quantitated as in Figure 1B. (**B**) Cell lysates were analyzed on immunoblots for expression of Flag-RNase L (full-length and truncated proteins) and normalized to β-actin levels. LNCaP cells overexpressing full-length or truncated RNase L mutants as above were treated with R1881 or mock-treated and (**C**) quantitative RT-PCR of mRNA levels of *PSA*, *ETV1* and *sGCα1* was determined and normalized to *GAPDH* mRNA levels. (**D**) PSA-luciferase promoter activity was determined 18 h after R1881 treatment and normalized to β-galactosidase levels. Data shown are mean values ± SD of experiments performed in triplicate from three independent experiments. Student's *t*-test was used to determine *p*-values. * $p < 0.01$, ** $p < 0.001$, # not significant, and compared to cells expressing vector alone and treated with R1881. Scale bar 10 μm; Magnification ×63.

2.2. The Effect of RNase L on AR Signaling Is Not Due to Altered AR Stability

Our results show that knockdown of endogenous RNase L increased AR signaling and over-expression of RNase L in LNCaP cells resulted in marked reduction of nuclear AR levels. We analyzed the subcellular distribution of AR in cells with endogenous levels of RNase L and FLNA, knockdown of either RNase L or FLNA and knockdown of both RNase L and FLNA by cellular fractionation following R1881 treatment. We validated the knockdown of RNase L and or FLNA and induction of AR expression in response to R1881 in immunoblots (Figure 3A). Cytosolic and nuclear extracts of cells with and without R1881 treatment were analyzed for AR protein levels. Consistent with our imaging experiments (Figure 1B), treatment of cells with R1881 resulted in translocation of AR from the cytosolic to the nuclear fraction. Quantitation of immunoblots for AR protein shows that depleting RNase L or FLNA increased AR nuclear-to-cytoplasmic ratio (N/C ratio), and depleting both resulted in further increase in nuclear AR levels (Figure 3B).

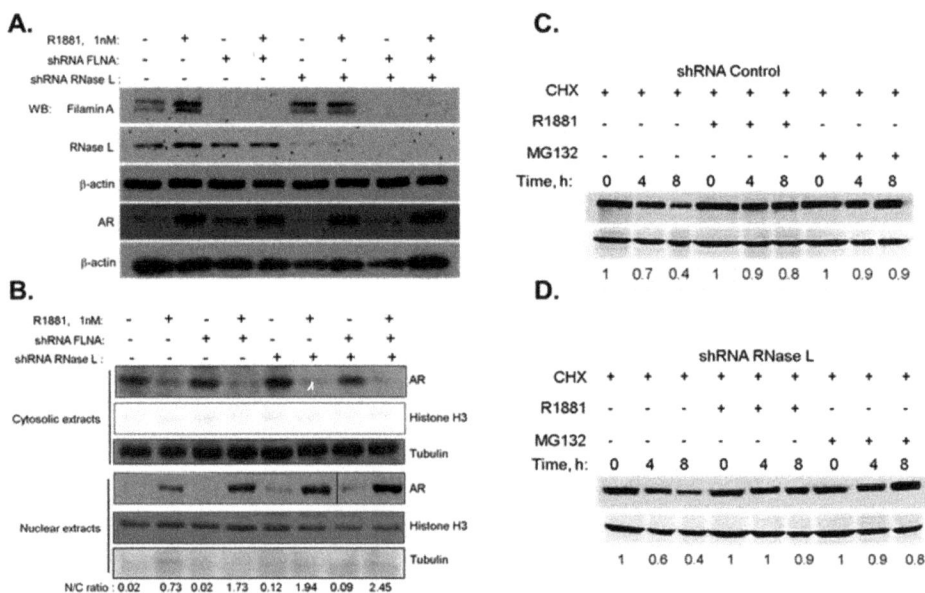

Figure 3. Increased AR localization in the nucleus in RNase L-depleted cells is not due to altered AR stability. LNCaP cells expressing shRNA to knockdown RNase L, FLNA or both and control shRNA were grown in 2% charcoal-stripped serum containing media and treated with vehicle or R1881 and (**A**) Cell lysates were analyzed on immunoblots for knockdown and AR expression levels. (**B**) Cells were fractionated and AR from nuclear (N) and cytosolic (C) extracts were analyzed by immunoblotting with anti-AR, histone H3 (marker for nuclear extract), α-tubulin (marker for cytosolic extract) antibodies. Nuclear-to-cytoplasmic (N/C) ratio of AR protein with or without R1881 was determined by densitometric analysis of band intensities using Image J software. LNCaP cells with control shRNA (**C**) or RNase L shRNA (**D**) in growth medium were treated with cycloheximide (CHX, 50 μg/mL) alone or combined with MG132 (20 μM) or R1881 (1 nM) for 4 h or 8 h. Cell lysates were prepared in 2% SDS and levels of AR were normalized on immunoblots with β-actin and relative changes in AR levels compared to CHX treatment for 0 h (band intensity set as 1) was determined. Similar results were observed in three independent experiments.

Due to the inhibitory effect of RNase L on AR nuclear accumulation, we tested the possibility that RNase L may affect AR protein stability. Control and RNase L shRNA expressing LNCaP cells were treated with cycloheximide (CHX) for the indicated times to block *de novo* protein translation and combined with either proteasome inhibitor, MG132, to block AR degradation or R1881 to induce AR protein levels. As demonstrated by other studies, AR protein levels declined in cells treated with CHX alone [35] and the reduced levels were comparable in cells depleted or expressing RNase L (Figure 3C,D). No differences in MG132 stabilized AR was observed in both cells suggesting that RNase L does not alter AR stability.

2.3. Hereditary Prostate Cancer 1 (HPC1)-Associated Mutants of RNase L Enhance AR Transcriptional Activity

To understand how mutations in RNase L contribute to HPC, we examined the effect of HPC1-associated RNase L mutants on androgen signaling. To further address if RNase L enzyme activity is required for androgen signaling, other mutations in RNase L were generated by site-directed mutagenesis of Flag-RNase L construct including R667A (nuclease-dead), K166E (reduced enzyme activity), K240/274N (defective 2–5A binding and lacks enzyme activity), Y312A (defective 2–5A binding and lacks enzyme activity). Expression of Flag-RNase L Wild type (WT) and mutant proteins is shown in immunoblots (Figure 4E). RNase L activity of the WT and mutant constructs was determined by rRNA cleavage in response to PolyI:C transfection in HeLa cells which lack detectable RNase L activity (Figure 4F) [36,37]. As endogenous RNase L may mask some of the effects of mutants we express, we used a knockdown/rescue approach in which endogenous RNase L levels were depleted in LNCaP cells with shRNA targeting the 3′-UTR (Figure 1E) and reconstituted with Flag-RNase L (WT) or RNase L mutants and stimulated with R1881 for 24 h. We then monitored the subcellular localization of AR by confocal microscopy (Figure 4A) and quantitated the amount of nuclear AR (Figure 4B). Expression of WT RNase L suppressed nuclear translocation of AR as we observed in Figure 2. However, cells expressing HPC1-associated mutants, R462Q and E265X showed increased AR in the nucleus. Since both mutants compromised RNase L enzyme activity, we tested other RNase L mutants which had reduced activity (K166E), lacked enzyme activity due to defect in binding the activator 2–5A (K240/274N and Y312A) or in the nuclease domain (R667A) to determine if enzyme activity of RNase L was required for androgen signaling. In contrast with other mutants, RNase L Y312A, which is defective in 2–5A binding and lacks enzyme activity, was able to suppress nuclear AR localization like WT RNase L (Figure 4A, arrows). We did not observe any change in AR subcellular localization in the absence of R1881. The increase in nuclear AR correlated with mRNA levels of AR target genes and AR-transcriptional activity (Figure 4C,D). Together, these results suggest that HPC1-associated RNase L mutants may contribute to HPC by regulating androgen signaling. Further, the effect of RNase L on androgen signaling appears to be independent of nucleolytic functions (Figure 4F).

2.4. Cell Migration Is Increased in Cells with Reduced Levels of RNase L

Our published observation that RNase L interacts with the actin-binding protein, FLNA, to regulate actin dynamics [3] suggests that RNase L may have additional roles in HPC besides regulating androgen signaling. The ability of RNase L to modulate actin cytoskeleton prompted us to explore whether RNase L affected cell migration in prostate cancer cells. To investigate the role of RNase L in prostate cancer cell motility, RNase L levels were knocked-down using shRNA in DU145, PC3 and LNCaP cells and compared to cells expressing endogenous RNase L levels (expressing control non-targeting shRNA) (Figure 5B, inset). Confluent monolayer of cells in serum-free media were scratched and replaced with growth media. Cell migration to close the wound was imaged (Figure 5A) and quantitated (Figure 5B) over time as indicated. In all three prostate cancer cells, depletion of RNase L had marked effect (1.5–2-fold) in enhancing cell migration and the effect was most significant at 24 h experimental endpoint. No significant difference in cell proliferation was observed between control and knockdown cells over the 24 h time course of the experiment (data not shown). The difference in cell migration was quantitated by transwell cell migration assays in response to serum (Figure 5C). Consistent with scratch wound healing assays, RNase L knockdown cells migrated significantly more (1.5–2-fold) through fibronectin-coated filters in response to serum.

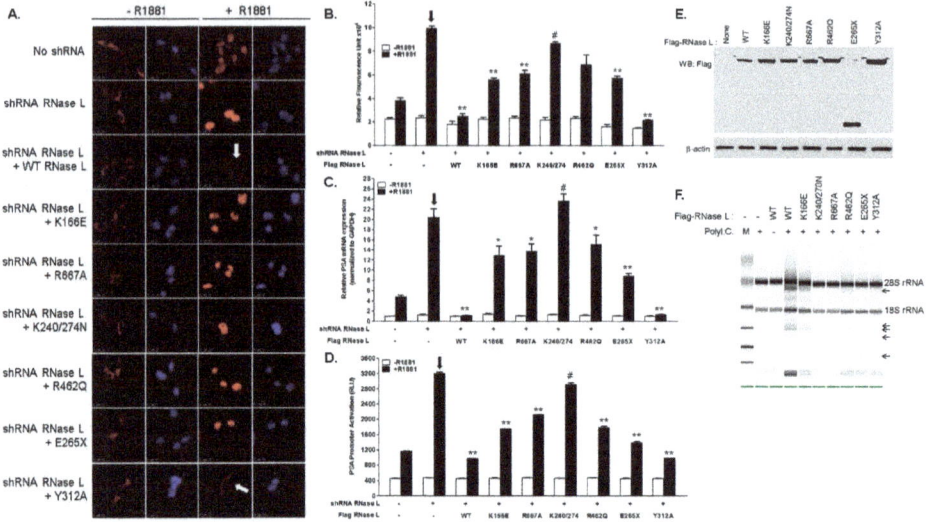

Figure 4. Hereditary Prostate Cancer 1-associated mutants of RNase L enhance AR transcriptional activity. (**A**) Endogenous RNase L-knockdown LNCaP cells were reconstituted with Flag-Wild-type (WT) or mutated RNase L as indicated and treated with vehicle (−R1881) or R1881 (1 nM) for 24 h. AR nuclear localization was analyzed under confocal microscope; and (**B**) quantitated by measuring fluorescence intensity using Image J software as described in Figure 1B. White arrows point to lack of nuclear localization in WT and RNase L Y312A mutant expressing cells. AR transcriptional activity was monitored in LNCaP cells reconstituted with WT and RNase L mutants treated with vehicle or R1881 by (**C**) quantitative RT-PCR of mRNA levels of *PSA*, *ETV1* and *sGCα1* normalized to *GAPDH* mRNA levels. (**D**) PSA-luciferase promoter activity 18 h after R1881 treatment and normalized to β-galactosidase levels. Data shown are mean values ± SD of experiments performed in triplicate from three independent experiments. Student's *t*-test was used to determine *p*-values. * $p < 0.01$, ** $p < 0.001$, # not significant, and compared to RNase L knockdown cells (shown with arrow on graphs) treated with R1881. (**E**) Cell lysates were analyzed for expression of Flag-RNase L (WT) and mutants as indicated using anti-Flag antibodies and normalized to β-actin levels. (**F**) HeLa cells were transfected with Flag-RNase L (WT) and RNase L mutants and enzyme activity was determined by monitoring rRNA cleavage (shown by arrows) as determined in RNA chips by transfecting cells with PolyI:C (2 µg/mL) and isolation of total RNA. Scale bar 10 µm, Magnification ×63.

To further demonstrate that the increased cell migration was due to lack of RNase L, we used WT and *Rnase l*$^{-/-}$ (RNase L KO) primary mouse embryonic fibroblasts (MEFs) and monitored wound closure by wound healing assays and cell migration in transwell assays. Migration of RNase L KO MEFs was enhanced 2–3-fold in response to serum or fibronectin compared to WT MEFs (Figure 5D–F). In addition, we reconstituted RNase L expression in DU145 or PC3 RNase L knockdown cells and compared wound closure in cells with endogenous levels of RNase L, depleted of RNase L or over-expressing RNase L (Figure 6A,C). Expression of RNase L is shown in immunoblots as inset in Figure 6B,D. At 24 h experimental endpoint wound closure was 47% in DU145 cells with endogenous RNase L (WT), 76% in RNase L-depleted cells and 29% in over-expressing cells. In PC3 cells, endogenous RNase L expressing cells (WT) showed 45%, RNase L-depleted cells showed 68%

and over-expressing cells showed 27% wound closure. To determine the consequence of RNase L activation on cell migration, DU145 cells were transfected with activator, 2–5A which causes RNase L dimerization, or mock transfected and wound closure was imaged and quantitated (Figure 6E,F). 2–5A treatment of cells decreased migration of DU145 cells by 48% compared to mock transfected cells. Taken together, these results demonstrate that RNase L inhibits cell migration in prostate cancer and primary cells. Furthermore, activation of RNase L by 2–5A, which results in conformational change and dimerization of RNase L, also suppresses cell migration.

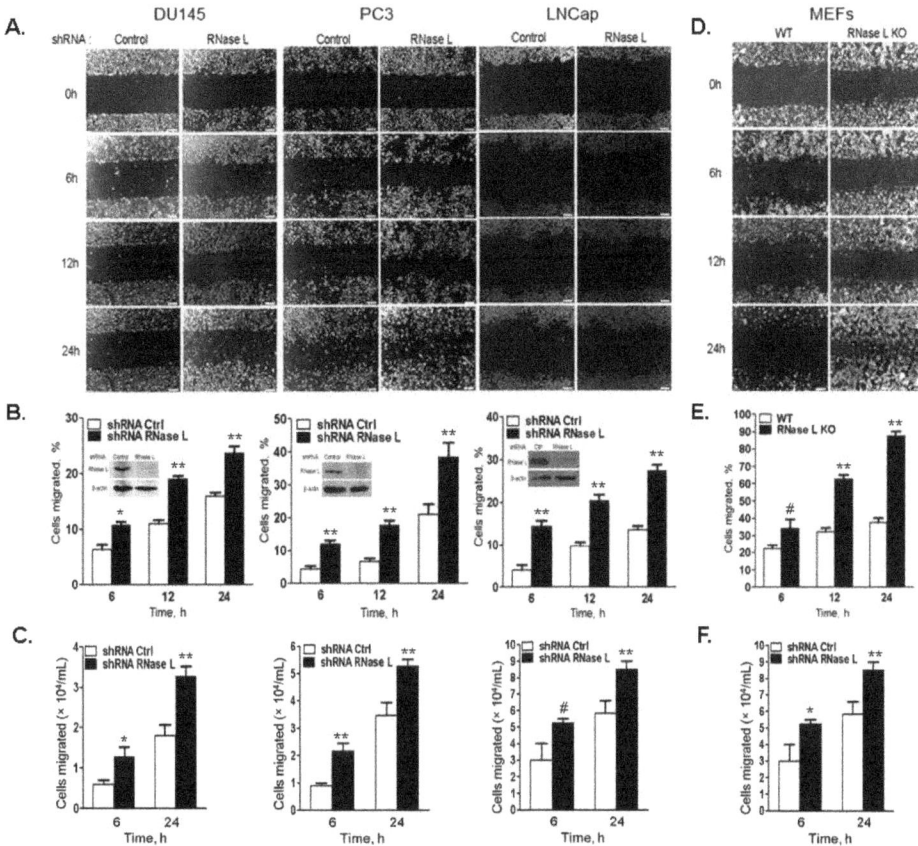

Figure 5. Cell migration is increased in cells with reduced RNase L levels. DU145, PC3 or LNCaP cells expressing control or RNase L shRNA (**A–C**) and WT or RNase L KO mouse embryonic fibroblasts (MEFs) (**D–F**) were grown to confluence. (**A,D**) The cell monolayer was scratched and wound closure was imaged under phase-contrast microscope at indicated times. (**B,E**) Cells migrated (%) to close the wound was quantitated by Image J software. Knock-down of RNase L on immunoblots is shown as inset in (**B**). (**C,F**) Transwell chambers coated with fibronectin were used to measure cell migration at 6 h and 24 h in response to growth medium with 10% serum by counting cells that migrated to the lower surface of the filters from experiments performed in triplicate. Data shown are mean values ± SD of experiments performed in triplicate from three independent experiments. Student's *t*-test was used to determine *p*-values. * $p < 0.01$, ** $p < 0.001$, # not significant, and compared to cells expressing control shRNA (**A–C**) and compared to WT MEFs (**D–F**). Scale bar 100 μm, Magnification ×10.

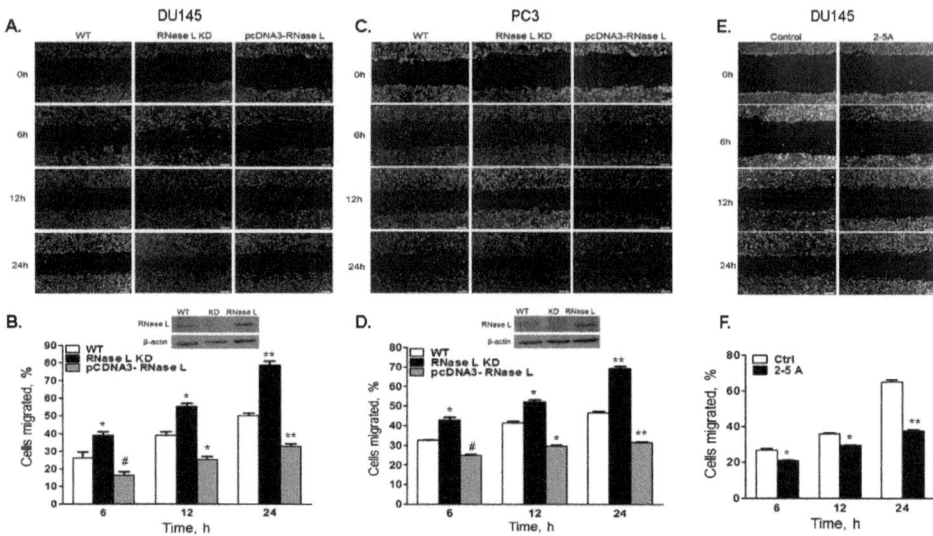

Figure 6. Over-expression or activation of RNase L inhibits cell migration. DU145 (**A,B**) or PC3 (**C,D**) cells expressing control shRNA and expressing endogenous levels of RNase L (WT), shRNA to knockdown RNase L (RNase L KD) or overexpressing RNase L after knockdown of endogenous RNase L (pcDNA3-RNase L) were grown to confluence. (**A,C**) Cell monolayers were scratched and wound closure was imaged under phase-contrast microscope at indicated times. (**B,D**) Cells migrated (%) to close the wound was quantitated by Image J software. RNase L protein levels were analyzed on immunoblots using anti-RNase L antibody and shown as inset in B and D. Data shown are mean values ± SD of experiments performed in triplicate from three independent experiments. Student's *t*-test was used to determine *p*-values. * $p < 0.01$, ** $p < 0.001$, # not significant, and compared to control cells expressing endogenous levels (WT) of RNase L. (**E**) DU145 cells were mock transfected (control) or transfected with 2–5A (10 μM) to activate RNase L. The monolayer was scratched and wound closure was imaged under phase-contrast microscope at indicated times. (**F**) Cells migrated (%) to close the wound was quantitated by Image J software. Data shown are mean values ± SD of experiments performed in triplicate from three independent experiments. Student's *t*-test was used to determine *p*-values. * $p < 0.01$, ** $p < 0.001$, and compared to mock-treated (control) cells (**E,F**). Scale bar 100 μm, Magnification ×10.

2.5. Hereditary Prostate Cancer-Associated Mutants of RNase L Promote Cell Migration

We have shown that RNase L suppresses AR signaling and cell migration. Because the most common RNase L mutations associated with HPC enhance AR signaling, we investigated whether these RNase L mutants also affect cell migration. To further explore if RNase L enzyme activity is required for cell migration, we also tested the RNase L mutants we generated that have reduced enzyme activity (K166E) or lack enzyme activity due to mutation in the nuclease domain (R667A) or defect in binding the activator, 2–5A (K240/274N, Y312A). Loss of heterozygosity of RNase L has been observed in HPC1 prostate tumors [8]. Therefore, we depleted RNase L in DU145 and PC3 cells and reconstituted with vector alone (none), Flag-RNase L (WT) or Flag-RNase L mutants as indicated (Figure 7A–F). Confluent monolayers of cells were scratched and wound closure was imaged (Figure 7A,C) and quantitated (Figure 7B,D). Differences in cell migration between cells expressing various RNase L mutants were also quantitated by transwell assays through fibronectin coated filters in response to serum (Figure 7E,F). Consistent with our observations, RNase L knockdown cells showed increased migration and expression of WT RNase L suppressed cell migration. HPC1-associated RNase L mutants, R462Q and E265X, showed increase in cell migration compared to cells expressing WT

RNase L (shown by arrows in Figure 7B,D–F). RNase L K166E, which has reduced enzyme activity, R667A, which is nuclease-dead, and K240/274N which lacks enzyme activity showed 1.5–2-fold increase in migration compared to WT expressing cells. In contrast with other RNase L mutants which lack enzyme function, the Y312A mutant which also lacks enzyme activity due to its inability to bind 2–5A, suppressed cell migration (shown by arrows in Figure 7B,D–F). This effect was comparable to WT RNase L indicating that different regions of RNase L, presumably through interacting proteins, may contribute to cell migration rather than enzyme activity.

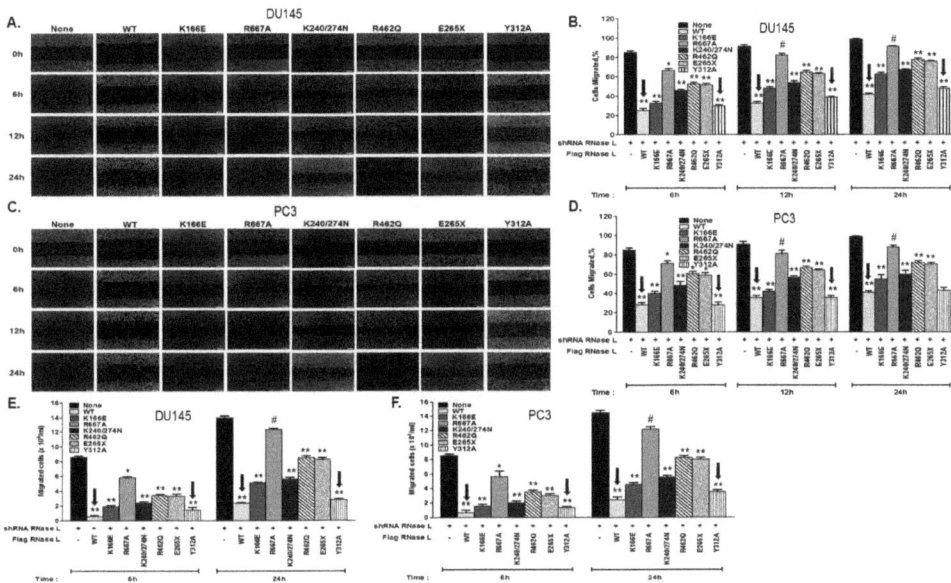

Figure 7. Hereditary prostate cancer-associated mutants of RNase L promote cell migration. Endogenous RNase L was knocked down in DU145 or PC3 cells and reconstituted with Flag vector (none), Flag-RNase L (WT) or Flag-RNase L mutants as indicated. (**A,C**) Cell monolayers were scratched and wound closure was imaged under phase-contrast microscope at indicated times. (**B,D**) Cells migrated (%) to close the wound was quantitated by Image J software. (**E,F**) Transwell chambers coated with fibronectin were used to measure cell migration at 6 h and 24 h in response to growth medium with 10% serum by counting cells that migrated to the lower surface of the filters from experiments performed in triplicate. Arrows are used to highlight RNase L mutants that suppress migration as WT RNase L. Data shown are mean values ± SD of experiments performed in triplicate from three independent experiments. Student's *t*-test was used to determine *p*-values. * $p < 0.01$, ** $p < 0.001$, # not significant, and compared to cells expressing shRNA to knockdown RNase L. Scale bar 100 μm, Magnification ×10.

2.6. RNase L-Depletion Increases Cell Attachment and Cell Spreading

Interaction of the cell with extracellular matrix (ECM) causes engagement of specific transmembrane receptors and signaling molecules which modulate cytoskeletal organization resulting in distinct cell morphology, attachment and migration [38]. To characterize the consequence of RNase L on cell attachment, RNase L was knocked-down using shRNA in DU145, PC3 or LNCaP cells and compared to control shRNA expressing cells. Cells were allowed to attach to dishes coated with fibronectin (FN), laminin (LN), collagen I (C I), collagen IV (C IV) or vitronectin (VN) for 1 h. Wells were then washed and attached cells were quantitated by staining. Knockdown of RNase L in all three prostate cancer cell lines tested showed increase in cell attachment to all the ECM substrates, and the difference was 2–2.5-fold greater in response to fibronectin (Figure 8A–C). We monitored DU145 and PC3 cells for both shape and cell spreading after incubation on FN substrates. Expression of RNase L appeared to result in the presence of more rounded cells which spread slowly. In contrast RNase L-depleted cells tend to spread rapidly and the cell area increased in a time-dependent manner (Figure 8D,E). Both the rate and extent of cell spreading were significantly lower at all the time points in RNase L expressing cells. These results suggest that RNase L inhibits cell attachment and cell spreading.

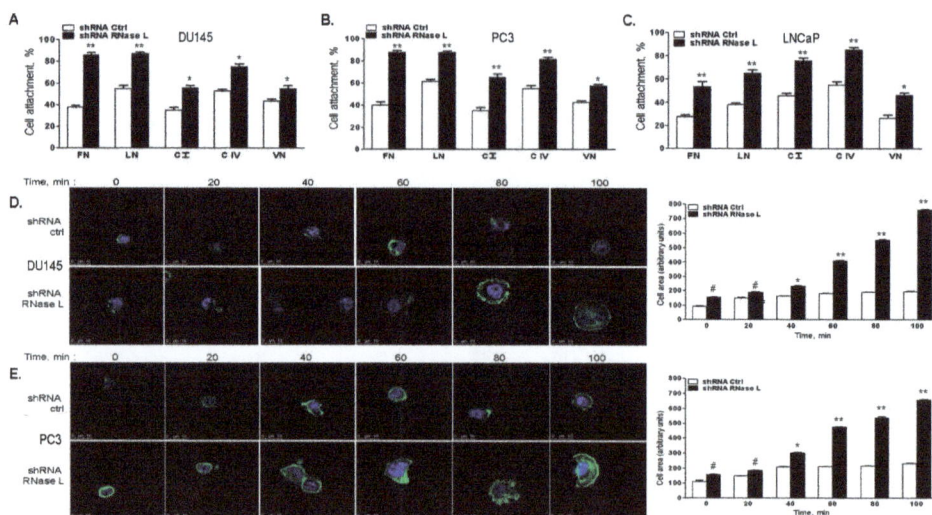

Figure 8. RNase L depletion increases cell attachment to extracellular matrix substrates and cell spreading. (**A**) DU145; (**B**) PC3; or (**C**) LNCaP cells expressing control shRNA or RNase L shRNA were allowed to attach to wells coated with 10 μg/mL each of fibronectin (FN), laminin (LN), collagen I (C I), collagen IV (C IV) or vitronectin (VN) for 1 h. Plates were washed and attached cells were stained. Cell attachment (%) was determined from means ± SD of experiments performed in triplicate from three separate experiments. (**D**) DU145; or (**E**) PC3 expressing control shRNA or RNase L shRNA were allowed to attach and spread on surfaces coated with fibronectin (10 μg/mL) at various time points, fixed and F-actin was labeled with Alexa 488-labeled phalloidin and imaged by confocal microscope. Areas of individual cells from at least 30 measurements were determined by Image J software. Data shown represent the mean of cell area ± standard error of mean (SEM) of three experiments performed in triplicate. Student's *t*-test was used to determine *p*-values. * *p* < 0.01, ** *p* < 0.001, # not significant, and compared to cells expressing control shRNA. Scale bar 10 μm, Magnification ×63.

2.7. Depletion of RNase L Promotes Integrin Activation and FAK-Src Signaling in Response to Fibronection

Integrins are transmembrane receptors that physically link ECM to intracellular actin cytoskeleton [39]. Trafficking and recycling of integrin β1 to the cell membrane and clustering is required for cell attachment, spreading and migration [40]. Clustering of integrins activates signaling events involving focal adhesion kinase (FAK) and Src [41,42]. Since RNase L regulates cell attachment and migration, we investigated if the effects were mediated by integrin-regulated Focal Adhesion Kinase-Sarcoma (FAK-Src) signaling pathways. To assess the role of RNase L in regulating integrin β1 expression and activation, prostate cancer cells depleted of RNase L or WT were allowed to spread on FN and cell suspensions were fixed and incubated with integrin β1 antibodies to label cell surface integrin β1. Flow cytometric analysis showed that cell surface integrin β1 was increased 1.5–3-fold following knockdown of RNase L (Figure 9A–C). No changes in protein levels of integrin β1 were observed between WT and RNase L knockdown cells. Consistent with roles in cell migration, attachment and spreading, RNase L is important for activation of integrin β1. Activation of RNase L with 2–5A which promotes dimerization, inhibits cell migration (Figure 6E,F) and we observed decrease in cell surface integrin β1 expression in DU145, PC3 and LNCaP cells transfected with 2–5A and this effect was not due to decrease in integrin β1 protein levels (Figure 10A–C).

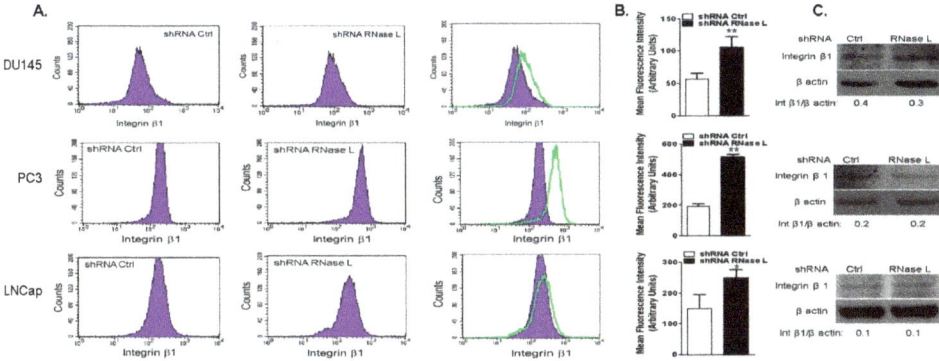

Figure 9. Cells with reduced levels of RNase L exhibit increased surface expression of integrin β1. DU145, PC3 or LNCaP cells expressing control shRNA or RNase L shRNA were plated on fibronectin-coated dishes and analyzed by flow cytometry for surface staining with antibodies against integrin β1 and Alexa-488 conjugated secondary antibodies. (**A**) Representative histograms; and (**B**) bar graphs for the mean fluorescence intensity of at least three independent experiments for integrin β1 are shown. Student's *t*-test was used to determine *p*-values. * $p < 0.01$, ** $p < 0.001$ and compared to cells expressing control shRNA. (**C**) Expression of integrin β1 in cell lysates was determined on immunoblots and normalized to levels of β-actin.

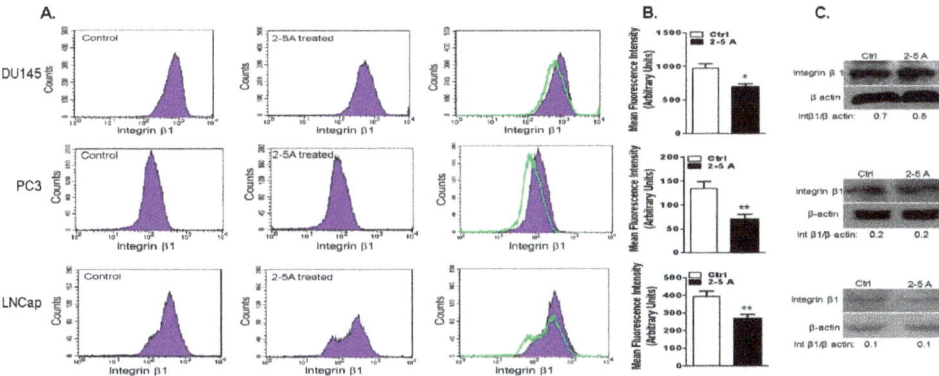

Figure 10. Activation of RNase L reduces surface expression of integrin β1. DU145, PC3 or LNCaP cells were mock-treated (control) or transfected with 2–5A (10 μM) to activate RNase L and plated on fibronectin-coated dishes and analyzed by flow cytometry for surface staining with antibodies against integrin β1 and Alexa-488 conjugated secondary antibodies. (**A**) Representative histograms; and (**B**) bar graphs for the mean fluorescence intensity of at least three independent experiments for integrin β1 are shown. Student's *t*-test was used to determine *p*-values. * $p < 0.01$, ** $p < 0.001$ and compared to mock-treated cells. (**C**) Expression of integrin β1 in cell lysates was determined on immunoblots and normalized to levels of β-actin.

FAK is recruited to sites of integrin clustering and activated by autophosphorylation at Y397 which in turn facilitates Src binding [43–45]. Phosphorylation of Src at Y416 can lead to formation of FAK-Src signaling complex which acts with downstream regulators including RhoGTPases to control cell shape and turnover of focal adhesion during cell migration [46,47]. We tested FAK activation after integrin β1 clustering in response to FN in PC3 cells with endogenous or knockdown of RNase L and in WT and RNase L KO MEFs. FAK Y397 phosphorylation was increased 43% in PC3 RNase L knockdown cells and 48% in RNase L KO MEFs compared to PC3 control or WT MEFs, respectively (Figure 11A,B). Src is activated by binding to pY397-FAK and we observed a corresponding 60% increase in pY416 Src in PC3 RNase L knockdown cells and 70% in RNase L KO MEFs compared to PC3 control or WT MEFs respectively (Figure 11A,B). In PC3 RNase L knockdown cells, we observed increase in basal levels of phospho-FAK and phospho-Src which increased further on FN stimulation.

To further demonstrate that Src activity is involved in cell migration, RNase L-depleted PC3 cells were treated with Src inhibitor, PP2, and monolayers were scratched and cell migration to close the wound was imaged and quantitated. In mock-treated cells 85% of wound closure was observed at 24 h compared to 43% in PP2-treated cells (Figure 11C,D). These results suggest that integrin clustering in response to FN in RNase L-depleted cells was significantly enhanced over control cells which in turn reflected in increased phosphorylation of FAK and Src which effects cell migration.

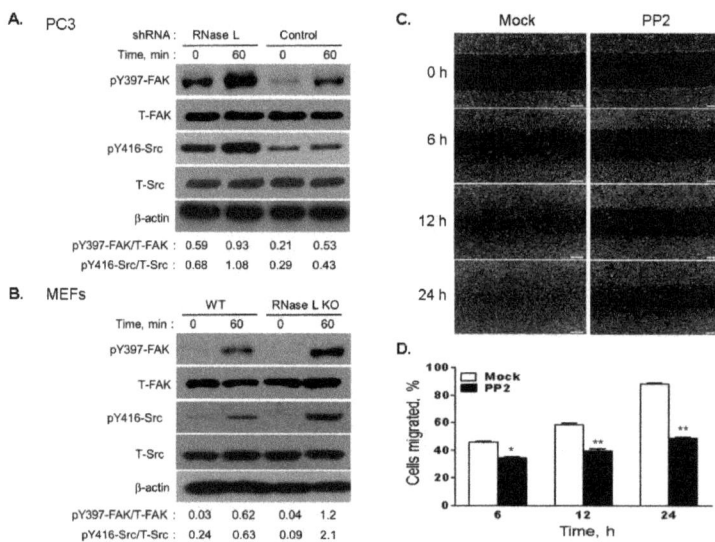

Figure 11. Increased FAK and Src phosphorylation in response to fibronectin in RNase L-depleted cells effects cell migration. (**A**) PC3 cells expressing control or RNase L shRNA; and (**B**) WT and *Rnase l$^{-/-}$* (RNase L KO) MEFs were serum starved and plated on fibronectin-coated plates for indicated times. Phospho-FAK (Y397) and phospho-Src (Y416) was detected in cell lysates using phospho-specific antibodies on immunoblots followed by Western blotting with anti-FAK, anti-Src antibodies and β-actin for loading control. (pY397)-FAK and (pY416)-Src intensity values were normalized to total-FAK and total-Src intensities respectively using Image J software. Representative immunoblots from experiments performed in triplicate are shown. (**C**) PC3 RNase L-depleted cells growing in confluent monolayers were pretreated with vehicle (mock) or Src inhibitor (PP2, 10 μM) for 1 h. Cell monolayers were scratched and wound closure was imaged under phase-contrast microscope at indicated times in growth medium containing PP2 inhibitor. (**D**) Cells migrated (%) to close the wound was quantitated by Image J software. Data shown are mean values $\pm$ SD of experiments performed in triplicate from three independent experiments. Student's *t*-test was used to determine *p*-values. * $p < 0.01$, ** $p < 0.001$, and compared to vehicle-treated (mock) cells. Scale bar 100 μm, Magnification $\times$10.

2.8. Increased Rac1 Activity Mediates Enhanced Cell Migration in RNase L-Depleted Cells

Signaling pathways triggered by integrins regulate FAK-Src activity and involve Rho GTPases which are crucial for remodeling cytoskeleton and cell mobility [48,49]. To investigate which of the Rho family, GTPases-RhoA, Cdc42 or Rac1, mediate RNase L-dependent migration, RNase L-depleted DU145 or PC3 cells were transfected with empty vector (mock), or dominant-negative forms of RhoA (T19N), Cdc42 (T17N) or Rac1 (T17N) [50]. Expression of the proteins on immunoblots is shown in Figure S1. Monolayers of cells were scratched and wound closure was imaged and quantitated at indicated times (Figure 12A–D). Expression of the dominant-negative Rac1 significantly inhibited migration of RNase L-depleted cells whereas RhoA (T19N) and Cdc42 (T17N) had marginal effects. To examine the effect of reduced RNase L levels on Rac1 activity in response to FN, we measured Ras-related C3 botulinum toxin substrate 1 (Rac1) activity by precipitating active GTP-bound Rac1 with GST-PAK-binding domain (GST-PBD) and estimating the abundance relative to total levels of Rac1 protein. In both DU145 (Figure 13A,B) and PC3 (Figure 13C,D), adhesion to FN stimulated a rapid and transient increase in Rac1 activity which peaked at 30 min and declined by 1 h. In contrast, RNase L knockdown cells had higher basal level of Rac1 which increased further on FN stimulation and was sustained at 1 h (Figure 13A–D). Our data support the observation that increased integrin and FAK-Src signaling in response to FN in RNase L knockdown cells contributes to increased Rac1 activity and cell migration.

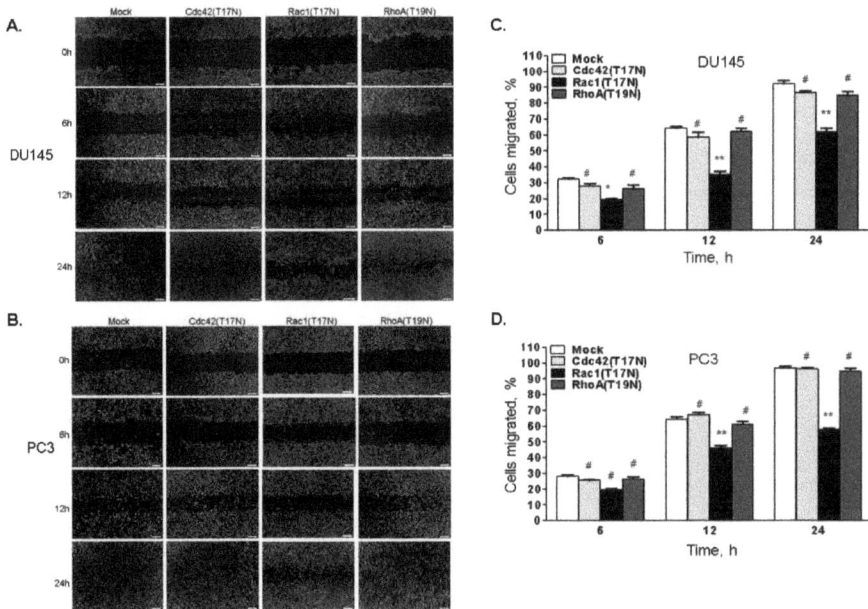

Figure 12. Ras-related C3 botulinum toxin substrate 1 (Rac1) mediates enhanced cell migration in RNase L-depleted cells. RNase L shRNA expressing (**A**) DU145 cells; or (**B**) PC3 cells were transfected with empty vector (mock), dominant negative Cdc42 (T17N), dominant negative Rac1 (T17N) or dominant negative RhoA (T19N) expressing plasmids. After 24 h, cell monolayers were scratched and wound closure was imaged under phase-contrast microscope at indicated times. (**C,D**) Cells migrated (%) to close the wound was quantitated by Image J software. Data shown are mean values ± SD of experiments performed in triplicate from three independent experiments. Student's *t*-test was used to determine *p*-values. * $p < 0.01$, ** $p < 0.001$, # not significant, and compared to RNase L-depleted cells mock transfected with empty vector. Scale bar 100 μm, Magnification ×10.

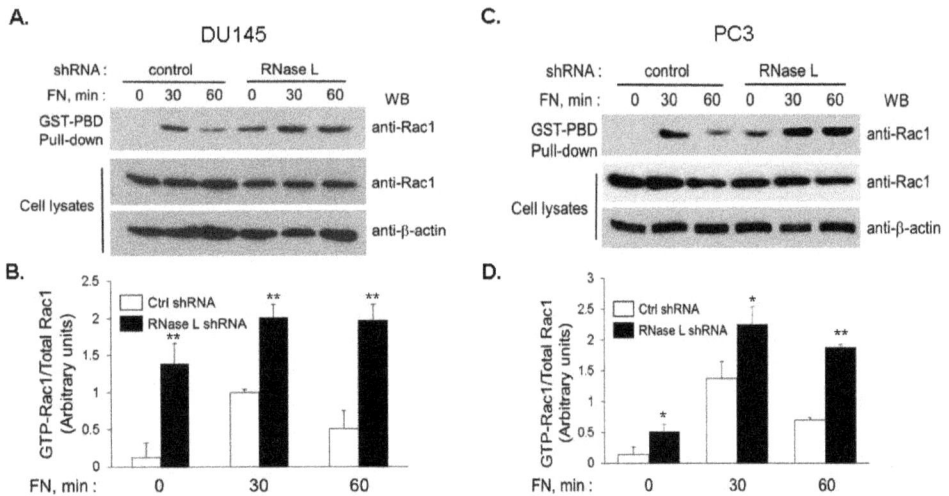

Figure 13. Rac1 activity in response to fibronectin is increased in RNase L-depleted cells. (**A**) DU145; or (**C**) PC3 cells expressing control or RNase L shRNA were plated on fibronectin-coated dishes for indicated times. Cell lysates were incubated with agarose-immobilized GST-PAK1 binding domain (GST-PBD) and co-precipitated proteins were subject to immunoblotting with anti-Rac1 antibodies to detect the amount of GTP-bound Rac1 (active Rac1) and compared to expression of Rac1 in cell lysates. Representative immunoblots from experiments performed in triplicate are shown. (**B,D**) Activity of Rac1 was quantitated by comparing the intensities of active Rac1 with those of total Rac1 in each lane using Image J software. Data are mean values ± SEM expressed as increase in Rac1 activity from three independent experiments. Student's *t*-test was used to determine *p*-values. * $p < 0.01$, ** $p < 0.001$, and compared to control shRNA expressing cells.

2.9. RNase L Regulates MMP-2 and MMP-9 Activities

Matrix metalloproteinases (MMPs) remodel the ECM and play a critical role in cell migration, invasion, tissue metastasis and impact tumor progression [51]. Elevated levels of MMP-2 and MMP-9 are observed in prostate cancer and correlate with increased metastasis [52,53]. To determine if activity of MMP-2 and -9 are regulated by RNase L, culture supernatants of control and RNase L knockdown DU145, PC3 and LNCaP cells were analyzed for activity of secreted MMP-2 and -9 by gelatin zymography. MMP-2 and -9 gelatinase activity, observed by cleared areas on gels, was higher in RNase L-depleted cells than in control cells (1.5–5-fold increase in MMP-2 activity compared to control cells and 1.5–2-fold increase in MMP-9 activity compared to control cells) (Figure 14A,B). Similar increase in MMP-2 and -9 activities was also observed in RNase L KO MEFs compared to WT MEFs (Figure S2). Activation of RNase L by transfecting 2–5A inhibited cell migration (Figure 6E,F) and we observed a decrease in MMP-2 and -9 activity in culture supernatants following 2–5A transfection compared to control cells (Figure 14A,C). Taken together, these results demonstrate that depleting RNase L leads to increased activity of MMP-2 and -9 in prostate cancer cells which correlates with increased cell migration. Silencing of RNase L or activation of RNase L did not affect expression of *MMP-9* mRNA (Figure S3).

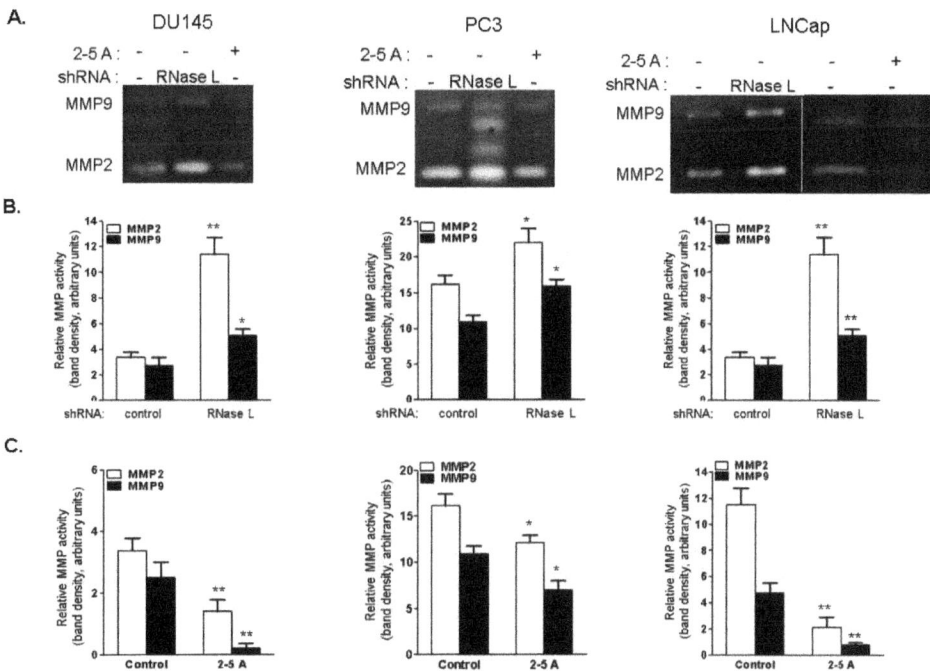

Figure 14. RNase L regulates matrix metalloproteinases (MMP) -2 and MMP-9 activities. (**A**) Gelatin zymography analysis of MMP-2 and MMP-9 activities in conditioned media harvested from DU145, PC3 or LNCaP cells expressing control or RNase L shRNA or transfected with 2–5A complexed with lipofectamine 2000 and added to cells to activate RNase L. Data shown are representative of three independent experiments. Quantitative analysis of MMP-2 and MMP-9 activities in (**B**) Cells expressing RNase L shRNA compared to control shRNA, and (**C**) Cells transfected with 2–5A to activate RNase L compared to control cells. Data shown are mean values ± SEM from three independent experiments. Student's *t*-test was used to determine *p*-values. * $p < 0.01$, ** $p < 0.001$, and compared to control shRNA expressing cells (**B**); and control cells (**C**).

3. Discussion

In the present study, we demonstrate the role of RNase L as a suppressor of AR signaling, cell migration and activity of matrix metalloproteinases identifying an unrecognized role of RNase L in prostate cancer. Importantly, the nuclease function of RNase L was not required for suppression. The relevance of the HPC1 (*RNASEL*) mutations in prostate cancer is poorly understood. We provide evidence that the most prevalent HPC1-associated mutations in RNase L, R462Q substitution and E265X truncation, enhance AR nuclear translocation, transcriptional activity and cell migration. This study not only identifies Androgen Receptor as a target of RNase L regulation, but also demonstrates that the effect of HPC1-associated mutations on AR signaling and cell migration is amenable to regulation by physiological conditions that exist in prostate cancer cells.

In androgen-treated cells, the interaction between RNase L, Filamin A and AR is disrupted and AR translocates to the nucleus to induce AR-responsive gene expression. In LNCaP cells lacking RNase L significantly higher nuclear AR was observed, similar to FLNA-depleted cells, and cells lacking both proteins showed further increase in nuclear AR which was reflected by increase in AR transcriptional activity, and expression of AR target genes (Figure 1). Increased nuclear AR was only observed in cells treated with R1881 indicating that RNase L effect on AR is ligand-dependent. As would be expected for a dynamic interaction, over-expression of full-length RNase L suppressed AR nuclear localization and AR nuclear activities. In other studies, over-expression of Filamin A inhibited AR transcriptional activity on androgen treatment indicating common underlying mechanisms [27]. Interestingly, the N-terminal fragment of RNase L which interacted with FLNA inhibited AR signaling similar to the full-length protein, while the C-terminal nuclease domain that fails to interact with FLNA, enhanced AR signaling (Figure 2) [3]. Analysis of subcellular AR distribution following R1881 treatment demonstrated increase in nuclear/cytoplasmic ratio in cells lacking RNase L or FLNA which was exacerbated when both proteins were lacking (Figure 3B). Taken together, our data show that RNase L, along with FLNA, may sequester AR in the cytoplasm and in response to R1881 the dissociation of the complex facilitates nuclear localization of AR. In the absence of the ligand, AR remains in the cytoplasm as a part of a multiprotein complex that includes HSP70 and HSP90 [54–56]. We have not ruled out the possibility that RNase L may alter AR interaction with HSP proteins. However, given that RNase L has been shown to regulate actin dynamics [3], interact with ECM and cytoskeletal proteins [24,26] and, importantly, inhibition of microtubule and cytoskeletal dynamics inhibits androgen-dependent AR nuclear translocation and AR transcriptional activity [30,31,57,58]; together these data suggest significant involvement of RNase L-regulated actin dynamics in AR translocation.

Sustained signaling through AR is a hallmark of castration-resistant prostate cancer (CRPC) and alternative splicing variants of AR (AR-Vs) that lack ligand-binding domain and constitutively active are reported to be upregulated [59–61]. It is established that AR-Vs upregulate transcription of canonical AR-responsive genes and other unique set of target genes [60,62]. Unlike full-length AR, transcriptional activities of the constitutively active AR-Vs are refractory to treatment with taxanes which do not block nuclear translocation of AR-V [63]. Future studies will address if RNase L impacts AR-V-mediated signaling activity.

In addition to transcriptional regulation, AR activity and abundance is regulated at the level of protein degradation and stability [64]. Inhibition of ubiquitin-proteasome degradation pathway has been reported to increase AR levels and several ubiquitin ligases bind to AR and regulate AR functions [65–67]. AR levels decreased to similar levels in WT and RNase L knockdown LNCaP cells (Figure 3) in the absence of ligand, R1881, when *de novo* protein translation is inhibited by cycloheximide treatment. Furthermore, inhibiting proteasome-mediated AR degradation by treating cells with MG132 did not result in altered stability of AR protein when RNase L was depleted. Thus, the inhibitory effect of RNase L on AR signaling does not appear to be due to the effect on AR degradation or stability. These observations support the notion that the final activity of AR in any given cell may eventually reflect balance and coordination of several regulators, including RNase L.

HPC1-associated mutations in RNase L have been studied in the context of RNase L enzyme activity and inducing apoptosis [12] as possible explanation for HPC, but do not address how the mutations contribute specifically to prostate cancer. RNase L R462Q mutant had reduced ability to dimerize into an active enzyme and had three-fold reduced activity and E265X produced a truncated protein which lacked the nuclease domain [12]. Since both mutants compromised RNase L enzyme activity, we raised the possibility that nuclease activity may be important. Recent description of RNase L structure allowed identification of residues critical for recognition of 2–5A, dimerization and nuclease function and we designed mutants based on the structural predictions and tested the ribonuclease activity (Figure 4E,F) [37,68]. In addition to the HPC1-associated mutants, we analyzed the RNase L mutants we generated that either lacked enzyme activity due to mutations in nuclease domain (R667A), or have reduced enzyme activity (K166E) or lack activity due to defects in binding activator, 2–5A (Y312A, K240/274N) for effects on AR signaling. The Y312A mutant lacks enzyme activity like the K240/274N and R667A mutations; however it can suppress AR signaling like WT RNase L indicating that enzyme activity may be dispensable (Figure 4). We conclude that in normal prostate cells RNase L is a negative regulator of AR signaling and loss of RNase L function in HPC can enhance AR signaling which is a hallmark of most prostate cancers.

RNase L regulates actin dynamics suggesting a possible role in cell migration [3]. Our results, consistent with other published data [21,22], shows that prostate cancer cells depleted of RNase L show greater migration in wound healing and transwell migration assays in response to fibronectin and serum (Figure 5). In reconstitution experiments, over-expression of RNase L suppressed cell migration compared to both endogenous levels and knockdown cells while activation of RNase L, which requires RNase L dimerization, inhibited cell migration (Figure 6). Unlike WT RNase L, HPC1-associated mutants, R462Q and E265X supported enhanced cell migration. Other RNase L mutants which have reduced activity (K166E) or lack activity (R667A and K240/274N) show enhanced cell migration. RNase L Y312A is an exception in that it lacks enzyme activity but suppresses cell migration like WT RNase L. The difference in the effect of K240/274N and Y312A may reflect the roles each of the residues play in contacting 2–5A versus disrupting domain interactions. K240 and K274 are critical residues in the P-loop motifs in the ankyrin repeats and may disrupt ankyrin repeat-protein kinase domain interaction and prevent 2–5A binding thereby affecting RNase activity [69]. It is possible that K240/274N mutant alters binding of proteins that contact RNase L through ankyrin repeats and the pseudokinase domains. Y312 is one of the residues that have been shown to provide direct contact with 2–5A and Y312A substitution may therefore lack activity while possibly retaining folding [37]. The precise mechanism by which these mutants affect interactions with other proteins will be addressed in future studies.

The mutations we have tested for AR signaling and cell migration span all the functional domains of RNase L. Mutations in the N-terminal ankyrin repeat domain, which serves as protein interaction domain, have variable effects on nuclear AR and cell migration and mutation in the C-terminal catalytic domain fails to suppress like WT. Biochemical studies showed that RNase L associated with the cytoskeleton assumes an inactive conformation and exists as a monomer while retaining the ability to bind with interacting proteins [24,70]. 2–5A binds to ankyrin repeat 2 and 4 and the pseudokinase domain facilitating dimerization and enzymatic activation. Activation of RNase L may cause dramatic conformational change and dimerization that induces its release from interacting proteins. Based on our results, we propose that the effect of RNase L on AR signaling and cell migration is mediated, in part, by protein-protein interactions and does not require enzymatic activity. Several proteins like Filamin A, LNX and or other cytoskeletal proteins may contribute to both these effects by interacting with RNase L [3,24,26]. Additional studies are required to investigate if the effect of the RNase L mutants is mediated by altered interaction with any of these interacting proteins. RNase L also inhibits cell attachment to ECM substrates and cell spreading on fibronectin. An earlier report showed that activity of RNase L downregulates transcripts involved in cell adhesion both transcriptionally and post-transcriptionally [22]. In contrast, we observed the suppressive effect of RNase L on cell migration in cells expressing endogenous RNase L in the absence of activation or with mutants that lacked enzyme activity.

We explored the underlying mechanisms for increased cell migration in RNase L-depleted cells by analyzing activation of integrin $\beta 1$ which is primarily involved in adhesion to fibronectin. Increased cell surface expression of integrin $\beta 1$ in RNase L-depleted cells corresponded with increase in cell migration and decrease correlated with activation of RNase L and inhibition of cell migration (Figures 9 and 10). Further, in androgen responsive LNCap cells, treatment with R1881 resulted in increase in cell surface expression of integrin $\beta 1$ and RNase L-depleted cells treated with R1881 showed a corresponding increase in integrin $\beta 1$ expression on cell surface (Figure S4). Activated integrins recruit and induce phosphorylation of FAK and Src which initiate signaling events to activate Rho GTPases to control cell shape and motility [41,42,47]. Accordingly, increase in integrin-stimulated FAK and Src phosphorylation was observed in RNase L-depleted cells and RNase L KO MEFs in response to fibronectin. Inhibiting Src activity with PP2 inhibitor in RNase L-depleted cells decreased cell migration compared to mock-treated cells demonstrating the involvement of the pathway for cell migration. Activity of Rho GTPases is required for cell migration and our testing of small Rho GTPases revealed that expression of dominant-negative form of Rac1 (T17N) significantly inhibited migration of RNase L-depleted cells, whereas dominant-negative Cdc42 (T17N) or RhoA (T19N) did not. Rac1 activity is regulated by Src in many cell types and elevated expression of Rac1 is observed in prostate cancer cells and tumors [71,72]. Our results show higher activity of Rac1 in response to fibronectin in RNase L-depleted cells compared to control cells although total levels of Rac1 protein were very similar. This is the first evidence of regulation of Rac1 activity by RNase L. Interestingly, the two proteins that interact with RNase L, namely, Filamin A and IQGAP1 are reported to regulate Rac1 activity [73]. Based on these observations and published data, RNase L appears to regulate cytoskeletal events, including AR signaling and cell migration, by virtue of its association and interaction with cytoskeletal and motor assembly proteins [26]. Apart from effects on cell migration, RNase L regulates activity of MMP-2 and MMP-9. Knockdown of RNase L increases gelatinase activity of both MMP-2 and -9 (Figure 14) which can enhance ECM remodeling and invasive potential of cells. Consistent with our observations, activation of RNase L inhibited integrin activation, cell migration as well as activity of MMP-2 and -9. MMP activities can be regulated at the level of transcription and post-transcriptionally by activators and inhibitors [74]. Levels of *MMP-9* mRNA did not change in knockdown cells and while not explored here, RNase L may affect MMP activity by regulating natural MMP inhibitors, tissue inhibitors of metalloproteinases (TIMPs).

Collectively, the results presented here identify RNase L as a suppressor of AR signaling, cell migration and MMP activity. Using RNase L mutants we show that nucleolytic activity is dispensable for both AR signaling and migration. HPC1-associated mutations in RNase L enhance AR signaling and cell migration and our study has identified a novel role of RNase L as a prostate cancer susceptibility gene. It is likely that RNase L mutations identified as risk factors in other types of cancers may affect cell migration and contribute to tumor development extending the antitumor role of RNase L beyond prostate cancer.

4. Materials and Methods

4.1. Chemicals, Reagents and Antibodies

Chemicals, unless indicated otherwise, were from Sigma Aldrich (St. Louis, MO, USA). Synthetic androgen R1881 was from Sigma Aldrich, extracellular matrices fibronectin, vitronectin, laminin, collagen I, collagen IV and Src inhibitor PP2 were from Millipore (Billerica, MA, USA). Cycloheximide (Sigma-Aldrich) and MG132 (EMD-Millipore, Billerica, MA, USA) were used at indicated concentrations. Antibodies to AR (N-20), Integrin β1 (M-106) and Filamin A were from Santa Cruz Biotechnology (Santa Cruz, CA, USA), Rac-1 (clone 102) was from BD Biosciences (San Jose, CA, USA). Total FAK, phospho-FAK (Y397), total Src, phospho-Src (Y416) were from Cell Signaling, Inc. (Danvers, MA, USA). RNase L monoclonal antibody was kindly provided by Robert Silverman (Cleveland Clinic). Antibodies to β-actin, monoclonal and polyclonal antibodies to Flag tag, Flag-M2 agarose beads were from Sigma Aldrich. Anti-mouse IgG and anti-rabbit IgG HRP linked secondary antibodies were from Cell Signaling, Inc. (Danvers, MA, USA) and ECL reagents were from GE Healthcare (Piscataway, NJ, USA) and Boston Bioproducts (Ashland, MA, USA). Alexa 488-labeled Phalloidin, and Alexa fluor 647 donkey anti-rabbit IgG were from Life Technologies, Carlsbad, CA, USA.

4.2. Cell Culture and Transfections

DU145, PC3 cells, WT and Rnasel$^{-/-}$ MEFs ((RNase L KO) primary and transformed with SV40 large T antigen, kindly provided by R.H. Silverman, Cleveland Clinic) and LNCaP cells (ATCC, Manassas, VA, USA) were grown in Roswell Park Memorial Institute (RPMI) 1640 supplemented with 100 μg/mL penicillin/streptomycin, 2 mM L-glutamine and 10% heat-inactivated fetal bovine serum (Sigma-Aldrich, St. Louis, MO, USA). For experiments involving androgen treatment, cells were transferred to phenol red-free medium supplemented with 2% charcoal-stripped serum (Hyclone, Logan, UT, USA) at least 24 h prior to addition of the synthetic androgen R1881 (1 nM) or ethanol vehicle (0.01%). Cells were maintained in 95% air, 5% CO_2 at 37 °C. RNase L-silencing (targeting the 3′-UTR), Filamin-A silencing, and non-silencing shRNAs were generated as suggested by the manufacturer using a GIPZ-lentiviral shRNA system and knock down cells or controls were selected with 1 µg/mL of puromycin as described previously [3] (Open Biosystems, Thermo Scientific, PA, USA). In some experiments, RNase L was knocked down using shRNA plasmid as described previously [16]. Knock-down of endogenous proteins was determined by western blotting. Transfection of 2–5A (10 µM) was performed using lipofectamine 2000 (Invitrogen, Thermo Fisher Scientific, Waltham, MA, USA) according to the manufacturer's protocol as described previously [75]. Briefly, cells were plated 1 day before transfection, so that the cells are 80%–90% confluent at the time of transfection. 2–5A was diluted into serum-free media and then mixed with lipofectamine 2000 reagent for 15 min before being added to cells in growth media. Preparation of 2–5A using ATP and recombinant 2–5A synthetase (a generous gift from Rune Hartmann, University of Aarhus, Aarhus, Denmark) has been described previously [75]. PolyI:C (2 µg/mL) was transfected into cells using Polyjet reagent (SignaGen Laboratories, Gaithersburg, MD, USA). Activity of RNase L in intact cells was determined in HeLa cells reconstituted with Flag-RNase L mutant constructs as described previously [75]. In experiments involving inhibitors, cells were preincubated with inhibitor for 1 h prior to treatment and then replaced with growth medium.

4.3. Plasmids

Plasmids Flag-RNase L, Flag-RNase L R667A, Flag-RNase L (1–335, ΔC), Flag-RNase L (386–741, ΔN) (kindly provided by Robert Silverman, Cleveland Clinic) were transfected using lipofectamine 2000 as per manufacturer's instructions. The Flag-RNase L mutants were constructed by site-directed mutagenesis using the primers listed in Table 1 and QuikChange Lightning Multi Site-Directed Mutagenesis Kit (Agilent Technologies, Santa Clara, CA, USA). The constructs were sequenced to confirm the mutations and expression confirmed by immunoblot analysis. The dominant negative Myc-tagged RhoA (T19N), Rac1 (T17N) and Cdc42 (T17N) eukaryotic expression constructs were described previously [50].

Table 1. List of primers used for mutagenesis.

Primer Name	Sequence of Primer
K274N R	5′-CAAGCAGCAGTGCTGT*A*TTGCCATCACTGTCTGTG-3′
K274N F	5′-CACAGACAGTGATGGCAAT*A*CAGCACTGCTGCTTG-3′
K240N R	5′-GGATCAGGGGAGT*A*TTCCCTCTTTCTCCCC-3′
K240N F	5′-GGGGAGAAAGAGGGAAT*A*CTCCCCTGATCC-3′
R667A R	5′-CAATGTGTTCTCCCAAATTC*G*CGATGAACTTTAGCAGATCAC-3′
R667A F	5′-GTGATCTGCTAAAGTTCATC*G*CGAATTTGGGAGAACACATTG-3′
R462Q R	5′-TAAATATAGATGACAGGACATTT*T*GGGCAAATTCATCTTCCTCATTT-3′
R462Q F	5′-AAATGAGGAAGATGAATTTGCCC*A*AAATGTCCTGTCATCTATATTTA-3′
E265X R	5′-CTGTGTCATTAATCT*A*TATGTGCTCTTGCTCCAGAAGC-3′
E265X F	5′-GCTTCTGGAGCAAGAGCACATAT*A*GATTAATGACACAG-3′
K166E F	5′-AGAGCGGCTGAGGGAGGGAGGGGCCACAG-3′
K166E R	5′-CTGTGGCCCCTCCCTCCCTCAGCCGCTCT-3′
Y312A R	5′-CAAGGGAATGGTCAGCATTCCGCCTCGCTGTCATAACAAGAT-3′
Y312A F	5′-ATCTTGTTATGACAGCGAGGCGGAATGCTGACCATTCCCTTG-3′

4.4. Co-Immunoprecipitation and Immunoblotting

LNCaP cells expressing Flag-RNase L plasmid were either not treated, treated with R1881 (1 nM) for 1 h or 24 h and harvested. Cells were washed with ice cold PBS and lysed in buffer containing 0.5% NP-40, 90 mM KCl, 5 mM magnesium acetate, 20 mM Tris, pH 7.5, 5 mM β mercaptoethanol, 0.1 M phenylmethylsulfonyl fluoride (PMSF), 0.2 mM sodium orthovanadate, 50 mM NaF, 10 mM glycerophosphate, protease inhibitor (Roche Diagnostics, Indianapolis, IN, USA) on ice for 20 min. The lysates were clarified by centrifugation at $10,000 \times g$ (at 4 °C for 20 min). Clarified cell lysates were precleared and mixed with control IgG or FlagM2-agarose beads and rotated end-to-end 1 h or overnight at 4 °C. The beads were collected and washed five times in lysis buffer. The immunoprecipitated proteins were dissociated by boiling in Laemmli sample buffer, separated on 10% SDS-polyacrylamide gels, transferred to nitrocellulose membrane (Biorad, Hercules, CA, USA) and subjected to immunoblotting. Membranes were probed with different primary antibodies according to the manufacturer's protocols. Membranes were washed with Tris Buffered Saline (TBS) with 1% Tween 20 and incubated with goat anti-mouse or goat anti-rabbit antibody tagged with horseradish peroxidase (Cell Signaling, Danvers, MA, USA) for 1 h. Proteins in the blots were detected by enhanced chemiluminesence (GE Healthcare). Cell extracts from LNCaP cells treated with cycloheximide (50 µg/mL) alone or combined with MG132 (20 µM) or R1881 (1 nM) were prepared in 2% sodium dodecyl sulphate (SDS) and subjected to immunoblotting. PC3 (control and RNase L knockdown) cells or WT and RNase LKO MEFs were serum starved for 16 h and plated on fibronectin coated dishes for 1 h. FAK and Src phosphorylation was detected by immunoblotting using anti-phospho FAK (Y397) or anti-phospho Src (Y416) antibodies (Cell Signaling, Danvers, MA, USA).

4.5. Immunofluorescence Assays

LNCaP cells (WT, RNase L knockdown, Filamin A knockdown or RNase L and Filamin A knockdown) on glass coverslips were treated with phenol red-free RPMI medium supplemented with 2% charcoal-stripped serum (Hyclone, Logan, UT, USA) at least 24 h prior to addition of the synthetic androgen R1881 (1 nM) or ethanol vehicle (0.01%) for 24 h. In some experiments, LNCaP RNase L-knockdown cells were reconstituted with Flag-RNase L (WT) or various Flag-RNase L mutants and plated on coverslips and treated as described above. Cells were fixed in 4% paraformaldehyde (Boston Bioproducts, MA, USA) for 15 min and permeabilized with 5% goat serum, 0.3% Triton-X-100 in PBS for 1 h. After washing and blocking in 1% Bovine Serum Albumin (BSA) in PBS, the cells were reacted with anti-AR antibody (N-20, Santa Cruz Biotechnology, CA, USA, 1:200 in 1% BSA, 0.3% Triton-X-100 in PBS) at 4 °C for 16 h followed by washing and incubation with fluorescent dye-conjugated secondary antibodies, Alexa-647 Goat anti-rabbit IgG (1:200, 1 h at 4 °C, Molecular probes, CA, USA). Cells were mounted in Vectashield with DAPI to stain the nucleus (Vector Labs, Burlingame, CA, USA). Fluorescence and confocal microscopy assessments were performed with Leica CS SP5 multi-photon laser scanning confocal microscope (Leica Microsystems, Weitzler, Germany) and quantitated using Image J software (National Institutes of Health). Images were processed using Adobe Photoshop CS4 (Adobe, San Jose, CA, USA). More than 10 cells (from at least three fields) were analyzed for each condition from three independent experiments.

4.6. Luciferase Reporter Gene Assays

LNCaP cells (WT, RNase L knockdown, Filamin A knockdown or RNase L and Filamin A knockdown) were transfected with AR-responsive PSA-luciferase [76] (1.0 µg), and plasmid pCH110 expressing β-galactosidase (0.1 µg) to normalize transfection efficiency in phenol red-free RPMI medium supplemented with 2% charcoal-stripped serum (Hyclone, Logan, UT, USA) at least 24 h prior to addition of the synthetic androgen R1881 (1 nM) or ethanol vehicle (0.01%). Cells were harvested in luciferase lysis buffer 24 h after treatment, and luciferase activity was determined using luciferase assay kit (Promega, Madison, WI, USA) and normalized to β-galactosidase levels. Experiments were performed in triplicate, and the results are representative of three independent experiments and shown as ± SD.

4.7. RNA Isolation and Quantitative Real Time Polymerase Chain Reaction

RNA was isolated using Trizol reagent (Invitrogen, Thermo Fisher Scientific,) as per the manufacturer's instructions and used for cDNA synthesis using random decamers and a RETROscript cDNA synthesis kit (Life Technologies; Thermo Fisher Scientific). Expression of androgen-responsive genes was determined by quantitative reverse transcription polymerase chain reaction (qRT-PCR) using

SYBR Green PCR Master Mix (Bio-Rad Laboratories Inc., Hercules, CA, USA) using the gene-specific primers and normalized to *GAPDH* expression. Primer sequences used are listed in Table 2 below. Experiments were performed in triplicate, and the results are representative of three independent experiments and shown as $\pm$ SD.

Table 2. List of primers for quantitative real-time polymerase chain reaction.

Primer Name	Sequence of Primer
PSA F	5'-GCAGCATTGAACCAGAGGAG-3'
PSA R	5'-CCCATGACGTGATACCCTGA-3'
sGCα1 F	5'-CTGCCTCATTTGCTTCATCA-3'
sGCα1 R	5'-TTGCCATGCTGAGCTGTTTA-3'
ETV1 F	5'-CACTGGGTCGTGGTACTCCT-3'
ETV1 R	5'-TACCCCATGGACCACAGATT-3'
MMP9 F	5'-GCCATTCACGTCGTCCTTAT-3'
MMP9 R	5'-TTGACAGCGACAAGAAGTGG-3'

4.8. Cell Fractionation

LNCaP cells (WT, RNase L knockdown, Filamin A knockdown or RNase L and Filamin A knockdown) were treated with R1881 or vehicle for 24 h and harvested. Twenty-five percent of the cells were saved as input, and the remaining portion was fractionated into nuclear and cytosolic fractions using Nuclear/Cytosol Fractionation Kit (MBL International Corp., Woburn, MA, USA). The fractions were then subjected to immunoblotting as described above to measure AR levels and quantitated using NIH Image J software.

4.9. Wound Healing Assay

Confluent monolayers of cells in 6-well plates were incubated in serum-free medium for 24 h. Monolayers were scratched with a micropipette tip and washed in phosphate buffer saline (PBS), and replaced with complete growth medium. Migration of cells to close the wound was monitored at indicated times and phase-contrast images were acquired on an Olympus IX81 inverted microscope using a 10× objective lens and a XM10 camera (Olympus, Tokyo, Japan). Wound closure was calculated from at least three independent experiments using NIH Image J software. Data shown are representative of three independent experiments and quantitation is shown as $\pm$ SD.

4.10. Cell Migration Assay

Transwell cell migration assays were performed using a modified Boyden chamber (Corning Inc., Corning, NY, USA) containing a fibronectin-coated polycarbonate membrane filter (6.5 mm diameter, 8 μm pore size) in growth medium. DU145, PC3 or LNCaP cells (2×10^5) cells (control or RNase L knockdown) and WT or RNase L KO MEFs were incubated in serum-free medium for 24 h and plated in the upper chamber and allowed to migrate for indicated times to lower chamber that contained growth medium with 10% fetal bovine serum (FBS). Non-migrated cells on the upper chamber were scraped with a cotton swab, and migrated cells on the bottom surface were trypsinized and counted with a hemocytometer. Experiments were performed in triplicate, and the results are representative of three independent experiments and shown as mean $\pm$ SEM.

4.11. Cell Spreading Assay

DU145 or PC3 (control or RNase L-knockdown) were plated on fibronectin-coated coverslips for indicated times up to 100 min. Attached and spread cells were fixed with 3.7% paraformaldehyde in PBS (Boston Bioproducts, Ashland, MA, USA) for 10 min and permeabilized with 0.1% Triton-X-100 for 5 min. F-actin was labeled with Alexa 488-labeled phalloidin (Life Technologies, CA, USA) and mounted in Vectashield with DAPI (4',6-diamidino-2-phenylindole, Vector Laboratories, Burlingame,

CA, USA). Cells were imaged by the use of a Leica TCS SP5 multiphoton laser scanning confocal microscope (Leica Microsystems, Weitzler, Germany), and cell area was quantitated using NIH Image J software. Experiments were repeated in triplicate with at least 30 measurements per time point for each experiment. Data are shown as mean area at indicated time points ± SEM.

4.12. Cell Attachment Assay

To quantitate cell attachment to different extracellular matrix substrates, 96-well plates were coated with 10 µg/mL each of fibronectin, vitronectin, laminin, collagen I or collagen IV (EMD Millipore, Billerica, MA, USA). Prior to use, the wells were treated with 1% BSA in 1× PBS (pH 7.4). Cells (2×10^5) in single cell suspension in serum-free medium were added per well in triplicate and incubated at 37 °C for 1 h. Wells were washed three times with PBS, fixed in 95% ethanol and stained with 0.1% crystal violet for 30 min at room temperature. The wells were washed extensively to remove excess stain. Cells were lysed in 0.2% Triton-X-100 and absorbance measured at 570 nm. Percent cell attachment was determined from three independent experiments performed in triplicate and shown as mean ± SEM.

4.13. Flow Cytometry and Analysis

DU145, PC3 or LNCaP cells (control or RNase L-knockdown) were grown in 100 mm dishes to 80%–90% confluence and then plated on dishes coated with 10 µg/mL fibronectin for 2 h. Cells were harvested and resuspended in ice-cold HEPES-buffer (20 mM HEPES, 125 mM NaCl, 45 mM glucose, 5 mM KCl, 0.1% albumin, pH 7.4) and incubated with anti-integrin β1 antibodies (Santa Cruz Biotechnologies, Santa Cruz, CA, USA) for 1 h. Isotype-specific antibodies were used as controls. Cells were washed three times in HEPES-buffer and incubated with Alexa-fluor-488 labeled secondary antibodies. Flow cytometry was performed using a FACSCalibur System (BectonDickinson, Heidelberg, Germany) equipped with Cell Quest software (Becton-Dickinson, San Jose, CA, USA). In some experiments, cells were transfected with 2–5A (10 µM) for 4 h and subject to flow cytometry analysis.

4.14. Rac Activity Assays

DU145 or PC3 cells (control and RNase L knockdown) were plated on dishes coated with 10 µg/mL fibronectin for 30 min or 1 h. Assays for GTP-bound Rac1 were performed as described [77]. Cells were lysed and precipitated using GST-PBD (PAK-binding domain) beads for pulling down active Rac1. Bead bound proteins (for active Rac1) and cell lysates (total Rac1) were resolved on a 13% SDS-PAGE gel and transferred onto nitrocellulose membranes and probed with primary antibodies against Rac1. The levels of active Rac1 were calculated by comparing the intensities of the active Rac1 bands with those of the total Rac1 bands in each lane using Image J software (National Institutes of Health). Data are expressed as fold increase in Rac1 activity over control and are representative of three independent experiments.

4.15. Gelatin Zymography

Gelatin zymography was performed under non-reducing conditions on 8% polyacrylamide gels copolymerized with 0.1% gelatin (Sigma-Aldrich, St. Louis, MO, USA). Activity of MMP-9 and -2 was determined in culture supernatants of control and RNase L knockdown cells (2×10^5 cells) treated or not with 2–5A (10 µM) as described previously [78]. Clear bands representing MMP-9 and -2 activities were imaged and quantitated using Image J software (National Institute of Health). Data shown are representative of three independent experiments and quantitation is shown mean as ± SD.

4.16. Statistical Analysis

All values are presented as mean ± SEM from at least three independent experiments or are representative of three independent experiments performed in triplicate and shown as mean ± SD.

Student's *t*-tests were used for determining statistical significance between groups. *p*-values are shown for all experiments and $p < 0.05$ was considered significant.

5. Conclusions

In this study, we demonstrate a role of RNase L as a suppressor of Androgen Receptor (AR) signaling, cell migration and matrix metalloproteinase activity. Using RNase L mutants, we show that its nucleolytic activity is dispensable for both AR signaling and migration. The most prevalent HPC1-associated mutations in RNase L, R462Q and E265X, enhanced AR signaling and cell migration and our studies identify a novel role of RNase L as a prostate cancer susceptibility gene.

Supplementary Materials: Supplementary materials can be found at www.mdpi.com/1422-0067/18/3/529/s1.

Acknowledgments: This work was supported by National Institutes of Health (NIH) Grants AI089518 (Krishnamurthy Malathi), AI119980-01A1 (Krishnamurthy Malathi), 1R21CA194776-01A1 (Rafael Garcia-Mata), 1R15CA199101-01A1 (Rafael Garcia-Mata), internal grants (Krishnamurthy Malathi) and startup funds from University of Toledo (Krishnamurthy Malathi). We thank Sushovita Mukherjee for data in the preliminary studies and technical help. We thank Robert Silverman (Cleveland Clinic) for RNase L KO MEFs, RNase L antibody and Flag-RNase L plasmids. We thank Douglas Leaman (University of Toledo) and Travis Taylor (University of Toledo) for valuable discussions through the course of this work.

Author Contributions: Krishnamurthy Malathi and Lirim Shemshedini conceived and designed the experiments; Shubham Dayal, Jun Zhou, Praveen Manivannan, Mohammad Adnan Siddiqui, Omaima Farid Ahmad, Matthew Clark and Krishnamurthy Malathi performed the experiments; Krishnamurthy Malathi, Rafael Garcia-Mata and Lirim Shemshedini analyzed the data; Rafael Garcia-Mata, Sahezeel Awadia contributed important reagents and materials; and Krishnamurthy Malathi wrote the paper.

Conflicts of Interest: The authors declare no conflict of interest.

References

1. Borden, E.C.; Sen, G.C.; Uze, G.; Silverman, R.H.; Ransohoff, R.M.; Foster, G.R.; Stark, G.R. Interferons at age 50: Past, current and future impact on biomedicine. *Nat. Rev. Drug Discov.* **2007**, *6*, 975–990. [CrossRef] [PubMed]

2. Silverman, R.H. Viral encounters with 2′,5′-oligoadenylate synthetase and RNase L during the interferon antiviral response. *J. Virol.* **2007**, *81*, 12720–12729. [CrossRef] [PubMed]

3. Malathi, K.; Siddiqui, M.A.; Dayal, S.; Naji, M.; Ezelle, H.J.; Zeng, C.; Zhou, A.; Hassel, B.A. RNase L interacts with filamin a to regulate actin dynamics and barrier function for viral entry. *mBio* **2014**, *5*, e02012. [CrossRef] [PubMed]

4. Al-Ahmadi, W.; Al-Haj, L.; Al-Mohanna, F.A.; Silverman, R.H.; Khabar, K.S. RNase L downmodulation of the rna-binding protein, hur, and cellular growth. *Oncogene* **2009**, *28*, 1782–1791. [CrossRef] [PubMed]

5. Zeng, C.; Yi, X.; Zipris, D.; Liu, H.; Zhang, L.; Zheng, Q.; Malathi, K.; Jin, G.; Zhou, A. RNase L contributes to experimentally induced type 1 diabetes onset in mice. *J. Endocrinol.* **2014**, *223*, 277–287. [CrossRef] [PubMed]

6. Fabre, O.; Salehzada, T.; Lambert, K.; Boo Seok, Y.; Zhou, A.; Mercier, J.; Bisbal, C. RNase L controls terminal adipocyte differentiation, lipids storage and insulin sensitivity via Chop10 mRNA regulation. *Cell Death Differ.* **2012**, *19*, 1470–1481. [CrossRef] [PubMed]

7. Ireland, D.D.; Stohlman, S.A.; Hinton, D.R.; Kapil, P.; Silverman, R.H.; Atkinson, R.A.; Bergmann, C.C. RNase L mediated protection from virus induced demyelination. *PLoS Pathog.* **2009**, *5*, e1000602. [CrossRef] [PubMed]

8. Carpten, J.; Nupponen, N.; Isaacs, S.; Sood, R.; Robbins, C.; Xu, J.; Faruque, M.; Moses, T.; Ewing, C.; Gillanders, E.; et al. Germline mutations in the ribonuclease l gene in families showing linkage with Hpc1. *Nat. Genet.* **2002**, *30*, 181–184. [CrossRef] [PubMed]

9. Casey, G.; Neville, P.J.; Plummer, S.J.; Xiang, Y.; Krumroy, L.M.; Klein, E.A.; Catalona, W.J.; Nupponen, N.; Carpten, J.D.; Trent, J.M.; et al. RNase L arg462gln variant is implicated in up to 13% of prostate cancer cases. *Nat. Genet.* **2002**, *32*, 581–583. [CrossRef] [PubMed]

10. Rennert, H.; Bercovich, D.; Hubert, A.; Abeliovich, D.; Rozovsky, U.; Bar-Shira, A.; Soloviov, S.; Schreiber, L.; Matzkin, H.; Rennert, G.; et al. A novel founder mutation in the *RNASEL* gene, 471delaaag, is associated with prostate cancer in Ashkenazi jews. *Am. J. Hum. Genet.* **2002**, *71*, 981–984. [CrossRef] [PubMed]

11. Rokman, A.; Ikonen, T.; Seppala, E.H.; Nupponen, N.; Autio, V.; Mononen, N.; Bailey-Wilson, J.; Trent, J.; Carpten, J.; Matikainen, M.P.; et al. Germline alterations of the *RNASEL* gene, a candidate *Hpc1* gene at 1q25, in patients and families with prostate cancer. *Am. J. Hum. Genet.* **2002**, *70*, 1299–1304. [CrossRef] [PubMed]

12. Xiang, Y.; Wang, Z.; Murakami, J.; Plummer, S.; Klein, E.A.; Carpten, J.D.; Trent, J.M.; Isaacs, W.B.; Casey, G.; Silverman, R.H. Effects of *RNASEL* mutations associated with prostate cancer on apoptosis induced by 2′,5′-oligoadenylates. *Cancer Res.* **2003**, *63*, 6795–6801. [PubMed]

13. Maier, C.; Haeusler, J.; Herkommer, K.; Vesovic, Z.; Hoegel, J.; Vogel, W.; Paiss, T. Mutation screening and association study of *RNASEL* as a prostate cancer susceptibility gene. *Br. J. Cancer* **2005**, *92*, 1159–1164. [CrossRef] [PubMed]

14. Orr-Urtreger, A.; Bar-Shira, A.; Bercovich, D.; Matarasso, N.; Rozovsky, U.; Rosner, S.; Soloviov, S.; Rennert, G.; Kadouri, L.; Hubert, A.; et al. *RNASEL* mutation screening and association study in Ashkenazi and non-Ashkenazi prostate cancer patients. *Cancer Epidemiol. Biomark. Prev.* **2006**, *15*, 474–479. [CrossRef] [PubMed]

15. Wiklund, F.; Jonsson, B.A.; Brookes, A.J.; Stromqvist, L.; Adolfsson, J.; Emanuelsson, M.; Adami, H.O.; Augustsson-Balter, K.; Gronberg, H. Genetic analysis of the *RNASEL* gene in hereditary, familial, and sporadic prostate cancer. *Clin. Cancer Res.* **2004**, *10*, 7150–7156. [CrossRef] [PubMed]

16. Malathi, K.; Paranjape, J.M.; Ganapathi, R.; Silverman, R.H. Hpc1/*RNASEL* mediates apoptosis of prostate cancer cells treated with 2′,5′-oligoadenylates, topoisomerase i inhibitors, and tumor necrosis factor-related apoptosis-inducing ligand. *Cancer Res.* **2004**, *64*, 9144–9151. [CrossRef] [PubMed]

17. Silverman, R.H. A scientific journey through the 2–5A/RNase L system. *Cytokine Growth Factor Rev.* **2007**, *18*, 381–388. [CrossRef] [PubMed]

18. Malathi, K.; Dong, B.; Gale, M., Jr.; Silverman, R.H. Small self-RNA generated by RNase L amplifies antiviral innate immunity. *Nature* **2007**, *448*, 816–819. [CrossRef] [PubMed]

19. Chakrabarti, A.; Banerjee, S.; Franchi, L.; Loo, Y.M.; Gale, M., Jr.; Nunez, G.; Silverman, R.H. RNase L activates the NLRP3 inflammasome during viral infections. *Cell Host Microbe* **2015**, *17*, 466–477. [CrossRef] [PubMed]

20. Siddiqui, M.A.; Mukherjee, S.; Manivannan, P.; Malathi, K. RNase L cleavage products promote switch from autophagy to apoptosis by caspase-mediated cleavage of Beclin-1. *Int. J. Mol. Sci.* **2015**, *16*, 17611–17636. [CrossRef] [PubMed]

21. Banerjee, S.; Li, G.; Li, Y.; Gaughan, C.; Baskar, D.; Parker, Y.; Lindner, D.J.; Weiss, S.R.; Silverman, R.H. RNase L is a negative regulator of cell migration. *Oncotarget* **2015**, *6*, 44360–44372.

22. Rath, S.; Donovan, J.; Whitney, G.; Chitrakar, A.; Wang, W.; Korennykh, A. Human RNase L tunes gene expression by selectively destabilizing the microrna-regulated transcriptome. *Proc. Natl. Acad. Sci. USA* **2015**, *112*, 15916–15921. [CrossRef] [PubMed]

23. Sato, A.; Naito, T.; Hiramoto, A.; Goda, K.; Omi, T.; Kitade, Y.; Sasaki, T.; Matsuda, A.; Fukushima, M.; Wataya, Y.; et al. Association of RNase L with a ras gtpase-activating-like protein IQGAP1 in mediating the apoptosis of a human cancer cell-line. *FEBS J.* **2010**, *277*, 4464–4473. [CrossRef] [PubMed]

24. Ezelle, H.J.; Malathi, K.; Hassel, B.A. The roles of RNase L in antimicrobial immunity and the cytoskeleton-associated innate response. *Int. J. Mol. Sci.* **2016**, *17*, E74. [CrossRef] [PubMed]

25. Bettoun, D.J.; Scafonas, A.; Rutledge, S.J.; Hodor, P.; Chen, O.; Gambone, C.; Vogel, R.; McElwee-Witmer, S.; Bai, C.; Freedman, L.; et al. Interaction between the androgen receptor and RNase L mediates a cross-talk between the interferon and androgen signaling pathways. *J. Biol. Chem.* **2005**, *280*, 38898–38901. [CrossRef] [PubMed]

26. Gupta, A.; Rath, P.C. Expression of mrna and protein-protein interaction of the antiviral endoribonuclease RNase L in mouse spleen. *Int. J. Biol. Macromol.* **2014**, *69*, 307–318. [CrossRef] [PubMed]

27. Loy, C.J.; Sim, K.S.; Yong, E.L. Filamin-a fragment localizes to the nucleus to regulate androgen receptor and coactivator functions. *Proc. Natl. Acad. Sci. USA* **2003**, *100*, 4562–4567. [CrossRef] [PubMed]

28. Mooso, B.A.; Vinall, R.L.; Tepper, C.G.; Savoy, R.M.; Cheung, J.P.; Singh, S.; Siddiqui, S.; Wang, Y.; Bedolla, R.G.; Martinez, A.; et al. Enhancing the effectiveness of androgen deprivation in prostate cancer by inducing Filamin A nuclear localization. *Endocr. Relat. Cancer* **2012**, *19*, 759–777. [CrossRef] [PubMed]

29. Savoy, R.M.; Ghosh, P.M. The dual role of filamin a in cancer: Can't live with (too much of) it, can't live without it. *Endocr. Relat. Cancer* **2013**, *20*, R341–356. [CrossRef] [PubMed]

30. Zhu, M.L.; Horbinski, C.M.; Garzotto, M.; Qian, D.Z.; Beer, T.M.; Kyprianou, N. Tubulin-targeting chemotherapy impairs androgen receptor activity in prostate cancer. *Cancer Res.* **2010**, *70*, 7992–8002. [CrossRef] [PubMed]

31. Darshan, M.S.; Loftus, M.S.; Thadani-Mulero, M.; Levy, B.P.; Escuin, D.; Zhou, X.K.; Gjyrezi, A.; Chanel-Vos, C.; Shen, R.; Tagawa, S.T.; et al. Taxane-induced blockade to nuclear accumulation of the androgen receptor predicts clinical responses in metastatic prostate cancer. *Cancer Res.* **2011**, *71*, 6019–6029. [CrossRef] [PubMed]

32. Ting, H.J.; Chang, C. Actin associated proteins function as androgen receptor coregulators: An implication of androgen receptor's roles in skeletal muscle. *J. Steroid Biochem. Mol. Biol.* **2008**, *111*, 157–163. [CrossRef] [PubMed]

33. Taplin, M.E.; Balk, S.P. Androgen receptor: A key molecule in the progression of prostate cancer to hormone independence. *J. Cell. Biochem.* **2004**, *91*, 483–490. [CrossRef] [PubMed]

34. Karantanos, T.; Evans, C.P.; Tombal, B.; Thompson, T.C.; Montironi, R.; Isaacs, W.B. Understanding the mechanisms of androgen deprivation resistance in prostate cancer at the molecular level. *Eur. Urol.* **2015**, *67*, 470–479. [CrossRef] [PubMed]

35. Lee, D.K.; Chang, C. Endocrine mechanisms of disease: Expression and degradation of androgen receptor: Mechanism and clinical implication. *J. Clin. Endocrinol. Metab.* **2003**, *88*, 4043–4054. [CrossRef] [PubMed]

36. Malathi, K.; Paranjape, J.M.; Bulanova, E.; Shim, M.; Guenther-Johnson, J.M.; Faber, P.W.; Eling, T.E.; Williams, B.R.; Silverman, R.H. A transcriptional signaling pathway in the IFN system mediated by 2′-5′-oligoadenylate activation of RNase L. *Proc. Natl. Acad. Sci. USA* **2005**, *102*, 14533–14538. [CrossRef]

37. Huang, H.; Zeqiraj, E.; Dong, B.; Jha, B.K.; Duffy, N.M.; Orlicky, S.; Thevakumaran, N.; Talukdar, M.; Pillon, M.C.; Ceccarelli, D.F.; et al. Dimeric structure of pseudokinase RNase L bound to 2–5A reveals a basis for interferon-induced antiviral activity. *Mol. Cell* **2014**, *53*, 221–234. [CrossRef] [PubMed]

38. Juliano, R.L.; Haskill, S. Signal transduction from the extracellular matrix. *J. Cell Biol.* **1993**, *120*, 577–585. [CrossRef] [PubMed]

39. Guo, W.; Giancotti, F.G. Integrin signalling during tumour progression. *Nat. Rev. Mol. Cell Biol.* **2004**, *5*, 816–826. [CrossRef] [PubMed]

40. Boudreau, N.J.; Jones, P.L. Extracellular matrix and integrin signalling: The shape of things to come. *Biochem. J.* **1999**, *339*, 481–488. [CrossRef] [PubMed]

41. Schlaepfer, D.D.; Hunter, T. Signal transduction from the extracellular matrix–a role for the focal adhesion protein-tyrosine kinase fak. *Cell Struct. Funct.* **1996**, *21*, 445–450. [CrossRef] [PubMed]

42. Mitra, S.K.; Schlaepfer, D.D. Integrin-regulated FAK-Src signaling in normal and cancer cells. *Curr. Opin. Cell Biol.* **2006**, *18*, 516–523. [CrossRef] [PubMed]

43. Hamadi, A.; Bouali, M.; Dontenwill, M.; Stoeckel, H.; Takeda, K.; Ronde, P. Regulation of focal adhesion dynamics and disassembly by phosphorylation of FAK at tyrosine 397. *J. Cell Sci.* **2005**, *118*, 4415–4425. [CrossRef] [PubMed]

44. Schlaepfer, D.D.; Mitra, S.K.; Ilic, D. Control of motile and invasive cell phenotypes by focal adhesion kinase. *Biochim. Biophys. Acta* **2004**, *1692*, 77–102. [CrossRef] [PubMed]

45. Arias-Salgado, E.G.; Lizano, S.; Sarkar, S.; Brugge, J.S.; Ginsberg, M.H.; Shattil, S.J. Src kinase activation by direct interaction with the integrin beta cytoplasmic domain. *Proc. Natl. Acad. Sci. USA* **2003**, *100*, 13298–13302. [CrossRef] [PubMed]

46. Palazzo, A.F.; Eng, C.H.; Schlaepfer, D.D.; Marcantonio, E.E.; Gundersen, G.G. Localized stabilization of microtubules by integrin- and FAK-facilitated rho signaling. *Science* **2004**, *303*, 836–839. [CrossRef] [PubMed]

47. Lawson, C.D.; Burridge, K. The on-off relationship of Rho and Rac during integrin-mediated adhesion and cell migration. *Small GTPases* **2014**, *5*, e27958. [CrossRef] [PubMed]

48. Nobes, C.D.; Hawkins, P.; Stephens, L.; Hall, A. Activation of the small GTP-binding proteins Rho and Rac by growth factor receptors. *J. Cell Sci.* **1995**, *108*, 225–233. [PubMed]

49. Hall, A. Small GTP-binding proteins and the regulation of the actin cytoskeleton. *Annu. Rev. Cell Biol.* **1994**, *10*, 31–54. [CrossRef] [PubMed]

50. Boulter, E.; Garcia-Mata, R.; Guilluy, C.; Dubash, A.; Rossi, G.; Brennwald, P.J.; Burridge, K. Regulation of Rho GTPase crosstalk, degradation and activity by Rhogdi1. *Nat. Cell Biol.* **2010**, *12*, 477–483. [CrossRef] [PubMed]

51. Kessenbrock, K.; Plaks, V.; Werb, Z. Matrix metalloproteinases: Regulators of the tumor microenvironment. *Cell* **2010**, *141*, 52–67. [CrossRef] [PubMed]
52. Morgia, G.; Falsaperla, M.; Malaponte, G.; Madonia, M.; Indelicato, M.; Travali, S.; Mazzarino, M.C. Matrix metalloproteinases as diagnostic (MMP-13) and prognostic (MMP-2, MMP-9) markers of prostate cancer. *Urol. Res.* **2005**, *33*, 44–50. [CrossRef] [PubMed]
53. Incorvaia, L.; Badalamenti, G.; Rini, G.; Arcara, C.; Fricano, S.; Sferrazza, C.; Di Trapani, D.; Gebbia, N.; Leto, G. MMP-2, MMP-9 and Activin A blood levels in patients with breast cancer or prostate cancer metastatic to the bone. *Anticancer Res.* **2007**, *27*, 1519–1525. [PubMed]
54. Fang, Y.; Fliss, A.E.; Robins, D.M.; Caplan, A.J. Hsp90 regulates androgen receptor hormone binding affinity in vivo. *J. Biol. Chem.* **1996**, *271*, 28697–28702. [CrossRef] [PubMed]
55. Schneider, C.; Sepp-Lorenzino, L.; Nimmesgern, E.; Ouerfelli, O.; Danishefsky, S.; Rosen, N.; Hartl, F.U. Pharmacologic shifting of a balance between protein refolding and degradation mediated by hsp90. *Proc. Natl. Acad. Sci. USA* **1996**, *93*, 14536–14541. [CrossRef] [PubMed]
56. He, B.; Bai, S.; Hnat, A.T.; Kalman, R.I.; Minges, J.T.; Patterson, C.; Wilson, E.M. An androgen receptor NH2-terminal conserved motif interacts with the cooh terminus of the hsp70-interacting protein (chip). *J. Biol. Chem.* **2004**, *279*, 30643–30653. [CrossRef] [PubMed]
57. Mistry, S.J.; Oh, W.K. New paradigms in microtubule-mediated endocrine signaling in prostate cancer. *Mol. Cancer Ther.* **2013**, *12*, 555–566. [CrossRef] [PubMed]
58. Yuan, X.; Cai, C.; Chen, S.; Chen, S.; Yu, Z.; Balk, S.P. Androgen receptor functions in castration-resistant prostate cancer and mechanisms of resistance to new agents targeting the androgen axis. *Oncogene* **2014**, *33*, 2815–2825. [CrossRef] [PubMed]
59. Dehm, S.M.; Schmidt, L.J.; Heemers, H.V.; Vessella, R.L.; Tindall, D.J. Splicing of a novel androgen receptor exon generates a constitutively active androgen receptor that mediates prostate cancer therapy resistance. *Cancer Res.* **2008**, *68*, 5469–5477. [CrossRef] [PubMed]
60. Guo, Z.; Yang, X.; Sun, F.; Jiang, R.; Linn, D.E.; Chen, H.; Chen, H.; Kong, X.; Melamed, J.; Tepper, C.G.; et al. A novel androgen receptor splice variant is up-regulated during prostate cancer progression and promotes androgen depletion-resistant growth. *Cancer Res.* **2009**, *69*, 2305–2313. [CrossRef] [PubMed]
61. Hu, R.; Dunn, T.A.; Wei, S.; Isharwal, S.; Veltri, R.W.; Humphreys, E.; Han, M.; Partin, A.W.; Vessella, R.L.; Isaacs, W.B.; et al. Ligand-independent androgen receptor variants derived from splicing of cryptic exons signify hormone-refractory prostate cancer. *Cancer Res.* **2009**, *69*, 16–22. [CrossRef] [PubMed]
62. Hu, R.; Lu, C.; Mostaghel, E.A.; Yegnasubramanian, S.; Gurel, M.; Tannahill, C.; Edwards, J.; Isaacs, W.B.; Nelson, P.S.; Bluemn, E.; et al. Distinct transcriptional programs mediated by the ligand-dependent full-length androgen receptor and its splice variants in castration-resistant prostate cancer. *Cancer Res.* **2012**, *72*, 3457–3462. [CrossRef] [PubMed]
63. Zhang, G.; Liu, X.; Li, J.; Ledet, E.; Alvarez, X.; Qi, Y.; Fu, X.; Sartor, O.; Dong, Y.; Zhang, H. Androgen receptor Splice variants circumvent ar blockade by microtubule-targeting agents. *Oncotarget* **2015**, *6*, 23358–23371. [CrossRef] [PubMed]
64. Gioeli, D.; Paschal, B.M. Post-translational modification of the androgen receptor. *Mol. Cell. Endocrinol.* **2012**, *352*, 70–78. [CrossRef] [PubMed]
65. Lin, H.K.; Wang, L.; Hu, Y.C.; Altuwaijri, S.; Chang, C. Phosphorylation-dependent ubiquitylation and degradation of androgen receptor by Akt require MDM2 E3 ligase. *EMBO J.* **2002**, *21*, 4037–4048. [CrossRef] [PubMed]
66. Qi, J.; Tripathi, M.; Mishra, R.; Sahgal, N.; Fazli, L.; Ettinger, S.; Placzek, W.J.; Claps, G.; Chung, L.W.; Bowtell, D.; et al. The e3 ubiquitin ligase siah2 contributes to castration-resistant prostate cancer by regulation of androgen receptor transcriptional activity. *Cancer Cell* **2013**, *23*, 332–346. [CrossRef] [PubMed]
67. Xu, K.; Shimelis, H.; Linn, D.E.; Jiang, R.; Yang, X.; Sun, F.; Guo, Z.; Chen, H.; Li, W.; Chen, H.; et al. Regulation of androgen receptor transcriptional activity and specificity by Rnf6-induced ubiquitination. *Cancer Cell* **2009**, *15*, 270–282. [CrossRef]
68. Han, Y.; Donovan, J.; Rath, S.; Whitney, G.; Chitrakar, A.; Korennykh, A. Structure of human RNase L reveals the basis for regulated RNA decay in the IFN response. *Science* **2014**, *343*, 1244–1248. [CrossRef] [PubMed]
69. Zhou, A.; Hassel, B.A.; Silverman, R.H. Expression cloning of 2–5a-dependent RNaase: A uniquely regulated mediator of interferon action. *Cell* **1993**, *72*, 753–765. [CrossRef]

70. Tnani, M.; Aliau, S.; Bayard, B. Localization of a molecular form of interferon-regulated RNase l in the cytoskeleton. *J. Interferon Cytokine Res.* **1998**, *18*, 361–368. [CrossRef] [PubMed]

71. Timpson, P.; Jones, G.E.; Frame, M.C.; Brunton, V.G. Coordination of cell polarization and migration by the rho family GTPases requires src tyrosine kinase activity. *Curr. Biol. CB* **2001**, *11*, 1836–1846. [CrossRef]

72. Engers, R.; Ziegler, S.; Mueller, M.; Walter, A.; Willers, R.; Gabbert, H.E. Prognostic relevance of increased Rac GTPase expression in prostate carcinomas. *Endocr. Relat. Cancer* **2007**, *14*, 245–256. [CrossRef] [PubMed]

73. Jacquemet, G.; Morgan, M.R.; Byron, A.; Humphries, J.D.; Choi, C.K.; Chen, C.S.; Caswell, P.T.; Humphries, M.J. Rac1 is deactivated at integrin activation sites through an Iqgap1-Filamin-A-Racgap1 pathway. *J. Cell Sci.* **2013**, *126*, 4121–4135. [CrossRef] [PubMed]

74. Gong, Y.; Chippada-Venkata, U.D.; Oh, W.K. Roles of matrix metalloproteinases and their natural inhibitors in prostate cancer progression. *Cancers* **2014**, *6*, 1298–1327. [CrossRef] [PubMed]

75. Siddiqui, M.A.; Malathi, K. Rnase L induces autophagy via c-Jun N-terminal kinase and double-stranded RNA-dependent protein kinase signaling pathways. *J. Biol. Chem.* **2012**, *287*, 43651–43664. [CrossRef] [PubMed]

76. Cai, C.; Hsieh, C.L.; Omwancha, J.; Zheng, Z.; Chen, S.Y.; Baert, J.L.; Shemshedini, L. Etv1 is a novel androgen receptor-regulated gene that mediates prostate cancer cell invasion. *Mol. Endocrinol.* **2007**, *21*, 1835–1846. [CrossRef] [PubMed]

77. Guilluy, C.; Dubash, A.D.; Garcia-Mata, R. Analysis of RhoA and Rho gef activity in whole cells and the cell nucleus. *Nat. Protoc.* **2011**, *6*, 2050–2060. [CrossRef] [PubMed]

78. Mukherjee, S.; Siddiqui, M.A.; Dayal, S.; Ayoub, Y.Z.; Malathi, K. Epigallocatechin-3-gallate suppresses proinflammatory cytokines and chemokines induced by toll-like receptor 9 agonists in prostate cancer cells. *J. Inflamm. Res.* **2014**, *7*, 89–101. [PubMed]

International Journal of
Molecular Sciences

MDPI

Article

SUMO-Specific Cysteine Protease 1 Promotes Epithelial Mesenchymal Transition of Prostate Cancer Cells via Regulating SMAD4 deSUMOylation

Xiaoyan Zhang [1], Hao Wang [1], Hua Wang [1], Fengjun Xiao [1], Prem Seth [2], Weidong Xu [2], Qinghua Jia [1], Chutse Wu [1], Yuefeng Yang [1,*] and Lisheng Wang [1,*]

[1] Department of Experimental Hematology, Beijing Institute of Radiation Medicine, Beijing 100850, China; zhangxy1120@aliyun.com (X.Z.); wang-home163@163.com (H.W.); wanghua@bmi.ac.cn (H.W.); xiaofjun@sina.com (F.X.); jjiaqh@hotmail.com (Q.J.); wuct@bmi.ac.cn (C.W.)

[2] Gene Therapy Program, Department of Medicine, NorthShore Research Institute, Evanston, IL 60201, USA; PSeth@northshore.org (P.S.); wdxuii@gmail.com (W.X.)

* Correspondence: yuefengyang1981@163.com (Y.Y.); lishengwang@ymail.com (L.W.); Tel.: +86-10-6693-1083 (Y.Y.); +86-10-6693-2041 (L.W.); Fax: +86-10-6815-8312 (Y.Y. & L.W.)

Academic Editor: Carsten Stephan
Received: 6 March 2017; Accepted: 7 April 2017; Published: 12 April 2017

Abstract: In advanced prostate cancer, small ubiquitin-like modifier (SUMO)-specific cysteine protease 1 (SENP1) is up-regulated. However, the role of SENP1 in regulating deSUMOylation of TGF-β/SMADs signaling is unknown. In this study, we developed a lentiviral vector, PLKO.1-shSENP1, to silence SENP1 in prostate cancer cells with high metastatic characteristics (PC3M). Likewise, we also created an adenovirus vector, Ad5/F11p-SENP1 to over-express SENP1 in prostate cancer cells with low metastatic potential (LNCaP). We showed that silencing of SENP1 promoted cellular apoptosis, and inhibited proliferation and migration of PC3M cells. Moreover, SENP1 silencing increased the SMAD4 expression at protein level, up-regulated E-cadherin and down-regulated Vimentin expression, indicating the inhibition of epithelial mesenchymal transition (EMT). Furthermore, SMAD4 interference abolished SENP1-mediated up-regulation of E-cadherin, suggesting that SENP1 regulated E-cadherin expression via SMAD4. SENP1 over-expression in LNCaP cells reduced SMAD4 protein, and promoted EMT via decreasing E-cadherin and increasing Vimentin. Moreover, down-regulation of SMAD4 and E-cadherin were blocked, after transfection with two SUMOylation sites mutated SMAD4, suggesting that SENP1 might reduce SMAD4 levels to regulate E-cadherin expression via deSUMOylation of SMAD4. In conclusion, SENP1 deSUMOylated SMAD4 to promote EMT via up-regulating E-cadherin in prostate cancer cells. Therefore, SENP1 is a potential target for treatment of advanced prostate cancer.

Keywords: SENP1; deSUMOylation; EMT; SMAD4; E-cadherin; prostate cancer

1. Introduction

Small ubiquitin-like modifier (SUMO) is an ubiquitin-like protein, and SUMOylation regulates many cellular events, including nuclear signaling, transcription activities, and DNA repair [1,2]. SUMOylation is a dynamic process, and can be reversed by SUMO-specific cysteine proteases (SENPs). One such protease SENP1 has been widely investigated in many cancers including prostate cancer, breast cancer and colon cancer [3,4]. Prostate cancer is the most commonly diagnosed cancer in United States and is also on rapid rise in China [5]. It has been reported that SENP1 is up-regulated in prostate cancer patients, and promotes both androgen receptors (ARs)-dependent and ARs-independent cell proliferation [6,7]. Importantly, high levels of SENP1 have been linked to advanced pathological stages, higher Gleason grade, positive lymph node status, and prostate specific antigen (PSA) recurrence [8].

Some reports have shown that SENP1 stabilizes hypoxia inducible factor 1 (HIF-1α) to promote tumor growth and metastasis, and increases vascular endothelial growth factor (VEGF) expression to increase angiogenesis in the tumors [9,10].

Transforming growth factor β (TGF-β), known as a pleiotropic cytokine in regulating various biological processes, plays dual roles in the cancer development and progression. Under physiological condition, TGF-β exerts biological activities, such as the inhibition of cell proliferation, and induction of cell apoptosis, via TGFβ/SMADs signaling pathways. However, during cancer progression, TGF-β/SMADs signaling-mediated growth inhibition is generally blocked, due to the loss and inactivation of the mother against decapentaplegic homolog (SMAD) molecules [11]. Among the SMADs, SMAD4 is an important tumor suppressor, which has also been recognized as a potential molecular maker for diagnosis of prostate cancer [12,13]. It has been reported that SUMOylation of SMAD4 increases protein expression and stability of SMAD4, which can enhance the transcriptional activities of SMAD4 target genes [14,15]. However, the role of SENP1 in regulating deSUMOylation of SMAD4 in prostate cancer is largely unknown.

In this study, we silenced SENP1 in PC3M cells, a prostate cancer cell line with high metastatic potential, and over-expressed SENP1 in LNCaP cells, prostate cancer cells with low metastatic phenotype. Using these transduced cells, we examined the biological characteristics, SMAD4 protein, and epithelial mesenchymal transitions (EMT) markers, including E-cadherin and Vimentin. Then, the role of SENP1-mediated SMAD4 deSUMOylation in regulating E-cadherin expression was analyzed by SMAD4 silencing, as well as by introducing SMAD4 mutations in the SUMOylation sites.

2. Results

2.1. SENP1 Silencing Induces Apoptosis, Inhibits Cell Growth and Migration in PC3M, an Androgen-Independent Prostate Cancer Cell Line

To study the effects of SENP1 on the biological effects of prostate cancer cells, we constructed a lentiviral vector expressing short hairpin RNA targeting SENP1, PLKO.1-shSENP1, and a control vector, PLKO.1-shScramble. Transduction of PC3M cells with PLKO.1-shSENP1 down-regulated the SENP1 expression, both at protein (Figure 1A) and mRNA level (Figure 1B). Then, the biological characteristics were analyzed in PC3M cells infected with lentiviral vectors. We found that PLKO.1-shSENP1-mediated SENP1 silencing induced cellular apoptosis (Figure 1C), inhibited cell proliferation (Figure 1D) and reduced cell migration (Figure 1E). These results suggest that SENP1 interference might be a potential therapeutic approach to inhibit tumor growth and prevent tumor metastasis.

2.2. SENP1 Interference Enhances TGF-β/Smads Signaling and Inhibits EMT in PC3M Cells

SMAD4 can be SUMOylated to regulate expression of TGF-β target genes. To test if SENP1 could deSUMOylate SMAD4 in prostate cancer cells, we analyzed SMAD4 expression in PC3M cells after infection with PLKO.1-shSENP1 or PLKO.1-shScramble. Interestingly, SENP1 silencing increased the expression of SMAD4 at the protein level (Figure 2A), but not at the mRNA level (Figure 2B), which suggested that SENP1 regulates the protein expression of SMAD4 at post-translational level. Furthermore, SENP1 interference increased E-cadherin protein, and reduced vimentin protein expression, which indicated the inhibition of EMT (Figure 2C,D). This is consistent with previous reports that TGF-β could promote the EMT in various tumor cells.

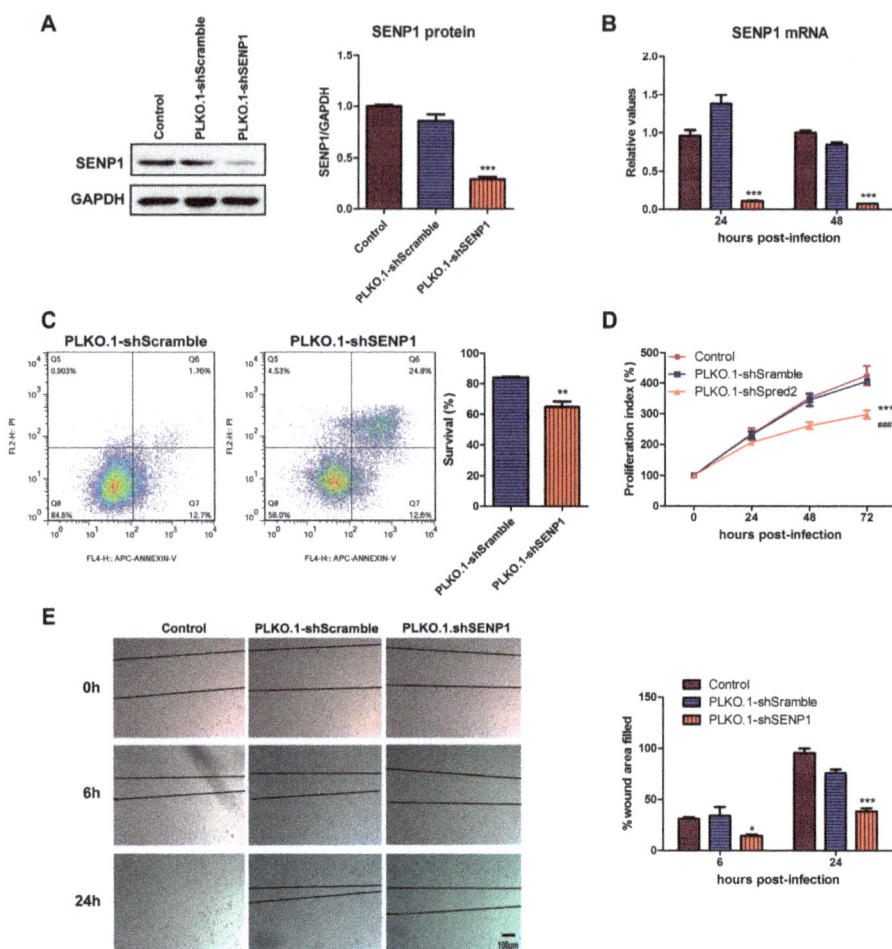

Figure 1. Effect of small ubiquitin-like modifier (SUMO)-specific cysteine protease 1 (SENP1) on the biological characteristics of PC3M prostate cancer cells. (**A,B**) Lentiviral vector mediated silencing of SENP1 in PC3M cells. PC3M cells were infected with 20 MOI (multiplicity of infection) of PLKO.1-shSENP1 or PLKO.1-shScramble. 48 h after infection, the SENP1 protein expression was detected by Western-blotting (**A**); At 24 and 48 h after infection, the total RNA was isolated and the mRNA expression of SENP1 was also analyzed by real-time reverse transcript polymerase chain reaction (RT-PCR) (**B**); (**C**) PLKO.1-shSENP1 induces apoptosis in PC3M cells. PC3M cells were infected with lentiviral vectors using 20 MOI. Forty-eight hour later, cells were collected, labeled with Annexin-V-APC and cellular apoptosis was analyzed by flow cytometry; (**D**) Proliferation of PC3M cells transduced with lentiviral vectors. PC3M cells were labeled with dye670, and then infected with lentiviral vectors. At indicated time points after infection, cells were collected and analyzed by flow cytometry. The proliferation index was calculated, using uninfected PC3M cells as control; (**E**) PLKO.1-shSENP1 inhibits the migration of PC3M cells. Confluent PC3M cells were scratched to generate wounds, at 24 h following infection with lentiviral vectors. 6 and 24 h later, the percentages of wound area filled were determined and analyzed. All the data were obtained from at least three independent experiments, and are shown as mean $\pm$ s.e.m. * $p < 0.05$, ** $p < 0.01$, *** $p < 0.001$, vs. PLKO.1-shScramble group; ### $p < 0.001$ vs. control group.

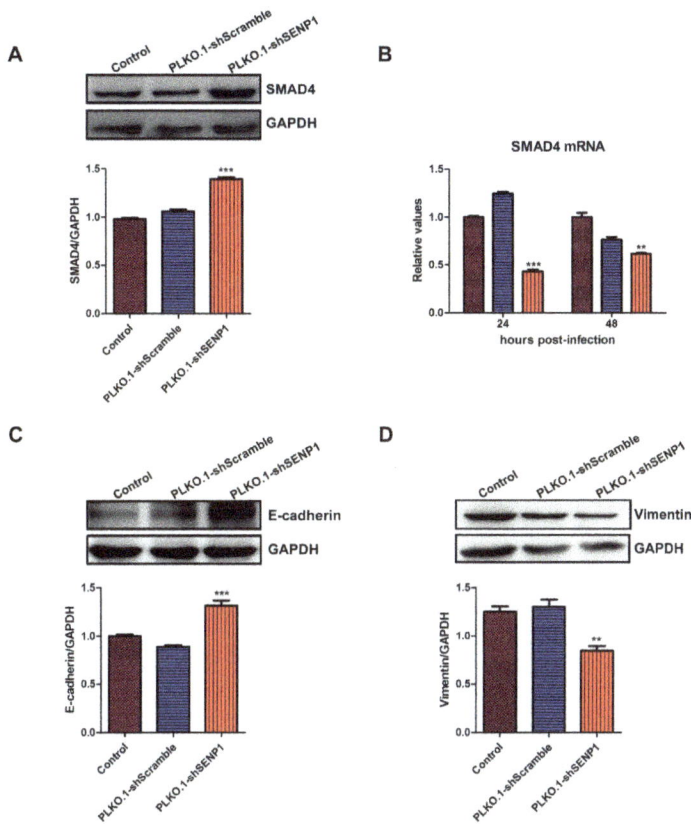

Figure 2. SENP1 interference enhances transforming growth factor (TGF-β)/SMADs signals, and inhibits epithelial mesenchymal transition (EMT) in PC3M cells. (**A**) PLKO.1-shSENP1 increases SMAD4 protein expression. PC3M cells were infected with 20 MOI PLKO.1-shSENP1 or PLKO.1-shScramble. 48 h later, cells were collected and SMAD4 protein was detected by Western-blotting; (**B**) SENP1 silencing decreased SMAD4 mRNA expression. At 24 and 48 h post-infection, cells were collected, and SMAD4 mRNA expression was detected by real-time RT-PCR; (**C,D**) SENP1 interference up-regulates E-cadherin protein, and reduces vimentin protein in PC3M cells. At 48h after infection with lentiviral vectors, protein expression of E-cadherin (**C**) and vimentin (**D**) was analyzed by Western-blotting as described above. All the data were obtained from at least three independent experiments, and are shown as mean ± s.e.m. ** $p < 0.01$, *** $p < 0.001$, vs. PLKO.1-shScramble group.

2.3. SENP1 Over-Expression Impairs TGF-β/Smads Signaling and Promotes EMT of Androgen-Dependent Prostate Cancer Cells, LNCaP

To further investigate the effects of SENP1 on TGF-β/SMADs signals and EMT markers, a chemic fiber modified replication deficiency adenovirus, Ad5/F11p.SENP1, and control adenovirus, Ad5/F11p.Null were constructed. In low endogenous SENP1 expressing prostate cancer cells, LNCaP, Ad5/F11p.SENP1 infection produced SENP1 protein efficiently (Figure 3A,B). Moreover, SENP1 over-expression reduced SMAD4 protein expression at 48 h after infection (Figure 3A,C). However, the mRNA expression of SMAD4 was up-regulated at 36 h and 48 h post-infection (Figure 3D), which again suggested that SENP1 regulated the protein expression at post-translation level, in consistent with the results in PC3M cells. Moreover, SENP1 down-regulated E-cadherin protein and increased vimentin protein in LNCaP cells, at 48 h after Ad5/F11p-SENP1 transduction, indicating that SENP1 promoted the

EMT of LNCaP cells (Figure 3E,F). Taken together, these studies suggest that in low-expressing SENP1 LNCaP cells, SENP1 over-expression down-regulated SMAD4 protein expression and promoted EMT of tumor cells.

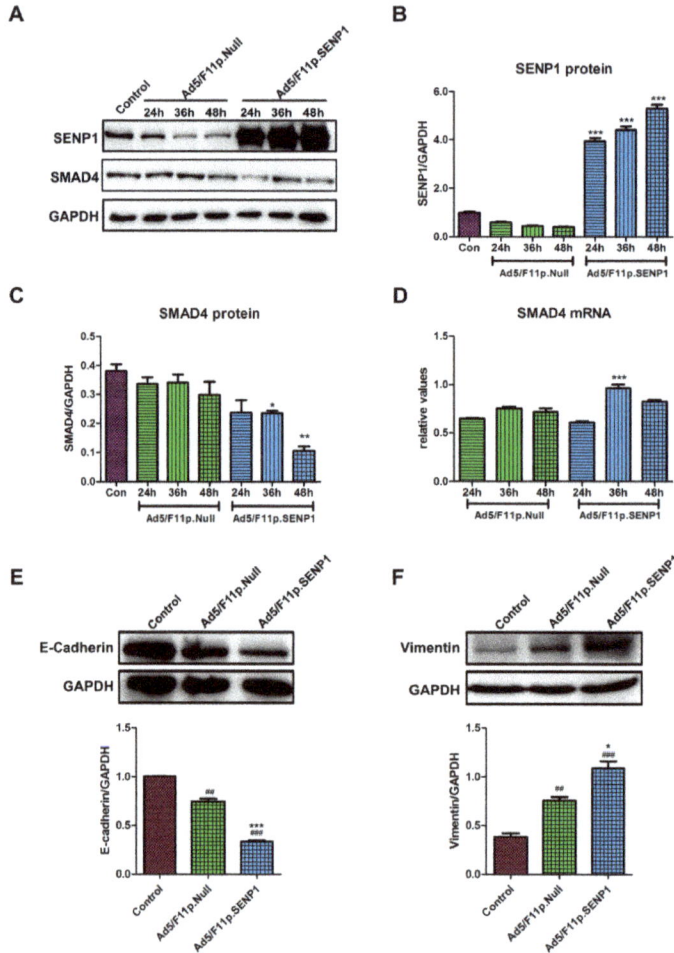

Figure 3. SENP1 over-expression decreases TGF-β/SMADs signals and promotes EMT of LNCaP cells. (**A–C**) SENP1 over-expression inhibits SMAD4 protein expression in LNCaP cells. LNCaP cells were infected with 10 MOI Ad5/F11p.SENP1 or Ad5/F11p.Null. At 24 h, 36 h and 48 h after infection, cells were collected and protein expression of SENP1 and SMAD4 was detected by Western-blotting (**A**), and the corresponding semi-quantitative results were shown in B and C respectively; (**D**) SENP1 increases SMAD4 mRNA expression in LNCaP cells. At 24 h, 36 h and 48 h after infection with adenoviruses, cells were collected, and SMAD4 mRNA expression was evaluated by real-time RT-PCR, and normalized by its expression in normal cultured LNCaP cells; (**E,F**) SENP1 reduces E-cadherin expression and promotes vimentin expression in LNCaP cells. 48 h after transduction with Ad5/F11p.SENP1 or Ad5/F11p.Null, the protein expression of E-cadherin (**E**) and vimentin (**F**) was detected by Western-blotting, and the semi-quantitative data are shown. All the data were obtained from at least three independent experiments, and are shown as mean ± s.e.m. * $p < 0.05$, ** $p < 0.01$, *** $p < 0.001$, vs. Ad5/F11p.Null group; ## $p < 0.01$, ### $p < 0.001$, vs. Ad5/F11p.Null group at the same time point.

2.4. SENP1 Regulates deSUMOylation of SMAD4 and Promotes EMT of Tumor Cells

Our studies described above have shown that SMAD4 is up-regulated by SENP1 silencing in PC3M cells, while it is down-regulated in SENP1 over-expressing LNCaP cells. Next, we investigated if SENP1 regulates the EMT of tumor cells via controlling SMAD4 expression. Specific short interfering RNAs targeting SMAD4 were transfected into PC3M cells for analyzing interference efficiency. Among these, siRNA621 was the most promising one, and was selected for SMAD4 knockdown. siRNA621 was transfected into PC3M cells or SENP1-silenced PC3M cells (Figure 4A,B). We found that siRNA621 transduction reduced the SMAD4 protein expression in PC3M cells. Moreover, siRNA621 prevented SENP1 interference-mediated up-regulation of SMAD4 protein (Figure 4A,C). Importantly, at 48 h after transduction, siRNA621 abolished SENP1 silence induced E-cadherin expression (Figure 4A,D), suggesting that SENP1 silencing could elevate E-cadherin protein levels via increasing SMAD4 protein expression. Interestingly, both SENP1 interference and SMAD4 silencing reduced Vimentin expression in PC3M cells (Figure 4A,E). These results indicated that SENP1 silencing down-regulated Vimentin expression by other mechanisms, but not via up-regulating SMAD4 protein. Importantly, our data also showed that SMAD4 silencing could partly recover the ability of migration in SENP1 silenced PC3M cells (Figure 4F,G). Therefore, SMAD4 plays pivotal roles in SENP1-mediated regulation of E-cadherin expression, EMT and migration in PC3M cells.

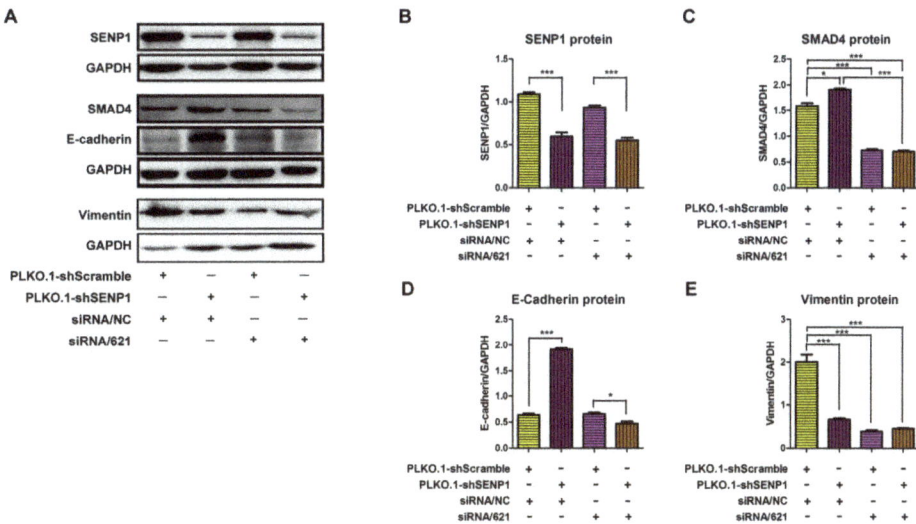

Figure 4. *Cont.*

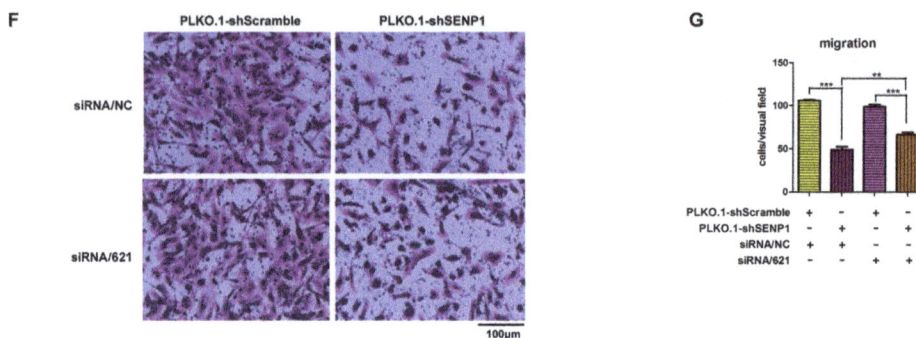

Figure 4. SMAD4 interference abolishes PLKO.1-shSENP1 induced E-cadherin expression in PC3M cells. PC3M cells were infected with 20 MOI PLKO.1-shScramble or PLKO.1-shSENP1. 24 h later, cells were transfected with siRNA targeting SMAD4 (siRNA/621) or control (siRNA/NC). After 48 h incubation, cells were collected and the protein expression of SENP1, SMAD4, E-cadherin and Vimentin was detected by Western-blotting (**A**). The semi-quantitative analysis of SENP1, SMAD4, E-cadherin and Vimentin was conducted, and are presented in (**B–E**), respectively. Moreover, the ability of migration was analyzed by transwell at 24 h after transfection with siRNAs. The representative images were shown in (**F**), and the statistical results were shown in (**G**). * $p < 0.05$, ** $p < 0.01$, *** $p < 0.001$, vs. corresponding group.

As described above, SENP1 could regulate SMAD4 protein expression at the post-translation level. Therefore, we investigated if SENP1 deSUMOylated SMAD4 in prostate cancer cells. In PC3M cells, we showed that PLKO.1-shSENP1 transfection could enhance the SUMOylation of SMAD4 via improving SUMO-1-SMAD4 complex (Figure 5A). To further confirm SENP1-mediated deSUMOylation of SMAD4, we constructed a plasmid vector pcDNA3.0-mutSMAD4 expressing SMAD4 with two mutation at SUMOylation sites, K113R and K159R; and using pcDNA3.0-SMAD4 as a control vector expressing wild type SMAD4. Both of these vectors were shown to produce SMAD4 protein in LNCaP cells (Figure 5A). Next, Ad5/F11p.SENP1 and Ad5/F11p.Null were transduced in SMAD4 over-expressing LNCaP cells (Figure 5B,C). We found that SENP1 could down-regulate SMAD4 protein level, both in control and wild SMAD4 over-expressing LNCaP cells. However, in mutated SMAD4 over-expressing LNCaP cells, SENP1 had no effect on SMAD4 expression, indicating that SENP1 could first deSUMOylate SMAD4 protein that leads to protein degradation (Figure 5B,D). Furthermore, SENP1 reduced E-cadherin expression both in the control and wild SMAD4 over-expressing LNCaP cells, but not in mutSMAD4 expressing LNCaP cells (Figure 5B,E). These results suggested that SENP1 could deSUMOylate and degrade SMAD4 protein to inhibit E-cadherin expression. Therefore, we propose that SENP1 up-regulation in prostate cancer cells deSUMOylates SMAD4, which in turn regulates E-cadherin, induces EMT of tumor cells and promotes tumor metastasis.

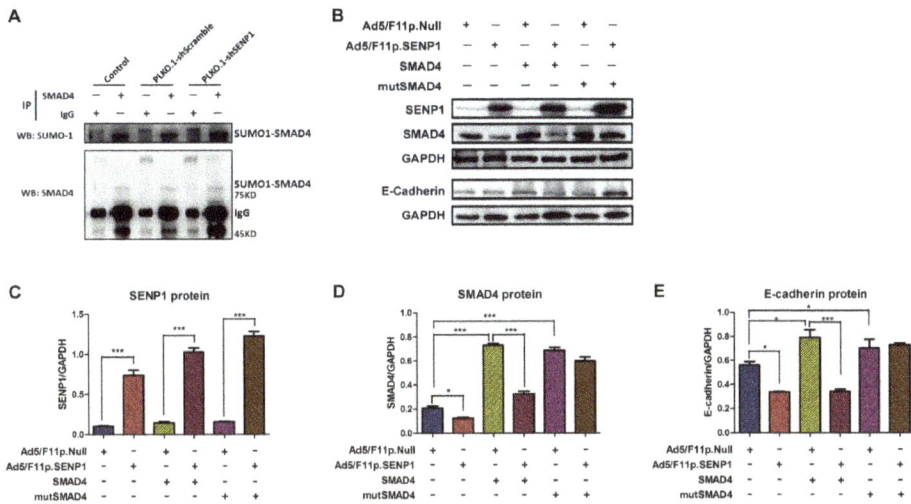

Figure 5. Mutations in SMAD4 at the SUMOylation sites of prevent SENP1-mediated degradation of SMAD4 and the down-regulation of E-cadherin. PC3M cells were transfected with lentiviral vectors, PLKO.1-shScramble and PLKO.1-shSENP1. 48 h later, cells were collected for immunoprecipitation analysis. The representative images were shown in (**A**). LNCaP cells were transfected with pcDNA3.0-SMAD4, pcDNA3.0-mutSMAD4 or pcDNA3.0 (control). 48 h later, cells were infected with 10 MOI Ad5/F11p.SENP1 or Ad5/F11p.Null. Cells were incubated for another 48 h, and then, the SENP1, SMAD4 and E-cadherin expression was detected by Western-blotting. The representative images are shown in (**B**); the semi-quantitative results of SENP1, SMAD4, and E-cadherin were analyzed and are presented in (**C**–**E**) respectively. All the data were obtained from at least three independent experiments, and are shown as mean ± s.e.m. * $p < 0.05$, *** $p < 0.001$, vs. corresponding group.

3. Discussion

In advanced stage of prostate cancer, a majority of patients develop distant metastasis, such as bone metastasis. Patients with metastases are insensitive to conventional treatments, including androgen-deprivation, chemotherapy and radiotherapy [16,17]. For bone metastatic patients, bisphosphonates and denosumab, a human monoclonal antibody against receptor activator of nuclear factor κ-B ligand (RANKL) could inhibit bone resorption and improve bone density to relieve pain and tumor-induced hypercalcemia [18,19]. However, the effective approaches to improve patients' overall survival are still lacking. Therefore, it is urgent to discover novel and more promising therapeutic targets for advanced and metastatic prostate cancer.

TGF-β is considered to be a potent growth inhibitor under physiological conditions, and inhibits tumor growth at the early stage of cancers. However, during the advanced stages, TGF-β is aberrantly activated and cross-talks with several cell signaling pathways, both canonical and non-canonical, to facilitate tumor growth and metastases [20]. For example, TGF-β can increase VEGF expression via Src/Fak/Akt signaling to promote angiogenesis, and activate HIFs through PI3K/Akt/mTOR signaling to regulate metabolism and growth of tumor cells. Therefore, several TGF-β inhibitors have been developed, and show effective anti-tumor responses in animal models [21,22]. Our group has also developed oncolytic adenoviruses expressing soluble TGF-β receptor II fusion human IgG Fc fragment (sTGFβRIIFc), which can block TGF-β signaling and inhibit tumor bone metastasis [23,24].

In addition to crosstalk with several cell signaling pathways, which promotes tumor growth and metastasis, the blockade of TGF-β/SMADs-mediated growth inhibition and transcriptional activities also plays pivotal roles to tumor progression [11,20]. SMAD4, which lies downstream of

SMAD2/3, is known to be an important tumor suppressor. It has been reported that down-regulation or inactivation of SMAD4 could promote the development and progression of prostate cancer, and SMAD4 has been emerged as a potential biomarker for diagnosis [12,13]. The SMAD4 levels are not only determined by transcription activities, but also influenced by post-translation modification, such as ubiquitinoylation, acetylization and SUMOylation [25,26]. Several reports have shown that SUMOylation could improve the stability and enhance the expression levels of SMAD4, which not only increases TGF-β-mediated transcription, but also negatively regulates ARs in prostate cancer [14,15]. However, SENP1, which could reverse the SUMOylation of protein, is significantly up-regulated in prostate cancers. In this study, we have shown that SENP1 silencing up-regulated SMAD4 in PC3M cells at protein level, but not at mRNA level, suggesting that SENP1 regulates SMAD4 expression at post-translation level. Moreover, in LNCaP cells, over-expression of SMAD4 protein with two mutations at SUMOylation sites could not be down-regulated by SENP1 over-expression. Therefore, we believe that SENP1 down-regulates SMAD4 protein through deSUMOylation.

EMT is an essential event during tumor metastasis to distant sites, and is associated with tumor cells to lose epithelial makers, such as E-cadherin, while acquiring mesenchymal makers, such as vimentin and N-cadherin. During advanced stages of cancers, TGF-β can induce EMT in tumor cells via both SMADs-dependent and -independent activation of EMT related transcription factors [27,28]. However, whether SENP1 participates in TGF-β induced EMT in prostate cancers has not been previously reported. In this study, we have shown that SENP1 silencing enhanced E-cadherin expression, while it inhibited vimentin expression in PC3M cells, indicating its role in the inhibition of EMT. We have also shown that the interference of SMAD4 could abolish this inhibitory effect of EMT. Taken together, these studies suggest that SENP1 silencing could inhibit the EMT of prostate cancer cells via up-regulating SMAD4 expression.

4. Materials and Methods

4.1. Cell Lines

Human androgen-independent prostate cancer cell, PC3M, was kindly provided by the Institute of Urology, Peking University (Beijing, China). Human androgen-dependent prostate cancer cell LNCaP was obtained from the Institute of Basic Medicine, Chinese Academy of Medicine Science (Beijing, China). Human embryonic kidney cell line HEK293 and 293T were purchased from American Type Culture Collection (ATCC, Manassas, VA, USA). All the cells were maintained in Dulbecco's Minimal Essential Medium (DMEM, Gibco, Grand Island, NY, USA) supplemented with 10% fetal calf serum (FCS, Logan, UT, USA).

4.2. Vectors

Lentiviral short hairpin RNA (shRNA) vector targeting SENP1 (PLKO.1-shSENP1) was constructed according to the protocol of PLKO.1-puro vector (Addgene, Cambridge, MA, USA). Briefly, the forward oligo, 5′TCGAGCGCCAGATTGAAGAACTCGAGTTCTGTTCTTCAATCTGGCGC TTTTTG3′ and reverse oligo, 5′GATCCAAAAAGCGCCAGATTGAAGAACAGAACTCGAGTTCTGT TCTTCAATCTGGCGC3′ were annealed and inserted into the PLKO.1-puro vector. Control vector PLKO.1-shScramble was also purchased from addgene. Lentiviruses were produced in 293T cells after co-transfection of PLKO.1-shSENP1 or PLKO.1-shScramble, packing plasmid psPAX2 and envelope plasmid pMD2.G using the phosphate co-precipitation kit (Promega, Madison, WI, USA). Viruses were purified and concentrated by PEG, followed by determination of viral titers on HT1080 cells.

A chemic fiber-modified and replication-deficient adenovirus-expressing SENP1, Ad5/ F11p.SENP1, and control vector, Ad5/F11p.Null, were constructed as previously described [29]. Wild type SMAD4 gene and mutated SMAD4 gene with two mutations (K113R and K159R) was synthesized (AuGCT, Beijing, China), and inserted into the multiple cloning sites to generate pcDNA3.0-SMAD4 and pcDNA3.0-mutSMAD4, respectively. siRNA621, targeting human SMAD4, and the corresponding control, siRNANC were purchased from GenePharma (Shanghai, China).

4.3. Biological Analysis of PC3M Cells after Infection with Lentiviruses

4.3.1. Apoptosis Analysis

Exponentially growing PC3M cells were seeded into 6-well plate at a density of 2.0×10^5 cells/well. 24 h later, cells were infected with PLKO.1-shSENP1 or PLKO.1-shScramble at 20 multiplicity of infection (MOI). After 24 h incubation, 1 mL of fresh complete culture media was added. 24 h later, cells were collected, labeled with APC conjugated Annexin-V and PI (Sungene Biotech Co., Ltd., Tianjin, China), and analyzed by flow cytometry on FACSCalibur (BD Bioscience, San Jose, CA, USA).

4.3.2. Proliferation Assay

Exponentially growing PC3M cells were collected and labeled with Dye eFluor® 670 (eBioscience, San Diego, CA, USA) according to the manufacturers' instructions. Labeled cells were plated into 6-well plates at a density of 2×10^5 cells per well. Next day, cells were infected with PLKO.1-shSENP1 and PLKO.1-shScramble at 20 MOI. At 0, 24, 48 and 72 h after infection, cells were collected and the fluorescence intensity of Dye 670 was measured by flow cytometry, and the proliferation index was calculated.

4.3.3. Migration Assay

Exponentially growing PC3M cells were seeded into 6-well plate at a density of 2×10^5 cells per well. Next day, cells were infected with 20 MOI of PLKO.1-shSENP1 and PLKO.1-shScramble, and the incubation continued for 24 h. Then, confluent cells were scratched to generate wounds, as described previously [30]. Cells were incubated for 0, 6, 24 h. Live cell images were taken, the distances between the two margins of the wounds were measured, and percentages of wound area filled were determined and analyzed as described previously [30].

For transwell assay, PC3M cells were infected with lentiviral vectors. 24 h later, cells were transfected with siRNAs. After another 24 h incubation, the migration ability of treated PC3M cells was examined by transwell. Briefly, the transwell insert was added to the well by merging the bottom of the insert into the medium in the lower compartment. Then, the cells were seeded into the transwell insert. After 20 h incubation, migrated cells were fixed, stained and counted.

4.4. Western-Blotting and Immunoprecipitation

48 h post-infection with PLKO.1-shSENP1 or PLKO.1-shScramble, the protein expressions of SENP1, SMAD4, E-cadherin and Vimentin in PC3M cells were detected by Western-blotting. Various antibodies used were rabbit anti-human SENP1 monoclonal antibody (Abcam, Cambridge, MA, USA), rabbit anti-human SMAD4 monoclonal antibody (Cell Signaling Technology, CST, Danvers, MA, USA), rabbit anti-human E-cadherin monoclonal antibody (Abcam), and rabbit anti-human Vimentin monoclonal antibody (Abcam). The relative expression levels were normalized by the glyceraldehyde-3-phosphate dehydrogenase (GAPDH). In LNCaP cells, the SENP1, SMAD4, E-cadherin and Vimentin expression were also detected by Western-blotting at 24, 36 and 48 h following infection with Ad5/F11p.SENP1 and Ad5/F11p.Null adenoviruses.

To analyze the role of SMAD4 in SENP1-mediated up-regulation of E-cadherin, PC3M cells were transduced with PLKO.1-shSENP1 and PLKO.1-shScramble. 24 h later, cells were transfected with siRNAs (siRNA621 or siRNANC), and the incubations continued for 48 h. Then, protein expression of SENP1, SMAD4, E-cadherin and Vimentin was analyzed by Western-blotting.

To investigate if SENP1 deSUMOylates SMAD4 to up-regulate E-cadherin, mutated SMAD4 expressing vector pcDNA3.0-mutSMAD4, wild type SMAD4 encoding vector pcDNA3.0-SMAD4 or control vector pcDNA3.0 were transfected into LNCaP cells. 24 h later, cells were infected with adenoviruses, Ad5/F11p.SENP1 or Ad5/F11p.Null. After 48 h incubation, the protein expression of SENP1, SMAD4 and E-cadherin were detected by Western-blotting.

At 48h after infection with PLKO.1-shScramble and PLKO.1-shSENP1, PC3M cells were used to conduct immunoprecipitation analsysis, using the primary antibodies anti-SMAD4 (CST), anti-SUMO-1 (CST) and anti-IgG, according to the manufacturer's instructions.

4.5. Real-Time Reverse Transcript Polymerase Chain Reaction (RT-PCR)

PC3M cells infected with lentiviruses were collected at 24 and 48 h after infection. LNCaP cells were obtained at 24, 36 and 48 h post-infection with adenoviruses. Then, total RNA was extracted, and cDNA was synthesized by using RevertAid First Strand cDNA Synthesis Kit (Thermo Scientific, Wilmington, DE, USA). The mRNA expressions of SENP1 and SMAD4 were quantified by using SYBR® Premix Ex Taq™ (Tli RNaseH Plus) (Takara, Shiga, Japan) on 7500 Fast Real-Time PCR System (Applied Biosystems/Life Technologies, Foster City, CA, USA). The relative expression levels were calculated by $2^{-\Delta Ct}$, using β-actin as the control. The primers for SENP1 and SMAD4 were as following, SENP1: forward, 5′ATCAGGCAGTGAAACGTTGGAC3′ and reverse, 5′GCA GGCTTCATTGTTTATCCCA3′; SMAD4: forward, 5′GGACTGCACCATACACCT3′ and reverse, 5′AATGGGCTGGAATGCAA3′; and β-actin: forward, 5′CATCCTCACCCTGAAGTACCC3′ and reverse, 5′AGCCTGGATAGCAACGTA CATG3′.

4.6. Statistical Analysis

Data are presented as mean ± s.e.m. and statistically analyzed by using GraphPad Prism software version 5 (GraphPad software, San Diego, CA, USA). One-way ANOVA followed by Bonferroni post hoc tests were performed to analyze multiple groups. Difference were considered significant at two sided $p < 0.05$.

5. Conclusions

Up-regulation of SENP1 promotes deSUMOylation of SMAD4 in prostate cancer cells, which reduces the SMAD4 levels and impairs TGF-β/SMADs-mediated transcription activities. SENP1 silencing leads to the increased expression of E-cadherin, and inhibition of EMT of tumor cells. We believe SENP1 silencing leads to the increased SMAD4 levels due to improved stability of SMAD4 in the tumor cells. Moreover, silencing of SENP1 promoted apoptosis, and inhibited the proliferation and migration of tumor cells. Therefore, SENP1 should be considered as a potential target for the treatment of advanced and metastatic prostate cancer.

Acknowledgments: This work was supported by Chinese National Natural Science Foundation of China (No. 81402558 and 81472396), and National High Technology Research and Development Program of China (863 Program) (No. 2012AA02A211 and SS2014AA020515).

Author Contributions: Yuefeng Yang, Lisheng Wang and Chutse Wu conceived and designed the experiments; Xiaoyan Zhang, Hao Wang, Qinghua Jia and Yuefeng Yang performed the experiments; Xiaoyan Zhang, Weidong Xu, Hua Wang analyzed the data; Fengjun Xiao contributed reagents/materials/analysis tools; Yuefeng Yang and Prem Seth wrote the paper.

Conflicts of Interest: The authors declare no conflict of interest.

References

1. Sutinen, P.; Malinen, M.; Heikkinen, S.; Palvimo, J.J. SUMOylation modulates the transcriptional activity of androgen receptor in a target gene and pathway selective manner. *Nucleic Acids Res.* **2014**, *42*, 8310–8319. [CrossRef] [PubMed]
2. Coleman, K.E.; Huang, T.T. How SUMOylation fine-tunes the fanconi anemia DNA repair pathway. *Front. Genet.* **2016**, *7*, 61. [CrossRef] [PubMed]
3. Bawa-Khalfe, T.; Yeh, E.T. SUMO losing balance: SUMO proteases disrupt SUMO homeostasis to facilitate cancer development and progression. *Genes Cancer* **2010**, *1*, 748–752. [CrossRef] [PubMed]

4. Wang, Z.; Jin, J.; Zhang, J.; Wang, L.; Cao, J. Depletion of SENP1 suppresses the proliferation and invasion of triple-negative breast cancer cells. *Oncol. Rep.* **2016**, *36*, 2071–2078. [CrossRef] [PubMed]
5. American Cancer Society Inc. Cancer Facts & Figures 2015. Available online: http://www.cancer.org/acs/groups/content/@editorial/documents/document/acspc-044552.pdf (accessed on 12 April 2017).
6. Kaikkonen, S.; Jaaskelainen, T.; Karvonen, U.; Rytinki, M.M.; Makkonen, H.; Gioeli, D.; Paschal, B.M.; Palvimo, J.J. SUMO-specific protease 1 (SENP1) reverses the hormone-augmented SUMOylation of androgen receptor and modulates gene responses in prostate cancer cells. *Mol. Endocrinol.* **2009**, *23*, 292–307. [CrossRef] [PubMed]
7. Bawa-Khalfe, T.; Cheng, J.; Lin, S.H.; Ittmann, M.M.; Yeh, E.T. SENP1 induces prostatic intraepithelial neoplasia through multiple mechanisms. *J. Biol. Chem.* **2010**, *285*, 25859–25866. [CrossRef] [PubMed]
8. Burdelski, C.; Menan, D.; Tsourlakis, M.C.; Kluth, M.; Hube-Magg, C.; Melling, N.; Minner, S.; Koop, C.; Graefen, M.; Heinzer, H.; et al. The prognostic value of SUMO1/Sentrin specific peptidase 1 (SENP1) in prostate cancer is limited to ERG-fusion positive tumors lacking PTEN deletion. *BMC Cancer* **2015**, *15*, 538. [CrossRef] [PubMed]
9. Zhou, F.; Dai, A.; Jiang, Y.; Tan, X.; Zhang, X. SENP1 enhances hypoxiainduced proliferation of rat pulmonary artery smooth muscle cells by regulating hypoxiainducible factor1α. *Mol. Med. Rep.* **2016**, *13*, 3482–3490. [PubMed]
10. Ao, Q.; Su, W.; Guo, S.; Cai, L.; Huang, L. SENP1 desensitizes hypoxic ovarian cancer cells to cisplatin by up-regulating HIF-1α. *Sci. Rep.* **2015**, *5*, 16396. [CrossRef] [PubMed]
11. Akhurst, R.J.; Derynck, R. TGF-β signaling in cancer—A double-edged sword. *Trends Cell Biol.* **2001**, *11*, S44–S51. [CrossRef]
12. Ding, Z.; Wu, C.J.; Chu, G.C.; Xiao, Y.; Ho, D.; Zhang, J.; Perry, S.R.; Labrot, E.S.; Wu, X.; Lis, R.; et al. SMAD4-dependent barrier constrains prostate cancer growth and metastatic progression. *Nature* **2011**, *470*, 269–273. [CrossRef] [PubMed]
13. Qin, J.; Wu, S.P.; Creighton, C.J.; Dai, F.; Xie, X.; Cheng, C.M.; Frolov, A.; Ayala, G.; Lin, X.; Feng, X.H.; et al. COUP-TFII inhibits TGF-β-induced growth barrier to promote prostate tumorigenesis. *Nature* **2013**, *493*, 236–240. [CrossRef] [PubMed]
14. Liang, M.; Melchior, F.; Feng, X.H.; Lin, X. Regulation of Smad4 sumoylation and transforming growth factor-β signaling by protein inhibitor of activated STAT1. *J. Biol. Chem.* **2004**, *279*, 22857–22865. [CrossRef] [PubMed]
15. Lin, X.; Liang, M.; Liang, Y.Y.; Brunicardi, F.C.; Melchior, F.; Feng, X.H. Activation of transforming growth factor-β signaling by SUMO-1 modification of tumor suppressor Smad4/DPC4. *J. Biol. Chem.* **2003**, *278*, 18714–18719. [CrossRef] [PubMed]
16. Harris, W.P.; Mostaghel, E.A.; Nelson, P.S.; Montgomery, B. Androgen deprivation therapy: Progress in understanding mechanisms of resistance and optimizing androgen depletion. *Nat. Clin. Pract. Urol.* **2009**, *6*, 76–85. [CrossRef] [PubMed]
17. Cannata, D.H.; Kirschenbaum, A.; Levine, A.C. Androgen deprivation therapy as primary treatment for prostate cancer. *J. Clin. Endocrinol. Metab.* **2012**, *97*, 360–365. [CrossRef] [PubMed]
18. Coleman, R. The use of bisphosphonates in cancer treatment. *Ann. N. Y. Acad. Sci.* **2011**, *1218*, 3–14. [CrossRef] [PubMed]
19. Smith, M.R.; Saad, F.; Coleman, R.; Shore, N.; Fizazi, K.; Tombal, B.; Miller, K.; Sieber, P.; Karsh, L.; Damiao, R.; et al. Denosumab and bone-metastasis-free survival in men with castration-resistant prostate cancer: Results of a phase 3, randomised, placebo-controlled trial. *Lancet* **2012**, *379*, 39–46. [CrossRef]
20. Zhu, M.L.; Partin, J.V.; Bruckheimer, E.M.; Strup, S.E.; Kyprianou, N. TGF-β signaling and androgen receptor status determine apoptotic cross-talk in human prostate cancer cells. *Prostate* **2008**, *68*, 287–295. [CrossRef] [PubMed]
21. Gallo-Oller, G.; Vollmann-Zwerenz, A.; Melendez, B.; Rey, J.A.; Hau, P.; Dotor, J.; Castresana, J.S. P144, a transforming growth factor β inhibitor peptide, generates antitumoral effects and modifies SMAD7 and SKI levels in human glioblastoma cell lines. *Cancer Lett.* **2016**, *381*, 67–75. [CrossRef] [PubMed]
22. Herbertz, S.; Sawyer, J.S.; Stauber, A.J.; Gueorguieva, I.; Driscoll, K.E.; Estrem, S.T.; Cleverly, A.L.; Desaiah, D.; Guba, S.C.; Benhadji, K.A.; et al. Clinical development of galunisertib (LY2157299 monohydrate), a small molecule inhibitor of transforming growth factor-β signaling pathway. *Drug Des. Dev. Ther.* **2015**, *9*, 4479–4499.

23. Xu, W.; Zhang, Z.; Yang, Y.; Hu, Z.; Wang, C.H.; Morgan, M.; Wu, Y.; Hutten, R.; Xiao, X.; Stock, S.; et al. Ad5/48 hexon oncolytic virus expressing sTGFβRIIFc produces reduced hepatic and systemic toxicities and inhibits prostate cancer bone metastases. *Mol. Ther. J. Am. Soc. Gene Ther.* **2014**, *22*, 1504–1517. [CrossRef] [PubMed]

24. Hu, Z.; Gerseny, H.; Zhang, Z.; Chen, Y.J.; Berg, A.; Stock, S.; Seth, P. Oncolytic adenovirus expressing soluble TGFβ receptor II-Fc-mediated inhibition of established bone metastases: A safe and effective systemic therapeutic approach for breast cancer. *Mol. Ther. J. Am. Soc. Gene Ther.* **2011**, *19*, 1609–1618. [CrossRef] [PubMed]

25. Long, J.; Wang, G.; He, D.; Liu, F. Repression of Smad4 transcriptional activity by SUMO modification. *Biochem. J.* **2004**, *379*, 23–29. [CrossRef] [PubMed]

26. Dupont, S.; Inui, M.; Newfeld, S.J. Regulation of TGF-β signal transduction by mono- and deubiquitylation of Smads. *FEBS Lett.* **2012**, *586*, 1913–1920. [CrossRef] [PubMed]

27. Ji, Q.; Liu, X.; Han, Z.; Zhou, L.; Sui, H.; Yan, L.; Jiang, H.; Ren, J.; Cai, J.; Li, Q. Resveratrol suppresses epithelial-to-mesenchymal transition in colorectal cancer through TGF-β1/Smads signaling pathway mediated Snail/E-cadherin expression. *BMC Cancer* **2015**, *15*, 97. [CrossRef] [PubMed]

28. Derynck, R.; Muthusamy, B.P.; Saeteurn, K.Y. Signaling pathway cooperation in TGF-β-induced epithelial-mesenchymal transition. *Curr. Opin. Cell Biol.* **2014**, *31*, 56–66. [CrossRef] [PubMed]

29. Hu, Z.B.; Wu, C.T.; Wang, H.; Zhang, Q.W.; Wang, L.; Wang, R.L.; Lu, Z.Z.; Wang, L.S. A simplified system for generating oncolytic adenovirus vector carrying one or two transgenes. *Cancer Gene Ther.* **2008**, *15*, 173–182. [CrossRef] [PubMed]

30. Xu, W.; Neill, T.; Yang, Y.; Hu, Z.; Cleveland, E.; Wu, Y.; Hutten, R.; Xiao, X.; Stock, S.R.; Shevrin, D.; et al. The systemic delivery of an oncolytic adenovirus expressing decorin inhibits bone metastasis in a mouse model of human prostate cancer. *Gene Ther.* **2015**, *22*, 247–256. [CrossRef] [PubMed]

International Journal of
Molecular Sciences

MDPI

Article

The Prostate Cancer Cells Resistant to Docetaxel as in vitro Model for Discovering MicroRNAs Predictive of the Onset of Docetaxel Resistance

Lorenzo Bascetta [1,†], Arianna Oliviero [1,†], Romina D'Aurizio [2], Monica Evangelista [1],
Alberto Mercatanti [1], Marco Pellegrini [2], Francesca Marrocolo [3], Sergio Bracarda [3,4]
and Milena Rizzo [1,4,*]

[1]   Non-Coding RNA Laboratory, Institute of Clinical Physiology (IFC), National Research Council (CNR),
      via G. Moruzzi 1, 56124 Pisa, Italy; lorenzobascetta@hotmail.it (L.B.); arianna90oli@hotmail.it (A.O.);
      m.evangelista@ifc.cnr.it (M.E.); alberto.mercatanti@ifc.cnr.it (A.M.)
[2]   Laboratory for Integrative System Medicine (LISM), Institute of Informatics and Telematics (IIT),
      National Research Council (CNR), via G. Moruzzi 1, 56124 Pisa, Italy; romina.daurizio@gmail.com (R.D.);
      marco.pellegrini@iit.cnr.it (M.P.)
[3]   Department of Oncology, San Donato Hospital, Azienda USL Toscana Sud-Est, via P. Nenni 20,
      52100 Arezzo, Italy; francesca.marrocolo@uslsudest.toscana.it (F.M.);
      sergio.bracarda@uslsudest.toscana.it (S.B.)
[4]   Istituto Toscano Tumori (ITT), via T. Alderotti 26/N, 50139 Firenze, Italy
*   Correspondence: milena.rizzo@ifc.cnr.it; Tel.: +39-050-3098; Fax: +39-050-3153327
†   These authors contributed equally to this work.

Received: 19 June 2017; Accepted: 5 July 2017; Published: 13 July 2017

Abstract: On the grounds that miRNAs present in the blood of prostate cancer (PCa) patients are
released in the growth medium by PCa cells, it is conceivable that PCa cells resistant to docetaxel
(DCT) (DCT^R) will release miRNAs that may be found in PCa patients under DCT therapy if resistant
PCa cells appear. We isolated DCT^R clones respectively from 22Rv1 and DU-145 PCa cell lines and
performed through next-generation sequencing (NGS) the miRNAs profiles of the released miRNAs.
The analysis of the NGS data identified 105 and 1 miRNAs which were differentially released in
the growth medium of the 22Rv1/DCT^R and DU-145/DCT^R clones, respectively. Using additional
filters, we selected 12 and 1 miRNA more released by all 22Rv1/DCT^R and DU-145/DCT^R clones,
respectively. Moreover, we showed that 6 of them were more represented in the growth medium of
the DCT^R cells than the ones of DCT-treated cells. We speculated that they have the pre-requisite
to be tested as predictive biomarkers of the DCT resistance in PCa patients under DCT therapy. We
propose the utilization of clones resistant to a given drug as in vitro model to identify the differentially
released miRNAs, which in perspective could be tested as predictive biomarkers of drug resistance in
tumor patients under therapy.

Keywords: prostate cancer cell lines; docetaxel resistance; circulating miRNAs; predictive biomarkers

---

1. Introduction

Prostate cancer (PCa) evolves into a condition known as castration-resistant prostate cancer
(CRPC) when the androgen deprivation therapy fails. At this stage, docetaxel (DCT) represents the gold
standard treatment [1]. Unfortunately, most patients develop resistance to the drug within a year from
the beginning of the therapy, and the disease progresses [2]. The identification of biomarkers capable
to predict in advance the onset of the DCT resistance could facilitate the shift towards alternative
effective therapies, thus increasing the life expectancy of the patients.

MicroRNAs (miRNAs), and more recently circulating miRNAs (c-miRNA), have been proposed as molecules with a diagnostic and prognostic value for PCa [3,4] but only few studies explored c-miRNAs as predictive biomarker of DCT resistance [5,6].

We have already reported that PCa cells in vitro specifically released in the growth medium the same miRNAs found in the peripheral blood of PCa patients [7]. Therefore, it is conceivable that PCa cells resistant to DCT will release in the growth medium miRNAs that, if detected in the blood of PCa patients, could reveal the presence of DCT-resistant cells. To approach the problem, we isolated PCa cells resistant to DCT and used the next-generation sequencing (NGS) technology and bioinformatics tools to identify miRNAs differentially released in the growth medium by DCT-resistant cells (DCT^R-miRNAs). We also discussed that DCT^R-miRNAs may represent potential predictive biomarkers of the DCT resistance suitable to be tested in PCa patients under DCT therapy.

2. Results

2.1. Isolation of DCT^R Clones from 22Rv1 and DU-145 Prostate Cancer Cell Lines

We exposed for three days either 22Rv1 or DU-145 PCa cells to increasing concentrations of DCT and observed a dose-dependent inhibition of cell proliferation which reached a plateau phase at 4 nM concentration and 3 nM, respectively (Figure 1A,C). We then grew continuously 22Rv1 cells and DU-145 cells respectively in 4 nM and 3 nM DCT-containing medium until we observed colonies presumably resistant to DCT. We isolated and expanded as clones 6 independent colonies from both 22Rv1 cells (22Rv1/DCT^R clones) and DU-145 cells (DU-145/DCT^R clones). Afterwards, we tested them for the DCT resistance. We treated cells with increasing DCT concentration and showed that 22Rv1/DCT^R and DU-145/DCT^R clones were more resistant to DCT with respect to parental cell lines (Figure 1B,D).

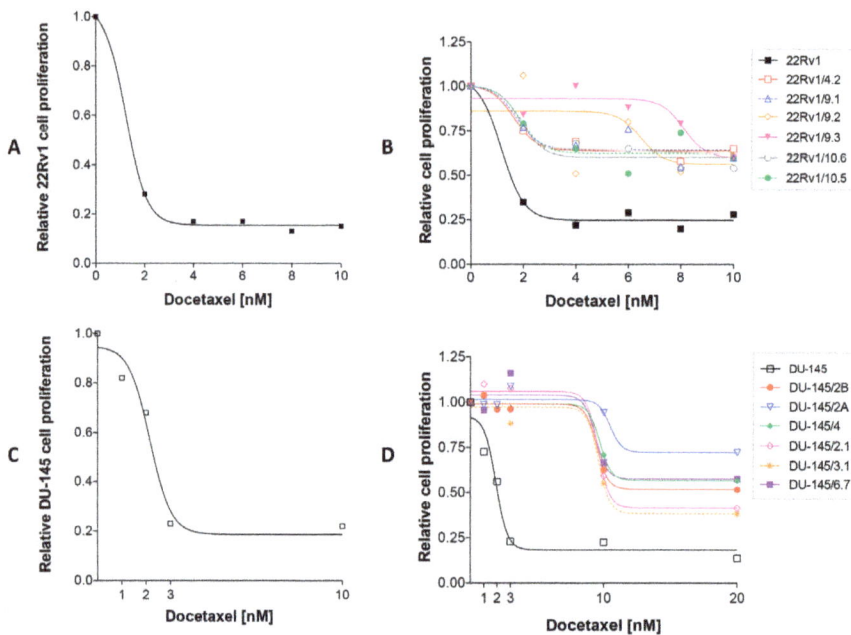

Figure 1. Effect of docetaxel (DCT) on DU-145 cell proliferation. Relative cell proliferation of 22Rv1 cells (**A**), 22Rv1/DCT^R clones (**B**), DU-145 cells (**C**) and DU-145/DCT^R clones (**D**) exposed to increasing concentrations of docetaxel for 72 h.

In addition, we showed that 22Rv1/DCT^R as well as DU-145/DCT^R clones presented expanded generation times (Figure 2A,B) and an acquired capability to extrude Rhodamine 123 (Figure 2C,D). These results, together with the upregulation of *MDR1* gene (Figure 2E,F), indicated that the drug export increase through the ABC transporters was at the basis of the DCT resistance in all clones. We tested also the expression of *BCL2* and *βIII tubulin* genes, whose upregulation is involved in the DCT resistance in PCa [8,9].

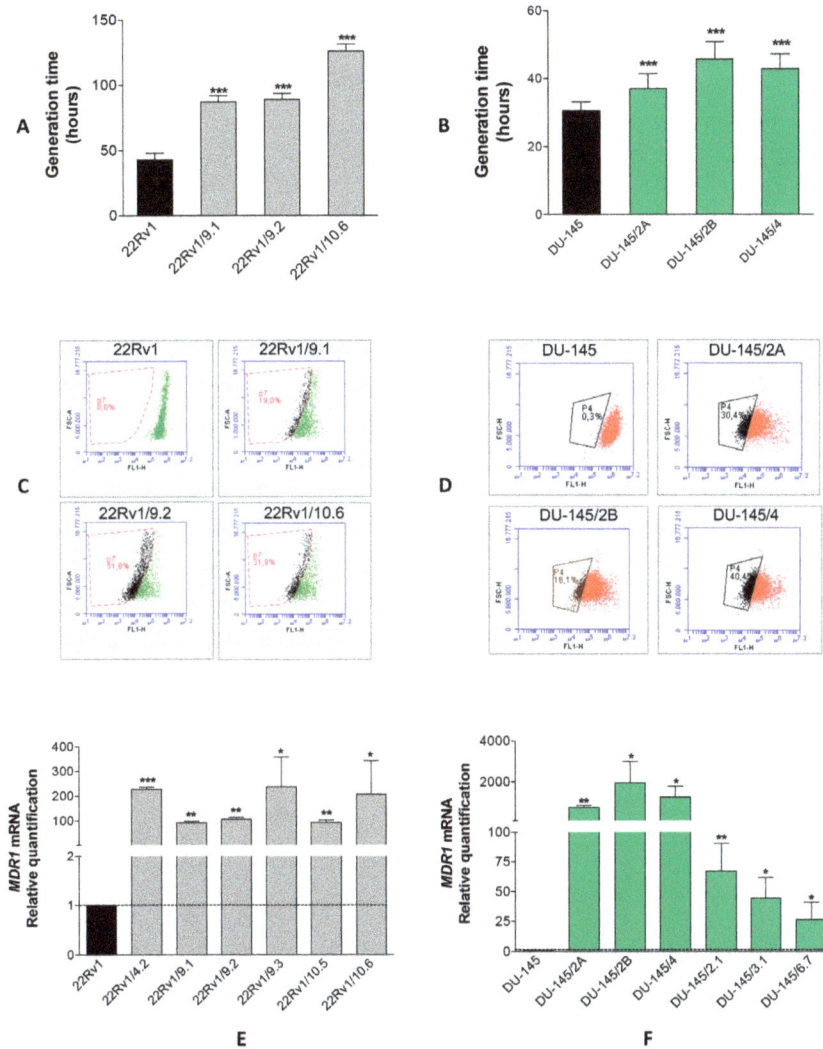

Figure 2. Characterization of DCT^R clones. Generation times (calculated dividing the hours in culture by the number of cell doublings) (**A,B**), percentage of cells which extruded Rh123 (**C,D**), *MDR1* mRNA relative quantification with qRT-PCR (**E,F**) in 22Rv1/DCT^R or DU-145/DCT^R clones with respect to parental cells (dotted line). Data were shown as mean ± SD from three independent experiments (* $p < 0.05$, ** $p < 0.01$, *** $p < 0.001$, unpaired *t*-test).

It is worth noting that while in 22Rv1/DCT^R clones the expression level of both *BCL2* (Figure 3A) and *βIII tubulin* (Figure 3B) genes did not change, these genes showed slight but significant variations in DU-145/DCT^R clones. In particular, *BCL2* was upregulated in DU-145/2B and DU-145/4 clones and downregulated in DU-145/2.1 and DU-145/6.7 clones (Figure 3B). In addition, *βIII tubulin* gene was upregulated in DU-145/2A clones and downregulated in DU-145/3.1 clones (Figure 3D).

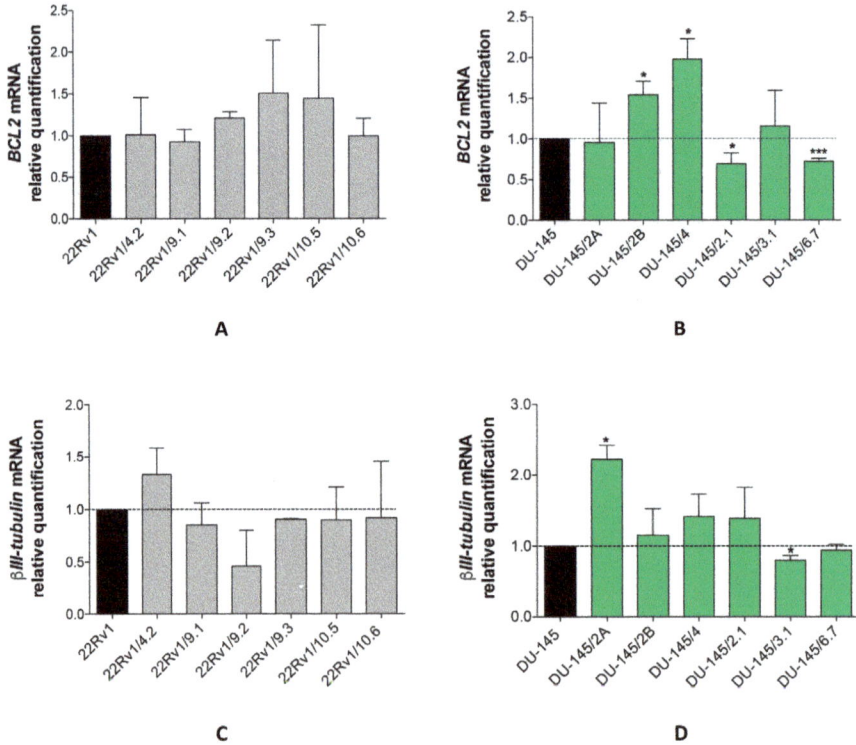

Figure 3. Characterization of DCT^R clones. *BCL2* (**A,B**) and *βIII tubulin* (**C,D**) mRNA relative quantification with qRT-PCR in 22Rv1/DCT^R or DU-145/DCT^R clones with respect to parental cells (dotted line). Data were shown as mean $\pm$ SD from three independent experiments (* $p < 0.05$, *** $p < 0.001$, unpaired *t*-test).

2.2. Identification of miRNAs Differentially Released by DCT^R Clones

We collected the growth medium of each 22Rv1/DCT^R and DU-145/DCT^R clone as well as the one of parental cells and performed the miRNAs expression profile using next-generation sequencing (small RNA-seq). We conducted the differential analysis to identify, for each PCa cell line, the miRNAs that were differentially released with a statistical significance from all DCT^R clones with respect to the corresponding parental cells. The analysis was performed using two different methods, i.e., DESeq2 and edgeR. For DU-145/DCT^R clones, the small RNA-seq experiment was conducted in two different runs. To test whether the experiment could suffer a batch effect on distribution of detected miRNAs, we performed a principal component analysis and observed a significant bias (Figure S1). Therefore, for DU-145/DCT^R clones, we use an additive model formula in the design formula of DESeq2 and edgeR. We identified 134 (with DESeq2) and 127 (with edgeR) miRNAs differentially released (FDR $\leq$ 0.01) by 22Rv1/DCT^R clones (Table S1) and 1 (with DESeq2) and 3 (with edgeR) miRNAs differentially released (FDR $\leq$ 0.01) by DU-145/DCT^R clones (Table S2). We then considered only the miRNAs identified by both algorithms which were 105 miRNAs for 22Rv1/DCT^R clones (44 more released miRNAs, 61 retained/less released miRNAs) and 1 more released miRNA for DU-145/DCT^R clones (Figure 4A). We then focused on miRNAs more released (red in Figure 4A) and we selected those whose average number of normalized reads (NR) in all 22Rv1/DCT^R (or DU-145/DCT^R) clones or in the corresponding parental cells was greater than 100 (Figure 4A, box i). Among them, we selected those with a difference greater than 10 NR ($\Delta > 10$) between the clone with the lowest value and the biological replicate of the parental cells with the highest value (Figure 4A, box ii). Thus, we identified 12 miRNAs for 22Rv1/DCT^R clones (Figure 4B) and 1 miRNA (miR-146a-5p) for DU-145/DCT^R clones (Figure 4C and Figure S2).

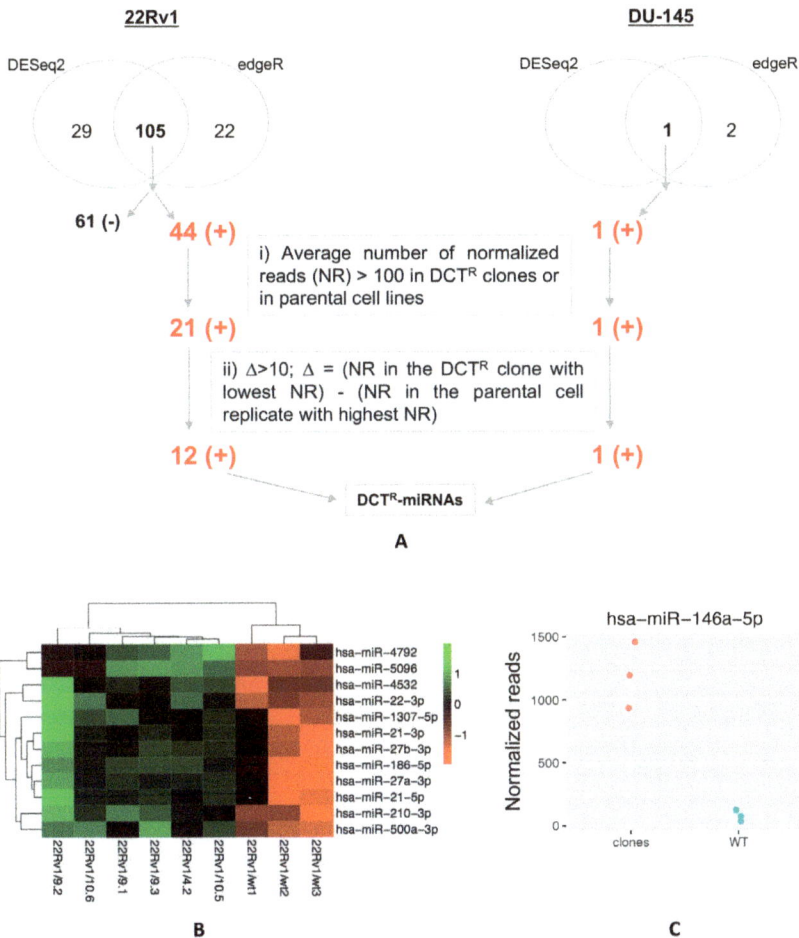

Figure 4. Extracellular miRNA profile of DCT^R clones. (**A**) Schematic representation of the selective criteria used to identify DCT^R-miRNAs in both 22Rv1/DCT^R and DU-145/DCT^R clones (+, more released; −, retained/less released); (**B**) Heatmap representing the result of DCT^R-miRNAs in 22Rv1/DCT^R clones and parental cells biological replicates (22Rv1WT1-3); (**C**) Normalized reads of miR-146a-5p in DU-145/DCT^R clones (red spots) and parental cells biological replicates (blue spots) of DU-145/2A, /2B and /4 clones (see Figure S2 for DU-145/2.1, /3.1 and /6.7 clones).

Finally, we validated the expression of these miRNAs (DCT^R-miRNAs) in the growth medium of both 22Rv1/DCT^R and DU-145/DCT^R clones by qRT-PCR. With the exception of miR-1307-5p (whose level showed a slight change), the levels of all DCT^R-miRNAs were higher in the growth medium of the DCT^R clones (Figure 5A,B) thus confirming the NGS data.

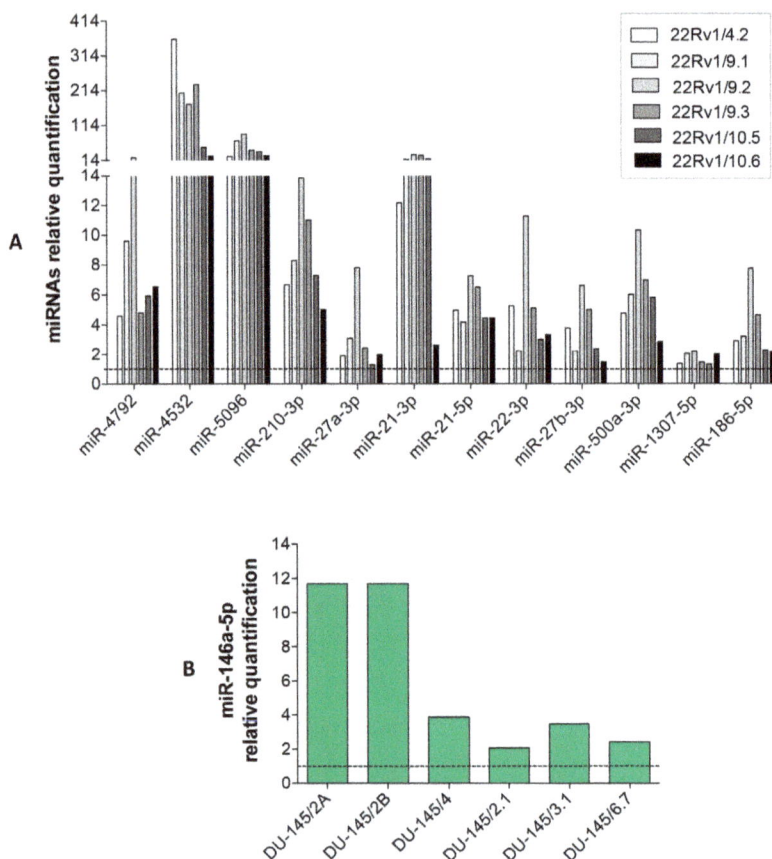

Figure 5. DCT^R-miRNAs validation. Relative quantification with qRT-PCR of DCT^R-miRNAs in the growth medium of 22Rv1/ (**A**) or DU-145/ (**B**) DCT^R clones with respect to parental cells (dotted line).

2.3. DCT^R-miRNAs as Possible Biomarkers of DCT Resistance

To evaluate whether the level of the DCT^R-miRNAs was affected by DCT, we measured by qRT-PCR the level of the 12 DCT^R-miRNAs in the growth medium of parental cells (22Rv1 or DU-145) treated for three days with DCT and we calculated the fold change of these miRNAs in comparison to that found in the growth medium of untreated cells (column B, Table 1). Then, we evaluated the mean fold change of the DCT^R-miRNAs (measured by qRT-PCR) in the growth medium of 22Rv1/DCT^R or DU-145/DCT^R clones with respect to that of the corresponding parental cells line (column A, Table 1). Finally, for each DCT^R-miRNAs, we calculate the ratio between the fold change value of DCT-treated versus untreated cells and the mean fold change of DCT^R-miRNA in 22Rv1/DCT^R or DU-145/DCT^R clones versus parental cells (column A/column B, Table 1). We observed that the level of 6 DCT^R-miRNAs was similar in cells either resistant to or treated with DCT, whereas the level of the remaining 6 DCT^R-miRNAs was higher (more than 2-fold) in cells resistant to DCT versus cells treated with DCT (Table 1, miRNAs underlined and in bold), suggesting that their differential release was specific of the DCT-resistant phenotype.

Table 1. DCT^R-miRNAs level in the growth medium of DCT^R clones or DCT-treated cells.

DCT^R/miRNAs	Column A *. DCT^R-miRNAs Level in DCT^R Clones Versus Parental Cells Growth Medium	Column B **. DCT^R-miRNAs Level in DCT-Treated Versus Untreated Cells Growth Medium	Column A/ Column B	Unpaired *t*-Test Pvalue (A Versus B)
miR-4792 ***	8.94 ± 2.75	1.69 ± 0.11	5.27	$p < 0.05$
miR-4532	175.20 ± 50.66	2.10 ± 0.48	83.42	$p < 0.01$
miR-5096	48.61 ± 10.00	1.28 ± 0.03	37.97	$p < 0.001$
miR-210-3p	8.80 ± 1.13	2.28 ± 0.19	3.86	$p < 0.001$
miR-27a-3p	3.48 ± 0.85	2.13 ± 0.2	1.63	NS
miR-21-3p	17.7 ± 4.12	3.9 ± 0.49	4.53	$p < 0.01$
miR-21-5p	5.18 ± 0.43	2.49 ± 0.39	2.07	$p < 0.001$
miR-22-3p	5.22 ± 1.21	3.86 ± 0.32	1.35	NS
miR-27b-3p	3.52 ± 0.77	4.29 ± 0.25	0.82	NS
miR-500a-3p	6.11 ± 1.01	3.78 ± 0.03	1.61	$p < 0.05$
miR-186-5p	3.82 ± 0.86	3.64 ± 0.43	1.05	NS
miR-146a-5p ****	5.87 ± 1.86	10.21 ± 1.46	0.57	NS

* Each value represents the mean ± SD of the DCT^R/miRNAs level in the DCT^R clones. ** Each value represents the mean ± SD of the DCT^R/miRNAs level in six biological replicates. *** DCT^R-miRNAs whose column A/column B value was higher than 2 are underlined and in bold. **** miR-146a-5p was the only miRNA selected from DU-145/DCT^R clones. NS, not significant.

2.4. Expression Level of Selected DCT^R-miRNAs in DCT^R Clones

To gain more insight into the differential release of the DCT^R-miRNAs by the DCT^R clones we measured the intracellular level of the DCT^R-miRNAs specifically associated with DCT-resistant phenotype (Table 1, underlined and in bold) in 22Rv1/DCT^R clones with respect to parental cell lines (Figure 6). We found that DCT^R-miRNAs were upregulated in 22Rv1/DCT^R clones (particularly miR-4532, miR-5096 and miR-210-3p) except for miR-21-3p and miR-21-5p whose level did not change significantly, suggesting that the release of DCT^R-miRNAs by DCT^R PCa cells could be mediated by both active or passive mechanisms.

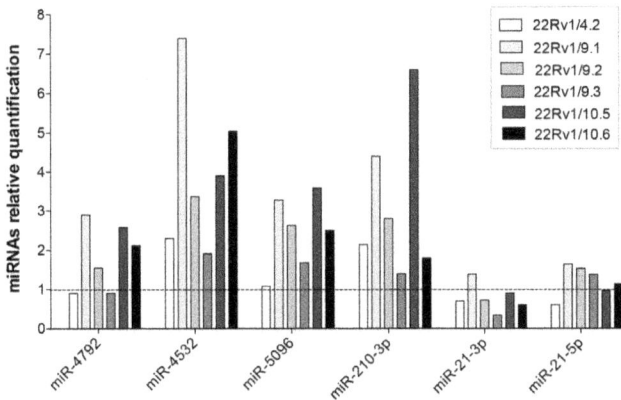

Figure 6. Intracellular level of selected DCT^R-miRNAs in 22Rv1/DCT^R clones. Relative quantification by qRT-PCR of DCT^R-miRNAs in the growth medium of 22Rv1/DCT^R clones with respect to parental cells (dotted line).

3. Discussion

Docetaxel (DCT) is the first line chemotherapy for patients who become insensitive to androgen deprivation therapy (castration-resistant prostate cancer, CRPC). Unfortunately, the DCT therapy frequently favors the development of DCT resistance in metastatic CRPC patients [2]. Hence, the discovery of biomarkers that indicate in advance the onset of DCT resistance could allow the early switch to other effective treatment options such as abiraterone acetate and enzalutamide [10,11]. While the detection of circulating miRNAs (c-miRNAs) in tumor patients is widely applied in clinical research [12–14], the detection of c-miRNAs in cancer patients under medical treatment is under-investigated, especially in PCa patients as confirmed by the poor availability of published data. So far, the change of miR-21 [5] and miR-210 [6] levels in the blood of PCa patients under DCT treatment have been associated with the clinical outcome. In addition, miR-141 [15,16], miR-146b-3p and miR-194 [16] have been shown to predict PCa clinical progression after therapies. Finally, in a recent work Lin et al. [17,18] demonstrated that miR-200a, miR-200b, miR-200c, miR-132, miR-375, miR-429 were associated with overall survival after DCT treatment.

In this work, we detected the miRNAs released by clones of 22Rv1 and DU-145 PCa cells resistant to DCT with the aim to identify those preferentially released by the DCT-resistant cells. The decision to isolate DCT^R clones from the androgen-dependent (22Rv1) and androgen-independent (DU-145) cell lines was due to the fact that DCT-based chemotherapy has become a therapeutic option not only for CRPC patients but also for those who have not yet reached the full androgen-independence [19]. In line with other data [20,21], we showed that the main molecular mechanism at the basis of the resistance of the DCT^R clones was the increase in the activity of the ABC membrane transporters, a family of ATP-dependent membrane-bound drug efflux pump that protects the tumor cells by cytotoxic drugs. In addition, we showed that both 22Rv1/DCT^R and DU-145/DCT^R clones had a generation time longer than those of the parental cell line. These data were in line with those obtained by Corcoran et al. [22]. We cannot exclude that the reduced growth rate of the DCT^R clones may contribute to the DCT-resistant phenotype.

Once the DCT^R clones were obtained, we detected miRNAs differentially released in the growth medium and selected those that passed several filters. First, we considered the miRNAs differentially released by all DCT^R clones derived from each cell line and this increased the probability that they could be those really released by DCT-resistant cells. Second, we focused on miRNAs that were more released and on these miRNAs we imposed a cut-off on normalized reads (NR > 100) as we believe that this level was adequate for an easier detection. Third, we selected miRNAs whose level were significantly different between DCT^R clones and parental cells by selecting not only the miRNAs more released by a |log2FC| > 1, but also the miRNAs whose level between the DCT^R clone with the lower NR and the parental line replicate with the highest NR was greater than 10 ($\Delta > 10$). At the end, we selected 12 DCT^R-miRNAs and 1 DCT^R-miRNA released by 22Rv1/DCT^R and DU-145/DCT^R clones respectively. In our opinion, the low abundance of DU-145/DCT^R-miRNAs in comparison with 22Rv1/DCT^R-miRNAs depends on the fact that DU-145/DCT^R clones are more heterogeneous than 22Rv1/DCT^R clones as suggested by the molecular data (*BLC2* and *βIII tubulin* expression levels) and the different levels of *MDR1* activation. Another observation is that 22Rv1/DCT^R and DU-145/DCT^R clones did not release the same miRNAs. This finding is not completely surprising given that DCT^R clones derived from different tumor contexts and therefore the resistant phenotype may be different. As DCT resistance in PCa patients may arise from different tumor context as well, this observation underlines the importance of using more PCa cell lines to have a higher chance of identifying miRNAs which could potentially predict DCT resistance.

We then examined whether DCT^R-miRNAs could be potential biomarkers of DCT resistance. In particular, we evaluated whether the release of the DCT^R-miRNAs was affected also by the DCT treatment. We obtained promising results from the comparison of the DCT^R-miRNAs released in the growth medium by DCT^R clones and by parental cells treated with DCT for 3 days. We showed that the release of miR-4792, miR-4532, miR-5096, miR-210-3p, miR-21-3p and miR-21-5p were higher in the growth medium of DCT^R clones than in DCT-treated cells thus rendering these miRNAs specific of DCT-resistant cells and then promising putative biomarkers of DCT resistance.

To be a good biomarker, a miRNA should be highly released before the onset of the DCT resistance. So far, many c-miRNAs diagnostic/prognostic of PCa have been identified, but it is still unclear if these c-miRNAs are useful in the management of PCa patients under DCT therapy. At the moment, only two c-miRNAs (miR-21, miR-210) have been associated to the outcome of DCT therapy [5,6] and, as such, have characteristics of predicting the DCT resistance of PCa patients. It has been shown that miR-21 level was elevated in CRPC patients, especially in those who developed resistance after DCT chemotherapy [5]. In the same way, it was demonstrated that the serum levels of miR-210 was higher in patients whose disease was resistant to DCT treatment as assessed by change in prostate specific antigen (PSA) [6]. It is worth noticing that both miR-21 and miR-210 belong to the 6 DCT^R-miRNAs that were highly released only by cells resistant to DCT, suggesting that our approach to isolate DCT^R PCa cells and to select the differentially released miRNAs not affected by DCT treatment could represent a manner to discover miRNAs to be tested as predictive biomarkers of the DCT resistance. As a consequence, the other DCT^R-miRNAs specifically released by DCT^R cells (miR-4792, miR-4532, miR-5096), and not yet tested in PCa patients, represent promising candidates to be tested as DCT predictive biomarkers in PCa patients.

Finally, we considered the intracellular level of the 6 DCT^R-miRNAs in DCT^R clones to gain more insight on the miRNA release mechanism. We found that not all DCT^R-miRNAs were upregulated in all 22Rv1/DCT^R clones suggesting that the release of these miRNAs by DCT-resistant PCa cells may be also mediated by an active mechanism. There are several reports in favor of either a specific or non-specific secretion of miRNAs by the cells (for review see [23,24]).

Among the miRNAs that were more upregulated (i.e., miR-210-3p, miR-4532 and miR-5096), miR-210 has been found to be transcriptionally activated by the hypoxia-inducible factor 1 α (HIF-1α) [25]. In particular, Cheng et al. [6] demonstrated that the treatment of PCa cells with hypoxic conditions determined the upregulation and the release of miR-210 in the growth medium. In addition, they demonstrated that high miR-210 serum level is associated with therapy (hormone therapy in combination with chemotherapy) resistance in CRPC patients. Given that it is known that tumor hypoxia is associated with therapy resistance [26], the author suggested that an increased hypoxia response signaling is present in a subset of CRPC patients, leading to increased miR-210 level and therapy resistance. So, the upregulation/release of miR-210-3p by 22Rv1/DCT^R clones is in line with these data: we can only speculate that an increased hypoxia signaling could be associated with DCT resistance mechanism of 22Rv1/DCT^R clones. However, miR-210 has a central role in promoting cancer-associated fibroblast (CAF) formation in PCa [27], hence in cancer progression, suggesting that it may act as a signal molecule in tumor microenvironment. Regarding miR-5096 and miR-4532, since they have been recently discovered, there are no reports on their biological functions. However, there are two studies that demonstrated that the transfer of miR-5096 through the gap junctions from glioma cells to astrocyte [28] or to human microvascular endothelial cells [29] promotes the glioma invasion with an unknown mechanism suggesting an oncogenic role of this miRNA, in line with its upregulation in 22Rv1/DCT^R clones.

In conclusion, we showed that DCT resistant PCa cells release in the growth medium miRNAs whose level change in PCa patients resistant to DCT therapy, suggesting that clones resistant to DCT represent a good in vitro model to identify potential DCT predictive miRNAs to be tested as c-miRNA in PCa patients during DCT treatment. In addition, we showed that not all miRNA specifically released by DCT^R clones were upregulated, suggesting that the miRNAs release may be mediated also by an active mechanism. In perspective, this in vitro approach for discovering miRNAs predictive of the onset of drug resistance may be applied to other tumor types as well as other anticancer drugs.

4. Materials and Methods

4.1. Cells and Culture Conditions

DU-145 and 22Rv1 cell lines were grown in RPMI 1640 medium added of 10% fetal bovine serum, 1% penicillin/streptomycin 2 mM and 1% L-glutammine 2 mM (Euroclone, Milan, Italy). Cells were incubated at 37 °C in a humidified atmosphere containing 6% CO_2. For DCT treatment, cells were grown for 24 h and treated with DCT (Taxotere, 20 mg/mL, Sanofi Aventis, Milan, Italy) for 3 days.

4.2. Dose Response Curves

2×10^5 cells were seeded per well (6 wells multiplate) and exposed to increasing concentration of DCT (Taxotere, 20 mg/mL, Sanofi Aventis). At specified time points, cells were fixed in 2% paraformaldehyde in PBS (Oxoid, Altrincham, Cheshire, UK), and subsequently stained with 0.1% crystal violet (SIGMA, St. Louis, MO, USA) dissolved in 20% methanol (SIGMA) and let dry at room temperature. Cells were then lysed with 10% acetic acid and the optical density (OD 590 nm) of the solution, detected with ChroMate Microtetraplate Reader apparatus (Awareness Technology, Westport, CT, USA), was used to measure cell proliferation.

4.3. Isolation and Characterization of DCT-Resistant (DCT^R) Clones

To obtain DCT^R clones, 5×10^5 cells/100 mm diameter dish were grown in presence of DCT of either 3 nM (22Rv1) or 4 nM (DU-145). Each 3–4 days, the medium was removed and replaced with fresh selective medium until colonies formation (30–50 days). Thereafter, independent colonies were isolated and expanded to clonal populations. To verify whether clones were still resistant to DCT, cells were seeded at density of 2×10^4 cells/cm^2 and 24 h later exposed at increasing concentrations of DCT (range 1–20 nM). After 72 h exposure, cell proliferation was determined as reported above. The generation times were calculated dividing the hours in culture by the number of cell doublings calculated as [lg$_2$ (final cell number/initial cell number)/lg$_2$ 2].

4.4. Rhodamine 123 Exclusion Assay

The assay was performed as previously described [30] with some modifications. Briefly, for each sample of DU-145 or 22Rv1 cells and DCT^R clones, 2 aliquots of 4×10^4 cells/mL were incubated at 37 °C with 1 μL of Rhodamine 123 (Rh123) (1 mM, SIGMA) and with or without 1 μL of Verapamil (50 mM, SIGMA). After 30 min, cells were harvested and centrifuged and each aliquot was split into two aliquots and analyzed with FACScalibur cytofluorimeter (BD Biosciences, San Jose, CA, USA) immediately (aliquot 1) and after 1 h (aliquot 2). Rh123 green was recorded on FL1 channel and the gate was set to exclude fluorescence of both aliquot 1 samples and verapamil minus aliquot 2 sample. The Rh123 extrusion was detected in the verapamil plus aliquot 2 sample and expressed as percentage of cells with a fluorescence different from the one excluded from the gate.

4.5. Total RNA Isolation

Intracellular total RNA was extracted with miRNeasy mini kit (Qiagen, Hilden, Germany) following the manufacturer's instructions. For NGS experiment, extracellular total RNA was isolated from 2 mL growth medium with QIAamp circulating nucleic acid kit (Qiagen), following the manufacturer's instructions. For qRT-PCR analysis, extracellular total RNA extraction was performed as previously described [7].

4.6. miRNAs and mRNA Quantification

miRNAs and mRNAs quantification were performed as previously described [7]. Transcripts values were normalized to those obtained from the amplification of the internal control (GAPDH,

HPRT1, β-actin for mRNAs and U6, sno-44, sno-55, sno-110 for intracellular miRNA and cel-miR-39 for extracellular miRNAs).

4.7. miRNA Profiling with Next-Generation Sequencing (NGS) Technology (Small RNA-Seq)

The small RNA libraries were constructed using TruSeq Small RNA kit (Illumina, San Diego, CA, USA) according to the manufacturer's suggestions. cDNA libraries were loaded at six-plex level of multiplexing (~4 million reads per samples) into a flow cell V3, and sequenced in a single-reads mode (50 bp) on a MiSeq sequencer (Illumina). Raw reads were analyzed as described in Barsanti et al. [31]. Briefly, raw sequences were de-multiplexed using the Illumina pipline CASAVA v.1.8.2. software (Illumina). FastQC was used for quality check and primary reads were trimmed of adapters sequence using Cutadapt v1.2.1 [32]. Remaining reads, with a minimum length of 17 bzp and maximum 35 bp after trimming, were clustered for unique hits and mapped to pre-miRNA sequences present into the miRBase (rel. 21) employing miRExpress tool v 2.1.3 [33].

4.8. Statistical Analysis

Data are expressed as mean ± SD of at least three independent experiments, and analyzed with Student's *t*-test.

The differential analysis was performed using two different statistical methods implemented in the two packages, edgeR [34] and DESeq2 [35] of the R Bioconductor repository. Exact *p*-values were adjusted for multiple testing according to Benjamini–Hochberg procedure. In order to control the batch effect of the two sequencing runs of the DU-145/DCT^R clones experiment, the analysis was corrected for baseline differences between the batches using an additive model in the design formula of both edgeR and DESeq2 approaches.

The hierarchical clustering was performed using the mean centered Log2 normalized reads, the Euclidean distance and complete agglomerative method.

Supplementary Materials: Supplementary materials can be found at www.mdpi.com/1422-0067/18/7/1512/s1.

Acknowledgments: We would like to thank Stefania Biagini (Fondazione Toscana Gabriele Monasterio, FTGM, Pisa) for giving us taxotere and Giuseppe Rainaldi for critical reading of the manuscript. This work was supported by the Istituto Toscano Tumori (grant 2010-Giuseppe Rainaldi; grant 2013-Milena Rizzo).

Author Contributions: Milena Rizzo conceived and designed the experiments. Lorenzo Bascetta, Monica Evangelista, Arianna Oliviero and Milena Rizzo performed the experiments. Milena Rizzo, Lorenzo Bascetta, Arianna Oliviero analyzed the data. Romina D'Aurizio and Alberto Mercatanti analyzed the NGS data. Milena Rizzo wrote the paper. Sergio Bracarda, Marco Pellegrini, Francesca Marrocolo revised the manuscript.

Conflicts of Interest: The authors declare no conflict of interest.

References

1. Schurko, B.; Oh, W.K. Docetaxel chemotherapy remains the standard of care in castration-resistant prostate cancer. *Nat. Clin. Pract. Oncol.* **2008**, *5*, 506–507. [CrossRef] [PubMed]
2. Petrylak, D.P.; Tangen, C.M.; Hussain, M.H.; Lara, P.N., Jr.; Jones, J.A.; Taplin, M.E.; Burch, P.A.; Berry, D.; Moinpour, C.; Kohli, M.; et al. Docetaxel and estramustine compared with mitoxantrone and prednisone for advanced refractory prostate cancer. *N. Engl. J. Med.* **2004**, *351*, 1513–1520. [CrossRef] [PubMed]
3. Endzelins, E.; Melne, V.; Kalnina, Z.; Lietuvietis, V.; Riekstina, U.; Llorente, A.; Line, A. Diagnostic, prognostic and predictive value of cell-free miRNAs in prostate cancer: A systematic review. *Mol. Cancer* **2016**, *15*, 41. [CrossRef] [PubMed]
4. Filella, X.; Foj, L. miRNAs as novel biomarkers in the management of prostate cancer. *Clin. Chem. Lab. Med.* **2017**, *55*, 715–736. [CrossRef] [PubMed]
5. Zhang, H.L.; Yang, L.F.; Zhu, Y.; Yao, X.D.; Zhang, S.L.; Dai, B.; Zhu, Y.P.; Shen, Y.J.; Shi, G.H.; Ye, D.W. Serum miRNA-21: Elevated levels in patients with metastatic hormone-refractory prostate cancer and potential predictive factor for the efficacy of docetaxel-based chemotherapy. *Prostate* **2011**, *71*, 326–331. [CrossRef] [PubMed]

6. Cheng, H.H.; Mitchell, P.S.; Kroh, E.M.; Dowell, A.E.; Chery, L.; Siddiqui, J.; Nelson, P.S.; Vessella, R.L.; Knudsen, B.S.; Chinnaiyan, A.M.; et al. Circulating microRNA profiling identifies a subset of metastatic prostate cancer patients with evidence of cancer-associated hypoxia. *PLoS ONE* **2013**, *8*, e69239. [CrossRef] [PubMed]

7. Lucotti, S.; Rainaldi, G.; Evangelista, M.; Rizzo, M. Fludarabine treatment favors the retention of miR-485-3p by prostate cancer cells: Implications for survival. *Mol. Cancer* **2013**, *12*, 52. [CrossRef] [PubMed]

8. Yamanaka, K.; Rocchi, P.; Miyake, H.; Fazli, L.; So, A.; Zangemeister-Wittke, U.; Gleave, M.E. Induction of apoptosis and enhancement of chemosensitivity in human prostate cancer LNCaP cells using bispecific antisense oligonucleotide targeting *Bcl-2* and *Bcl-xL* genes. *BJU Int.* **2006**, *97*, 1300–1308. [CrossRef] [PubMed]

9. Terry, S.; Ploussard, G.; Allory, Y.; Nicolaiew, N.; Boissiere-Michot, F.; Maille, P.; Kheuang, L.; Coppolani, E.; Ali, A.; Bibeau, F.; et al. Increased expression of class III beta-tubulin in castration-resistant human prostate cancer. *Br. J. Cancer* **2009**, *101*, 951–956. [CrossRef] [PubMed]

10. Sartor, O.; Gillessen, S. Treatment sequencing in metastatic castrate-resistant prostate cancer. *Asian J. Androl.* **2014**, *16*, 426–431. [CrossRef] [PubMed]

11. Hathaway, A.R.; Baker, M.K.; Sonpavde, G. Emerging agents for the therapy of advanced prostate cancer. *Future Oncol.* **2015**, *11*, 2775–2787. [CrossRef] [PubMed]

12. Cheng, G. Circulating miRNAs: Roles in cancer diagnosis, prognosis and therapy. *Adv. Drug Deliv. Rev.* **2015**, *81*, 75–93. [CrossRef] [PubMed]

13. Larrea, E.; Sole, C.; Manterola, L.; Goicoechea, I.; Armesto, M.; Arestin, M.; Caffarel, M.M.; Araujo, A.M.; Araiz, M.; Fernandez-Mercado, M.; et al. New concepts in cancer biomarkers: Circulating miRNAs in liquid biopsies. *Int. J. Mol. Sci.* **2016**, *17*, 627. [CrossRef] [PubMed]

14. Armand-Labit, V.; Pradines, A. Circulating cell-free microRNAs as clinical cancer biomarkers. *Biomol. Concepts* **2017**, *8*, 61–81. [CrossRef] [PubMed]

15. Gonzales, J.C.; Fink, L.M.; Goodman, O.B., Jr.; Symanowski, J.T.; Vogelzang, N.J.; Ward, D.C. Comparison of circulating MicroRNA 141 to circulating tumor cells, lactate dehydrogenase, and prostate-specific antigen for determining treatment response in patients with metastatic prostate cancer. *Clin. Genitourin. Cancer* **2011**, *9*, 39–45. [CrossRef] [PubMed]

16. Selth, L.A.; Townley, S.L.; Bert, A.G.; Stricker, P.D.; Sutherland, P.D.; Horvath, L.G.; Goodall, G.J.; Butler, L.M.; Tilley, W.D. Circulating microRNAs predict biochemical recurrence in prostate cancer patients. *Br. J. Cancer* **2013**, *109*, 641–650. [CrossRef] [PubMed]

17. Lin, H.M.; Castillo, L.; Mahon, K.L.; Chiam, K.; Lee, B.Y.; Nguyen, Q.; Boyer, M.J.; Stockler, M.R.; Pavlakis, N.; Marx, G.; et al. Circulating microRNAs are associated with docetaxel chemotherapy outcome in castration-resistant prostate cancer. *Br. J. Cancer* **2014**, *110*, 2462–2471. [CrossRef] [PubMed]

18. Lin, H.M.; Mahon, K.L.; Spielman, C.; Gurney, H.; Mallesara, G.; Stockler, M.R.; Bastick, P.; Briscoe, K.; Marx, G.; Swarbrick, A.; et al. Phase 2 study of circulating microRNA biomarkers in castration-resistant prostate cancer. *Br. J. Cancer* **2017**, *116*, 1002–1011. [CrossRef] [PubMed]

19. James, N.D.; Sydes, M.R.; Clarke, N.W.; Mason, M.D.; Dearnaley, D.P.; Spears, M.R.; Ritchie, A.W.; Parker, C.C.; Russell, J.M.; Attard, G.; et al. Addition of docetaxel, zoledronic acid, or both to first-line long-term hormone therapy in prostate cancer (STAMPEDE): Survival results from an adaptive, multiarm, multistage, platform randomised controlled trial. *Lancet* **2016**, *387*, 1163–1177. [CrossRef]

20. O'Neill, A.J.; Prencipe, M.; Dowling, C.; Fan, Y.; Mulrane, L.; Gallagher, W.M.; O'Connor, D.; O'Connor, R.; Devery, A.; Corcoran, C.; et al. Characterisation and manipulation of docetaxel resistant prostate cancer cell lines. *Mol. Cancer* **2011**, *10*, 126. [CrossRef] [PubMed]

21. Zhu, Y.; Liu, C.; Nadiminty, N.; Lou, W.; Tummala, R.; Evans, C.P.; Gao, A.C. Inhibition of ABCB1 expression overcomes acquired docetaxel resistance in prostate cancer. *Mol. Cancer Ther.* **2013**, *12*, 1829–1836. [CrossRef] [PubMed]

22. Corcoran, C.; Rani, S.; O'Brien, K.; O'Neill, A.; Prencipe, M.; Sheikh, R.; Webb, G.; McDermott, R.; Watson, W.; Crown, J.; et al. Docetaxel-resistance in prostate cancer: Evaluating associated phenotypic changes and potential for resistance transfer via exosomes. *PLoS ONE* **2012**, *7*, e50999. [CrossRef] [PubMed]

23. Makarova, J.A.; Shkurnikov, M.U.; Wicklein, D.; Lange, T.; Samatov, T.R.; Turchinovich, A.A.; Tonevitsky, A.G. Intracellular and extracellular microRNA: An update on localization and biological role. *Prog. Histochem. Cytochem.* **2016**, *51*, 33–49. [CrossRef] [PubMed]

24. Turchinovich, A.; Tonevitsky, A.G.; Burwinkel, B. Extracellular miRNA: A collision of two paradigms. *Trends Biochem. Sci.* **2016**, *41*, 883–892. [CrossRef] [PubMed]
25. Huang, X.; Le, Q.T.; Giaccia, A.J. MiR-210-micromanager of the hypoxia pathway. *Trends Mol. Med.* **2010**, *16*, 230–237. [CrossRef] [PubMed]
26. Bertout, J.A.; Patel, S.A.; Simon, M.C. The impact of O_2 availability on human cancer. *Nat. Rev. Cancer* **2008**, *8*, 967–975. [CrossRef] [PubMed]
27. Taddei, M.L.; Cavallini, L.; Comito, G.; Giannoni, E.; Folini, M.; Marini, A.; Gandellini, P.; Morandi, A.; Pintus, G.; Raspollini, M.R.; et al. Senescent stroma promotes prostate cancer progression: The role of miR-210. *Mol. Oncol.* **2014**, *8*, 1729–1746. [CrossRef] [PubMed]
28. Hong, X.; Sin, W.C.; Harris, A.L.; Naus, C.C. Gap junctions modulate glioma invasion by direct transfer of microRNA. *Oncotarget* **2015**, *6*, 15566. [CrossRef] [PubMed]
29. Thuringer, D.; Boucher, J.; Jego, G.; Pernet, N.; Cronier, L.; Hammann, A.; Solary, E.; Garrido, C. Transfer of functional microRNAs between glioblastoma and microvascular endothelial cells through gap junctions. *Oncotarget* **2016**, *7*, 73925–73934. [CrossRef] [PubMed]
30. Touil, Y.; Zuliani, T.; Wolowczuk, I.; Kuranda, K.; Prochazkova, J.; Andrieux, J.; Le Roy, H.; Mortier, L.; Vandomme, J.; Jouy, N.; et al. The PI3K/AKT signaling pathway controls the quiescence of the low-Rhodamine123-retention cell compartment enriched for melanoma stem cell activity. *Stem Cells* **2013**, *31*, 641–651. [CrossRef] [PubMed]
31. Barsanti, C.; Trivella, M.G.; D'Aurizio, R.; El Baroudi, M.; Baumgart, M.; Groth, M.; Caruso, R.; Verde, A.; Botta, L.; Cozzi, L.; et al. Differential regulation of microRNAs in end-stage failing hearts is associated with left ventricular assist device unloading. *BioMed Res. Int.* **2015**, *2015*, 592512. [CrossRef] [PubMed]
32. Martin, M. Cutadapt removes adapter sequences from high-throughput sequencing reads. *EMBnet. J.* **2011**, *17*, 10–12. [CrossRef]
33. Wang, W.C.; Lin, F.M.; Chang, W.C.; Lin, K.Y.; Huang, H.D.; Lin, N.S. miRExpress: Analyzing high-throughput sequencing data for profiling microRNA expression. *BMC Bioinform.* **2009**, *10*, 328. [CrossRef] [PubMed]
34. McCarthy, D.J.; Chen, Y.; Smyth, G.K. Differential expression analysis of multifactor RNA-Seq experiments with respect to biological variation. *Nucleic Acids Res.* **2012**, *40*, 4288–4297. [CrossRef] [PubMed]
35. Love, M.I.; Huber, W.; Anders, S. Moderated estimation of fold change and dispersion for RNA-seq data with DESeq2. *Genome Biol.* **2014**, *15*, 550. [CrossRef] [PubMed]

International Journal of
Molecular Sciences

MDPI

Article

Cancer/Testis Antigens: "Smart" Biomarkers for Diagnosis and Prognosis of Prostate and Other Cancers

Prakash Kulkarni [1,*] and Vladimir N. Uversky [2,3,*]

1   Institute for Bioscience and Biotechnology Research, University of Maryland, Rockville, MD 20850, USA
2   Department of Molecular Medicine, Morsani College of Medicine, University of South Florida, Tampa,
    FL 33612, USA
3   Laboratory of New methods in Biology, Institute for Biological Instrumentation,
    Russian Academy of Sciences, Pushchino 142290, Moscow Region, Russia
*   Correspondence: pkulkar4@ibbr.umd.edu (P.K.); vuversky@health.usf.edu (V.N.U.);
    Tel.: +1-240-314-6122 (P.K.); +1-813-974-5816 (V.N.U.); Fax: +1-813-974-7357 (V.N.U.)

Academic Editor: Carsten Stephan
Received: 25 February 2017; Accepted: 27 March 2017; Published: 31 March 2017

Abstract: A clinical dilemma in the management of prostate cancer (PCa) is to distinguish men with aggressive disease who need definitive treatment from men who may not require immediate intervention. Accurate prediction of disease behavior is critical because radical treatment is associated with high morbidity. Here, we highlight the cancer/testis antigens (CTAs) as potential PCa biomarkers. The CTAs are a group of proteins that are typically restricted to the testis in the normal adult but are aberrantly expressed in several types of cancers. Interestingly, >90% of CTAs are predicted to belong to the realm of intrinsically disordered proteins (IDPs), which do not have unique structures and exist as highly dynamic conformational ensembles, but are known to play important roles in several biological processes. Using prostate-associated gene 4 (PAGE4) as an example of a disordered CTA, we highlight how IDP conformational dynamics may regulate phenotypic heterogeneity in PCa cells, and how it may be exploited both as a potential biomarker as well as a promising therapeutic target in PCa. We also discuss how in addition to intrinsic disorder and post-translational modifications, structural and functional variability induced in the CTAs by alternate splicing represents an important feature that might have different roles in different cancers. Although it is clear that significant additional work needs to be done in the outlined direction, this novel concept emphasizing (multi)functionality as an important trait in selecting a biomarker underscoring the theranostic potential of CTAs that is latent in their structure (or, more appropriately, the lack thereof), and casts them as next generation or "smart" biomarker candidates.

Keywords: biomarkers; prostate cancer; cancer/testis antigens; intrinsically disordered protein; prostate-associated gene 4 (PAGE4); nucleolar protein 4 (NOL4); centrosomal protein of 55 kDa (CEP55)

1. Introduction

Prostate cancer (PCa) is one of the most prevalent forms of cancer in older men over the age of 50. Worldwide, >1 million men are diagnosed with PCa each year and more than 300,000 die of the disease. Current US statistics show that one in five or six men will be diagnosed with PCa during their lifetime. In fact, by extrapolating statistical data from the past 40 years (1973–2013) it is estimated that in little over 100 years, one in two men will develop the disease [1]. Although these numbers appear daunting, only a fraction of those diagnosed have forms of the disease that can be considered to be "lethal" in nature.

As is true for other types of cancer, early diagnosis is believed to be crucial for the selection of the most successful and suitable PCa treatment strategy. Therefore, it follows that regular screening of men over the age 50 may be a logical thing to do. However, even today, there is no reliable test other than prostate-specific antigen (PSA), and there is no unanimous opinion in the medical community regarding the benefits of PSA screening [2,3]. Those who advocate regular screening believe that early diagnosis and treatment of PCa offers men a better chance to address the disease. On the other hand, urologists who recommend against regular screening note that because PCa is typically slow growing, the side effects of treatment would likely outweigh any benefit that might be derived from detecting the disease at a stage when it is unlikely to cause problems. Consistently, in 2012, the United States Preventive Services Task Force (USPSTF), an independent panel of experts in primary care and prevention that systematically reviews the evidence of effectiveness and develops recommendations for clinical preventive services, recommended against PCa screening in adult men of all ages. Indeed, in a roundtable discussion organized by Lee et al. to analyze studies of screening in two large randomized trials, it became obvious that the benefits of screening may not occur for 10 or more years after screening given the long natural history of the disease and that, perhaps as many as 1000 men will need to be screened and about 50 will need to be treated to save one life from PCa [4].

Many factors, including an increase in the aging population and widespread screening for PSA have contributed to the rise in the diagnoses of men who present early-stage (low or intermediate Gleason scores (GS)) or "low-risk" disease. While immediate treatment is recommended for patients with high GS ($\geq$8), the appropriate treatment for patients with low GS ($\leq$6) or intermediate GS (=7) remains ambiguous. Patients with low-risk disease are typically recommended the "wait and watch" or "active surveillance" protocol but are routinely monitored including repeat biopsies with the intention of avoiding treatment unless there is evidence of disease progression [5–7]. It is, therefore, not surprising that a staggering number of biopsies—>1 million in the US alone—are performed every year adding to the burgeoning healthcare cost and undesirable risk of serious complications requiring hospitalization [8].

While the intent of the active surveillance protocol is to minimize over-treatment, the concern is that active surveillance may miss the opportunity for early intervention of tumors that are seemingly low risk but that are actually aggressive. Indeed, despite the cautious approach, up to 40% of patients enrolled in active surveillance develop full-blown PCa [6]. Thus, a clinical dilemma today in the management of PCa is to distinguish men with aggressive disease who need definitive treatment from men whose disease does not require such intervention. Furthermore, accurate prediction of disease behavior is critical because radical treatment is associated with high morbidity.

Currently, at the time of diagnosis, most PCa cases present as localized disease and are preferentially treated by radical prostatectomy or radiation therapy with curative intent. During the last decade, a significant shift towards localized, well-differentiated tumors at radical prostatectomy (so-called stage migration) has occurred [9,10] perhaps due to the widespread use of PSA screening or a change in PCa biology [11], although, the latter would seem less likely. Therefore nowadays, PCa detected by PSA alone is often characterized by small-size and low-grade tumors in relatively younger male populations. In fact, it is reported that around 30% of cancers treated with radical prostatectomy in the US are "insignificant" tumors [12]. On the other hand, nearly 30% patients are reported to experience an isolated increase in serum PSA with long-term follow-up [13–19]. Therefore, it is important for physicians and patients to know the likelihood of disease progression following radical prostatectomy. Considered together, it follows that there is a critical need to identify reliable biomarkers that may be used for better diagnosis as well as to distinguish most of the low-GS tumors that will remain indolent from the few that are truly aggressive to better treat and manage PCa.

The advent of advanced technologies and sophisticated bioinformatics algorithms has fueled the discovery of novel biomarkers that include serum-, urine- and tissue-based assays that may supplement PSA testing, or even replace it over time (reviewed in [20–28]). These include extracellular vesicles [29], long noncoding RNAs [30,31], microRNAs [32–34] and circulating tumor cells (CTCs) [35,36], among many others. These advances have provided new insights into the individual patient's tumor biology, and several biomarkers with specific indications for disease diagnosis, prediction and prognosis, as well as risk stratification of aggressive PCa at the time of diagnosis are now commercially available.

For example, Decipher™ (GenomeDx Biosciences, Vancouver, Canada) is a tool based on 22 genes that evaluates the risk of adverse outcomes (metastasis) after radical prostatectomy [37], while Oncotype DX® (Genomics Health, Redwood City, CA, USA) was developed for use with fixed paraffin-embedded (FFPE) diagnostic prostate needle biopsies and measures expression of 12 cancer-related genes representing four biological pathways and five reference genes to calculate the Genomic Prostate Score (GPS) [38]. This assay has been analytically and subsequently clinically validated as a predictor of aggressive disease [38]. Prolaris (Myriad Genetics, Salth Lake City, UT, USA), on the other hand, is a 46-gene prognostic test that quantitatively determines the risk of recurrence in patients who have undergone prostatectomy. The assay measures the expression of 31 cell cycle progression (CCP) genes and 15 housekeeping genes that act as internal controls and normalization standards in each patient sample. The assay is also performed on FFPE samples and the results are reported as a numerical score along with accompanying interpretive information [39]. Of note, since the expression of CCP genes is likely to represent a fundamental aspect of tumor biology, the rationale for selecting these genes for prediction of outcome in PCa is based on a common biological function of the individual genes in this panel. Interestingly, the other two genomics tests also include genes associated with cell proliferation. Finally, ProMark® (Metamark Corp., Waltham, MA, USA) s based on a multiplexed proteomics assay [40] and predicts PCa aggressiveness in patients found with similar features to Oncotype DX®. These biomarkers can be helpful for post-biopsy decision-making in low-risk patients and post-radical prostatectomy in selected risk groups. These biomarkers that are intended to be used in combination with the accepted clinical criteria (i.e., GS, PSA, clinical stage) to stratify PCa according to biological aggressiveness and direct initial patient management have gained considerable popularity; however, additional studies are needed to investigate the clinical benefit of these new technologies, the financial ramifications and how they should be utilized in clinics.

2. Cancer/Testis Antigens (CTAs) as Novel PCa Biomarkers

The cancer/testis antigens (CTAs) are a group of proteins that are typically restricted to the testis in the normal adult but are aberrantly expressed in several types of cancers [41]. To date, ~250 genes encoding CTAs have been identified [42] that can be broadly divided into two groups: CT-X antigens located on the X chromosome and non-X CTAs located on various autosomes. Furthermore, members of the CT-X antigens, in particular, are typically associated with advanced disease characterized by poorer outcomes in several types of cancers, including PCa [43–48]. Because of these intriguing expression patterns, the CTAs serve as unique biomarkers for cancer diagnosis/prognosis.

A systematic study by Suyama et al. [48] using a custom DNA microarray revealed that several CT-X antigens from melanoma-associated antigen A/chondrosarcoma-associate gene (MAGE-A/CSAG) subfamilies are coordinately upregulated in castrate-resistant PCa, but not in primary PCa. Interestingly, however, the CT-X antigen prostate-associated gene 4 (PAGE4) was found to be highly upregulated in primary PCa but silent in castrate-resistant PCa, thereby raising the possibility that CTA-based "gene signature" could potentially be developed to distinguish men with aggressive PCa who need treatment from men with indolent disease not requiring immediate intervention [48].

To test this possibility, Shiraishi et al. [49] devised a multiplex real-time polymerase chain reaction (PCR) assay. From a panel of 22 CTAs that showed differential expression, they selected a subpanel of 5 CTAs that included 4 non-X CT antigens (centrosomal protein of 55 kDa—CEP55, NUF2, lymphokine-activated killer T-cell-originated protein kinase—PBK and the dual specificity protein kinase TTK) and the CT-X antigen, PAGE4 [49]. The authors found that while the non-X CTAs were upregulated, the CT-X antigen, PAGE4, was downregulated in patients with recurrent PCa after radical prostatectomy (Figure 1). Kaplan-Meier curves revealed that higher levels of expression of CEP55 and NUF2 were significantly correlated with shorter biochemical recurrence-free time [49]. In contrast, higher expression of PAGE4 was significantly correlated with longer biochemical recurrence-free time (Figure 2). Further, with the exception of TTK, the other CTAs were significantly correlated with prostatectomy GS, but none were correlated with age, preoperative PSA and tumor stage [49]. It is important note that, like in the case of the genomics tests that include several cell cycle progression genes, the five CTAs used in the study by Shiraishi et al. [49] are also associated with the cell cycle and proliferation. In fact, some of the CTAs are common to both gene sets highlighting the potential of this CTA panel.

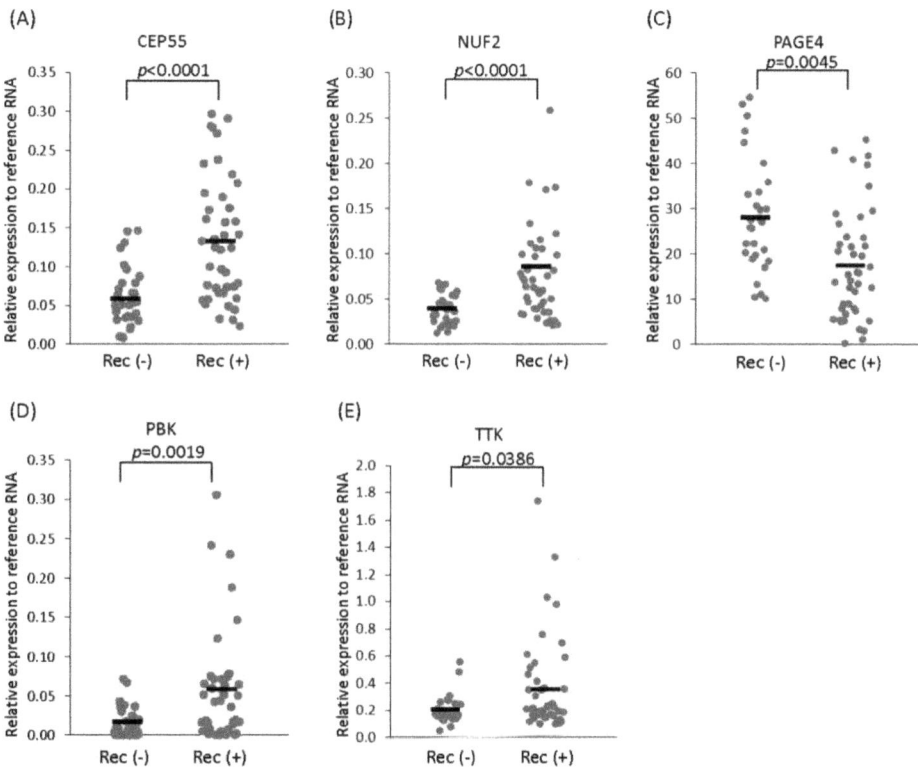

Figure 1. Cancer/Testis Antigen (CTA) expression in recurrent and non-recurrent prostate cancer. CTA expression in clinically localized prostate cancer with recurrence (Rec (+)) (*n* = 43) and without recurrence (Rec (−)) (*n* = 29). (**A**) centrosomal protein of 55 kDa (CEP55); (**B**) NDC80 kinetohore complex component NUF2; (**C**) prostate-associated gene 4 (PAGE4); (**D**) lymphokine-activated killer T-cell-originated protein kinase (PBK); (**E**) the dual specificity protein kinase TTK. Reproduced with permission from ref. [49].

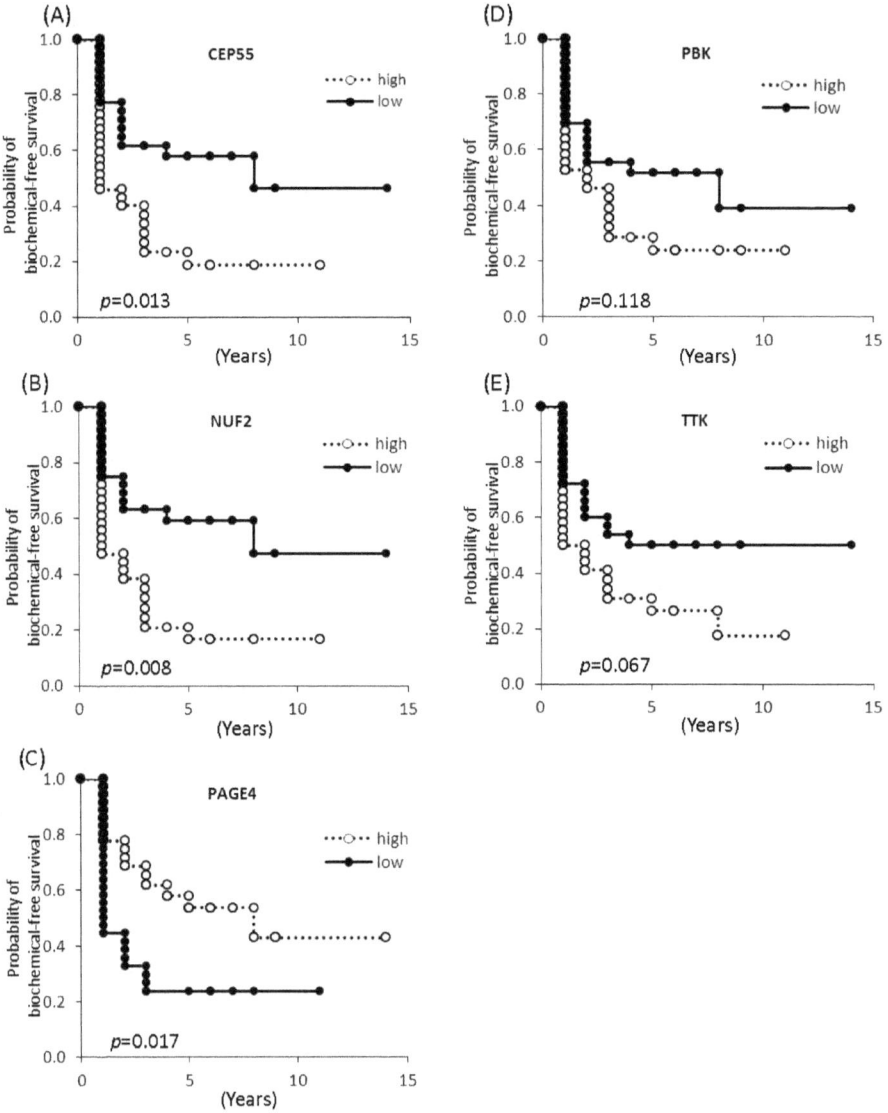

Figure 2. Kaplan-Meier analyses. Kaplan-Meier curves showing biochemical recurrence-free survival against time after radical prostatectomy stratified by the mRNA expression of (**A**) CEP55; (**B**) NUF2; (**C**) PAGE4; (**D**) PBK; and (**E**) TTK (high versus low groups dichotomized by median value). Reproduced with permission from ref. [49].

Despite the promise however, there are some limitations to the study by Shiraishi et al. [49]. First, the sample number was limited (*n* = 72), and they were not derived from patients who were consecutively and prospectively recruited for this study. Second, a high-risk cohort was used as a result of selection of specimens with large-volume tumors appropriate for frozen tissue collection, not reflecting contemporary, newly screened radical prostatectomy population. Third, there was no significant difference in the CT-X antigens (synovial sarcoma antigen X (SSX), synovial sarcoma X breakpoint 2 (SSX2), chondrosarcoma-associated gene 2/3 protein (CSAG2), melanoma-associated antigen 2 (MAGE-A2) and melanoma-associated antigen 12 (MAGE-A12)) between patients with or without recurrence [49]. However, von Boehmer et al. [50] observed that the CT-X antigen melanoma-associated antigen C2/Cancer-testis antigen 10 (MAGE-C2/CT10) may be a predictor of biochemical recurrence after radical prostatectomy, even though its expression was detected only in 3.3% of primary PCa samples.

More recently, the same research group employed the nCounter Gene Expression Assay (NanoString Technologies, Seattle, WA, USA) instead of quantitative multiplex PCR to evaluate the CTA gene signature in PCa patients [51]. The nCounter Analysis System utilizes a novel digital technology that is based on direct multiplexed measurement of gene expression and offers high levels of precision and sensitivity (<1 copy per cell). The technology uses molecular "barcodes" and single molecule imaging to detect and count hundreds of unique transcripts in a single reaction. Each color-coded barcode is attached to a single target-specific probe corresponding to a gene of interest. Mixed together with controls, they form a multiplexed CodeSet. The assay does not rely on enzymes for processing or amplification and enables highly sensitive detection and quantification of gene expression from a wide variety of sample types including direct measurement from purified total RNA, cell and tissue lysates, RNA extracted from FFPE samples and blood without globin mitigation.

The nCounter Analysis System is an integrated system comprised of a fully-automated assay and is designed to provide a sensitive, reproducible, quantitative and highly multiplexed method (up to 800 transcripts in one tube) with a wide dynamic range with superior gene expression quantification results when compared to real-time PCR without RNA purification, cDNA preparation, or amplification [51]. Of the 22 CTAs selected initially by Shiraishi et al. [49], Takahashi et al. [51] found that, in mRNA samples extracted from surgical samples, at least 5 CTAs (CEP55, NUF2, TTK, PBK, and PAGE4) appeared to be differentially expressed between metastatic and localized PCa both by quantitative PCR and the nanowire technology. As expected, CEP55 ($p < 0.01$), PBK ($p < 0.01$), NUF2 ($p < 0.01$) and sperm-associated antigen 4 (SPAG4) ($p < 0.01$) were significantly upregulated and PAGE4 ($p < 0.01$) was downregulated in metastatic PCa compared to localized disease. Further, using this assay FFPE samples, the authors found that RNA expression levels of the CTAs CSAG2 and nucleolar protein 4 (NOL4) were significantly higher in men with GS 8–10 disease than those with GS $\leq 4 + 3$ disease [51]. By contrast, the RNA expression level of PAGE4 was lower in men with GS 8–10 disease than those with GS ≤ 6 disease. Notwithstanding the slight disparity in the CTAs that appear to discriminate disease progression, this study further demonstrated the potential of the CTAs as PCa biomarkers [51] using achieved samples.

3. A Vast Majority of CTAs Are Predicted to Be Intrinsically Disordered

A bioinformatics study by Rajagopalan et al. [52] discovered that a majority of CTAs (>90%) are predicted to be intrinsically disordered proteins (IDPs). IDPs and hybrid proteins containing ordered domains and intrinsically disordered protein regions (IDPRs) are biologically active proteins that correspondingly lack rigid 3D structure either along their entire length or in localized regions, at least under physiological conditions in vitro [53–55]. Indeed, computational studies revealed that the per-proteome amounts of IDPs/IDPRs are high and increase with the increase in the organism complexity [56–58]. Indeed, all the CTAs selected by Shiraishi et al. [49] and Takahashi et al. [51] are predicted to be IDPs or as hybrid proteins containing long IDPRs (Figure 3 and Table 1). Despite the lack of unique structures, many IDPs/IDPRs can transition from disordered to ordered state upon binding to various targets [59]. The structural plasticity and conformational adaptability of IDPs/IDPRs, their ability to react and change easily and quickly in response to the changes in their environment, their capability to fold under the variety of conditions [53–55,60–69] combined with their binding promiscuity and unique capability to fold differently while interacting with different binding partners [66,70] define a wide set of functions exerted by IDPs/IDPRs in different biological systems. These same features determine the broad participation of IDPs/IDPRs in various biological processes [59,71,72] where they are involved in numerous signaling processes [73,74], regulation of different cellular pathways [75–80], cell protection [81], protein protection [82,83], cellular homeostasis [84,85] and cell cycle regulation [86–90]. Thus, biological activities of many IDPs/IDPRs are known to be precisely and tightly controlled and regulated by extensive posttranslational modifications (PTMs), such as phosphorylation, acetylation, glycosylation, etc. [59,91–93] and by alternative splicing (AS) [94–96].

Furthermore, IDPs/IDPRs are often associated with dosage sensitivity and are frequently engaged in highly promiscuous interactions, especially when concentrations of these proteins are increased [97]. Importantly, IDPs/IDPRs can interact with numerous binding partners of different natures, and many of these proteins are known to serve as essential hubs within various protein-protein interaction networks [68,98–102], where intrinsic disorder and related disorder-to-order transitions could enable one protein to interact with multiple partners (one-to-many signaling) or to enable multiple partners to bind to one protein (many-to-one signaling) [103].

Consistent with these observations, several CTAs are also predicted to bind DNA, and their forced expression appears to increase cell growth implying a potential dosage-sensitive function [52]. Furthermore, the CTAs appear to often occupy "hub" positions in protein-regulatory networks that typically adopt a "scale-free" power law distribution. Thus, the observations by Rajagopalan et al. [52] provide a novel perspective on the CTAs, implicating them in integrating and interpreting information in altered physiological states in a dosage-sensitive manner (see [52] and references therein). Considered together, these observations emphasizing the functional role of CTAs together with the development of a biomarker panel based on functionality (e.g., cell cycle progression), underscore the potential of CTAs to differentiate and discern diseased states of the prostate that is latent in their structure or lack thereof.

4. The Functional Role of Intrinsic Disorder in CTAs as Biomarkers

Here, using PAGE4 as an example, we illustrate how intrinsic disorder in CTAs may cast them as "next generation" biomarker candidates. More specifically, we discuss how conformational dynamics of this intrinsically disordered CTA may regulate the phenotype of the PCa cell and how the functionality of this IDP may be exploited as a potential biomarker in PCa.

PCa that is androgen-dependent is responsive to androgen-ablation therapy (ADT), the first line of treatment against advanced PCa, as well as an adjuvant to local treatment of high-risk disease. Although most patients initially respond to ADT, they eventually progress to a hormone-refractory state, which can prove fatal (reviewed in [104]). Yet the mechanism(s) underlying hormone resistance in PCa remains quite elusive. However, contrary to conventional wisdom, a new treatment paradigm called Bipolar ADT or BAT [105] is being tested where chemotherapy patients cycle through ADT followed by a supra-physiological dose of androgen. The results of this pilot study indicate that BAT may be more beneficial than ADT alone [105].

PAGE4 is a highly (~100%) intrinsically disordered CTA (Figure 3 and Table 1) that bears the hallmarks of a proto-oncogene; thus, it is highly expressed in the fetal human prostate, is undetectable in the normal adult gland, but is aberrantly expressed in androgen-dependent primary PCa and not in androgen-independent metastatic disease [106–108]. PAGE4 is also a strong potentiator of the transcription factor, AP-1, which is implicated in PCa [109] and is phosphorylated by two kinases namely, homeodomain-interacting protein kinase 1 (HIPK1) and CDC-like kinase 2 (CLK2). HIPK1 phosphorylates PAGE4 at predominantly T51, which is critical for its transcriptional activity [110]. In contrast, CLK2 is responsible for hyperphosphorylation of PAGE4 at multiple S/T residues. Furthermore, while HIPK1-phosphorylated PAGE4 potentiates AP-1, CLK2-phosphorylated PAGE4 attenuates its activity. Consistently, biophysical measurements indicate that HIPK1-phosphorylated PAGE4 exhibits a relatively compact conformational ensemble that binds AP-1 [111], while hyperphosphorylated PAGE4 is more expanded, resembles a random coil and is characterized by the diminished affinity for AP-1 [112].

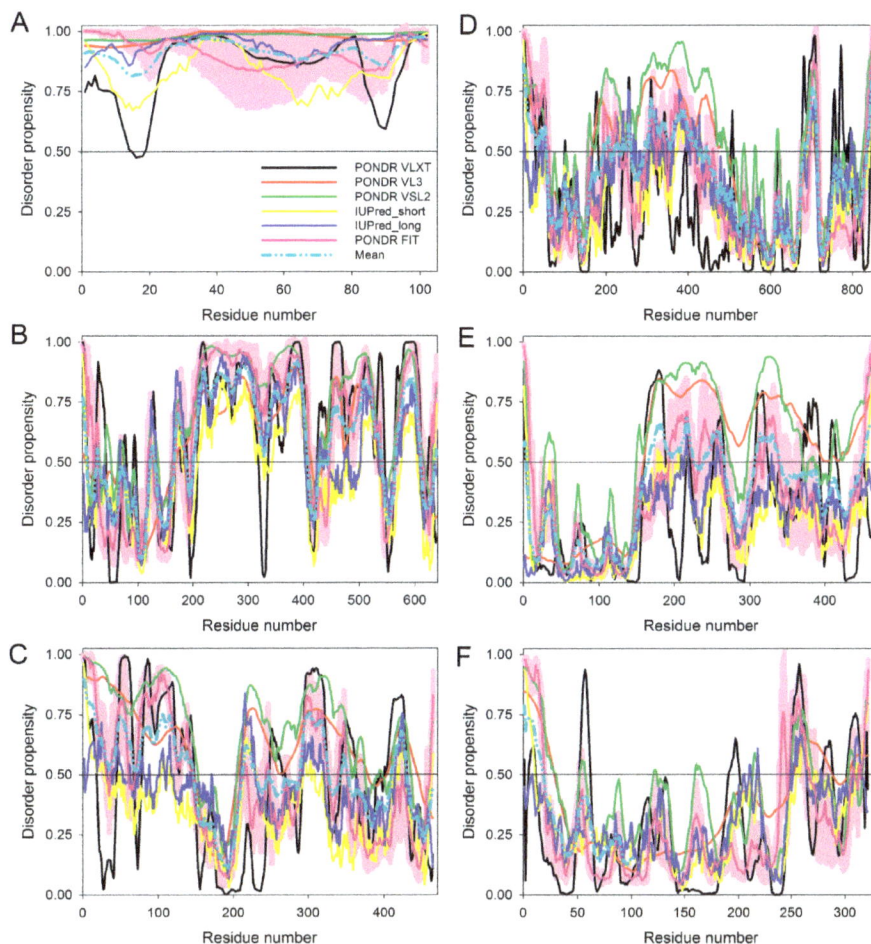

Figure 3. Variability of predicted intrinsic disorder levels and peculiarities of intrinsic disorder distributions within amino acid sequences of several CATs, PAGE4 (**A**, UniProt ID: O60829), nucleolar protein 4—NOL4 (**B**, UniProt ID: O94818), CEP55 (**C**, UniProt ID: Q53EZ4), TTK (**D**, UniProt ID: P33981), NUF2 (**E**, UniProt ID: Q9BZD4) and PBK (**F**, UniProt ID: Q96KB5). Intrinsic disorder profiles for query proteins generated by PONDR® VLXT [113], PONDR® VL3 [114], PONDR® VSL2 [114,115], PONDR® FIT [116], IUPred_short and IUPred_long [117] are shown by black, red, green, pink, yellow and blue lines, respectively. Cyan dash-dot-dotted lines show the mean disorder propensity calculated by averaging disorder profiles of individual predictors. Light pink shadow around the PONDR® FIT curve shows error distribution. In these analyses, the predicted intrinsic disorder scores above 0.5 are considered to correspond to the disordered residues/regions, whereas regions with the disorder scores between 0.2 and 0.5 are considered flexible.

Table 1. Intrinsic disorder-related characterization of some human Cancer/Testis Antigens that can be used as the prostate cancer biomarkers.

Protein	UniProt ID	Protein Length (N_{AIBS}) [a]	PONDR-FIT (%) [b]	MobiDB Consensus (%) [c]	Location (Length) of Long Disordered Regions [d]	Location (Length) of AIBSs [e]	N_{int} [f]
PAGE4, P antigen family member 4	O60829	102 (4/64.7)	100.00	100.00	1–102	1–8 (8) 14–31 (18) 53–81 (29) 92–102 (11)	N.P.
NOL4, Nucleolar protein 4	O94818	638 (13/40.1)	65.98	51.72	201–261 (61) 273–295 (23) 343–403 (61) 503–535 (33) 576–603 (28)	142–162 (21) 184–209 (26) 219–283 (65) 296–323 (28) 329–343 (15) 354–364 (11) 372–378 (7) 407–423 (17) 438–445 (8) 461–466 (6) 486–496 (11) 540–569 (30) 609–628 (20)	9
CEP55, centrosomal protein of 55 kDa	Q53EZ4	464 (8/12.1)	42.24	26.29	1–26 (26)	32–38 (7) 60–65 (6) 158–163 (6) 195–203 (9) 236–243 (8) 280–285 (6) 331–338 (8) 457–462 (6)	102
TTK, Dual specificity protein kinase TTK	P33981	857 (8/7.7)	28.70	23.64	371–410 (40) 836–857 (22)	238–244 (7) 271–279 (9) 292–297 (6) 347–356 (10) 362–368 (7) 407–412 (6) 811–819 (9) 824–835 (12)	72
NUF2, Kinetochore protein Nuf2	Q9BZD4	464 (1/3.0)	23.28	5.17	N.P.	278–291 (14)	48
PBK, Lymphokine-activated killer T-cell-originated protein kinase	Q96KB5	322 (3/8.7)	16.77	13.35	N.P.	268–275 (8) 290–300 (11) 314–322 (9)	30

[a] N_{AIBS} (A/B) represents the number of potential disorder-based binding sites identified by the ANCHOR algorithm (AIBS, A) and the percentage of residues involved in disorder-based interactions (B). [b] Content of disordered residues (i.e., residues with the disorder propensity ≥0.5) in a protein based on the PONDR-FIT disorder prediction. [c] Content of predicted disordered residues in a protein based on the MobiDB consensus score. [d] Information on long disordered regions (i.e., disordered regions of at least 10 residues) was obtained based on the MobiDB consensus profile. [e] AIBSs are potential disorder-based binding sites identified by the ANCHOR algorithm. [f] N_{int}, number of interactions as found using the BioGRID server [118]; N.P.: not present.

AP-1 can negatively regulate AR activity [119,120], and AR inhibits CLK2 expression [112]. Furthermore, cells resistant to ADT often have enhanced AR activity (AR protein expression can increase >25 fold), suggesting a positive correlation between ADT resistance and AR activity [121]. Based on these interactions, Kulkarni et al. constructed a circuit representing the PAGE4/AP-1/AR/ CLK2 interactions that drives non-genetic phenotypic heterogeneity in PCa cells and developed a mathematical model to represent the dynamics of this circuit [112]. The model predicts that this circuit can display sustained or damped oscillations; i.e., androgen dependence of a cell need not be a fixed state, but can vary temporally. Thus, the model suggests that an isogenic population of PCa cells displays a continuum of phenotypes with varying androgen-dependence. These cells can reversibly switch between androgen-dependent and androgen-independent states, without any specific genetic perturbation [112]. These findings appear to explain why BAT treatment appears more beneficial than ADT alone.

If this model is correct, then it suggests that higher levels of PAGE4 could be used as an indicator of a better PCa prognosis. Indeed, as noted by Shiraishi et al. [49], PAGE4 expression is significantly correlated with longer biochemical recurrence-free time (Figure 2). Furthermore, Sampson et al. [107] observed that in hormone-naive PCa, the median survival of patients with tumors expressing high PAGE4 levels was 8.2 years compared with 3.1 years for patients with tumors expressing negative/low levels of PAGE4 lending further credence to the model (Figure 4). Taken together, the work by Kulkarni et al. demonstrates a plausible functional link between IDP conformational dynamics and state switching in cancer [112]. Therefore, in theory, PAGE4 and its various phosphorylated variants represent novel biomarkers, as well as therapeutic targets to treat and manage PCa. For example, the detection of high levels of HIPK1-phosphorylated PAGE4 may imply that it can potentiate AP-1 and thus render the cells androgen sensitive and such patients may benefit from ADT alone. On the other hand, high levels of hyperphosphorylated PAGE4 would imply that the cells are heading towards an androgen-resistant state and thus, patients may benefit from BAT. Finally, pharmacologically targeting PAGE4 may also emerge as a viable option to treat PCa, especially low-risk disease.

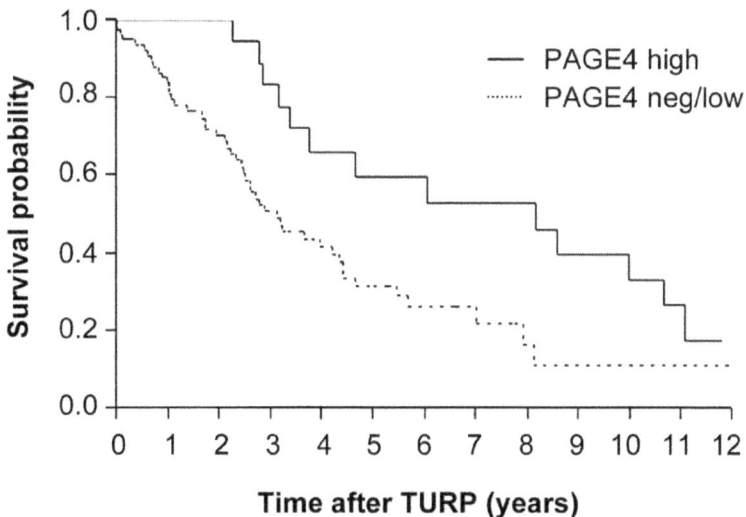

Figure 4. PAGE4 levels correlate with survival of patients with hormonenaive PCa. Overall survival of patients with hormone-naive PCa after transurethral resection of the prostate (TURP) for local advanced obstructive PCa stratified for high versus negative/low (neg/low) epithelial PAGE4 levels on the advanced PCa tissue microarray (TMA) (third quartile of mean epithelial PAGE4 intensity was set as the cut-off level). Reproduced with permission from ref. [107].

Curiously, two other highly disordered CTAs, NOL4 and CEP55, were shown to be associated with different types of cancer. For example, aberrant methylation of CpG islands in the NOL4 gene promoter was shown to be associated with cervical [122] and head and neck squamous cell carcinoma (HNSCC) [123]. These studies showed that NOL4 is methylated in 85% of cervical cancers [122] and in 91% HNSCC samples [123] and therefore the analysis of the epigenetic alteration of this gene can be used for early detection and risk prediction of cancers. Furthermore, NOL4 was recently shown to be one of the 20 aberrantly expressed genes in the most common and the most lethal primary brain tumor, glioblastoma (GBM) [124]. Although the exact biological function of NOL4 protein is not known as of yet, recent analysis revealed that different AS variants of mouse NOL4 (canonical NOL4-L, NOL4-S that lacks the N-terminal tail of NOL4-L and NOL4-SΔ, a NOL4-S with missed nuclear localization signal (NLS)) differently regulate the transactivation activities of the transcription factors Mlr1 (Mblk-1-related protein-1, where Mblk stands for mushroom body large-type kenyon cell-specific protein) and Mlr2 [125]. According to UniProt [126], human NOL4 (UniProt ID: O94818) also might exist in 4 isoforms generated by AS, such as canonical form containing the full-length polypeptide chain, isoform-2 missing 413–514 region, isoform-3, where residues 1–87 (MESERDMYRQ...KQVLYVPVKT) are changed to a shorter sequence MADLMQETFLHHA and isoform-4 with missing N-terminal residues 1–285. Analysis of the functional disorder profile generated by the D^2P^2 platform [127] (see Figure 5A) suggests that the disorder-based functionality of this protein that includes the peculiarities of the PTM distribution and presence of the molecular recognition features (which are specific binding sites that undergo disorder-to-order transition at binding to biological partners) is dramatically affected by AS, providing further support to the important idea that functionality of NOL4 can be regulated by AS (see also Table 1).

Like NOL4, CEP55 is also known to be expressed in various cancers [128,129], being barely detectable in normal tissues except for testis and thymus [130]. In fact, enhanced levels of this protein can be found in breast carcinoma, colorectal carcinoma and lung carcinoma tissues [130], as well as in human gastric carcinoma [131], urinary bladder transitional cell carcinoma [132] and in lung and liver cancers [129]. This protein is also induced at all stages of cervical cancer [133]. Furthermore, in breast cancer, CEP55 is one of the 16 genes, genomic alterations of which may be involved in tumorigenesis and in the processes of invasion and progression of disease [134]. CEP55-derived peptides were shown to serve as suitable candidates for the vaccine therapy of colorectal carcinoma [135]. Aberrant expression levels of the CEP55 genes are known to serve as prognostic marker of the estrogen receptor (ER) positive breast cancer [136]. In HNSCC, genomic instability and malignant transformation might involve CEP55 activation by aberrantly upregulated Forkhead box protein M1 (FOXM1) [137]. In gastric cancer, CEP55 plays a role in the induction of cell transformation in the RAC-alpha serine/threonine-protein kinase (AKT) signaling pathway-dependent manner [131]. CEP55 regulates cytokinesis via interaction with the peptidyl-prolyl isomerase Pin1 followed by the Polo-like kinase 1 (Plk1)-mediated phosphorylation of CEP55 needed for the function of this protein during cytokinesis.

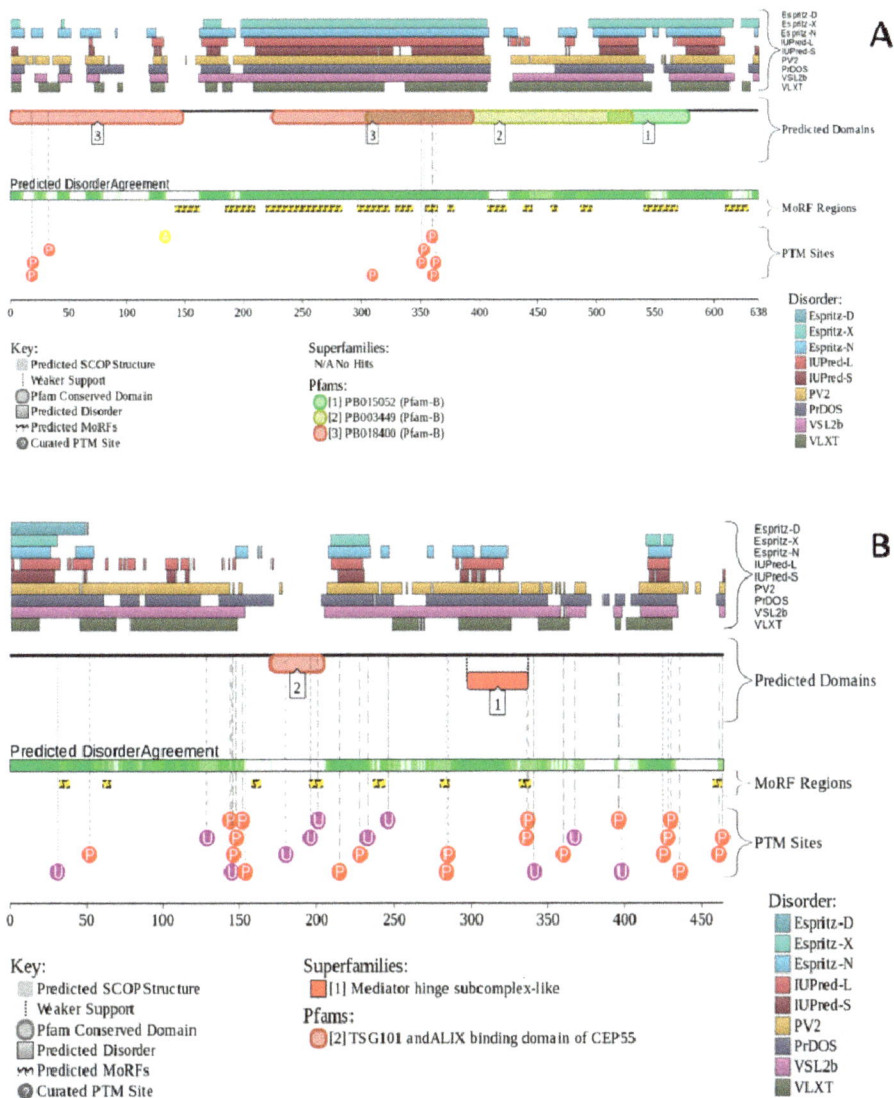

Figure 5. Intrinsic disorder propensity and some important disorder-related functional information generated for human NOL4 (**A**) and CEP55 (**B**) by the D^2P^2 database [127]. The D^2P^2 is a database of predicted disorder for a large library of proteins from completely sequenced genomes [127]. D^2P^2 database uses outputs of IUPred [117], PONDR® VLXT [113], PrDOS [138], PONDR® VSL2B [114,115], PV2 [127] and ESpritz [139] and is further supplemented by data concerning location of various curated posttranslational modifications and predicted disorder-based protein binding sites. Here, the green-and-white bar in the middle of the plot shows the predicted disorder agreement between nine predictors, with green parts corresponding to disordered regions by consensus. Yellow bar shows the location of the predicted disorder-based binding sites (molecular recognition features, MoRFs which are predicted by ANCHOR algorithm [140,141]), whereas colored circles at the bottom of the plot show location of various posttranslational modifications (PTMs).

In fact, pathologic levels of Pin1 being associated with tumorigenesis [142] and with Plk1 activity being needed for the negative regulation of the CEP55 function in cytokinesis [143]. In the BRCA2-dependent manner, CEP55 forms CEP55-ALIX (ALG-2 interacting protein X, also known as programmed cell death 6 interacting protein) and CEP55-TSG101 (another component of the ESCRT-1 (endosomal sorting complex required for transport-1) complex) complexes during abscission, whereas cancer-associated mutations in BRCA2 disrupts these interactions leading to the enhanced cytokinetic defects [144].

CEP55 is known to homodimerize, likely via its coiled-coil domains that are also responsible for protein-protein interactions, and can directly interact with centrosome components [128]. In agreement with this hypothesis, and with the emphasized ability of CEP55 to be engaged in interaction with the ALIX (which is a protein associated with the ESCRT), structural analysis revealed that the 160–217 region of CEP55 forms a non-canonical coiled-coil dimer that binds the Pro-rich sequence of ALIX (residues 797–809) [145]. Although no structural information is available for the remaining parts of human CEP55, Figures 3D and 5B and Table 1 show that this protein is predicted to contain high levels of intrinsic disorder. Furthermore, human CEP55 (UniProt ID: Q53EZ4) is expected to have two AS-generated isoforms [126], a canonical full-length form and an isoform-2 with the missing 401–464 region and the 389–400 region NQITQLESLKQL being changed to KNNTVGILETAS. Figure 5B and Table 1 show that alternative splicing causes elimination of several phosphorylation sites and one MoRF in human CEP55. In other words, it is likely that similar to NOL4, the physiological and pathological functionalities of human CEP55 can be modulated by AS.

5. Conclusions and Future Directions

Preliminary evidence in the literature indicates PAGE4 protein is detected in serum [146]. Although the authors evaluated PAGE4 as a biomarker to discern symptomatic and asymptomatic benign prostate hypertrophy (BPH), it is plausible that serum PAGE4 levels could discern PCa from normal and hence substitute for PSA given that, in the adult human male, PAGE4 is remarkably prostate-specific marker and is undetectable in the normal adult prostate [108,147]. Furthermore, it is even conceivable that a minimally invasive test could potentially also be developed to discern "good" (organ-confined/androgen-dependent disease) and "bad" (metastatic/androgen -independent disease) PCa given the positive correlation between PAGE4 and biochemical recurrence-free survival following radical prostatectomy. Additionally, monoclonal antibodies against the differentially phosphorylated forms of PAGE4 could be explored as novel tools to discern any correlation with disease prognosis. With advances in technology, estimating the levels of PAGE4 RNA and/or protein in CTCs using single-cell transcriptome (RNA-Seq) and single-cell westerns, respectively could be developed as minimally invasive tests for diagnosis and/or disease prognosis.

Although the corresponding data on the differential involvement of different AS isoforms of NOL4 and CEP55 in cancer are lacking at the moment, it is tempting to conjecture that in addition to intrinsic disorder and PTMs, structural and functional variability induced in proteins by AS represents an important feature that might have different roles in different cancers. In fact, AS was indicated as one of the cellular mechanisms (such as chromosomal translocations, altered expression, PTMs, aberrant proteolytic degradation and defective trafficking) that might cause pathogenic transformations in IDPs [148]. Furthermore, the indicated structural plasticity and multifunctionality of PAGE4, NOL4 and CEP55 are in line with the proteoform concept, according to which a functional protein product of a single gene exists in different molecular forms generated by genetic variations, alternative splicing and PTMs [149], as well as by intrinsic conformational plasticity and as a result of protein functioning [150].

Therefore, as opposed to current practice wherein any analyte such as a protein(s), RNA (structural, messenger, small interfering, long non-coding), DNA and its genetic and/or epigenetic modifications, metabolite(s) or circulating tumor cells themselves are selected as biomarkers merely based on their potential for disease diagnostics or prognostics, here we emphasize functionality as an additional

Int. J. Mol. Sci. **2017**, *18*, 740

trait in selecting a biomarker. For example, the standard biomarker for PCa is PSA, which is a kallikrein protease, whose function in the disease remains poorly understood. By contrast, PAGE4, which is a remarkably prostate-specific cancer/testis antigen in the adult male, is an IDP. Therefore, when overexpressed, PAGE4 can engage in promiscuous interactions resulting in pathological changes, that is, it is dosage-sensitive (see [52] and cross references therein). Furthermore, PAGE4 is a putative proto-oncogene that also appears to contribute to phenotypic heterogeneity in PCa cells due to its conformational plasticity. In other words, PAGE4 not only serves as a biomarker but also represents a therapeutic target (a theranostic). Therefore, PAGE4 and other examples of CTAs discussed here, by virtue of their functionality (for example, cell cycle progression), represent a set of "smart" biomarkers.

Considered together, these observations and considerations support an important notion: the analysis of the protein expression levels in biological fluids may not be the optimal focus of clinical proteomic research and that novel proteomic approaches are needed for the discovery of structure- and function-based next generation or smart biomarkers [151,152].

Since data presented in Figures 3 and 5 and Table 1 are the results of computational analyses used to show that some CTAs (PAGE4, NOL4 and CEP55) could be IDPs, this raises a legitimate question of whether any current biological methods can be utilized to confirm that these putative IDPs are really intrinsically disordered in cancer cells. Although earlier on there was some skepticism about the existence of disorder in proteins in the crowded cellular environment, this has been refuted by several studies that demonstrate that IDPs remain disordered in vivo both in bacterial and mammalian cells using in-cell NMR [153–155]. Clearly, conducting detailed structural and functional characterization of CTAs in vitro and in vivo represents an important future direction in this field. Another important question is related to the existence of the PAGE4-AR (androgen receptor) axis, namely, are the expression levels of PAGE4 as a PCa biomarker associated with the AR expression in the tissue specimens collected from PCa patients? Unfortunately, currently there are no direct data correlating PAGE4 and AR levels. However, as indicated in [112], one might suspect that there is an inverse correlation between the two, since PAGE4 is downregulated in metastatic disease, whereas AR is known to be upregulated at protein and/or mRNA level. Obviously, finding an exact answer to this question constitutes a very important subject for future research. Finally, it would be important to know if there is a link between PAGE4 and resistance to anti-cancer drugs, such as abiraterone or enzalutamide. Although we are not aware of any publication addressing this issue, and do not have corresponding data, we suspect an inverse correlation, since PAGE4 is downregulated in androgen-independent PCa cells. Again, careful analysis of this subject should be conducted in the future.

Author Contributions: Prakash Kulkarni and Vladimir N. Uversky conceived idea, analyzed literature data and wrote article.

Conflicts of Interest: The authors declare no conflict of interest.

References

1. Pollock, P.A.; Ludgate, A.; Wassersug, R.J. In 2124, half of all men can count on developing prostate cancer. *Curr. Oncol.* **2015**, *22*, 10–12. [CrossRef] [PubMed]

2. Dubben, H.H. Prostate-cancer screening. *N. Engl. J. Med.* **2009**, *361*, 204, author reply 204–206. [PubMed]

3. Barry, M.J. Screening for prostate cancer—The controversy that refuses to die. *N. Engl. J. Med.* **2009**, *360*, 1351–1354. [CrossRef] [PubMed]

4. Lee, T.H.; Kantoff, P.W.; McNaughton-Collins, M.F. Screening for prostate cancer. *N. Engl. J. Med.* **2009**, *360*, e18. [CrossRef] [PubMed]

5. Hayes, J.H.; Ollendorf, D.A.; Pearson, S.D.; Barry, M.J.; Kantoff, P.W.; Stewart, S.T.; Bhatnagar, V.; Sweeney, C.J.; Stahl, J.E.; McMahon, P.M. Active surveillance compared with initial treatment for men with low-risk prostate cancer: A decision analysis. *JAMA* **2010**, *304*, 2373–2380. [CrossRef] [PubMed]

6. Cooperberg, M.R.; Carroll, P.R.; Klotz, L. Active surveillance for prostate cancer: Progress and promise. *J. Clin. Oncol.* **2011**, *29*, 3669–3676. [CrossRef] [PubMed]

7. Tosoian, J.J.; Trock, B.J.; Landis, P.; Feng, Z.; Epstein, J.I.; Partin, A.W.; Walsh, P.C.; Carter, H.B. Active surveillance program for prostate cancer: An update of the Johns Hopkins experience. *J. Clin. Oncol.* **2011**, *29*, 2185–2190. [CrossRef] [PubMed]

8. Loeb, S.; Carter, H.B.; Berndt, S.I.; Ricker, W.; Schaeffer, E.M. Complications after prostate biopsy: Data from seer-medicare. *J. Urol.* **2011**, *186*, 1830–1834. [CrossRef] [PubMed]

9. Stephenson, R.A.; Stanford, J.L. Population-based prostate cancer trends in the united states: Patterns of change in the era of prostate-specific antigen. *World J. Urol.* **1997**, *15*, 331–335. [CrossRef] [PubMed]

10. Galper, S.L.; Chen, M.H.; Catalona, W.J.; Roehl, K.A.; Richie, J.P.; D'Amico, A.V. Evidence to support a continued stage migration and decrease in prostate cancer specific mortality. *J. Urol.* **2006**, *175*, 907–912. [CrossRef]

11. Polascik, T.J.; Oesterling, J.E.; Partin, A.W. Prostate specific antigen: A decade of discovery—what we have learned and where we are going. *J. Urol.* **1999**, *162*, 293–306. [CrossRef]

12. Pound, C.R.; Partin, A.W.; Eisenberger, M.A.; Chan, D.W.; Pearson, J.D.; Walsh, P.C. Natural history of progression after PSA elevation following radical prostatectomy. *JAMA* **1999**, *281*, 1591–1597. [CrossRef] [PubMed]

13. Catalona, W.J.; Smith, D.S. 5-year tumor recurrence rates after anatomical radical retropubic prostatectomy for prostate cancer. *J. Urol.* **1994**, *152*, 1837–1842. [PubMed]

14. Ohori, M.; Goad, J.R.; Wheeler, T.M.; Eastham, J.A.; Thompson, T.C.; Scardino, P.T. Can radical prostatectomy alter the progression of poorly differentiated prostate cancer? *J. Urol.* **1994**, *152*, 1843–1849. [PubMed]

15. Trapasso, J.G.; deKernion, J.B.; Smith, R.B.; Dorey, F. The incidence and significance of detectable levels of serum prostate specific antigen after radical prostatectomy. *J. Urol.* **1994**, *152*, 1821–1825. [PubMed]

16. Zincke, H.; Oesterling, J.E.; Blute, M.L.; Bergstralh, E.J.; Myers, R.P.; Barrett, D.M. Long-term (15 years) results after radical prostatectomy for clinically localized (stage T2C or lower) prostate cancer. *J. Urol.* **1994**, *152*, 1850–1857. [PubMed]

17. Pound, C.R.; Partin, A.W.; Epstein, J.I.; Walsh, P.C. Prostate-specific antigen after anatomic radical retropubic prostatectomy. Patterns of recurrence and cancer control. *Urol. Clin. N. Am.* **1997**, *24*, 395–406. [CrossRef]

18. Catalona, W.J.; Smith, D.S. Cancer recurrence and survival rates after anatomic radical retropubic prostatectomy for prostate cancer: Intermediate-term results. *J. Urol.* **1998**, *160*, 2428–2434. [CrossRef]

19. Han, M.; Partin, A.W.; Pound, C.R.; Epstein, J.I.; Walsh, P.C. Long-term biochemical disease-free and cancer-specific survival following anatomic radical retropubic prostatectomy. The 15-year Johns Hopkins experience. *Urol. Clin. N. Am.* **2001**, *28*, 555–565. [CrossRef]

20. Agell, L.; Hernandez, S.; Nonell, L.; Lorenzo, M.; Puigdecanet, E.; de Muga, S.; Juanpere, N.; Bermudo, R.; Fernandez, P.L.; Lorente, J.A.; et al. A 12-gene expression signature is associated with aggressive histological in prostate cancer: SEC14L1 and TCEB1 genes are potential markers of progression. *Am. J. Pathol.* **2012**, *181*, 1585–1594. [CrossRef] [PubMed]

21. Artibani, W. Landmarks in prostate cancer diagnosis: The biomarkers. *BJU Int.* **2012**, *110*, 8–13. [CrossRef] [PubMed]

22. Irshad, S.; Bansal, M.; Castillo-Martin, M.; Zheng, T.; Aytes, A.; Wenske, S.; Le Magnen, C.; Guarnieri, P.; Sumazin, P.; Benson, M.C.; et al. A molecular signature predictive of indolent prostate cancer. *Sci. Transl. Med.* **2013**, *5*, 202ra122. [CrossRef] [PubMed]

23. Klein, E.A.; Cooperberg, M.R.; Magi-Galluzzi, C.; Simko, J.P.; Falzarano, S.M.; Maddala, T.; Chan, J.M.; Li, J.; Cowan, J.E.; Tsiatis, A.C.; et al. A 17-gene assay to predict prostate cancer aggressiveness in the context of Gleason grade heterogeneity, tumor multifocality and biopsy undersampling. *Eur. Urol.* **2014**, *66*, 550–560. [CrossRef] [PubMed]

24. Qu, M.; Ren, S.C.; Sun, Y.H. Current early diagnostic biomarkers of prostate cancer. *Asian J. Androl.* **2014**, *16*, 549–554. [PubMed]

25. Sartori, D.A.; Chan, D.W. Biomarkers in prostate cancer: What's new? *Curr. Opin. Oncol.* **2014**, *26*, 259–264. [CrossRef] [PubMed]

26. Kelly, R.S.; Vander Heiden, M.G.; Giovannucci, E.; Mucci, L.A. Metabolomic biomarkers of prostate cancer: Prediction, diagnosis, progression, prognosis and recurrence. *Cancer Epidemiol. Biomark. Prev.* **2016**, *25*, 887–906. [CrossRef] [PubMed]

27. Lima, A.R.; Bastos Mde, L.; Carvalho, M.; Guedes de Pinho, P. Biomarker discovery in human prostate cancer: An update in metabolomics studies. *Transl. Oncol.* **2016**, *9*, 357–370. [CrossRef] [PubMed]

28. Tonry, C.L.; Leacy, E.; Raso, C.; Finn, S.P.; Armstrong, J.; Pennington, S.R. The role of proteomics in biomarker development for improved patient diagnosis and clinical decision making in prostate cancer. *Diagnostics* **2016**, *6*. [CrossRef] [PubMed]

29. Nawaz, M.; Camussi, G.; Valadi, H.; Nazarenko, I.; Ekstrom, K.; Wang, X.; Principe, S.; Shah, N.; Ashraf, N.M.; Fatima, F.; et al. The emerging role of extracellular vesicles as biomarkers for urogenital cancers. *Nat. Rev. Urol.* **2014**, *11*, 688–701. [CrossRef] [PubMed]

30. Mouraviev, V.; Lee, B.; Patel, V.; Albala, D.; Johansen, T.E.; Partin, A.; Ross, A.; Perera, R.J. Clinical prospects of long noncoding RNAs as novel biomarkers and therapeutic targets in prostate cancer. *Prostate Cancer Prostatic Dis.* **2015**, *19*, 14–20. [CrossRef] [PubMed]

31. Martens-Uzunova, E.S.; Bottcher, R.; Croce, C.M.; Jenster, G.; Visakorpi, T.; Calin, G.A. Long noncoding RNA in prostate, bladder and kidney cancer. *Eur. Urol.* **2014**, *65*, 1140–1151. [CrossRef] [PubMed]

32. Fendler, A.; Stephan, C.; Yousef, G.M.; Kristiansen, G.; Jung, K. The translational potential of microRNAs as biofluid markers of urological tumours. *Nat. Rev. Urol.* **2016**, *13*, 734–752. [CrossRef] [PubMed]

33. Kumar, B.; Lupold, S.E. Microrna expression and function in prostate cancer: A review of current knowledge and opportunities for discovery. *Asian J. Androl.* **2016**, *18*, 559–567. [PubMed]

34. Fabris, L.; Ceder, Y.; Chinnaiyan, A.M.; Jenster, G.W.; Sorensen, K.D.; Tomlins, S.; Visakorpi, T.; Calin, G.A. The potential of microRNAs as prostate cancer biomarkers. *Eur. Urol.* **2016**, *70*, 312–322. [CrossRef] [PubMed]

35. Gorin, M.A.; Verdone, J.E.; van der Toom, E.; Bivalacqua, T.J.; Allaf, M.E.; Pienta, K.J. Circulating tumour cells as biomarkers of prostate, bladder and kidney cancer. *Nat. Rev. Urol.* **2016**, *14*, 90–97. [CrossRef] [PubMed]

36. Hugen, C.M.; Zainfeld, D.E.; Goldkorn, A. Circulating tumor cells in genitourinary malignancies: An evolving path to precision medicine. *Front. Oncol* **2017**, *7*, 6. [CrossRef] [PubMed]

37. Erho, N.; Crisan, A.; Vergara, I.A.; Mitra, A.P.; Ghadessi, M.; Buerki, C.; Bergstralh, E.J.; Kollmeyer, T.; Fink, S.; Haddad, Z.; et al. Discovery and validation of a prostate cancer genomic classifier that predicts early metastasis following radical prostatectomy. *PLoS ONE* **2013**, *8*, e66855. [CrossRef] [PubMed]

38. Knezevic, D.; Goddard, A.D.; Natraj, N.; Cherbavaz, D.B.; Clark-Langone, K.M.; Snable, J.; Watson, D.; Falzarano, S.M.; Magi-Galluzzi, C.; Klein, E.A.; et al. Analytical validation of the oncotype Dx prostate cancer assay—A clinical RT-PCR assay optimized for prostate needle biopsies. *BMC Genom.* **2013**, *14*, 690. [CrossRef] [PubMed]

39. Cuzick, J.; Stone, S.; Fisher, G.; Yang, Z.H.; North, B.V.; Berney, D.M.; Beltran, L.; Greenberg, D.; Moller, H.; Reid, J.E.; et al. Validation of an RNA cell cycle progression score for predicting death from prostate cancer in a conservatively managed needle biopsy cohort. *Br. J. Cancer* **2015**, *113*, 382–389. [CrossRef] [PubMed]

40. Blume-Jensen, P.; Berman, D.M.; Rimm, D.L.; Shipitsin, M.; Putzi, M.; Nifong, T.P.; Small, C.; Choudhury, S.; Capela, T.; Coupal, L.; et al. Development and clinical validation of an in situ biopsy-based multimarker assay for risk stratification in prostate cancer. *Clin. Cancer Res.* **2015**, *21*, 2591–2600. [CrossRef] [PubMed]

41. Scanlan, M.J.; Simpson, A.J.; Old, L.J. The cancer/testis genes: Review, standardization and commentary. *Cancer Immun.* **2004**, *4*, 1. [PubMed]

42. Almeida, L.G.; Sakabe, N.J.; deOliveira, A.R.; Silva, M.C.; Mundstein, A.S.; Cohen, T.; Chen, Y.T.; Chua, R.; Gurung, S.; Gnjatic, S.; et al. Ctdatabase: A knowledge-base of high-throughput and curated data on cancer-testis antigens. *Nucleic Acids Res.* **2009**, *37*, D816–D819. [CrossRef] [PubMed]

43. Gure, A.O.; Chua, R.; Williamson, B.; Gonen, M.; Ferrera, C.A.; Gnjatic, S.; Ritter, G.; Simpson, A.J.; Chen, Y.T.; Old, L.J.; et al. Cancer-testis genes are coordinately expressed and are markers of poor outcome in non-small cell lung cancer. *Clin. Cancer Res.* **2005**, *11*, 8055–8062. [CrossRef] [PubMed]

44. Velazquez, E.F.; Jungbluth, A.A.; Yancovitz, M.; Gnjatic, S.; Adams, S.; O'Neill, D.; Zavilevich, K.; Albukh, T.; Christos, P.; Mazumdar, M.; et al. Expression of the cancer/testis antigen NY-ESO-1 in primary and metastatic malignant melanoma (mm)—Correlation with prognostic factors. *Cancer Immun.* **2007**, *7*, 11. [PubMed]

45. Andrade, V.C.; Vettore, A.L.; Felix, R.S.; Almeida, M.S.; Carvalho, F.; Oliveira, J.S.; Chauffaille, M.L.; Andriolo, A.; Caballero, O.L.; Zago, M.A.; et al. Prognostic impact of cancer/testis antigen expression in advanced stage multiple myeloma patients. *Cancer Immun.* **2008**, *8*, 2. [PubMed]

46. Napoletano, C.; Bellati, F.; Tarquini, E.; Tomao, F.; Taurino, F.; Spagnoli, G.; Rughetti, A.; Muzii, L.; Nuti, M.; Benedetti Panici, P. MAGE-A and NY-ESO-1 expression in cervical cancer: Prognostic factors and effects of chemotherapy. *Am. J. Obstet. Gynecol.* **2008**, *198*, e91–e97. [CrossRef] [PubMed]

47. Grigoriadis, A.; Caballero, O.L.; Hoek, K.S.; da Silva, L.; Chen, Y.T.; Shin, S.J.; Jungbluth, A.A.; Miller, L.D.; Clouston, D.; Cebon, J.; et al. CT-X antigen expression in human breast cancer. *Proc. Natl. Acad. Sci. USA* **2009**, *106*, 13493–13498. [CrossRef] [PubMed]
48. Suyama, T.; Shiraishi, T.; Zeng, Y.; Yu, W.; Parekh, N.; Vessella, R.L.; Luo, J.; Getzenberg, R.H.; Kulkarni, P. Expression of cancer/testis antigens in prostate cancer is associated with disease progression. *Prostate* **2010**, *70*, 1778–1787. [CrossRef] [PubMed]
49. Shiraishi, T.; Terada, N.; Zeng, Y.; Suyama, T.; Luo, J.; Trock, B.; Kulkarni, P.; Getzenberg, R.H. Cancer/testis antigens as potential predictors of biochemical recurrence of prostate cancer following radical prostatectomy. *J. Transl. Med.* **2011**, *9*, 153. [CrossRef] [PubMed]
50. Von Boehmer, L.; Keller, L.; Mortezavi, A.; Provenzano, M.; Sais, G.; Hermanns, T.; Sulser, T.; Jungbluth, A.A.; Old, L.J.; Kristiansen, G.; et al. MAGE-C2/CT10 protein expression is an independent predictor of recurrence in prostate cancer. *PLoS ONE* **2011**, *6*, e21366. [CrossRef] [PubMed]
51. Takahashi, S.; Shiraishi, T.; Miles, N.; Trock, B.J.; Kulkarni, P.; Getzenberg, R.H. Nanowire analysis of cancer-testis antigens as biomarkers of aggressive prostate cancer. *Urology* **2015**, *85*, e701–e707. [CrossRef] [PubMed]
52. Rajagopalan, K.; Mooney, S.M.; Parekh, N.; Getzenberg, R.H.; Kulkarni, P. A majority of the cancer/testis antigens are intrinsically disordered proteins. *J. Cell. Biochem.* **2011**, *112*, 3256–3267. [CrossRef] [PubMed]
53. Wright, P.E.; Dyson, H.J. Intrinsically unstructured proteins: Re-assessing the protein structure-function paradigm. *J. Mol. Biol.* **1999**, *293*, 321–331. [CrossRef] [PubMed]
54. Uversky, V.N.; Gillespie, J.R.; Fink, A.L. Why are "natively unfolded" proteins unstructured under physiologic conditions? *Proteins* **2000**, *41*, 415–427. [CrossRef]
55. Dunker, A.K.; Lawson, J.D.; Brown, C.J.; Williams, R.M.; Romero, P.; Oh, J.S.; Oldfield, C.J.; Campen, A.M.; Ratliff, C.M.; Hipps, K.W.; et al. Intrinsically disordered protein. *J. Mol. Graph. Model.* **2001**, *19*, 26–59. [CrossRef]
56. Xue, B.; Dunker, A.K.; Uversky, V.N. Orderly order in protein intrinsic disorder distribution: Disorder in 3500 proteomes from viruses and the three domains of life. *J. Biomol. Struct. Dyn.* **2012**, *30*, 137–149. [CrossRef] [PubMed]
57. Peng, Z.; Yan, J.; Fan, X.; Mizianty, M.J.; Xue, B.; Wang, K.; Hu, G.; Uversky, V.N.; Kurgan, L. Exceptionally abundant exceptions: Comprehensive characterization of intrinsic disorder in all domains of life. *Cell. Mol. Life Sci.* **2015**, *72*, 137–151. [CrossRef] [PubMed]
58. Walsh, I.; Giollo, M.; Di Domenico, T.; Ferrari, C.; Zimmermann, O.; Tosatto, S.C. Comprehensive large-scale assessment of intrinsic protein disorder. *Bioinformatics* **2015**, *31*, 201–208. [CrossRef] [PubMed]
59. Uversky, V.N.; Dunker, A.K. Understanding protein non-folding. *Biochim. Biophys. Acta* **2010**, *1804*, 1231–1264. [CrossRef] [PubMed]
60. Iakoucheva, L.M.; Brown, C.J.; Lawson, J.D.; Obradovic, Z.; Dunker, A.K. Intrinsic disorder in cell-signaling and cancer-associated proteins. *J. Mol. Biol.* **2002**, *323*, 573–584. [CrossRef]
61. Uversky, V.N. Natively unfolded proteins: A point where biology waits for physics. *Protein Sci.* **2002**, *11*, 739–756. [CrossRef] [PubMed]
62. Uversky, V.N. What does it mean to be natively unfolded? *Eur. J. Biochem.* **2002**, *269*, 2–12. [CrossRef] [PubMed]
63. Dunker, A.K.; Brown, C.J.; Lawson, J.D.; Iakoucheva, L.M.; Obradovic, Z. Intrinsic disorder and protein function. *Biochemistry* **2002**, *41*, 6573–6582. [CrossRef] [PubMed]
64. Tompa, P. Intrinsically unstructured proteins. *Trends Biochem. Sci.* **2002**, *27*, 527–533. [CrossRef]
65. Dyson, H.J.; Wright, P.E. Insights into the structure and dynamics of unfolded proteins from nuclear magnetic resonance. *Adv. Protein Chem.* **2002**, *62*, 311–340. [PubMed]
66. Dyson, H.J.; Wright, P.E. Intrinsically unstructured proteins and their functions. *Nat. Rev. Mol. Cell Biol.* **2005**, *6*, 197–208. [CrossRef] [PubMed]
67. Fink, A.L. Natively unfolded proteins. *Curr. Opin. Struct. Biol.* **2005**, *15*, 35–41. [CrossRef] [PubMed]
68. Uversky, V.N.; Oldfield, C.J.; Dunker, A.K. Showing your ID: Intrinsic disorder as an ID for recognition, regulation and cell signaling. *J. Mol. Recognit.* **2005**, *18*, 343–384. [CrossRef] [PubMed]
69. Dunker, A.K.; Cortese, M.S.; Romero, P.; Iakoucheva, L.M.; Uversky, V.N. Flexible nets. The roles of intrinsic disorder in protein interaction networks. *FEBS J.* **2005**, *272*, 5129–5148. [CrossRef] [PubMed]

70. Oldfield, C.J.; Meng, J.; Yang, J.Y.; Yang, M.Q.; Uversky, V.N.; Dunker, A.K. Flexible nets: Disorder and induced fit in the associations of p53 and 14-3-3 with their partners. *BMC Genom.* **2008**, *9* (Suppl. S1), S1. [CrossRef] [PubMed]

71. Ferreon, A.C.; Ferreon, J.C.; Wright, P.E.; Deniz, A.A. Modulation of allostery by protein intrinsic disorder. *Nature* **2013**, *498*, 390–394. [CrossRef] [PubMed]

72. Cozzetto, D.; Jones, D.T. The contribution of intrinsic disorder prediction to the elucidation of protein function. *Curr. Opin. Struct. Biol.* **2013**, *23*, 467–472. [CrossRef] [PubMed]

73. Mitrea, D.M.; Kriwacki, R.W. Regulated unfolding of proteins in signaling. *FEBS Lett.* **2013**, *587*, 1081–1088. [CrossRef] [PubMed]

74. Follis, A.V.; Llambi, F.; Ou, L.; Baran, K.; Green, D.R.; Kriwacki, R.W. The DNA-binding domain mediates both nuclear and cytosolic functions of p53. *Nat. Struct. Mol. Biol.* **2014**, *21*, 535–543. [CrossRef] [PubMed]

75. Galea, C.A.; Wang, Y.; Sivakolundu, S.G.; Kriwacki, R.W. Regulation of cell division by intrinsically unstructured proteins: Intrinsic flexibility, modularity and signaling conduits. *Biochemistry* **2008**, *47*, 7598–7609. [CrossRef] [PubMed]

76. Wang, Y.; Fisher, J.C.; Mathew, R.; Ou, L.; Otieno, S.; Sublet, J.; Xiao, L.; Chen, J.; Roussel, M.F.; Kriwacki, R.W. Intrinsic disorder mediates the diverse regulatory functions of the cdk inhibitor p21. *Nat. Chem. Biol.* **2011**, *7*, 214–221. [CrossRef] [PubMed]

77. Ou, L.; Waddell, M.B.; Kriwacki, R.W. Mechanism of cell cycle entry mediated by the intrinsically disordered protein p27(kip1). *ACS Chem. Biol.* **2012**, *7*, 678–682. [CrossRef] [PubMed]

78. Follis, A.V.; Galea, C.A.; Kriwacki, R.W. Intrinsic protein flexibility in regulation of cell proliferation: Advantages for signaling and opportunities for novel therapeutics. *Adv. Exp. Med. Biol.* **2012**, *725*, 27–49. [PubMed]

79. Moldoveanu, T.; Grace, C.R.; Llambi, F.; Nourse, A.; Fitzgerald, P.; Gehring, K.; Kriwacki, R.W.; Green, D.R. Bid-induced structural changes in BAK promote apoptosis. *Nat. Struct. Mol. Biol.* **2013**, *20*, 589–597. [CrossRef] [PubMed]

80. Frye, J.J.; Brown, N.G.; Petzold, G.; Watson, E.R.; Grace, C.R.; Nourse, A.; Jarvis, M.A.; Kriwacki, R.W.; Peters, J.M.; Stark, H.; et al. Electron microscopy structure of human APC/C(CDH1)-EMI1 reveals multimodal mechanism of E3 ligase shutdown. *Nat. Struct. Mol. Biol.* **2013**, *20*, 827–835. [CrossRef] [PubMed]

81. Mei, Y.; Su, M.; Soni, G.; Salem, S.; Colbert, C.L.; Sinha, S.C. Intrinsically disordered regions in autophagy proteins. *Proteins* **2014**, *82*, 565–578. [CrossRef] [PubMed]

82. Chakrabortee, S.; Tripathi, R.; Watson, M.; Schierle, G.S.; Kurniawan, D.P.; Kaminski, C.F.; Wise, M.J.; Tunnacliffe, A. Intrinsically disordered proteins as molecular shields. *Mol. BioSyst.* **2012**, *8*, 210–219. [CrossRef] [PubMed]

83. De Jonge, N.; Garcia-Pino, A.; Buts, L.; Haesaerts, S.; Charlier, D.; Zangger, K.; Wyns, L.; de Greve, H.; Loris, R. Rejuvenation of CCDB-poisoned gyrase by an intrinsically disordered protein domain. *Mol. Cell* **2009**, *35*, 154–163. [CrossRef] [PubMed]

84. Norholm, A.B.; Hendus-Altenburger, R.; Bjerre, G.; Kjaergaard, M.; Pedersen, S.F.; Kragelund, B.B. The intracellular distal tail of the Na+/H+ exchanger NHE1 is intrinsically disordered: Implications for NHE1 trafficking. *Biochemistry* **2011**, *50*, 3469–3480. [CrossRef] [PubMed]

85. Follis, A.V.; Chipuk, J.E.; Fisher, J.C.; Yun, M.K.; Grace, C.R.; Nourse, A.; Baran, K.; Ou, L.; Min, L.; White, S.W.; et al. Puma binding induces partial unfolding within Bcl-XL to disrupt p53 binding and promote apoptosis. *Nat. Chem. Biol.* **2013**, *9*, 163–168. [CrossRef] [PubMed]

86. Borriello, A.; Cucciolla, V.; Oliva, A.; Zappia, V.; Della Ragione, F. P27kip1 metabolism: A fascinating labyrinth. *Cell Cycle* **2007**, *6*, 1053–1061. [CrossRef] [PubMed]

87. Barberis, M. Sic1 as a timer of clb cyclin waves in the yeast cell cycle—Design principle of not just an inhibitor. *FEBS J.* **2012**, *279*, 3386–3410. [CrossRef] [PubMed]

88. Mitrea, D.M.; Yoon, M.K.; Ou, L.; Kriwacki, R.W. Disorder-function relationships for the cell cycle regulatory proteins p21 and p27. *Biol. Chem.* **2012**, *393*, 259–274. [CrossRef] [PubMed]

89. Yoon, M.K.; Mitrea, D.M.; Ou, L.; Kriwacki, R.W. Cell cycle regulation by the intrinsically disordered proteins p21 and p27. *Biochem. Soc. Trans.* **2012**, *40*, 981–988. [CrossRef] [PubMed]

90. Cianfanelli, V.; De Zio, D.; Di Bartolomeo, S.; Nazio, F.; Strappazzon, F.; Cecconi, F. AMBRA1 at a glance. *J. Cell Sci.* **2015**, *128*, 2003–2008. [CrossRef] [PubMed]

91. Collins, M.O.; Yu, L.; Campuzano, I.; Grant, S.G.; Choudhary, J.S. Phosphoproteomic analysis of the mouse brain cytosol reveals a predominance of protein phosphorylation in regions of intrinsic sequence disorder. *Mol. Cell. Proteom.* **2008**, *7*, 1331–1348. [CrossRef] [PubMed]

92. Pejaver, V.; Hsu, W.L.; Xin, F.; Dunker, A.K.; Uversky, V.N.; Radivojac, P. The structural and functional signatures of proteins that undergo multiple events of post-translational modification. *Protein Sci.* **2014**, *23*, 1077–1093. [CrossRef] [PubMed]

93. Kurotani, A.; Tokmakov, A.A.; Kuroda, Y.; Fukami, Y.; Shinozaki, K.; Sakurai, T. Correlations between predicted protein disorder and post-translational modifications in plants. *Bioinformatics* **2014**, *30*, 1095–1103. [CrossRef] [PubMed]

94. Romero, P.R.; Zaidi, S.; Fang, Y.Y.; Uversky, V.N.; Radivojac, P.; Oldfield, C.J.; Cortese, M.S.; Sickmeier, M.; LeGall, T.; Obradovic, Z.; et al. Alternative splicing in concert with protein intrinsic disorder enables increased functional diversity in multicellular organisms. *Proc. Natl. Acad. Sci. USA* **2006**, *103*, 8390–8395. [CrossRef]

95. Buljan, M.; Chalancon, G.; Dunker, A.K.; Bateman, A.; Balaji, S.; Fuxreiter, M.; Babu, M.M. Alternative splicing of intrinsically disordered regions and rewiring of protein interactions. *Curr. Opin. Struct. Biol.* **2013**, *23*, 443–450. [CrossRef] [PubMed]

96. Buljan, M.; Chalancon, G.; Eustermann, S.; Wagner, G.P.; Fuxreiter, M.; Bateman, A.; Babu, M.M. Tissue-specific splicing of disordered segments that embed binding motifs rewires protein interaction networks. *Mol. Cell* **2012**, *46*, 871–883. [CrossRef] [PubMed]

97. Vavouri, T.; Semple, J.I.; Garcia-Verdugo, R.; Lehner, B. Intrinsic protein disorder and interaction promiscuity are widely associated with dosage sensitivity. *Cell* **2009**, *138*, 198–208. [CrossRef] [PubMed]

98. Patil, A.; Nakamura, H. Disordered domains and high surface charge confer hubs with the ability to interact with multiple proteins in interaction networks. *FEBS Lett.* **2006**, *580*, 2041–2045. [CrossRef] [PubMed]

99. Haynes, C.; Oldfield, C.J.; Ji, F.; Klitgord, N.; Cusick, M.E.; Radivojac, P.; Uversky, V.N.; Vidal, M.; Iakoucheva, L.M. Intrinsic disorder is a common feature of hub proteins from four eukaryotic interactomes. *PLoS Comput. Biol.* **2006**, *2*, e100. [CrossRef] [PubMed]

100. Ekman, D.; Light, S.; Bjorklund, A.K.; Elofsson, A. What properties characterize the hub proteins of the protein-protein interaction network of saccharomyces cerevisiae? *Genome Biol.* **2006**, *7*, R45. [CrossRef]

101. Dosztanyi, Z.; Chen, J.; Dunker, A.K.; Simon, I.; Tompa, P. Disorder and sequence repeats in hub proteins and their implications for network evolution. *J. Proteome Res.* **2006**, *5*, 2985–2995. [CrossRef] [PubMed]

102. Singh, G.P.; Ganapathi, M.; Sandhu, K.S.; Dash, D. Intrinsic unstructuredness and abundance of pest motifs in eukaryotic proteomes. *Proteins* **2006**, *62*, 309–315. [CrossRef] [PubMed]

103. Dunker, A.K.; Garner, E.; Guilliot, S.; Romero, P.; Albrecht, K.; Hart, J.; Obradovic, Z.; Kissinger, C.; Villafranca, J.E. Protein disorder and the evolution of molecular recognition: Theory, predictions and observations. *Pac. Symp. Biocomput.* **1998**, 473–484.

104. Karantanos, T.; Corn, P.G.; Thompson, T.C. Prostate cancer progression after androgen deprivation therapy: Mechanisms of castrate resistance and novel therapeutic approaches. *Oncogene* **2013**, *32*, 5501–5511. [CrossRef] [PubMed]

105. Schweizer, M.T.; Antonarakis, E.S.; Wang, H.; Ajiboye, A.S.; Spitz, A.; Cao, H.; Luo, J.; Haffner, M.C.; Yegnasubramanian, S.; Carducci, M.A.; et al. Effect of bipolar androgen therapy for asymptomatic men with castration-resistant prostate cancer: Results from a pilot clinical study. *Sci. Transl. Med.* **2015**, *7*, 269ra262. [CrossRef] [PubMed]

106. Zeng, Y.; He, Y.; Yang, F.; Mooney, S.M.; Getzenberg, R.H.; Orban, J.; Kulkarni, P. The cancer/testis antigen prostate-associated gene 4 (PAGE4) is a highly intrinsically disordered protein. *J. Biol. Chem.* **2011**, *286*, 13985–13994. [CrossRef] [PubMed]

107. Sampson, N.; Ruiz, C.; Zenzmaier, C.; Bubendorf, L.; Berger, P. PAGE4 positivity is associated with attenuated ar signaling and predicts patient survival in hormone-naive prostate cancer. *Am. J. Pathol.* **2012**, *181*, 1443–1454. [CrossRef] [PubMed]

108. Zeng, Y.; Gao, D.; Kim, J.J.; Shiraishi, T.; Terada, N.; Kakehi, Y.; Kong, C.; Getzenberg, R.H.; Kulkarni, P. Prostate-associated gene 4 (PAGE4) protects cells against stress by elevating p21 and suppressing reactive oxygen species production. *Am. J. Clin. Exp. Urol.* **2013**, *1*, 39–52. [PubMed]

109. Rajagopalan, K.; Qiu, R.; Mooney, S.M.; Rao, S.; Shiraishi, T.; Sacho, E.; Huang, H.; Shapiro, E.; Weninger, K.R.; Kulkarni, P. The stress-response protein prostate-associated gene 4, interacts with c-Jun and potentiates its transactivation. *Biochim. Biophys. Acta* **2014**, *1842*, 154–163. [CrossRef] [PubMed]

110. Mooney, S.M.; Qiu, R.; Kim, J.J.; Sacho, E.J.; Rajagopalan, K.; Johng, D.; Shiraishi, T.; Kulkarni, P.; Weninger, K.R. Cancer/testis antigen PAGE4, a regulator of c-jun transactivation, is phosphorylated by homeodomain-interacting protein kinase 1, a component of the stress-response pathway. *Biochemistry* **2014**, *53*, 1670–1679. [CrossRef] [PubMed]

111. He, Y.; Chen, Y.; Mooney, S.M.; Rajagopalan, K.; Bhargava, A.; Sacho, E.; Weninger, K.; Bryan, P.N.; Kulkarni, P.; Orban, J. Phosphorylation-induced conformational ensemble switching in an intrinsically disordered cancer/testis antigen. *J. Biol Chem.* **2015**, *290*, 25090–25102. [CrossRef] [PubMed]

112. Kulkarni, P.; Jolly, M.K.; Jia, D.; Mooney, S.M.; Bhargava, A.; Kagohara, L.T.; Chen, Y.; Hao, P.; He, Y.; Veltri, R.W.; et al. Phosphorylation-induced conformational dynamics in an intrinsically disordered protein and potential role in phenotypic heterogeneity. *Proc. Natl. Acad. Sci. USA* **2017**. [CrossRef] [PubMed]

113. Romero, P.; Obradovic, Z.; Li, X.; Garner, E.C.; Brown, C.J.; Dunker, A.K. Sequence complexity of disordered protein. *Proteins* **2001**, *42*, 38–48. [CrossRef]

114. Peng, K.; Radivojac, P.; Vucetic, S.; Dunker, A.K.; Obradovic, Z. Length-dependent prediction of protein intrinsic disorder. *BMC Bioinform.* **2006**, *7*, 208. [CrossRef] [PubMed]

115. Obradovic, Z.; Peng, K.; Vucetic, S.; Radivojac, P.; Dunker, A.K. Exploiting heterogeneous sequence properties improves prediction of protein disorder. *Proteins* **2005**, *61*, 176–182. [CrossRef] [PubMed]

116. Xue, B.; Dunbrack, R.L.; Williams, R.W.; Dunker, A.K.; Uversky, V.N. PONDR-FIT: A meta-predictor of intrinsically disordered amino acids. *Biochim. Biophys. Acta* **2010**, *1804*, 996–1010. [CrossRef] [PubMed]

117. Dosztanyi, Z.; Csizmok, V.; Tompa, P.; Simon, I. IUPred: Web server for the prediction of intrinsically unstructured regions of proteins based on estimated energy content. *Bioinformatics* **2005**, *21*, 3433–3434. [CrossRef] [PubMed]

118. Chatr-Aryamontri, A.; Oughtred, R.; Boucher, L.; Rust, J.; Chang, C.; Kolas, N.K.; O'Donnell, L.; Oster, S.; Theesfeld, C.; Sellam, A.; et al. The BioGRID interaction database: 2017 Update. *Nucleic Acids Res.* **2017**, *45*, 369–379. [CrossRef] [PubMed]

119. Sato, N.; Sadar, M.D.; Bruchovsky, N.; Saatcioglu, F.; Rennie, P.S.; Sato, S.; Lange, P.H.; Gleave, M.E. Androgenic induction of prostate-specific antigen gene is repressed by protein-protein interaction between the androgen receptor and AP-1/c-Jun in the human prostate cancer cell line lncap. *J. Biol. Chem.* **1997**, *272*, 17485–17494. [CrossRef] [PubMed]

120. Tillman, K.; Oberfield, J.L.; Shen, X.Q.; Bubulya, A.; Shemshedini, L. C-fos dimerization with c-Jun represses c-Jun enhancement of androgen receptor transactivation. *Endocrine* **1998**, *9*, 193–200. [CrossRef]

121. Isaacs, J.T.; D'Antonio, J.M.; Chen, S.; Antony, L.; Dalrymple, S.P.; Ndikuyeze, G.H.; Luo, J.; Denmeade, S.R. Adaptive auto-regulation of androgen receptor provides a paradigm shifting rationale for bipolar androgen therapy (BAT) for castrate resistant human prostate cancer. *Prostate* **2012**, *72*, 1491–1505. [CrossRef] [PubMed]

122. Wang, S.S.; Smiraglia, D.J.; Wu, Y.Z.; Ghosh, S.; Rader, J.S.; Cho, K.R.; Bonfiglio, T.A.; Nayar, R.; Plass, C.; Sherman, M.E. Identification of novel methylation markers in cervical cancer using restriction landmark genomic scanning. *Cancer Res.* **2008**, *68*, 2489–2497. [CrossRef] [PubMed]

123. Demokan, S.; Chuang, A.Y.; Pattani, K.M.; Sidransky, D.; Koch, W.; Califano, J.A. Validation of nucleolar protein 4 as a novel methylated tumor suppressor gene in head and neck cancer. *Oncol. Rep.* **2014**, *31*, 1014–1020. [PubMed]

124. Stangeland, B.; Mughal, A.A.; Grieg, Z.; Sandberg, C.J.; Joel, M.; Nygard, S.; Meling, T.; Murrell, W.; Vik Mo, E.O.; Langmoen, I.A. Combined expressional analysis, bioinformatics and targeted proteomics identify new potential therapeutic targets in glioblastoma stem cells. *Oncotarget* **2015**, *6*, 26192–26215. [CrossRef] [PubMed]

125. Takayanagi-Kiya, S.; Misawa-Hojo, K.; Kiya, T.; Kunieda, T.; Kubo, T. Splicing variants of NOL4 differentially regulate the transcription activity of MLR1 and MLR2 in cultured cells. *Zool. Sci.* **2014**, *31*, 735–740. [CrossRef] [PubMed]

126. Pundir, S.; Martin, M.J.; O'Donovan, C. Uniprot protein knowledgebase. *Methods Mol. Biol.* **2017**, *1558*, 41–55. [PubMed]

127. Oates, M.E.; Romero, P.; Ishida, T.; Ghalwash, M.; Mizianty, M.J.; Xue, B.; Dosztanyi, Z.; Uversky, V.N.; Obradovic, Z.; Kurgan, L.; et al. D(2)p(2): Database of disordered protein predictions. *Nucleic Acids Res.* **2013**, *41*, D508–D516. [CrossRef] [PubMed]

128. Martinez-Garay, I.; Rustom, A.; Gerdes, H.H.; Kutsche, K. The novel centrosomal associated protein CEP55 is present in the spindle midzone and the midbody. *Genomics* **2006**, *87*, 243–253. [CrossRef] [PubMed]

129. Jeffery, J.; Sinha, D.; Srihari, S.; Kalimutho, M.; Khanna, K.K. Beyond cytokinesis: The emerging roles of CEP55 in tumorigenesis. *Oncogene* **2016**, *35*, 683–690. [CrossRef] [PubMed]

130. Inoda, S.; Hirohashi, Y.; Torigoe, T.; Nakatsugawa, M.; Kiriyama, K.; Nakazawa, E.; Harada, K.; Takasu, H.; Tamura, Y.; Kamiguchi, K.; et al. CEP55/C10ORF3, a tumor antigen derived from a centrosome residing protein in breast carcinoma. *J. Immunother.* **2009**, *32*, 474–485. [CrossRef] [PubMed]

131. Tao, J.; Zhi, X.; Tian, Y.; Li, Z.; Zhu, Y.; Wang, W.; Xie, K.; Tang, J.; Zhang, X.; Wang, L.; et al. CEP55 contributes to human gastric carcinoma by regulating cell proliferation. *Tumour Biol.* **2014**, *35*, 4389–4399. [CrossRef] [PubMed]

132. Singh, P.K.; Srivastava, A.K.; Rath, S.K.; Dalela, D.; Goel, M.M.; Bhatt, M.L. Expression and clinical significance of centrosomal protein 55 (CEP55) in human urinary bladder transitional cell carcinoma. *Immunobiology* **2015**, *220*, 103–108. [CrossRef] [PubMed]

133. Koch, M.; Wiese, M. Gene expression signatures of angiocidin and darapladib treatment connect to therapy options in cervical cancer. *J. Cancer Res. Clin. Oncol.* **2013**, *139*, 259–267. [CrossRef] [PubMed]

134. Colak, D.; Nofal, A.; Albakheet, A.; Nirmal, M.; Jeprel, H.; Eldali, A.; Al-Tweigeri, T.; Tulbah, A.; Ajarim, D.; Malik, O.A.; et al. Age-specific gene expression signatures for breast tumors and cross-species conserved potential cancer progression markers in young women. *PLoS ONE* **2013**, *8*, e63204. [CrossRef] [PubMed]

135. Inoda, S.; Morita, R.; Hirohashi, Y.; Torigoe, T.; Asanuma, H.; Nakazawa, E.; Nakatsugawa, M.; Tamura, Y.; Kamiguchi, K.; Tsuruma, T.; et al. The feasibility of CEP55/C10ORF3 derived peptide vaccine therapy for colorectal carcinoma. *Exp. Mol. Pathol.* **2010**, *90*, 55–60. [CrossRef] [PubMed]

136. Martin, K.J.; Patrick, D.R.; Bissell, M.J.; Fournier, M.V. Prognostic breast cancer signature identified from 3D culture model accurately predicts clinical outcome across independent datasets. *PLoS ONE* **2008**, *3*, e2994. [CrossRef] [PubMed]

137. Gemenetzidis, E.; Bose, A.; Riaz, A.M.; Chaplin, T.; Young, B.D.; Ali, M.; Sugden, D.; Thurlow, J.K.; Cheong, S.C.; Teo, S.H.; et al. FOXM1 upregulation is an early event in human squamous cell carcinoma and it is enhanced by nicotine during malignant transformation. *PLoS ONE* **2009**, *4*, e4849. [CrossRef] [PubMed]

138. Ishida, T.; Kinoshita, K. Prdos: Prediction of disordered protein regions from amino acid sequence. *Nucleic Acids Res.* **2007**, *35*, W460–W464. [CrossRef] [PubMed]

139. Walsh, I.; Martin, A.J.; Di Domenico, T.; Tosatto, S.C. Espritz: Accurate and fast prediction of protein disorder. *Bioinformatics* **2012**, *28*, 503–509. [CrossRef] [PubMed]

140. Meszaros, B.; Simon, I.; Dosztanyi, Z. Prediction of protein binding regions in disordered proteins. *PLoS Comput. Biol.* **2009**, *5*, e1000376. [CrossRef] [PubMed]

141. Dosztanyi, Z.; Meszaros, B.; Simon, I. ANCHOR: Web server for predicting protein binding regions in disordered proteins. *Bioinformatics* **2009**, *25*, 2745–2746. [CrossRef] [PubMed]

142. van der Horst, A.; Khanna, K.K. The peptidyl-prolyl isomerase pin1 regulates cytokinesis through CEP55. *Cancer Res.* **2009**, *69*, 6651–6659. [CrossRef] [PubMed]

143. Bastos, R.N.; Barr, F.A. Plk1 negatively regulates CEP55 recruitment to the midbody to ensure orderly abscission. *J. Cell Biol.* **2010**, *191*, 751–760. [CrossRef] [PubMed]

144. Mondal, G.; Rowley, M.; Guidugli, L.; Wu, J.; Pankratz, V.S.; Couch, F.J. BRCA2 localization to the midbody by filamin a regulates CEP55 signaling and completion of cytokinesis. *Dev. Cell* **2012**, *23*, 137–152. [CrossRef] [PubMed]

145. Lee, H.H.; Elia, N.; Ghirlando, R.; Lippincott-Schwartz, J.; Hurley, J.H. Midbody targeting of the escrt machinery by a noncanonical coiled coil in CEP55. *Science* **2008**, *322*, 576–580. [CrossRef] [PubMed]

146. Cannon, G.W.; Mullins, C.; Lucia, M.S.; Hayward, S.W.; Lin, V.; Liu, B.C.; Slawin, K.; Rubin, M.A.; Getzenberg, R.H. A preliminary study of JM-27: A serum marker that can specifically identify men with symptomatic benign prostatic hyperplasia. *J. Urol.* **2007**, *177*, 610–614, discussion 614. [CrossRef] [PubMed]

147. Prakash, K.; Pirozzi, G.; Elashoff, M.; Munger, W.; Waga, I.; Dhir, R.; Kakehi, Y.; Getzenberg, R.H. Symptomatic and asymptomatic benign prostatic hyperplasia: Molecular differentiation by using microarrays. *Proc. Natl. Acad. Sci. USA* **2002**, *99*, 7598–7603. [CrossRef] [PubMed]

148. Uversky, V.N. Wrecked regulation of intrinsically disordered proteins in diseases: Pathogenicity of deregulated regulators. *Front. Mol. Biosci.* **2014**, *1*, 6. [CrossRef] [PubMed]

149. Smith, L.M.; Kelleher, N.L. Proteoform: A single term describing protein complexity. *Nat. Methods* **2013**, *10*, 186–187. [CrossRef] [PubMed]

150. Uversky, V.N. P53 proteoforms and intrinsic disorder: An illustration of the protein structure-function continuum concept. *Int. J. Mol. Sci.* **2016**, *17*, 1874. [CrossRef] [PubMed]

151. Zaslavsky, B.Y.; Uversky, V.N.; Chait, A. Analytical applications of partitioning in aqueous two-phase systems: Exploring protein structural changes and protein-partner interactions in vitro and in vivo by solvent interaction analysis method. *Biochim. Biophys. Acta* **2016**, *1864*, 622–644. [CrossRef] [PubMed]

152. Zaslavsky, B.Y.; Uversky, V.N.; Chait, A. Solvent interaction analysis as a proteomic approach to structure-based biomarker discovery and clinical diagnostics. *Expert Rev. Proteom.* **2016**, *13*, 9–17. [CrossRef] [PubMed]

153. Felli, I.C.; Gonnelli, L.; Pierattelli, R. In-cell ^{13}C-NMR spectroscopy for the study of intrinsically disordered proteins. *Nat. Protoc.* **2014**, *9*, 2005–2016. [CrossRef] [PubMed]

154. Theillet, F.X.; Binolfi, A.; Frembgen-Kesner, T.; Hingorani, K.; Sarkar, M.; Kyne, C.; Li, C.; Crowley, P.B.; Gierasch, L.; Pielak, G.J.; et al. Physicochemical properties of cells and their effects on intrinsically disordered proteins (IDPS). *Chem. Rev.* **2014**, *114*, 6661–6714. [CrossRef] [PubMed]

155. Ikeya, T.; Hanashima, T.; Hosoya, S.; Shimazaki, M.; Ikeda, S.; Mishima, M.; Guntert, P.; Ito, Y. Improved in-cell structure determination of proteins at near-physiological concentration. *Sci. Rep.* **2016**, *6*, 38312. [CrossRef] [PubMed]

International Journal of
Molecular Sciences

MDPI

Article

Wisteria floribunda Agglutinin and Its Reactive-Glycan-Carrying Prostate-Specific Antigen as a Novel Diagnostic and Prognostic Marker of Prostate Cancer

Kazuhisa Hagiwara [1], Yuki Tobisawa [1], Takatoshi Kaya [2], Tomonori Kaneko [2], Shingo Hatakeyama [1], Kazuyuki Mori [1], Yasuhiro Hashimoto [3], Takuya Koie [1], Yoshihiko Suda [2], Chikara Ohyama [1,3] and Tohru Yoneyama [1,3,*]

[1]   Department of Urology, Hirosaki University Graduate School of Medicine, Hirosaki 036-8562, Japan;
      hagiwara.kazuhisa@gmail.com (K.H.); tobisawa@hirosaki-u.ac.jp (Y.T.); shingoh@hirosaki-u.ac.jp (S.H.);
      moribio@hirosaki-u.ac.jp (K.M.); goodwin@hirosaki-u.ac.jp (T.K.); coyama@hirosaki-u.ac.jp (C.O.)
[2]   Corporate R&D Headquarters, Konica Minolta, Inc., Hino-shi, Tokyo 191-8511, Japan;
      takatoshi.kaya@konicaminolta.com (T.K.); tomonori.kaneko1@konicaminolta.com (T.K.);
      yoshihiko.suda@konicaminolta.com (Y.S.)
[3]   Department of Advanced Transplant and Regenerative Medicine,
      Hirosaki University Graduate School of Medicine, Hirosaki 036-8562, Japan; bikkuri@opal.plala.or.jp
*   Correspondence: tohruyon@hirosaki-u.ac.jp; Tel.: +81-172-39-5091

Academic Editor: Carsten Stephan
Received: 28 December 2016; Accepted: 19 January 2017; Published: 26 January 2017

Abstract: *Wisteria floribunda* agglutinin (WFA) preferably binds to LacdiNAc glycans, and its reactivity is associated with tumor progression. The aim of this study to examine whether the serum LacdiNAc carrying prostate-specific antigen–glycosylation isomer (PSA-Gi) and WFA-reactivity of tumor tissue can be applied as a diagnostic and prognostic marker of prostate cancer (PCa). Between 2007 and 2016, serum PSA-Gi levels before prostate biopsy (Pbx) were measured in 184 biopsy-proven benign prostatic hyperplasia patients and 244 PCa patients using an automated lectin-antibody immunoassay. WFA-reactivity on tumor was analyzed in 260 radical prostatectomy (RP) patients. Diagnostic and prognostic performance of serum PSA-Gi was evaluated using area under the receiver-operator characteristic curve (AUC). Prognostic performance of WFA-reactivity on tumor was evaluated via Cox proportional hazards regression analysis and nomogram. The AUC of serum PSA-Gi detecting PCa and predicting Pbx Grade Group (GG) 3 and GG $\geq$ 3 after RP was much higher than those of conventional PSA. Multivariate analysis showed that WFA-reactivity on prostate tumor was an independent risk factor of PSA recurrence. The nomogram was a strong model for predicting PSA-free survival provability with a c-index $\geq$0.7. Serum PSA-Gi levels and WFA-reactivity on prostate tumor may be a novel diagnostic and pre- and post-operative prognostic biomarkers of PCa, respectively.

Keywords: prostate-specific antigen; *N*-glycan; LacdiNAc; *Wisteria floribunda* agglutinin (WFA) lectin; biomarker

1. Introduction

Prostate cancer (PCa) is a common cancer in men worldwide [1,2]. The most important issues regarding PCa is overdiagnosis and overtreatment [3,4]. Although the majority of patients diagnosed as clinically localized PCa, 30%–40% of patients who receive aggressive treatment such as radical prostatectomy (RP) experience biochemical recurrence [5,6]. Although, active surveillance (AS) is also proposed for low-risk PCa patients who meet the Prostate Cancer Research International Active Surveillance (PRIAS) criteria, 10%–30% of AS patients experience extraprostatic extension, and 42%–80% of AS patients experience an upgrade of the Gleason score after RP (ope GS) [7–10]. Pre-operative prostate-specific antigen (PSA) levels and biopsy GS are also powerful indicators of biological outcomes after RP [11]. Nevertheless, these indicators are not sufficient to prevent the overtreatment of PCa, and there is a need for more accurate diagnostic and prognostic indicators to select an appropriate treatment option for localized PCa.

N- and *O*-glycosylation plays important roles in disease progression. The nonreducing terminal GalNAcβ1-4GlcNAc-(LacdiNAc) structure is found in *N*- and *O*-glycans of many mammalian glycoproteins though in very small amounts [12]. *Wisteria floribunda* agglutinin (WFA) is a good probe for LacdiNAc glycan [12]. Several researchers reported about LacdiNAc expression in cancer using WFA. They stated that LacdiNAc in *N*-glycans significantly decreases during progression of human breast cancer [13,14]. In contrast, the enhanced expression of LacdiNAc has been shown to be associated with the progression of human prostate, ovarian, colon, and liver cancers [12,15–17]. Therefore, the quantification of LacdiNAc glycan carrying glycoproteins or tissue-specific expression of LacdiNAc glycan detected by the WFA has shown promise as cancer glycobiomarkers [17–19]. In particular, regarding PCa, there are only three papers about LacdiNAc distribution in prostate biopsy (Pbx) and RP specimens using WFA [15,16,20], and they did not report the relation between WFA-reactivity in tissues and PCa prognosis. Although there are only a few reports including our group's about PCa-associated aberrant LacdiNAc carrying PSA-glycosylation isomer (PSA-Gi) (Figure 1) [21,22], we demonstrate a pilot study of serum PSA-Gi as a diagnostic biomarker by using an automated two-step WFA–anti-PSA antibody sandwich immunoassay using high-sensitivity surface plasmon field-enhanced fluorescence spectrometry (SPFS) (Figure 2) [22]. Therefore, in this study, we retrospectively evaluated diagnostic and pre-operative prognostic performance of serum PSA-Gi and examined the association between WFA-reactivity on PCa tissues and PSA recurrence after RP.

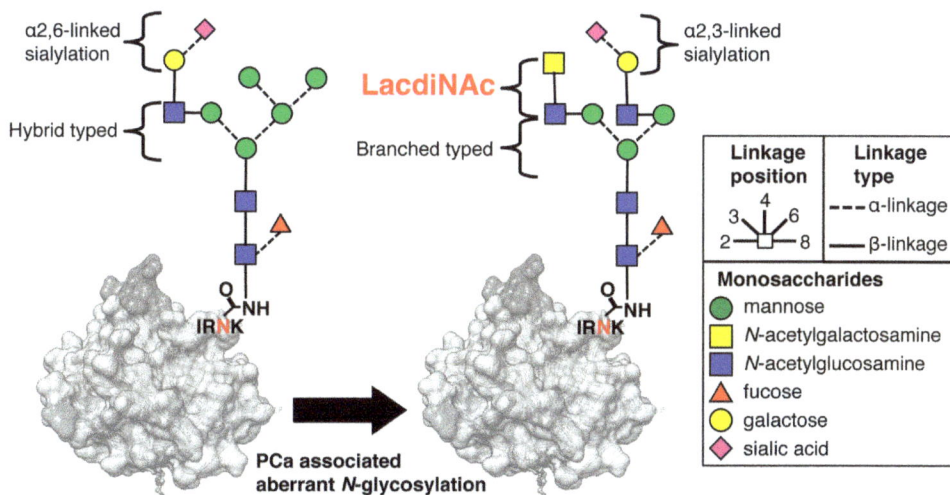

Figure 1. Prostate cancer (PCa)-associated aberrant *N*-glycosylation of prostate-specific antigen (PSA). PSA derived from PCa serum and culture supernatant of LNCaP carries *Wisteria floribunda* agglutinin (WFA)-reactive LacdiNAc glycans; this is not the case for PSA derived from benign prostatic hyperplasia (BPH) serum. PCa-associated aberrant LacdiNAc carrying PSA glycosylation isomer designated as PSA–glycosylation isomer (PSA-Gi) [21]. Carbon linkage positions are denoted by the bond position drawn on each monosachharide. IRNK indicate *N*-glycosylation site of PSA.

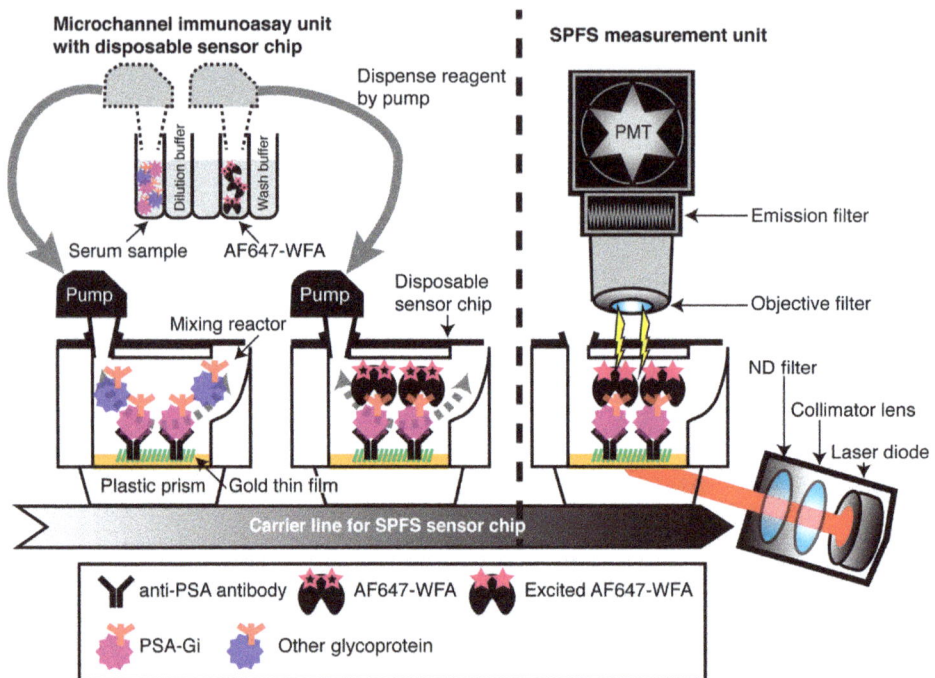

Figure 2. The schematic representation of serum PSA-Gi detection using a two-step surface plasmon field-enhanced fluorescence spectrometry (SPFS)-based WFA lectin-anti-PSA antibody immunoassay. Gray line arrows indicated that reagent dispense from reagent container to mixing reactor using pump. Gray dotted line arrows indicated mixing the content of mixing reactor by pump.

2. Results

2.1. Diagnostic Performance of Serum PSA-Gi before Pbx Much Superior to Total PSA

Serum PSA-Gi levels before Pbx was measured in patients with benign prostatic hyperplasia (BPH) (n = 184) or PCa (n = 244) to evaluate diagnostic performance. Patients' characteristics in the BPH and PCa groups are shown in Table 1. Serum PSA-Gi levels in the both total PSA range ≤20 ng/mL (Figure 3a,b) and ≤10 ng/mL (Figure 3d,e) were significantly higher in patients with PCa (median: 0.1680 U/mL and median: 0.1140 U/mL, respectively) than in patients with BPH (median: 0.0715 U/mL and median: 0.0670 U/mL, respectively), $p < 0.0001$. The area under the receiver-operator characteristic curve (AUC) of PSA-Gi predicting PCa in any concentration range of total PSA (0.795, 95% CI; 0.753–0.837 and 0.752, 95% CI; 0.690–0.813, respectively) was much higher than those of PSA-Gi/total PSA (0.734, 95% CI; 0.686–0.782 and 0.718, 95% CI; 0.659–0.779, respectively) and total PSA (0.638, 95% CI; 0.586–0.691 and 0.550, 95% CI; 0.483–0.618, respectively) (Table 2, Figure 3c,f). At the cutoff PSA-Gi levels (0.0495 U/mL) for the prediction of PCa, the specificity at 90% sensitivity was 36.8%—much higher than the specificity of total PSA (18.8%). Furthermore, we found that higher PSA-Gi levels (≥0.1140 U/mL) in patients with BPH at first Pbx moderately predicted a diagnosis of PCa within 1–4 years after the first Pbx (Figure 3a,d). The nonparametric spearman correlation coefficient between the PSA-Gi level in BPH and total PSA in BPH was 0.3294 (95% CI, 0.1989–0.4559, $p < 0.0001$) and that between the PSA-Gi level in PCa and total PSA in PCa was 0.4613 (95% CI, 0.3531–0.5573, $p < 0.0001$) (Figure 3g). This means the PSA-Gi level was positively correlated with total PSA in BPH and PCa patients.

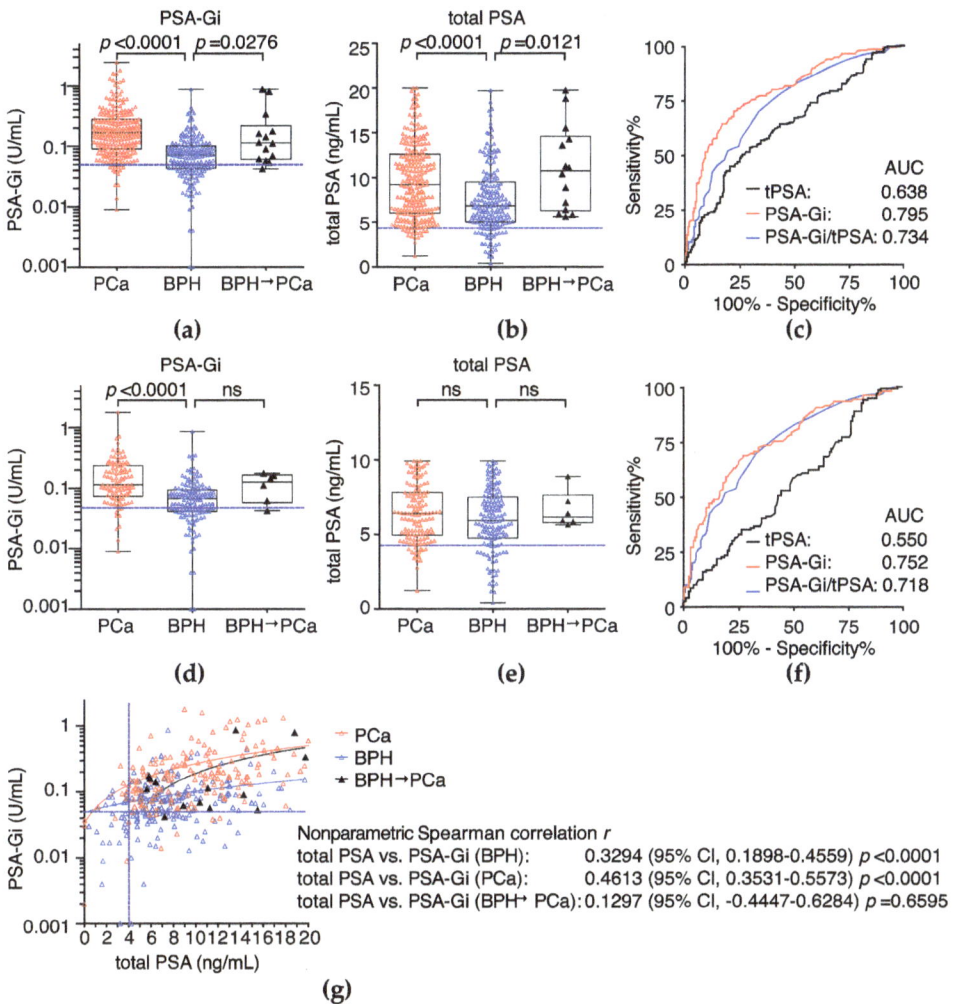

Figure 3. Serum levels of the PSA-Gi at Pbx in the patients who diagnosed as BPH or PCa by an SPFS-based lectin-antibody immunoassay. (**a**) PSA-Gi and (**b**) total PSA levels in patients with a diagnosis of BPH or PCa at a total PSA ≤ 20 ng/mL; (**c**) receiver-operator characteristic (ROC) curve analysis of total PSA, PSA-Gi, and PSA-Gi/total PSA in patients who had a diagnosis of BPH or PCa at a total PSA ≤ 20 ng/mL. The areas under the ROC curve (AUCs) for the prediction of PCa of PSA-Gi, total PSA, and PSA-Gi/total PSA were 0.795, 0.638, and 0.734, respectively; (**d**) PSA-Gi and (**e**) total PSA levels in patients with BPH or PCa at total PSA ≤ 10 ng/mL; (**f**) ROC curve analysis of total PSA, PSA-Gi, and PSA-Gi/total PSA in patients with BPH or PCa at a total PSA ≤ 10 ng/mL. The AUCs for the prediction of PCa by means of PSA-Gi, total PSA, and PSA-Gi/total PSA were 0.752, 0.550, and 0.718, respectively; (**g**) correlation between PSA-Gi and total PSA. Correlation coefficient was analyzed by non-parametric Spearman's *r*-test. (**a–g**) The cutoff level at 90% sensitivity of PSA-Gi and/or total PSA is presented as a blue dotted line.

Table 1. Characteristics of BPH patients and PCa patients.

Characteristics $n = 442$	BPH [a] 184	PCa [b] 244	BPH-> PCa 14	p ([a] vs. [b])
Age, median (range)	69 (30–87)	68 (44–85)	69 (52–80)	ns [1]
PSA [2], ng/mL, median (range)	6.8 (0.4–19.7)	9.0 (1.2–62.6)	6.3 (5.9–19.7)	<0.001
PSA-Gi, U/mL, median (range)	0.0715 (0.001–0.86)	0.165 (0.002–2.43)	0.113 (0.04–0.87)	<0.001
PSA-Gi/total PSA, U/ng, median (range)	0.0100 (0.00–0.1150)	0.0200 (0.002–0.1980)	0.0135 (0.003–0.0640)	<0.001
Clinical T stage, n (%)		$n = 244$		
cT1		144	(59.3)	
cT2		46	(18.5)	
cT3		55	(22.2)	
Pbx GS [3], n (%) Pbx GG [4]		$n = 244$		
3 + 3 [1]		6	(2.4)	
3 + 4 [2]		79	(32.4)	
4 + 3 [3]		29	(11.9)	
4 + 4 [4]		30	(12.3)	
3 + 5 [4]		3	(1.2)	
4 + 5 [5]		72	(29.5)	
5 + 4 [5]		20	(8.2)	
5 + 5 [5]		5	(2.0)	
Pathological T stage, n (%)		$n = 92$	$n = 8$	
pT1		4	(4.3)	0	(0)
pT2		53	(57.6)	5	(62.5)
pT3		38	(41.3)	3	(37.5)
Ope GS [5], n (%) Ope GG [6]		$n = 92$	$n = 8$	
3 + 3 [1]		1	(1.1)	
3 + 4 [2]		13	(14.1)	2	(25.0)
4 + 3 [3]		14	(15.2)	
3 + 5 [4]		3	(3.2)	1	(12.5)
4 + 4 [4]		9	(9.8)	1	(12.5)
5 + 3 [4]		1	(1.1)	
4 + 5 [5]		37	(40.2)	3	(37.5)
5 + 4 [5]		12	(13.0)	1	(12.5)
5 + 5 [5]		2	(2.2)	

[1] not significantly difference; [2] total PSA; [3] prostate biopsy Gleason score; [4] prostate biopsy grade group; [5] Gleason score after radical prostatectomy; [6] grade group after radical prostatectomy. Pbx: prostate biopsy; [a] BPH; [b] PCa.

Table 2. Comparison of areas under the receiver-operator characteristic curve (AUCs) of PSA, PSA-Gi, and PSA-Gi/total PSA for the detection of PCa.

Test Name	PSA Range	AUC	95% CI	p (vs. [a])	p (vs. [b])	p (vs. [c])
Total PSA [a]	-	0.638	0.586–0.691	-	<0.0001	0.0376
PSA-Gi [b]	20 ng/mL	0.795	0.753–0.837	<0.0001	-	0.0003
PSA-Gi/total PSA[c]	-	0.734	0.586–0.691	0.0376	0.0003	-
Total PSA [a]	-	0.550	0.483–0.618	-	<0.0001	<0.0001
PSA-Gi [b]	10 ng/mL	0.752	0.690–0.813	<0.0001	-	0.567
PSA-Gi/total PSA [c]	-	0.719	0.659–0.779	0.0009	0.0009	-

[a] Total PSA test; [b] PSA-Gi test; [c] PSA-Gi/total PSA test.

2.2. Serum PSA-Gi before Pbx Can Discriminate between Pbx Grade Group 2 and 3

Serum PSA-Gi levels before Pbx was measured in 244 PCa patients to evaluate the pre-operative predictor for a prostate biopsy. PSA-Gi levels were significantly correlated with Pbx grade group (GG) [23] (Figure 4a,b). Although total PSA could not discriminate between Pbx GG 2 and 3, serum PSA-Gi levels were significantly higher at ope GG 3 (median: 0.2500 U/mL, $p = 0.0118$) than at ope GG 2 (median: 0.1280 U/mL, Figure 4a,b). The AUC of PSA-Gi predicting Pbx GG 3 tumors was 0.649 (95% CI, 0.5221–0.7735) in contrast to the total PSA AUC of 0.520 (95% CI, 0.4091–0.6312; $p = 0.162$; Figure 4c). At the cutoff PSA-Gi level (0.1930 U/mL) for the prediction of GG 3 tumors at

Pbx, sensitivity was 57.1%, and specificity was 80.8%—muchhigher than the specificity of the total PSA test (47.4%).

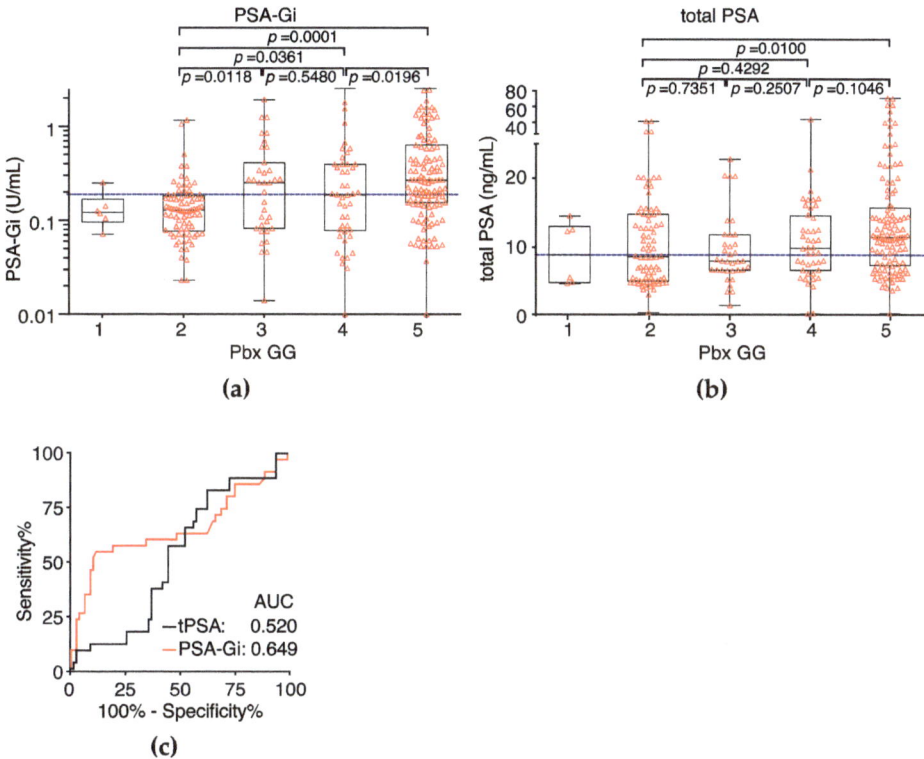

Figure 4. The serum PSA-Gi levels at Pbx in PCa patients who underwent radical prostatectomy (RP). (a) PSA-Gi levels before Pbx among PCa patients classified by the Pbx grade group (Pbx GG); (b) total PSA level before Pbx of PCa patients classified by the Pbx GG. Cutoff levels at 57.1% sensitivity of PSA-Gi and/or total PSA is presented as a blue dotted line; (c) ROC curve analysis of total PSA and PSA-Gi in PCa patients with Pbx GG 2 and Pbx GG 3. The AUCs for the prediction of patients with Pbx GG 3 of PSA-Gi and total PSA were 0.649 and 0.520, respectively.

2.3. Serum PSA-Gi before Pbx Can Discriminate between Ope Grade Group ≤2 and ≥3

Serum PSA-Gi levels before Pbx was measured in 92 PCa patients who underwent RP to evaluate the pre-operative prognostic performance. PSA-Gi levels were moderately correlated with grade group after RP (ope GG) [23] (Figure 5a,b). Although total PSA could not discriminate tumors with ope GG ≥ 3, serum PSA-Gi levels was significantly higher at ope GG ≥ 3 (median: 0.1885 U/mL, $p = 0.0068$) than at ope GG ≤ 2 (median: 0.0985 U/mL, Figure 5c,d). The AUC of PSA-Gi predicting ope GG ≥ 3 tumors was 0.724 (95% CI, 0.603–0.845) in contrast to the total PSA AUC of 0.618 (95% CI, 0.442–0.794; $p = 0.202$; Figure 5e). Furthermore, the PSA-Gi levels tended to be higher in patients with a GG upgrade from 2 at Pbx to ope GG ≥ 3 and were associated with a GG downgrade from ≥3 at Pbx to ope GG ≤ 2 (Figure 5f,g). At the cutoff PSA-Gi level (0.1445 U/mL) for the prediction of GG ≥ 3 tumors, sensitivity was 60.3%, and specificity was 78.6%—much higher than the specificity of the total PSA test (50.0%).

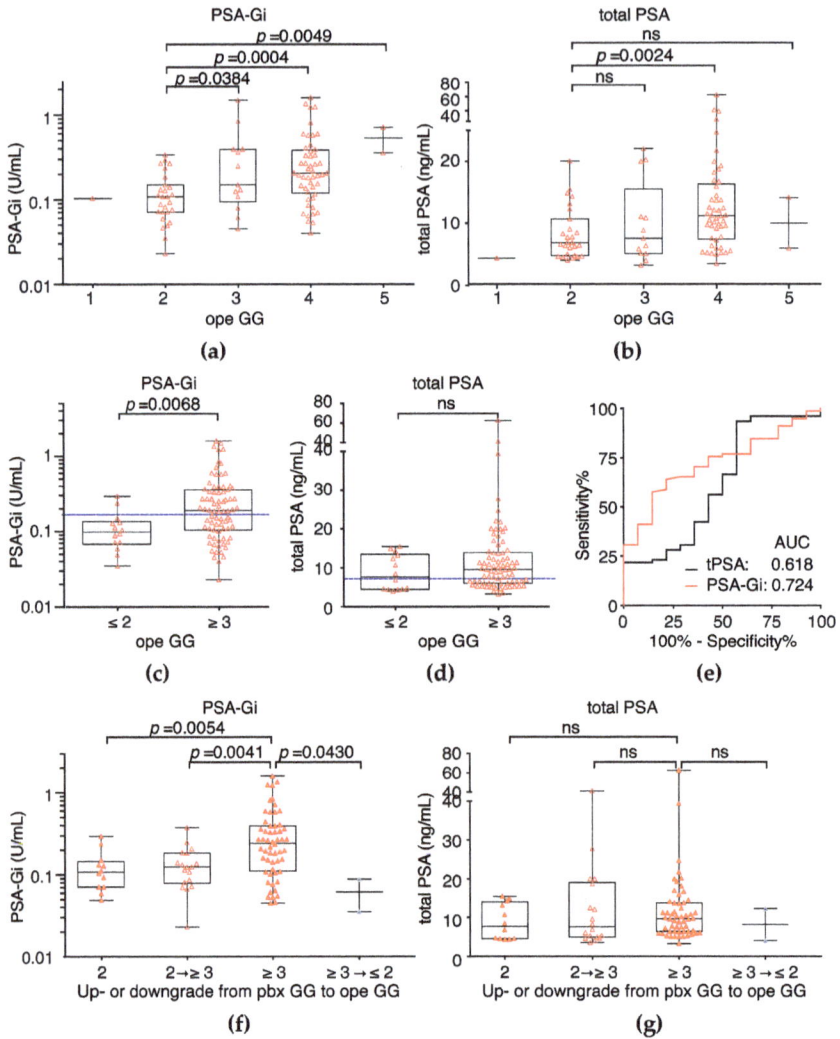

Figure 5. The serum PSA-Gi levels at Pbx in PCa patients who underwent RP. (**a**) PSA-Gi levels before Pbx among PCa patients classified by the grade group after RP (ope GG); (**b**) total PSA level before Pbx of PCa patients classified by the ope GG; (**c,d**) PSA-Gi and total PSA levels before Pbx between patients with ope GG ≤ 2 and ope GG ≥ 3. Cutoff levels at 60% sensitivity of PSA-Gi and/or total PSA is presented as a blue dotted line; (**e**) ROC curve analysis of total PSA and PSA-Gi in PCa patients with ope GG ≤ 2 and ope GG ≥ 3. The AUCs for the prediction of patients with ope GG ≥ 3 of PSA-Gi and total PSA were 0.724 and 0.618, respectively; (**f,g**) PSA-Gi and total PSA levels in patients with a GG upgrade from 2 at Pbx to ope GG ≥ 3 and a GG downgrade from ≥3 at Pbx to ope GG ≤ 2.

2.4. Tumors Strongly and Moderately Positive for WFA Is an Independent Risk Factor of PSA Recurrence

Immunohistochemical staining of RP specimens by WFA was performed to examine the association between WFA-reactivity of tumor site and clinicopathological status. Patients' characteristics in the 260 RP patients are shown in Table 3. WFA-reactive glycan was expressed in both benign prostate glands and tumors. On the basis of the reciprocal intensity of a tumor site [24], the WFA-reactivity was classified into three groups: weakly positive (median 78.5, range 74–85), moderately positive (median 98.5, range 86–104), and strongly positive (median 132, range 105–170; Figures 6a and A1). When collated with these criteria, tumors strongly and moderately positive for WFA were significantly associated with a higher ope GS, pathological stage (≥pT3), and perineural invasion (pn)-positive status (Figure 6b and Table 3). As shown in Figure 5c, patients with tumors strongly and moderately positive for WFA had a much shorter period of PSA recurrence after RP than patients with tumors weakly positive for WFA (log-rank test, *p* = 0.0044). Multivariate Cox regression analysis revealed that WFA-reactivity was an independent risk factor of PSA recurrence (Table 4) and developed post-operative nomogram including WFA-reactivity, age, grade group, pT, RM, and pn status for prediction of PSA-free survival provability (Figure 6d). The c-index of nomogram was 0.754 (95% CI, 0.697–0.812) [25].

Table 3. Characteristics of PCa patients who underwent RP categorized by WFA-reactivity.

Characteristics		Weakly Positive [a]		Moderately Positive [b]		Strongly Positive [c]		*p* [a] vs. [b + c]
n, Total = 260		51		95		112		
Age, median (range)		68 (48–75)		68 (56–76)		68 (52–78)		0.555
PSA [1], ng/mL, median (range)		7.5 (2.3–18.4)		7.4 (0.6–27.6)		7.5 (0.5–35.9)		0.473
Pathological T stage, *n* (%)								0.008 [2]
pT2, *n* = 163		41	(26.4)	48	(29.4)	72	(44.2)	0.002
pT3, *n* = 96		10	(10.4)	47	(49.0)	39	(40.6)	0.002
pT4, *n* = 1		0	(0)	0	(0)	1	(100)	0.612
Ope GS [3], *n* (%)	Ope GG [4]							0.045 [2]
3 + 3, *n* = 11	Ope GG [1]	5	(45.4)	3	(27.3)	3	(27.3)	0.035
3 + 4, *n* = 112	Ope GG [2]	28	(26.5)	34	(27.9)	50	(44.6)	0.108
4 + 3, *n* = 63	Ope GG [3]	13	(19.3)	28	(45.2)	22	(35.5)	0.955
4 + 4, *n* = 9	Ope GG [4]	2	(22.3)	3	(33.3)	4	(44.4)	0.889
3 + 5, *n* = 9	Ope GG [4]	1	(11.1)	3	(33.3)	5	(55.6)	0.482
4 + 5, *n* = 42	Ope GG [5]	4	(9.5)	17	(40.5)	21	(50.0)	0.056
5 + 4, *n* = 14	Ope GG [5]	0	(0)	7	(50.0)	7	(50.0)	0.052
pn [5], *n* (%)								
pn−, *n* = 56		21	(37.5)	18	(32.1)	17	(30.4)	<0.001
pn+, *n* = 204		32	(15.7)	77	(37.7)	95	(46.6)	<0.001
RM [6], *n* (%)								
RM−, *n* = 188		43	(22.9)	65	(34.6)	80	(42.5)	0.108
RM+, *n* = 72		10	(13.9)	30	(41.7)	32	(44.4)	0.108
PSA failure, *n* (%)								
−, *n* = 194		49	(25.3)	66	(34.0)	79	(40.7)	<0.001
+, *n* = 66		4	(6.1)	29	(43.9)	33	(50.0)	<0.001

[1] total PSA; [2] χ^2 test; [3] Ope GS, Gleason score after radical prostatectomy; [4] Ope GG, grade group after radical prostatectomy; [5] pn, perineural invasion; [6] RM, resection margin; [a] weakly positive; [b] moderately positive; [c] strongly positive.

Table 4. Multivariate analysis to determine an independent predictor of PSA recurrence.

Variable	Hazard Ratio	Standard Error	*p*
Age	1.046	0.027	0.099
WFA-reactivity	2.831	0.529	0.049
pT [1]	1.589	0.336	0.168
Grade group	1.246	0.099	0.027
RM [2]	2.424	0.319	0.006
pn [3]	1.715	0.447	0.227

[1] pathological T stage; [2] resection margin; [3] perineural invasion.

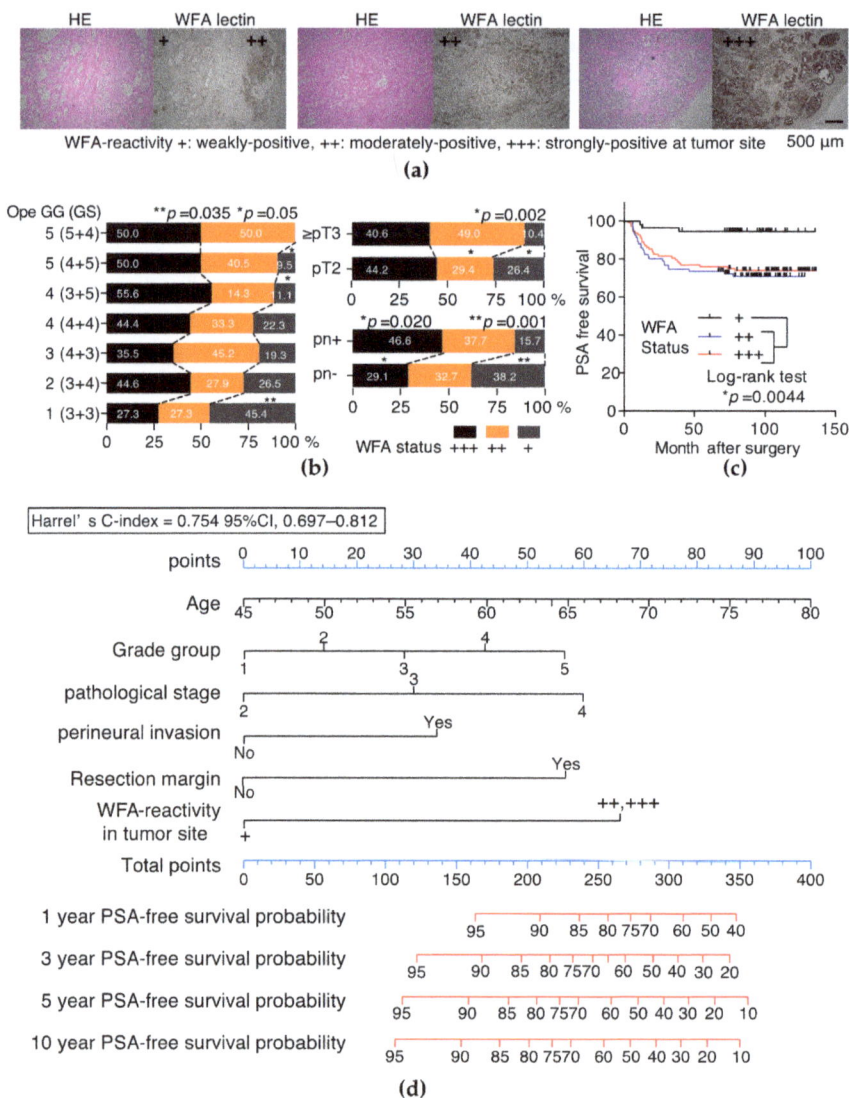

Figure 6. Immunohistochemical analysis of RP specimens using WFA lectin and post-operative nomogram predicting PSA-free survival probability. (**a**) Representative hematoxylin-eosin (HE) staining and WFA reactive-glycan expression of tumors of RP specimens. WFA-reactivity was classified into three groups: weakly positive, moderately positive, and strongly positive at a tumor site, respectively. Scale bar indicated 500 μm; (**b**) association between with WFA-reactive glycan expression and ope GG (ope GS), pathological stage, and perineural invasion-status; (**c**) PSA-free survival was evaluated using Kaplan–Meier curves and differences between three groups were assessed using the log-rank test. Patients with tumors strongly or moderately positive for WFA had a much shorter period of PSA recurrence after RP than did patients with tumors weakly positive for WFA; (**d**) Cox hazard regression analysis-based post-operative nomogram predicting PSA-free survival probability after RP. The c-index (0.754, 95% CI, 0.697–0.812), which is similar to the area under a receiver operating characteristic curve, was used to estimate the discrimination ability of the nomogram [25].

3. Discussion

One of the most important problems with PCa is overdiagnosis [3]. PSA-based screening has become controversial due to false positive results of total PSA in the PSA gray zone [4]. Overtreatment is also a major problem among certain segments of PCa patients [3] such as localized PCa and active surveillance patients [7–10]. Current biomarkers are not sufficient to prevent the overtreatment of PCa. Several serum-based testing (Phi, %p2PSA, and 4KScore), urine-based testing (PCA3) and MRI imaging has shown promising results in terms of diagnosis, localization, risk stratification, and staging of clinically significant PCa [26,27]. However, these promising biomarkers and imaging data are not yet cost-effective enough for routine clinical practice [28]. Therefore, there is a need for more accurate and cost-effective diagnostic and prognostic biomarkers. PCa-associated aberrant glycosylation of PSA is one of the candidate biomarkers. Fukushima et al. demonstrated that PSA derived from PCa serum and culture supernatant of LNCaP carries WFA-reactive LacdiNAc glycans; this is not the case for PSA derived from BPH serum [21] (Figure 1).

In the present study, we evaluated the diagnostic and pre-operative prognostic performance of WFA-reactive glycan-carrying PSA-Gi by using an SPFS-based automated immunoassay system [22]. We demonstrated that the AUC of PSA-Gi predicting PCa was much higher than that of the total PSA and PSA-Gi/total PSA (Figure 3c,f). We also demonstrated that a higher PSA-Gi level in BPH patients was moderately associated with a diagnosis of PCa within 1–4 years after first biopsy (Figure 3a,d). These results suggested that the diagnostic performance of a PSA-Gi single marker was much superior to conventional total PSA.

Furthermore, we showed that PSA-Gi before Pbx significantly higher in patients with Pbx GG 3 than that of patients with Pbx GG 2 and specificity for prediction of Pbx GG 3 was much higher than PSA (Figure 4a–c). This suggests that PSA-Gi can discriminate between GG 2 and GG 3 tumors and may be used as a predictor for a prostate biopsy to discriminate between non-aggressive and aggressive tumors in the active surveillance program. We also showed that the AUC of PSA-Gi predicting ope GG $\geq$ 3 tumors was higher than that of the total PSA and specificity for prediction of ope GG $\geq$ 3 was much higher than PSA (Figure 5e). The PSA-Gi levels before Pbx tends to be higher in patients with GG upgraded from 2 at Pbx to ope GG $\geq$ 3. A similar result was reported that pre-operative fucosylated haptoglobin (Fuc-Hpt) levels is significantly higher in patients with GS $\geq$ 7 than those with GS $\geq$ 6 [29]. Nevertheless, the serum Fuc-Hpt levels is also higher in patients with pancreatic, ovarian, and hepatocellular cancers [30,31]. In addition, Li et al. reported that the serum fucosylated PSA (Fuc-PSA) levels is significantly higher in patients with GS $\geq$ 7 than those with GS $\geq$ 6 [32]. It is well-known that PSA is a prostate-specific protein, and aberrant glycosylation of PSA including Fuc-PSA and PSA-Gi was thus found to be a more specific glycobiomarker of PCa than Fuc-Hpt. Although our sample size is small and retrospective, these results suggest that aberrant glycosylation of PSA is associated with PCa aggressiveness. Stark et al. demonstrated that GG 3 tumors are associated with a three-fold increase in lethal PCa compared with GG 2 tumors in RP specimens [33]. More recently, Epstein et al. also demonstrated that there are large differences in 5-year recurrence rates between both the GG 2 and GG 3 in a large multi-institutional surgical cohort and hazard ratios for GG 3 disease were generally threefold higher than for GG 2 [34]. Therefore, discrimination between GG 2 and GG 3 is an important task for the reduction of overtreatment of PCa. Thus, our PSA-Gi may be a promising pre-operative prognostic biomarker predicting Pbx GG 3 tumors and ope GG $\geq$ 3 tumors, particularly in very low-risk PCa patients who have met PRIAS criteria and PCa patients at an intermediate risk.

Moreover, we examined WFA-reactivity of prostate tumors showed that tumors strongly and moderately positive for WFA are significantly associated with higher ope GG, pT, and pn-positive status (Figure 6b) and worse PSA-free survival as compared to patients with weakly positive tumors for WFA (Figure 6c). Cox regression analysis here provided WFA-reactivity in tumors was an independent risk factor of PSA recurrence (Table 4). Thus, nomogram developed in this study including WFA-reactivity in the tumor site combined with clinocopathological parameters seemed to be a strong model for predicting PSA-free survival provability with a c-index (0.754) (Figure 6d). Further internal and external validation study was required for the evaluation of predictive performance in this nomogram.

Our results reveal that serum PSA-Gi levels before Pbx is useful for the discrimination of PCa as well as Pbx GG 3 and ope GG $\geq$ 3 patients and the WFA-reactivity of tumors is also useful for the prediction of PSA recurrence. Thus, both PSA-Gi and WFA-reactivity of tumors may reduce overdiagnosis and overtreatment of PCa.

4. Materials and Methods

This study was performed in accordance with the ethical standards of the Declaration of Helsinki and was approved by the Ethics Committee of Hirosaki University Graduate School of Medicine ("The Study about Carbohydrate Structure Change in Urological Disease"; approval number: 2014-195; approval date: 22 December 2014). Informed consent was obtained from all patients.

4.1. Serum Samples from Patients with BPH and PCa

A total of 442 patients with benign prostatic hyperplasia (BPH) and PCa were treated at our hospital between June 2007 and August 2016. Serum samples from patients with BPH (n = 184), PCa (n = 244 of whom 92 patients underwent RP), or PCa who diagnosed as BPH at first Pbx (n = 14) were obtained before the first Pbx. The final diagnoses of BPH or PCa were confirmed using the histopathological findings of prostate biopsies. Staging and grading information of the tumors for RP patients was obtained from medical charts. The grade group of prostate biopsy and prostatectomy specimens were evaluated according to the International Society of Urological Pathology (ISUP) guidelines [23]. Patient demographics are shown in Table 1. All samples were stored at $-80\ ^{\circ}$C until use.

4.2. Detection of Serum PSA-Gi and Total PSA

The serum PSA-Gi was detected by using an SPFS-based two-step WFA–anti-PSA antibody sandwich immunoassay with a disposable sensor chip as described previously [22]. The system was developed by Konica Minolta Inc. (Figure 1). Two-step sandwich SPFS immunoassays of PSA-Gi were carried out automatically by moving a cylindrical pump between the anti-total-PSA monoclonal antibody (No. 72, Mikuri Immunological Laboratories Co., Ltd., Osaka, Japan) immobilized on a thin gold film in a disposable sensor chip and a reagent container in a self-developed assay machine. The reagent container already contained a number of separate reagents, including wash buffer (TBS 0.05% Tween 20, 10× TBS (Nippon Gene Co., Ltd., Tokyo, Japan) and polysorbate 20 (MP Biomedicals, LLC., Santa Ana, CA, USA)), AF647-WFA (WFA (vector laboratories, Inc., Burlingame, CA, USA) labeled using an Alexa Fluor 647-labeling kit (A20186, Thermo Fisher Scientific Inc., Waltham, MA, USA)) and the sample for measurement. The 20 µL of serum was diluted by 100 µL of a PBS-based dilution buffe. Then the 100 µL diluted serum samples and AF647-WFA solution (10.0 µg/mL in 1% BSA in PBS) were allowed to react for 10 min, and unreacted lectins were removed with washing buffer (four washes) after the WFA lectin reaction. After four washes, the final washing buffer was kept for SPFS optical measurement in the microchannel of each disposable sensor chips. After the final washing step, AF647 in the microchannel of disposable sensor chips were sequentially excited by laser light, which was applied on the backside of a thin gold film through the plastic prism. The laser light was already p-polarized and collimated by the internal laser diode system. A laser diode (635 nm, 0.95 mW; Edmund Optics Japan, Ltd., Tokyo, Japan) was used as a light source with a Neutral Density filter (AND20C-10 (10%), Sigmakoki Co., Ltd., Saitama, Japan). The fluorescent signal of AF647 that passed through the emission filter (DIF-BP-1 (half width: 668 ± 5 nm), Optical Coatings Japan, Tokyo, Japan) was detected by a photomultiplier tube (H7421-40, Hamamatsu Photonics K.K., Shizuoka, Japan), which was located at the end of a light-converging optical system (numerical aperture, NA = 0.6; Edmund Optics Japan Ltd., Tokyo, Japan). All assays were conducted automatically at 25 °C; four immunoassays were carried out simultaneously. Standard PSA-Gi sample was obtained from culture supernatant of LNCaP cells (RCB2144, RIKEN Bio-resource Center through the National Bio-Resource Project of the MEXT, Tsukuba, Japan), as reported previously [22]. In brief, LNCaP cells were cultured in the RPMI 1640 medium (Thermo Fisher Scientific Inc., Waltham, MA, USA) supplemented with 10% fetal calf serum (FCS) at 5% CO_2 at 37 °C. PSA secreted into the medium by the human PCa cell line, LNCaP cells, was used as a standard material of PSA-Gi in this study. The standard PSA-Gi concentration in the medium of the human LNCaP cell line was measured by WFA agarose column chromatography combined with a total-PSA enzyme-linked immunosorbent assay, as reported previously [22]. Fifty-five percent of total PSA in the medium of the LNCaP cell line possessed PSA-Gi (data not shown) [22]. Serum total PSA was measured by Architect i1000 system (Abbott Japan, Tokyo, Japan) and special reagents for total PSA (Abbott Japan) in a PSA range from 0.001 to 100 ng/mL.

4.3. Immunohistochemical Analysis of RP Specimens by WFA

A total of 260 paraffin-embedded RP specimens were obtained from PCa patients who underwent RP without neoadjuvant therapy between June 2007 and August 2016 in Hirosaki University Hospital. Patient demographics are shown in Table 3. Staging and grading information regarding the tumors and patient follow-up have been described previously [35]. In brief, PSA recurrence after RP was defined by two consecutive PSA values of >0.2 ng/mL with a 1-month interval and after a postoperative decrease below the detection limit (<0.001 ng/mL). Time zero was defined as the day of surgical treatment. Patients with constantly undetectable PSA levels (<0.001 ng/mL as the detection limit) after surgery were considered as patients without biochemical recurrence. Follow-up intervals were calculated from the date of the operation to the last recorded follow-up. Information on patients with PCa and tumor characteristics was obtained from medical charts. The grade group of prostate biopsy and prostatectomy specimens were evaluated according to the International Society of Urological Pathology (ISUP) guidelines [23]. Deparaffinized RP specimens were incubated with the biotinylated-WFA (Vector Laboratories, Burlingame, CA, USA) in PBS containing 5% of bovine serum albumin (1:500 dilution) at 4 °C, overnight. Biotinylated-WFA was detected by Vectastain Elite ABC kit (Vector Laboratories). WFA-reactivity was classified into three groups according to the reciprocal intensity scale as described previously [24]. Representative images of each Gleason grade tumor are shown in Figure A1.

4.4. Statistical Analysis

All calculations for clinical data were performed in the SPSS software, ver. 21.0 (SPSS, Inc., Chicago, IL, USA) and in GraphPad Prism 6.03 (GraphPad Software, San Diego, CA, USA). Intergroup differences were statistically analyzed by a Student's *t*-test for normally distributed variables or by the Mann–Whitney *U*-test for non-normally distributed models. Data with $p < 0.05$ were considered significant. ROC curves developed using the library "rms" in R (available on: http://www.r-project.org/) [25] and the statistical difference of AUCs were calculated by the same program. The χ^2 test was used to analyze the association of the WFA-reactivity status with clinicopathological parameters. PSA-free survival was evaluated using Kaplan–Meier curves, and differences between groups were assessed by the log-rank test. Multivariate test by Cox proportional hazards regression analysis was performed to detect significant and independent parameters with which PSA recurrence after RP can be predicted. Post-operative nomogram predicting PSA-free survival provability after RP was developed using the library "rms" in R (available on: http://www.r-project.org/), and the c-index was also calculated by same program [25].

5. Conclusions

At present, the majority of promising markers such as Phi, 4KScore, and tissue-based markers [26] are used in multiplex testing to improve diagnostic and prognostic accuracy. PSA-Gi is used as a single marker and yields results comparable to the diagnostic and prognostic performance of multiplex markers. PCA3 was also a promising urine marker for repeat biopsy decision-making [26]. However, there are a few cumbersome procedures for sample handling for avoiding RNA degradation. In this

Int. J. Mol. Sci. **2017**, *18*, 261

study, although we used frozen serum samples stored from 2007 to 2016, diagnostic and prognostic performance of PSA-Gi was substantially superior to total PSA. The serum sample handling of PSA-Gi was almost the same as the PSA test. Therefore, serum PSA-Gi is a promising pre-operative marker for detecting PCa and assessing the aggressiveness of PCa and has an advantage of cost-effectiveness and sample handling for routine clinical practice. Furthermore, the nomogram developed in this study is also a promising predictive tool for determining PSA-free survival probability. Larger clinical trials are warranted to confirm our findings.

Acknowledgments: All of the authors thank Katsuko Yamashita, Ph.D. for providing basically information about PSA-Gi and Yukie Nishizawa, Kaname Higuchi, and Satomi Sakamoto, technical assistant of Hirosaki University Graduate School of Medicine, for their invaluable help with sample collection and patient data management. This study was supported by the Japan Agency for Medical Research and Development-SENTAN KEISOKU BUNSEKIGIJYUTU KAIHATSU program (AMED)-SENTAN project from Japan Agency for Medical Research and Development (AMED), and also supported by the Japan Society for the Promotion of Science (JSPS) KAKENHI grant no. 15K15579 and grant no. 15H02563.

Author Contributions: Tohru Yoneyama and Yuki Tobisawa conceived and designed the experiments; Kazuhisa Hagiwara, Takatoshi Kaya, Tomonori Kaneko, and Tohru Yoneyama performed the experiments; Takatoshi Kaya and Tohru Yoneyama analyzed the data; Takatoshi Kaya, Tomonori Kaneko, and Yoshihiko Suda developed the SPFS-based automated immunoassay system; Shingo Hatakeyama, Yasuhiro Hashimoto, Takuya Koie, and Chikara Ohyama diagnosed BPH and PCa, and performed biopsy and radical prostatectomy; Chikara Ohyama and Yoshihiko Suda obtained funding. Chikara Ohyama and Tohru Yoneyama supervised. Yuki Tobisawa and Kazuyuki Mori critically revised the manuscript for intellectual content; Tohru Yoneyama wrote the paper.

Conflicts of Interest: The authors declare no conflict of interest.

Abbreviations

SPFS	surface plasmon field-enhanced fluorescence spectrometry
WFA	*Wisteria floribunda* agglutinin
PSA	prostate-specific antigen
PCa	prostate cancer
BPH	benign prostatic hyperplasia
LacdiNAc	GalNAcβ1-4GlcNAc-
Gal	galactose
Man	mannose
Fuc	fucose
Sia	sialic acid
GalNAc	*N*-acetylgalactosamine
GlcNAc	*N*-acetylglucosamine
Pbx GS	prostate biopsy Gleason Score
Pbx GG	prostate biopsy grade group
cT	clinical T stage
pT	pathological T stage
Ope GS	gleason score after radical prostatectomy
Ope GG	grade group after radical prostatectomy
RP	radical prostatectomy
pn	perineural invasion
RM	resection margin

Appendix A

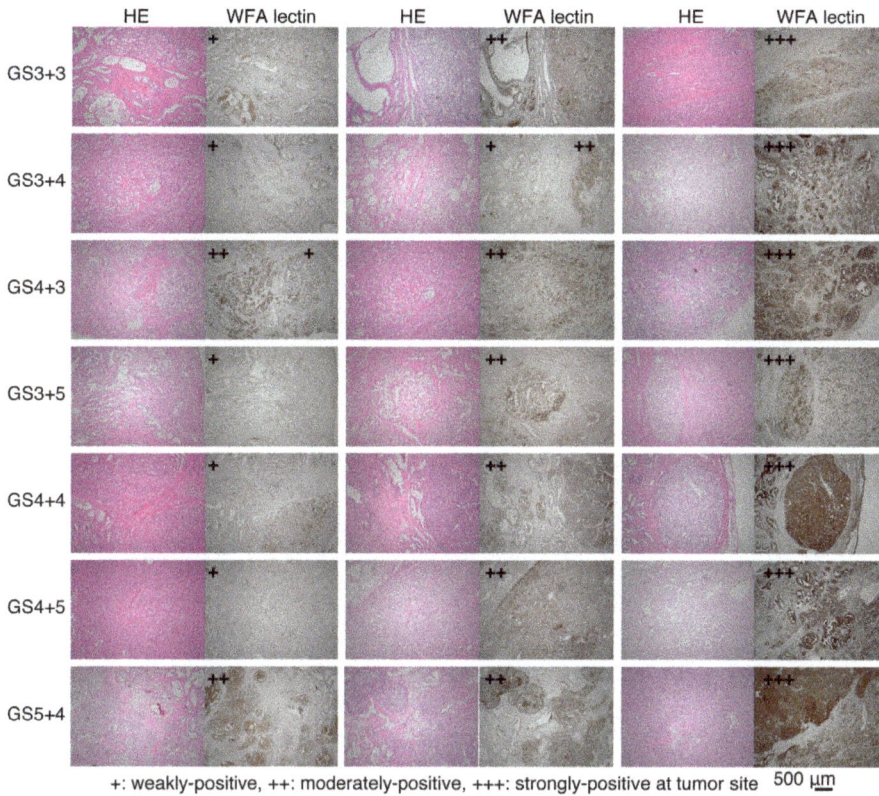

+: weakly-positive, ++: moderately-positive, +++: strongly-positive at tumor site   500 μm

Figure A1. Representative hematoxylin-eosin (HE) staining and WFA reactive-glycan expression of tumors of each Gleason grade among RP specimens. WFA staining intensity was classified into three groups: weakly positive, moderately positive, and strongly positive at a tumor site, respectively. Scale bar indicated 500 μm.

References

1. Torre, L.A.; Bray, F.; Siegel, R.L.; Ferlay, J.; Lortet-Tieulent, J.; Jemal, A. Global cancer statistics, 2012. *CA: Cancer J. Clin.* **2015**, *65*, 87–108. [CrossRef] [PubMed]
2. Ferlay, J.; Steliarova-Foucher, E.; Lortet-Tieulent, J.; Rosso, S.; Coebergh, J.W.; Comber, H.; Forman, D.; Bray, F. Cancer incidence and mortality patterns in Europe: Estimates for 40 countries in 2012. *Eur. J. Cancer* **2013**, *49*, 1374–1403. [CrossRef] [PubMed]
3. Klotz, L. Prostate cancer overdiagnosis and overtreatment. *Curr. Opin. Endocrinol. Diabetes Obes.* **2013**, *20*, 204–209. [CrossRef] [PubMed]
4. Kim, E.H.; Andriole, G.L. Prostate-specific antigen-based screening: Controversy and guidelines. *BMC Med.* **2015**, *13*, 61. [CrossRef] [PubMed]
5. Ward, J.F.; Blute, M.L.; Slezak, J.; Bergstralh, E.J.; Zincke, H. The long-term clinical impact of biochemical recurrence of prostate cancer 5 or more years after radical prostatectomy. *J. Urol.* **2003**, *170*, 1872–1876. [CrossRef] [PubMed]

6. Powell, I.J.; Tangen, C.M.; Miller, G.J.; Lowe, B.A.; Haas, G.; Carroll, P.R.; Osswald, M.B.; deVere White, R.; Thompson, I.M., Jr.; Crawford, E.D. Neoadjuvant therapy before radical prostatectomy for clinical T3/T4 carcinoma of the prostate: 5-Year followup, phase II Southwest Oncology Group Study 9109. *J. Urol.* **2002**, *168*, 2016–2019. [CrossRef]

7. Mitsuzuka, K.; Narita, S.; Koie, T.; Kaiho, Y.; Tsuchiya, N.; Yoneyama, T.; Kakoi, N.; Kawamura, S.; Tochigi, T.; Habuchi, T.; et al. Pathological and biochemical outcomes after radical prostatectomy in men with low-risk prostate cancer meeting the Prostate Cancer International: Active surveillance criteria. *Br. J. Urol. Int.* **2013**, *111*, 914–920. [CrossRef] [PubMed]

8. Satkunasivam, R.; Kulkarni, G.S.; Zlotta, A.R.; Kalnin, R.; Trachtenberg, J.; Fleshner, N.E.; Hamilton, R.J.; Jewett, M.A.; Finelli, A. Pathological, oncologic and functional outcomes of radical prostatectomy following active surveillance. *J. Urol.* **2013**, *190*, 91–95. [CrossRef] [PubMed]

9. Maurice, M.J.; Sundi, D.; Schaeffer, E.M.; Abouassaly, R. Risk of pathological upgrading and upstaging among men with low-risk prostate cancer varies by race: Results from the National Cancer Data Base. *J. Urol.* **2016**. [CrossRef] [PubMed]

10. Sussman, R.; Staff, I.; Tortora, J.; Champagne, A.; Meraney, A.; Kesler, S.S.; Wagner, J.R. Impact of active surveillance on pathology and nerve sparing status. *Can. J. Urol.* **2014**, *21*, 7299–7304. [PubMed]

11. Kattan, M.W.; Eastham, J.A.; Stapleton, A.M.; Wheeler, T.M.; Scardino, P.T. A preoperative nomogram for disease recurrence following radical prostatectomy for prostate cancer. *J. Natl. Cancer Inst.* **1998**, *90*, 766–771. [CrossRef] [PubMed]

12. Haji-Ghassemi, O.; Gilbert, M.; Spence, J.; Schur, M.J.; Parker, M.J.; Jenkins, M.L.; Burke, J.E.; van Faassen, H.; Young, N.M.; Evans, S.V. Molecular basis for recognition of the cancer glycobiomarker LacdiNAc (GalNAc(β1–4)GlcNAc) by *Wisteria floribunda* agglutinin. *J. Biol. Chem.* **2016**, *291*, 24085–24095. [CrossRef] [PubMed]

13. Kitamura, N.; Guo, S.; Sato, T.; Hiraizumi, S.; Taka, J.; Ikekita, M.; Sawada, S.; Fujisawa, H.; Furukawa, K. Prognostic significance of reduced expression of β-N-acetylgalactosaminylated N-linked oligosaccharides in human breast cancer. *Int. J. Cancer* **2003**, *105*, 533–541. [CrossRef] [PubMed]

14. Hirano, K.; Matsuda, A.; Kuji, R.; Nakandakari, S.; Shirai, T.; Furukawa, K. Enhanced expression of the β4-N-acetylgalactosaminyltransferase 4 gene impairs tumor growth of human breast cancer cells. *Biochem. Biophys. Res. Commun.* **2015**, *461*, 80–85. [CrossRef] [PubMed]

15. McMahon, R.F.; McWilliam, L.J.; Mosley, S. Evaluation of three techniques for differential diagnosis of prostatic needle biopsy specimens. *J. Clin. Pathol.* **1992**, *45*, 1094–1098. [CrossRef] [PubMed]

16. McMahon, R.F.; McWilliam, L.J.; Clarke, N.W.; George, N.J. Altered saccharide sequences in two groups of patients with metastatic prostatic carcinoma. *Br. J. Urol.* **1994**, *74*, 80–85. [CrossRef] [PubMed]

17. Yamaguchi, T.; Yokoyama, Y.; Ebata, T.; Matsuda, A.; Kuno, A.; Ikehara, Y.; Shoda, J.; Narimatsu, H.; Nagino, M. Verification of WFA-sialylated MUC1 as a sensitive biliary biomarker for human biliary tract cancer. *Ann. Surg. Oncol.* **2016**, *23*, 671–677. [CrossRef] [PubMed]

18. Iio, E.; Ocho, M.; Togayachi, A.; Nojima, M.; Kuno, A.; Ikehara, Y.; Hasegawa, I.; Yatsuhashi, H.; Yamasaki, K.; Shimada, N.; et al. A novel glycobiomarker, *Wisteria floribunda* agglutinin macrophage colony-stimulating factor receptor, for predicting carcinogenesis of liver cirrhosis. *Int. J. Cancer* **2016**, *138*, 1462–1471. [CrossRef] [PubMed]

19. Sogabe, M.; Nozaki, H.; Tanaka, N.; Kubota, T.; Kaji, H.; Kuno, A.; Togayachi, A.; Gotoh, M.; Nakanishi, H.; Nakanishi, T.; et al. Novel glycobiomarker for ovarian cancer that detects clear cell carcinoma. *J. Proteome Res.* **2014**, *13*, 1624–1635. [CrossRef] [PubMed]

20. Khabaz, M.N.; McClure, J.; McClure, S.; Stoddart, R.W. Glycophenotype of prostatic carcinomas. *Folia Histochem. Cytobiol.* **2010**, *48*, 637–645. [PubMed]

21. Fukushima, K.; Satoh, T.; Baba, S.; Yamashita, K. α1,2-Fucosylated and β-N-acetylgalactosaminylated prostate-specific antigen as an efficient marker of prostatic cancer. *Glycobiology* **2010**, *20*, 452–460. [CrossRef] [PubMed]

22. Kaya, T.; Kaneko, T.; Kojima, S.; Nakamura, Y.; Ide, Y.; Ishida, K.; Suda, Y.; Yamashita, K. High-sensitivity immunoassay with surface plasmon field-enhanced fluorescence spectroscopy using a plastic sensor chip: Application to quantitative analysis of total prostate-specific antigen and GalNAcβ1–4GlcNAc-linked prostate-specific antigen for prostate cancer diagnosis. *Anal. Chem.* **2015**, *87*, 1797–1803. [PubMed]

23. Humphrey, P.A.; Moch, H.; Cubilla, A.L.; Ulbright, T.M.; Reuter, V.E. The 2016 WHO classification of tumours of the urinary system and male genital organs-part B: Prostate and bladder tumours. *Eur. Urol.* **2016**, *70*, 106–119. [CrossRef] [PubMed]

24. Nguyen, D. Quantifying chromogen intensity in immunohistochemistry via reciprocal intensity. *Protoc. Exchange* **2013**. [CrossRef]

25. Harrell, F.E., Jr.; Lee, K.L.; Mark, D.B. Multivariable prognostic models: Issues in developing models, evaluating assumptions and adequacy, and measuring and reducing errors. *Stat. Med.* **1996**, *15*, 361–387. [CrossRef]

26. Hatakeyama, S.; Yoneyama, T.; Tobisawa, Y.; Ohyama, C. Recent progress and perspectives on prostate cancer biomarkers. *Int. J. Clin. Oncol.* **2016**. [CrossRef] [PubMed]

27. Haider, M.A.; Yao, X.; Loblaw, A.; Finelli, A. Multiparametric magnetic resonance imaging in the diagnosis of prostate cancer: A systematic review. *Clin. Oncol.* **2016**, *28*, 550–567. [CrossRef] [PubMed]

28. Nicholson, A.; Mahon, J.; Boland, A.; Beale, S.; Dwan, K.; Fleeman, N.; Hockenhull, J.; Dundar, Y. The clinical effectiveness and cost-effectiveness of the PROGENSA (R) prostate cancer antigen 3 assay and the prostate health index in the diagnosis of prostate cancer: A systematic review and economic evaluation. *Health Technol. Assess.* **2015**, *19*, 1–191. [CrossRef] [PubMed]

29. Fujita, K.; Shimomura, M.; Uemura, M.; Nakata, W.; Sato, M.; Nagahara, A.; Nakai, Y.; Takamatsu, S.; Miyoshi, E.; Nonomura, N. Serum fucosylated haptoglobin as a novel prognostic biomarker predicting high-Gleason prostate cancer. *Prostate* **2014**, *74*, 1052–1058. [CrossRef] [PubMed]

30. Okuyama, N.; Ide, Y.; Nakano, M.; Nakagawa, T.; Yamanaka, K.; Moriwaki, K.; Murata, K.; Ohigashi, H.; Yokoyama, S.; Eguchi, H.; et al. Fucosylated haptoglobin is a novel marker for pancreatic cancer: A detailed analysis of the oligosaccharide structure and a possible mechanism for fucosylation. *Int. J. Cancer* **2006**, *118*, 2803–2808. [CrossRef] [PubMed]

31. Thompson, S.; Dargan, E.; Turner, G.A. Increased fucosylation and other carbohydrate changes in haptoglobin in ovarian cancer. *Cancer Lett.* **1992**, *66*, 43–48. [CrossRef]

32. Li, Q.K.; Chen, L.; Ao, M.H.; Chiu, J.H.; Zhang, Z.; Zhang, H.; Chan, D.W. Serum fucosylated prostate-specific antigen (PSA) improves the differentiation of aggressive from non-aggressive prostate cancers. *Theranostics* **2015**, *5*, 267–276. [CrossRef] [PubMed]

33. Stark, J.R.; Perner, S.; Stampfer, M.J.; Sinnott, J.A.; Finn, S.; Eisenstein, A.S.; Ma, J.; Fiorentino, M.; Kurth, T.; Loda, M.; et al. Gleason score and lethal prostate cancer: Does 3 + 4 = 4 + 3? *J. Clin. Oncol.* **2009**, *27*, 3459–3464. [CrossRef] [PubMed]

34. Epstein, J.I.; Zelefsky, M.J.; Sjoberg, D.D.; Nelson, J.B.; Egevad, L.; Magi-Galluzzi, C.; Vickers, A.J.; Parwani, A.V.; Reuter, V.E.; Fine, S.W.; et al. A contemporary prostate cancer grading system: A validated alternative to the gleason score. *Eur. Urol.* **2016**, *69*, 428–435. [CrossRef] [PubMed]

35. Koie, T.; Ohyama, C.; Hatakeyama, S.; Imai, A.; Yoneyama, T.; Hashimoto, Y.; Yoneyama, T.; Tobisawa, Y.; Hosogoe, S.; Yamamoto, H.; et al. Significance of preoperative butyrylcholinesterase as an independent predictor of biochemical recurrence-free survival in patients with prostate cancer treated with radical prostatectomy. *Int. J. Clin. Oncol.* **2016**, *21*, 379–383. [CrossRef] [PubMed]

International Journal of
Molecular Sciences

MDPI

Article

An Automated Micro-Total Immunoassay System for Measuring Cancer-Associated α2,3-linked Sialyl N-Glycan-Carrying Prostate-Specific Antigen May Improve the Accuracy of Prostate Cancer Diagnosis

Tomokazu Ishikawa [1,2], Tohru Yoneyama [1,3,*], Yuki Tobisawa [1], Shingo Hatakeyama [1], Tatsuo Kurosawa [2], Kenji Nakamura [2], Shintaro Narita [4], Koji Mitsuzuka [5], Wilhelmina Duivenvoorden [6], Jehonathan H. Pinthus [6], Yasuhiro Hashimoto [3], Takuya Koie [1], Tomonori Habuchi [4], Yoichi Arai [5] and Chikara Ohyama [1,3]

[1] Department of Urology, Hirosaki University Graduate School of Medicine, Hirosaki 036-8562, Japan;
    ishikawa.tomokazu@wako-chem.co.jp (T.I.); tobisawa@hirosaki-u.ac.jp (Y.T.);
    shingoh@hirosaki-u.ac.jp (S.H.); goodwin@hirosaki-u.ac.jp (T.K.); coyama@hirosaki-u.ac.jp (C.O.)
[2] Diagnostics Research Laboratories, Wako Pure Chemical Industries, Hyogo 661-0963, Japan;
    kurosawa.tatsuo@wako-chem.co.jp (T.K.); nakamura.kenji@wako-chem.co.jp (K.N.)
[3] Department of Advanced Transplant and Regenerative Medicine, Hirosaki University Graduate School of
    Medicine, Hirosaki 036-8562, Japan; bikkuri@opal.plala.or.jp
[4] Department of Urology, Akita University Graduate School of Medicine, Akita 010-8543, Japan;
    narishin@doc.med.akita-u.ac.jp (S.N.); thabuchi@doc.med.akita-u.ac.jp (T.H.)
[5] Department of Urology, Tohoku University Graduate School of Medicine, Sendai 980-8574, Japan;
    mitsuzuka@uro.med.tohoku.ac.jp (K.M.); yarai@uro.med.tohoku.ac.jp (Y.A.)
[6] Department of Surgery, McMaster University, Hamilton, ON L8S4L8, Canada; duiven@mcmaster.ca (W.D.);
    pinthusj@hhsc.ca (J.H.P.)
* Correspondence: tohruyon@hirosaki-u.ac.jp; Tel.: +81-172-39-5091

Academic Editor: Carsten Stephan
Received: 1 February 2017; Accepted: 18 February 2017; Published: 22 February 2017

Abstract: The low specificity of the prostate-specific antigen (PSA) for early detection of prostate cancer (PCa) is a major issue worldwide. The aim of this study to examine whether the serum PCa-associated α2,3-linked sialyl N-glycan-carrying PSA (S2,3PSA) ratio measured by automated micro-total immunoassay systems (µTAS system) can be applied as a diagnostic marker of PCa. The µTAS system can utilize affinity-based separation involving noncovalent interaction between the immunocomplex of S2,3PSA and *Maackia amurensis* lectin to simultaneously determine concentrations of free PSA and S2,3PSA. To validate quantitative performance, both recombinant S2,3PSA and benign-associated α2,6-linked sialyl N-glycan-carrying PSA (S2,6PSA) purified from culture supernatant of PSA cDNA transiently-transfected Chinese hamster ovary (CHO)-K1 cells were used as standard protein. Between 2007 and 2016, fifty patients with biopsy-proven PCa were pair-matched for age and PSA levels, with the same number of benign prostatic hyperplasia (BPH) patients used to validate the diagnostic performance of serum S2,3PSA ratio. A recombinant S2,3PSA- and S2,6PSA-spiked sample was clearly discriminated by µTAS system. Limit of detection of S2,3PSA was 0.05 ng/mL and coefficient variation was less than 3.1%. The area under the curve (AUC) for detection of PCa for the S2,3PSA ratio (%S2,3PSA) with cutoff value 43.85% (AUC; 0.8340) was much superior to total PSA (AUC; 0.5062) using validation sample set. Although the present results are preliminary, the newly developed µTAS platform for measuring %S2,3PSA can achieve the required assay performance specifications for use in the practical and clinical setting and may improve the accuracy of PCa diagnosis. Additional validation studies are warranted.

Int. J. Mol. Sci. **2017**, *18*, 470

Keywords: prostate-specific antigen; α2,3-linked sialyl *N*-glycan; *Maackia amurensis* lectin (MAA) lectin; biomarker

1. Introduction

Serum prostate-specific antigen (PSA) is widely used as a powerful biomarker for detecting prostate cancer (PCa) [1,2]. However, PSA-based PCa screening has resulted in over-diagnosis, leading to unnecessary prostate biopsies and overtreatment of indolent cancer [3–5]. Therefore, novel, more specific screening methods are urgently needed, especially for young healthy men. Of the various molecular isoforms of PSA, proPSA is one of the most promising potential biomarkers [6–10]. A multi-biomarker test for identifying aggressive prostate cancer that combines total PSA, free PSA, intact PSA, and human kallikrein-2 has been developed [11,12]. In contrast, we focused on the cancer-associated glycan alterations that have frequently been observed during carcinogenesis [13,14]. More importantly, some glycans, for example α-fetoprotein [15] and human chorionic gonadotropin [16], have been found to have specific cancer-associated carbohydrate alterations compared with their normal counterparts. The glycoprotein PSA has an *N*-glycosylation site on its 45th amino acid from the N-terminus. In a prior study, we cleaved a PSA-specific sequence (Ile-Arg-Asn-Lys, IRNK) that includes the glycosylated-Asn (N) and performed an intensive structural analysis of the glycan profile of PSA using matrix-assisted laser desorption/ionization time-of-flight (MALDI-TOF) mass spectrometry [17]. This resulted in our identification of a PCa-associated aberrant glycosylation of PSA, which produces α2,3-linked sialyl *N*-glycan that is readily observed on free PSA (S2,3PSA), whereas α2,6-linked sialyl *N*-glycan on free PSA (S2,6PSA) is exclusive to benign prostatic hyperplasia (BPH) [18] (Figure 1).

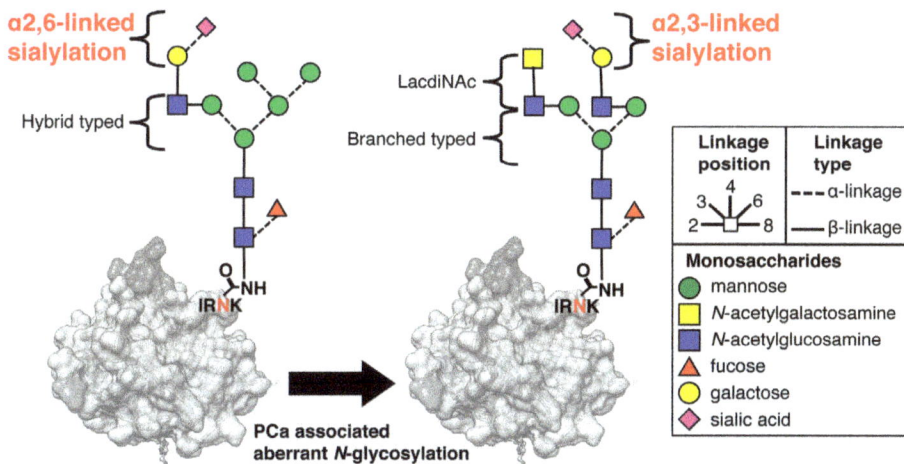

Figure 1. Prostate cancer-associated aberrant glycosylation of *N*-glycan on prostate-specific antigen (PSA). In normal PSA, the terminal sialic acids link to galactose residues with an α2,6 linkage whereas in prostate cancer (PCa)-associated PSA, the linkage between the terminal sialic acid and galactose residues is an α2,3 linkage [18]. Carbon linkage positions are denoted by the bond position drawn on each monosaccharide. Ile-Arg-Asn-Lys, (IRNK): N-glycosylation site of PSA.

We have also previously developed a magnetic bead-based S2,3PSA assay (Luminex method) that more accurately diagnoses early PCa than the conventional PSA test [19]. However, this method cannot measure benign S2,6PSA and does not have enough quantitative capability for clinical application. With the aim of overcoming these issues, we investigated using microfluidic technology, also referred to as micro-total analysis systems (µTAS system) [20–23], to quantitate both S2,6PSA and S2,3PSA in patient serum samples with a one-time measurement. µTAS immunoassays incorporating an automated platform can achieve the required assay performance specifications in clinical laboratories more efficiently than existing methodologies. In this study, we developed an automated µTAS system for measuring serum S2,3PSA ratio (%S2,3PSA) in clinical settings.

2. Results

2.1. Preparation of FLAG-Tag-Fused Recombinat S2,3 and S2,6 PSA in Chinese Hamster Ovary (CHO) Cells

To construct S2,3PSA and S2,6PSA recombinant standard protein, FLAG-tag-fused human PSA cDNA (PSA–FLAG) was transiently transfected into CHO cells. Results of immunoblotting of PSA–FLAG-transfected CHO culture supernatant are shown in Figure 2a. Using an *Agrocybe cylindracea* (ACG) lectin affinity column and gel filtration column chromatography, S2,3PSA and S2,6PSA containing asialo-type *N*-glycan carrying recombinant PSA standard protein was purified from culture supernatants of human PSA expressed by CHO cells (Figure 2b,c). Lectin array analysis then revealed the putative glycan structure of purified S2,3 and S2,6PSA recombinant standard (Figure 2d). Signal intensity of recombinant S2,3PSA against *Mackkia amurensis* lectin (MAL) that preferentially bound the sialic acid α2,3-linked galactose structure was much higher than that of recombinant S2,6PSA. Signal intensity of recombinant S2,3PSA against *Sambucus Nigra* lectin (SNA) and *Sambucus Sieboldiana* lectin (SSA) that preferentially bound the sialic acid α2,6-linked galactose structure was much lower than that of recombinant S2,6PSA. These results suggest that the culture supernatant of PSA–FLAG-transfected CHO cells contained both α2,3- and α2,6-linked sialyl *N*-glycan-carrying recombinant PSA and could be utilized as a standard protein of µTAS system measuring S2,3PSA ratio (%S2,3PSA).

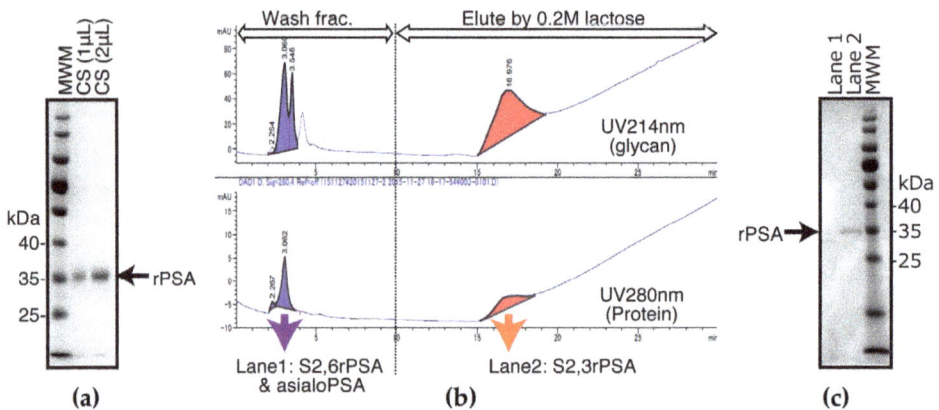

Figure 2. *Cont.*

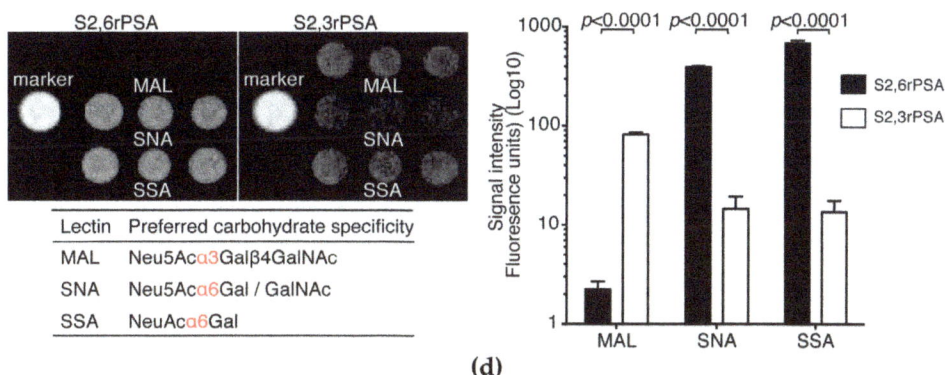

Figure 2. Preparation of α2,3-linked sialyl *N*-glycan-carrying PSA (S2,3PSA) and α2,6-linked sialyl *N*-glycan-carrying PSA (S2,6PSA) in Chinese hamster ovary (CHO)-K1 cells transfected with PSA–FLAG. CHO-K1 cells were transfected with FLAG-tag-fused human PSA cDNA (PSA–FLAG). (**a**) Culture supernatant (CS) with serum free media were blotted onto polyvinylidene fluoride (PVDF) membrane and probed with anti-FLAG antibodies; (**b**) Chromatograms obtained using *Agrocybe cylindracea* (ACG) lectin column chromatography for α2,3-linked sialyl *N*-glycan-carrying recombinant PSA (S2,3rPSA) collected from 0.2 M lactose eluted fraction and other glyco-isoform free PSA containing α2,6-linked sialyl *N*-glycan (S2,6rPSA) or asialo-type *N*-glycan collected from washed fraction. S2,3rPSA and S2,6PSA was further purified by gel filtration column chromatography; (**c**) Washed fraction (lane 1) and 0.2 M lactose-eluted fraction (lane 2) (indicating chromatogram) (**b**) were blotted onto PVDF membrane and probed with anti-FLAG antibodies; (**d**) Lectin array profiling of S2,3rPSA and S2,6PSA proteins. Purified S2,3rPSA and S2,6rPSA was applied to the lectin array, including triplicate spots. The glass slides were scanned using a GlycoStationReader1200 (Glycotechnica, Sapporo, Japan). MWM: molecular weight markers; MAL: *Mackkia amurensis* lectin; SNA: *Sambucus Nigra* lectin; SSA: *Sambucus sieboldiana* lectin.

2.2. S2,6PSA and S2,3PSA Separation in a Microfluidic Channel Filled with Electrophoresis Leading Buffer Containing Maackia Amurensis Lectin

Simultaneous quantitative measurement of S2,6PSA and S2,3PSA was achieved by performing microchip capillary electrophoresis and liquid-phase binding assay (LBA) on a μTAS Wako i30 auto analyzer with microfluidic chip IO6 (Wako Pure Chemical Industries, Hyogo, Japan) for electrokinetic analyte transport assay (EATA) (Figure 3a–c). Total assay time of μTAS system measuring %S2,3PSA ratio was 9 min, substantially shorter than the previously developed Luminex method (4 h). The reaction time during transport of DNA conjugate and bound free PSA analyte through the chip channels is short (25–50 s), and the binding kinetics are greatly increased by isotachophoresis (ITP) stacking and concentration of the reactants (Figure 3b) [24,25]. Figure 3c shows a typical electropherogram of S2,6PSA and S2,3PSA in a microfluidic channel separated by lectin affinity electrophoresis. Because S2,3PSA is reactive to *Mackkia amurensis* lectin (MAA), electrophoretic migration of the sandwich complex of DNA–labeled anti-total PSA Fab′ (DNA–Fab′), S2,3PSA, and HiLyte Fluor 647–labeled anti-free PSA Fab′ (HiLyte–Fab′) in the separation channel is slower than that of the immunocomplex of DNA–Fab′, S2,6PSA, and HiLyte–Fab′.

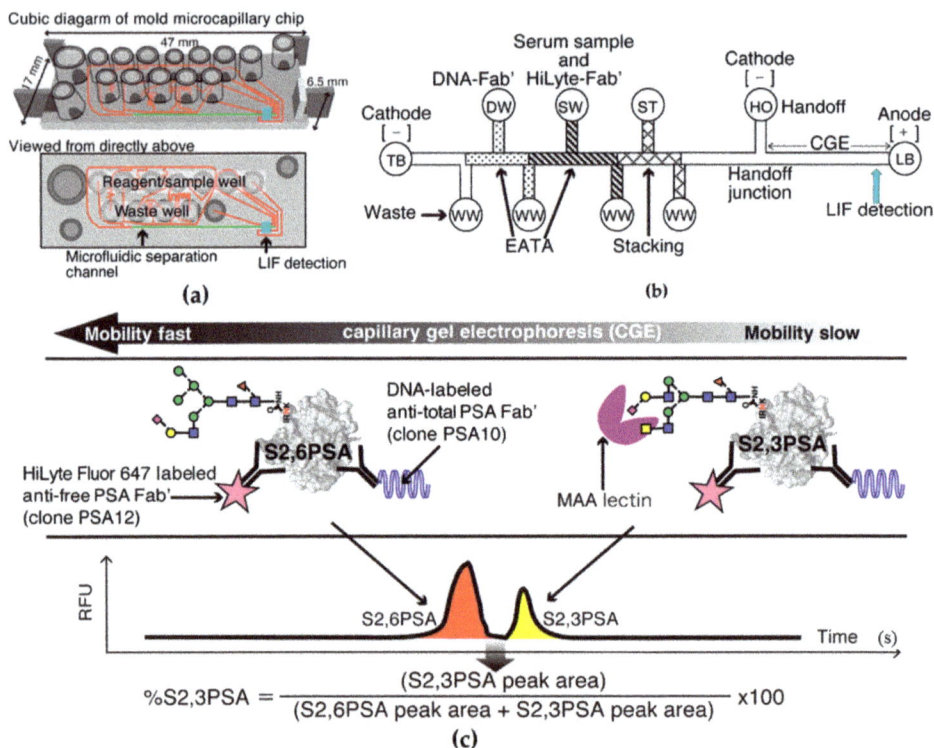

Figure 3. %S2,3PSA test using micro-total immunoassay systems (μTAS)-based microcapillary electrophoresis. (**a**) Molded plastic chip with top wells for reagents and precision microchannels on the bottom. Red line indicates microchannels. Highlight green line indicates microcapillary separation channel. Fluorescence signals are detected through the thin film closing the channels on the bottom of the chip; (**b**) Schematic diagram of a chip showing the electrokinetic analyte transport assay (EATA) method. Waste wells (WWs), trailing buffer (TB) well, leading buffer (LB) well, handoff (HO), DNA–labeled anti-total PSA Fab′ (DNA–Fab′) (DW), serum sample and HiLyte Fluor 647–labeled anti-free PSA Fab′ (HiLyte–Fab′) mixture (SW), and stacking buffer (ST) wells are shown. Vacuum applied to the WWs loads the reagents and sample-to-chip channel segments and voltage applied between cathode and anode mixes the sample and reagents for the binding reaction. Switching the voltage from the TB well to the HO well switches from isotachophoresis (ITP) stacking mode to capillary gel electrophoresis (CGE) mode. Separation of S2,3PSA from other glyco-isoform of free PSA occurs in the *Mackkia amurensis* lectin (MAA)-filled CGE separation channel prior to laser-induced fluorescence (LIF) detection. (**c**) Typical electropherogram of S2,6PSA and S2,3PSA peak separation. Fluorescent markers have been coelectrophoresed to identify the PSA protein peak positions (data not shown). RFU: relative fluorescence units.

When MAA was eliminated from leading buffer (LB) in the separation channel, both S2,6PSA and S2,3PSA immunocomplex co-migrated to the same position (Figure 4a). Furthermore, using affinity-based separation with MAA lectin against aberrant glycosylation, we also confirmed a capture efficiency of applied S2,3rPSA standard of almost 100% in this assay (Figure 4b,c).

Purified S2,6 and S2,3rPSA standard without MAA lectin

(a)

Purified S2,6rPSA standard

(b)

Purified S2,3rPSA standard

(c)

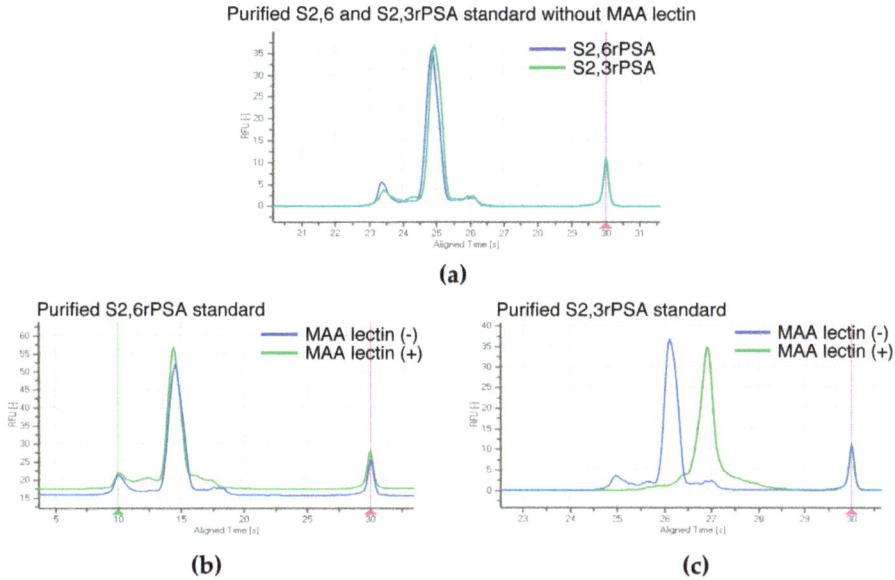

Figure 4. Capture efficiency of applied S2,3PSA standard in μTAS assay. (**a**) Electropherogram of S2,6rPSA and S2,3rPSA without MAA lectin. S2,6rPSA (**b**) and S2,3rPSA (**c**) standard were applied to the detection of peak separation respectively in a microfluidic channel filled with leading buffer (LB) containing MAA lectin (+) or not (−). The capture efficiency was evaluated by the peak mobility shift based on the specific interaction with lectin reactivity.

2.3. Assay Linearity and Sensitivity of %S2,3PSA Test by μTAS System

To demonstrate that our assay is robust for determining %S2,3PSA, recombinant standard samples were serially diluted with control PSA protein and %S2,3PSA determined at a constant concentration of free PSA. As shown in Figure 5a,b, there was a linear relationship between percentage of S2,3PSA and fluorescence intensities in the two prepared samples tested. The assay's sensitivity was determined by testing samples of buffer spiked with serially diluted S2,3PSA with the means ± 2 standard deviations (SD) being calculated for five replicates. As shown in Figure 5c, 0.05 ng/mL S2,3PSA was clearly detectable over the zero sample, there being no overlap of the 2SD range with zero. The reproducibility of the peak area detection for the 0.05 ng/mL level was within 15% coefficient of variation (CV) for %S2,3PSA, indicating that the limits of detection of quantitation of the assay are 0.05 ng/mL.

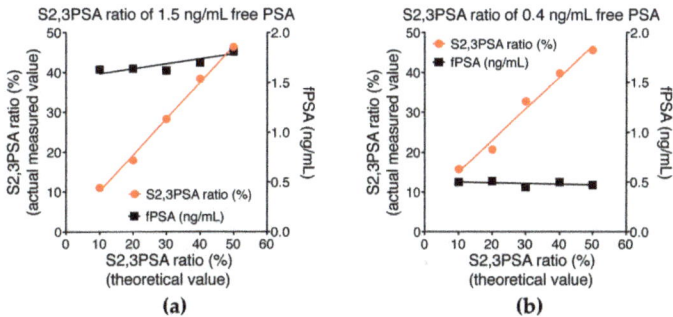

(a)

(b)

Figure 5. *Cont.*

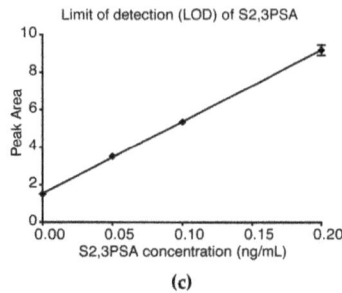

(c)

Figure 5. Assay linearity for measurement of S2,3PSA ratio (%S2,3PSA) and limit of detection of S2,3PSA measurement. (**a**) Sample was a prepared sample, the original sample having an PSA concentration of 1.5 ng/mL and %S2,3PSA of 50.0%; (**b**) Sample was prepared by spiking with PSA (0.4 ng/mL) and %S2,3PSA at 50%. The two samples were serially diluted by control PSA containing S2,6 sialylation or asialo at the same concentration. PSA concentrations and %S2,3PSA were determined by the μTAS system; (**c**) The limit of detection (LOD) was 0.05 ng/mL, this level showing no overlap between the 2 standard deviation (SD) ranges for S2,3PSA concentration and negative control.

2.4. Assay Reproducibility of %S2,3PSA Test Using the μTAS System

Assay reproducibility was examined by measuring two samples at each concentration with total of free PSA and %S2,3PSA, using 10 replicate measurements of each sample. We confirmed the assay's very good reproducibility. The CV was calculated to within 3% for free PSA and 4% for %S2,3PSA for all ranges tested (Table 1).

Table 1. Assay reproducibility of S2,3PSA test.

Sample Number	1.0 ng/mL Free PSA		5.0 ng/mL Free PSA	
	Free PSA	%S2,3PSA	Free PSA	%S2,3PSA
1	1.04	50.5	5.06	38.1
2	0.99	50.7	5.08	38.1
3	1.04	49.4	5.00	38.3
4	1.05	50.1	5.07	37.9
5	1.04	47.7	5.05	37.7
6	1.03	50.4	5.24	37.7
7	1.13	50.3	5.13	38.3
8	1.10	46.8	5.01	37.9
9	1.05	47.1	5.04	38.1
10	1.01	48.5	5.21	37.9
Ave. [1]	1.04	49.1	5.09	38.0
SD [2]	0.03	1.51	0.08	0.21
CV [3]	2.8%	3.1%	1.6%	0.6%

[1] Ave.: average; [2] SD, standard deviation; [3] CV: coefficient variation.

2.5. Validation of %S2,3PSA Test

Relevant clinical details of patients from whom samples were obtained are shown in Table 2. In this study, age and PSA level of 100 matched serum samples were assessed, comprising 50 from patients with PCa and 50 from BPH cases (Table 2, Figure 6a). Figure 6c shows that %S2,3PSA was significantly higher in patients with PCa than in patients with BPH ($p < 0.0001$). Total PSA level was not significantly different between the groups (Figure 6d). Receiver operating characteristic curve analyses were then used to compare the diagnostic potential of total PSA and %S2,3PSA (Figure 6e). The area under the curve (AUC) showed that results of conventional PSA testing did not differ between patients with BPH and PCa (AUC 0.5062, 95% CI 0.3922–0.6202), whereas there was a good separation for %S2,3PSA (AUC 0.8340, 95% CI 0.7555–0.9125, $p < 0.0001$) without any correlation between each assay (Figure 6b). The optimum cutoff point giving high specificity (72.0%) at 80% sensitivity was determined to be 42.20% of %S2,3PSA, with positive and negative predictive values of 75.5% and 78.7%, respectively, and much superior total PSA specificity (14.0%) at 80% sensitivity was determined to be 4.45 ng/mL of total PSA, with positive and negative predictive values of 48.2% and 41.2%, respectively. In this validation sample set, 54% ($n = 27/50$) of patients with PCa were found to be in the low Grade Group (GG) [26], with GG 1 (Gleason Score, GS 3 + 3) and GG 2 (GS 3 + 4) tumors. In addition, 46% ($n = 23/50$) of patients with PCa were found to have a greater than GG 3 (GS 4 + 3) tumor by prostate biopsy (Pbx) (Table 2). Figure 6g,f shows the total PSA and %S2,3PSA ratio of PCa patients classified by the prostate biopsy grade group (Pbx GG). The %S2,3PSA and total PSA level showed no significant difference between Pbx GG ≤ 2 and Pbx GG ≥ 3. In the case of greater than 50% S2,3PSA ratio, %S2,3PSA of Pbx GG ≥ 3 patients (range 50.7%–71.7%)(47%, $n = 11/23$) was significantly higher than that of Pbx GG ≤ 2 patients (range 50.3%–52.6%) (26%, $n = 7/27$) ($p = 0.0019$).

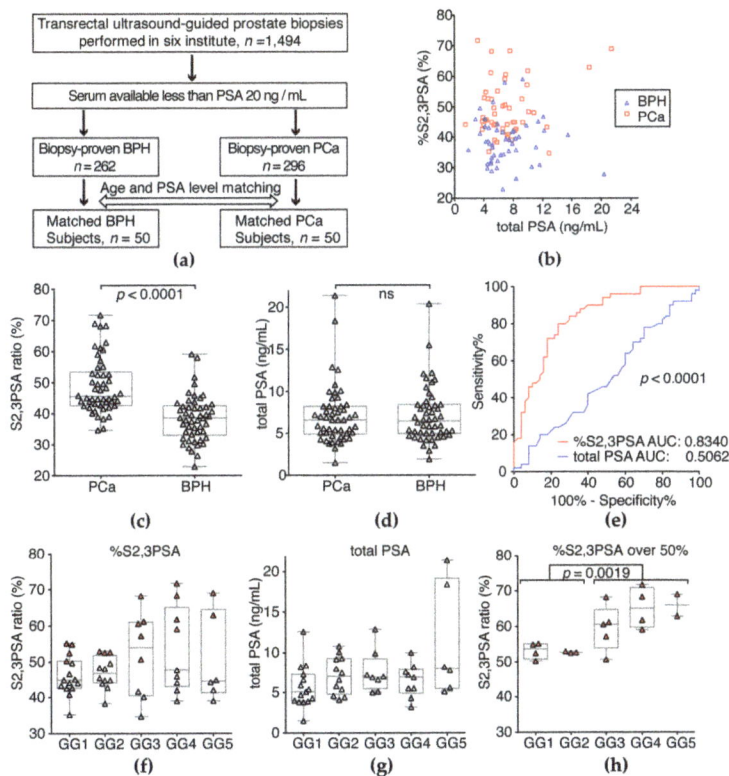

Figure 6. Serum %S2,3PSA and total PSA level in the validation sample set. (**a**) Fifty patients with PCa were pair-matched for age and PSA level with the same number of patients with BPH; (**b**) Correlations between %S2,3PSA and total PSA value in patient serum samples; (**c**) Serum %S2,3PSA was significantly higher in patients with PCa than BPH ($p < 0.0001$). The grand mean represents the overall mean of the Y variable; (**d**) Serum total PSA was not significantly different between PCa and BPH (**e**) Receiver operating characteristics (ROC) curve analysis for detection of PCa showed that the areas under the curve (AUC) for %S2,3PSA and conventional total PSA tests were 0.8340 and 0.5062, respectively; (**f**) %S2,3PSA among PCa patients classified by the Pbx GG; (**g**) total PSA level of PCa patients classified by the Pbx GG. Red triangle indicates %S2,3PSA over 50%; (**h**) Comparison between Pbx GG ≤ 2 and Pbx GG ≥ 3 patients that had an %S2,3PSA ratio over 50% ($p = 0.0019$).

Table 2. Age and total PSA BPH and PCa patients in the validation study.

Characteristics	BPH [a]	PCa [b]	p ([a] vs. [b])
$n = 100$	50	50	
Age, median (range)	66.5 (51–85)	67 (51–86)	ns [1]
PSA [2], ng/mL, median (range)	6.45 (1.9–20.4)	6.6 (1.5–21.4)	ns [1]
%S2,3PSA, median (range)	38.55 (22.9–59.1)	45.70 (34.7–71.7)	<0.0001
pbx GG [3]		n (%)	
GG 1		15 (30%)	
GG 2		12 (24%)	
GG 3		8 (16%)	
GG 4		9 (18%)	
GG 5		6 (12%)	

[1] no significant difference; [2] total PSA; [3] pbx GG: prostate biopsy grade group; [a] benign prostatic hyperplasia; [b] prostate cancer.

3. Discussion

The use of aberrant PSA glycosylation for early detection of prostate cancer has been reported [27–30]. In particular, we have previously identified that the terminal *N*-glycan structure of PSA from patients with PCa is rich in sialic acid α2,3-linked to the galactose residue, whereas the terminal *N*-glycan structures of PSA from seminal plasma are exclusively α2,6-linked [17]. We have also developed a magnetic bead-based S2,3PSA assay (Luminex method, [19]) that more accurately diagnoses early PCa than conventional PSA testing; however, this method is not sufficiently versatile and does not have enough quantitative capability for clinical application. To overcome these issues, we here developed an automated microcapillary electrophoresis-based immunoassay system (μTAS system) to be able to measure serum %S2,3PSA in clinical settings.

In this study, we developed a μTAS system based on the principles of EATA and LBA, precision-injection molded microfluidic plastic chips, and instrumentation optimized for running microfluidic chips and demonstrated that this EATA method can utilize affinity-based separation involving noncovalent interaction between the immunocomplex of α2,3-linked sialylated PSA glyco-isoform and MAA to simultaneously determine concentrations of total free PSA and its glyco-isoform, S2,3PSA. We have previously demonstrated that the migration speed of an immunocomplex of DNA–Fab in a microfluidic channel can be controlled by changing the length of the DNA fragment [31]; capillary gel electrophoresis (CGE) is also known to be capable of separating molecules on the basis of size and charge.

In this study, we established an assay standard with recombinant S2,3PSA protein expressed by CHO cells and a standard curve for the quantitative measurement of %S2,3PSA according to assay linearity (Figure 5a,b). We also confirmed a capture efficiency of applied S2,3PSA standard of almost 100% in this assay (Figure 4b,c). By using specific lectin MAA to recognize α2,3-linked sialylation directly, the calculating analyte percentage excluded an unspecified factor contributed by different glyco-isoforms like asialo-type glycan as compared to the indirect assay with other lectin-based approaches [32]. These observations suggest that our assay strategy allows the reliable determination of the percentage ratio of S2,3PSA without loss of capturing the specific PSA glyco-isoform.

In terms of assay performance, the sensitivity for detection of S2,3PSA was 0.05 ng/mL using the μTAS immunoassay system. As noted in previous studies of EATA methods [33], up to 140-fold concentrations of sample and reagent mixtures can be achieved by ITP stacking, greatly enhancing assay sensitivity and binding kinetics. The newly developed μTAS system has higher analytical sensitivity than our previous methods [18,19] and %S2,3PSA can be determined at lower free PSA concentrations. This higher analytical sensitivity enables low CVs (<4%) for a precise percentage ratio of S2,3PSA and should improve the clinical utility of %S2,3PSA in patients with small tumors. Clinical studies employing the μTAS system are currently underway; the results will be published separately.

In addition to its high sensitivity, the µTAS in %S2,3PSA assay also shortens time to first result from the 4 h required for our previous assay using the Luminex method, [19] to less than 10 min. We achieved this improvement by combining the LBA principle and microchip capillary electrophoresis. Because LBA does not rely on diffusion of antigen to antibody immobilized on the surfaces of a solid phase, the immunoreaction is quick and stoichiometric. Furthermore, in the on-chip system, electrophoresis can be performed quickly by applying higher voltage under better thermal control; the principle of EATA results in a concentration of reactants that shortens binding and separation times [25,33].

In the present study, a novel assay system for measuring %S2,3PSA with a µTAS system discriminated patients with PCa from patients with BPH with 72.0% specificity at 80.0% sensitivity with a α2,3-sialic acid percentage cut-off of 42.20%, as indicated by an AUC of 0.8340 which was significantly higher than that for conventional PSA testing (0.5062, $p < 0.0001$ Figure 6e). According to some reports [10–12], combination assay containing multiple biomarkers such as the 4K test or Prostate Health Index would be applicable for improving diagnostic accuracy. In contrast, since %S2,3PSA test is a single biomarker assay based on the measurement of aberrant glycosylation, this simple assay strategy would achieve easy-to-use clinical application with a higher accuracy of diagnosis of PCa. Additionally, %S2,3PSA of Pbx GG $\geq$ 3 patients (range 50.7%–71.7%) was significantly higher than that of Pbx GG $\leq$ 2 patients (range 50.3%–52.6%) in the case of an over-50% S2,3PSA ratio ($p = 0.0019$) (Figure 6h). Although sample size was small, this result suggests that an over-50% S2,3PSA ratio holds the promise to discriminate between GG 2 and GG 3 tumors and may be used as a predictor for a prostate biopsy to discriminate between non-aggressive and aggressive tumors in the active surveillance program. To address this issue, we would need larger cohort study in the future.

4. Materials and Methods

4.1. Immunoassay Reagents

DNA (245-bp) and HiLyte Fluor 647 dye (AnaSpec, San Jose, CA, USA) were coupled to Fab′ prepared from anti-total PSA and anti-free PSA monoclonal antibodies (clones PSA10 and PSA12, respectively) that had been selected from our panel of mouse immunoglobulin G antibodies. These antibodies were digested with pepsin, followed by reduction of F(ab′)2 using 50 mM 2-aminoethanethiol to form Fab′. Each Fab′ was purified by Diol 200 gel filtration column chromatography (Wako Pure Chemical Industries, Osaka, Japan).

DNA-labeled anti-total PSA Fab′ (clone PSA10) was prepared as follows, a 245-bp DNA fragment was amplified by polymerase chain reaction using lambda DNA as a template and 5′ amine-modified primer as a forward primer to provide the amino group needed to couple the DNA fragment to anti-total PSA (clone PSA10) Fab′ antibody via bifunctional linker. Gene Taq polymerase (Wako Pure Chemical Industries) was used as polymerase. The DNA fragment was purified by using Diol-200 gel filtration and DEAE ion exchange chromatography on a DEAE-P5 W column (Tosoh, Tokyo, Japan). Next, 10 µM of the 245-bp DNA fragment was reacted with 10 mM N-(6-maleimidocaproyloxy)succinimide (EMCS) linker (Wako Pure Chemical Industries), a bifunctional linker having both maleimide and succinimide groups, at 37 °C for 30 min in 50 mM phosphate-buffered saline (PBS, pH 7.5). The linker-modified, 245-bp DNA fragment was purified by Diol-200 gel filtration, concentrated to 10 µM, and reacted with 500 µM PSA10 Fab′ at 4 °C for 3 h in 50 mM PBS (pH 6.5). To remove DNA-coupled Fab′ that had more than one DNA fragment, the conjugate was further purified by both Diol-200 gel filtration and DEAE column chromatography.

HiLyte Fluor 647 labeled anti-free PSA Fab′ (clone PSA12) was prepared as follows, anti-free PSA monoclonal antibody (clone PSA12) Fab′ was reacted with 1 mM HiLyte Fluor 647 C2 maleimide (Wako Pure Chemical Industries) at 4 °C for 2 h in 50 mM PBS (pH 7.5), after which 1 mM N-ethylmaleimide was added to the reaction mixture to block free thiol groups on the Fab′ molecule after the labeling reaction. Excess unreacted HiLyte Fluor 647 C2 maleimide was removed by Diol-200 gel filtration.

Electrophoresis leading buffer (LB) and trailing buffer (TB) for isotachophoresis (ITP) and capillary electrophoresis (CE) were formulated by adding nonionic surfactants to block the plastic chip surface and facilitate chip filling. The composition of the LB was 4.5% (w/v) polyethylene glycol (PEG8000), 3% (w/v) glycerol, 75 mM Tris–HCl (pH 7.5), 10 mM NaCl, 6.0% (w/v) dextran sulfate, 0.01% bovine serum albumin (BSA), and 4 mg/mL *Maackia amurensis* lectin (MAA). The composition of the sample buffer (SB) was 5.0% (w/v) PEG20000, 3% (w/v) glycerol, 75 mM Tris–HCl (pH 7.5), 150 mM NaCl, 0.01% BSA, and 10 mM 2-(*N*-morpholino)ethanesulfonic acid (MES). The TB consisted of 2.0% (w/v) PEG20000, 3% (w/v) glycerol, 75 mM Tris, 0.01% BSA, and 125 mM Hepes. The stacking buffer (ST) was composed of 2.0% (w/v) PEG20000, 3% (w/v) glycerol, 75 mM Tris–HCl (pH 7.5), and 0.01% BSA.

4.2. Microfluidic Electrophoresis Assay

Simultaneous quantitative measurement of S2,6PSA and S2,3PSA was achieved by performing microchip capillary electrophoresis and liquid-phase binding assay (LBA) on a μTAS Wako i30 auto analyzer with microfluidic chip IO6 (Figure 2a,b; Wako Pure Chemical Industries) for electrokinetic analyte transport assay (EATA). The details of the EATA method have been described by us previously [24] and are shown in Figure 2b. In brief, HyLite–Fab' was mixed with the serum sample in SB and the resulting mixture loaded into a sample well. DNA–Fab', the LB containing MAA, and the TB were also loaded into their designated wells, and the ST and focusing dye solutions into the ST well and focusing dye solution well, respectively. After all buffers and sample had been loaded, positive pressure (+20 psi) was applied to all wells to equilibrate all zones. Next, an electrical field was applied from the cathode to the anode to initiate ITP stacking of the DNA–Fab'. The stacked DNA–Fab' migrated from the DNA–Fab' zone into the sample + HiLyte–Fab' zone to initiate formation of the sandwich immunocomplex of DNA–Fab', PSA, and HiLyte–Fab'. The resulting immunocomplex was then further transported to the separation zone by ITP. During the ITP step, unreacted HiLyte–Fab' was left behind in the serum sample and HiLyte–Fab' zone. When the boundary of TB and LB reached the handoff junction, the electrical field was automatically switched from cathode–anode wells to handoff–anode wells and capillary gel electrophoresis (CGE) started to both separate the remaining noise components from the immunocomplex and S2,3PSA from S2,6PSA by affinity electrophoresis in an MAA-containing separation gel. Fluorescence signals from laser-induced fluorescence (LIF) detection were analyzed by software developed to use internal fluorescent markers to align and identify the peaks for S2,6PSA and S2,3PSA. The analyte peaks were integrated for peak area, which was then used to quantitate their ratio, as shown in Figure 2c.

4.3. Prostate Biopsy and Serum Samples

Between June 2007 and June 2016, 1494 transrectal ultrasound-guided prostate biopsies were performed in response to detection in regional PCa screening programs of PSA concentrations of ≥4.0 ng/mL or palpable prostate nodules in Hirosaki University Hospital, Akita University hospital, Tohoku University Hospital and McMaster University (Juravinski Hospital, Hamilton, ON, Canada). Serum samples were obtained from all patients at the time of biopsy and stored at −80 °C until use. The final diagnoses were established by histopathological examination of prostate biopsies. The grade group of prostate biopsy specimens were evaluated according to the International Society of Urological Pathology (ISUP) guidelines [26]. This study was performed in accordance with the ethical standards of the Declaration of Helsinki and was approved by the Ethics Committees of all participating institute including Hirosaki University Graduate School of Medicine ("The Study about Carbohydrate Structure Change in Urological Disease"; approval number: 2014–195). Informed consent was obtained from all patients. Table 2 shows relevant clinical characteristics of the study subjects. To evaluate diagnostic performance of serum S2,3PSA ratio, fifty patients with biopsy-proven PCa were pair-matched for age and PSA level with the same number of BPH patients selected from our serum bank.

4.4. Forced Expression of FLAG-Tag-Fused S2,3 and S2,6PSA in Chinese Hamster Ovary (CHO)-K1 Cells

CHO-K1 cells were obtained from the American Type Culture Collection and grown in Ham's F12 Nutrient Mixture medium supplemented with penicillin, streptomycin, and 10% fetal bovine serum at 37 °C with 5% CO_2. FLAG-tag (N-DYKDDDDK-C)-fused human PSA (kallikrein-3, KLK3) cDNA was amplified from RNA isolated from the prostate of a patient with benign prostatic hyperplasia using the primers hPSA-F1 5′-CCCAAGCTTACCACCTGCAC-3′ and hPSA-FLAG-Xho-R1 5′-TTTCTCGAGCTACTTGTCATCGTCGTCCTTGTAATCAGCGGGGTTGGCCACGATGGT-3′ and subcloned into the PCaG-Neo vector (Wako Pure Chemical Industries). The PSA–FLAG vector was then transiently transfected into CHO-K1 cells. After transfection, recombinant PSA was purified using a FLAG-tag system (Sigma, St. Louis, MO, USA) from serum free media. S2,3PSA and S2,6PSA containing asialo-type standard protein was purified utilizing an ACG lectin column and gel filtration column chromatography.

4.5. Lectin Microarray

One hundred microliters of purified S2,6PSA and S2,3PSA (31.25–2000 ng/mL) were applied to a lectin array (LecChipver1.0; GlycoTechnica, Sapporo, Japan), including triplicate spots of lectins in each of seven divided incubation baths on glass slides [34]. After incubation at 20 °C for 17 h, the reaction solution was discarded and the glass slides scanned using a GlycoStation Reader1200 (GlycoTechnica). Abbreviation of lectins are as follows: MAL, *maackia amurensis* lectin; SNA, *sambucus nigra* lectin; and SSA, *Sambucus sieboldiana* lectin.

4.6. Statistical Analysis

All calculations for clinical data were performed in the SPSS software, ver. 21.0 (SPSS, Inc., Chicago, IL, USA) and in GraphPad Prism 6.03 (GraphPad Software, San Diego, CA, USA). Intergroup differences were statistically analyzed by a Student's *t*-test for normally distributed variables or by the Mann–Whitney *U*-test for non-normally distributed models. Data with $p < 0.05$ were considered significant. Receiver operating characteristics (ROC) curves were developed using the library "rms" in R (available on: http://www.r-project.org/) [25] and the statistical difference of AUCs were calculated by the same program.

5. Conclusions

In conclusion, although the present results are preliminary, they suggest that the newly developed serum %S2,3PSA test may have superior diagnostic accuracy to currently available tests. Additionally, we believe that the %S2,3PSA assay with µTAS system has great potential for further application in the clinical laboratory when rapid and quantitative testing is required. Larger-scale studies are warranted to confirm these findings.

Acknowledgments: All of the authors thank Yukie Nishizawa, Kaname Higuchi, and Satomi Sakamoto, technical assistant of Hirosaki University Graduate School of Medicine, for their invaluable help with sample collection and patient data management. This study was supported by the by the Japan Society for the Promotion of Science (JSPS) KAKENHI grant no. 15K15579 and grant no. 15H02563.

Author Contributions: Tohru Yoneyama and Yuki Tobisawa conceived and designed the experiments; Tomokazu Ishikawa performed the experiments; Tohru Yoneyama analyzed the data; Tomokazu Ishikawa, Tatsuo Kurosawa, and Kenji Nakamura developed the automated µTAS-based immunoassay system; Shingo Hatakeyama, Yasuhiro Hashimoto, Takuya Koie, Shintaro Narita, Koji Mitsuzuka, Wilhelmina Duivenvoorden, Jehonathan H. Pinthus, Tomonori Habuchi, Yoichi Arai, and Chikara Ohyama diagnosed BPH and PCa, and performed biopsies; Chikara Ohyama obtained funding. Chikara Ohyama and Tohru Yoneyama supervised. Yuki Tobisawa critically revised the manuscript for intellectual content; Tomokazu Ishikawa and Tohru Yoneyama wrote the paper.

Conflicts of Interest: The authors declare no conflict of interest.

Abbreviations

µTAS	micro-total analysis system
MAA	*maackia amurensis* lectin
EATA	electrokinetic analyte transport assay
LBA	liquid-phase binding assay
C.V.	coefficient of variation
ITP	isotachophoresis
LIF	laser-induced fluorescence
LB	leading buffer
TB	trailing buffer
CE	capillary electrophoresis
SB	sample buffer
ST	stacking buffer
CGE	capillary gel electrophoresis
LOD	limit of detection
PSA	prostate-specific antigen
PCa	prostate cancer
BPH	benign prostatic hyperplasia
Pbx GG	prostate biopsy grade group
LacdiNAc	GalNAcβ1-4GlcNAc-
Gal	galactose
Man	mannose
Fuc	fucose
Sia	sialic acid
GalNAc	N-acetylgalactosamine
GlcNAc	N-acetylglucosamine

References

1. Schroder, F.H.; Hugosson, J.; Roobol, M.J.; Tammela, T.L.; Ciatto, S.; Nelen, V.; Kwiatkowski, M.; Lujan, M.; Lilja, H.; Zappa, M.; et al. Screening and prostate-cancer mortality in a randomized European study. *N. Engl. J. Med.* **2009**, *360*, 1320–1328. [CrossRef] [PubMed]

2. Hugosson, J.; Carlsson, S.; Aus, G.; Bergdahl, S.; Khatami, A.; Lodding, P.; Pihl, C.G.; Stranne, J.; Holmberg, E.; Lilja, H. Mortality results from the Goteborg randomised population-based prostate-cancer screening trial. *Lancet Oncol.* **2010**, *11*, 725–732. [CrossRef]

3. Loeb, S.; Catalona, W.J. Prostate-specific antigen in clinical practice. *Cancer Lett.* **2007**, *249*, 30–39. [CrossRef] [PubMed]

4. Ito, K.; Ichinose, Y.; Kubota, Y.; Imai, K.; Yamanaka, H. Clinicopathological features of prostate cancer detected by transrectal ultrasonography-guided systematic six-sextant biopsy. *Int. J. Urol.* **1997**, *4*, 474–479. [CrossRef] [PubMed]

5. Ito, K.; Ohi, M.; Yamamoto, T.; Miyamoto, S.; Kurokawa, K.; Fukabori, Y.; Suzuki, K.; Yamanaka, H. The diagnostic accuracy of the age-adjusted and prostate volume-adjusted biopsy method in males with prostate specific antigen levels of 4.1–10.0 ng/mL. *Cancer* **2002**, *95*, 2112–2119. [CrossRef] [PubMed]

6. Balk, S.P.; Ko, Y.J.; Bubley, G.J. Biology of prostate-specific antigen. *J. Clin. Oncol.* **2003**, *21*, 383–391. [CrossRef] [PubMed]

7. Lazzeri, M.; Haese, A.; de la Taille, A.; Palou Redorta, J.; McNicholas, T.; Lughezzani, G.; Scattoni, V.; Bini, V.; Freschi, M.; Sussman, A.; et al. Serum isoform [−2]proPSA derivatives significantly improve prediction of prostate cancer at initial biopsy in a total PSA range of 2–10 ng/mL: A multicentric European study. *Eur. Urol.* **2013**, *63*, 986–994. [CrossRef] [PubMed]

8. Loeb, S.; Sanda, M.G.; Broyles, D.L.; Shin, S.S.; Bangma, C.H.; Wei, J.T.; Partin, A.W.; Klee, G.G.; Slawin, K.M.; Marks, L.S.; et al. The prostate health index selectively identifies clinically significant prostate cancer. *J. Urol.* **2015**, *193*, 1163–1169. [CrossRef] [PubMed]

9. Foley, R.W.; Gorman, L.; Sharifi, N.; Murphy, K.; Moore, H.; Tuzova, A.V.; Perry, A.S.; Murphy, T.B.; Lundon, D.J.; Watson, R.W. Improving multivariable prostate cancer risk assessment using the Prostate Health Index. *BJU Int.* **2016**, *117*, 409–417. [CrossRef] [PubMed]

10. Fossati, N.; Lazzeri, M.; Haese, A.; McNicholas, T.; de la Taille, A.; Buffi, N.M.; Lughezzani, G.; Gadda, G.M.; Lista, G.; Larcher, A.; et al. Clinical performance of serum isoform [−2]proPSA (p2PSA), and its derivatives %p2PSA and the Prostate Health Index, in men aged <60 years: Results from a multicentric European study. *BJU Int.* **2015**, *115*, 913–920. [PubMed]

11. Punnen, S.; Pavan, N.; Parekh, D.J. Finding the Wolf in Sheep's Clothing: The 4Kscore Is a Novel Blood Test That Can Accurately Identify the Risk of Aggressive Prostate Cancer. *Rev. Urol.* **2015**, *17*, 3–13. [PubMed]

12. Vedder, M.M.; de Bekker-Grob, E.W.; Lilja, H.G.; Vickers, A.J.; van Leenders, G.J.; Steyerberg, E.W.; Roobol, M.J. The added value of percentage of free to total prostate-specific antigen, PCA3, and a kallikrein panel to the ERSPC risk calculator for prostate cancer in prescreened men. *Eur. Urol.* **2014**, *66*, 1109–1115. [CrossRef] [PubMed]

13. Fukuda, M. Possible roles of tumor-associated carbohydrate antigens. *Cancer Res.* **1996**, *56*, 2237–2244. [PubMed]

14. Hakomori, S. Glycosylation defining cancer malignancy: New wine in an old bottle. *Proc. Natl. Acad. Sci. USA* **2002**, *99*, 10231–10233. [CrossRef] [PubMed]

15. Yamashita, K.; Taketa, K.; Nishi, S.; Fukushima, K.; Ohkura, T. Sugar chains of human cord serum α-fetoprotein: Characteristics of N-linked sugar chains of glycoproteins produced in human liver and hepatocellular carcinomas. *Cancer Res.* **1993**, *53*, 2970–2975. [PubMed]

16. Amano, J.; Nishimura, R.; Mochizuki, M.; Kobata, A. Comparative study of the mucin-type sugar chains of human chorionic gonadotropin present in the urine of patients with trophoblastic diseases and healthy pregnant women. *J. Biol. Chem.* **1988**, *263*, 1157–1165. [PubMed]

17. Tajiri, M.; Ohyama, C.; Wada, Y. Oligosaccharide profiles of the prostate specific antigen in free and complexed forms from the prostate cancer patient serum and in seminal plasma: A glycopeptide approach. *Glycobiology* **2008**, *18*, 2–8. [CrossRef] [PubMed]

18. Ohyama, C.; Hosono, M.; Nitta, K.; Oh-eda, M.; Yoshikawa, K.; Habuchi, T.; Arai, Y.; Fukuda, M. Carbohydrate structure and differential binding of prostate specific antigen to Maackia amurensis lectin between prostate cancer and benign prostate hypertrophy. *Glycobiology* **2004**, *14*, 671–679. [CrossRef] [PubMed]

19. Yoneyama, T.; Ohyama, C.; Hatakeyama, S.; Narita, S.; Habuchi, T.; Koie, T.; Mori, K.; Hidari, K.I.; Yamaguchi, M.; Suzuki, T.; et al. Measurement of aberrant glycosylation of prostate specific antigen can improve specificity in early detection of prostate cancer. *Biochem. Biophys. Res. Commun.* **2014**, *448*, 390–396. [CrossRef] [PubMed]

20. Jacobson, S.C.; Ramsey, J.M. Microchip electrophoresis with sample stacking. *Electrophoresis* **1995**, *16*, 481–486. [CrossRef] [PubMed]

21. Koutny, L.B.; Schmalzing, D.; Taylor, T.A.; Fuchs, M. Microchip electrophoretic immunoassay for serum cortisol. *Anal. Chem.* **1996**, *68*, 18–22. [CrossRef] [PubMed]

22. Chiem, N.; Harrison, D.J. Microchip-based capillary electrophoresis for immunoassays: Analysis of monoclonal antibodies and theophylline. *Anal. Chem.* **1997**, *69*, 373–378. [CrossRef] [PubMed]

23. Reyes, D.R.; Iossifidis, D.; Auroux, P.-A.; Manz, A. Micro total analysis systems. 1. Introduction, theory, and technology. *Anal. Chem.* **2002**, *74*, 2623–2636. [CrossRef] [PubMed]

24. Kawabata, T.; Wada, H.G.; Watanabe, M.; Satomura, S. Electrokinetic analyte transport assay for α-fetoprotein immunoassay integrates mixing, reaction and separation on-chip. *Electrophoresis* **2008**, *29*, 1399–1406. [CrossRef] [PubMed]

25. Park, C.C.; Kazakova, I.; Kawabata, T.; Spaid, M.; Chien, R.L.; Wada, H.G.; Satomura, S. Controlling data quality and reproducibility of a high-sensitivity immunoassay using isotachophoresis in a microchip. *Anal. Chem.* **2008**, *80*, 808–814. [CrossRef] [PubMed]

26. Humphrey, P.A.; Moch, H.; Cubilla, A.L.; Ulbright, T.M.; Reuter, V.E. The 2016 WHO Classification of Tumours of the Urinary System and Male Genital Organs-Part B: Prostate and Bladder Tumours. *Eur. Urol.* **2016**, *70*, 106–119. [CrossRef] [PubMed]

27. Vermassen, T.; Speeckaert, M.M.; Lumen, N.; Rottey, S.; Delanghe, J.R. Glycosylation of prostate specific antigen and its potential diagnostic applications. *Clin. Chim. Acta* **2012**, *413*, 1500–1505. [CrossRef] [PubMed]

28. Drake, R.R.; Jones, E.E.; Powers, T.W.; Nyalwidhe, J.O. Altered glycosylation in prostate cancer. *Adv. Cancer Res.* **2015**, *126*, 345–382. [PubMed]

29. Sarrats, A.; Comet, J.; Tabares, G.; Ramirez, M.; Aleixandre, R.N.; de Llorens, R.; Peracaula, R. Differential percentage of serum prostate-specific antigen subforms suggests a new way to improve prostate cancer diagnosis. *Prostate* **2010**, *70*, 1–9. [CrossRef] [PubMed]

30. Sarrats, A.; Saldova, R.; Comet, J.; O'Donoghue, N.; de Llorens, R.; Rudd, P.M.; Peracaula, R. Glycan characterization of PSA 2-DE subforms from serum and seminal plasma. *Omics* **2010**, *14*, 465–474. [CrossRef] [PubMed]

31. Kawabata, T.; Watanabe, M.; Nakamura, K.; Satomura, S. Liquid-phase binding assay of α-fetoprotein using DNA-coupled antibody and capillary chip electrophoresis. *Anal. Chem.* **2005**, *77*, 5579–5582. [CrossRef] [PubMed]

32. Llop, E.; Ferrer-Batalle, M.; Barrabes, S.; Guerrero, P.E.; Ramirez, M.; Saldova, R.; Rudd, P.M.; Aleixandre, R.N.; Comet, J.; de Llorens, R.; et al. Improvement of Prostate Cancer Diagnosis by Detecting PSA Glycosylation-Specific Changes. *Theranostics* **2016**, *6*, 1190–1204. [CrossRef] [PubMed]

33. Kagebayashi, C.; Yamaguchi, I.; Akinaga, A.; Kitano, H.; Yokoyama, K.; Satomura, M.; Kurosawa, T.; Watanabe, M.; Kawabata, T.; Chang, W.; et al. Automated immunoassay system for AFP-L3% using on-chip electrokinetic reaction and separation by affinity electrophoresis. *Anal. Biochem.* **2009**, *388*, 306–311. [CrossRef] [PubMed]

34. Hirabayashi, J.; Yamada, M.; Kuno, A.; Tateno, H. Lectin microarrays: Concept, principle and applications. *Chem. Soc. Rev.* **2013**, *42*, 4443–4458. [CrossRef] [PubMed]

International Journal of
Molecular Sciences

MDPI

Article

Does the Prostate Health Index Depend on Tumor Volume?—A Study on 196 Patients after Radical Prostatectomy

Frank Friedersdorff [1,*], Britt Groß [1], Andreas Maxeiner [1], Klaus Jung [1,2], Kurt Miller [1], Carsten Stephan [1,2], Jonas Busch [1,†] and Ergin Kilic [3,†]

1   Department of Urology, Charité University Hospital, 10098 Berlin, Germany; grossbritt@yahoo.de (B.G.); andres.maxeiner@charite.de (A.M.); klaus.jung@charite.de (K.J.); kurt.miller@charite.de (K.M.); carsten.stephan@charite.de (C.S.); Jonas.busch@charite.de (J.B.)
2   Berlin Institute for Urologic Research, 10115 Berlin, Germany
3   Department of Pathology, Charité University Hospital, 10098 Berlin, Germany; ergin.kilic@charite.de
*   Correspondence: frank.friedersdorff@charite.de; Tel.: +49-30-450-515-052; Fax: +49-30-450-515-904
†   These authors contributed equally to this work.

Academic Editors: Jack A. Schalken and William Chi-shing Cho
Received: 27 October 2016; Accepted: 4 February 2017; Published: 24 February 2017

Abstract: The Prostate Health Index (PHI) has been used increasingly in the context of prostate cancer (PCa) diagnostics since 2010. Previous studies have shown an association between PHI and a tumor volume of >0.5 cm^3. The aim of this study was to investigate the correlation between PHI and tumor volume as well as the Gleason score. A total of 196 selected patients with prostate cancer treated with radical prostatectomy at our institution were included in our study. The tumor volume was calculated and preoperative serum parameters total prostate-specific antigen (tPSA), free PSA (fPSA), [−2]proPSA, and PHI were evaluated. The association between the pathological findings such as Gleason score, pathological T-stage (pT stage), and tumor volume were evaluated. We further used logistic regression and Cox proportional hazard regression analyses for assessing the association between tumor volume and PHI and for predicting biochemical recurrence. With an area under the curve (AUC) of 0.79, PHI is the most accurate predictor of a tumor volumes >0.5 cm^3. Moreover, PHI correlates significantly with the tumor volume ($r = 0.588$), which is significantly different ($p = 0.008$) from the correlation of the Gleason score with tumor volume ($r = 0.385$). PHI correlates more strongly with the tumor volume than does the Gleason score. Using PHI improves the prediction of larger tumor volume and subsequently clinically significant cancer.

Keywords: prostate cancer; Prostate Health Index; tumor volume

1. Introduction

Prostate cancer (PCa) is still one of the most frequent illnesses in men. Since the introduction of screening with prostate-specific antigen (PSA) in the early 1990s, the incidence of PCa has risen sharply, though the age at first diagnosis has been shifting toward younger patients with less advanced stages of PCa [1]. The number of deaths from PCa has been reduced in recent years, though not proportionally with the rise in incidence.

The degree of tumor expansion is the most important factor for both PCa prognosis and therapy [2]. The PCa tumor volume has a substantial influence on the course of the illness. Growth exceeding the capsule and lymph node metastases occurs mainly with larger tumors [3–5]. There is an approximately proportional relationship between the tumor volume and the stage of differentiation. Tumors that are smaller than 0.5 cm^3 are classified as "insignificant", since the tumor grows so slowly that there is a high probability that it will never reach a significant size before the death of the patient [6]. The probability of the existence of a tumor larger than 0.5 cm^3 can be calculated, for example, with a nomogram [7].

The Prostate Health Index (PHI) is currently one of the most promising new markers of PCa. Initial studies have shown that PHI preferentially detects aggressive carcinomas and, due to its high specificity, can reduce the number of unnecessarily performed biopsies [8,9]. Furthermore, tumors larger than 0.5 cm^3 have significantly higher PHI values than tumors with smaller volumes [10–12]. However, it is still not known whether PHI correlates continuously with tumor volume. A marker that is able to reliably predict the tumor volume would be advantageous for therapeutic decision-making. PHI and [−2]proPSA are currently the best available serum parameters for PCa detection [9]. At 90% sensitivity, the specificity of PHI for PCa detection is on average 31.6% [9]. The use of PHI in the framework of PCa screening reduces the use of biopsies by 15%–41%, in comparison to classic screening with only total prostate-specific antigen (tPSA) measurement [8,13]. Moreover, there is an association between PHI and carcinomas with unfavorable prognostic characteristics [12,14].

The use of PHI appears quite promising in patients with diagnosed PCa who are following an active surveillance strategy. Approximately 30%–37% of these tumors advance to stages requiring intervention [15,16]. With a PHI score >43, there is a 3.6 times higher risk for disease progression, according to a study by Hirama et al. [15]. Carcinomas with an elevated tendency for deterioration can be identified earlier this way and subjected to a definitive therapy. Overall, the integration of PHI into active surveillance appears to be also economically useful. The increased expenditures from this additional blood test would be outweighed since the reduction of unnecessary biopsies, doctor visits, and laboratory tests would reduce the costs in comparison to examinations based only on tPSA [17,18].

The aim of the present study was to determine if there is a continuous correlation between PHI and tumor volume over the entire range of values commonly seen.

2. Results

2.1. Clinicopathological Characteristics of the Study Cohort

All clinicopathological characteristics are summarized in Table 1. One-third of the patients (*n* = 65) had a well-differentiated tumor (Gleason < 7). A total of 131 patients had a poorly differentiated tumor (Gleason ≥ 7). Within this group, a Gleason 7 tumor was found in 57.6% of the patients and 9.7% of the patients had a Gleason 8 or 9, Table 1). About 80% of the patients had a locally confined tumor (pT ≤ 2c). An "insignificant" carcinoma (≤0.5 cm^3, range 0.03 to 0.48 cm^3) was found in 39 specimens (20%) and a "significant" tumor (>0.5 cm^3, range 0.57 to 22.8 cm^3) in 157 cases (80%).

Table 1. Clinicopathological data (median and range) of the study cohort.

Variable	Median (Range)	Mean ± S.D.
Age (years)	71 (49–85)	70 ± 7
Prostate volume (cm^3)	36.7 (9.2–264)	42.9 ± 27.1
Tumor volume (cm^3)	1.57 (0.03–22.8)	2.76 ± 3.47
Percentage of tumor volume	4.3 (0.02–80.4)	7.8 ± 10.6
	Number (Percent)	
Pathological tumor stage		
≤2c	154 (78.6)	
3a	29 (14.8)	
3b	11 (5.6)	
4	2 (1.0)	
Gleason Score		
<7	65 (33.2)	
7a (3 + 4)	77 (39.2)	
7b (4 + 3)	36 (18.4)	
≥8	18 (9.2)	
Resection margin status		
R0	143 (73.0)	
R1	50 (25.5)	
Rx	3 (1.5)	

2.2. PSA Parameters in Relation to Clinicopatholological Factors and Tumor Volume

In Table 2, the data of all four PSA derivatives and the prostate and tumor volumes are compiled in relation to the pathological Gleason score (well and poorly differentiated tumors) and the pathological stage (locally limited and advanced tumors). While percent free PSA (%fPSA) did not differ between the two tumor staging groups nor prostate volume both between the Gleason and staging groups, all other parameters were significantly different. The highest significance levela of $p < 0.0001$ were reached for the tumor volume and the percentage of tumor as well as for [−2]proPSA and PHI. In addition, Table 3 shows that all four serum parameters had significantly higher (tPSA, [−2]proPSA, PHI) or lower (%fPSA) values in patients with tumor volumes larger than 0.5 cm³, the so-called significant tumors. Additional details are summarized in the Supplementary Table S1.

Table 2. Prostate volume, tumor volume, and prostate-specific antigen (PSA) analytes (median and range) in dependence on Gleason score and pathological tumor stage.

Variable	All Patients	Gleason Score <7 ≥7		p-Value	pT Stage ≤2c ≥3		p-Value
	(*n* = 196)	(*n* = 65)	(*n* = 131)		(*n* = 154)	(*n* = 42)	
Prostate volume (cm³)	36.7 (9.2–264)	36.8 (17.6–167)	36.1 (9.2–264)	0.568	35.3 (14.7–167)	39.8 (9.2–264)	0.295
Tumor volume (cm³)	1.58 (0.03–22.8)	0.66 (0.03–10.9)	2.26 (0.04–22.8)	<0.0001	1.21 (0.03–18.6)	4.36 (0.87–22.8)	<0.0001
Percentage of tumor (%)	4.3 (0.02–80.4)	1.4 (0.02–40.3)	6.6 (0.1–80.4)	<0.0001	3.1 (0.1–80.4)	10.0 (3.1–80.4)	<0.0001
tPSA (ng/mL)	5.0 (0.7–61.6)	4.2 (0.7–17.7)	5.4 (0.7–61.6)	<0.001	4.6 (0.7–61.6)	5.6 (1.8–32.6)	0.005
%fPSA (%)	12.3 (4.0–76.6)	14.4 (4.9–36.9)	11.3 (4.0–76.6)	<0.001	12.7 (4.0–76.6)	11.5 (4.9–28.2)	0.154
[−2]proPSA (pg/mL)	12.3 (2.3–117)	10.0 (2.3–46.7)	14.2 (3.2–117)	<0.0001	11.7 (2.3–117)	17.4 (3.7–58.2)	0.0005
Prostate Health Index (PHI)	47.6 (9.3–228)	37.5 (10.5–102)	55.3 (9.3–228)	<0.0001	43.4 (9.3–211)	67.1 (22.3–228)	<0.0001

Statistical significances were calculated using the Mann–Whitney *U*-test.

Table 3. PSA parameters (medians and ranges) in dependence on the tumor volume.

Variable	Tumor Volume		p-Value
	≤0.5 cm³ (*n* = 39)	>0.5 cm³ (*n* = 157)	
tPSA (ng/mL)	2.8 (0.7–10.8)	5.4 (0.7–61.6)	<0.0001
%fPSA	15.8 (6.0–35.0)	11.7 (4.0–76.6)	0.0006
[−2]proPSA (pg/mL)	9.4 (2.3–32.0)	13.6 (3.2–117)	<0.0001
PHI	32.9 (16.1–66.4)	53.7 (9.3–228)	<0.0001

Statistical significance calculated using the Mann–Whitney *U*-test.

The further associations between the clinicopathological variables and the PSA analytes were examined using correlation calculations and receiver-operating characteristi (ROC) curve analyses. Table 4 shows the correlations between the serum PSA parameters and the clinicopathological variables. PHI showed a significantly higher correlation to tumor volume than Gleason score (0.588 vs. 0.385; $p = 0.008$). Moreover, [−2]proPSA showed a significantly higher correlation with tumor volume than Gleason score (0.659 vs. 0.385, $p = 0.0002$). Interestingly, [−2]proPSA correlated significantly more with tumor volume than pT stage (0.659 vs. 0.522; $p = 0.037$). The correlation comparison between PHI and [−2]proPSA showed no difference ($p = 0.25$). PHI correlated significantly better with tumor volume than PSA (0.588 vs. 0.363; $p = 0.004$). The scatter plots of the PHI values in relation to the tumor volume as well as to the Gleason score categories and pT classifications are displayed in Figure 1A–C.

The areas under the ROC curves (AUCs) of the individual serum parameters are presented in Table 5. At 0.79, PHI reaches the greatest AUC value in differentiating tumor volumes ≤0.5 cm³ from those >0.5 cm³. In general, PHI is the best parameter to predict aggressive PCa with Gleason score (GS) ≥ 7 and locally advanced stages with ≥pT3a. Furthermore, PHI reached a significantly higher AUC than that of the pT stage in predicting a tumor volume of >0.5 cm³ (0.79 vs. 0.69; $p = 0.04$). Thus, to show the mutual influence and interdependence between all these variables, we performed a logistic regression analysis as a multivariable approach. For that purpose, the categorized Gleason score and pT stage in combination with PHI were analyzed to differentiate between the two tumor volume categories. All three variables showed a significant odds ratio (Gleason score: 6.61 with a 95% confidence interval of 2.72 to 16.1), p-value < 0.001; pT stage: (1.60 (1.01–2.55), p = 0.047; PHI: 1.04 (1.02–1.07, p = 0.002)). This combination distinctly increased the AUC value to 0.87 (0.82–0.91) indicating the improved prediction between the insignificant and significant tumor volume.

Table 4. Pearson correlation of the serum parameters and the pathology findings.

Parameter	Tumor Volume (cm³)	Tumor (%)	Gleason Score (≤6, 7a, 7b, ≥8)	pT Stage (≤2c, 3a, 3b, 4)
tPSA	0.363 (<0.0001)	0.287 (<0.0001)	0.153 (0.032)	0.237 (0.0008)
%fPSA	−0.101 (0.158)	−0.145 (0.043)	−0.157 (0.028)	−0.071 (0.324)
[−2]proPSA	0.659 (<0.0001)	0.389 (<0.0001)	0.246 (0.0005)	0.344 (<0.0001)
PHI	0.588 (<0.0001)	0.478 (<0.0001)	0.309 (<0.0001)	0.317 (<0.0001)
Gleason Score	0.385 (<0.0001)	0.373 (<0.0001)	-	0.397 (<0.0001)
pT Stage	0.522 (<0.0001)	0.486 (<0.0001)	0.397 (<0.0001)	-

The table presents the r values, with the p-values in parentheses.

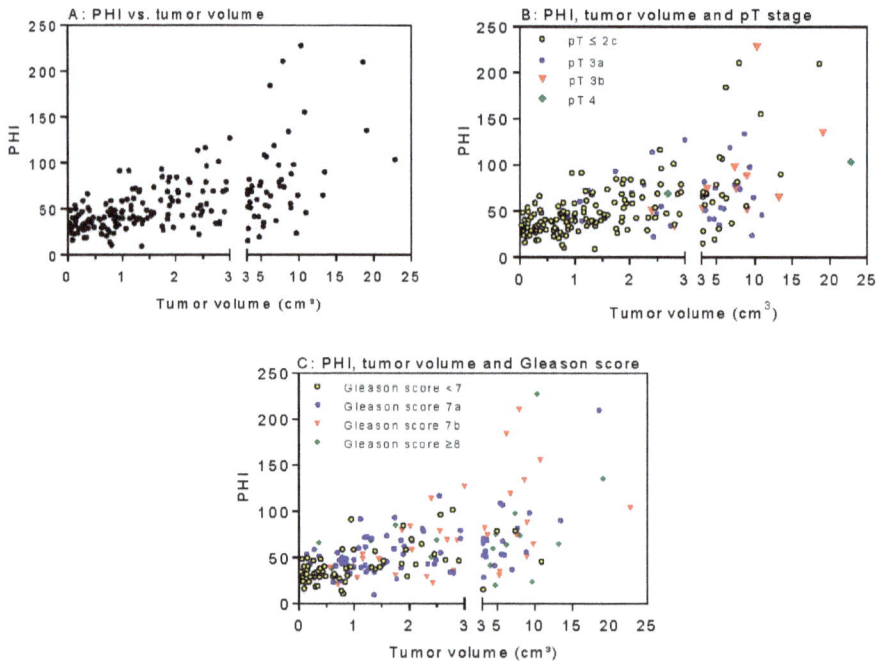

Figure 1. PHI values in relation to tumor volume, pT stage and Gleason score as indicated in (**A–C**).

Table 5. Calculated areas under the receiver operating characteristic (ROC) curves of the serum parameters (left column) to predict tumors with particular characteristics (top row). The 95% confidence intervals of the area under the curves (AUCs) are given in parentheses.

Parameter	Gleason Score ≥ 7 vs. <7	pT Stage ≥3a vs. ≤2c	Tumor Volume >0.5 vs. ≤0.5 cm³
tPSA	0.64 (0.56–0.73)	0.64 (0.55–0.73)	0.74 (0.65–0.83)
%fPSA	0.64 (0.56–0.72)	0.57 (0.48–0.67)	0.68 (0.58–0.77)
[−2]proPSA	0.66 (0.59–0.74)	0.68 (0.58–0.77)	0.72 (0.64–0.81)
PHI	0.72 (0.65–0.80)	0.70 (0.62–80.0)	0.79 (0.72–0.86)

2.3. PHI and Tumor Volume as Independent Predictors of Biochemical Relapse

To evaluate the relationships between the conventional clinicopathological variables (Gleason score, pathological tumor stage, resection status) and age, tumor value, and PHI as predictors of the biochemical recurrence, we performed Cox regression analyses (Table 6). Both tumor volume and PHI were statistically significant predictors of a biochemical recurrence (follow-up time: 37 months (95% confidence interval 31 to 44)) in the univariate Cox regression model but also remained, despite the limited number of patients and events ($n = 25$) and the independent factors in the multivariable model ($p = 0.048$ and 0.0009).

Table 6. Cox proportional hazard regression analyses of tumor volume and PHI in relation to conventional clinicopathological variables for predicting biochemical recurrence after radical prostatectomy.

Variable [a]	Univariable Analysis		Multivariable Analysis [b]	
	Hazard Ratio (95% CI [c])	*p*-Value	Hazard Ratio (95% CI)	*p*-Value
Age (continuous)	1.05 (0.98–1.12)	0.159	not included	
pT stage	4.27 (1.91–9.56)	0.0004	3.92 (1.60–9.56)	0.003
Gleason Score	1.97 (1.22–3.18)	0.006	1.91 (1.01–3.57)	0.042
Margin resection status	3.34 (1.52–7.35)	0.003	1.65 (0.67–4.01)	0.273
Tumor volume	1.12 (1.01–1.24)	0.047	0.83 (0.68–0.99)	0.048
PHI	1.02 (1.01–1.03)	<0.0001	1.02 (1.01–1.03)	0.0009

[a] Age and tumor volume were used with their continuous values, the pathological factors as categorized data as indicated in the previous tables; [b] The multivariable analysis included all variables with *p*-values < 0.10 obtained in the univariable analysis; [c] CI = confidence interval.

3. Discussion

The use of PHI as a marker with high specificity for aggressive PCa and tumors with a low grade of differentiation has been proven [10,11,19]. In the present study, PHI correlates significantly with the tumor volume ($r = 0.588$). This is significantly stronger ($p = 0.008$) than the correlation of the Gleason score with tumor volume ($r = 0.385$). The hypothesis was supported by the significantly higher PHI value for tumor volumes >0.5 cm^3 [10–12,14]. Furthermore, it seems plausible that the PHI values rise with increasing carcinoma volumes, since there is a proven relationship between tumor size and tumor differentiation [4,5,12,20]. Previous studies have shown that PHI can predict a tumor volume >0.5 cm^3 even more precisely (AUC: 0.72–0.94) than Gleason score ≥ 7 (AUC: 0.64–0.74) or a pT-stage $\geq$ 3a (AUC: 0.72–0.85) [10,11,14,21,22]. The strong correlation between PHI and tumor volume found in this study describes a further partial aspect of the association of this marker with aggressive carcinomas.

PHI as a preoperative parameter shows significant explanatory power to predict a significant tumor volume in the prostatectomy preparation. In the present study, PHI reached a test strength (AUC) of 0.79 in the detection of clinical significant tumors. The portion of substantial tumors in our analysis was 80%. Comparable results for the test strength of PHI were reported in the work by Guazzoni et al. [11]. In 350 preparations, PHI detected tumors >0.5 cm^3 with an AUC of 0.8. In that study, the portion of substantial tumors was 93%. Somewhat lower AUC values were found in the work from Tallon et al. [21]. In the framework of that study, 154 prostate preparations were examined. A volume >0.5 cm^3 was found in 88% of the study participants. The calculated AUC value of PHI for the detection of substantial tumors was 0.72. In that study and in the present study, T3 tumors were included. In a study by Ferro et al., the AUC of PHI for the detection of substantial tumors was 0.94 [10]. The more advantageous results of that study might be explainable by the different composition of their study sample. On the one hand, that study limited itself to locally limited tumors (T1–T2); on the other hand, a comparatively smaller portion of the carcinomas (17%) had a substantial size ($\geq$ 0.5 cm^3). In the most recent study on 135 Asian patients by Chiu et al. [12], PHI reached a comparable high AUC of 0.82 to predict a tumor volume of >0.5 cm^3, but only about half (52.3%) of all patients had a tumor volume >0.5 cm^3. Since the portion of significant tumor volumes was

80% in our present study, comparability seems to be given more with the studies by Guazzoni et al. and Tallon et al. [11,21]. Furthermore, there is an association between the serum marker PHI and the pathological Gleason score. In our present study, PHI was the best parameter to detect tumors with a Gleason score ≥7 (AUC: 0.72) while the other parameters tPSA, %fPSA, and [−2]proPSA showed comparatively smaller AUCs of 0.64, 0.64, and 0.66, respectively. In other studies, the calculated AUC of PHI for the detection of tumors with a Gleason score ≥7 was between 0.67 and 0.74 [11,19,22]. In the work by Ferro et al. [10], an exceptionally high-test strength (AUC: 0.83) of PHI was found for the detection of tumors with a Gleason score ≥7. In addition, in this case, the clearly advantageous results from the research group of Ferro et al. [10] might be due to a different composition of the study group. Chiu et al. [12] provided only data for prediction of pathological Gleason score ≥7 or pathological pT3 stage and reached an AUC of exactly 0.8.

The relationship between PHI and biopsy Gleason score has already been investigated in numerous studies. According to a meta-analysis from 2014, PHI detects carcinomas with a Gleason ≥7 with an AUC of 0.90 [19]. That value is clearly above the calculated test strength of the present study for the detection of low differentiated tumors (AUC: 0.72). One possible source of such a large difference in results is that the biopsy Gleason score and the pathological Gleason score differ in many cases. The tissue samples of the core needle biopsy harbor a comparatively higher risk to incorrectly evaluate the Gleason grade more favorably than it actually is, since this technique of examination is inherently not capable of imaging the entire PCa. Since the Gleason score can differ according to the methods chosen, that meta-analysis is only conditionally comparable with the present study. Moreover, the processing of a prostatectomy preparation is the more precise method to characterize a PCa and, thus, to estimate the relation of Gleason score and PHI.

There is an association between PHI and the extensiveness of the local tumor. In the present study, PHI detected tumors with growth exceeding the capsule with an AUC of 0.70. The AUCs of other parameters were lower (tPSA: 0.64, %fPSA: 0.57, [−2]proPSA: 0.68). In other studies, the AUC for the detection of T3 tumors was reported as 0.69–0.72 [11,20], which is comparable to our study. In the study by Ferro et al. [10], PHI detected tumors exceeding the capsule with an AUC of 0.85.

In the present study, it is apparent that PHI shows the highest AUCs for the prediction of the tumor characteristics (0.79 for tumor volume, 0.72 for Gleason score ≥7, 0.70 for pT ≤ 3a). In other studies, it has been shown that PHI predicts the tumor volume more precisely than the Gleason score of the T-stage does [10–12,14,21]. Consequently, it can be assumed that the levels of PSA, fPSA, and [−2]proPSA (the components of the PHI) are influenced not only by the cell differentiation but likewise by the amount of carcinoma cells (tumor volume). According to the results of this study, it is conceivable that the tumor volume is the strongest determining factor. This is also supported by the high Pearson's correlation factors of 0.588 and 0.478 for PHI with the tumor volume and the percentage of tumor, respectively. The correlation of PHI with the Gleason score (r = 0.309) and pT stage (r = 0.317) is weaker in our cohort. All these correlation data were unfortunately not provided by all previous studies. Furthermore, tumor volume (borderline significance with p = 0.048) and PHI (p = 0.0009) remained as independent factors in multivariable analysis to predict a biochemical recurrence (Table 6) beside the known pathological data pT stage and Gleason score. For example, the resection margin status could not reach significance (p = 0.27) in this Cox proportional hazard regression model, which emphasizes the importance of tumor volume and PHI.

The tumor volume is an important predictor of the further disease course and, correspondingly, should be kept in mind in planning therapy including focal therapy. Our method very precisely estimates tumor volume and has not been described before. Other earlier studies do not focus on methodological aspects. In the present work, the median tumor volume was 1.58 cm^3, and the median PHI was 47.6. The results of this work show that a high PHI value is most likely connected with a large tumor volume, which could be a contraindication for a therapy approach of active surveillance. The integration of PHI into active surveillance could reduce the frequency of biopsies. Larger intervals between follow-up biopsies would be a clear advantage for the patient. Nevertheless, large-scale studies about this are needed in order to rule out the possibility that a lower frequency of biopsies in combination with regular PHI determinations leads to a worse outcome for the patient. The use of PHI in the preparation of a prostatectomy is likewise conceivable. High PHI values indicate a large tumor volume, which in turn makes the possibility of performing a nerve-sparing operation more improbable. Correspondingly, it could be advantageous to advise the patient of this aspect before operating. Future studies might be able to answer the question of whether there is a PHI limit value, above which a nerve-sparing operation becomes very improbable. This could lead to a more transparent risk prognosis before surgery.

The serum markers PHI and [−2]proPSA are accurate predictors of the tumor volume in prostatectomy preparations. Further, it was shown in the AUC analysis that PHI shows a stronger association to tumor volume than the established predictive parameters of the Gleason score and T-stage. It is important to integrate PHI into active surveillance of PCa patients, since an estimate of the tumor volume, as an important predictor of the illness progression, could support an adequate therapy decision.

4. Materials and Methods

4.1. Study Population

The study was approved by the Charité Ethics Committee. All patients provided written informed consent for this research study. Preoperative serum samples (tPSA, free PSA (fPSA), [−2]proPSA) were collected from 460 men who underwent radical prostatectomy between 2001 and 2014. The surgical approach was retroperitoneal, laparoscopic, or da Vinci based. The exclusion criteria were incomplete ($n = 233$) or missing pathological sections ($n = 24$), lack of appropriate serum samples ($n = 2$), or use of neoadjuvant therapies ($n = 5$) so that finally 196 patients were included in the analysis.

The study groups (Table 1) were defined as follows: carcinomas with a pathological GS of 4–6 were assembled into the GS < 7 group (well differentiated, $n = 65$) and carcinomas with a GS of 7–10 ($n = 131$) were assembled into the GS ≥ 7 group (poorly differentiated). The stage "locally limited" ($n = 154$) included tumors with pathological T-stage (pT stage) 2a, 2b, and 2c. The stage "locally advanced" included tumors in the pT-stages 3a, 3b, and 4 ($n = 42$). The volume of the tumors was divided into two categories: "insignificant" (≤0.5 cm^3, $n = 39$) and "significant" (>0.5 cm^3, $n = 157$) according to the Epstein criteria [23].

The prognostic potential of the tumor volume in relation to the other clinicopathological parameters including PHI was assessed to predict tumor recurrence after radical prostatectomy. The conventional criterion of biochemical recurrence of increased PSA after surgery was used. Biochemical recurrence was defined as the first postoperative PSA value of >0.1 ng/mL following a nadir PSA level after surgery and confirmed by persistent increased PSA values >0.1 ng/mL. The interval between radical prostatectomy and biochemical recurrence was calculated in months as time to the development of this event, while patients without biochemical recurrence were censored at the last follow-up visit.

4.2. Pathology Assessment

After the macroscopic assessment, the preparations were weighed and measured in three planes (apico-basal, horizontal, ventro-dorsal). The prostates were sliced first medially along the urethra and then the apex and base were cut off. Both mid-pieces, the apex, and the base were each divided into approximately 5 mm thick tissue blocks and then embedded in paraffin. The entire paraffin blocks were sliced by machine into 2–4 µm thin sections and then mounted onto specimen slides. They were then stained with hematoxyline and eosin (H&E). The prostatectomy preparations of this study had a mean (range) of 28 (10–63) sections; in total, over 5500 H&E sections were examined microscopically.

4.3. Serological Diagnosis and Volumetric Analysis

The preoperative determination of serum tPSA, fPSA, and [−2]proPSA was performed on the fully automated immunoassay device Access® (Beckman Coulter, Brea, CA, USA) as described before [24]. PHI was calculated according to the equation [−2]proPSA/fPSA × $\sqrt{PSA}$.

All tissue sections were examined under a microscope, and carcinomas were outlined in color. The appraisal of the preparation was performed together with an experienced pathologist (EK).

As Figure 2 illustrates, carcinomas were surrounded with a hand-drawn line. The surfaces of the prostate and the surface of the tumor on the H&E sections were identified by means of a gridwork screen (3 mm × 3 mm). The gridwork screen was placed on the slides, and the surfaces were enumerated. The boxes were evaluated as "1", "0.5", and "0" when they were covered with clearly more than 50%, approximately with 50%, or clearly less than 50% by tumor, respectively. For each H&E section, the number of boxes for prostate and for cancer were recorded separately. Periprostatic soft tissue and intraprostatic parts of the seminal vesicle were excluded from the evaluation. The volume of the prostate was calculated with the aid of the ellipsis formula ($\pi/6$ × height × width × depth). The percentage portion of the carcinoma was calculated as follows: %tumor = (total number of tumor boxes/total number of prostate boxes) × 100. The tumor volume was then calculated with the following formula: tumor volume = (%tumor × prostate volume)/100.

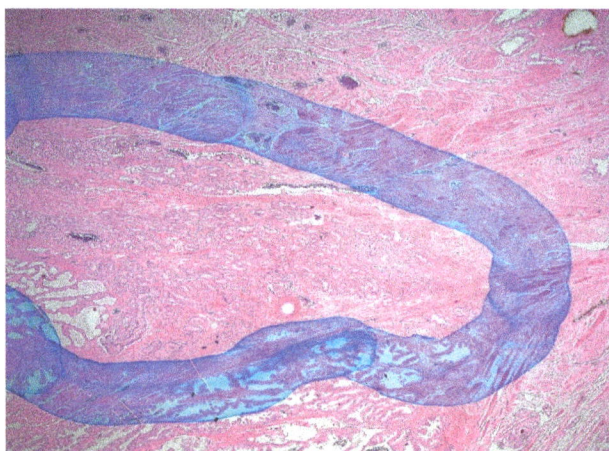

Figure 2. Outlining of the prostate carcinoma. (Hematoxyline-eosin stain; magnification: 20×).

4.4. Statistical Analysis

All statistical analysis was performed with SPSS 22.0 (IBM Corporation; Armonk, NY, USA) and MedCalc 16.8.4 (MedCalc software bvba, Ostend, Belgium). Several tests were performed (Mann–Whitney *U*-test, Kruskal–Wallis test, Pearson correlation coefficient, logistic regression, and Cox hazard regression). ROC analysis was used for estimating the AUCs. A *p*-value < 0.05 was considered statistically significant.

Supplementary Materials: Supplementary materials can be found at www.mdpi.com/1422-0067/18/3/488/s1.

Acknowledgments: We thank Silke Rabenhorst and Sabine Becker for their valuable technical assistance.

Author Contributions: This manuscript is part of the doctoral thesis of Britt Groß. Frank Friedersdorff and Britt Groß designed the study and drafted the manuscript; Andreas Maxeiner, Carsten Stephan, and Jonas Busch provided the clinical specimens for this study; Klaus Jung, Carsten Stephan, and Kurt Miller supervised the research, analyzed the data, and edited the paper; Britt Groß and Klaus Jung performed the statistical analysis and the data processing; Ergin Kilic and Britt Groß performed the pathological analysis. All authors read and approved the final manuscript.

Conflicts of Interest: The authors declare no conflict of interest.

References

1. The American Cancer Society. Cancer Facts & Figures 2015. Available online: Http://www.Cancer.Org/ research/cancerfactsstatistics (accessed on 15 May 2015).

2. European Association of Urology. Guidelines on Prostate-Cancer 2015. Available online: Http://uroweb. Org/guideline/prostate-cancer/ (accessed on 4 June 2015).

3. Epstein, J.I.; Amin, M.; Boccon-Gibod, L.; Egevad, L.; Humphrey, P.A.; Mikuz, G.; Newling, D.; Nilsson, S.; Sakr, W.; Srigley, J.R.; et al. Prognostic factors and reporting of prostate carcinoma in radical prostatectomy and pelvic lymphadenectomy specimens. *Scand. J. Urol. Nephrol. Suppl.* **2005**, *216*, 34–63. [CrossRef] [PubMed]

4. McNeal, J.E.; Villers, A.A.; Redwine, E.A.; Freiha, F.S.; Stamey, T.A. Capsular penetration in prostate cancer. Significance for natural history and treatment. *Am. J. Surg. Pathol.* **1990**, *14*, 240–247. [CrossRef] [PubMed]

5. McNeal, J.E.; Villers, A.A.; Redwine, E.A.; Freiha, F.S.; Stamey, T.A. Histologic differentiation, cancer volume, and pelvic lymph node metastasis in adenocarcinoma of the prostate. *Cancer* **1990**, *66*, 1225–1233. [PubMed]

6. Stamey, T.A.; Freiha, F.S.; McNeal, J.E.; Redwine, E.A.; Whittemore, A.S.; Schmid, H.P. Localized prostate cancer. Relationship of tumor volume to clinical significance for treatment of prostate cancer. *Cancer* **1993**, *71*, 933–938. [CrossRef]

7. Kattan, M.W.; Eastham, J.A.; Wheeler, T.M.; Maru, N.; Scardino, P.T.; Erbersdobler, A.; Graefen, M.; Huland, H.; Koh, H.; Shariat, S.; et al. Counseling men with prostate cancer: A nomogram for predicting the presence of small, moderately differentiated, confined tumors. *J. Urol.* **2003**, *170*, 1792–1797. [CrossRef] [PubMed]

8. De la Calle, C.; Patil, D.; Wei, J.T.; Scherr, D.S.; Sokoll, L.; Chan, D.W.; Siddiqui, J.; Mosquera, J.M.; Rubin, M.A.; Sanda, M.G. Multicenter evaluation of the prostate health index to detect aggressive prostate cancer in biopsy naive men. *J. Urol.* **2015**, *194*, 65–72. [CrossRef] [PubMed]

9. Filella, X.; Gimenez, N. Evaluation of [−2]proPSA and prostate health index (PHI) for the detection of prostate cancer: A systematic review and meta-analysis. *Clin. Chem. Lab. Med.* **2013**, *51*, 729–739. [CrossRef] [PubMed]

10. Ferro, M.; Lucarelli, G.; Bruzzese, D.; Perdona, S.; Mazzarella, C.; Perruolo, G.; Marino, A.; Cosimato, V.; Giorgio, E.; Tagliamonte, V.; et al. Improving the prediction of pathologic outcomes in patients undergoing radical prostatectomy: The value of prostate cancer antigen 3 (PCa3), prostate health index (PHI) and sarcosine. *Anticancer Res.* **2015**, *35*, 1017–1023. [PubMed]

11. Guazzoni, G.; Lazzeri, M.; Nava, L.; Lughezzani, G.; Larcher, A.; Scattoni, V.; Gadda, G.M.; Bini, V.; Cestari, A.; Buffi, N.M.; et al. Preoperative prostate-specific antigen isoform p2PSA and its derivatives, %p2PSA and prostate health index, predict pathologic outcomes in patients undergoing radical prostatectomy for prostate cancer. *Eur. Urol.* **2012**, *61*, 455–466. [CrossRef] [PubMed]

12. Chiu, P.K.; Lai, F.M.; Teoh, J.Y.; Lee, W.M.; Yee, C.H.; Chan, E.S.; Hou, S.M.; Ng, C.F. Prostate health index and %p2PSA predict aggressive prostate cancer pathology in Chinese patients undergoing radical prostatectomy. *Ann. Surg. Oncol.* **2016**, *23*, 2707–2714. [CrossRef] [PubMed]

13. Lazzeri, M.; Haese, A.; de la Taille, A.; Palou Redorta, J.; McNicholas, T.; Lughezzani, G.; Scattoni, V.; Bini, V.; Freschi, M.; Sussman, A.; et al. Serum isoform [−2]proPSA derivatives significantly improve prediction of prostate cancer at initial biopsy in a total PSA range of 2–10 ng/mL: A multicentric European study. *Eur. Urol.* **2013**, *63*, 986–994. [CrossRef] [PubMed]

14. Cantiello, F.; Russo, G.I.; Ferro, M.; Cicione, A.; Cimino, S.; Favilla, V.; Perdona, S.; Bottero, D.; Terracciano, D.; de Cobelli, O.; et al. Prognostic accuracy of prostate health index and urinary prostate cancer antigen 3 in predicting pathologic features after radical prostatectomy. *Urol. Oncol.* **2015**, *33*, 163.e115–163.e123. [CrossRef] [PubMed]

15. Hirama, H.; Sugimoto, M.; Ito, K.; Shiraishi, T.; Kakehi, Y. The impact of baseline [−2]proPSA-related indices on the prediction of pathological reclassification at 1 year during active surveillance for low-risk prostate cancer: The Japanese multicenter study cohort. *J. Cancer Res. Clin. Oncol.* **2014**, *140*, 257–263. [CrossRef] [PubMed]

16. Klotz, L.; Zhang, L.; Lam, A.; Nam, R.; Mamedov, A.; Loblaw, A. Clinical results of long-term follow-up of a large, active surveillance cohort with localized prostate cancer. *J. Clin. Oncol.* **2010**, *28*, 126–131. [CrossRef] [PubMed]

17. Nichol, M.B.; Wu, J.; An, J.J.; Huang, J.; Denham, D.; Frencher, S.; Jacobsen, S.J. Budget impact analysis of a new prostate cancer risk index for prostate cancer detection. *Prostate Cancer Prostatic Dis.* **2011**, *14*, 253–261. [CrossRef] [PubMed]

18. Nichol, M.B.; Wu, J.; Huang, J.; Denham, D.; Frencher, S.K.; Jacobsen, S.J. Cost-effectiveness of prostate health index for prostate cancer detection. *BJU Int.* **2012**, *110*, 353–362. [CrossRef] [PubMed]

19. Wang, W.; Wang, M.; Wang, L.; Adams, T.S.; Tian, Y.; Xu, J. Diagnostic ability of %p2PSA and prostate health index for aggressive prostate cancer: A meta-analysis. *Sci. Rep.* **2014**, *4*, 5012. [CrossRef] [PubMed]

20. Stamey, T.A.; Yemoto, C.M.; McNeal, J.E.; Sigal, B.M.; Johnstone, I.M. Prostate cancer is highly predictable: A prognostic equation based on all morphological variables in radical prostatectomy specimens. *J. Urol.* **2000**, *163*, 1155–1160. [CrossRef]

21. Tallon, L.; Luangphakdy, D.; Ruffion, A.; Colombel, M.; Devonec, M.; Champetier, D.; Paparel, P.; Decaussin-Petrucci, M.; Perrin, P.; Vlaeminck-Guillem, V. Comparative evaluation of urinary PCa3 and tmprss2: ERG scores and serum PHI in predicting prostate cancer aggressiveness. *Int. J. Mol. Sci.* **2014**, *15*, 13299–13316. [CrossRef] [PubMed]

22. Fossati, N.; Buffi, N.M.; Haese, A.; Stephan, C.; Larcher, A.; McNicholas, T.; de la Taille, A.; Freschi, M.; Lughezzani, G.; Abrate, A.; et al. Preoperative prostate-specific antigen isoform p2PSA and its derivatives, %p2PSA and prostate health index, predict pathologic outcomes in patients undergoing radical prostatectomy for prostate cancer: Results from a multicentric European prospective study. *Eur. Urol.* **2015**, *68*, 132–138. [CrossRef] [PubMed]

23. Kryvenko, O.N.; Epstein, J.I. Definition of insignificant tumor volume of Gleason score 3 + 3 = 6 (grade group 1) prostate cancer at radical prostatectomy: Is it time to increase the threshold? *J. Urol.* **2016**, *196*, 1664–1669. [CrossRef] [PubMed]

24. Stephan, C.; Vincendeau, S.; Houlgatte, A.; Cammann, H.; Jung, K.; Semjonow, A. Multicenter evaluation of [-2]proprostate-specific antigen and the prostate health index for detecting prostate cancer. *Clin. Chem.* **2013**, *59*, 306–314. [CrossRef] [PubMed]

International Journal of
Molecular Sciences

MDPI

Article

Comparative Study of Blood-Based Biomarkers, α2,3-Sialic Acid PSA and PHI, for High-Risk Prostate Cancer Detection

Montserrat Ferrer-Batallé [1,2,†], Esther Llop [1,2,†], Manel Ramírez [2,3], Rosa Núria Aleixandre [2,3], Marc Saez [4,5], Josep Comet [2,3], Rafael de Llorens [1,3,*] and Rosa Peracaula [1,3,*]

[1]   Biochemistry and Molecular Biology Unit, Department of Biology, University of Girona, 17003 Girona, Spain; montserrat.ferrer@udg.edu (M.F.-B.); esther.llop@udg.edu (E.L.)
[2]   Girona Biomedical Research Institute (IDIBGI), 17190 Salt (Girona), Spain; jramirez.girona.ics@gencat.cat (M.R.); rnaleixandre@infonegocio.com (R.N.A.); 28547jcb@comb.cat (J.C.)
[3]   Catalan Health Institute, University Hospital of Girona Dr. Josep Trueta, 17007 Girona, Spain
[4]   Research Group on Statistics, Econometrics and Health (GRECS), University of Girona, 17003 Girona, Spain; marc.saez@udg.edu
[5]   CIBER of Epidemiology and Public Health (CIBERESP), 28029 Madrid, Spain
*   Correspondence: rafael.llorens@udg.edu (R.d.L.); rosa.peracaula@udg.edu (R.P.); Tel.: +34-972-418-370 (R.d.L. & R.P.)
†   These authors contributed equally to this work.

Academic Editor: Carsten Stephan
Received: 21 March 2017; Accepted: 12 April 2017; Published: 17 April 2017

Abstract: Prostate Specific Antigen (PSA) is the most commonly used serum marker for prostate cancer (PCa), although it is not specific and sensitive enough to allow the differential diagnosis of the more aggressive tumors. For that, new diagnostic methods are being developed, such as PCA-3, PSA isoforms that have resulted in the 4K score or the Prostate Health Index (PHI), and PSA glycoforms. In the present study, we have compared the PHI with our recently developed PSA glycoform assay, based on the determination of the α2,3-sialic acid percentage of serum PSA (% α2,3-SA), in a cohort of 79 patients, which include 50 PCa of different grades and 29 benign prostate hyperplasia (BPH) patients. The % α2,3-SA could distinguish high-risk PCa patients from the rest of patients better than the PHI (area under the curve (AUC) of 0.971 vs. 0.840), although the PHI correlated better with the Gleason score than the % α2,3-SA. The combination of both markers increased the AUC up to 0.985 resulting in 100% sensitivity and 94.7% specificity to differentiate high-risk PCa from the other low and intermediate-risk PCa and BPH patients. These results suggest that both serum markers complement each other and offer an improved diagnostic tool to identify high-risk PCa, which is an important requirement for guiding treatment decisions.

Keywords: diagnosis; glycosylation; prostate cancer; prostate specific antigen; proPSA; PHI; α2,3-sialic acid

1. Introduction

Prostate cancer (PCa) is an important problem in public health and a major disease that affects men's health worldwide. It was the most commonly diagnosed male neoplasia in western countries and Japan last year. It is expected that around one of each six men will be diagnosed with PCa during his life. In addition, as the number of older people increase, the incidence of the disease will raise dramatically in the coming decades [1].

The serum marker Prostate Specific Antigen (PSA), adopted in the early 1990's, has been the widely used and preferred assay for prostate diseases, including PCa, with important levels of success, and represents the gold standard marker as an essential tool for urologists [1,2]. In addition, since PCa has a long natural history, the PSA assay predicts a prostate pathology decades before a confirmatory diagnostic [3]. This means that a majority of men diagnosed with PCa could be detected at early stages and with localized prostate cancer. Epidemiologic studies indicate an important and continuous decrease in prostate cancer mortality since the application of the PSA screening test [4].

However, the PSA test presents some limitations. It is organ specific but not cancer specific [5]. Serum PSA levels could also be elevated in benign prostate hyperplasia (BPH), prostatitis, and prostate manipulations (as DRE and bicycling), and cannot discriminate between aggressive and non-aggressive cancers. PSA assays present a high rate of false positives that leads to over-diagnosis, unnecessary biopsies, and over-treatments [6]. Actually only 25% of men biopsied after an elevated PSA level have PCa, and many of these cancers are slow growing, with no impact in the patient's life [7].

New non-invasive biomarkers with greater sensitivity and specificity that are capable of distinguishing aggressive tumors from indolent ones are required [5]. To improve the specificity of the PSA as a biomarker, different strategies using several PSA isoforms (ratio free PSA/total PSA, PSA density and velocity, proPSA forms, 4K score, and Prostate Health Index (PHI)) have been developed [7], recently including, PSA glycoforms [8–10].

Regarding proPSA forms, these were first identified in the serum of patients with prostate cancer in 1997 [11]. ProPSAs were preferentially elevated in the peripheral zone of prostatic tissue containing cancer, whilst remaining largely undetectable in the transitional zone of the prostate [12]. These proPSA forms comprised the complete sequence of inactive zymogen $[-7]$proPSA, and also shorter forms as $[-5]$ $[-4]$ and $[-2]$proPSA. $[-2]$proPSA was present in the sera of prostate cancer patients and it was a more specific serum marker that could improve a PSA assay. $[-7]$ and $[-5]$proPSA did not give adequate results as biomarkers [13]. The interesting positive results of the $[-2]$proPSA detection moved Beckman & Coulter Inc. (Brea, CA, USA), in partnership with the NCI Early Detection Research Network, to develop a mathematical algorithm with $[-2]$proPSA, tPSA, and fPSA serum levels, the so called Prostate Health Index: PHI = $([-2]$proPSA/fPSA$) \times \sqrt{\text{tPSA}}$ [14]. PHI received the FDA approval in 2012 [7]. Several works, including numerous international multicenter studies, have indicated that PHI score outperforms its individual components for the prediction of overall and high-grade prostate cancer [6,15–17]. PHI score has a high diagnostic accuracy rate and may be useful as a tumor marker in predicting patients harboring more aggressive disease. PHI also predicts the likelihood of progression during active surveillance. PHI score has been reported to correlate with PSA serum levels and Gleason scores. Nowadays, PHI has regulatory approval in more than 50 countries worldwide and is now being incorporated into prostate cancer guidelines for early prostate cancer detection and risk stratification [18]. However, others studies do not completely agree with these results and indicated that when the goal is to detect at least 95% of the aggressive tumors, PHI does not seem to be much more effective than the %fPSA and the PSA density [19].

To address the problem of the discovery of new non-invasive PCa markers that can predict PCa aggressiveness, several authors have determined the glycosylation pattern of PSA from healthy donors, PCa cancer cell lines, and PCa serum patients, and have shown specific changes in the PSA core fucosylation and sialylation levels in PCa patients [8,10,20–27]. In this regard, we have developed a methodology to quantify the ratio of core fucosylation of serum PSA and the percentage of $\alpha2,3$-sialic acid of serum PSA and have shown a decrease in the content of core fucose and an increase in $\alpha2,3$-sialic acid of PSA N-glycans in patients with high-risk PCa [9]. In particular, the percentage of $\alpha2,3$-sialic acid of PSA was increased in the high-risk PCa patients compared with low or intermediate-risk PCa and BPH patients and gave an AUC of 0.971, with 85.7% sensitivity and 95.3% specificity. Interestingly, the percentage of $\alpha2,3$-sialic acid of PSA also correlated with the Gleason score of PCa patients.

With the aim of searching for new serum markers that could assist in the identification of the aggressive prostate cancers, the present study compared the potential of PHI and the percentage of $\alpha2,3$-sialic of PSA, alone and in combination, to identify high risk PCa cancer in a cohort of 79 patients' serum samples.

2. Results

2.1. Clinical and Pathological Characteristics of the Patients

A cohort of 79 serum samples containing 29 BPH and 50 PCa samples was used for the study of the two blood-based biomarkers, PHI and the percentage of $\alpha 2,3$-sialic of PSA. PCa staging was determined according to the International Union Against Cancer (IUAC) and patients were classified in high-risk ($N = 22$), intermediate-risk ($N = 21$) and low-risk ($N = 7$). Clinical data of the subjects included in this study are summarized in Table 1.

The seven low-risk PCa patients had tPSA levels below 10 ng/mL and Gleason scores $\leq$6. The 21 intermediate-risk PCa group comprised five patients with a Gleason score of six and clinical stage >pT2a; 15 patients with a Gleason score of seven and one subject presenting a focal Gleason score of eight. Their tPSA levels were between 3.73 and 12.42 ng/mL. The 22 high-risk PCa included 18 with a Gleason score $\geq$8, two with a Gleason score of seven and metastasis, and two other subjects with an undetermined Gleason score who also presented metastasis. Data corresponding to the age and total and free PSA values of all groups of patients are shown in Table 1.

Evaluation of the clinical outcome of the PCa patients showed a PCa recurrence one year after treatment of 0%, 4.8%, and 59% in the low, intermediate, and high-risk PCa groups respectively. Data of the five-year relapse-free survival was reported for all patients in the low-risk group being 100%. However, this information was not available for all patients in the other two groups. The five-year relapse-free survival was 95% for the intermediate-risk group corresponding to 20 out of the 21 patients, and it was 40% in the high-risk group corresponding to 15 out of the 22 patients.

2.2. Analysis of $\alpha 2,3$-Sialic Acid PSA in Serum Samples

For the analysis of percentage of $\alpha 2,3$-sialic acid PSA, 0.75 mL of each serum were required. First, the serum samples were treated with ethanolamine, in order to release PSA from its complex with $\alpha 1$-antichymotrypsin. Then, total PSA from the serum samples was immunoprecipitated and loaded into a SNA lectin column. This lectin chromatography, which binds to $\alpha 2,6$-sialylated glycoconjugates, allows for the separation of $\alpha 2,3$-sialylated from $\alpha 2,6$-sialylated PSA glycoforms [9]. After the lectin chromatography, free PSA in the unbound ($\alpha 2,3$-sialylated PSA) and bound fractions ($\alpha 2,6$-sialylated PSA) was measured, and from these data the percentage of fPSA in both fractions was calculated. The percentage of the unbound fraction corresponded to the percentage of $\alpha 2,3$-sialic acid PSA.

The potential of the percentage of $\alpha 2,3$-sialic acid PSA as a blood biomarker for aggressive PCa was assessed in the cohort of sera (29 BPH, seven low-risk, 21 intermediate-risk and 22 high-risk PCa). Three different PCa serum samples, containing different values of tPSA (12.87, 23.08, and 40.61 ng/mL) were repeatedly analyzed in the different batches of samples in order to calculate the inter-assay variation of the method that was lower than 12%.

The plot of the percentage of $\alpha 2,3$-sialylated PSA is represented against the concentration of the total PSA of each sample (Figure 1A) and in the four groups (Figure 1B). A significant increase of percentage of $\alpha 2,3$-sialylated PSA in the group of high-risk PCa patients (26.8–61.4%) compared with the other three groups, intermediate-risk PCa (12.7–35.5%; $p < 0.001$), low-risk PCa (12.3–29.9%; $p = 0.006$), and BPH (10.9–33.5%; $p < 0.001$) was shown. However, no significant differences were found between BPH and low and intermediate-risk PCa patients. The correlation of $\alpha 2,3$-sialylated PSA values of the samples with their corresponding tPSA levels was tested and resulted to be non-significant in any of the BPH and PCa groups. Both parameters were then independent, indicating that a high or a low percentage of $\alpha 2,3$-sialylated PSA could be found in sera with either low or high tPSA levels in any group of patients (Figure 1A).

Table 1. Clinical and pathological characteristics of the patients.

Pathology	Cases	N	PCa Recurrence, 1 Year	Gleason Score	N	Age Average	Range	tPSA ng/mL	±SD	Range	fPSA ng/mL	±SD	Range
BPH		29				63.24	44–76	7.59	2.39	3.89–14.47	1.26	0.53	0.30–2.28
PCa N = 50	Low-risk	7	0%	Gleason 5	1	84		2.45			0.27		
				Gleason 6	6	66.2	61–74	4.91	1.38	2.64–6.33	0.88	0.22	0.61–1.14
	Intermediate risk	21	4.8%	Gleason 6	5	56	47–75	5.79	3.73	3.73–12.42	0.53	0.33	0.19–0.97
				Gleason 7	15	65.2	46–78	6.61	1.84	5.13–10.39	0.65	0.29	0.58–1.36
				Gleason 8 focal	1	70		7.16			1.76		
	High-risk	22	59%	Gleason 7/metastasis	2	76	69–83	12.08	2.85	10.07–14.1	1.93	1.63	0.78–3.09
				Gleason 8	10	65.5	51–83	16.23	11.93	1.96–40.61	1.58	1.46	0.35–5.29
				Gleason 9	7	67.8	49–79	14.81	5.22	4.34–18.77	3.28	2.18	0.7–7.09
				Gleason 10	1	67		87.51			12.77		
				Gleason ND */metastasis	2	75	67–83	7.28	3.65	4.7–9.86	1.37	1.22	0.51–2.23

* ND: not determined; SD: Standard deviation.

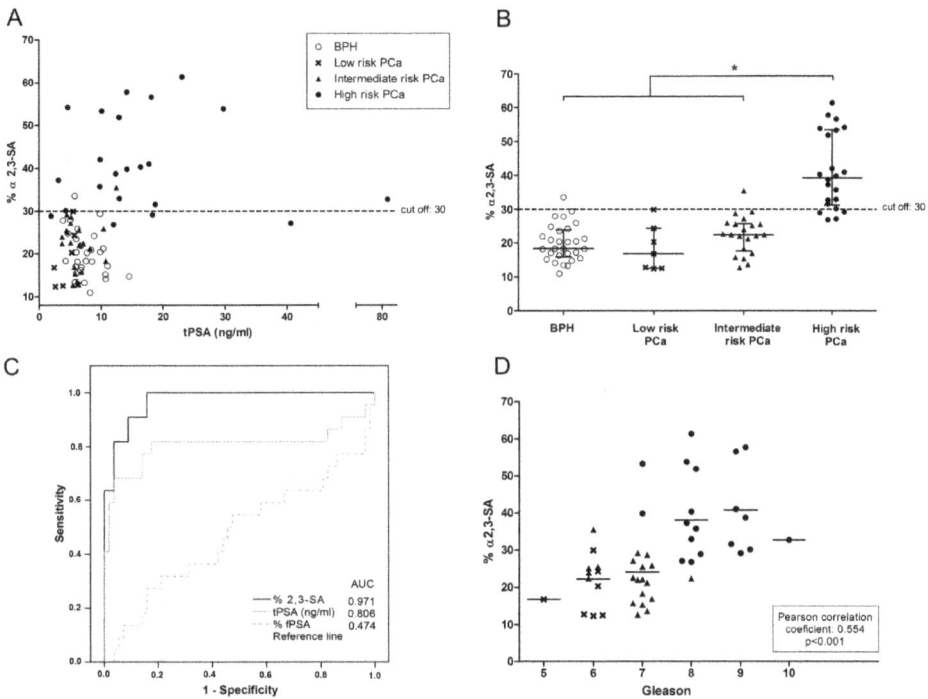

Figure 1. α2,3-SA percentage of Prostate Specific Antigen (PSA) (% α2,3-SA) of the cohort of 79 serum samples. Benign Prostate Hyperplasia (BPH) samples are represented with an open circle (o), low risk PCa with a cross (×), intermediate risk PCa with a filled triangle (▲) and high risk PCa with a filled circle (●). (**A**) Representation of % α2,3-SA against tPSA serum levels; dotted line (- - -) shows the cutoff value for discriminating high risk PCa samples from the other three groups; (**B**) Representation of % α2,3-SA against the pathology. The center line indicates the median, and the top and bottom lines, the 75th and 25th percentiles, respectively; (**C**) Representation of the Receiver operating characteristic (ROC) curves for % α2,3-SA, tPSA, and %fPSA; (**D**) Correlation plot of % α2,3-SA from the PCa serum samples with their Gleason score. The mean of % α2,3-SA of each Gleason score is shown with a horizontal line (-).

In order to compare the performance of PSA α2,3-sialic acid percentage with that of tPSA and the %fPSA values, the Receiver operating characteristic (ROC) curves of these three parameters were compared (Figure 1C). The ROC assay showed that % α2,3-sialic acid had the highest performance and could separate high-risk PCa patients from BPH, low, or intermediate-risk prostate cancers with 81.8% sensitivity and 96.5% specificity with a cutoff of 30%, resulting in an AUC of 0.97. In addition, this biomarker, which is based on the detection of specific PSA glycoforms, significantly correlated with the Gleason score of the tumor (correlation coefficient 0.554, $p < 0.001$) (Figure 1D), which highlights its potential as a marker for aggressive PCa.

2.3. Prostate Health Index (PHI) Score Analysis of Serum Samples

For this analysis, patients' sera were analyzed for total PSA (tPSA), free PSA (fPSA), and [−2]proPSA. Then the Prostate Health Index (PHI) score was calculated [PHI = ([−2]proPSA /fPSA) × $\sqrt{\text{tPSA}}$]. This methodology was used to analyze the cohort of serum samples tested previously for α2,3-sialic acid percentage of PSA.

The plot of the PHI score is shown against the concentration of total serum PSA of each sample (Figure 2A) and in the four groups (Figure 2B). There was a significant increase of PHI score in the group of high-risk PCa patients compared with the other two groups, low-risk PCa ($p = 0.006$) and BPH ($p < 0.001$). The intermediate-risk PCa group showed also a significant increase of PHI compared with low-risk PCa ($p = 0.006$) and BPH ($p = 0.022$). No significant differences were found between high-risk PCa patients and intermediate-risk PCa neither between BPH and low-risk PCa patients.

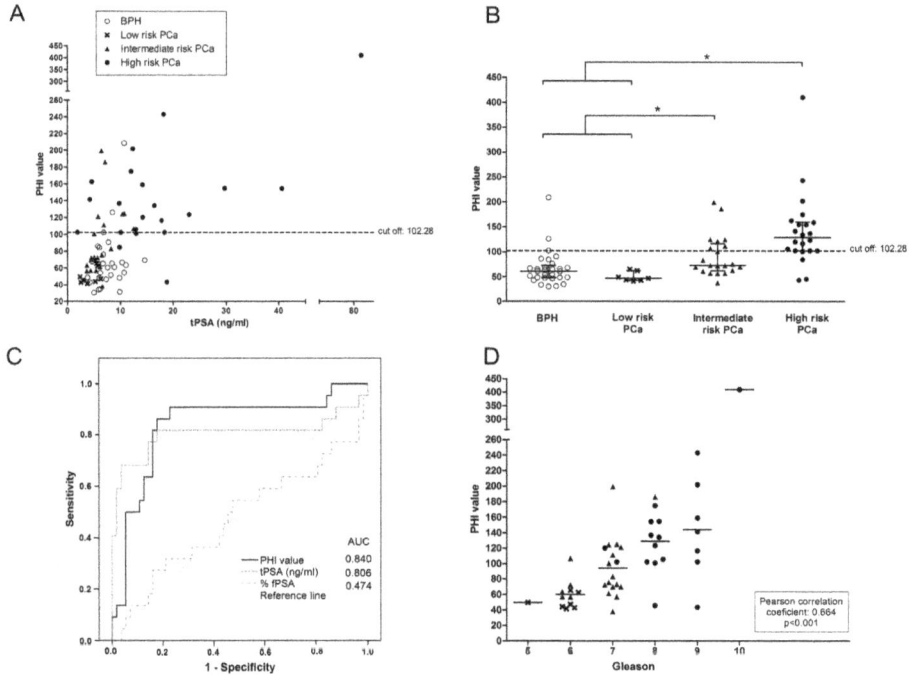

Figure 2. Prostate Health Index (PHI) values of the cohort of 79 serum samples. BPH samples are represented with an open circle (o), low risk PCa with a cross (×), intermediate risk PCa with a filled triangle (▲) and high risk PCa with a filled circle (●). (**A**) Representation of PHI value against tPSA serum levels; dotted line (- - -) shows the cutoff value for discriminating high risk PCa samples from the other three groups; (**B**) Representation of PHI value against the pathology. The center line indicates the median, and the top and bottom lines, the 75th and 25th percentiles, respectively; (**C**) Representation of the ROC curves for the PHI value, tPSA and %fPSA (**D**) Correlation plot of PHI value of the PCa samples with their Gleason score. The mean PHI value of each Gleason score is shown with a horizontal line (-).

PHI values correlated with the tPSA levels of the sample in the high-risk PCa group (correlation coefficient 0.758, $p < 0.001$), while there was no correlation for the other individual groups.

ROC analysis of the PHI score gave an AUC of 0.840 to discriminate high-risk PCa patients from the other groups, BPH and low- and intermediate-risk PCa. With a PHI cutoff of 102.28, the sensitivity was 81.8% and the specificity was 84.2%. The performance of the PHI score was higher than that of tPSA and %fPSA (Figure 2C). PHI score values showed a significant correlation with the Gleason score of the prostate tumor tissues (correlation coefficient of 0.664; $p < 0.001$) (Figure 2D).

Since PHI values of the high risk group were dependent on tPSA values, a subcohort of patients with tPSA levels lower than 13 ng/mL ($N = 67$, 28 BPH, seven low-risk, 21 intermediate-risk and 11 high-risk PCa) was evaluated. This subcohort reduced basically the number of high-risk PCa patients, which had high levels of tPSA. In this subcohort, there was no correlation of PHI values and tPSA levels within the high-risk group. The AUC of PHI in this subcohort for identifying high-risk PCa was 0.81, slightly lower than when analyzing the whole cohort.

When PHI was assayed to discriminate PCa from BPH, the AUC was of 0.735, sensitivity of 84% and specificity 45%, with a cutoff of 55.7. The diagnostic performance of PHI was higher than tPSA (AUC of 0.506) and %fPSA (AUC of 0.632), in agreement with bibliographic studies. In the subcohort of patients with tPSA levels lower than 13 ng/mL (N = 67, 28 BPH, seven low-risk, 21 intermediate-risk and 11 high-risk PCa), PHI performance for PCa diagnosing (AUC of 0.694) was still higher than tPSA (AUC of 0.382) and %fPSA (AUC of 0.630).

2.4. Combinatorial Analysis of PHI and α2,3-Sialic Acid PSA

In order to assess the performance of the combination of PHI and α2,3-sialic acid PSA, the R statistic package was used. The combination of both biomarkers showed a high performance to differentiate the high-risk PCa group from the other groups with an AUC of 0.985, much higher than PHI alone (Figure 3A,C). The combination of PHI and α2,3-sialic acid PSA also correlated with the Gleason score of the PCa patients and interestingly the two high-risk PCa patients with GS = 7 were classified correctly and were differentiated from 14 out of 15 patients of GS = 7 of the intermediate-risk PCa group (Figure 3B).

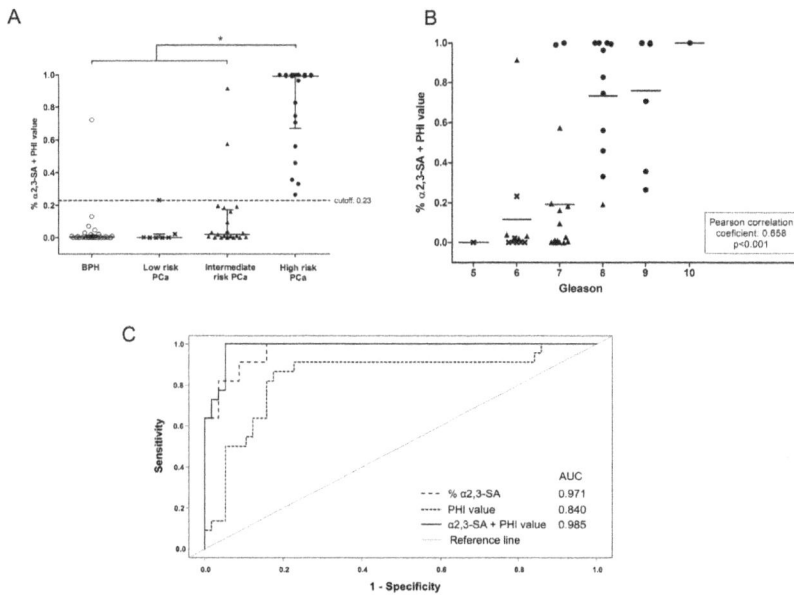

Figure 3. % α2,3-SA and PHI combination of the cohort of 79 serum samples. BPH samples are represented with an open circle (o), low risk PCa with a cross (×), intermediate risk PCa with a filled triangle (▲) and high risk PCa with a filled circle (●). (**A**) Representation of % α2,3-SA and PHI combination values against the pathology. The center line indicates the median, and the top and bottom lines, the 75th and 25th percentiles, respectively; dotted line (- - -) shows the cutoff value for discriminating high risk PCa samples from the other three groups; (**B**) Correlation plot of % α2,3-SA and PHI combination of the PCa samples with their Gleason score. The mean % α2,3-SA and PHI combination value of each Gleason score is shown with a horizontal line (-); (**C**) ROC curves for the diagnosis of high-risk PCa versus low- and intermediate-risk PCa and BPH. Diagnostic performance of % α2,3-SA and PHI combination (solid line) compared with PHI (dotted line) and % α2,3-SA (dashed line).

With the aim of implementing the combination of PHI and % α2,3-SA in clinics, an algorithm that includes both variables was developed. This consisted of a generalized lineal model (GLM) with a binomial response. After the introduction of PHI and α2,3-sialic acid percentage values, the GLM allowed to classify the patients as high-risk PCa with 100% sensitivity and 94.7% specificity. The cutoff for PHI score was 65.4 and for α2,3-sialic acid percentage of PSA was 29.94%. The model calculates the probability of a patient to be diagnosed as high-risk PCa or not (either low and intermediate-risk PCa or BPH). For a probability equal to, or higher than 23.2% (that corresponds to the point with maximum sensitivity and specificity) the patient will be classified as high-risk PCa with a sensitivity of 100% and a specificity of 94.7%. For a probability lower than 23.2% the patient will be classified either as a low-or intermediate-risk PCa, or a BPH. The probability for each patient is calculated with the following function using the patient values of PHI and α2,3-sialic acid percentage of PSA (% α2,3-SA), where β_0, β_1 and β_2 are parameters estimated by the model:

$$\text{Prob(High} - \text{riskPCa)} = \frac{e^{(\beta_0 + \beta_1 \text{PHI} + \beta_2 \% \alpha 2,3 - \text{SA})}}{1 + e^{(\beta_0 + \beta_1 \text{PHI} + \beta_2 \% \alpha 2,3 - \text{SA})}}.$$

3. Discussion

New generation of tumor markers for PCa diagnosis should be able to discriminate between patients with aggressive tumors and those without cancer or low aggressive tumors. Thus, the skills required for the new generation of markers of PCa are high sensitivity and specificity for aggressive tumors. This way, an unnecessary biopsy in men who do not have an aggressive or asymptomatic PCa could be avoided [19,28]. Early diagnosis of PCa frequently, involves the over-detection of non-aggressive tumors.

In the next future, PCa diagnosis and prognosis will probably depend on panels of biomarkers that will allow a more accurate prediction of PCa presence, stage and aggressiveness, so they will be key factors in a clinician making decisions. These markers could include serum non-invasive markers, as well as imaging markers, such as multi-parametric prostate magnetic resonance (mpMRI), which has also been proposed as a means to avoid the incidental detection of low-grade cancers [29–31].

PHI is a simple and affordable blood test that could be used as part of a multivariable approach to screening. In this sense, PHI has shown good performance for PCa diagnosis [16]. Our results are in agreement with the reported data and have shown that PHI identifies PCa from BPH with an AUC of 0.735 with higher performance than tPSA (AUC = 0.506) and %fPSA (AUC = 0.632). Since PHI has been recommended for PSA levels between 4–10 ng/mL, we examined PHI performance in the subcohort with levels of tPSA lower than 13 ng/mL and the AUC decreased to 0.694, but was still higher than tPSA (AUC = 0.382) and %fPSA (AUC = 0.630).

However, the performance of PHI in identifying high-risk PCa from the non-aggressive PCa and BPHs is much higher than for identifying PCa from BPH in both the whole cohort and the subcohort, which can be explained because PHI correlates with the Gleason score, as has also been described previously by other studies [32].

The potential of % α2,3-SA to identify high-risk PCa has been confirmed in this study. The AUC was 0.97 with a cutoff of 30%, as previously described. Interestingly, % α2,3-SA performance was not influenced by the tPSA levels of the samples, and had the same performance in the subcohort of tPSA levels lower than 13 ng/mL.

% α2,3-SA test identifies PSA glycoforms containing α2,3-sialic acid, which have been linked to PCa aggressiveness [9,10,33]. PHI score comprises other PSA isoforms linked to PCa, namely [−2]proPSA, fPSA and tPSA. In this work, we have assessed whether these different PSA forms could complement each other to better identify high-risk PCa. The combination of both markers, % α2,3-SA and PHI, has given the best performance to identify high-risk PCa, with an AUC of 0.985 (100% sensitivity, 94% specificity), although larger independent cohorts are required to validate these promising results. In this regard, the methodology to determine the percentage of α2,3-sialic acid of PSA is currently being implemented to make it more automated so that it could be used in a clinical setting.

These results highlight that the future of prostate cancer diagnosis might rely on the combination of a panel of markers based on PSA forms that can give accurate molecular diagnosis and staging and indicate the likelihood of aggressive behavior.

4. Materials and Methods

4.1. Serum Samples

The study population included 79 patients (29 BPH and 50 PCa) from Hospital Universitari Dr. Josep Trueta (Girona, Spain) between 2006 and 2013. The study was approved by the Hospital Ethics Committee (Refs. 169.06 and 023.10) and all patients provided written informed consent before being enrolled. Patients' sera were collected and stored at −80 °C. Urology and Pathology units from Hospital Universitari Dr. J. Trueta (Girona, Spain) performed the diagnosis using Transrectal Ultrasound-guided biopsy and/or adenomectomy/prostatectomy followed by pathological analysis.

The 29 BPH patients of the study (age range 44–76 years old) had a medical follow-up for a minimum of 2 years. 24 BPH patients had, at least, two negative biopsies with no evidence of high-grade Prostatic Intraepithelial Neoplasia (PIN). The 5 BPH left were subjected to prostate surgery (adenomectomy or prostate transurethral resection) and confirmed not to have prostate cancer by the Pathology Unit.

The 50 PCa patients of the study (age range 46–84 years old) were graded according to the Tumor-Node-Metastasis (TNM) classification following the general guidelines of the European Association of Urology. PCa patients were treatment naïve when serum samples were collected, except one PCa patient of the high-risk group, who was receiving hormonal therapy. High-risk PCa group comprised 22 patients with Gleason scores ≥ 8 (4 + 4) and/or with metastasis. The low-risk PCa group included 7 patients with Gleason scores of ≤ 6 (3 + 3), tPSA levels <10 ng/mL and clinical stage $\leq$pT2a. The group of intermediate-risk patients was comprised of 21 patients that did not meet the above criteria. They had Gleason scores of 7 (3 + 4 or 4 + 3) and 6 (3 + 3) and also included a patient with focal Gleason 8, tPSA levels <10 ng/mL and clinical stage pT2a considering his 10-year relapse-free survival.

The average of tPSA serum levels for BPH patients was 7.59 ng/mL (range, 3.89 to 14.47 ng/mL). The average of tPSA for the PCa groups was: 17.83 ng/ml (range, 1.96 to 87.51 ng/mL) for high-risk PCa patients, 6.44 ng/mL (range, 3.73 to 12.42 ng/mL) for intermediate-risk PCa patients, and 4.56 ng/mL (range, 2.45 to 6.33 ng/mL) for low-risk PCa patients.

4.2. Analysis of α2,3-Sialic Acid of Serum PSA

The determination of % α2,3-sialic acid of PSA was performed using a previously published method [9]. Briefly, ethanolamine 5 M was added to 0.75 mL of each serum sample to a final concentration of 1 M to release the PSA complexed to α1-antichymotrypsin. Total PSA was immunopurified using the Access Hybritech PSA assay Kit (Beckman Coulter, Brea, CA, USA). Amicon Ultra-0.5 3K Centrifugal Filter Devices (Millipore, Cork, Ireland) were used for desalting and concentrating the immunopurified tPSA samples up to a final volume of 40 μL. Samples were then applied to a lectin chromatography using *Sambucus nigra* (SNA)-agarose lectin (Vector Laboratories, Inc., Burlingame, CA, USA). Eluted unbound and bound chromatographic fractions were collected by centrifugation and quantification of free PSA of these fractions was performed using the Roche ELECSYS platform and used to determine the percentages of fPSA in the unbound fraction, corresponding to α2,3-sialic acid PSA, and in the bound fractions, which correspond to α2,6-sialic acid PSA.

4.3. Quantification of tPSA, fPSA and [−2]proPSA

Patient sera were analyzed for total PSA (tPSA), free PSA (fPSA), and [−2]proPSA on the Beckman Coulter Access 2 analyzer using WHO-standard-calibration. The Prostate Health Index (PHI) score was then calculated [PHI = ([−2]proPSA/fPSA) × $\sqrt{\text{tPSA}}$]. Assays kits used were: Hybritech total PSA assay kit (Beckman Coulter, Fullerton, CA, USA; cat. no. 37200; Lot no. 523610), Hybritech free PSA assay kit (Beckman Coulter, Fullerton, CA, USA; cat. no. 37210; Lot no. 570228) and Hybritech p2PSA assay kit (Beckman Coulter, Fullerton, CA, USA; cat. no. P090026; Lot no. 527739). Assays were performed according to the instructions of their manufacturer and calibration and control materials used in each assay where the ones recommended by the manufacturer.

4.4. Statistics

Statistical analyses of both PHI and % α2,3-SA as PCa biomarkers were performed using IBM SPSS Statistics 23 for Windows and graphics were generated with SPSS software and GraphPad Prism 5 (GraphPad Software, Inc., La Jolla, CA, USA).

Patients were classified into four groups (BPH, low-risk PCa, intermediate-risk PCa, and high-risk PCa) and Shapiro-Wilk and Levene's tests were used to assess the normality and homoscedasticity of variables. Differences of % α2,3-SA and PHI value between groups were analyzed using a Mann–Whitney U test. Receiver operating characteristic (ROC) curves were analyzed for tPSA, fPSA, % α2,3-SA, and PHI for distinguishing between high-risk PCa from the group of low-risk PCa, intermediate-risk PCa, and BPH, and also for distinguishing between PCa from BPH.

Bivariate regression (Pearson correlation) was used to analyze the correlation of % α2,3-SA and PHI with either the Gleason score or the tPSA levels.

To combine PHI and % α2,3-SA, a logistic regression was performed, in which the response variable corresponded to the probability that the event of interest was a high-risk PCa (variable taking the value 1) or the group comprising low- and intermediate-risk PCa and BPH (variable taking the value 0). An R statistical package was used to develop a generalized lineal model (GLM) with binomial response. The construction and the comparison of the AUC of the ROC curves were performed using the Epi [34,35] and pROC libraries [36].

In all these analyses, $p < 0.05$ was considered statistically significant.

Acknowledgments: This work was supported by the Spanish Ministerio de Economia y Competitividad (CDTI grant IDI20130186 and grant BIO 2015-66356-R), by the Generalitat de Catalunya, Spain (grant 2014 SGR 229), and by Roche Diagnostics (Barcelona, Spain). We thank Mireia Lopez-Siles for her support with GraphPad Prism 5.

Author Contributions: Rafael de Llorens and Rosa Peracaula conceived and designed the experiments; Montserrat Ferrer-Batallé, Esther Llop and Manel Ramírez performed the experiments; Montserrat Ferrer-Batallé, Esther Llop, Marc Saez, Josep Comet, Rafael de Llorens and Rosa Peracaula analyzed the data; Manel Ramírez and Rosa Núria Aleixandre contributed reagents/materials/analysis tools; Montserrat Ferrer-Batallé, Esther Llop, Rafael de Llorens and Rosa Peracaula wrote the paper. All of the authors read and approved the final manuscript.

Conflicts of Interest: The authors have filed a patent: "In vitro method for prostate cancer diagnosis". PCT/ES2016/070781, Priority date: 6 November 2015.

Abbreviations

PCa	Prostate cancer
PSA	Prostate Specific Antigen
PHI	Prostate Health Index
BPH	Benign Prostate Hyperplasia
AUC	Area Under the Curve
% α2,3-SA	Percentage of α2,3 sialic acid of PSA
DRE	Digital rectal examination
tPSA	Total PSA
fPSA	Free PSA
FDA	Food and Drug Administration
%fPSA	Free-to-total prostate-specific antigen ratio
PIN	Prostatic Intraepithelial Neoplasia

TNM	Tumor-Node-Metastasis
ACT	α1-Antichymotrypsin
SNA	*Sambucus nigra* lectin
CV	Coefficient of variation
mpMRI	Multi-parametric magnetic resonance imaging

References

1. Shoag, J.E.; Schlegel, P.N.; Hu, J.C. Prostate-Specific Antigen Screening: Time to Change the Dominant Forces on the Pendulum. *J. Clin. Oncol.* **2016**. [CrossRef] [PubMed]
2. Schmid, M.; Trinh, Q.D.; Graefen, M.; Fisch, M.; Chun, F.K.; Hansen, J. The role of biomarkers in the assessment of prostate cancer risk prior to prostate biopsy: Which markers matter and how should they be used? *World J. Urol.* **2014**, *32*, 871–880. [CrossRef] [PubMed]
3. Sohn, E. Screening: Diagnostic dilemma. *Nature* **2015**, *528*, S120–S122. [CrossRef] [PubMed]
4. Roobol, M. Perspective: Enforce the clinical guidelines. *Nature* **2015**, *528*, S123. [CrossRef] [PubMed]
5. Heidegger, I.; Klocker, H.; Steiner, E.; Skradski, V.; Ladurner, M.; Pichler, R.; Schafer, G.; Horninger, W.; Bektic, J. [−2]proPSA is an early marker for prostate cancer aggressiveness. *Prostate Cancer Prostatic Dis.* **2014**, *17*, 70–74. [CrossRef] [PubMed]
6. Hatakeyama, S.; Yoneyama, T.; Tobisawa, Y.; Ohyama, C. Recent progress and perspectives on prostate cancer biomarkers. *Int. J. Clin. Oncol.* **2016**, *22*, 214–221. [CrossRef] [PubMed]
7. Crawford, E.D.; Denes, B.S.; Ventil, K.H.; Shore, N. Prostate cancer: Incorporating genomic biomarkers in prostate cancer decisions. *Clin. Pract.* **2014**, *11*, 605–612. [CrossRef]
8. Li, Q.K.; Chen, L.; Ao, M.H.; Chiu, J.H.; Zhang, Z.; Zhang, H.; Chan, D.W. Serum fucosylated prostate-specific antigen (PSA) improves the differentiation of aggressive from non-aggressive prostate cancers. *Theranostics* **2015**, *5*, 267–276. [CrossRef] [PubMed]
9. Llop, E.; Ferrer-Batallé, M.; Barrabés, S.; Guerrero, P.; Ramírez, M.; Saldova, R.; Rudd, P.; Aleixandre, R.; Comet, J.; de Llorens, R.; et al. Improvement of Prostate Cancer Diagnosis by Detecting PSA Glycosylation-Specific Changes. *Theranostics* **2016**, *6*, 1190–1204. [CrossRef] [PubMed]
10. Ishikawa, T.; Yoneyama, T.; Tobisawa, Y.; Hatakeyama, S.; Kurosawa, T.; Nakamura, K.; Narita, S.; Mitsuzuka, K.; Duivenvoorden, W.; Pinthus, J.H.; et al. An Automated Micro-Total Immunoassay System for Measuring Cancer-Associated α2,3-linked Sialyl *N*-Glycan-Carrying Prostate-Specific Antigen May Improve the Accuracy of Prostate Cancer Diagnosis. *Int. J. Mol. Sci.* **2017**, *18*, 470. [CrossRef] [PubMed]
11. Sartori, D.A.; Chan, D.W. Biomarkers in prostate cancer: What's new? *Curr. Opin. Oncol.* **2014**, *26*, 259–264. [CrossRef] [PubMed]
12. Mikolajczyk, S.D.; Millar, L.S.; Wang, T.J.; Rittenhouse, H.G.; Marks, L.S.; Song, W.; Wheeler, T.M.; Slawin, K.M. A precursor form of prostate-specific antigen is more highly elevated in prostate cancer compared with benign transition zone prostate tissue. *Cancer Res.* **2000**, *60*, 756–759. [PubMed]
13. Stephan, C.; Meyer, H.A.; Paul, E.M.; Kristiansen, G.; Loening, S.A.; Lein, M.; Jung, K. Serum (-5, -7) proPSA for distinguishing stage and grade of prostate cancer. *Anticancer Res.* **2007**, *27*, 1833–1836. [PubMed]
14. Hori, S.; Blanchet, J.S.; McLoughlin, J. From prostate-specific antigen (PSA) to precursor PSA (proPSA) isoforms: A review of the emerging role of proPSAs in the detection and management of early prostate cancer. *BJU Int.* **2013**, *112*, 717–728. [CrossRef] [PubMed]
15. Lazzeri, M.; Lughezzani, G.; Haese, A.; McNicholas, T.; de la Taille, A.; Buffi, N.M.; Cardone, P.; Hurle, R.; Casale, P.; Bini, V.; et al. Clinical performance of prostate health index in men with tPSA >10 ng/ml: Results from a multicentric European study. *Urol. Oncol.* **2016**, *34*, 415. [CrossRef] [PubMed]
16. Loeb, S.; Catalona, W.J. The Prostate Health Index: A new test for the detection of prostate cancer. *Ther. Adv. Urol.* **2014**, *6*, 74–77. [CrossRef] [PubMed]
17. Wang, W.; Wang, M.; Wang, L.; Adams, T.S.; Tian, Y.; Xu, J. Diagnostic ability of %p2PSA and prostate health index for aggressive prostate cancer: A meta-analysis. *Sci. Rep.* **2014**, *4*, 5012. [CrossRef] [PubMed]
18. Loeb, S. Time to replace prostate-specific antigen (PSA) with the Prostate Health Index (PHI)? Yet more evidence that the phi consistently outperforms PSA across diverse populations. *BJU Int.* **2015**, *115*, 500. [CrossRef] [PubMed]

19. Morote, J.; Celma, A.; Planas, J.; Placer, J.; Ferrer, R.; de Torres, I.; Pacciuci, R.; Olivan, M. Diagnostic accuracy of prostate health index to identify aggressive prostate cancer. An Institutional validation study. *Actas Urol. Esp.* **2016**, *40*, 378–385. [CrossRef] [PubMed]

20. Peracaula, R.; Tabares, G.; Royle, L.; Harvey, D.J.; Dwek, R.A.; Rudd, P.M.; de Llorens, R. Altered glycosylation pattern allows the distinction between prostate-specific antigen (PSA) from normal and tumor origins. *Glycobiology* **2003**, *13*, 457–470. [CrossRef] [PubMed]

21. Ohyama, C.; Hosono, M.; Nitta, K.; Oh-eda, M.; Yoshikawa, K.; Habuchi, T.; Arai, Y.; Fukuda, M. Carbohydrate structure and differential binding of prostate specific antigen to Maackia amurensis lectin between prostate cancer and benign prostate hypertrophy. *Glycobiology* **2004**, *14*, 671–679. [CrossRef] [PubMed]

22. Tabares, G.; Radcliffe, C.M.; Barrabes, S.; Ramirez, M.; Aleixandre, R.N.; Hoesel, W.; Dwek, R.A.; Rudd, P.M.; Peracaula, R.; de Llorens, R. Different glycan structures in prostate-specific antigen from prostate cancer sera in relation to seminal plasma PSA. *Glycobiology* **2006**, *16*, 132–145. [CrossRef] [PubMed]

23. Tajiri, M.; Ohyama, C.; Wada, Y. Oligosaccharide profiles of the prostate specific antigen in free and complexed forms from the prostate cancer patient serum and in seminal plasma: A glycopeptide approach. *Glycobiology* **2008**, *18*, 2–8. [CrossRef] [PubMed]

24. Meany, D.L.; Zhang, Z.; Sokoll, L.J.; Zhang, H.; Chan, D.W. Glycoproteomics for prostate cancer detection: Changes in serum PSA glycosylation patterns. *J. Proteom. Res.* **2009**, *8*, 613–619. [CrossRef] [PubMed]

25. Sarrats, A.; Saldova, R.; Comet, J.; O'Donoghue, N.; de Llorens, R.; Rudd, P.M.; Peracaula, R. Glycan characterization of PSA 2-DE subforms from serum and seminal plasma. *OMICS* **2010**, *14*, 465–474. [CrossRef] [PubMed]

26. Sarrats, A.; Comet, J.; Tabares, G.; Ramirez, M.; Aleixandre, R.N.; de Llorens, R.; Peracaula, R. Differential percentage of serum prostate-specific antigen subforms suggests a new way to improve prostate cancer diagnosis. *Prostate* **2010**, *70*, 1–9. [CrossRef] [PubMed]

27. Yoneyama, T.; Ohyama, C.; Hatakeyama, S.; Narita, S.; Habuchi, T.; Koie, T.; Mori, K.; Hidari, K.I.; Yamaguchi, M.; Suzuki, T.; et al. Measurement of aberrant glycosylation of prostate specific antigen can improve specificity in early detection of prostate cancer. *Biochem. Biophys. Res. Commun.* **2014**, *448*, 390–396.

28. Mohammed, A.A. Biomarkers in prostate cancer: New era and prospective. *Med. Oncol.* **2014**, *31*, 140.

29. Leapman, M.S.; Carroll, P.R. What is the best way not to treat prostate cancer? *Urol. Oncol.* **2017**, *35*, 42–50.

30. Vilanova, J.C.; Barcelo-Vidal, C.; Comet, J.; Boada, M.; Barcelo, J.; Ferrer, J.; Albanell, J. Usefulness of prebiopsy multifunctional and morphologic MRI combined with free-to-total prostate-specific antigen ratio in the detection of prostate cancer. *AJR Am. J. Roentgenol.* **2011**, *196*, W715–W722. [CrossRef] [PubMed]

31. Polascik, T.J.; Passoni, N.M.; Villers, A.; Choyke, P.L. Modernizing the diagnostic and decision-making pathway for prostate cancer. *Clin. Cancer Res.* **2014**, *20*, 6254–6257. [CrossRef] [PubMed]

32. Stephan, C.; Vincendeau, S.; Houlgatte, A.; Cammann, H.; Jung, K.; Semjonow, A. Multicenter evaluation of [-2]proprostate-specific antigen and the prostate health index for detecting prostate cancer. *Clin. Chem.* **2013**, *59*, 306–314. [CrossRef] [PubMed]

33. Kosanovic, M.M.; Jankovic, M.M. Sialylation and fucosylation of cancer-associated prostate specific antigen. *J. BUON* **2005**, *10*, 247–250. [PubMed]

34. Hills, M.; Cartensen, B.; Plummer, M. Follow-Up with the Epi Package. 2009. Available online: http://bendixcarstensen.com/Epi/Follow-up.pdf (accessed on 20 February 2017).

35. Cartensen, B.; Plummer, M.; Laara, E.; Hills, M. Epi: A Package for Statistical Analysis in Epidemiology, R Package Version 2.10. 2017. Available online: https://cran.r-project.org/web/packages/Epi/Epi.pdf (accessed on 20 February 2017).

36. Robin, X.; Turck, M.; Hainard, A.; Tiberti, N.; Lisacek, F.; Sanchez, J.C.; Müller, M.; Siegert, S. Display and Analyze ROC Curves, Package "pROC" Version 1.9.1. 2017. Available online: https://cran.r-project.org/web/packages/pROC/pROC.pdf (accessed on 20 February 2017).

International Journal of
Molecular Sciences

MDPI

Article

A Panel of MicroRNAs as Diagnostic Biomarkers for the Identification of Prostate Cancer

Rhonda Daniel [1,†], Qianni Wu [1,†], Vernell Williams [2], Gene Clark [1], Georgi Guruli [3] and Zendra Zehner [1,*]

[1]   Department of Biochemistry and Molecular Biology, VCU Medical Center and the Massey Cancer Center,
      Virginia Commonwealth University, Richmond, VA 23298-0614, USA; danielr@vcu.edu (R.D.);
      wuq3@vcu.edu (Q.W.); clarkgc@mymail.vcu.edu (G.C.)
[2]   Molecular Diagnostic Laboratory, Department of Pathology, VCU Health System,
      Virginia Commonwealth University, Richmond, VA 23298-0248, USA; Vernell.Williamson@vcuhealth.org
[3]   Division of Urology, VCU Medical Center and the Massey Cancer Center,
      Virginia Commonwealth University, Richmond, VA 23298-0037, USA; Georgi.guruli@vcuhealth.org
*   Correspondence: zendra.zehner@vcuhealth.org; Tel.: +1-804-304-2411
†   These authors contributed equally to this work.

Received: 4 April 2017; Accepted: 13 June 2017; Published: 16 June 2017

Abstract: Prostate cancer is the most common non-cutaneous cancer among men; yet, current diagnostic methods are insufficient, and more reliable diagnostic markers need to be developed. One answer that can bridge this gap may lie in microRNAs. These small RNA molecules impact protein expression at the translational level, regulating important cellular pathways, the dysregulation of which can exert tumorigenic effects contributing to cancer. In this study, high throughput sequencing of small RNAs extracted from blood from 28 prostate cancer patients at initial stages of diagnosis and prior to treatment was used to identify microRNAs that could be utilized as diagnostic biomarkers for prostate cancer compared to 12 healthy controls. In addition, a group of four microRNAs (miR-1468-3p, miR-146a-5p, miR-1538 and miR-197-3p) was identified as normalization standards for subsequent qRT-PCR confirmation. qRT-PCR analysis corroborated microRNA sequencing results for the seven top dysregulated microRNAs. The abundance of four microRNAs (miR-127-3p, miR-204-5p, miR-329-3p and miR-487b-3p) was upregulated in blood, whereas the levels of three microRNAs (miR-32-5p, miR-20a-5p and miR-454-3p) were downregulated. Data analysis of the receiver operating curves for these selected microRNAs exhibited a better correlation with prostate cancer than PSA (prostate-specific antigen), the current gold standard for prostate cancer detection. In summary, a panel of seven microRNAs is proposed, many of which have prostate-specific targets, which may represent a significant improvement over current testing methods.

Keywords: microRNA; high throughput RNA sequencing; small RNA sequencing; qRT-PCR; prostate cancer; PSA

1. Introduction

Prostate Cancer (PCa) is the most common non-cutaneous cancer among men, yet current diagnostic methods are insufficient at detecting this disease, and more reliable biomarkers need to be developed. Currently, the prostate-specific antigen (PSA) is used as a diagnostic marker for PCa; however, many factors have been found to elevate PSA levels. Age, infection, trauma, ejaculation, urinary retention, instrumentation, certain medications and even bike riding can lead to false positive diagnoses, generating unnecessary concern and over-treatment with dire outcomes for the patient [1–4]. Even worse are the chances of false negative diagnoses, which result in PCa remaining undetected until its later stages. Therefore, although the use of the PSA level has had its clinical advantages, it has failed to sufficiently bridge the gap to accurately diagnose disease or distinguish indolent from aggressive disease. One answer that might close this gap and enable more efficient diagnoses may lie in microRNAs (miRs) [5].

Small RNAs play an extremely important role in gene regulation. Their function in the suppression of unwanted genetic materials is vital to the proper operation of the cell. Small RNAs fall into three classifications: microRNAs, siRNA and PIWI-interacting RNAs (piRNA), the most dominating of which are microRNAs [6]. MicroRNAs are small non-coding RNA molecules (18–22 nts in length) that are evolutionarily conserved and associated with the Argonaute family of proteins. These microRNAs function at the translational level through silencing mechanisms to regulate gene expression.

MicroRNAs have been shown to be significantly altered throughout the course of disease progression [7]. This is especially true in cancer where abnormal cell growth and angiogenesis are critical for tumorigenesis to occur. The loss of microRNAs that suppress the translation of oncogenes, termed tumor suppressors, has been shown to contribute to the development and progression of many cancers [7]. These microRNAs are primarily responsible for controlling apoptotic pathways and cell cycle checkpoints [7].

Since the discovery of microRNAs, many research groups have analyzed blood in hopes of establishing a correlation to disease. Mitchell et al. first reported that PCa cells released microRNAs into the bloodstream in protective capsules, the content of which could be monitored by PCR-based methods [8]. Schultz et al. studied whole blood for the identification of microRNAs that could be used as biomarkers for the detection of pancreatic cancer [9]. By confirmatory qRT-PCR, they found 38 microRNAs dysregulated and were able to identify two diagnostic microRNA panels that could distinguish between patients with pancreatic cancer from healthy controls [9]. Another study compared microRNA levels between plasma and serum from PCa patients by measuring four microRNAs: hsa-miR-15b, hsa-miR-16, hsa-miR-19b and hsa-miR-24. Interestingly, they found a strong correlation in the microRNA content of these two types of body fluids supporting either serum or plasma as a sufficient source of material for disease studies [8]. Using qRT-PCR, Cochetti et al. suggested a panel of serum microRNAs that could distinguish PCa from benign prostatic hyperplasia in age-matched patients with elevated PSA levels [10]. Thus, the use of blood, serum or plasma as a worthwhile source of material to diagnose disease is well documented [8–10].

To date, a number of studies have used PCR technology to identify microRNAs that could be used as relevant biomarkers to diagnose PCa [11,12]. Certainly, these are important studies, but for the most part, they have used preformed panels of microRNA arrays or focused qRT-PCR assays for specific microRNAs suggested from studying a wide range of different cancers and then applied to PCa. By this approach, only predetermined, known microRNAs are being evaluated. In an effort to widen the scope of microRNA candidates, high throughput sequencing (HTS), also referred to as deep sequencing or RNA sequencing, would better evaluate all possible microRNAs, as well as permitting the discovery of new, novel microRNAs. Keller et al. used HTS of whole blood samples collected with PAXgene blood tubes to study microRNA profiles in lung cancer patients [13]. However, in this case, samples were pooled prior to sequencing, thereby preventing an analysis of microRNA dysregulation across individual samples. To our knowledge, only two reports have used HTS to identify microRNAs diagnostic for PCa. In one case, HTS was used to compare the microRNA content of prostate tumors to adjacent tumor-free margins with the discovery of a loss of miR-143 and miR-145 expression in tumor tissues [14]. A second report applied HTS to exosomal material isolated from blood and found miR-1290 and miR-375 as prognostic markers for castration-resistant prostate cancer (CRPC) [15]. However, in this case, these microRNAs would be useful for identifying late stage prostate cancers.

To better define microRNAs that could be used to more accurately predict PCa at early, not later stages of disease, in this pilot study, we have used HTS of blood from PCa patients at initial stages of diagnosis and before undergoing treatment compared to healthy controls. Moreover, samples were analyzed individually rather than as a pool so that variability between patients or control samples could be followed. RNA sequencing results were also analyzed to identify normalization microRNAs that could be used as endogenous controls for subsequent qRT-PCR analyses. Confirmatory qRT-PCR was then used to corroborate HTS results for the top seven dysregulated microRNAs. Data analysis of the area under the curve (AUC) of the receiver operating curves (ROC) for these selected microRNAs exhibited a better correlation with prostate cancer (AUC range = 0.819–0.950) than the reported value for PSA (AUC 0.678 comparing PCa to non-cancer) [16]. In summary, a panel of seven microRNAs is proposed, many of which have prostate-specific targets, which upon follow-up confirmatory studies could represent a significant improvement over current testing methods.

2. Results

2.1. High Throughput Sequencing Results

A summary of the characteristics and pathological data for patients (n = 28) and controls (n = 12) selected for this study is compared in Table 1. Data for each individual can be found in Appendix A Table A1. Blood was retrieved from patients at early stages of diagnosis and prior to treatment. For most cases, age, ethnicity, PSA and Gleason scores obtained from biopsy were reported. The Gleason score was obtained by microscopic analysis by a trained pathologist and is the combined score of the most common and second most abundant cell type based on cell morphology. When the Gleason score or PSA were not available, it is designated as unknown. Although some mix of ethnicity was obtained, Caucasian was most prevalent with no ethnicity or age recorded for nine individuals. Low Gleason scores of G6 and G7 and PSA values ranging from 3.4–22 predominated, since samples were taken from patients at early stages of diagnosis. Although Gleason scores were not reported for four patients, elevated PSA levels including the high of 22 was found within this group supporting their inclusion to analyze as many samples as possible in this pilot study. Every effort was made to select a control group that had no evidence of PCa either for the individual or within the family. PCa being predominately a disease of the elderly, the average age of the patient group did exceed that of the controls, but since all data were analyzed as individuals, we could subsequently evaluate differences within each group. In this case, we did not find notable discrepancies in data within either the patient or control group due to age or the group of four with elevated PSA values, but unknown Gleason scores, further supporting their inclusion in this study.

Table 1. Characteristics and pathological data of patients and controls involved in the study.

Characteristics	PCa (n = 28)	Controls (n = 12)
AGE (years)		
Range Age	55–92 (n = 19)	23–91 (n = 12)
Mean Age	65.9	50
Unknown	9	0
ETHINICITY (race)		
Caucasian	12	9
African American	6	1
Asian/Hawaiian	1	1
Unknown	9	0
PSA (Prostate Specificity Antigen)		
Range	3.2–22 (n = 19)	–
Mean	7.39	–
Elevated	n = 3	–
Unknown	n = 6	–
PATHOLOGY (Gleason Score)		
G6	n = 9	–
G7	n = 11	–
G8	n = 2	–
G9	n = 2	–
Unknown	n = 4	–

Blood was collected and small RNAs extracted from individual patient and control samples as described in the Materials and Methods. The HTS data revealed that among the 2588 microRNAs present in the miRBase (mature 21 June 2014) [17], about 550 were found at detectable levels in the samples tested. To better refine this list of potential candidates, *p*-values were adjusted using the Benjamini-Hochberg method to yield a False Detection Rate, or FDR value. The FDR value indicates the possible false detection rate using a generalized linearization model. This method is considered a Type 1 error expansion multiple comparison model that reduces the risk of rejecting a true null hypothesis. In order to include as many positive hits as possible in the HTS screening, a cutoff FDR value of <0.2 was selected. An FDR value of 0.2 would mean that 20% of selected microRNAs may be false positives. Since all HTS results would be subsequently confirmed by qRT-PCR, it was felt that lowering the stringency to include more potential microRNA candidates for future confirmation was acceptable at this initial stage. In fact, lowering the stringency of this selection generated a list of 10 possible dysregulated microRNAs for future study (Table 2). Subsequently, miR-5582-3p and miR-543 were dropped because there were no manufactured primers readily available in the market, and their abundance was low. In addition, miR-500b-3p was also dropped due to its low abundance. Thus, seven microRNAs were chosen for future analysis.

Table 2. HTS differential expression analysis the top 10 dysregulated miRNA candidates.

MicroRNA	LogCPM	LogFC	*p*-Value	FDR
miR-5582-3p	0.861	−2.477	2.36×10^{-6}	0.001
miR-32-5p	2.436	−2.036	1.23×10^{-5}	0.003
miR-500b-3p	1.141	2.035	5.77×10^{-4}	0.105
miR-329-3p	1.760	2.096	1.11×10^{-3}	0.132
miR-487b-3p	0.854	2.596	1.20×10^{-3}	0.132
miR-454-3p	5.239	−0.933	1.50×10^{-3}	0.138
miR-204-5p	1.646	1.781	1.93×10^{-3}	0.151
miR-20a-5p	9.049	−1.085	2.97×10^{-3}	0.167
miR-127-3p	4.337	1.433	3.16×10^{-3}	0.167
miR-543	3.353	1.359	3.48×10^{-3}	0.167

LogFC = Log2 comparing patients to controls. FDR = false detection rate.

During the bioinformatics analysis, the HTS total reads for patients and controls were not significantly different from each other, suggesting that blood from normal and patient groups contained similar amounts of total microRNA (Figure 1). This similarity increased the confidence of dysregulation, as it could be confirmed that the differential expression of certain microRNAs was not due to differences in library size.

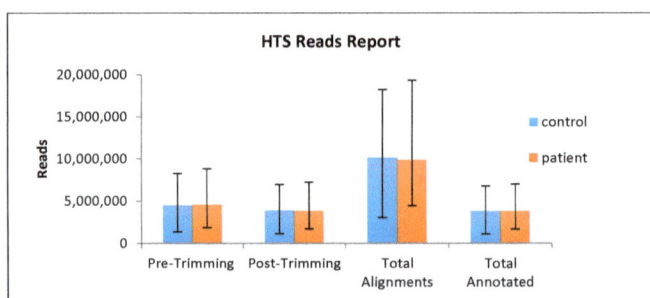

Figure 1. Analysis of HTS reads in blood from PCa patients and controls. HTS was performed on a total number of 40 samples (28 patients and 12 controls) with an RNA concentration of 100 ng/μL. Total reads are shown before and after the Partek Flow® (St. Louis, MO, USA) process.

Processed raw reads were further normalized using the Trimmed Mean of M-values (TMM) method provided by the Edge R program [18]. Based on the hypothesis that most genes are not differentially expressed, the TMM method generates a scaling factor applied to library sizes, which attempts to minimize the intra-group variation in gene expression. This normalization method can further minimize the effect of technical variations caused by sequencing depth and batch variation. The TMM method is a very powerful method when varying library size, and high-count genes can exist [19]. Compared to the commonly-used normalization methods of Total Counts (TC) or Reads Per Kilobase per Million mapped reads (RPKM), the TMM method is more reliable, because it not only normalizes the library size, but also takes into account the effect of RNA composition [18]. The effectiveness of the TMM method in normalizing the microRNA sequencing results was later confirmed via qRT-PCR.

The seven dysregulated microRNAs indicated by HTS differential expression analysis showed great differences in normalized reads between control and patient groups (Figure 2a–g). According to the HTS data, four microRNAs were upregulated (miR-127-3p, miR-204-5p, miR-329-3p and miR-487b-3p) in patients' blood samples (Figure 2a–d), while three microRNAs (miR-32-5p, miR-20a-5p and miR-454-3p) were downregulated (Figure 2e–g).

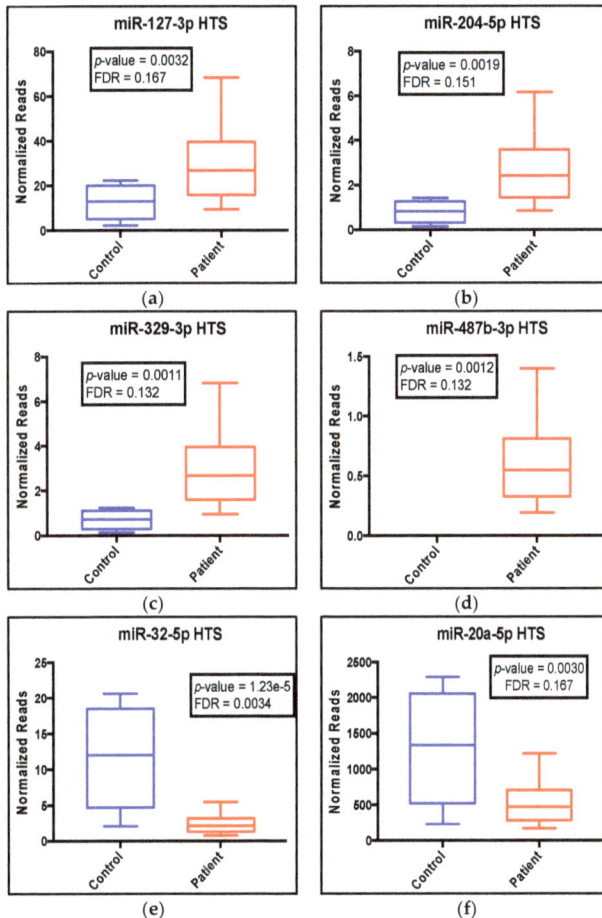

Figure 2. *Cont.*

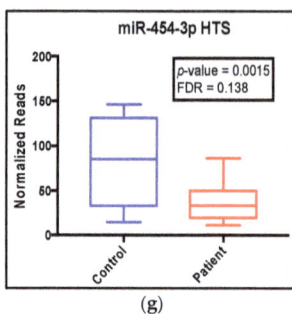

(g)

Figure 2. HTS data show dysregulation of seven microRNAs in blood from PCa patients (red) compared to controls (blue). (**a–g**) Box plots of the top seven dysregulated microRNAs as indicated on each panel are based on Edge R differential expression analysis. *p*-Values and FDR values were generated by Edge R using the generalized linear method. Box-and-whiskers graphs were plotted using Prism. The minimum, the 25th percentile, the median, the 75th percentile and the maximum are shown on each box plot as the bottom to the top lines, respectively. An FDR < 0.2 was considered significant.

2.2. Identification of MicroRNAs as Normalization Standards for qRT-PCR

In order to confirm HTS data by qRT-PCR, a normalization method needed to be developed to ensure that microRNA dysregulation was due to true biological variation and not technical error. Ideally, microRNA normalizers should exhibit small standard deviations and display similar expression levels to the dysregulated microRNAs under study. To this end, the concentration of small RNAs in each sample was determined from bioanalyzer results, set to a constant amount, and the Cq value determined by qRT-PCR. Results were analyzed using the NormFinder program, which scrutinizes intra- and inter-group variations to determine which microRNA candidates are best suited for normalization using an algorithm to calculate a stability value for each microRNA, i.e., the lower the value, the lower the variation.

The NormFinder program selected eight microRNAs as exhibiting stable expression patterns; miR-146a-5p, miR-1538, miR-197-3p, miR-1468-3p, miR-26b-5p, miR-296-5p, miR-1248 and miR-23a-3p (Figure 3a; raw data Appendix A Table 2). Due to limiting amounts of material, the search for potential microRNA normalizers was initially monitored in a subset of samples, i.e., eight patients and eight controls. Of these, the first four candidates, which exhibited the closest stability values (ranging from 0.009–0.0016) with minimal differences in expression between control and patient samples, were subsequently analyzed in a fuller spectrum of samples (26 patients and 10 controls). Unfortunately, two samples from each group of HTS data had to be dropped from further analysis due to a lack of material (Figure 3b; raw data Appendix A Table 3). NormFinder suggested that the single best candidate was miR-146a-5p. However, when there is no obvious single, outstanding normalization candidate, NormFinder suggests using a combination of microRNAs to increase reliability and produce less intra- and inter-group variability. Since the top four microRNAs (miR-146a-5p, miR-1538, miR-197-3p, miR-1468-3p) all showed very close stability values, the Cq value of each was compared to each other, as well as the geometric mean of the top two candidates (miR-146-5p and miR-1538) or all four top candidates together (Figure 3b). The top two candidates showed decreased intra-group variation, especially for the patient group (Figure 3b). However, the geometric mean of all four candidates together showed even smaller intra- and inter-variations within and between both control and patient groups (Figure 3b). Therefore, these four microRNAs were selected as a group of normalizers to be used for downstream qRT-PCR analyses.

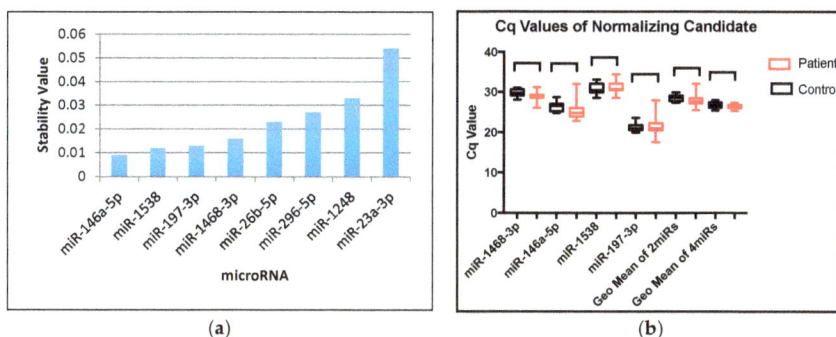

(a)　　　　　　　　　　　　　(b)

Figure 3. Analysis by the NormFinder program identified four microRNAs as the best normalization candidates for qRT-PCR studies. (**a**) Eight stably-expressed microRNAs (miR-146a-5p, miR-1538, miR-197-3p, miR-1468-3p, miR-26b-5p, miR-296-5p, miR-1248 and miR-23a-3p) suggested by HTS data were confirmed by qRT-PCR in triplicate (eight controls and eight patients). The small RNA concentration for each sample was normalized to roughly 0.012 ng/μL in each reaction. A stability value was generated for each candidate by the NormFinder program, where the lower the value, the better; (**b**) The Cq value of the top four microRNA candidates (miR-1468-3p, miR-146a-5p, miR-1538, miR-197-3p) was subsequently evaluated in 26 patients and 10 controls and plotted individually as box plots versus the geometric (Geo) mean of two candidates (miR-146a-5p and miR-1538) or four candidates (miR-146a-5p, miR-1538, miR-197-3p and miR-1468-3p) as analyzed in triplicate by qRT-PCR.

2.3. Validation of HTS Data by qRT-PCR Analysis

The elucidation of valid microRNA normalizers permitted further analysis of HTS results via qRT-PCR. The individual dot plots of dCq (ΔCq) values are shown for the seven dysregulated miRs suggested by HTS data (Figure 4). According to the qRT-PCR results, miR-127-3p, miR-204-5p, miR-329-3p and miR-487b-3p were all upregulated in patients compared to controls (Figure 4a–d), while miR-32-5p, miR-20a-5p and miR-454-3p were downregulated (Figure 4e–g) The differences in expression in control versus patient samples was calculated for each microRNA as the –ddCq (Log2 fold change) and shown in Figure 4h. Raw and normalized Cq values for controls and patients are included in Appendix A Tables 3 and 4, respectively. Thus, the qRT PCR results agreed with the HTS data, confirming that all seven microRNAs were dysregulated in PCa patients.

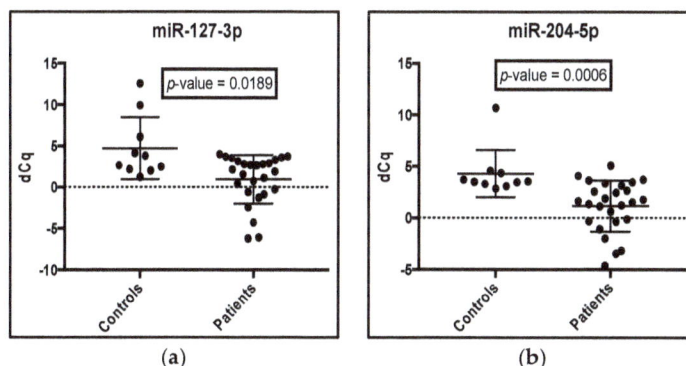

(a)　　　　　　　　　　　　　(b)

Figure 4. *Cont.*

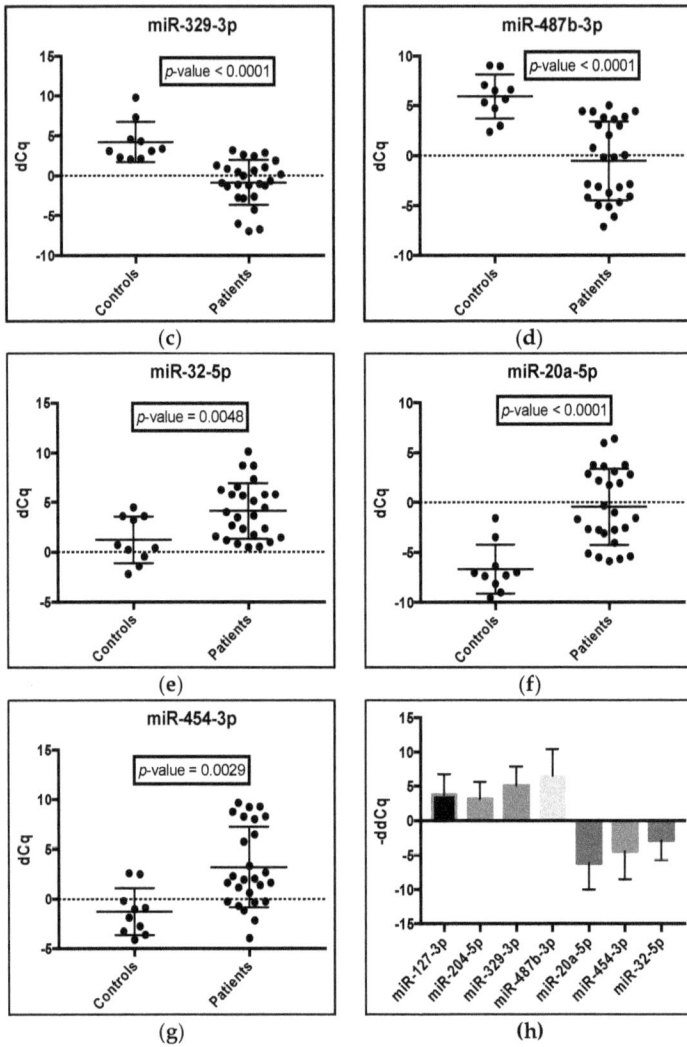

Figure 4. Confirmatory qRT-PCR results for dysregulated miRNA candidates suggested by HTS data. (**a–g**) A comparison between normalized Cq values (dCq) from qRT-PCR analysis of blood from patients and controls and plotted as dot blots. qRT-PCR was performed on 36 samples (10 controls and 26 patients) in triplicate. Samples were adjusted to the same small RNA concentration (0.012 ng/μL) per reaction. Raw Cq values were normalized by subtracting the geometric mean Cq value of the top four normalization candidates (miR-146a-5p, miR-1538, miR-197-3p and miR-1468-3p) suggested by the NormFinder program from individual Cq values to generate dCq. A p-value was obtained using the Mann–Whitney nonparametric test assuming that data do not follow a Gaussian distribution. A p-value < 0.05 was considered significant. The minimum, median and maximum values are shown as respective lines from the bottom to the top; (**h**) The −ddCq values of the seven dysregulated microRNAs are shown. The −ddCq for each candidate was obtained by taking the mean of the normalized dCq of all controls minus the normalized dCq of each patient sample. This value equals the fold change on a Log2 scale.

In order to further assess whether the seven microRNAs could serve as good biomarkers, Receiver Operator Curves (ROC) were drawn based on the qRT-PCR data. ROC analysis demonstrates the trade-off between sensitivity and specificity where a good biomarker should display both high sensitivity and high specificity [20]. The ROC curve for each microRNA is shown in Figure 5. In ROC analysis, the Area Under the Curve (AUC) quantifies the biomarker potential for each candidate where the higher the AUC value, the better a candidate microRNA is at distinguishing PCa patients from controls. Via ROC analysis, the currently used PCa biomarker, PSA, has a reported AUC value of 0.678 for distinguishing PCa from no cancer [16]. The seven microRNAs identified in our study exhibited a respectable range of AUC values from 0.7538 for miR-127-3p up to 0.9462 for miR-329-3p, all significantly better than that reported for PSA, with *p*-values ranging from 1.9435×10^{-6} to 0.0094 (Figure 5a–g).

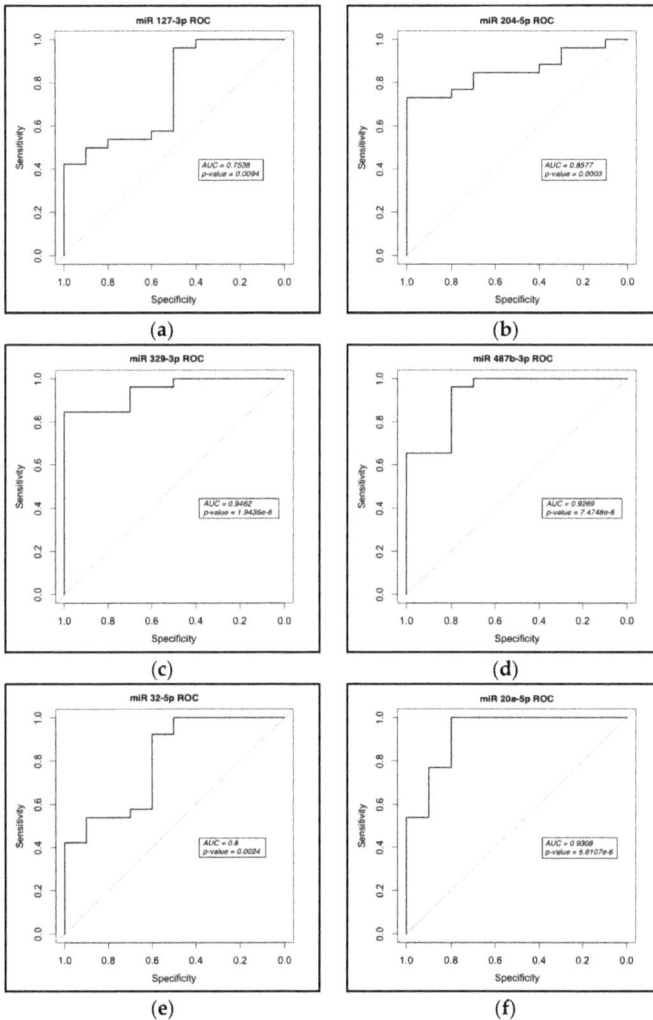

Figure 5. *Cont.*

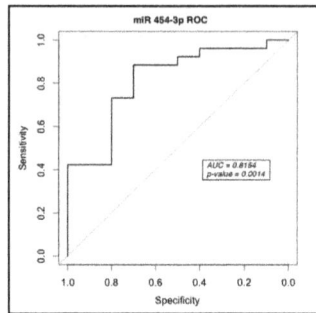

(g)

Figure 5. Receiver operator curves for dysregulated microRNAs. (**a–g**) Analysis was performed based on the qRT-PCR results in triplicate of individual microRNAs as indicated on each graph and plotted as sensitivity versus specificity. An AUC > 0.5 is considered significant.

2.4. Comparison of Blood Results to TCGA Database

The expression of our panel of blood microRNAs was compared to expression levels in tumor tissue by our analysis of data in The Cancer Genome Atlas (TCGA) database. Although the TCGA microRNA sequencing data were annotated with the stem-loop transcripts instead of the mature strands, all seven microRNAs from our study derive from the major expressed mature strand of their stem-loop precursor based on data in miRBase [17]. Therefore, the expression of these seven mature microRNAs is directly proportional to the abundance of their stem-loop precursors. The mature miR-127-3p, miR-204-5p, miR-487b-3p, miR-32-5p, miR-20a-5p and miR-454-3p are derived from precursors miR-127, miR-204, miR-487b, miR-32, miR-20a and miR-454, respectively. The mature miR-329-3p was derived from two precursors, miR-329-1 and miR-329-2.

The expression of each precursor in PCa tissues compared to their disease-free matched margins showed significant dysregulation, and the direction of dysregulation agreed with the literature results (Figure 6). However, the pattern of dysregulation for each microRNA in tumor tissue was opposite to that pattern observed in our blood samples. For example, miR-127, miR-204, miR-329-1, miR-329-2 and miR-487b were all upregulated in PCa tissue, which suggested that their major mature strands (miR-127-3p, miR-204-5p, miR-329-3p and miR-487b-3p) were also upregulated. However, these four microRNAs were shown to be downregulated in our blood samples. The inverse correlation was observed for the three microRNAs (miR-32-5p, miR-20a-5p and miR-454-3p) that are downregulated in blood. Again, their precursor transcripts and presumably major, mature microRNA products were upregulated in the TCGA tissue data. A comparison of the fold changes between our HTS blood data versus that from the TCGA database are included in the Appendix A (Table 5).

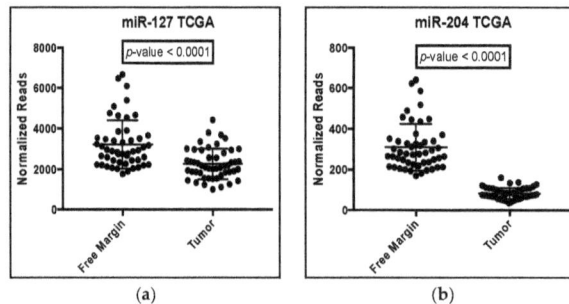

(a)

(b)

Figure 6. *Cont.*

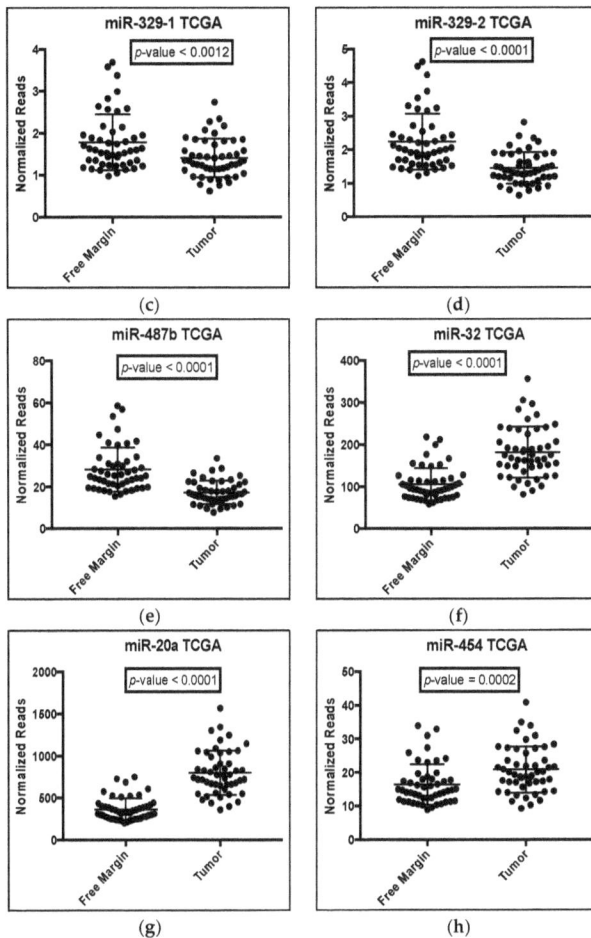

Figure 6. An analysis of microRNA sequencing results from the TGCA matched tissue database for the seven microRNA candidates. (**a–h**) Reads for each microRNA candidate as indicated were normalized using the Edge R TMM method and plotted as a dot blot with a line (bottom to top, respectively) representing the minimum, median (or mean) and maximum value for the tumor versus the disease-free matched tissue (free-margin) from PCa patients. A *p*-value was obtained using the Mann–Whitney nonparametric test assuming that that data do not follow a Gaussian distribution. A $p < 0.05$ was considered significant. TCGA, The Cancer Genome Atlas.

3. Discussion

Analysis of HTS sequencing results suggested a panel of seven microRNAs that could be useful in diagnosing PCa in blood. Previously, the lack of reliable microRNA standards for normalization across different samples had been detrimental to subsequent qRT-PCR validation studies. In some studies, snRNAs have been used for this purpose, but since these are not normally secreted and are not produced by pathways that correlate with microRNA synthesis, their use as normalizers for complex body fluids such as blood is questionable. A review of our HTS data selected four microRNAs (miR-197-3p or -5p, miR-1538, miR-1468-3p and miR-146a-5p) that were consistently expressed across all patient and control samples and could be used as reliable normalization standards for future qRT-PCR studies. Kirschner et al. also found miR-146a to be stably expressed in plasma and serum, not affected by hemolysis, in agreement with our results in blood [21]. Moreover, the enhanced geometric mean of these microRNAs was shown to be significantly better than any single microRNA alone.

With a proven group of normalization standards, it was important to confirm HTS results via qRT-PCR. Significantly, results from these two very different methodologies agreed well, further supporting the validity of our approach. All of the microRNAs with low *p*- and FDR-values via HTS data showed significant *p*-values with qRT-PCR and notable AUC values upon ROC analysis. This agreement was encouraging because thus far, investigators had been determining diagnostic microRNAs by screening of pre-selected microRNA arrays, which represented only a small subset of microRNAs from the database of >2580 total microRNAs [17]. This approach is limited to analyzing only those microRNAs that have already been shown to be dysregulated in some disease and then selected for analyzing PCa. However, HTS permits the identification of all possible diagnostic miRNAs, both known and perhaps novel, expanding the spectrum of microRNA candidates evaluated. Interestingly, a few novel microRNA species were identified, but these always turned out to be a single report; thus, their relevance as a "new" molecule warranting further verification was hard to justify in this pilot study due to their low abundance. More importantly, HTS results were validated by qRT-PCR for all seven candidates generating ROC curves with individual AUC values better than PSA (AUC = 0.678), the current gold standard for diagnosing PCa [16].

Another value of our suggested panel is that four microRNAs are upregulated in blood, whereas three are downregulated. This result means that each group can serve as an additional internal control for each other, thereby further serving to verify the accuracy of results, i.e., they do not all go up or all go down. Constructing a diagnostic panel with only downregulated microRNAs is always hard to justify; however, by pairing the loss of three microRNAs with an increase in the other four allows for greater diagnostic confidence.

In some previous studies, PCa samples were pooled in order to obtain sufficient material for subsequent analysis [22,23]. This approach prevents an analysis of variability across individual samples and blocks any correlation to the stage of disease when Gleason scores are available. Not only is it important to diagnose PCa, but eventually to identify biomarkers that could serve to stage disease and, more importantly, discern indolent from aggressive disease, thereby impacting subsequent treatment options. Thus, the fact that valid HTS data could be acquired from individual samples without requiring pooling might enable a correlation between microRNA and tumor stage in future studies. Samples analyzed here were predominately from lower Gleason-scored patients (Table 1: 9 or 11 patients with a G6 or G7 score respectively with only 2 samples scored as G8 or G9). Thus, our panel is more diagnostic for early detection and, if these patients could be followed, might elucidate microRNAs useful for separating indolent from more aggressive disease. In any case, more patient samples are needed particularly with higher Gleason scores to determine microRNAs that could identify later stages of disease as proposed for miR-1290 and miR-375 in CRPC [15].

3.1. Literature Review of Diagnostic Panel of Dysregulated MicroRNAs in Cancer

A brief review of these diagnostic seven microRNAs, their Chromosomal (Chr) location and known targets was carried out to determine if their dysregulation might support a functional role in prostate tumorigenesis.

3.1.1. miR-127-3p, miR-204-5p, miR-329-3p and miR-487b-3p as Tumor Suppressors

miR-127-3p (Chr 14) is situated near a cluster of microRNAs (has-miR-431, hsa-miR-433, hsa-miR-432 and hsa-miR-136) susceptible to epigenetic silencing [24]. It has been shown to target BCL6 and is downregulated in breast cancer tissue where overexpression of miR-127-3p or depletion of BCL6 supported its role as a tumor suppressor [25]. In addition, BCL6 plays an important role in cell proliferation by suppressing transcription of the anti-apoptotic *BCL-XL* gene and the adhesion molecule VCAM [26,27].

miR-204-5p (Chr 9) is highly downregulated in many tumor types including breast, kidney and prostate [28]. The absence of miR-204-5p led to a decrease in Kir7.1 proteins, which connect TGF-BR2 and maintain potassium homeostasis, thereby playing a crucial role in maintaining epithelial barrier function and cell physiology [28]. miR-204-5p has been shown to suppress the growth, migration and invasion of endometrial carcinomas by binding to TrkB mRNA and interfering with JAK2 and STAT3 phosphorylation [29].

miR-329-3p (Chr14) is part of an extensive microRNA cluster containing over 40 microRNAs. Yang et al. found miR-329-3p to be downregulated in metastatic, neuroblastoma tumor tissue compared to the primary tumor [30]. One promising target for miR-329-3p is KDM1A, which has been shown to be significantly upregulated in the androgen-dependent LnCaP prostate cell line [30,31]. Upon depletion of KDMA1 using siRNA, VEGF-A expression was also decreased, which in turn blocked androgen-induced VEGF-A, PSA and Tmprss2 expression, suggesting a role for miR-329-3p as a tumor suppressor.

miR-487-3p (Chr 14 within the same microRNA cluster as miR-329) has been found to be downregulated in neuroblastomas and in PCa [32,33]. Moreover, 10 microRNAs from this cluster were found to be significantly downregulated in PCa as Gleason scores increased, thereby playing an important role in regulating proliferation, apoptosis, migration and invasion in metastatic PCa cells [32]. An interesting predicted target for miR-487b-3p is ALDH1A3, aldehyde dehydrogenase 1A3, an enzyme known to be upregulated four-fold in the LnCaP PCa cell lines [34] when exposed to the androgen Dihydrotestosterone (DHT).

3.1.2. miR-32-5p, miR-20a-5p and miR-454-3p as OncomiRs

miR-32-5p (Chr 9) has been found to be an androgen-regulated microRNA that targets BTG2 [35]. Its overexpression has been shown to block apoptosis and promote PCa in CRPC. Furthermore, this microRNA was discovered to be regulated by DHT and displays putative upstream androgen receptor-binding sites (ARBS).

miR-20a-5p (Chr 13) is part of the miR-17–92 cluster, which plays an important role in cell cycle progression, proliferation, apoptosis and other cellular processes [36]. One of the most studied targets of miR-20a is the E2F family, particularly E2F2 and E2F3 [37]. The overexpression of miR-20a-5p in the PC3 PCa cell line was shown to regulate the cell cycle via targeting of E2F2 and E2F3 mRNAs [36]. In addition, this microRNA also targets several cyclin-dependent kinases, including p21 and p57, which halt cell cycle progression. Finally, another notable target is FasI, which promotes cell death [37]. Thus, the major targets of miR-20a-5p promote tumorigenesis and angiogenesis by blocking cell cycle checkpoints [36].

miR-454-3p is located on Chr 17 in the first intronic region of its host gene *SKA2* (Spindle and Kinetochore-Associated Complex Subunit 2). SKA2 is essential for proper chromosome segregation. During the cell cycle, both SKA2 and miR-454-3p have been shown to be upregulated. miR-454-3p targets the tumor suppressor gene, *BTG1* (B cell Translocation Gene 1), which plays an important role in cell cycle progression and is involved in the stress response [38]. This anti-proliferative gene is expressed at its highest concentration during the G0/G1 phases of the cell cycle and is then downregulated when the cell progresses through the G1 phase. In renal carcinoma cells, an increase in miR-454-3p displayed a marked decrease in BTG1 via a direct interaction with the 3′-UTR of BTG1 mRNA [38].

3.1.3. Summary of the Literature Review for Targets of Panel MicroRNAs

A summary of these results is shown in Table 3. The four upregulated microRNAs in patient blood (miR-127-3p, miR-329-3p, miR-487b-3p and miR-204-5p) cumulatively target BCL6, TrkB, KDM1A and ALDH1A3, all of which have been shown to be important regulators in PCa [24–34]. Since these proteins exert oncogenic effects in prostate tissue, their regulators are viewed as tumor suppressors, the loss of which could contribute to tumorigenesis. On the other hand, the three downregulated microRNAs in patient blood (miR-20a-5p, miR-32-5p and miR-454-3p) have been shown to target the tumor suppressor proteins E2F2/3, BTG2 and BTG1, respectively [35–38]. Although they were downregulated in our patient blood samples, they have been shown to be oncomiRs in tumor tissue, the retention of which could promote tumor progression. A review of this literature supports how these microRNAs could play a role in PCa progression.

Table 3. Summary of targets and their role in prostate cancer for the miRNA panel.

MicroRNA	Validated Target	Possible Role in Cancer
miR-127-3p	BCL6 [24–27]	Tumor suppressor
miR-204-5p	TrkB [28,29]	Tumor suppressor
miR-329-3p	KDMA1 [30,31]	Tumor suppressor
miR-487b-3p	ALDH1A3 [32–34] *	Tumor suppressor
miR-32-5p	BTG2 [35]	OncomiR
miR-20a-5p	E2F family, P21, p57 [36,37]	OncomiR
miR-454-3p	BTG1 [38]	OncomiR

* A predicted target.

Interestingly, three of the microRNAs in our panel belong to the same mega cluster on Chr 14. A post-review of our data did note differential expression for several additional members from this cluster. However, due to their low abundance, slightly higher FDR values and limited budget, they were not included in subsequent qRT-PCR confirmatory studies. Analysis of the HTS data showed that five mega cluster members (miR-654-5p, miR 654-3p, miR-493-3p, miR-493-5p and 433-5p) were present in the top 50 dysregulated microRNAs ranking 17th–59th from the top (Appendix A Table 6). A review of the TGCA data showed that all were downregulated to different degrees in tumor tissue, fitting with their loss as tumor suppressors. Interestingly, one of these microRNAs, miR-433-3p, has been shown to target CREB (cAMP Response Element Binding protein), a nuclear transcription factor shown to be involved in tumor initiation, progression and metastasis [39]. Sun et al. showed that overexpression of miR-433-3p could counteract the effects of CREB. Studies have shown that the microRNAs in this Chr 14 cluster are downregulated through unknown mechanisms. If increases in these microRNAs are found in blood, it is possible to hypothesize that the expression of these microRNAs is not just being turned off at the transcriptional level, but that they are being shuttled out of the tumor cell and into the blood as a survival and growth mechanism for the developing tumor.

3.2. Relevance of Comparing Blood HTS and qRT-PCR Data to the TCGA Database

An unanticipated discovery from this study was the inverse relationship between blood and tumor microRNA expression levels (Figures 2 and 4 compared to Figure 6). Since all of our blood microRNAs displayed an inverse expression level with our analysis of data from the TCGA database, we propose that it is possible that tumors are retaining oncomiRs for the purpose of driving tumorigenesis and angiogenesis, and therefore, less of these oncomiRs are released into the blood (Figure 7). Conversely, tumor suppressors block tumor growth and may need to be disposed of to enhance tumorigenesis and ultimately metastasis; hence, the increase in blood levels of these microRNAs. If this were the case for only one or two members of our diagnostic panel of seven microRNAs, perhaps not, but the recurrence for all seven candidates lends credibility to this hypothesis. Moreover, a review of their known targets and the roles they could play in PCa further supports this idea (Table 3).

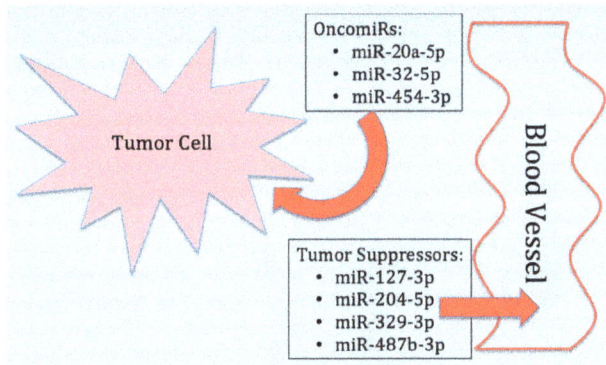

Figure 7. Relationship between microRNA content in prostate tumor cell derived from the TCGA database versus our analysis of blood in prostate cancer patients. The tumor cell retains oncomiRs, but disposes of tumor suppressors to enhance tumorigenesis.

It has been shown that cancer cells secrete vesicles containing not only mature microRNAs to modify their environment for future metastasis, but the entire processing machinery (dicer, RISC with premiR) to ensure that once taken up by the target cell, the microRNA is efficiently processed and actively moved into the translational silencing mechanism of target mRNAs [40]. It is proposed that if only the mature microRNA were delivered to a secondary site, it might not be as efficient in modifying translation within the target cell. Thus, the preferential cellular export of certain microRNAs as "hormomirs" may function to modulate gene expression at secondary sites, thereby affecting disease pathology [41,42]. With this in mind, it is not much of a stretch to propose a developing tumor wants to dispose of compromising microRNAs that could restrict its growth; thus, excluding tumor suppressor microRNAs. Concomitantly, holding onto an oncomiR to quickly modulate the proteome is fast and efficient, faster than modifying gene expression at the transcriptional level. In support of this hypothesis, Selth et al. found a similar inverse relationship between the loss of miR-146b-3p expression in the prostate tumor with a concomitant increase in circulation [41]. Conversely, in the same study, a direct correlation between increased expression of miR-194 in the tumor and in circulation was noted, suggesting that microRNAs may vary in their expression and patterns of secretion. Since overexpression of miR-194 blocks cell proliferation, induces apoptosis, caspase-3/-9 activities and p53/p21 signaling while suppressing PI3K/AKT/FoxO3a signaling, it is difficult to understand how a tumor could tolerate an increase in the expression of this microRNA, which was not discussed [43]. Another study found miR-194 to be decreased in prostate tumors, befitting its function as a tumor suppressor [44]. On the other hand, miR-1 and miR-133a have been shown to be increased in serum in response to acute myocardial infarction where the levels of both are reduced in the infarcted myocardial tissue [45]. Thus, some evidence for an inverse correlation between tissue expression and microRNAs in circulation exists, but at this time, additional studies in PCa with more patient samples and expansion to other cancers and disease states need to be completed to determine the overall merits of this hypothesis.

3.3. Comparison of HTS Data to Screens of MicroRNA Panels and qRT-PCR Analysis

To date, the elucidation of microRNAs to identify PCa has been mostly generated from screens of preformed microRNA panels or qRT-PCR assays for microRNAs already shown to be involved in cancer. In both cases, the decision as to what microRNAs should be surveyed has already been made rather than using a technology like HTS, which uses no preselection and permits the identification of any potential microRNA. Cochetti et al. chose 23 microRNAs from an in silico survey of predicted target genes to analyze serum from PCa patients [10]. A review of a variety of such studies using serum or plasma has suggested miR-141, -21, -200b, -375, -221, -26a, -195, -15b, -16, -19b, -24, -451 or let-7i as biomarkers to distinguish PCa patients from healthy individuals [5,8,14,41,42]. Some of these microRNAs have also been proposed as biomarkers for other cancers not being unique to PCa (miR-141, -21, -16, -451) or are heavily influenced by hemolysis (miR-15b, -16, -451), making their utility for PCa diagnosis debatable [8,42]. Interestingly, we did not find any of these microRNAs to be significantly altered in our HTS data. In part, this could be due to differences in analyzing serum or plasma versus blood, a very different source of body fluid, as well as the use of HTS data as the starting point for investigation.

HTS has been applied to a very limited number of studies for identifying microRNAs diagnostic for PCa. Szczyrba et al. looked at microRNA profiles of prostate carcinoma compared to normal tissue and cell lines [14]. Here, the loss of miR-143 and miR-145 targeting myosin VI (MYO6) was suggested as a diagnostic marker for prostate carcinoma. However, we did not find these two microRNAs to be dysregulated in blood. Of more significance to our study was the reported HTS data of blood exosomal material isolated from CRPC patients [15]. Here, as well, a group of normalizing RNAs (miR-301a/e-5p, miR-99a-5p, let-7c, miR-125a-5p, miR-16-5 and RNU6B) was proposed for subsequent qRT-PCR analysis. Since RNU6B is not a secreted microRNA, it is doubtful that it should have been included in this analysis, but discounting this snRNA, the rest would be useful normalizers for exosomal material. Interestingly, we found a completely different group of normalizing microRNAs with no overlap of this group, supporting how different an exosomal pool might be to the blood samples analyzed here. Upon normalization, these authors proposed miR-1290 and miR-375 as prognostic markers for CRPC. Since this was only for CRPC patients, perhaps it is not surprising that we did not find these two microRNAs in our analysis, since we are not focused on CRPC. Perhaps our panel of seven microRNAs would be better suited for identifying early stages of prostate cancer, rather than this later stage. Again, additional studies for CRPC versus patients that have not progressed to later stage disease are needed to clarify this hypothesis.

3.4. Limitations to This Pilot Study

This study was meant as a pilot study and, as such, suffers from some limitations, which should be addressed. First, to obtain sufficient material for HTS analysis, blood was used as the initial source of small RNAs. Thus, small RNAs will be contaminated with cellular microRNAs, not just circulating microRNAs. However, circulating microRNAs can come in many forms as exosomes, microvesicles, apoptotic bodies or bound to HDL, argonaute 2 or RNA-binding proteins, such as nucleophosmin 1 [42]. At this time, it is not clear which of these forms or combinations thereof would be the most diagnostic as biomarkers for PCa. Thus, focusing on some particles (exosomes or microvesicles) at the exclusion of the others might not be the most relevant source. Isolation of small RNAs from whole blood rather than purification of a subset of these particles seemed like a more inclusive starting point. More importantly, it was assumed that control samples will contain the same contaminates as patient samples, and thus, contaminating cellular microRNAs should cancel out and not be found amongst the dysregulated microRNAs, if samples are handled consistently. In support of this premise, it was reassuring that microRNAs known to reflect WBC (miRs-15- and -230), RBCs affected by hemolysis (miR-16-485-3p, -532-3p, -15b, -16 and -451), RBCs not-affected by hemolysis (miR-1274b, -142-3p and 146a), myeloid (miR-7a, -223, -197 and 574-3p) or lymphoid (miR-150) cells were not found in our panel of dysregulated microRNAs [21,41,42,46]. Our final panel of seven microRNAs was unique amongst those proposed in the literature, perhaps due to the fact that they were compiled from HTS data rather than screens of predetermined miR array panels or primer sets. Second, PSA values were not reported for control samples, making it impossible to obtain an AUC value for this cohort. Thus, an AUC value for PSA was taken from the literature [16], and it could be higher for our control group with its younger age, a problem encountered by other such studies [42]. However, all seven panel members yielded individual AUC values considerably better than the reported value for PSA, which when used as a panel should be stronger than any single microRNA. In fact, in a review by Selth, it was proposed that "no single analyte is likely to achieve the desired level of diagnostic or prognostic accuracy for PCa ... requiring a signature of multiple microRNAs rather than a single miR", as proposed here [42]. Third, validation of HTS results by qRT-PCR was on the same cohort of samples analyzed by HTS. In a future study, a third larger cohort should be evaluated independently to better validate panel members as relevant biomarkers. Finally, these samples did not span the spectrum of higher Gleason scores, but reflect earlier stages of PCa. Thus, at this time, they are not useful for staging or separating indolent from aggressive disease, an important future correlate. Nevertheless, it is proposed that despite these limitations, this initial pilot study does present new novel microRNAs that have not been previously suggested, which warrant inclusion in future studies sampling larger cohorts.

Here, we have proposed a panel of seven microRNAs generated from HTS data rather than pre-judged screens of microRNA arrays proposed from other cancers with unknown relevance to PCa. The same criticism exists for data generated by qRT-PCR studies, since again, only a subset of chosen total microRNAs is being investigated. We propose that HTS data confirmed by qRT-PCR analysis are a worthwhile approach for deducing biomarkers for PCa. Certainly, our ROC curves and AUC values appear superior compared to the current PSA gold standard [16]. As a group, the value of a panel of diagnostic microRNAs is substantial. However, additional studies with more extensive patient sampling are required to determine their future usefulness in not only identifying PCa, but ultimately staging prostate cancer and separating indolent versus aggressive disease. This will require sampling, preferably individually, of a vast number of patient and control samples, but our initial study certainly justifies the merits of future investigation.

4. Materials and Methods

4.1. Sample Extraction and HTS Sequencing

Whole blood samples from patients and controls were obtained from the Nelson Urology Clinic, VCU (Virginia Commonwealth University) Medical Center and Mcguire Veterans Hospital following approval by the ethics committee (IRB Panel D Approval #HM14344). All patients provided written consent. Blood samples were taken prior to treatment, radiotherapy or prostatectomy. In most cases, age, ethnicity, PSA values and Gleason scores from biopsies were provided (Table 1). Controls were carefully selected to not have any history of PCa either as an individual or within the family, and written consent was obtained. A complete analysis of information provided to us for each individual is included in Appendix A Table A1. Samples were collected in PAXgene blood tubes (PreAnalytiX, Qiagen/BD, Franklin Lakes, NJ, USA), which contain a manufacture's additive to stabilize RNA. The total RNA, including microRNAs in each sample, was extracted using a corresponding PAXgene blood miRNA kit following the manufacture's protocol, which removes DNA and results in the purification of pure RNA. The quality and concentration of small RNAs ranging from 10–40 nts were measured using the small RNA Chip Assay (Agilent) based on the manufacturer's instructions. A total number of 40 samples (12 controls and 28 patients) with a small RNA concentration >100 ng/μL was selected for HTS microRNA sequencing using the Illumina® TruSeq Small RNA Library Preparation kit (New England Biolabs, Ipswich, MA, USA) and HiSeq 2500 system (Illumina, San Diego CA, USA) according to the manufacturer's protocol.

4.2. Bioinformatics Analysis

The raw deep sequencing data were processed using Flow® v 3.0 (Partek Incorporated, St. Louis, MO, USA). The adapter sequence "AGATCGGAAGAGCACACGTCT" (TruSeq Adapter, Index 7), frequently detected from all reads, was removed from both the 5′- and 3′-ends. A second trimming was performed to further eliminate bases at both ends with a Phred quality score lower than the average 35, indicating a probability that every 1 in 5000 bases was incorrect; accuracy of 99.95%. The minimum read length detected by the program was changed from 25 to 16 nts in order to include all possible microRNA reads in a suitable range. The trimmed data were aligned to the human genome (GRCh38) with only 1 seed mismatch allowed. The three best alignments satisfying such criteria were reported for each read using Bowtie 1.0. The parameters used by Bowtie while performing the alignment were as follows: alignment mod = quality limit, seed mismatch limit = 1, seed length = 28 and quality limit = 70, both strand alignment and alignments reported per read = 3. The aligned reads were annotated with miRBase (mature 21, Version 2). The differential expression analysis was conducted on the annotated sequencing reads exported from Partek Flow® using Edge R (Version 3.12) (Roswell Park Cancer Institute, Buffalo, NY, USA) [18,47]. The reads were normalized using the default "Trimmed Mean of M values" (TMM method) algorithm, which aims at minimizing the effect of sequencing depth and RNA composition [47].

4.3. Quantitative Real Time-Polymerase Chain Reaction

The search for microRNA candidates serving as good qRT-PCR endogenous controls was attempted using NormFinder software [48]. The microRNAs showing a relative high abundance and minimal intergroup variation suggested by the HTS data were selected for exploratory qRT-PCR on 16 samples (8 controls, 8 patients) in triplicate (Appendix A Table 2). Each RNA sample (3 ngs) was converted to cDNA (final volume 20 µL) using the qScript™ synthesis kit (Quanta Biosciences Inc., Gaithersburg, MD, USA) following the manufacturer's protocol and qPCR conducted as previously described [49]. Briefly, the cDNA was diluted with RNase-free water 1:1 (v/v), and 2 µL were used for each PCR reaction run in triplicate. Each PCR reaction was scaled down to 6.25 µL SYBR® Green Master Mix, 0.25 µL primer, 4.0 µL RNase-free H_2O for the purpose of saving reagents without compromising the results. qRT-PCR was conducted in an Applied Biosystems 7500 real-time PCR instrument (Life Technologies, Foster City, CA, USA) using the following conditions: 50 °C for 2 min, followed by 40 cycles at 95 °C for 15 s, 60 °C for 15 s and 70 °C for 30 s. Data were collected at 70 °C and analyzed using SDS software v1.3.1 (Life Technologies), using automatic threshold and baseline settings. Negative amplification controls and DNase-treated controls were routinely included for each microRNA and did not impact analysis. PCR efficiency for each microRNA primer set was tested and found to be within the acceptable range (80–110%). The Cq values of each candidate microRNA were imported into the NormFinder program, which generated stability values for each candidate after evaluating both intra- and inter-group variations. Lower stability values suggested higher consistency of a microRNA across different samples and groups. A combination of the four microRNAs with the lowest stability values showed a greater stability and consistency. Therefore, a normalization factor for each plate was determined by taking the geometric mean of the Cq values of the four microRNAs.

Next qRT-PCR was performed on dysregulated candidate microRNAs using 10 controls and 26 patients (raw and normalized values in Appendix A Tables 3 and 4, respectively) analyzed in triplicate (SD < 0.2). Unfortunately, material from two patient and control samples evaluated by HTS was found to be insufficient for confirmatory qRT-PCR analysis. For all other samples, the protocol was the same as described above. Raw Cq values were normalized by subtracting the geometric mean Cq value of the top four normalization candidates (miR-146a-5p, miR-1538, miR-197-3p and miR-1468-3p) from individual Cq values to generate dCq. A *p*-value was obtained using the Mann–Whitney nonparametric test assuming that data do not follow a Gaussian distribution. A *p*-value <0.05 was considered significant. The −ddCq value for each candidate was obtained by taking the mean of the normalized dCq for all controls minus the normalized dCq of each patient sample. The −ddCq values were equivalent to Log2 fold change, as the fold change was calculated by 2^{-ddCq}.

4.4. The Cancer Genome Atlas Tissue Data Analysis

Illumina HiSeq level 3 miRNA sequencing data of 50 prostate tumor tissue samples and their matched normal margins were selected and downloaded from The Cancer Genome Atlas database (TCGA database) for analysis. The reads were annotated with stem-loop transcripts of each microRNA [50]. The raw reads were normalized in the same way as described above using the Edge R software [18].

4.5. Statistical Analysis

Differential expression analysis was conducted on the normalized next generation sequencing reads for both blood samples and TCGA tissue data using EdgeR [18]. As the distribution of microRNA sequencing reads remains unclear, the dispersion of reads was estimated via the Cox–Reid profile adjusted likelihood method default in Edge R [51]. The reads matrix was fitted to a generalized linear model and a likelihood ratio test was performed on the fitted data. The *p*-value was adjusted to the number of comparisons (equal to total number of microRNAs detected in the sequencing) using the Benjamini–Hochberg method, which yields a False Discovery Rate (FDR) to minimize Type I error [52]. In order to maximize the screening results, a FDR value smaller than 0.2 was considered significant in this experiment.

For dysregulated microRNA PCR results, the normalized dCq values of each candidate between control and patient groups were compared using the Mann–Whitney nonparametric test assuming that the data do not follow a Gaussian distribution on Prism (GraphPad Software Inc., Version 6, 2015). A *p*-value lower than 0.05 was considered significant. ROC curves were generated for candidate microRNAs showing a statistically-significant difference between two groups [20]. The ROC was obtained by plotting sensitivity against specificity using the pROC package (Version 1.8). An area greater than 0.5 under the curve (AUC) suggests the diagnostic potential of each microRNA candidate.

5. Conclusions

In summary, we propose a group of four microRNAs (miR-146a-5p, miR-1538, miR197-3p and miR-1468-5p) that could be used as normalization standards for the comparative analysis of blood samples at least for PCa and perhaps other cancers, as well. In addition, a panel of seven microRNAs (miR-127-3p, miR-204-5p, miR-329-3p, miR-487b-3p, miR-32-5p, miR-20a-5p and miR-454-3p) might be useful for diagnosing PCa dependent on further validation. Individual members of this panel display better diagnostic capabilities than PSA alone and as a group are superior.

Acknowledgments: The authors acknowledge the VCU Urology Clinic and Mcguire Veterans Administration Hospital for providing patient samples utilized in this research. In addition, we are grateful to the Massey Cancer Center for supplying funds for this pilot project. No funds were reserved for covering the costs to publish in open access from this source. HTS data have been submitted to the GEO Database as submission GSE97901.

Author Contributions: Rhonda Daniel, Qianni Wu, Gene Clark, Georgi Guruli and Zendra Zehner conceived of and designed the experiments. Rhonda Daniel performed the experiments. Rhonda Daniel, Qianni Wu, Vernell Williams, Gene Clark and Zendra Zehner analyzed the data. Georgi Guruli and Vernell Williams contributed reagents, materials and analysis tools. Rhonda Daniel, Qianni.Wu, Vernell Williams and Zendra Zehner wrote the manuscript. All authors reviewed and approved of the final manuscript.

Conflicts of Interest: The authors declare no conflict of interest. The founding sponsors had no role in the design of the study; in the collection, analyses or interpretation of data; in the writing of the manuscript; nor in the decision to publish the results.

Abbreviations

ALDH1A3	Aldehyde Dehydrogenase 1A3
ARBS	Androgen Receptor-Binging Sites
AUC	Area Under the Curve
BCL6	B-Cell Lymphoma 6 Protein
BCL-XL	B-Cell Lymphoma-Extra Large
BTG1	B Cell Translocation Gene 1
Chr	Chromosome
CPM	Counts Per Million
CREB	Cyclic-AMP Response Element Binding Protein
CRPC	Castration-Resistant Prostate Cancer
Cq	Quantification Cycle
FC	Fold Change
FDR	False Detection Rate
hsa	A Human microRNA
HTS	High Throughput Sequencing
JAK2	Janus Kinase 2
KDM1A	Lysine Demethylase 1A
LnCaP	Androgen-Sensitive Human Prostate Adenocarcinoma Cells derived from the left supraclavicular lymph node metastasis from a 50-year old caucasian male
miR	MicroRNA
PIWI	P-Element-Induced Wimpy Testis
PSA	Prostate-Stimulating Antigen
PCa	Cancer of the Prostate
qRT-PCR	Quantitative Reverse Transcription Polymerase Chain Reaction
RPKM	Reads Per Kilobase of transcript per Million mapped reads
SKA2	Spindle and Kinetochore-Associated Complex Subunit 2
siRNA	Small interfering RNA
snRNA	Small nuclear RNA
STAT3	Signal Transducer and Activator of Transcription 3
TGF-BR2	Transforming Growth Factor Beta Receptor, Type II
TC	Total Counts
TCGA	The Cancer Genome Atlas
TMM	Trim Mean of M values
Tmprss2	Transmembrane protease, serine 2
TRKB	Tropomyosin Receptor Kinase B
UTR	Untranslated Region
VEGF-A	Vascular Endothelial Growth Factor A
VCAM	Vascular Cell Adhesion Protein

Appendix A

Table A1. Characteristics and pathological data of individual patients and controls involved in the study.

Sample Name	Sample Type	Gleason Score	Age	PSA	Race/Ethnicity	HTS Sample	Confirmatory PCR Sample	NormFinder Sample	Number in Appendix
Z1B-SEQ 1	Patient	G9	92	-	Caucasian	yes	yes	yes	Sample 36
Z2B-SEQ 1	Patient	G7	-	-	-	yes	yes	yes	Sample 26
Z3B-SEQ 1	Patient	G7	-	-	-	yes	yes	yes	Sample 27
Z4B-SEQ 1	Patient	G7	-	-	-	yes	yes	yes	Sample 28
Z5B-SEQ 1	Patient	G9	-	-	-	yes	yes	yes	Sample 35
Z6B-SEQ 1	Patient	-	-	Elevated	-	yes	yes		Sample 11
Z7B-SEQ 1	Patient	-	-	22	-	yes	yes		sample 12
Z8B-SEQ 1	Patient	-	-	Elevated	-	yes	yes		Sample 13
091714-SEQ 2	Control	Control	51	-	Caucasian	yes	no		-
Case100-SEQ 2	Patient	G7	61	7.78	African American	yes	yes	yes	Sample 24
10212014FAM-SEQ 2	Control	Control	62	-	-	yes	yes	yes	Sample 10
12172014a-SEQ 2	Patient	G7	-	-	-	yes	no		-
12172014b-SEQ 2	Patient	-	-	Elevated	-	yes	no		-
12192014-SEQ 2	Control	Control	51	-	Caucasian	yes	yes	yes	Sample 8
Case18-SEQ 2	Patient	G6	68	6.8	African American	yes	yes	yes	Sample 14
Case20-SEQ 2	Patient	G6	64	4.15	African American	yes	yes	yes	Sample 18
Case31-SEQ 2	Patient	G6	67	12.47	Caucasian	yes	yes	yes	Sample 15
Case33-SEQ 2	Patient	G6	63	3.2	Caucasian	yes	yes		Sample 21
Case36-SEQ 2	Patient	G6	55	10.73	Caucasian	yes	yes	yes	Sample 22
Case40-SEQ 2	Patient	G7	66	3.42	Caucasian	yes	yes		Sample 29
Case56-SEQ 2	Patient	G7	68	4.57	Caucasian	yes	yes		Sample 30
Case72-SEQ 2	Patient	G8	63	4.77	Caucasian	yes	yes		Sample 34
Case85-SEQ 2	Patient	G7	65	6.1	Caucasian	yes	yes	yes	Sample 23
Case9-SEQ 2	Patient	G6	55	8.82	Asian/Hawaiian	yes	yes		Sample 20
Z9B-SEQ 2	Control	Control	54	-	Caucasian	yes	yes	yes	Sample 7
Z10B-SEQ 2	Control	Control	39	-	African American	yes	yes		Sample 6
03242015-SEQ 3	Control	Control	46	-	Caucasian	yes	yes	yes	Sample 4
03262015-SEQ 3	Control	Control	43	-	Caucasian	yes	yes	yes	Sample 3
04062015-SEQ 3	Control	Control	91	-	Caucasian	yes	yes	yes	Sample 9
04242015-SEQ 3	Control	Control	59	-	Caucasian	yes	yes	yes	Sample 5
04282015-SEQ 3	Control	Control	55	-	Asian/Hawaiian	yes	no		-
Case16-SEQ 3	Patient	G8	76	7.99	Caucasian	yes	yes	yes	Sample 33
Case51-SEQ 3	Patient	G7	64	6.06	Caucasian	yes	yes	yes	Sample 25
Case 6-SEQ 3	Patient	G6	58	11	African American	yes	yes	yes	Sample 16
Case82-SEQ 3	Patient	G7	62	8.36	African American	yes	yes		Sample 31
Case83-SEQ 3	Patient	G6	66	4.21	Caucasian	yes	yes		Sample 19
Case89-SEQ 3	Patient	G7	66	3.5	Caucasian	yes	yes		Sample 32
Case8-SEQ 3	Patient	G6	73	4.4	African American	yes	yes		Sample 17
Z11B-SEQ 3	Control	Control	24	-	Caucasian	yes	yes		Sample 2
Z12B-SEQ 3	Control	Control	23	-	Caucasian	yes	yes	yes	Sample 1

Table 2. Raw Cq values of exploratory qRT-PCR analysis of blood samples from controls ($n = 8$) and patients ($n = 8$) to identify normalizer microRNAs for confirmatory qRT-PCR analysis. Cq values are the average of triplicates (SD $\leq$ 0.2).

Sample	Sample Type	miR197-3p	miR26b-5p	miR296-5p	miR23a-3p	miR146a-5p	miR 1248	miR1468-3p	miR1538
7	Control	22.74829	21.628092	26.089388	18.800253	27.064646	19.802492	30.705496	31.02948
10	Control	21.136293	25.414007	32.061626	18.889477	26.500956	16.915016	30.051926	29.690506
1	Control	22.46137	19.360636	24.926224	17.68174	25.192787	17.260515	29.683975	30.3445345
8	Control	29.555853	27.222218	31.271189	19.254599	29.127176	21.902388	33.016678	32.38471
5	Control	20.826368	23.358286	28.328314	19.247988	26.861567	16.149588	26.136414	28.535395
9	Control	22.516693	23.018911	27.688248	21.737234	28.745064	18.812151	29.16825	29.95835
3	Control	22.004297	21.192694	26.759714	17.74209	25.819382	15.312261	26.210001	29.330034
4	Control	20.42111	21.1323	27.237703	29.127176	26.609123	16.747427	26.76587	28.660437
22	Patient	21.886065	23.961235	30.295868	17.754295	26.563072	16.856455	28.191488	29.235413
14	Patient	28.050713	27.764153	32.219086	21.700003	28.97425	22.934767	33.360752	32.80031
23	Patient	20.232058	20.535894	28.087671	15.586144	25.002638	20.58262	27.846542	28.899343
24	Patient	20.934057	20.589144	26.781439	16.453804	24.992987	15.296409	28.57466	29.60646
15	Patient	20.324524	21.27265	28.345781	23.688972	31.2935	16.856455	31.182531	32.48879
25	Patient	22.465195	22.984718	26.560272	21.267275	27.856596	20.140722	28.645343	30.621424
16	Patient	21.8991	23.907438	28.562735	21.182196	27.760199	18.519064	27.570486	30.55234
33	Patient	19.746403	22.80751	28.66214	18.871817	27.803125	17.341208	27.578966	28.894281

Table 3. Raw Cq values of confirmatory qRT-PCR analysis of blood samples from controls ($n = 10$) and patients ($n = 26$). Cq values are the average of triplicates (SD $\leq$ 0.2).

Number	Sample Type	miR127-3p	miR1468-3p	miR146a-5p	miR1538	miR197-3p	miR204-5p	miR20a-5p	miR32-5p	miR329-3p	miR454-3p	miR487b-3p
1	Control	36.908432	30.543045	25.765848	32.149563	20.931528	30.270754	17.96527	25.546413	36.75606	22.844046	36.028534
2	Control	37.981623	28.062347	24.830622	30.136032	19.910437	36.117193	15.91111	23.228275	32.757341	22.673347	34.425587
3	Control	28.144186	28.446253	25.333971	30.515898	20.662188	30.312696	18.96527	29.557442	29.08623	25.073465	31.28196
4	Control	28.527613	29.671013	26.744165	30.476606	20.475235	30.221922	19.232796	29.752726	29.638098	26.315996	32.190907
5	Control	30.057663	29.518211	26.824263	28.535395	20.943573	30.793854	22.753603	26.935396	30.549356	28.72201	33.296204
6	Control	30.720682	29.358604	24.906527	31.398394	21.780424	29.43325	20.239492	30.17968	29.987643	25.529074	33.19684
7	Control	30.600016	31.048105	27.055967	33.107254	21.905685	31.476599	20.894003	27.493944	30.090225	24.6573	32.67403
8	Control	33.510212	30.512245	28.752623	29.60646	21.71274	30.86007	19.256899	31.885645	29.467866	23.78831	30.392588
9	Control	29.977875	29.733969	27.0654	29.95835	23.657553	30.979445	20.109362	27.893682	32.036114	25.58172	29.871424
10	Control	28.420685	30.848099	26.421408	32.04854	20.955032	30.269693	25.617834	27.432236	29.530313	29.78074	33.72329
11	Patient	25.514578	28.478271	23.705612	30.452873	22.588583	27.87504	23.345776	28.470451	23.393488	25.4003	25.936775
12	Patient	28.76194	28.892448	24.489096	31.246782	21.041815	29.743967	23.559605	27.84227	26.779068	27.28158	29.112656
13	Patient	28.005022	28.854507	23.039087	32.001001	21.060934	28.308046	23.194078	26.389101	27.779022	27.499535	29.67564
14	Patient	23.12572	30.165638	25.97425	34.46028	20.858582	23.945423	24.633371	33.6321	24.79482	28.784933	31.840466
15	Patient	25.970835	28.054586	22.90155	32.48879	26.455523	27.135048	25.562134	35.949183	27.743675	23.302767	29.316404
16	Patient	30.06621	31.15792	24.450277	31.151573	21.628485	23.59674	20.898695	32.535934	26.933802	28.414148	23.654408
17	Patient	29.233658	29.118433	24.842314	31.093117	22.109777	29.91673	28.335602	27.745428	29.466345	24.386953	30.998627
18	Patient	24.328703	28.683714	29.701113	30.66745	19.658997	25.6975	29.903137	30.435247	25.765816	26.52373	29.824835
19	Patient	21.014902	29.878317	31.95202	32.125797	17.540724	25.09019	26.061022	34.43073	22.8402	32.833908	26.911285
20	Patient	26.311941	28.65756	22.998682	32.747137	25.313272	26.886636	26.856186	27.684212	24.394753	33.664633	20.6597
21	Patient	26.961014	26.03805	22.823648	28.527018	27.915817	25.870775	30.0119	29.701286	26.257238	25.049133	20.115797
22	Patient	27.535292	28.895426	26.545149	30.66663	20.694304	29.86962	24.83243	35.122456	27.294931	29.74388	21.432922
23	Patient	29.402014	29.05626	25.683632	28.899343	20.198837	28.838037	22.598902	32.26222	28.878447	27.629982	22.840466
24	Patient	26.196487	29.45417	24.96972	32.38471	20.566633	29.10848	20.786528	28.78074	29.102547	35.674986	22.73802
25	Patient	29.829214	29.245398	25.854036	30.621424	20.477509	27.870811	29.884722	27.046827	20.211077	34.532679	23.396152
26	Patient	29.333616	29.284616	25.595312	30.323542	20.69581	29.889395	29.074312	27.739935	28.63606	34.9503	31.2132
27	Patient	26.915873	28.553482	24.359283	29.934687	23.699072	28.018053	28.724783	29.167513	25.274988	35.780039	27.273096
28	Patient	29.36677	29.126135	25.857595	31.09918	20.901865	27.674309	29.272202	27.905403	25.156372	34.727425	30.325487
29	Patient	30.076767	27.860743	24.468689	32.051258	21.122679	27.960495	20.559242	31.840103	25.185661	25.716452	21.408869
30	Patient	29.482193	28.605219	24.295252	30.460472	21.434538	21.330612	29.717953	29.96257	25.31598	25.70342	21.854279
31	Patient	29.213333	29.811007	26.022947	30.650206	20.445127	27.530617	21.284222	31.529474	25.304756	28.699488	30.818052
32	Patient	27.923819	28.06331	22.805868	29.62851	25.487047	30.44813	28.31951	27.352371	25.182838	34.379427	29.983652
33	Patient	29.104956	30.537256	28.080553	31.083038	20.618134	27.849699	21.825888	23.01379	20.256021	27.8463	23.03477
34	Patient	29.28003	28.26641	24.822348	30.753214	19.842566	26.921324	21.533236	30.028503	26.876923	35.232162	25.622961
35	Patient	28.177826	28.581133	23.973612	30.277636	20.03753	27.943693	31.369062	35.510693	26.46978	28.047266	20.259644
36	Patient	20.407839	28.817247	26.935812	30.680155	21.042847	31.687258	33.005196	32.28391	19.864021	28.640589	23.454853

Table 4. Normalized Cq values for confirmatory qRT-PCR analysis of blood samples from controls ($n = 10$) and patients ($n = 26$). Cq values are the average of triplicates (SD $\leq$ 0.2).

Number	Sample Type	miR127-3p	miR204-5p	miR20a-5p	miR32-5p	miR329-3p	miR454-3p	miR487b-3p
1	Control	9.932081552	3.294403552	-9.011080448	-1.429937448	9.779709552	-4.132304448	9.05218355
2	Control	12.55318916	10.68875916	-9.51732384	-2.20015884	7.32890716	-2.75498684	8.99715316
3	Control	2.181017005	4.349527005	-6.997898995	3.594273005	3.123061005	-0.889703995	5.318791005
4	Control	2.000546379	3.694855379	-7.294270621	3.225659379	3.111031379	-0.211070621	5.663840379
5	Control	3.829786882	4.565977882	-3.474273118	0.707519882	4.321479882	2.494133882	7.068327882
6	Control	4.128401402	2.840969402	-6.352788598	3.587399402	3.395362402	-1.063206598	6.604559402
7	Control	2.662065759	3.538648759	-7.043947241	-0.444006241	2.152274759	-3.280650241	4.736079759
8	Control	6.106260714	3.456118714	-8.147052286	4.481693714	2.063914714	-3.615641286	2.988636714
9	Control	2.496480815	3.498050815	-7.372032185	0.412287815	4.554719815	-1.899674185	2.390029815
10	Control	1.220601618	3.069609618	-1.582249382	0.232152618	2.330229618	2.580656618	6.523206618
11	Patient	-0.590230547	1.770231453	-2.759032547	2.365642453	-2.711320547	-0.704508547	-0.168033547
12	Patient	2.645655116	3.627682116	-2.556679884	1.725985116	0.662783116	1.165295116	2.996371116
13	Patient	2.133020459	2.436044459	-2.677920459	0.517090459	1.907020459	1.627533459	3.803638459
14	Patient	-4.268868334	-3.449165334	-2.761217334	6.237511666	-2.599768334	1.390344666	4.445877666
15	Patient	-1.289403891	-0.125190891	-1.698104891	8.688944109	0.483436109	-3.957471891	2.056165109
16	Patient	3.299833617	-3.169636383	-5.867681383	5.769557617	0.167425617	1.64777617	-3.111968383
17	Patient	2.67829958	3.36137158	1.78024358	1.19006958	2.91098658	-2.16840542	4.44326858
18	Patient	-2.442123688	-1.073226688	3.132310312	3.664420312	-1.005010688	-0.247096688	3.054008312
19	Patient	-6.067611542	-1.992323542	-1.021491542	7.348216458	-4.242313542	5.751394458	-0.171228542
20	Patient	-0.875350601	-0.300755601	-0.331105601	0.496920399	-2.792538601	6.477341399	-7.121321601
21	Patient	0.732417073	-0.357821927	3.783303073	3.472668073	0.028641073	-1.179463927	-6.112799927
22	Patient	1.121376344	3.455704344	-1.581485656	8.708540344	0.881015344	3.329964344	-4.980993656
23	Patient	3.711214401	3.147237401	-3.091897599	6.571420401	3.187647401	1.939182401	-2.850333599
24	Patient	-0.259009595	2.652983405	-5.668968595	2.325243405	2.647050405	9.219489405	-3.717476595
25	Patient	3.588714461	1.630311461	3.64422461	0.806327461	-6.029422539	8.292179461	-2.844347539
26	Patient	3.144817167	3.700596167	2.885513167	1.551136167	2.467261167	8.761501167	5.024401167
27	Patient	0.412117905	1.514297905	2.221027905	2.663757905	-1.228767095	9.276283905	0.769340905
28	Patient	2.915212989	1.222751989	2.820644989	1.453845989	-1.295185011	8.275867989	3.873929989
29	Patient	4.012268775	1.895996775	-5.505256225	5.775604775	-0.878837225	-0.348046225	-4.655629225
30	Patient	3.528189418	-4.623391582	3.763949418	4.008566418	-0.638023582	-0.250583582	-4.099724582
31	Patient	2.808155968	1.125439968	-5.120955032	5.124296968	-1.100421032	2.294310968	4.412874968
32	Patient	1.55724467	4.08155567	1.95293567	0.98579667	-1.18373633	8.01285267	3.61707767
33	Patient	1.877809959	0.622552959	-5.401258041	5.786643959	-6.971125041	0.619153959	-4.192376041
34	Patient	3.700034457	1.341328457	-4.046759543	4.448507457	1.296927457	9.652166457	0.042965457
35	Patient	2.785951344	2.551818344	5.977187344	10.11881834	1.07790534	2.655391344	-5.132230656
36	Patient	-6.198563209	5.080855791	6.398793791	5.677507791	-6.742381209	2.034186791	-3.151549209

Table 5. Comparison of fold change in blood HTS data to tissue HTS data from the TCGA database. Fold change was calculated in the same method as panel members using the Edge R program.

MicroRNA	Fold Change in Blood *	Fold Change in TCGA **
hsa-miR-32-5p	−4.11	1.71
hsa-miR-329-3p	4.29	−1.27
hsa-miR-487b-3p	6.06	−1.66
hsa-miR-454-3p	−1.91	1.26
hsa-miR-204-5p	3.43	−3.81
hsa-miR-20a-5p	−2.11	2.19
hsa-miR-127-3p	2.69	−1.43

* FDR < 0.2; ** FDR < 0.05.

Table 6. Chromosome 14 q32.31 dysregulated microRNAs in our analysis of blood samples compared to data from the TCGA tissue database. In addition to our proposed panel members (miR-329-3p, miR-487b-3p and miR-127-3p), subsequent analysis of HTS data uncovered five other microRNAs from this locus (miR-654-5p, miR-654-3p, miR-493-3p, miR-493-5p and miR-433-3p) to be upregulated in blood. Since these microRNAs were not within the top ten potential candidates, they were not carried forth for qRT-PCR confirmation. However, analysis of the TCGA database did confirm these microRNAs to be tumor suppressors lost in tumor tissue compared to matched tumor-free margins, fitting with our model that tumors may strive to get rid of tumor suppressors in order to progress. Fold change is shown in a Log2 scale and was calculated with the same method as panel members using the edge R program.

miRNA	Dysregulation Ranking in Blood	Log Fold Change in Blood	Log Fold Change in TCGA
miR-329-3p	4	2.10	−0.34
miR-487b-3p	5	2.60	−0.73
miR-127-3p	9	1.43	−0.52
miR-654-5p	17	1.96	−0.61
miR-654-3p	36	1.08	−0.61
miR-493-3p	37	1.74	−0.58
miR-493-5p	59	1.29	−0.58
miR-433-3p	43	1.30	−0.16

References

1. Barry, M.J. Clinical practice. Prostate-specific-antigen testing for early diagnosis of prostate cancer. *N. Engl. J. Med.* **2001**, *344*, 1373–1377. [CrossRef] [PubMed]
2. Tchetgen, M.-B.; Song, J.T.; Strawderman, M.; Jacobsen, S.J.; Oesterling, J.E. Ejaculation increases the serum prostate-specific antigen concentration. *Urology* **1996**, *47*, 511–516. [CrossRef]
3. Herschman, J.D.; Smith, D.S.; Catalona, W.J. Effect of ejaculation on serum total and free prostate-specific antigen concentrations. *Urology* **1997**, *50*, 239–243. [CrossRef]
4. Nadler, R.B.; Humphrey, P.A.; Smith, D.S.; Catalona, W.J.; Ratliff, T.L. Effect of inflammation and benign prostatic hyperplasia on elevated serum prostate specific antigen levels. *J. Urol.* **1995**, *154*, 407–413. [CrossRef]
5. Endzeliņš, E.; Melne, V.; Kalniņa, Z.; Lietuvietiss, V.; Riekstia, U.; Llorente, A.; Line, A. Diagnostic, prognostic and predictive value of cell-free miRNAs in prostate cancer: A systematic review. *Mol. Cancer* **2016**, *15*, 41–54. [CrossRef] [PubMed]
6. Ha, M.; Kim, V.N. Regulation of microRNA biogenesis. *Nat. Rev. Mol. Cell Biol.* **2014**, *15*, 509–524. [CrossRef] [PubMed]
7. O'Donnell, K.A.; Mendell, J.T. Dysregulation of microRNAs in human malignancy. In *MicroRNAs: From Basic Science to Disease Biology*, 1st ed.; Krishnarao, A., Ed.; Cambridge University Press: Cambridge, UK; New York, NY, USA, 2008; pp. 295–306.
8. Mitchell, P.S.; Parkin, R.K.; Kroh, E.M.; Fritz, B.R.; Wyman, S.K.; Pogosova-Agadjanyan, E.L.; Peterson, A.; Noteboom, J.; O'Briant, K.C.; Allen, A.; et al. Circulating microRNAs as stable blood-based markers for cancer detection. *Proc. Natl. Acad. Sci. USA* **2008**, *105*, 10513–10518. [CrossRef] [PubMed]

9. Schultz, N.A.; Dehlendorff, C.; Jensen, B.V.; Bjerregaard, J.K.; Nielsen, K.R.; Bojesen, S.E.; Calatayud, D.; Nielsen, S.E.; Yilmaz, M.; Hollander, N.H.; et al. MicroRNA biomarkers in whole blood for detection of pancreatic cancer. *JAMA* **2014**, *311*, 392–404. [CrossRef] [PubMed]

10. Cochetti, G.; Pol, G.; Guelfi, G.; Boni, A.; Egidi, M.G.; Mearini, E. Different levels of serum microRNAs in prostate cancer and benign prostatic hyperplasia: Evaluation of potential diagnostic and prognostic role. *OncoTargets Ther.* **2016**, *9*, 7545–7553. [CrossRef] [PubMed]

11. Lodes, M.J.; Caraballo, M.; Suciu, D.; Munro, S.; Kumar, A.; Anderson, B. Detection of cancer with serum miRNAs on an oligonucleotide microarray. *PLoS ONE* **2009**, *4*, e6229. [CrossRef] [PubMed]

12. Roberts, M.J.; Richards, R.S.; Chow, C.W.; Doi, S.A.; Schirra, H.J.; Buck, M.; Samaratunga, H.; Perry-Keene, J.; Payton, A.; Yaxley, J.; et al. Prostate-based biofluids for the detection of prostate cancer: A comparative study of the diagnostic performance of cell-sourced RNA biomarkers. *Prostate Int.* **2016**, *4*, 97–102. [CrossRef] [PubMed]

13. Keller, A.; Leidinger, P.; Messe, E.; Haas, J.; Backes, C.; Rasche, L.; Behrens, J.R.; Pfuhl, C.; Wakonig, K.; GieB, R.M.; et al. Next-generation sequencing identifies altered whole blood microRNAs in neuromyelitis optica spectrum disorder which may permit discrimination from multiple sclerosis. *J. Neuroinflamm.* **2015**, *12*, 196–208. [CrossRef] [PubMed]

14. Szczyrba, J.; Löprich, E.; Wach, S.; Jung, V.; Unteregger, G.; Barth, S.; Grobholz, R.; Wieland, W.; Stohr, R.; Hartmann, A.; et al. The MicroRNA profile of prostate carcinoma obtained by deep sequencing. *Mol. Cancer Res.* **2010**, *8*, 529–538. [CrossRef] [PubMed]

15. Huang, X.; Yuan, T.; Liang, M.; Du, M.; Xia, S.; Dittmar, R.; Wang, D.; See, W.; Costello, B.A.; Quevedo, F.; et al. Exosomal miR-1290 and miR-375 as prognostic markers in castration-resistant prostate cancer. *Eur. Urol.* **2015**, *67*, 33–41. [CrossRef] [PubMed]

16. Thompson, I.M.; Ankerst, D.P.; Chi, C.; Lucia, M.S.; Goodman, P.J.; Crowley, J.J.; Parnes, H.L.; Coltman, C.A. Operating characteristics of prostate-specific antigen in men with an initial PSA level of 3.0 ng/mL or lower. *JAMA* **2005**, *294*, 66–70. [CrossRef] [PubMed]

17. Sam, G.; Russell, J.G.; Stijn, D.; Alex, B.; Anton, J.E. MiRBase: MicroRNA sequences, targets and gene nomenclature. *Nucleic Acids Res.* **2006**, *34* (Suppl. S1), D140–D144. [CrossRef]

18. Chen, Y.; McCarthy, D.; Robinson, M.; Smyth, G. *EdgeR: Differential Expression Analysis of Digital Gene Expression User's Guide*; Bioconductor, Roswell Park Cancer Institute: Buffalo, NY, USA, 2014.

19. Dillies, M.-A.; Rau, A.; Aubert, J.; Hennequet-Antier, C.; Jeanmougin, M.; Servant, N.; Keime, C.; Marot, G.; Castel, D.; Estelle, J.; et al. A comprehensive evaluation of normalization methods for Illumina high-throughput RNA sequencing data analysis. *Brief Bioinform.* **2012**. [CrossRef]

20. Centor, R.M. Signal detectability: The use of ROC curves and their analyses. *Med. Decis. Mak.* **1991**, *11*, 102–106. [CrossRef] [PubMed]

21. Kirschner, M.B.; Edelman, J.J.B.; Kao, S.C.-H.; Vallely, M.P.; van Zandwijk, N.; Reid, G. The impact of hemolysis on cell-free microRNA biomarkers. *Front. Genet.* **2013**, *4*, 1–13. [CrossRef] [PubMed]

22. Watahiki, A.; Macfarlane, R.J.; Gleave, M.E.; Crea, F.; Wang, Y.; Helgason, C.D. Plasma miRNAs as biomarkers to identify patients with castration-resistant metastatic prostate cancer. *Int. J. Mol. Sci.* **2013**, *14*, 7757–7770. [CrossRef] [PubMed]

23. Cheng, H.H.; Mitchell, P.S.; Kroh, E.M.; Dowell, A.E.; Chery, L.; Siddiqui, J.; Nelson, P.S.; Vessella, R.L.; Knudsen, B.S.; Chinnaiyan, A.M.; et al. Circulating microRNA profiling identifies a subset of metastatic prostate cancer patients with evidence of cancer-associated hypoxia. *PLoS ONE* **2013**, *8*, e69239. [CrossRef] [PubMed]

24. Lopez-Serra, P.; Esteller, M. DNA methylation-associated silencing of tumor-suppressor microRNAs in cancer. *Oncogene* **2012**, *31*, 1609–1622. [CrossRef] [PubMed]

25. Chen, J.; Wang, M.; Guo, M.; Xie, Y.; Cong, Y.-S. MiR-127 Regulates Cell Proliferation and Senescence by Targeting BCL6. *PLoS ONE* **2013**, *8*, e80266. [CrossRef]

26. Tang, T.T.; Dowbenko, D.; Jackson, A.; Toney, L.; Lewin, D.A.; Dent, A.L.; Lasky, L.A. The Forkhead transcription factor AFX activates apoptosis by induction of the Bcl-6 transcriptional repressor. *J. Biol. Chem.* **2002**, *277*, 14255–14265. [CrossRef] [PubMed]

27. Mencarelli, A.; Renga, B.; Distrutti, E.; Fiorucci, S. Antiatherosclerotic effect of farnesoid X receptor. *Am. J. Physiol. Heart Circ. Physiol.* **2009**, *296*, 272–281. [CrossRef] [PubMed]

28. Wang, F.E.; Zhang, C.; Maminishkis, A.; Dong, L.; Zhi, C.; Li, R.; Zhao, J.; Majerciak, V.; Gaur, A.B.; Chen, S.; et al. MicroRNA-204/211 alters epithelial physiology. *FASEB J.* **2010**, *24*, 1552–1571. [CrossRef] [PubMed]

29. Bao, W.; Wang, H.H.; Tian, F.J.; He, X.Y.; Wang, J.Y.; Zhang, H.J.; Wang, L.H.; Wan, X.P. A TrkB-STAT3-miR-204–5p regulatory circuitry controls proliferation and invasion of endometrial carcinoma cells. *Mol. Cancer* **2013**, *12*, 155. [CrossRef] [PubMed]

30. Yang, H.; Li, Q.; Zhao, W.; Yian, D.; Zhao, H.; Zhou, Y. MiR-329 suppresses the growth and motility of neuroblastoma by targeting KDM1A. *FEBS Lett.* **2014**, *588*, 192–197. [CrossRef] [PubMed]

31. Kashyap, V.; Ahmad, S.; Nilsson, E.M.; Helczynski, L.; Kenna, S.; Persson, J.L.; Gudas, L.J.; Mongan, N.P. The lysine specific demethylase-1 (LSD1/KDM1A) regulates VEGF-A expression in prostate cancer. *Mol. Oncol.* **2013**, *7*, 555–566. [CrossRef] [PubMed]

32. Formosa, A.; Markert, E.K.; Lena, A.M.; Italiano, D.; Finazzi-Ago, E.; Levine, A.J.; Dernardini, S.; Garabadgiu, A.V.; Melino, G.; Candi, E. MicroRNAs, miR-154, miR-299-5p, miR-376a, miR-376c, miR-377, miR-381, miR-487b, miR-485-3p, miR-495 and miR-654-3p, mapped to the 14q32.31 locus, regulate proliferation, apoptosis, migration and invasion in metastatic prostate cancer cells. *Oncogene* **2014**, *33*, 5173–5182. [CrossRef] [PubMed]

33. Gattolliat, C.H.; Thomas, L.; Ciafrè, S.A.; Meurice, G.; Le Teuff, G.; Job, B.; Richon, C.; Combaret, B.; Dessen, P.; Valteau-Couanet, D.; et al. Expression of miR-487b and miR-410 encoded by 14q32.31 locus is a prognostic marker in neuroblastoma. *Br. J. Cancer* **2011**, *105*, 1352–1361. [CrossRef] [PubMed]

34. Le Magnen, C.; Bubendorf, L.; Rentsch, C.A.; Mengus, C.; Gsponer, J.; Zellweger, T.; Rieken, M.; Thaimann, G.N.; Cecchini, M.G.; Germann, M.; et al. Characterization and clinical relevance of ALDH bright populations in prostate cancer. *Clin. Cancer Res.* **2013**, *19*, 5361–5371. [CrossRef] [PubMed]

35. Jalava, S.E.; Urbanucci, A.; Latonen, L.; Waltering, K.K.; Sahu, B.; Janne, O.A.; Seppaia, J.; Lahdesmaki, H.; Tammela, T.L.; Visakorpi, T. Androgen-regulated miR-32 targets BTG2 and is overexpressed in castration-resistant prostate cancer. *Oncogene* **2012**, *31*, 4460–4471. [CrossRef] [PubMed]

36. Mogilyansky, E.; Rigoutsos, I. The miR-17/92 cluster: A comprehensive update on its genomics, genetics, functions and increasingly important and numerous roles in health and disease. *Cell Death Differ.* **2013**, *20*, 1603–1614. [CrossRef] [PubMed]

37. Pesta, M.; Klecka, J.; Kulda, V.; Topolcan, O.; Hora, M.; Eret, V.; Ludvikova, M.; Babjuk, M.; Novak, K.; Stoiz, J.; et al. Importance of miR-20a expression in prostate cancer tissue. *Anticancer Res.* **2010**, *30*, 3579–3583. [PubMed]

38. Wu, X.; Ding, N.; Hu, W.; He, J.; Xu, S.; Pei, H.; Hua, J.; Zhou, G.; Wang, J. Down-regulation of BTG1 by miR-454-3p enhances cellular radiosensitivity in renal carcinoma cells. *Radiat. Oncol.* **2014**, *9*, 179. [CrossRef] [PubMed]

39. Sun, S.; Wang, S.; Xu, X.; Di, H.; Du, J.; Xu, B.; Qang, W.; Wang, J. MiR-433-3p suppresses cell growth and enhances chemosensitivity by targeting CREB in human glioma. *Oncotarget* **2017**, *8*, 5057–5068. [CrossRef] [PubMed]

40. Melo, S.A.; Sugimoto, H.; O'Connell, J.T. Cancer Exosomes perform cell-independent microRNA biogenesis and promote tumorgenesis. *Cancer Cell* **2014**, *26*, 707–721. [CrossRef] [PubMed]

41. Selth, L.A.; Townley, S.L.; Bert, A.G.; Stricker, P.D.; Sutherland, P.D.; Horvath, L.G.; Goodall, G.J.; Butler, L.M.; Tilley, W.D. Circulating microRNAs predict biochemical recurrence in prostate cancer patients. *Br. J. Cancer* **2013**, *109*, 641–650. [CrossRef] [PubMed]

42. Selth, L.A.; Tilley, W.D.; Butler, W.D. Circulating microRNAs: Macro-utility as markers of prostate cancer? *Endocr. Relat. Cancer* **2012**, *19*, R99–R113. [CrossRef] [PubMed]

43. Bai, M.; Zhang, M.; Long, F.; Yu, N.; Zeng, A.; Zhao, R. Circulating microRNA-194 regulates human melanoma cells via PI3K/AKT/FoxO3a and p53/p21 signaling pathways. *Oncol. Rep.* **2017**, *37*, 2702–2710. [CrossRef] [PubMed]

44. Volinia, S.; Calin, G.A.; Liu, C.G.; Ambs, S.; Cimmino, A.J.; Petrocca, F.; Visone, R.; Lanza, G.; Scarpa, A.; Vecchione, A.; et al. A microRNA expression signature of human solid tumors defines cancer gene targets. *Proc. Natl. Acad. Sci. USA* **2006**, *103*, 2257–2261. [CrossRef] [PubMed]

45. Kuwabara, Y.; Ono, K.; Horie, T.; Nishi, H.; Nagao, K.; Kinoshita, M.; Watanabe, S.; Baba, O.; Kojima, Y.; Shizuta, S.; et al. Increased microRNA-1 and microRNA-133a levels in serum of patients with cardiovascular disease indicate myocardial damage. *Circ. Cardiovasc. Genet.* **2011**, *4*, 446–454. [CrossRef] [PubMed]

46. Pritchard, C.C.; Kroh, E.; Wood, B.; Arroyo, J.D.; Dougherty, K.J.; Miyaji, M.M.; Tait, J.F.; Tewari, M. Blood cell origin of circulating microRNAs; a cautionary note for cancer biomarker studies. *Cancer Prev. Res.* **2012**, *5*, 492–497. [CrossRef] [PubMed]

47. Robinson, M.D.; Oshlack, A. A scaling normalization method for differential expression analysis of RNA-seq data. *Genome Biol.* **2010**, *11*, R25. [CrossRef] [PubMed]

48. Andersen, C.L.; Ledet-Jensen, J.; Ørntoft, T. Normalization of real-time quantitative RT-PCR data: A model based variance estimation approach to identify genes suited for normalization—Applied to bladder—And colon-cancer data-sets. *Cancer Res.* **2004**, *64*, 5245–5250. [CrossRef] [PubMed]

49. Seashols-Williams, S.; Lewis, C.; Calloway, C.; Peace, N.; Harrison, A.; Hayes-Nash, C.; Fleming, S.; Wu, Q.; Zehner, Z.E. High-throughput miRNA sequencing and identification of biomarkers for forensically relevant biological fluids. *Electrophoresis* **2016**, *37*, 2780–2788. [CrossRef] [PubMed]

50. Chu, A.; Robertson, G.; Brooks, D.; Mungall, A.J.; Birol, I.; Coope, R.; Marra, M.A. Large-scale profiling of microRNAs for The Cancer Genome Atlas. *Nucleic Acids Res.* **2016**, *44*, e3. [CrossRef] [PubMed]

51. McCarthy, D.J.; Chen, Y.; Smyth, G.K. Differential expression analysis of multifactor RNA-Seq experiments with respect to biological variation. *Nucleic Acids Res.* **2012**, *40*, 4288–4297. [CrossRef] [PubMed]

52. Benjamini, Y.; Hochberg, Y. Controlling the false discovery rate: A practical and powerful approach to multiple testing. *J. R. Statist. Soc.* **1995**, *57*, 289–300.

International Journal of
Molecular Sciences

MDPI

Article

Inherited Variants in Wnt Pathway Genes Influence Outcomes of Prostate Cancer Patients Receiving Androgen Deprivation Therapy

Jiun-Hung Geng [1,2], Victor C. Lin [3,4], Chia-Cheng Yu [5,6,7], Chao-Yuan Huang [8,9], Hsin-Ling Yin [10,11], Ta-Yuan Chang [12], Te-Ling Lu [13], Shu-Pin Huang [1,14,15,*] and Bo-Ying Bao [13,16,17,*]

[1]   Department of Urology, Kaohsiung Medical University Hospital, Kaohsiung 807, Taiwan;
     u9001090@hotmail.com
[2]   Department of Urology, Kaohsiung Municipal Hsiao-Kang Hospital, Kaohsiung 812, Taiwan
[3]   Department of Urology, E-Da Hospital, Kaohsiung 824, Taiwan; victorlin0098@yahoo.com.tw
[4]   School of Medicine for International Students, I-Shou University, Kaohsiung 840, Taiwan
[5]   Division of Urology, Department of Surgery, Kaohsiung Veterans General Hospital, Kaohsiung 813, Taiwan;
     ccyu@vghks.gov.tw
[6]   Department of Urology, School of Medicine, National Yang-Ming University, Taipei 112, Taiwan
[7]   Department of Pharmacy, Tajen University, Pingtung 907, Taiwan
[8]   Department of Urology, National Taiwan University Hospital, College of Medicine,
     National Taiwan University, Taipei 100, Taiwan; cyhuang0909@ntu.edu.tw
[9]   Department of Urology, National Taiwan University Hospital Hsin-Chu Branch, Hsinchu 300, Taiwan
[10] Department of Pathology, Kaohsiung Medical University Hospital, Kaohsiung 807, Taiwan;
     schoolyin@gmail.com
[11] Department of Pathology, Faculty of Medicine, College of Medicine, Kaohsiung Medical University,
     Kaohsiung 807, Taiwan
[12] Department of Occupational Safety and Health, China Medical University, Taichung 404, Taiwan;
     tychang@mail.cmu.edu.tw
[13] Department of Pharmacy, China Medical University, Taichung 404, Taiwan; lutl@mail.cmu.edu.tw
[14] Department of Urology, Faculty of Medicine, College of Medicine, Kaohsiung Medical University,
     Kaohsiung 807, Taiwan
[15] Graduate Institute of Medicine, College of Medicine, Kaohsiung Medical University, Kaohsiung 807, Taiwan
[16] Sex Hormone Research Center, China Medical University Hospital, Taichung 404, Taiwan
[17] Department of Nursing, Asia University, Taichung 413, Taiwan
*   Correspondences: shpihu@yahoo.com.tw (S.-P.H.); bao@mail.cmu.edu.tw (B.-Y.B.);
     Tel.: +886-7-312-1101 (ext. 6694) (S.-P.H.); +886-4-2205-3366 (ext. 5126) (B.-B.Y.)

Academic Editor: Carsten Stephan
Received: 26 September 2016; Accepted: 21 November 2016; Published: 26 November 2016

Abstract: Aberrant Wnt signaling has been associated with many types of cancer. However, the association of inherited Wnt pathway variants with clinical outcomes in prostate cancer patients receiving androgen deprivation therapy (ADT) has not been determined. Here, we comprehensively studied the contribution of common single nucleotide polymorphisms (SNPs) in Wnt pathway genes to the clinical outcomes of 465 advanced prostate cancer patients treated with ADT. Two SNPs, *adenomatous polyposis coli* (*APC*) rs2707765 and rs497844, were significantly ($p \leq 0.009$ and $q \leq 0.043$) associated with both prostate cancer progression and all-cause mortality, even after multivariate analyses and multiple testing correction. Patients with a greater number of favorable alleles had a longer time to disease progression and better overall survival during ADT (p for trend ≤ 0.003). Additional, cDNA array and in silico analyses of prostate cancer tissue suggested that rs2707765 affects *APC* expression, which in turn is correlated with tumor aggressiveness and patient prognosis. This study identifies the influence of inherited variants in the Wnt pathway on the efficacy of ADT and highlights a preclinical rationale for using *APC* as a prognostic marker in advanced prostate cancer.

Int. J. Mol. Sci. **2016**, *17*, 1970

Keywords: prostate cancer; androgen deprivation therapy; outcomes; genetic variation; Wnt pathway

1. Introduction

Prostate cancer is the most common type of cancer in men worldwide, with 1.1 million cases and 307,000 related deaths in 2012. Most cases of prostate cancer are diagnosed and treated while the disease is localized. However, 10%–20% present with advanced-stage or metastatic prostate cancer, and others develop disseminated disease after definitive treatment [1]. Once an advanced-stage or metastatic prostate cancer is diagnosed, androgen deprivation therapy (ADT), which can be accomplished with either bilateral surgical or medical castration, is the standard systemic therapy [2,3]. Although the initial response rate of prostate cancer to ADT has been reported to be as high as 80%, ADT is not curative for prostate cancer, and many patients receiving ADT progress to castration-resistant prostate cancer (CRPC) within 18–30 months [4]. Once CRPC develops, a patient's life expectancy is approximately 16–18 months [5]. Several parameters, including the prostate-specific antigen (PSA) doubling time, PSA nadir, PSA level at ADT initiation, stage and Gleason score, have been reported as useful prognostic predictors of disease progression or survival in patients receiving ADT. However, the predictive ability of these parameters remains limited and might be improved by incorporating information on genetic variants, which may help to estimate disease progression and identify novel therapeutic targets.

The Wnt pathway is an evolutionarily-conserved signal transduction pathway that governs embryonic growth by directing processes, such as cell fate decisions, proliferation, neural patterning, polarity, migration and apoptosis [6,7]. The accumulated evidence has demonstrated that Wnt pathway aberrations are frequently associated with tumor development and the progression of many types of cancer, including colorectal cancer, hepatocellular cancer, breast cancer [8–15] and prostate cancer [16–20]. The best-known example, familial adenomatous polyposis, is an autosomal, dominantly inherited disease characterized by the development of polyps in the colon and rectum. It is most frequently caused by a mutation of the gene *adenomatous polyposis coli* (*APC*) [21,22]. In addition, mutations in β-catenin (*CTNNB1*) and *APC* have been identified in sporadic colorectal cancers and various other tumor types [15]. Although oncogenic mutations in Wnt pathway genes are rare in prostate cancer, increased expression levels of Wnt family proteins have been observed in clinical prostate cancer samples [18,23]. Additionally, the treatment of LNCaP human prostate cancer cells with conditioned medium containing the growth factor Wnt3a significantly enhanced cell growth in the absence of androgens, demonstrating that the Wnt pathway may represent a novel mechanism contributing to prostate cancer progression [24]. Therefore, it is reasonable to hypothesize that genetic variants in the Wnt signaling pathway could influence prostate cancer progression and might also contribute to the development of CRPC. In this study, we comprehensively evaluated the prognostic significance of 17 tagged single nucleotide polymorphisms (SNPs) in three key genes related to the Wnt pathway, *WNT1*, *APC* and *CTNNB1*, with regards to disease progression and all-cause mortality (ACM) in a cohort of prostate cancer patients treated with ADT.

2. Results

The demographic and clinicopathologic characteristics of the study participants are summarized in Table S1. With a median follow-up time of 92 months, 429 (92.3%) prostate cancer patients experienced disease progression after ADT. A total of 143 (30.8%) patients died at a median follow-up of 65 months. The clinical stage, Gleason score, PSA at ADT initiation, PSA nadir, time to PSA nadir and treatment modalities were associated with both disease progression and ACM ($p \leq 0.004$). The age at diagnosis was only correlated with the disease progression ($p < 0.001$).

Seventeen tagging SNPs in three core Wnt pathway genes, namely *CTNNB1*, *APC* and *WNT1*, were analyzed in this study. Details of the SNPs and their associations with disease progression and ACM during ADT are shown in Table S2. Seven *APC* SNPs were associated with disease progression or ACM with a nominal $p < 0.05$, according to a multivariate Cox model adjusted for age, clinical stage, Gleason score, PSA at ADT initiation, PSA nadir, time to PSA nadir and treatment modality. After adjusting for the false discovery rate (FDR) at the <0.05 level, four SNPs, namely *APC* rs3846716, rs2707765, rs41115 and rs497844, remained significantly associated with time to progression ($q \leq 0.012$; Table 1). Intriguingly, *APC* rs2707765 and rs497844 correlated significantly with a decreased risk of both disease progression and ACM ($q \leq 0.043$; Table 2). A gene-dosage effect on disease progression and ACM was found when two genetic loci of interest were analyzed in combination, with the hazards ratio decreasing as the number of protective alleles increased (p for trend ≤ 0.003; Table 3 and Figure 1).

Table 1. Association between htSNPs in *APC* and disease progression in prostate cancer patients treated with ADT.

SNP	Location	Allele	Event [a]	No Event [a]	Best Model	HR (95% CI) [b]	p [b]	q
rs3846716	5′ upstream	G>A	295/117/12	22/11/2	Dominant	0.67 (0.53–0.84)	<0.001	**0.009**
rs2289485	Intron	T>G	371/54/2	30/5/1	Additive	0.74 (0.56–0.99)	0.040	0.093
rs2707765	Intron	G>C	210/176/38	16/15/4	Additive	0.78 (0.67–0.91)	0.002	**0.009**
rs2431238	Intron	C>T	343/76/2	30/4/1	Additive	0.77 (0.60–0.99)	0.044	0.093
rs41115	Thr1493Thr	T>C	300/114/14	25/9/2	Dominant	0.71 (0.57–0.89)	0.003	**0.012**
rs497844	3′ downstream	C>T	301/107/15	24/10/1	Dominant	0.68 (0.54–0.85)	0.001	**0.009**

Abbreviations: htSNP, haplotype tagging single nucleotide polymorphism; *APC*, *adenomatous polyposis coli*; ADT, androgen deprivation therapy; HR, hazards ratio; CI, confidence interval; *p*, *p*-value; *q*, *q*-value. [a] The number represents major allele homozygotes, heterozygotes and minor allele homozygotes, respectively; [b] Multivariate Cox models adjusted for age, clinical stage, Gleason score, prostate-specific antigen (PSA) at ADT initiation, PSA nadir, time to PSA nadir and treatment modality. The major allele homozygotes were considered as the reference group, with a fixed HR = 1.00. *p* < 0.05 is in boldface.

Table 2. Association between htSNPs in *APC* and ACM in prostate cancer patients treated with ADT.

SNP	Location	Allele	Event [a]	No Event [a]	Best Model	HR (95% CI) [b]	p [b]	q
rs3846716	5′ upstream	G>A	106/32/2	211/96/12	Additive	0.65 (0.44–0.95)	0.026	0.055
rs2289485	Intron	T>G	129/12/0	272/47/3	Additive	0.46 (0.25–0.86)	0.014	0.054
rs2707765	Intron	G>C	77/54/9	149/137/33	Additive	0.68 (0.51–0.91)	0.009	**0.043**
rs17134945	Intron	A>G	130/12/0	276/43/3	Additive	0.52 (0.28–0.95)	0.033	0.057
rs41115	Thr1493Thr	T>C	108/31/3	217/92/13	Dominant	0.62 (0.41–0.95)	0.026	0.055
rs497844	3′ downstream	C>T	110/27/3	215/90/13	Dominant	0.51 (0.33–0.80)	0.003	**0.043**

Abbreviations: ACM, all-cause mortality. [a] The number represents major allele homozygotes, heterozygotes and minor allele homozygotes, respectively; [b] Multivariate Cox models adjusted for age, clinical stage, Gleason score, PSA at ADT initiation, PSA nadir, time to PSA nadir and treatment modality. The major allele homozygotes were considered as the reference group, with a fixed HR = 1.00. *p* < 0.05 is in boldface.

As a preliminary assessment of the putative functional roles of these SNPs, we investigated whether rs2707765 and rs497844 were associated with the differential expression of *APC*. The Genotype-Tissue Expression (GTEx) database revealed a significant trend toward increased *APC* mRNA expression in the prostate tissues of rs2707765 protective allele (C) carriers ($p = 0.018$; Figure 2, left) and rs497844 protective allele (T) carriers ($p = 0.21$; Figure 2, right). We used HaploReg to annotate these risk-associated SNPs and discover their potential causal link with the disease. rs2707765, rs497844 and several SNPs in strong linkage disequilibrium with them ($r^2 > 0.8$) (e.g., rs2431514 and rs2464803 for rs2707765 and rs712671 and rs455469 for rs497844) are situated within a locus showing promoter and enhancer histone marks, DNase hypersensitivity peaks and transcription factor binding in several cell lines (Tables S3 and S4). Together, rs2707765 and rs497844 might affect transcription factor binding, increase *APC* mRNA expression and reduce the aggressiveness of prostate cancer.

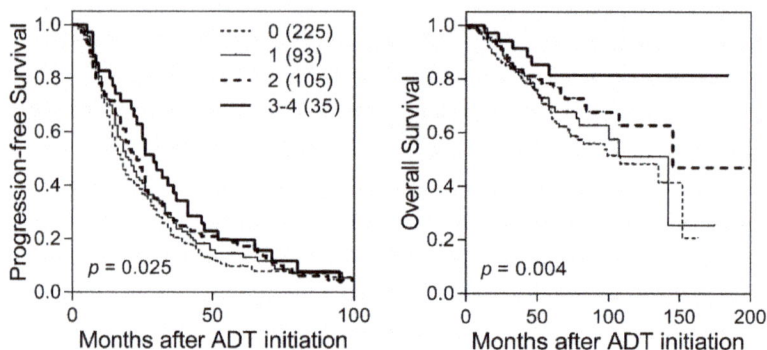

Figure 1. Kaplan-Meier curves of time to progression (**left** panel) and all-cause mortality (ACM) (**right** panel) during androgen deprivation therapy (ADT) for patients with 0, 1, 2 or 3–4 protective alleles at *APC* rs2707765 and rs497844. The protective alleles refer to C in rs2707765 and T in rs497844. The more protective alleles a prostate cancer patient carries, the longer their time to progression and ACM. Numbers in parentheses indicate the number of patients.

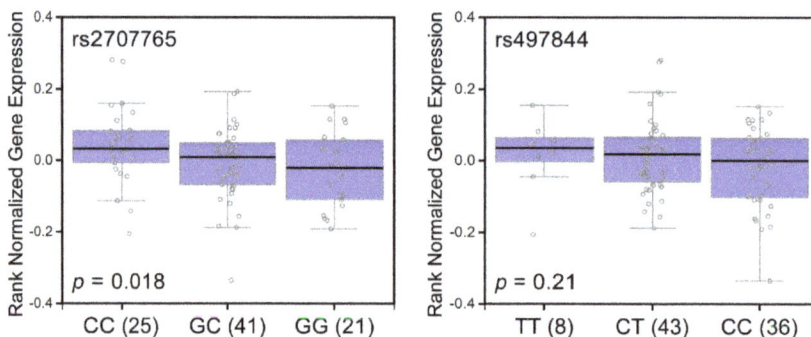

Figure 2. Correlation of rs2707765 and rs497844 genotypes and *APC* gene expression in prostate tissues. Boxplots represent *APC* mRNA expression according to the rs2707765 (**left** panel) and rs497844 (**right** panel) genotypes in 87 human prostate tissues (Genotype-Tissue Expression (GTEx) dataset). There is a trend toward increased *APC* mRNA expression in the prostate tissues of rs2707765 protective allele C carriers and rs497844 protective allele T carriers. The numbers in parentheses indicate the number of cases.

To determine whether *APC* expression is linked to prostate cancer progression, we performed a quantitative real-time polymerase chain reaction (qRT-PCR) analysis of *APC* transcripts using a prostate cancer complementary DNA (cDNA) array containing 48 tissue samples. A significant downregulation of *APC* expression was observed in cancer and more advanced stage cancer samples ($p \leq 0.048$; Figure 3A). To further confirm the relationship between *APC* expression and prostate cancer aggressiveness, we performed an in silico evaluation using publicly-available Memorial Sloan-Kettering Cancer Center Prostate Oncogenome datasets [25]. Our analysis revealed that more clinically-advanced prostate cancers, such as those with metastasis, had significantly lower levels of *APC* expression ($p < 0.001$; Figure 3B). Follow-up of this cohort established that *APC* homozygous deletion, mutation and mRNA downregulation were strongly associated with a worse recurrence-free survival ($p < 0.001$, Figure 3B).

Table 3. Cumulative effect of predictive markers on disease progression and ACM in prostate cancer patients treated with ADT.

No. of Protective Alleles [a]	Overall Patients, *n*	Disease Progression				ACM			
		Events, *n*	Median, *mo*	HR (95% CI) [b]	*p* [b]	Events, *n*	Median, *mo*	HR (95% CI) [b]	*p* [b]
0	225	109	17	1.00	-	77	108	1.00	-
1	93	86	20	0.91 (0.70–1.18)	0.467	30	142	0.76 (0.48–1.20)	0.240
2	105	96	23	0.72 (0.55–0.93)	**0.012**	27	145	0.60 (0.37–0.97)	**0.037**
3–4	35	32	29	0.56 (0.38–0.83)	**0.004**	6	NR [c]	0.36 (0.16–0.84)	**0.018**
Trend	-	-	-	0.84 (0.76–0.93)	**<0.001**	-	-	0.75 (0.62–0.90)	**0.003**

Abbreviations: mo, month. [a] Protective alleles refer to C in rs2707765 and T in rs497844; [b] Multivariate Cox models adjusted for age, clinical stage, Gleason score, PSA at ADT initiation, PSA nadir, time to PSA nadir and treatment modality; [c] NR means the median time-to-event has not been reached. $p < 0.05$ is in boldface.

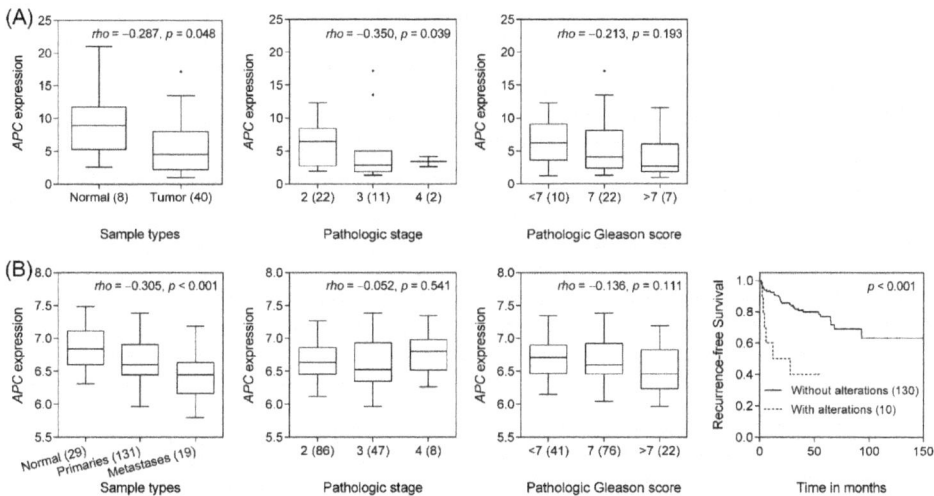

Figure 3. Correlation of *APC* mRNA expression with prostate cancer progression. (**A**) *APC* mRNA expression in 40 prostate cancers and eight normal human prostate tissue specimens, as determined by qRT-PCR, indicates that *APC* is downregulated in the cancer tissues, especially in those from patients with more advanced stages of cancer; (**B**) The associations between *APC* expression and prostate cancer aggressiveness were analyzed using an independent set of Memorial Sloan-Kettering Cancer Center Prostate Oncogenome data. Primary and metastatic prostate cancers display significantly lower *APC* mRNA expression. *rho*, Spearman's rank correlation coefficient. Data points beyond the quartiles are outliers. Kaplan-Meier curves of recurrence-free survival according to the alterations in *APC* are shown. Patients were dichotomized with or without *APC* homozygous deletion, mutation and mRNA downregulation. Numbers in parentheses indicate the number of patients.

3. Discussion

In this hypothesis-driven association study, we identified two SNPs in *APC*, namely rs2707765 and rs497844, as being associated with the efficacy of ADT. Notably, the relationships of these SNPs with disease progression and ACM persisted despite controlling for known prognostic factors and multiple comparison testing. This suggests that the patient genotype adds information beyond conventional factors. Moreover, rs2707765 affects *APC* expression, and a lower *APC* expression correlated with more aggressive cancer and a worse clinical outcome. These results support the connection between the Wnt pathway and prostate cancer progression.

APC is a multifunctional protein involved in a variety of cellular processes. Its best-known role is as a negative regulator of the Wnt pathway, but recent evidence also suggests that it has a role in regulating microtubule dynamics in mitosis. APC was found to associate with the plus-ends of microtubules that interact with kinetochores and to be required for the spindle checkpoint to prevent chromosome missegregation [26]. Many tumors exhibit chromosomal instability, which is probably caused by chromosome missegregation [27]. *APC* mutant mouse embryonic stem cells were reported to have multipolar spindles, indicating that *APC* mutations could lead to chromosomal instability through defective mitosis [28,29]. Therefore, it was postulated that mutations in *APC* lead to chromosomal instability and induce aberrant Wnt signaling, thus contributing to tumor progression. While *APC* is highly relevant with respect to chromosomal aberrations in human colorectal cancers [21,22], mutations in *APC* are rare in human prostate cancer [30]. Many studies have focused on *APC* hypermethylation, which might cause *APC* downregulation [31,32] and has been associated with clinicopathological features of tumor aggressiveness. In a systematic review and meta-analysis of studies on tissue samples, *APC* promoter hypermethylation was found to correlate with more advanced stages of prostate cancer [33]. Functional annotations derived from the Encyclopedia of DNA Elements (ENCODE) and Roadmap Epigenomics data indicate that rs2707765 and rs497844, as well as their linked SNPs, coincide with open chromatin regions that likely correspond to promoters or enhancers of *APC* (Tables S3 and S4). In addition, rs2707765 and rs497844 are predicted to influence the binding of several transcription factors, such as fetal Alz-50 reactive clone 1 (FAC1), activating transcription factor 3 (ATF3) and the E2 factor (E2F) family of transcription factors. These data were consistent with our findings that the protective alleles, rs2707765 C and rs497844 T, tend to have a higher expression of *APC* and a better response to therapy. Interestingly, our previous study also demonstrated a potential prognostic role of *APC* rs3846716 (although it did not reach the predefined significance threshold of $q < 0.05$ in ACM) with regards to post-radical prostatectomy recurrence for localized prostate cancer [34]. Taken together, these results highlight the importance of *APC* throughout all stages of cancer treatment.

The present study has several strengths. This is the first study to have assessed whether genetic variants in the Wnt pathway influence clinical outcomes for advanced prostate cancer patients receiving ADT. Using a cDNA array and after validation with a published dataset, we have also presented additional evidence for a role of *APC* in prostate cancer, namely that downregulated APC gene expression was associated with more aggressive cancer and a poor prognosis. The haplotype-tagging SNP approach ensured the exhaustive and detailed coverage of susceptibility alleles across the whole *APC* gene region. Moreover, complete clinical information and follow-up data allowed us to control for potential confounding factors. However, our study population comprised homogeneous Han Chinese individuals, and therefore, the findings might be less applicable to other ethnic groups. There are many important Wnt pathway genes that have not been investigated in the present study. Thus, further validation in larger interethnic cohorts and additional functional studies will be necessary to elucidate the underlying biological mechanisms.

4. Materials and Methods

4.1. Patient Selection and Data Collection

Our study population included prostate cancer patients treated from 1995–2009 at three medical centers in Taiwan: Kaohsiung Medical University Hospital (Kaohsiung, Taiwan), Kaohsiung Veterans General Hospital (Kaohsiung, Taiwan) and National Taiwan University Hospital (Taipei, Taiwan), as described previously [35]. The clinical characteristics at diagnosis were compiled from medical records, and patients were followed-up prospectively to evaluate whether the tested genetic variations could represent prognostic factors of clinical outcomes during ADT. After excluding patients with inadequate clinicopathological characteristics or follow-up periods, 465 patients remained in the cohort. Written informed consent was obtained from all participants, and the study was approved by the Institutional Review Board of Kaohsiung Medical University Hospital (#KMUHIRB-2013132; 21 January 2014; Kaohsiung, Taiwan).

Treatment outcomes, disease progression and ACM were measured and updated recently during a prospective follow-up in 2014. The PSA nadir was defined as the lowest PSA value achieved at any time during ADT treatment [36,37]. The time to PSA nadir was defined as the duration from the initiation of ADT to the PSA nadir [38]. Disease progression was defined as the presence of at least two consecutive increases in the PSA value more than one week apart that exceeded the PSA nadir [39]. The use of secondary hormone treatment for increasing PSA was also considered to be a progression event. The time to progression was defined as the duration from the start of ADT until disease progression. In this study, patients underwent regular monthly follow-up evaluations and PSA testing at three-month intervals. Causes of death were determined from the official cause of death registry provided by the Department of Health, Executive Yuan, Taipei, Taiwan. ACM was defined as the interval from ADT initiation to death from any cause.

4.2. Single Nucleotide Polymorphism (SNP) Selection and Genotyping

We used a tagging SNP approach to select genetic variants for the investigation of the genetic variability in three core genes of the Wnt pathway, namely *WNT1*, *APC* and *CTNNB1*. Eighteen tagged SNPs were selected from Han Chinese in Beijing HapMap data using Tagger with $r^2 \geq 0.8$ and a minor-allele frequency ≥ 0.2 [40,41]. Genomic DNA was extracted from the peripheral blood using QIAamp DNA Blood Mini Kit (Qiagen, Venlo, The Netherlands) following the manufacturer's instruction, and stored at $-80\ °C$. Genotyping was performed using iPLEX® matrix-assisted laser desorption/ionization time-of-flight mass-spectrometry technology (Agena Bioscience, Hamburg, Germany) at the National Center for Genome Medicine, Academia Sinica, Taipei, Taiwan, as described previously [42]. The average genotype call rate for genotyped SNPs was 99.7%, and the average concordance rate was 100% among five duplicated control samples. One SNP, rs1798802, significantly deviated from the Hardy-Weinberg equilibrium ($p < 0.05$) and was removed, leaving 17 SNPs for further statistical analysis.

4.3. Human Tissue cDNA Array and TaqMan® qRT-PCR Analysis

The expression levels of *APC* and *β-actin* (*ACTB*) were measured using the TissueScan™ human prostate cancer cDNA array II, including 8 normal and 40 prostate cancer samples (OriGene Technologies, Rockville, MD, USA). qRT-PCR was performed using prevalidated TaqMan® gene expression assays (Applied Biosystems, Foster City, CA, USA) for *APC* (Hs01568269_m1) and *ACTB* (Hs01060665_g1), on a 7500 real-time PCR system (Applied Biosystems, Foster City, CA, USA) according to the manufacturer's instructions. The quantification of each sample was normalized using the house-keeping gene *ACTB*.

4.4. Statistical Analysis

The clinical characteristics of the study population were summarized as numbers and percentages of patients or median values with interquartile ranges. Associations of clinical characteristics with disease progression and ACM were estimated using the log-rank test or Cox regression analysis. The median follow-up time and 95% confidence intervals of disease progression and ACM were assessed using the reverse Kaplan-Meier method. Associations of genotyped SNPs with the time to progression and ACM were assessed using multivariate Cox models adjusted for known prognostic factors, including age, clinical stage, Gleason score, PSA at ADT initiation, PSA nadir, time to PSA nadir and treatment modality. For each SNP, three genetic models (dominant, recessive and additive) of inheritance were assessed. The model with the strongest likelihood was considered to be the best model for each SNP. Only additive and dominant models were assessed if the variant homozygotes were observed in <0.05 of the study population. While testing the selected 17 SNPs, the FDR was calculated

Int. J. Mol. Sci. **2016**, *17*, 1970

to determine the degree to which the tests for association were prone to yielding false positive results, using the q-values method [43]. q-Values and two-sided p-values < 0.05 were considered to be statistically significant. SPSS software, Version 22.0.0 (IBM, Armonk, NY, USA), was used for the statistical analyses.

4.5. Bioinformatics Analysis

HaploReg Version 4.1 [44] was used to assess the evidence of regulatory potential for the associated SNPs. We used the GTEx data to examine the correlations between SNPs and gene expression levels in human prostate tissues [45]. To further validate our findings, the association between *APC* expression and prostate cancer aggressiveness was analyzed using data from the Memorial Sloan-Kettering Cancer Center Prostate Oncogenome [25].

5. Conclusions

We comprehensively evaluated haplotype-tagging SNPs in core genes of the Wnt pathway and provided the first evidence of an association between *APC* gene variants and outcomes in advanced prostate cancer patients. Our findings might influence personalized prostate cancer management by identifying patients who would not benefit from ADT.

Supplementary Materials: Supplementary materials can be found at www.mdpi.com/1422-0067/17/12/1970/s1.

Acknowledgments: This work was supported by the Ministry of Science and Technology of Taiwan (Grant Numbers 102-2628-B-039-005-MY3, 103-2314-B-037-060, 104-2314-B-650-006, 104-2314-B-037-052-MY3 and 105-2314-B-650-003-MY3), the Kaohsiung Medical University Hospital (Grant Number KMUH103-3R43), the E-Da Hospital (Grant Numbers EDPJ104059 and EDPJ105054) and the China Medical University (Grant Number CMU105-S-42). The funders had no role in study design, data collection and analysis, decision to publish nor the preparation of the manuscript. We thank Chao-Shih Chen for data analysis and the National Center for Genome Medicine, Ministry of Science and Technology of Taiwan, for technical support. The results published here are based in part on data generated by the GTEx, HaploReg, ENCODE and Roadmap Epigenomics projects.

Author Contributions: Jiun-Hung Geng, Victor C. Lin, Chia-Cheng Yu, Chao-Yuan Huang, Shu-Pin Huang and Bo-Ying Bao conceived of and designed the experiments. Hsin-Ling Yin, Ta-Yuan Chang and Te-Ling Lu performed the experiments and analyzed the data. All authors wrote, reviewed and approved the submission of the paper.

Conflicts of Interest: The authors declare no conflict of interest.

References

1. Djulbegovic, M.; Beyth, R.J.; Neuberger, M.M.; Stoffs, T.L.; Vieweg, J.; Djulbegovic, B.; Dahm, P. Screening for prostate cancer: Systematic review and meta-analysis of randomised controlled trials. *BMJ* **2010**, *341*, c4543. [CrossRef] [PubMed]
2. Harris, W.P.; Mostaghel, E.A.; Nelson, P.S.; Montgomery, B. Androgen deprivation therapy: Progress in understanding mechanisms of resistance and optimizing androgen depletion. *Nat. Clin. Pract. Urol.* **2009**, *6*, 76–85. [CrossRef] [PubMed]
3. Horwitz, E.M. Prostate cancer: Optimizing the duration of androgen deprivation therapy. *Nat. Rev. Urol.* **2009**, *6*, 527–529. [CrossRef] [PubMed]
4. Walczak, J.R.; Carducci, M.A. Prostate cancer: A practical approach to current management of recurrent disease. *Mayo Clin. Proc.* **2007**, *82*, 243–249. [CrossRef]
5. Pienta, K.J.; Bradley, D. Mechanisms underlying the development of androgen-independent prostate cancer. *Clin. Cancer Res.* **2006**, *12*, 1665–1671. [CrossRef] [PubMed]
6. Moon, R.T.; Miller, J.R. The APC tumor suppressor protein in development and cancer. *Trends Genet.* **1997**, *13*, 256–258. [CrossRef]
7. Spink, K.E.; Polakis, P.; Weis, W.I. Structural basis of the axin-adenomatous polyposis coli interaction. *EMBO J.* **2000**, *19*, 2270–2279. [CrossRef] [PubMed]
8. Logan, C.Y.; Nusse, R. The Wnt signaling pathway in development and disease. *Annu. Rev. Cell Dev. Biol.* **2004**, *20*, 781–810. [CrossRef] [PubMed]

9. Katoh, M. Wnt/pcp signaling pathway and human cancer (review). *Oncol. Rep.* **2005**, *14*, 1583–1588. [CrossRef] [PubMed]

10. Brennan, K.R.; Brown, A.M. Wnt proteins in mammary development and cancer. *J. Mammary Gland Biol. Neoplasia* **2004**, *9*, 119–131. [CrossRef] [PubMed]

11. Brown, A.M. Wnt signaling in breast cancer: Have we come full circle? *Breast Cancer Res.* **2001**, *3*, 351–355. [CrossRef] [PubMed]

12. Howe, L.R.; Brown, A.M. Wnt signaling and breast cancer. *Cancer Biol. Ther.* **2004**, *3*, 36–41. [CrossRef] [PubMed]

13. Park, J.Y.; Park, W.S.; Nam, S.W.; Kim, S.Y.; Lee, S.H.; Yoo, N.J.; Lee, J.Y.; Park, C.K. Mutations of β-catenin and AXIN I genes are a late event in human hepatocellular carcinogenesis. *Liver Int.* **2005**, *25*, 70–76. [CrossRef] [PubMed]

14. Takahashi, M.; Wakabayashi, K. Gene mutations and altered gene expression in azoxymethane-induced colon carcinogenesis in rodents. *Cancer Sci.* **2004**, *95*, 475–480. [CrossRef] [PubMed]

15. Clevers, H. Wnt breakers in colon cancer. *Cancer Cell* **2004**, *5*, 5–6. [CrossRef]

16. Mulholland, D.J.; Read, J.T.; Rennie, P.S.; Cox, M.E.; Nelson, C.C. Functional localization and competition between the androgen receptor and T-cell factor for nuclear β-catenin: A means for inhibition of the Tcf signaling axis. *Oncogene* **2003**, *22*, 5602–5613. [CrossRef] [PubMed]

17. Yardy, G.W.; Brewster, S.F. Wnt signalling and prostate cancer. *Prostate Cancer Prostatic Dis.* **2005**, *8*, 119–126. [CrossRef] [PubMed]

18. Verras, M.; Sun, Z. Roles and regulation of wnt signaling and β-catenin in prostate cancer. *Cancer Lett.* **2006**, *237*, 22–32. [CrossRef] [PubMed]

19. Bierie, B.; Nozawa, M.; Renou, J.P.; Shillingford, J.M.; Morgan, F.; Oka, T.; Taketo, M.M.; Cardiff, R.D.; Miyoshi, K.; Wagner, K.U.; et al. Activation of β-catenin in prostate epithelium induces hyperplasias and squamous transdifferentiation. *Oncogene* **2003**, *22*, 3875–3887. [CrossRef] [PubMed]

20. Chesire, D.R.; Isaacs, W.B. β-catenin signaling in prostate cancer: An early perspective. *Endocr. Relat. Cancer* **2003**, *10*, 537–560. [CrossRef] [PubMed]

21. Kinzler, K.W.; Nilbert, M.C.; Su, L.K.; Vogelstein, B.; Bryan, T.M.; Levy, D.B.; Smith, K.J.; Preisinger, A.C.; Hedge, P.; McKechnie, D.; et al. Identification of fap locus genes from chromosome 5q21. *Science* **1991**, *253*, 661–665. [CrossRef] [PubMed]

22. Nishisho, I.; Nakamura, Y.; Miyoshi, Y.; Miki, Y.; Ando, H.; Horii, A.; Koyama, K.; Utsunomiya, J.; Baba, S.; Hedge, P.; et al. Mutations of chromosome 5q21 genes in fap and colorectal cancer patients. *Science* **1991**, *253*, 665–669. [CrossRef] [PubMed]

23. Chen, G.; Shukeir, N.; Potti, A.; Sircar, K.; Aprikian, A.; Goltzman, D.; Rabbani, S.A. Up-regulation of Wnt-1 and β-catenin production in patients with advanced metastatic prostate carcinoma: Potential pathogenetic and prognostic implications. *Cancer* **2004**, *101*, 1345–1356. [CrossRef] [PubMed]

24. Verras, M.; Brown, J.; Li, X.; Nusse, R.; Sun, Z. Wnt3a growth factor induces androgen receptor-mediated transcription and enhances cell growth in human prostate cancer cells. *Cancer Res.* **2004**, *64*, 8860–8866. [CrossRef] [PubMed]

25. Hieronymus, H.; Schultz, N.; Gopalan, A.; Carver, B.S.; Chang, M.T.; Xiao, Y.; Heguy, A.; Huberman, K.; Bernstein, M.; Assel, M.; et al. Copy number alteration burden predicts prostate cancer relapse. *Proc. Natl. Acad. Sci. USA* **2014**, *111*, 11139–11144. [CrossRef] [PubMed]

26. Nathke, I.S. The adenomatous polyposis coli protein: The achilles heel of the gut epithelium. *Annu. Rev. Cell Dev. Biol.* **2004**, *20*, 337–366. [CrossRef] [PubMed]

27. Lengauer, C.; Kinzler, K.W.; Vogelstein, B. Genetic instabilities in human cancers. *Nature* **1998**, *396*, 643–649. [CrossRef] [PubMed]

28. Fodde, R.; Kuipers, J.; Rosenberg, C.; Smits, R.; Kielman, M.; Gaspar, C.; van Es, J.H.; Breukel, C.; Wiegant, J.; Giles, R.H.; et al. Mutations in the APC tumour suppressor gene cause chromosomal instability. *Nat. Cell Biol.* **2001**, *3*, 433–438. [CrossRef] [PubMed]

29. Kaplan, K.B.; Burds, A.A.; Swedlow, J.R.; Bekir, S.S.; Sorger, P.K.; Nathke, I.S. A role for the adenomatous polyposis coli protein in chromosome segregation. *Nat. Cell Biol.* **2001**, *3*, 429–432. [CrossRef] [PubMed]

30. Kypta, R.M.; Waxman, J. Wnt/β-catenin signalling in prostate cancer. *Nat. Rev. Urol.* **2012**, *9*, 418–428. [CrossRef] [PubMed]

31. Cho, N.Y.; Kim, J.H.; Moon, K.C.; Kang, G.H. Genomic hypomethylation and cpg island hypermethylation in prostatic intraepithelial neoplasm. *Virchows Arch.* **2009**, *454*, 17–23. [CrossRef] [PubMed]

32. Costa, V.L.; Henrique, R.; Jeronimo, C. Epigenetic markers for molecular detection of prostate cancer. *Dis. Markers* **2007**, *23*, 31–41. [CrossRef] [PubMed]

33. Chen, Y.; Li, J.; Yu, X.; Li, S.; Zhang, X.; Mo, Z.; Hu, Y. APC gene hypermethylation and prostate cancer: A systematic review and meta-analysis. *Eur. J. Hum. Genet.* **2013**, *21*, 929–935. [CrossRef] [PubMed]

34. Huang, S.P.; Ting, W.C.; Chen, L.M.; Huang, L.C.; Liu, C.C.; Chen, C.W.; Hsieh, C.J.; Yang, W.H.; Chang, T.Y.; Lee, H.Z.; et al. Association analysis of Wnt pathway genes on prostate-specific antigen recurrence after radical prostatectomy. *Ann. Surg. Oncol.* **2010**, *17*, 312–322. [CrossRef] [PubMed]

35. Bao, B.Y.; Pao, J.B.; Huang, C.N.; Pu, Y.S.; Chang, T.Y.; Lan, Y.H.; Lu, T.L.; Lee, H.Z.; Juang, S.H.; Chen, L.M.; et al. Polymorphisms inside micrornas and microrna target sites predict clinical outcomes in prostate cancer patients receiving androgen-deprivation therapy. *Clin. Cancer Res.* **2011**, *17*, 928–936. [CrossRef] [PubMed]

36. Stewart, A.J.; Scher, H.I.; Chen, M.H.; McLeod, D.G.; Carroll, P.R.; Moul, J.W.; D'Amico, A.V. Prostate-specific antigen nadir and cancer-specific mortality following hormonal therapy for prostate-specific antigen failure. *J. Clin. Oncol.* **2005**, *23*, 6556–6560. [CrossRef] [PubMed]

37. Kwak, C.; Jeong, S.J.; Park, M.S.; Lee, E.; Lee, S.E. Prognostic significance of the nadir prostate specific antigen level after hormone therapy for prostate cancer. *J. Urol.* **2002**, *168*, 995–1000. [CrossRef]

38. Choueiri, T.K.; Xie, W.; D'Amico, A.V.; Ross, R.W.; Hu, J.C.; Pomerantz, M.; Regan, M.M.; Taplin, M.E.; Kantoff, P.W.; Sartor, O.; et al. Time to prostate-specific antigen nadir independently predicts overall survival in patients who have metastatic hormone-sensitive prostate cancer treated with androgen-deprivation therapy. *Cancer* **2009**, *115*, 981–987. [CrossRef] [PubMed]

39. Ross, R.W.; Oh, W.K.; Xie, W.; Pomerantz, M.; Nakabayashi, M.; Sartor, O.; Taplin, M.E.; Regan, M.M.; Kantoff, P.W.; Freedman, M. Inherited variation in the androgen pathway is associated with the efficacy of androgen-deprivation therapy in men with prostate cancer. *J. Clin. Oncol.* **2008**, *26*, 842–847. [CrossRef] [PubMed]

40. De Bakker, P.I.; Yelensky, R.; Pe'er, I.; Gabriel, S.B.; Daly, M.J.; Altshuler, D. Efficiency and power in genetic association studies. *Nat. Genet.* **2005**, *37*, 1217–1223. [CrossRef] [PubMed]

41. Frazer, K.A.; Ballinger, D.G.; Cox, D.R.; Hinds, D.A.; Stuve, L.L.; Gibbs, R.A.; Belmont, J.W.; Boudreau, A.; Hardenbol, P.; Leal, S.M.; et al. A second generation human haplotype map of over 3.1 million SNPs. *Nature* **2007**, *449*, 851–861. [CrossRef] [PubMed]

42. Huang, S.P.; Huang, L.C.; Ting, W.C.; Chen, L.M.; Chang, T.Y.; Lu, T.L.; Lan, Y.H.; Liu, C.C.; Yang, W.H.; Lee, H.Z.; et al. Prognostic significance of prostate cancer susceptibility variants on prostate-specific antigen recurrence after radical prostatectomy. *Cancer Epidemiol. Biomark. Prev.* **2009**, *18*, 3068–3074. [CrossRef] [PubMed]

43. Storey, J.D.; Tibshirani, R. Statistical significance for genomewide studies. *Proc. Natl. Acad. Sci. USA* **2003**, *100*, 9440–9445. [CrossRef] [PubMed]

44. Ward, L.D.; Kellis, M. Haploreg: A resource for exploring chromatin states, conservation, and regulatory motif alterations within sets of genetically linked variants. *Nucleic Acids Res.* **2012**, *40*, D930–D934. [CrossRef] [PubMed]

45. Lonsdale, J.; Thomas, J.; Salvatore, M.; Phillips, R.; Lo, E.; Shad, S.; Hasz, R.; Walters, G.; Garcia, F.; Young, N.; et al. The genotype-tissue expression (GTEx) project. *Nat. Genet.* **2013**, *45*, 580–585. [CrossRef] [PubMed]

International Journal of
Molecular Sciences

MDPI

Article

Early Prediction of Therapy Response to Abiraterone Acetate Using PSA Subforms in Patients with Castration Resistant Prostate Cancer

Katrin Schlack [1,*], **Laura-Maria Krabbe** [1,2], **Manfred Fobker** [3], **Andres Jan Schrader** [1], **Axel Semjonow** [1] and **Martin Boegemann** [1]

[1] Department of Urology, Prostate Center, University Hospital Muenster, Albert-Schweitzer-Campus 1, GB A1, Muenster D-48149, Germany; laura-maria.krabbe@ukmuenster.de (L.-M.K.); andresjan.schrader@ukmuenster.de (A.J.S.); axel.semjonow@ukmuenster.de (A.S.); martin.boegemann@ukmuenster.de (M.B.)

[2] Department of Urology, University of Texas Southwestern Medical Center, Dallas, TX 75390-9110, USA

[3] Center for Laboratory Medicine, University Hospital Muenster, Albert-Schweitzer-Campus 1, GB A1, Muenster D-48149, Germany; manfred.fobker@ukmuenster.de

* Correspondence: katrin.schlack@ukmuenster.de; Tel.: +49-251-83-44600

Academic Editor: Carsten Stephan
Received: 9 August 2016; Accepted: 1 September 2016; Published: 9 September 2016

Abstract: The purpose of this study was to evaluate the prognostic ability of early changes of total prostate specific antigen (tPSA), free PSA (fPSA), [−2]proPSA and the Prostate Health Index (PHI) following initiation of Abiraterone-therapy in men with castration resistant prostate cancer (mCRPC). In 25 patients, PSA-subforms were analyzed before and at 8–12 weeks under therapy as prognosticators of progression-free-survival (PFS) and overall survival (OS). Comparing patients with a PFS < vs. ≥12 months by using Mann–Whitney–Wilcoxon Tests, the relative-median-change of tPSA (−0.1% vs. −86.8%; $p = 0.02$), fPSA (12.1% vs. −55.3%; $p = 0.03$) and [−2]proPSA (8.1% vs. −59.3%; $p = 0.05$) differed significantly. For men with ≤ vs. >15 months of OS there was a non-significant trend for a difference in the relative-median-change of fPSA (17.0% vs. −46.3%; $p = 0.06$). In Kaplan–Meier analyses, declining fPSA and [−2]proPSA were associated with a longer median PFS (13 months, 95% confidence interval (CI): 9.6–16.4 vs. 10 months, 95% CI: 3.5–16.5; $p = 0.11$), respectively. Correspondingly, decreasing fPSA and [−2]proPSA values indicated an OS of 32 months (95% CI: not reached (NR)) compared to 21 months in men with rising values (95% CI: 7.7–34.3; $p = 0.14$), respectively. We concluded that the addition of fPSA- and [−2]proPSA-changes to tPSA-information might be further studied as potential markers of early Abiraterone response in mCRPC patients.

Keywords: mCRPC; surrogate biomarker; abiraterone acetate; prognosticators; prostate cancer; [−2]proPSA; fPSA; PHI; tPSA

1. Introduction

During the past few years, there has been significant progress in treatment options for metastatic castration resistant prostate cancer (mCRPC). After prostate cancer develops, resistance to androgen-deprivation therapy (ADT), patients can be treated with different agents including chemotherapy, next generation ADT drugs, Sipuleucel-T or Radium-223 [1–8]. Amongst these, Abiraterone acetate (Abiraterone), a selective CYP450 17A1 inhibitor, is available in the pre- and post-chemotherapy setting and is broadly accepted as a standard of care due to its life-prolonging potential and generally low toxicity [1,2,9].

A challenging aspect for clinicians in this setting is the early determination of treatment success, particularly the early identification of response or failure in patients whose lack of symptoms makes a clinical decision difficult. Except a decline in total prostate specific antigen (tPSA), there are no easily available prognostic biomarkers besides clinical parameters to determine therapeutic success. Therefore, the decision whether to continue or to change a therapy regime is mainly based on clinical factors (e.g., eastern cooperative oncology group performance status (ECOG)) and tPSA, which is still the most commonly used therapy control marker in mCRPC.

However, there are several limitations concerning tPSA, especially in mCRPC. The parameter does not always decline right away even though patients might still benefit from therapy. Thus, in the registration trials, tPSA could not be validated as an independent prognosticator for therapy response [1,2]. Furthermore, an early rise of tPSA during the first 12 weeks of therapy followed by a delayed decline (PSA-flare) during Abiraterone-therapy was recently found to be a prognosticator for an improved median progression free survival (PFS) [10]. PSA-flare occurs in about 10% of the patients treated with Abiraterone. A larger proportion of patients with initial increase of tPSA (nearly 50%), however, will suffer a continuous rise of tPSA over time followed by clinical progression [10]. In clinical routines, the differentiation between PSA-flare and continuous rise is difficult, resulting in either early, sometimes premature, interruption of therapy or waiting for the potential benefit of a delayed tPSA-decline with, in many cases, a treatment beyond true progression. Therefore, additional biomarkers to distinguish between true progression and a delayed tPSA response are desperately needed for these situations.

Several biomarkers, including circulating tumor cells (CTCs) and lactate dehydrogenase (LDH), have recently been discussed as response-indicators [11,12].

CTC counts at baseline are considered to be prognostic for the duration of treatment with a resulting shorter time of treatment for ≥ 5 CTCs [11]. However, the sensitivity of detection of CTCs is limited in patients with low tumor load and differs depending on the site of metastasis [13]. Additionally, CTC-assays are expensive and not broadly available.

Similar data show that the combination of ≥ 5 CTC count and an elevated LDH level after 12 weeks of therapy may be a surrogate for poor overall survival (OS) [12]. However, for broad acceptance and use, more easily available as well as more sensitive, validated and non-expensive assays are required [14].

In 1991, the low specificity of PSA for the detection of prostate cancer was improved by the identification of two major molecular subforms of PSA, an unbound proportion (free PSA, fPSA) and a form complexed with the protease inhibitor alpha-1-antichimotrypsin [15]. It was shown that the chance of detecting prostate cancer was lower with increasing proportion of fPSA (%fPSA) [16,17]. Following the detection of precursor forms of fPSA (proPSA), further investigation concluded that proPSA was associated with prostate cancer with a higher probability of cancer as the percentage of proPSA in fPSA increases [18,19]. Within the proPSA fraction, further discrimination could be made by the detection of truncated forms which are more resistant to activation to mature PSA, with [−2]proPSA being the most consistent of these [20]. According to recent investigations, the ratio of [−2]proPSA to fPSA (%[−2]proPSA) and the Prostate Health Index (PHI = ([−2]proPSA/fPSA) × $\sqrt{\text{tPSA}}$) seem to be the strongest predictors of prostate cancer compared to tPSA and its subforms [21,22]. In addition, there was evidence for a correlation with more aggressive variants of prostate cancer since the variables increased with increasing Gleason-Score [23,24].

In this study, we aimed to investigate whether PSA subforms could be helpful for predicting treatment response and prognosticating survival outcomes in patients with mCRPC treated with Abiraterone. When added to tPSA in the setting of organ confined prostate cancer, the Supplementary Materials of fPSA, [−2]proPSA and PHI help to diagnose and identify more aggressive forms of prostate cancer [25,26]. Therefore, we assumed that these markers could also be useful to facilitate decision making in the treatment of mCRPC.

2. Results

2.1. Characteristics of the Study Group

For the patients alive at the time of analysis, the median follow-up was 25 months (Interquartile range (IQR) 14.5–28.0) in July 2015. The median time on Abiraterone-therapy was 13 months (IQR 10.5–19.0) with six (24%) patients on therapy at the time of last data acquisition. No dose modifications were necessary for any patients. Descriptive characteristics of the cohort are given in Table 1. The median age of the patients was 71 years (IQR 63–74). Bone metastases were present in 76%, lymphonodal metastases in 64% and visceral metastases in 8% of patients at the beginning of Abiraterone-therapy. An unfavorable Gleason-Score of ≥8 at initial diagnosis of PCa was found in 60%. Median baseline levels were 61.7 ng/mL for tPSA (IQR 29.0–299.5), 11.2 ng/mL for fPSA (IQR 5.2–29.1), 485.3 pg/mL for [−2]proPSA (IQR 272.8–1107.9) and 327.3 for PHI (IQR 212.5–612.0). No patient showed elevated liver enzyme- or creatinine-concentrations during the course of the study.

Antiresorptive therapy (Zoledronic acid or Denosumab) was administered in 12 patients (48%). All of these remained on a stable dose of the drug during the course of Abiraterone-therapy.

Table 1. Baseline characteristics of patients with mCRPC under therapy with Abiraterone IQR, interquartile range; ECOG, eastern cooperative oncology group performance status.

Variable	Number
Patients (*n*) (%)	25 (100)
Age, median (years) (IQR)	71.0 (62.5–74.0)
Median follow-up (months) (IQR)	25.0 (14.5–28.0)
Median duration of therapy (months) (IQR)	13.0 (10.5–19.0)
Presence of lymphnode metastases (*n*) (%)	16 (64.0)
Presence of bone metastases (*n*) (%)	19 (76.0)
Presence of visceral metastases (*n*) (%)	2 (8.0)
Pre chemotherapy (*n*) (%)	20 (80.0)
Post chemotherapy (*n*) (%)	5 (20.0)
Antiresorptive therapy (*n*) (%)	12 (48.0)
Zoledronic acid (*n*) (%)	7 (28.0)
Denosumab (*n*) (%)	5 (20.0)
ECOG (all) (*n*) (%)	
0–1	22 (88.0)
>1	3 (12.0)
Gleason Score ≥ 8 (*n*) (%)	15 (60.0)
Median tPSA baseline (ng/mL) (IQR)	61.7 (29.0–299.5)
Median fPSA baseline (ng/mL) (IQR)	11.2 (5.2–29.1)
Median [−2]proPSA baseline (pg/mL) (IQR)	485.3 (272.8–1107.9)
Median PHI baseline (IQR)	327.3 (212.5–612.0)
Best clinical outcome (*n*) (%)	
Complete remission	1 (4.0)
Partial remission	10 (40.0)
Stable disease	10 (40.0)
Progressive disease	4 (16.0)
tPSA decrease ≥ 50% (*n*) (%)	12 (48.0)
tPSA decrease ≥ 90% (*n*) (%)	7 (28.0)
Patients died (*n*) (%)	11 (44.0)

2.2. Value of PSA Subforms as Prognostic Markers

Alterations of the baseline values are presented as relative change of median values in percent (Figure 1). The definition of progressive disease is explained in paragraph 4 (Materials and Methods). For patients with shorter response to therapy (PFS < 12 months and OS $\leq$ 15 months) the median fPSA and [−2]proPSA increased at 8–12 weeks. In contrast to these two subforms, the median tPSA showed stable values at the same time for these patients. The other parameters (PHI, %[−2]proPSA and %fPSA) seem to play a minor role, considering the scope of our study, since changes did not differ between the groups (Table 2).

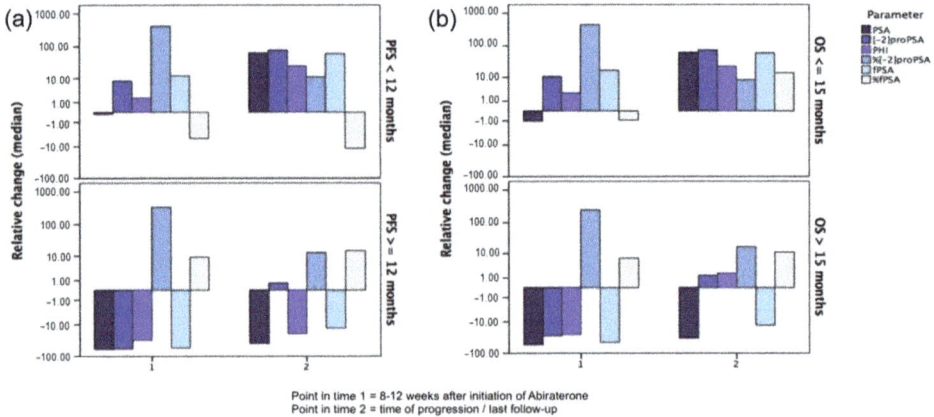

Point in time 1 = 8-12 weeks after initiation of Abiraterone
Point in time 2 = time of progression / last follow-up

Figure 1. Relative median changes of PSA subforms at 8–12 weeks and at progression or last follow-up compared to baseline for (**a**) PFS < 12 months vs. ≥12 months; and (**b**) OS ≤ 15 months vs. >15 months.

Table 2. Relative changes of median PSA-subform values (%) in 25 patients at 8–12 weeks for OS and PFS OS, overall survival; PFS, progression free survival.

Variable	OS ≤ 15 Months	OS > 15 Months	*p*
tPSA	−1.1 (−54.5–66.3)	−55.1 (−74.6–13.2)	0.20
[−2]proPSA	10.5 (−50.7–62.0)	−29.1 (−66.9–19.9)	0.15
PHI	2.6 (−42.9–39.0)	−25.9 (−41.4–19.0)	0.32
%[−2]proPSA	449.0 (260.6–588.4)	242.1 (149.5–577.7)	0.27
fPSA	17.0 (−41.8–50.3)	−46.3 (−72.0–8.1)	0.06
%fPSA	−0.9 (−19.6–28.1)	7.0 (−27.6–41.3)	0.61
	PFS < 12 Months	**PFS ≥ 12 Months**	
tPSA	−0.1 (−0.1–71.3)	−86.8 (−60.9–9.4)	0.02
[−2]proPSA	8.1 (−21.2–60.1)	−59.3 (−74.0–18.5)	0.05
PHI	1.7 (−24.4–63.6)	−32.1 (−47.8–4.1)	0.12
%[−2]proPSA	416.0 (159.8–754.2)	335.5 (163.2–550.7)	0.73
fPSA	12.1 (−22.7–43.8)	−55.3 (−83.3–13.6)	0.03
%fPSA	−5.2 (−28.3–16.7)	9.1 (−17.3–42.0)	0.15

In patients with < vs. ≥12 months of PFS, the relative change of median tPSA values (−0.1% vs. −86.8%; *p* = 0.02 (Mann–Whitney–Wilcoxon Test)) at 8–12 weeks compared to baseline differed significantly. Similar results were shown for decreasing [−2]proPSA- (8.1% vs. −59.3%; *p* = 0.05) and fPSA values (12.1% vs. −55.3%; *p* = 0.03), respectively.

For men with ≤ vs. >15 months of OS, there was a non-significant trend for a difference in relative changes of median fPSA values (17.0% vs. −46.3%; *p* = 0.06). Considering OS, no conclusive change of median values was seen for the other parameters (tPSA, [−2]proPSA, PHI, %[−2]proPSA and %fPSA).

In univariate Cox-regression analysis, rising [−2]proPSA and fPSA were non-significant prognosticators of shorter PFS (Hazard ratio (HR): 2.0, (95% confidence interval (CI): 0.8–4.8); p = 0.14 for both parameters). Analogously, the results for OS showed a non-significant trend for rising [−2]proPSA and fPSA regarding the prognostication of poor OS (HR: 2.5 (95% CI: 0.7–9.1); p = 0.16 for both parameters). In this analysis, PHI also showed a trend towards shortened PFS in case of an early increase (HR: 2.0 (95% CI: 0.9–5.0); p = 0.11). For OS, however, PHI did not show any trend for shortened survival (HR: 1.5 (95% CI: 0.4–5.4); p = 0.49). Of note, rising tPSA changes did not prove to be a prognosticator for worse PFS (HR: 1.6 (95% CI: 0.6–4.0); p = 0.29) or OS (HR: 1.3 (95% CI: 0.4–4.5); p = 0.72) (Table 3).

Table 3. Changes of median PSA-subform values in univariate analysis OS, overall survival; PFS, progression; HR, hazard ratio; CI, confidence interval.

Univariate Analysis for OS			Univariate Analysis for PFS		
Variable	HR (95% CI)	p	Variable	HR (95% CI)	p
tPSA increase yes vs. no	1.3 (0.4–4.5)	0.72	tPSA increase yes vs. no	1.6 (0.6–4.0)	0.29
fPSA increase yes vs. no	2.5 (0.7–9.1)	0.16	fPSA increase yes vs. no	2.0 (0.8–4.8)	0.14
[−2]proPSA increase yes vs. no	2.5 (0.7–9.1)	0.16	[−2]proPSA increase yes vs. no	2.0 (0.8–4.8)	0.14
PHI increase yes vs. no	1.5 (0.4–5.4)	0.49	PHI increase yes vs. no	2.0 (0.9–5.0)	0.11
%[−2]proPSA increase yes vs. no	23.1 (0–∞)	0.57	%[−2]proPSA increase yes vs. no	1.8 (0.2–13.5)	0.57
%fPSA increase no vs. yes	0.4 (0.1–1.7)	0.23	%fPSA increase no vs. yes	1.2 (0.5–2.8)	0.73

2.3. Kaplan–MEIER Survival Analysis

The Kaplan–Meier analyses for PFS and OS are presented in Figures 2–4. Here, the strongest marker for longer PFS in Mann–Whitney–Wilcoxon comparison of median values, declining fPSA, also showed better survival outcomes. A declining fPSA value at 8–12 weeks was associated with a median PFS of 13 months (95% CI: 9.6–16.4 months) compared to 10 months (95% CI: 3.5–16.5 months) in patients with increasing values (log-rank p = 0.11).

Correspondingly, in the analysis of OS, a decreasing fPSA at 8–12 weeks of therapy indicated a median OS of 32 months (95% CI: not reached (NR)) compared to 21 months (95% CI: 7.7–34.3 months) in men with rising fPSA (p = 0.14) (Figure 2).

The Kaplan–Meier analysis of PFS and OS with declining and rising [−2]proPSA showed similar results (Figure 3).

In contrast, in the analysis of PFS for tPSA, there was less difference between rising and declining values. A tPSA decrease was associated with a median PFS of 12 months (95% CI: 8.4–15.6 months) compared to 10 months (95% CI: 7.1–13.0 months) in the case of increasing values (p = 0.26). Correspondingly, the evaluation of OS did not show differences between tPSA values in both groups with 32 months (95% CI: NR) vs. 28 months (95% CI: 17.2–38.8 months, p = 0.72) (Figure 4).

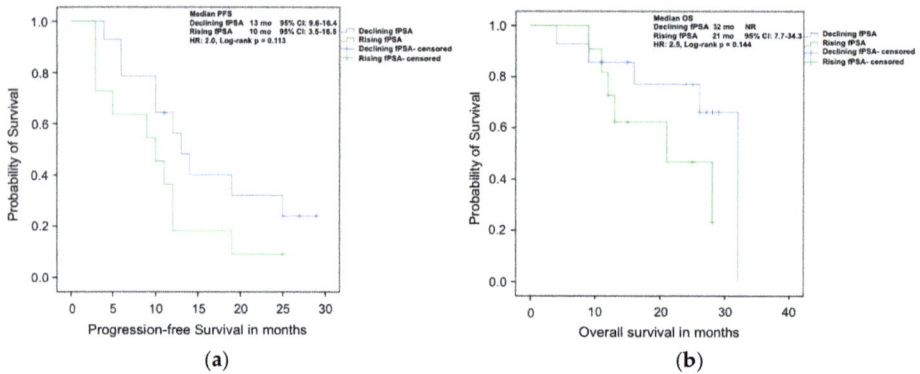

Figure 2. Kaplan–Meier analyses of (**a**) progression free survival; and (**b**) overall survival of mCRPC patients treated with Abiraterone with rising or declining fPSA values after 8–12 weeks of therapy.

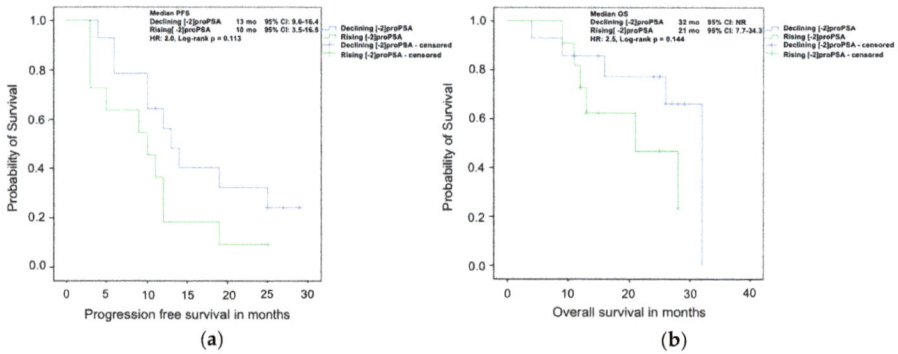

Figure 3. Kaplan–Meier analyses of (**a**) progression free survival; and (**b**) overall survival of mCRPC patients treated with Abiraterone with rising or declining [−2]proPSA values after 8–12 weeks of therapy.

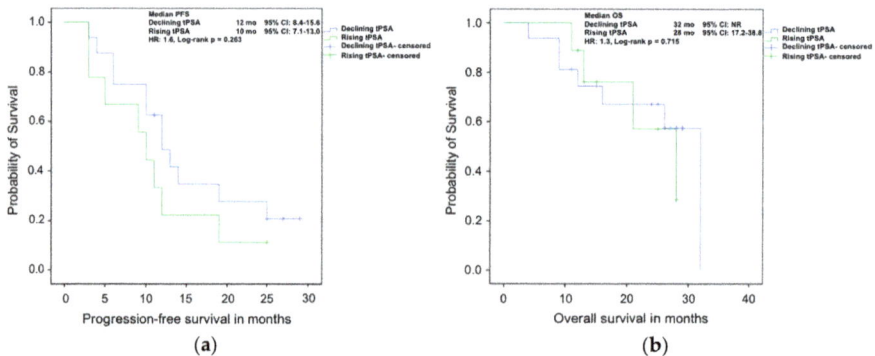

Figure 4. Kaplan–Meier analyses of (**a**) progression free survival; and (**b**) overall survival of mCRPC patients treated with Abiraterone with rising or declining tPSA values after 8–12 weeks of therapy.

3. Discussion

To our knowledge, the relevance of fPSA and [−2]proPSA as well as PHI in patients with mCRPC under Abiraterone has not been investigated thus far. Standard monitoring of therapy is mainly based on the measurement of tPSA as well as radiological criteria and the general condition of the patients like the assessment of ECOG performance status [27].

Among the group of patients with mCRPC, a large proportion of up to 90% shows bone metastases [28]. During early imaging, especially in the first three to six months of therapy, an early "bone-flare" can occur in a relevant amount of patients, which can lead to incorrect classification as progressive disease (PD) [29,30].

The most common parameter being used for determination of treatment success under any therapy for mCRPC is tPSA. Here, comparable to the above-mentioned "bone-flare", a "PSA-flare" can occur under therapy with Abiraterone [10]. These common phenomena of everyday practice leave the burden to decide whether a therapy should be terminated or continued to the clinician. This dilemma holds especially true in asymptomatic or oligosymptomatic patients, since lack of clinical information can render decision making in some cases impossible. Therefore, in the case of biochemical progression only, current guidelines recommend continuing treatment until unequivocal information on progression is evident [31].

The data concerning monitoring of response or progression at an early stage of treatment are contradictory. PSA-flare has been reported to occur under other therapies than Abiraterone, like Docetaxel [32] and could be caused by a delayed response or rapid cell destruction with release of tPSA to the circulation [10]. Under treatment with Abiraterone, immediate or delayed tPSA-responses are common, but there are patients who will not show a tPSA-decline at all [10,33]. In our cohort, as well, a PSA-flare was seen in three patients, which is in line with the previously described 10% in other cohorts [10]. These three patients showed various duration of response to Abiraterone and stayed under treatment for nine, 26 and 38 months. Thus, the early differentiation between "PSA-flare" and true PD is difficult and remains mostly unresolved. This clinical dilemma can be even more complex if a possible "bone-flare" is suspected [29,30].

Although a significant tPSA decline of at least 50% from baseline was found in a large proportion of the patients treated within the registration trials for Abiraterone (COU-AA-301 and COU-AA-302), tPSA-changes were not found to be independent prognosticators for survival outcomes [1,2]. These non-straightforward findings are in line with our results. In our cohort, the median change of tPSA at 8–12 weeks of Abiraterone-therapy was a prognosticator for a PFS < 12 vs. ≥12 months ($p = 0.02$) with a decline resulting in a longer PFS, but the univariate cox-regression analysis showed no significant association (HR 1.6, $p = 0.29$); this may be due to our relatively small cohort. For this reason, no multivariate analysis was performed.

Some alternative biomarkers were recently evaluated. In contrast to tPSA, CTCs do not show a "flare" phenomenon and therefore could be ideal as a surrogate for OS [34]. In this context, several studies reported that the post-therapy CTC enumeration could be considered to be an independent prognostic parameter for OS [9,35]. However, due to costs, a relevant rate of falsely negative results and a lack of easily available assays the broad use of CTC enumeration remains limited [34].

Additionally, few retrospective studies report a trend for baseline LDH values above the range of normal limits being inversely associated with survival outcomes [12,35]. However, LDH is a very unspecific biomarker of cell necrosis in many tumor types including mCRPC and probably only reflects tumor burden as well as cell turnover as a marker of cancer aggressiveness [34].

Because of this lack of broadly available surrogate markers, we analyzed the potential of readily accessible biomarkers that are already being used in different settings within prostate cancer.

Until now, the role of fPSA is limited to the diagnosis of prostate cancer [19]. A lower proportion of %fPSA can be found in men with prostate cancer compared to patients with benign prostate enlargement [36]. Clinicians aim to avoid unnecessary biopsies by the measurement of fPSA, especially in cases with a mild tPSA elevation of 4–10 ng/mL. Additionally, a correlation between fPSA and the aggressiveness of the prostate cancers was shown [37].

More recent investigations demonstrated that this correlation with aggressiveness is even higher for the subform-proportion of %[−2]proPSA and PHI [23,24]. In addition, both parameters are currently considered to be the strongest predictors for presence of prostate cancer [21–23,38] with a higher specificity compared to fPSA and tPSA [22].

Due to the correlation between PSA subforms and the aggressiveness of the tumor we assumed that these parameters could play a role as surrogate biomarkers under therapy of mCRPC as well. Our results underline the non-straightforward data regarding early rising tPSA measurements with no validity concerning therapy outcomes. In contrast, we were able to show a trend towards rising [−2]proPSA, fPSA and, less importantly, PHI in patients with shorter OS and PFS at an early stage of treatment. Compared to tPSA, we need to consider that tPSA was taken into account for the definition of PD during treatment with Abiraterone. This might be an explanation for the less obvious separation of the curves during Kaplan–Meier analyses when comparing tPSA to fPSA and [−2]proPSA. However, our data suggest that fPSA and [−2]proPSA might be further studied as potential parameters for the prognostication of therapy response under early treatment with Abiraterone in case of non-straightforward development of PSA-values. The relative changes of both, median [−2]proPSA and fPSA values, were shown to be almost equal.

Our study is limited due to its relatively small cohort of only 25 patients recruited in a single center. Further larger studies are needed to explore the potential prognostic and predictive impact of fPSA and [−2]proPSA in mCRPC treated with Abiraterone.

4. Materials and Methods

4.1. Patients

Twenty-five patients with mCRPC presenting at the Prostate Center of the University Hospital Muenster (Münster, Germany), and receiving Abiraterone-therapy between March 2012 and July 2014, were reviewed. They had given written informed consent before participating and ethics committee-approval was granted for this study.

All patients presented with confirmed mCRPC, defined by prostate cancer working group 2 (PCWG2) criteria [31], in either pre- or post-chemotherapy setting (20 of them received Abiraterone in a pre- and 5 in a post-chemotherapy setting) suitable for Abiraterone-treatment. The pre-chemotherapy patients were asymptomatic or oligo-symptomatic, not requiring opiates and had a pain level of no more than 3 out of 10 on the numeric-rating-scale. Patients receiving Abiraterone in the post-chemotherapy setting had previously documented PD either on or after docetaxel chemotherapy. One of the 25 patients had received Enzalutamide previously. All patients in our study were either on a stable dose of an antiresorptive agent (i.e., Zoledronic acid or Denosumab) at least three months prior to the start of Abiraterone and during the whole treatment phase or did not receive antiresorptive medication at all. The patients were not systematically followed up with 12-weekly imaging when no PD was suspected, but when biochemical progression was evident, or when clinical progression occurred, even when there was no biochemical progression. CT-, MRI- or PSMA-PET-CT-scans of thorax, abdomen and pelvis were used for the evaluation of soft tissue metastases. Bone scans were done to acquire additional information on baseline bone metastases. PD was defined by RECIST 1.1 criteria [39] for cross-sectional imaging and by PCWG2 criteria for bone scans [31].

A physician with large expertise in the treatment of mCRPC assessed the current response status i.e., complete remission (CR), partial remission (PR), stable disease (SD) or PD at each visit. For defining PD, deterioration of general condition, worsening of pain or laboratory constellations (rising tPSA) and imaging were taken into account, according to PCWG2 criteria. Rising tPSA alone was not sufficient to define progression. The results of PSA-subform analysis, which was done in a retrospective setting, were not taken into consideration for therapeutic decisions.

All patients presented the day before the start of Abiraterone-therapy to have blood drawn for baseline analysis, two and four weeks after initiation of therapy and every four weeks thereafter.

4.2. Analysis of PSA Subforms

Blood samples were allowed to clot at room temperature and then centrifuged at $1600\times g$ for 15 min at 4 °C. Taking into account the stability issue of [−2]proPSA, serum samples were frozen at −80 °C within 3 h after blood draw [40]. After thawing of the frozen samples at room temperature, samples were processed immediately and used to determine tPSA, fPSA, [−2]proPSA on the Access II instrument, WHO-calibrated (Beckman Coulter, Krefeld, Germany) for the timepoints 'baseline', '8–12 weeks under therapy' and 'progression'. Furthermore, %fPSA, %[−2]proPSA and PHI were calculated.

We carried out the analyses according to the manufacturers' instructions on the AccessII-instrument (Beckman-Coulter, Krefeld, Germany)—batch numbers: tPSA (470206), fPSA (470202) and [−2]proPSA (434426). The results for PSA-subforms exceeded the measuring range of the instrument in many cases, with [−2]proPSA values up to 8000 pg/mL. Therefore, dilution was necessary. Due to a nonlinear response with serial dilution of the proPSA kit diluent, we used the LowCross-Buffer® (CANDOR Bioscience GmbH, Wangen, Germany, lot number 100A766f) as sample diluent in order to reduce nonspecific binding and cross-reactivities of the antibodies and matrix effects from blood serum. This procedure helped to minimize interferences in this assay and improves the reliability of the results. When dilution of serum samples was necessary due to very high [−2]proPSA values, all the other parameters were measured within the same diluted sample.

4.3. Statistical Methods

The descriptive statistics are reported as medians with IQR for continuous variables and as populations and frequencies for categorical variables. We performed the Mann–Whitney–Wilcoxon Test to determine the significance of the differences between categorical and continuous variables, respectively. We used the Kaplan–Meier-estimates for survival analyses. Univariate analysis of the different biomarkers was done with Cox-regression-models. Hazard ratios are given with 95% confidence intervals. All reported *p*-values are two-sided and statistical significance was assumed with a $p \leq 0.05$. SPSS-Statistics V.22 (IBM Inc., Armonk, NY, USA) was used for statistical assessment.

5. Conclusions

When added to tPSA information, which is traditionally used as therapy response indicator, changes of fPSA and [−2]proPSA at 8–12 weeks of Abiraterone-therapy may be a promising therapy control marker to help physicians make meaningful decisions on whether to stop or to continue Abiraterone-therapy in men with mCRPC but need to be further studied. Due to its better availability, the measurement of fPSA seems to be more feasible than [−2]proPSA.

Supplementary Materials: Supplementary materials can be found at www.mdpi.com/1422-0067/17/9/1520/s1.

Acknowledgments: The authors acknowledge the free-of-charge supply of PSA reagents by Beckman Coulter (Krefeld, Germany), and the assistance of Hilla Bürgel, Barbara Kloke and Beate Pepping-Schefers from the Prostate Center Muenster (Münster, Germany) in archiving and handling of the samples. We additionally thank Maria Eveslage of the Institute of Biostatistics and Clinical Research (Münster, Germany) for support regarding the statistical analysis.

Author Contributions: Katrin Schlack participated in the design of the study, helped in acquisition of data, in statistical analysis, and drafting of the manuscript; Laura-Maria Krabbe participated in statistical analysis and critical revision of the manuscript; Manfred Fobker participated in optimizing the assay for [−2]proPSA in very high levels and critical revision of the manuscript; Andres Jan Schrader participated in critical revision of the manuscript; Axel Semjonow participated in interpretation of serial dilution experiments, statistical analysis and critical revision of the manuscript; Martin Boegemann conceived of the study and participated in its design and coordination, performed part of the statistical analysis, and helped to draft the manuscript.

Conflicts of Interest: The authors declare no conflict of interest.

Abbreviations

mCRPC	metastatic Castration Resistant Prostate Cancer
ADT	Androgen-Deprivation Therapy
PSA	Prostate Specific Antigen
tPSA	total PSA
ECOG	Eastern Cooperative Oncology Group performance status
PFS	Progression Free Survival
CTC	Circulating Tumor Cells
LDH	Lactate Dehydrogenase
OS	Overall Survival
fPSA	free PSA
PHI	Prostate Health Index
IQR	Interquartile Range
HR	Hazard Ratio
CI	Confidence Interval
NR	Not Reached
PD	Progressive Disease
PCWG2	Prostate Cancer Working Group 2
CR	Complete Remission
PR	Partial Remission
SD	Stable Disease

References

1. Ryan, C.J.; Smith, M.R.; Fizazi, K.; Saad, F.; Mulders, P.F.; Sternberg, C.N.; Miller, K.; Logothetis, C.J.; Shore, N.D.; Small, E.J.; et al. Abiraterone acetate plus prednisone versus placebo plus prednisone in chemotherapy-naive men with metastatic castration-resistant prostate cancer (COU-AA-302): Final overall survival analysis of a randomised, double-blind, placebo-controlled phase 3 study. *Lancet Oncol.* **2015**, *16*, 152–160. [CrossRef]
2. Logothetis, C.J.; Basch, E.; Molina, A.; Fizazi, K.; North, S.A.; Chi, K.N.; Jones, R.J.; Goodman, O.B.; Mainwaring, P.N.; Sternberg, C.N.; et al. Effect of abiraterone acetate and prednisone compared with placebo and prednisone on pain control and skeletal-related events in patients with metastatic castration-resistant prostate cancer: Exploratory analysis of data from the COU-AA-301 randomised trial. *Lancet Oncol.* **2012**, *13*, 1210–1217. [PubMed]
3. Kantoff, P.W.; Higano, C.S.; Shore, N.D.; Berger, E.R.; Small, E.J.; Penson, D.F.; Redfern, C.H.; Ferrari, A.C.; Dreicer, R.; Sims, R.B.; et al. Sipuleucel-t immunotherapy for castration-resistant prostate cancer. *N. Engl. J. Med.* **2010**, *363*, 411–422. [CrossRef] [PubMed]
4. Loriot, Y.; Miller, K.; Sternberg, C.N.; Fizazi, K.; de Bono, J.S.; Chowdhury, S.; Higano, C.S.; Noonberg, S.; Holmstrom, S.; Mansbach, H.; et al. Effect of enzalutamide on health-related quality of life, pain, and skeletal-related events in asymptomatic and minimally symptomatic, chemotherapy-naive patients with metastatic castration-resistant prostate cancer (prevail): Results from a randomised, phase 3 trial. *Lancet Oncol.* **2015**, *16*, 509–521. [PubMed]

5. Fizazi, K.; Scher, H.I.; Miller, K.; Basch, E.; Sternberg, C.N.; Cella, D.; Forer, D.; Hirmand, M.; de Bono, J.S. Effect of enzalutamide on time to first skeletal-related event, pain, and quality of life in men with castration-resistant prostate cancer: Results from the randomised, phase 3 affirm trial. *Lancet Oncol.* **2014**, *15*, 1147–1156. [CrossRef]

6. Berthold, D.R.; Pond, G.R.; Soban, F.; de Wit, R.; Eisenberger, M.; Tannock, I.F. Docetaxel plus prednisone or mitoxantrone plus prednisone for advanced prostate cancer: Updated survival in the TAX 327 study. *J. Clin. Oncol.* **2008**, *26*, 242–245. [CrossRef] [PubMed]

7. Petrylak, D.P.; Tangen, C.M.; Hussain, M.H.; Lara, P.N., Jr.; Jones, J.A.; Taplin, M.E.; Burch, P.A.; Berry, D.; Moinpour, C.; Kohli, M.; et al. Docetaxel and estramustine compared with mitoxantrone and prednisone for advanced refractory prostate cancer. *N. Engl. J. Med.* **2004**, *351*, 1513–1520. [CrossRef] [PubMed]

8. Sartor, O.; Coleman, R.; Nilsson, S.; Heinrich, D.; Helle, S.I.; O'Sullivan, J.M.; Fossa, S.D.; Chodacki, A.; Wiechno, P.; Logue, J.; et al. Effect of radium-223 dichloride on symptomatic skeletal events in patients with castration-resistant prostate cancer and bone metastases: Results from a phase 3, double-blind, randomised trial. *Lancet Oncol.* **2014**, *15*, 738–746. [CrossRef]

9. De Bono, J.S.; Logothetis, C.J.; Molina, A.; Fizazi, K.; North, S.; Chu, L.; Chi, K.N.; Jones, R.J.; Goodman, O.B., Jr.; Saad, F.; et al. Abiraterone and increased survival in metastatic prostate cancer. *N. Engl. J. Med.* **2011**, *364*, 1995–2005. [CrossRef] [PubMed]

10. Burgio, S.L.; Conteduca, V.; Rudnas, B.; Carrozza, F.; Campadelli, E.; Bianchi, E.; Fabbri, P.; Montanari, M.; Carretta, E.; Menna, C.; et al. PSA flare with abiraterone in patients with metastatic castration-resistant prostate cancer. *Clin. Genitourin. Cancer* **2015**, *13*, 39–43. [CrossRef] [PubMed]

11. Danila, D.C.; Morris, M.J.; de Bono, J.S.; Ryan, C.J.; Denmeade, S.R.; Smith, M.R.; Taplin, M.E.; Bubley, G.J.; Kheoh, T.; Haqq, C.; et al. Phase II multicenter study of abiraterone acetate plus prednisone therapy in patients with docetaxel-treated castration-resistant prostate cancer. *J. Clin. Oncol.* **2010**, *28*, 1496–1501. [CrossRef] [PubMed]

12. Scher, H.I.; Heller, G.; Molina, A.; Attard, G.; Danila, D.C.; Jia, X.; Peng, W.; Sandhu, S.K.; Olmos, D.; Riisnaes, R.; et al. Circulating tumor cell biomarker panel as an individual-level surrogate for survival in metastatic castration-resistant prostate cancer. *J. Clin. Oncol.* **2015**, *33*, 1348–1355. [CrossRef] [PubMed]

13. Danila, D.C.; Heller, G.; Gignac, G.A.; Gonzalez-Espinoza, R.; Anand, A.; Tanaka, E.; Lilja, H.; Schwartz, L.; Larson, S.; Fleisher, M.; et al. Circulating tumor cell number and prognosis in progressive castration-resistant prostate cancer. *Clin. Cancer Res.* **2007**, *13*, 7053–7058. [CrossRef] [PubMed]

14. Danila, D.C.; Fleisher, M.; Scher, H.I. Circulating tumor cells as biomarkers in prostate cancer. *Clin. Cancer Res.* **2011**, *17*, 3903–3912. [CrossRef] [PubMed]

15. Lilja, H.; Christensson, A.; Dahlen, U.; Matikainen, M.T.; Nilsson, O.; Pettersson, K.; Lovgren, T. Prostate-specific antigen in serum occurs predominantly in complex with α 1-antichymotrypsin. *Clin. Chem.* **1991**, *37*, 1618–1625. [PubMed]

16. Lilja, H.; Stenman, U.H. Successful separation between benign prostatic hyperplasia and prostate cancer by measurement of free and complexed psa. *Cancer Treat. Res.* **1996**, *88*, 93–101. [PubMed]

17. Stenman, U.H.; Leinonen, J.; Alfthan, H.; Rannikko, S.; Tuhkanen, K.; Alfthan, O. A complex between prostate-specific antigen and alpha 1-antichymotrypsin is the major form of prostate-specific antigen in serum of patients with prostatic cancer: Assay of the complex improves clinical sensitivity for cancer. *Cancer Res.* **1991**, *51*, 222–226. [PubMed]

18. Mikolajczyk, S.D.; Rittenhouse, H.G. Pro psa: A more cancer specific form of prostate specific antigen for the early detection of prostate cancer. *Keio J. Med.* **2003**, *52*, 86–91. [CrossRef] [PubMed]

19. Tosoian, J.; Loeb, S. Psa and beyond: The past, present, and future of investigative biomarkers for prostate cancer. *Sci. World J.* **2010**, *10*, 1919–1931. [CrossRef] [PubMed]

20. Mikolajczyk, S.D.; Marks, L.S.; Partin, A.W.; Rittenhouse, H.G. Free prostate-specific antigen in serum is becoming more complex. *Urology* **2002**, *59*, 797–802. [CrossRef]

21. Lazzeri, M.; Briganti, A.; Scattoni, V.; Lughezzani, G.; Larcher, A.; Gadda, G.M.; Lista, G.; Cestari, A.; Buffi, N.; Bini, V.; et al. Serum index test %[−2]proPSA and prostate health index are more accurate than prostate specific antigen and %fPSA in predicting a positive repeat prostate biopsy. *J. Urol.* **2012**, *188*, 1137–1143. [CrossRef] [PubMed]

22. Catalona, W.J.; Partin, A.W.; Sanda, M.G.; Wei, J.T.; Klee, G.G.; Bangma, C.H.; Slawin, K.M.; Marks, L.S.; Loeb, S.; Broyles, D.L.; et al. A multicenter study of [−2]pro-prostate specific antigen combined with prostate specific antigen and free prostate specific antigen for prostate cancer detection in the 2.0 to 10.0 ng/mL prostate specific antigen range. *J. Urol.* **2011**, *185*, 1650–1655. [CrossRef] [PubMed]

23. Lazzeri, M.; Haese, A.; Abrate, A.; de la Taille, A.; Redorta, J.P.; McNicholas, T.; Lughezzani, G.; Lista, G.; Larcher, A.; Bini, V.; et al. Clinical performance of serum prostate-specific antigen isoform [−2]proPSA (p2PSA) and its derivatives, %p2PSA and the prostate health index (PHI), in men with a family history of prostate cancer: Results from a multicentre european study, the prometheus project. *BJU Int.* **2013**, *112*, 313–321. [PubMed]

24. Stephan, C.; Jung, K.; Semjonow, A.; Schulze-Forster, K.; Cammann, H.; Hu, X.; Meyer, H.A.; Bogemann, M.; Miller, K.; Friedersdorff, F. Comparative assessment of urinary prostate cancer antigen 3 and TMPRSS2: Erg gene fusion with the serum [−2]proprostate-specific antigen-based prostate health index for detection of prostate cancer. *Clin. Chem.* **2013**, *59*, 280–288. [CrossRef] [PubMed]

25. Fossati, N.; Buffi, N.M.; Haese, A.; Stephan, C.; Larcher, A.; McNicholas, T.; de la Taille, A.; Freschi, M.; Lughezzani, G.; Abrate, A.; et al. Preoperative prostate-specific antigen isoform p2PSA and its derivatives, %p2PSA and prostate health index, predict pathologic outcomes in patients undergoing radical prostatectomy for prostate cancer: Results from a multicentric european prospective study. *Eur. Urol.* **2015**, *68*, 132–138. [CrossRef] [PubMed]

26. Fossati, N.; Lazzeri, M.; Haese, A.; McNicholas, T.; de la Taille, A.; Buffi, N.M.; Lughezzani, G.; Gadda, G.M.; Lista, G.; Larcher, A.; et al. Clinical performance of serum isoform [−2]proPSA (p2PSA), and its derivatives %p2PSA and the prostate health index, in men aged <60 years: Results from a multicentric european study. *BJU Int.* **2015**, *115*, 913–920. [PubMed]

27. Woo, H.H.; Begbie, S.; Gogna, K.; Mainwaring, P.N.; Murphy, D.G.; Parnis, F.; Steer, C.; Davis, I.D. Multidisciplinary consensus: A practical guide for the integration of abiraterone into clinical practice. *Asia Pac. J. Clin. Oncol.* **2014**, *10*, 228–236. [CrossRef] [PubMed]

28. Roghmann, F.; Antczak, C.; McKay, R.R.; Choueiri, T.; Hu, J.C.; Kibel, A.S.; Kim, S.P.; Kowalczyk, K.J.; Menon, M.; Nguyen, P.L.; et al. The burden of skeletal-related events in patients with prostate cancer and bone metastasis. *Urol. Oncol.* **2015**, *33*, e9–e18. [CrossRef] [PubMed]

29. Ryan, C.J.; Shah, S.; Efstathiou, E.; Smith, M.R.; Taplin, M.E.; Bubley, G.J.; Logothetis, C.J.; Kheoh, T.; Kilian, C.; Haqq, C.M.; et al. Phase II study of abiraterone acetate in chemotherapy-naive metastatic castration-resistant prostate cancer displaying bone flare discordant with serologic response. *Clin. Cancer Res.* **2011**, *17*, 4854–4861. [CrossRef] [PubMed]

30. Johns, W.D.; Garnick, M.B.; Kaplan, W.D. Leuprolide therapy for prostate cancer: An association with scintigraphic "flare" on bone scan. *Clin. Nucl. Med.* **1990**, *15*, 485–487. [CrossRef] [PubMed]

31. Scher, H.I.; Halabi, S.; Tannock, I.; Morris, M.; Sternberg, C.N.; Carducci, M.A.; Eisenberger, M.A.; Higano, C.; Bubley, G.J.; Dreicer, R.; et al. Design and end points of clinical trials for patients with progressive prostate cancer and castrate levels of testosterone: Recommendations of the prostate cancer clinical trials working group. *J. Clin. Oncol.* **2008**, *26*, 1148–1159. [CrossRef] [PubMed]

32. Berthold, D.R.; Pond, G.R.; Roessner, M.; de Wit, R.; Eisenberger, M.; Tannock, A.I. TAX-327 investigators Treatment of hormone-refractory prostate cancer with docetaxel or mitoxantrone: Relationships between prostate-specific antigen, pain, and quality of life response and survival in the TAX-327 study. *Clin. Cancer Res.* **2008**, *14*, 2763–2767. [CrossRef] [PubMed]

33. Leibowitz-Amit, R.; Templeton, A.J.; Omlin, A.; Pezaro, C.; Atenafu, E.G.; Keizman, D.; Vera-Badillo, F.; Seah, J.A.; Attard, G.; Knox, J.J.; et al. Clinical variables associated with psa response to abiraterone acetate in patients with metastatic castration-resistant prostate cancer. *Ann. Oncol.* **2014**, *25*, 657–662. [CrossRef] [PubMed]

34. Armstrong, A.J.; Eisenberger, M.A.; Halabi, S.; Oudard, S.; Nanus, D.M.; Petrylak, D.P.; Sartor, A.O.; Scher, H.I. Biomarkers in the management and treatment of men with metastatic castration-resistant prostate cancer. *Eur. Urol.* **2012**, *61*, 549–559. [CrossRef] [PubMed]

35. Scher, H.I.; Jia, X.; de Bono, J.S.; Fleisher, M.; Pienta, K.J.; Raghavan, D.; Heller, G. Circulating tumour cells as prognostic markers in progressive, castration-resistant prostate cancer: A reanalysis of IMMC38 trial data. *Lancet Oncol.* **2009**, *10*, 233–239. [CrossRef]

36. De Angelis, G.; Rittenhouse, H.G.; Mikolajczyk, S.D.; Blair Shamel, L.; Semjonow, A. Twenty years of PSA: From prostate antigen to tumor marker. *Rev. Urol.* **2007**, *9*, 113–123. [PubMed]

37. Catalona, W.J.; Partin, A.W.; Slawin, K.M.; Brawer, M.K.; Flanigan, R.C.; Patel, A.; Richie, J.P.; deKernion, J.B.; Walsh, P.C.; Scardino, P.T.; et al. Use of the percentage of free prostate-specific antigen to enhance differentiation of prostate cancer from benign prostatic disease: A prospective multicenter clinical trial. *JAMA* **1998**, *279*, 1542–1547. [CrossRef] [PubMed]

38. Abrate, A.; Lughezzani, G.; Gadda, G.M.; Lista, G.; Kinzikeeva, E.; Fossati, N.; Larcher, A.; Dell'Oglio, P.; Mistretta, F.; Buffi, N.; et al. Clinical use of [−2]proPSA (p2PSA) and its derivatives (%p2PSA and prostate health index) for the detection of prostate cancer: A review of the literature. *Korean J. Urol.* **2014**, *55*, 436–445. [CrossRef] [PubMed]

39. Eisenhauer, E.A.; Therasse, P.; Bogaerts, J.; Schwartz, L.H.; Sargent, D.; Ford, R.; Dancey, J.; Arbuck, S.; Gwyther, S.; Mooney, M.; et al. New response evaluation criteria in solid tumours: Revised recist guideline (version 1.1). *Eur. J. Cancer* **2009**, *45*, 228–247. [CrossRef] [PubMed]

40. Semjonow, A.; Kopke, T.; Eltze, E.; Pepping-Schefers, B.; Burgel, H.; Darte, C. Pre-analytical in vitro stability of [−2]proPSA in blood and serum. *Clin. Biochem.* **2010**, *43*, 926–928. [CrossRef] [PubMed]

International Journal of
Molecular Sciences

MDPI

Article

The Role of the Neutrophil to Lymphocyte Ratio for Survival Outcomes in Patients with Metastatic Castration-Resistant Prostate Cancer Treated with Abiraterone

Martin Boegemann [1,†,*], Katrin Schlack [1,†], Stefan Thomes [1], Julie Steinestel [1], Kambiz Rahbar [2], Axel Semjonow [1], Andres Jan Schrader [1], Martin Aringer [3] and Laura-Maria Krabbe [1,4]

[1] Department of Urology, Prostate Center, University Hospital Muenster, Albert-Schweitzer-Campus 1, GB A1, D-48149 Muenster, Germany; katrin.schlack@ukmuenster.de (K.S.); s_thom09@uni-muenster.de (S.T.); Julie.steinestel@ukmuenster.de (J.S.); axel.semjonow@ukmuenster.de (A.S.); andresjan.schrader@ukmuenster.de (A.J.S.); laura-maria.krabbe@ukmuenster.de (L.-M.K.)
[2] Department of Nuclear Medicine, Muenster University Medical Center, Albert-Schweitzer-Campus 1, GB A1, D-48149 Muenster, Germany; kambiz.rahbar@ukmuenster.de
[3] Department of Rheumatology, Dresden University Medical Center, Fetscherstraße 74, D-01307 Dresden, Germany; martin.aringer@uniklinikum-dresden.de
[4] Department of Urology, University of Texas Southwestern Medical Center, Dallas, TX 75390-9110, USA
[*] Correspondence: martin.boegemann@ukmuenster.de; Tel.: +49-(0)-251-834-4600
[†] These authors contributed equally to this work.

Academic Editor: Carsten Stephan
Received: 26 November 2016; Accepted: 7 February 2017; Published: 11 February 2017

Abstract: The purpose of this study was to examine the prognostic capability of baseline neutrophil-to-lymphocyte-ratio (NLR) and NLR-change under Abiraterone in metastatic castration-resistant prostate cancer patients. The impact of baseline NLR and change after eight weeks of treatment on progression-free survival (PFS) and overall survival (OS) was analyzed using Kaplan-Meier-estimates and Cox-regression. 79 men with baseline NLR <5 and 17 with NLR >5 were analyzed. In baseline analysis of PFS NLR >5 was associated with non-significantly shorter median PFS (five versus 10 months) (HR: 1.6 (95%CI:0.9–2.8); $p = 0.11$). After multivariate adjustment (MVA), ECOG > 0–1, baseline LDH>upper limit of normal (UNL) and presence of visceral metastases were independent prognosticators. For OS, NLR >5 was associated with shorter survival (seven versus 19 months) (HR: 2.3 (95%CI:1.3–4.0); $p < 0.01$). In MVA, ECOG > 0–1 and baseline LDH > UNL remained independent prognosticators. After 8 weeks of Abiraterone NLR-change to <5 prognosticated worse PFS (five versus 12 months) (HR: 4.1 (95%CI:1.1–15.8); $p = 0.04$). MVA showed a trend towards worse PFS for NLR-change to <5 ($p = 0.11$). NLR-change to <5 led to non-significant shorter median OS (seven versus 16 months) (HR: 2.3 (95%CI:0.7–7.1); $p = 0.15$). MVA showed non-significant difference for OS. We concluded baseline NLR <5 is associated with improved survival. In contrast, in patients with baseline NLR >5, NLR-change to <5 after eight weeks of Abiraterone was associated with worse survival and should be interpreted carefully.

Keywords: neutrophil-to-lymphocyte ratio; castration-resistant prostate cancer; prognostic biomarker; treatment response; abiraterone

1. Introduction

Lately, several agents were approved for treatment of metastatic castration-resistant prostate cancer (mCRPC). These include, prior to Docetaxel chemotherapy, Abiraterone [1], Enzalutamide [2] and Sipuleucel-T [3], or after Docetaxel chemotherapy, Abiraterone [4], Enzalutamide [5] and Cabazitaxel [6], or when no chemotherapy is indicated in bone-predominant mCRPC without visceral metastases, radium-223 [7]. All of these medications prolong overall survival (OS). However, there is a shortage of available biomarkers to predict response to each therapy and no data concerning the optimal sequence of therapies. Therefore, the decision which medication to choose is largely based on empirical data, experience, or matters like drug availability or the possibility of reimbursement. Further, early imaging, especially in bone mCRPC can be misleading due to the bone flare phenomenon which can represent a response to treatment, although suggesting progressive disease (PD) [8].

Inflammation is recognized to be a relevant driver of cancer progression [9]. The neutrophil-to-lymphocyte ratio (NLR) as an indicator of cancer-related inflammation has been shown to be predictive for treatment response or even suggested as a marker of therapy control in various tumour entities, for example in renal cell cancer, oesophageal cancer, gastrointestinal cancer, and many others [10–13]. In mCRPC patients treated with Ketoconazole, a NLR cut-off of <3 was identified for differentiation between patients with longer or shorter progression-free survival (PFS) [14]. In the SUN-1120-trial on Sunitinib vs. Placebo in mCRPC Sonpavde found that a NLR > 2.5 and even more so >5 was prognostic of poor OS [15]. And in a post hoc analysis of men treated with Docetaxel within the TAX-327 and VENICE-trials a NLR >than the median of 2, 2.1 was associated with worse survival [16]. A retrospective study on mCRPC patients treated with Abiraterone identified a baseline NLR of <5 in combination with the extent of metastatic spread to be associated with prostate-specific antigen (PSA) response, and overall survival [17].

We hypothesize that in mCRPC patients treated with Abiraterone, the NLR might not only be useable as a predictor of treatment response at baseline, but also as a marker of early treatment response.

2. Results

2.1. Patient Characteristics

The descriptive characteristics of the cohort are presented in Table 1. The study cohort consisted of 96 mCRPC patients, of whom 52 were treated within the pre- and 44 in the post-Docetaxel setting. The median follow-up was 20.0 months (interquartile range (IQR): 11.0–28.0). The median time of Abiraterone treatment was 10.0 months (IQR: 6.0–14.3). No dose modification was necessary for any patient. The median age of the patients was 70.0 years (IQR: 62.0–76.3). Visceral metastases were present in 28.1% at the beginning of Abiraterone treatment. An unfavourable Gleason-Score of ≥ 8 was seen in 59.8% of the men. The median baseline values of PSA, alkaline phosphatase (ALP) and LDH (lactate dehydrogenase) were 134 ng/mL (IQR: 45–349), 126 U/L (IQR: 86–296) and 251 U/L (IQR: 210–358), respectively.

At baseline, a NLR of <5 was present in 79 (82.3%) of the patients. Compared with men with a NLR >5, patients with a NLR <5 had a significantly lower median baseline PSA-level (91 vs. 224 ng/mL; Mann-Whitney U-test: $p = 0.04$) and lower baseline median LDH-levels (240 vs. 308 U/L; $p = 0.04$). Further, the baseline LDH was elevated above the upper limit of normal (ULN) at baseline in a larger proportion in the men with a NLR >5 (88.2% vs. 59.5%; χ^2-test: $p = 0.03$). The other baseline characteristics were not statistically different. Bone protective medication (Denosumab or Zoledronic acid) was well balanced between both groups (58.8% vs. 59.5%; χ^2-test: $p = 0.96$) (Table 1).

According to prostate cancer working group 2 (PCWG2)-criteria, PSA-response is defined as a PSA decline of $\geq 50\%$ and was seen in 35.4% of the patients with a baseline NLR >5 compared with 53.2% of the men with a baseline NLR <5 (χ^2-test: $p = 0.18$). A decline of $\geq 90\%$ was found in 5.9% vs. 27.8% (χ^2-test: $p = 0.06$).

Table 1. Characteristics of patients with metastatic castration-resistant prostate cancer (mCRPC) on Abiraterone with a neutrophil-to-lymphocyte-ratio (NLR) <5 and >5.

Variable	All	NLR <5	NLR >5	p
Patients (*n*) (%)	96	79 (82.3)	17 (17.7)	
Median NLR Baseline (ng/mL) (IQR)	3.2 (2.5–4.5)			
Age, median (years) (IQR)	70.0 (63.0–76.3)	70.0 (62.0–76.0)	70.0 (65.0–78.0)	0.60
Median follow-up	20.0 (11.0–28.0)	17.5 (11.0–26.8)	29.0 (29.0–29.0)	0.28
Median duration AA therapy (months) (IQR)	10.0 (6.0–14.3)	11.0 (7.0–15.0)	7.0 (5.0–11.5)	0.08
GS ≥8 (*n*) (%)	55 (59.8)	46 (60.5)	9 (56.3)	0.75
Lnn. Metastases (*n*) (%)	61 (63.5)	50 (63.3)	11 (64.7)	0.91
Visceral Metastases (*n*) (%)	27 (28.1)	24 (30.4)	3 (17.6)	0.38
Bone Metastases (*n*) (%)	85 (88.5)	69 (87.3)	16 (94.1)	0.68
Pre CTX (*n*) (%)	52 (54.2)	44 (55.7)	8 (47.1)	0.52
Post CTX (*n*) (%)	44 (45.8)	35 (44.3)	9 (52.9)	
Patients died (*n*) (%)	67 (69.8)	51 (64.6)	16 (94.1)	0.02
ECOG (all) (*n*) (%)				
0	19 (20.0)	16 (20.5)	3 (17.6)	
1	57 (60.0)	47 (60.3)	10 (58.8)	0.91
2	19 (20.0)	15 (19.2)	4 (23.5)	
Antiresorptive therapy (*n*) (%)	57 (59.4)	47 (59.5)	10 (58.8)	0.96
Zoledronic acid (*n*) (%)	40 (41.7)	33 (41.8)	7 (41.2)	0.96
Denosumab (*n*) (%)	19 (19.8)	15 (19.0)	4 (23.5)	0.67
Best clinical outcome (*n*) (%)				
CR	1 (1.1)	1 (1.3)	0 (0)	
PR	54 (56.8)	46 (59.0)	8 (47.1)	0.61
SD	27 (28.4)	20 (25.6)	7 (41.2)	
PD	13 (13.7)	11 (14.1)	2 (11.8)	
PSA red. ≥50% (*n*) (%)	48 (50.0)	42 (53.2)	6 (35.3)	0.18
PSA red. ≥90% (*n*) (%)	23 (24.0)	22 (27.8)	1 (5.9)	0.06
Median PSA Baseline (ng/mL) (IQR)	134 (45–349)	91 (35–334)	224 (107–589)	0.04
Median LDH Baseline (U/L) (IQR)	251 (210–358)	240 (205–357)	308 (245–386)	0.04
LDH BL >UNL (*n*) (%)	62 (64.6)	47 (59.5)	15 (88.2)	0.03
Median ALP Baseline (U/L) (IQR)	126 (86–296)	126 (91–297)	103 (77–266)	0.43

Abbreviations: NLR: neutrophil-to-lymphocyte ratio; IQR: interquartile range; Lnn: lymphonodal; CTX: chemotherapy; CR: complete remission; PR: partial remission; SD: stable disease; PD: progressive disease; ECOG: Easter collaborative Oncology Group; GS: Gleason score; PSA: prostate specific antigen; LDH: lactate dehydrogenase; ALP: alkaline phosphatase; BL: baseline; UNL: upper normal limit.

2.2. Prognostication of Survival Outcomes at Baseline

Results of univariate and multivariate analyses for PFS and OS based on baseline NLR (>5 vs. <5) are displayed in Tables 2–4.

Table 2. Univariate analysis for baseline biomarkers for progression-free survival in (a) 96 mCRPC-patients prior to Abiraterone therapy and (b) in 17 patients with a baseline NLR >5 after eight weeks of therapy.

(a) Univariate Analysis PFS at Baseline			(b) Univariate Analysis of PFS in *n* = 17 Patients with Baseline NLR > 5 after Eight Weeks of Abiraterone		
Variable	HR (95%CI)	p	Variable	HR (95%CI)	p
ECOG			ECOG		
0–1	1 (reference)	<0.01	0–1	1 (reference)	0.57
2	2.9 (1.7–4.9)		2	1.4 (0.4–4.8)	
LDH baseline >UNL			LDH baseline >UNL	1 (reference)	
No	1 (reference)	<0.01	No		0.52
Yes	2.6 (1.6–4.3)		Yes	2.0 (0.3–15.4)	
Visceral metastases			Visceral metastases		
No	1 (reference)	0.02	No	1 (reference)	0.85
Yes	2.2 (1.4–3.5)		Yes	0.9 (0.2–3.3)	
ALP baseline >UNL			ALP baseline >UNL		
No	1 (reference)	0.13	No	1 (reference)	0.55
Yes	1.4 (0.9–2.1)		Yes	1.4 (0.5–3.9)	

Table 2. *Cont.*

(a) Univariate Analysis PFS at Baseline			(b) Univariate Analysis of PFS in *n* = 17 Patients with Baseline NLR > 5 after Eight Weeks of Abiraterone		
Variable	HR (95%CI)	*p*	Variable	HR (95%CI)	*p*
NLR			NLR		
<5	1 (reference)	0.11	No change	1 (reference)	0.04
>5	1.6 (0.9–2.8)		Change to <5	4.1 (1.1–15.8)	
Abiraterone			PSA decline ≥50%		
Pre-Docetaxel	1 (reference)	0.10	Yes	1 (reference)	0.14
Post-Docetaxel	1.4 (0.9–2.2)		No	2.7 (0.7–10.2)	
Gleason Score			Abiraterone		
<8	1 (reference)	0.10	Pre-Docetaxel	1 (reference)	0.41
≥8	1.5 (0.9–2.3)		Post-Docetaxel	1.6 (0.5–4.6)	
Lymphonodal metastases			Gleason Score		
No	1 (reference)	0.77	<8	1 (reference)	0.63
Yes	0.9 (0.6–1.5)		≥8	1.3 (0.4–3.9)	
Bone Metastases			Lymphonodal metastases		
No	1 (reference)	0.82	No	1 (reference)	0.41
Yes	1.1 (0.5–2.2)		Yes	0.6 (0.2–1.9)	

Abbreviations: PFS: progression free survival; HR: hazard ratio; 95%CI: 95% confidence interval; ECOG: Eastern Collaborative Oncology Group; LDH: lactate dehydrogenase; UNL: upper normal limit; PSA: prostate-specific antigen; NLR: neutrophil-to-lymphocyte ratio; ALP: alkaline phosphatase; OS: overall survival.

Table 3. Univariate analysis for baseline biomarkers for overall survival in (a) 96 mCRPC-patients prior to Abiraterone therapy and (b) in 17 patients with a baseline NLR >5 after eight weeks of therapy.

(a) Univariate Analysis OS at Baseline			(b) Univariate Analysis of OS in *n* = 17 Patients with Baseline NLR >5 after Eight Weeks of Abiraterone		
Variable	HR (95%CI)	*p*	Variable	HR (95%CI)	*p*
ECOG			ECOG		
0–1	1 (reference)	<0.01	0–1	1 (reference)	0.60
2	3.4 (1.9–6.0)		2	1.4 (0.3–4.5)	
LDH baseline >UNL			LDH baseline >UNL		
No	1 (reference)	<0.01	No	1 (reference)	0.65
Yes	3.1 (1.7–5.8)		Yes	0.7 (0.2–3.3)	
Visceral metastases			Visceral metastases		
No	1 (reference)	0.35	No	1 (reference)	0.76
Yes	1.3 (0.8–2.2)		Yes	0.8 (0.2–2.9)	
ALP baseline >UNL			ALP baseline >UNL		
No	1 (reference)	<0.01	No	1 (reference)	0.83
Yes	1.9 (1.2–3.2)		Yes	0.9 (0.3–2.5)	
NLR			NLR		
<5	1 (reference)	<0.01	No change	1 (reference)	0.15
>5	2.3 (1.3–4.0)		Change to <5	2.3 (0.7–7.1)	
Abiraterone			PSA decline ≥50%		
Pre-Docetaxel	1 (reference)	0.06	Yes	1 (reference)	0.08
Post-Docetaxel	1.6 (1.0–2.6)		No	2.9 (0.9–9.4)	
Gleason Score			Abiraterone		
<8	1 (reference)	0.67	Pre-Docetaxel	1 (reference)	0.38
≥8	1.1 (0.7–1.9)		Post-Docetaxel	1.6 (0.6–4.6)	
Lymphonodal metastases			Gleason Score		
No	1 (reference)	0.35	<8	1 (reference)	0.93
Yes	0.8 (0.5–1.3)		≥8	1.1 (0.4–3.0)	
Bone metastases			Lymphonodal metastases		
No	1 (reference)	0.62	No	1 (reference)	0.64
Yes	1.3 (0.5–3.2)		Yes	0.8 (0.3–2.2)	

Abbreviations: PFS: progression-free survival; HR: hazard ratio; 95%CI: 95% confidence interval; ECOG: Eastern Collaborative Oncology Group; LDH: lactate dehydrogenase; UNL: upper normal limit; PSA: prostate-specific antigen; NLR: neutrophil-to-lymphocyte ratio; ALP: alkaline phosphatase; OS: overall survival.

Table 4. Multivariate analysis for significant baseline biomarkers for (a) progression-free and (b) overall survival in 96 mCRPC-patients prior to Abiraterone therapy and (c) progression-free survival and (d) overall survival in 17 patients with a baseline NLR >5 after eight weeks of therapy.

(a) Multivariate Analysis PFS at Baseline			(c) Multivariate analysis of PFS in *n* =17 Patients with Baseline NLR >5 after Eight Weeks of Abiraterone		
Variable	HR (95%CI)	*p*	Variable	HR (95%CI)	*p*
ECOG			NLR		
0–1	1 (reference)	0.01	No change	1 (reference)	0.11
2	2.6 (1.5–4.5)		Change to <5	3.4 (0.8–15.2)	
LDH baseline >UNL			PSA decline ≥50%		
No	1 (reference)	0.01	Yes	1 (reference)	0.14
Yes	2.2 (1.3–3.8)		No	1.5 (0.3–6.4)	
Visceral metastases					
No	1 (reference)	0.04			
Yes	1.7 (1.0–2.9)				
ALP baseline >UNL					
No	1 (reference)	0.25			
Yes	1.3 (0.8–2.1)				
NLR					
<5	1 (reference)	0.71			
>5	1.1 (0.6–2.0)				
(b) Multivariate analysis OS at baseline			(d) Multivariate Analysis of OS in *n* = 17 Patients with Baseline NLR >5 after Eight Weeks of Abiraterone		
Variable	HR (95%CI)	*p*	Variable	HR (95%CI)	*p*
ECOG			NLR		
0–1	1 (reference)	<0.01	No change	1 (reference)	0.47
2	3.0 (1.6–5.5)		Change to <5	1.6 (0.5–5.6)	
LDH baseline >UNL			PSA decline ≥50%		
No	1 (reference)	0.01	Yes	1 (reference)	0.25
Yes	2.4 (1.2–4.6)		No	2.2 (0.6–8.1)	
Visceral metastases					
No	1 (reference)	0.56			
Yes	1.2 (0.7–2.1)				
ALP baseline >UNL					
No	1 (reference)	0.05			
Yes	1.7 (1.0–2.9)				
NLR					
<5	1 (reference)	0.10			
>5	1.7 (0.9–3.0)				

Abbreviations: PFS: progression free survival; HR: hazard ratio; 95%CI: 95% confidence interval; ECOG: Eastern Collaborative Oncology Group; LDH: lactate dehydrogenase; UNL: upper normal limit; PSA: prostate-specific antigen; NLR: neutrophil-to-lymphocyte ratio; ALP: alkaline phosphatase; OS: overall survival.

In univariate analysis (Table 2), baseline LDH >ULN (hazard ratio (HR): 2.6 (95% confidence interval (95%CI): 1.6–4.3); $p < 0.01$), Eastern Collaborative Oncology Group performance status (ECOG) ≥ 2 (HR: 2.9 (95%CI: 1.7–4.9); $p < 0.01$), and the presence of visceral metastases (HR: 2.2 (95%CI: 1.4–3.5); $p = 0.02$) were associated with worse PFS.

At baseline there was a trend for improved median PFS of patients with a NLR <5 (10 months (95%CI: 8.1–11.9)) compared with men with a NLR >5 (five months (95%CI: 1.0–9.0), log-rank $p = 0.09$) (Figure 1a).

After adjusting for the parameters significant in univariate analysis and clinically-important variables, ECOG ≥ 2 (HR: 2.6 (95%CI: 1.5–4.5); $p = 0.01$), baseline LDH >UNL (HR: 2.2 (95%CI: 1.3–3.8); $p = 0.01$) and the presence of visceral metastases (HR: 1.7 (95%CI: 1.0–2.9); $p = 0.04$) remained independent prognosticators of poor PFS (Table 4).

The univariate analysis for OS showed that a baseline NLR >5 (HR: 2.3 (95%CI: 1.3–4.0); $p < 0.01$), ECOG ≥ 2 (HR: 3.4 (95%CI: 1.9–6.0); $p < 0.01$), baseline LDH >UNL (HR: 3.1 (95%CI: 1.7–5.8); $p < 0.01$) and baseline ALP >UNL (HR: 1.9 (95%CI: 1.2–3.2); $p < 0.01$) were associated with worse OS (Table 3).

In Kaplan-Meier analysis the median OS (Figure 1c) of men with a baseline NLR <5 (19 months (95%CI 13.0–25.0) was significantly longer than in patients with a NLR >5 (seven months (95%CI: 4.0–10.0), log-rank $p = 0.04$).

In multivariate analysis of OS (Table 4), only ECOG ≥ 2 (HR: 3.0 (95%CI: 1.6–5.5); $p < 0.01$) and baseline LDH >UNL (HR: 2.4 (95%CI: 1.2–4.6); $p = 0.01$) remained independent prognosticators of worse survival. For NLR >5, OS was reduced but not significantly (HR: 1.7 (95%CI: 0.9–3.0); $p = 0.10$).

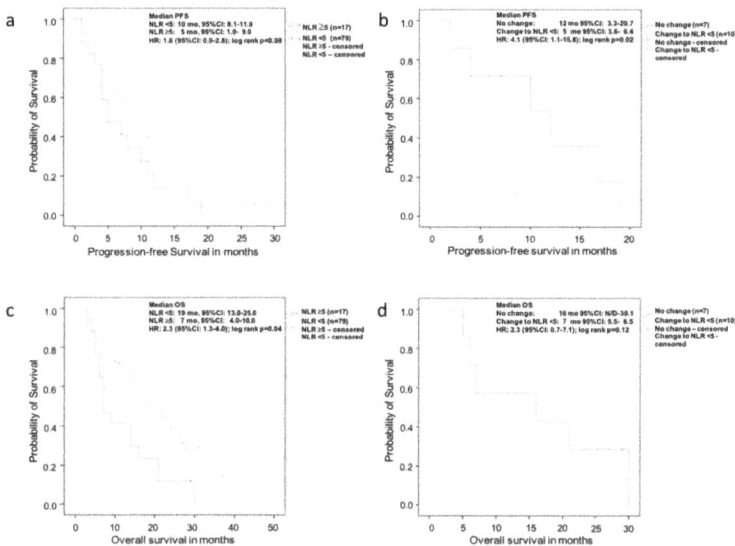

Figure 1. Kaplan-Meier analysis for progression-free survival probability of patients with mCRPC under therapy with Abiraterone (**a**) with baseline neutrophil-to-lymphocyte ratio (NLR) < 5 vs. ≥ 5; (**b**) Patients with or without change of NLR to <5 after eight weeks of therapy and (**c**) overall survival probability with baseline NLR < 5 vs. ≥ 5; and (**d**) patients with or without change of NLR to <5 after eight weeks of therapy.

2.3. Prognostication of Survival Outcomes with Change of Neutrophil-to-Lymphocyte Ratio (NLR) to <5 in Patients with a Baseline NLR > 5 vs. no Change after Eight Weeks of Abiraterone

Seventeen men (17.7%) of the whole cohort had a NLR >5 at baseline. In the patients with a change of NLR from >5 to <5 after eight weeks of treatment with Abiraterone ($n = 10$) the neutrophils declined in all but one of the cases in which the lymphocytes increased. Seven patients remained with an NLR of >5 after eight weeks of therapy.

For PFS the univariate Cox-regression analysis (Table 2) showed that change of NLR to <5 at eight weeks of Abiraterone therapy was associated with worse survival (HR: 4.1 (95%CI: 1.1–15.8); $p = 0.04$). All other variables, including decline of PSA $\geq 50\%$, were no significant prognosticators.

In the Kaplan-Meier analysis of PFS (Figure 1b), equally, the median survival was in favour of no change of NLR (12 months (95%CI 3.3–20.7) compared to change to NLR <5 (five months (95%CI 3.6–6.4), log-rank $p = 0.02$)).

After adjusting for decline of PSA $\geq 50\%$, there was only a trend for prognostication of worse PFS for change of NLR to <5 (HR: 3.4 (95%CI: 0.8–15.2); $p = 0.11$) (Table 4).

The univariate analysis of OS (Table 3) showed a trend for change of NLR to <5 (HR: 2.3 (95%CI: 0.7–7.1); $p = 0.15$) and no decline of PSA $\geq$50% (HR: 2.9 (95%CI: 0.9–9.4); $p = 0.08$) towards poor survival.

In Kaplan-Meier analysis of OS (Figure 1d) the change of NLR to <5 was associated with a shorter survival of 7 months (95%CI 5.5–8.5) compared with 16 months (95%CI N/D-39.1) in patients with no change of NLR. However, this difference was not statistically significant (log-rank $p = 0.12$).

In multivariate Cox-regression analysis (Table 4) there was no significant difference for the prognostication of worse OS in patients with change of NLR to <5 and no decline of PSA $\geq$50% ($p = 0.47$ and 0.25, respectively).

In Kaplan-Meier analysis of the 79 patients with an initial NLR of <5 at baseline (Figure 2) there was no difference between the subgroup of patients with change of NLR to >5 and in the subgroup with no change of NLR for both PFS and OS (log rank $p = 0.21$, respectively).

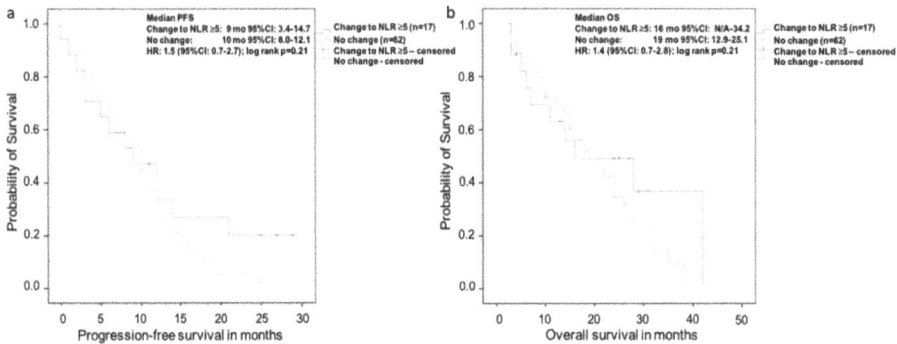

Figure 2. Kaplan-Meier analysis of the subgroup of patients with a NLR <5 at baseline for (**a**) overall survival (OS) and (**b**) progression-free survival (PFS). There was no difference of median survival in the subgroup of patients with a NLR <5 at baseline and change of NLR to >5 vs. no change after eight weeks of Abiraterone therapy (log rank $p = 0.21$). There equally was no difference of median PFS for change of NLR to >5 vs. no change after eight weeks of Abiraterone therapy (log rank $p = 0.21$).

3. Discussion

The NLR was found to be prognostic for survival outcomes and recurrence rate in various tumour entities [18,19]. For prostate cancer, especially in mCRPC, the NLR was equally shown to be prognostic for survival [10,17]. In a study on mCRPC patients treated with Abiraterone, Leibowitz-Amit et al. found that a composite score of baseline NLR <5 and extent of metastatic spread was associated with PSA-response [17]. Our results support these findings. In our cohort of mCRPC patients treated with Abiraterone the baseline NLR with a cut-off of <5 without addition of metastatic spread equally was a prognosticator for prolonged OS. However, after adjustment for other important variables in multivariate analysis there was only a trend towards statistical significance regarding survival. Maybe the limited number of patients with a baseline NLR >5 ($n = 17$) included in this study, precluded reaching significant values in multivariate analyses for survival endpoints.

The underlying immunologic mechanisms are still unclear, but most likely the interactions between tumour cells and immunologic cells and mediators undergo specific changes [9]. An elevated (expected unfavourable) NLR can be caused by rising neutrophils or declining lymphocytes or both. Assuming that a high baseline NLR is prognostic for outcomes under given treatments it would be reasonable to presume that a change of an elevated (expected unfavourable) NLR to a lower (potentially favourable) NLR under treatment could be useful to identify patients with exceptionally good and lasting responses to therapy.

So far, there is very little evidence on NLR as a therapy control marker. For mCRPC there is only one study by Lorente et al., which addressed this matter [10]. Here, in a post hoc analysis of the large registration trial of Cabazitaxel vs. Mitoxantrone, a baseline NLR of <3 was identified to be associated with better survival, better PSA-response and better objective response rate according to RECIST-criteria. Our study is the first to address change of NLR as a treatment response marker in patients treated with Abiraterone. Interestingly, in that study the conversion of an unfavourable NLR of >3 to a favourable NLR < 3 under therapy was independently associated with an improved OS. This result was in line with the expectable. In contrast and most surprisingly, in our study we found that a change of a supposedly unfavourable NLR > 5 to the expected to be favourable NLR <5 after eight weeks of Abiraterone therapy in mCRPC patients was associated with both worse PFS and OS. The reasons for these findings are unclear. One possible explanation could be the immunomodulating effect of Prednisone, which has to be administered to prevent mineral corticoid excess under Abiraterone therapy. In our study, all patients received the recommended dose of 5 mg Prednisone bi-daily with the start of Abiraterone, but no patient had prior prednisone therapy. In the trial of Lorente et al. [10] 45% of the patients had already been on a stable dose of a corticosteroid before initiation of Cabazitaxel-/Mitoxantron-therapy and, thus, at the time when the baseline NLR was determined. The subgroup analysis of patients with or without baseline Prednisone use showed that the patients with Prednisone use had a significantly higher NLR of 3.9 vs. 2.9 ($p < 0.001$) and patients with NLR > 3 were more likely to receive baseline Prednisone ($p = 0.016$). However, Lorente et al. tested for confounding variables and showed that the NLR < 3 was predictive of survival with or without corticosteroid use. This basically supports the role of corticosteroids for elevating the NLR in general. In our study all patients had been treated with Prednisone starting with initiation of Abiraterone therapy. This difference could explain our unexpected findings. Prednisone physiologically leads to an increase of neutrophils first by about 20% by demarginalisation of the intravascular pool, and consecutively, by a greater increase caused by mobilization of the bone marrow neutrophil reserve [20]. Naturally, this effect would be evident for all patients treated with Prednisone and Abiraterone. In the subset of patients with a change of NLR from >5 to <5 after eight weeks of Abiraterone therapy in our study the neutrophils declined in almost all of the cases. An explanation for this may be that these patients had a higher tumour load especially of bone metastases thus leading to a depletion of functioning bone marrow with the production of neutrophils most likely being impaired by bone marrow insufficiency as a result of tumour progression under Abiraterone. The immunomodulating effect of Prednisone under Abiraterone may have a stronger impact due to increasingly depleted bone marrow reserves in case of disease progression.

Recently there were findings that PSA change within four weeks after the start of therapy may be prognostic for survival outcomes. For example, Facchini et al. showed that PSA-change after 15 days of Abiraterone was a surrogate for PFS and OS [21]. Therefore, for future studies, research on change of NLR and the possibility of complement of PSA and change of NLR in this very early setting may be promising.

Our study is limited by problems inherent to the retrospective approach and a relatively small sample size. This holds especially true for the subset of patients with change of NLR to <5. Therefore, our findings need to be validated by prospective studies. The single parameter NLR cut-off of less than, or greater than five needs to be validated as well. We did not do imaging in a per protocol routine fashion and, therefore, could not calculate radiographic PFS.

The underlying effects of the NLR and changes of NLR in relationship to cancer are still unclear and especially our findings concerning unexpected changes under therapy with Prednisone/Abiraterone need to be elucidated.

4. Materials and Methods

4.1. Patients

After ethics committee-approval (2007-467-f-S), we retrospectively reviewed all patients with mCRPC presenting at the Department of Urology in the Muenster University Medical Center between December 2009 and July 2015 and receiving Abiraterone to analyse the impact of the NLR on progression-free and overall survival. All patients gave written informed consent before participating. We performed the study according to the declaration of Helsinki.

The study cohort consisted of confirmed mCRPC patients as defined by PCWG2-criteria [22] in pre- or post-chemotherapy setting. All patients met the requirements for Abiraterone-treatment. The men receiving Abiraterone prior to Docetaxel all had to be asymptomatic or oligo-symptomatic with no need to use opiates and degree of pain of no more than three out of 10 on the numeric-rating-scale. Patients receiving Abiraterone after Docetaxel all had progressive disease on or after chemotherapy. During the reviewed timeframe a total of 96 patients presented with evaluable NLR-data of which 52 patients were treated prior to, and 44 after, chemotherapy, respectively. Five patients (4.8%) had received Enzalutamide prior to Abiraterone. All patients included in the analysis were on a stable dose of anti-bone resorptive medication (Denosumab or Zoledronic acid) at least three months prior to the start of Abiraterone, and during the whole treatment phase, or did not receive antiresorptive medication at all.

The day before start of Abiraterone therapy all patients had blood drawn for baseline analysis of PSA, neutrophils, lymphocytes, and serum chemistry. The analysis of PSA was repeated two and four weeks after initiation of therapy, and four weekly thereafter. Neutrophils and lymphocyte measurements were repeated after eight weeks. PSA-progression was defined according to PCWG2-criteria, i.e., PSA-progression at the date that a 25% or greater increase and an absolute increase of 2 ng/mL or more from the nadir is documented, which is confirmed by a second value obtained three or more weeks later [22].

For documentation of baseline values and changes of soft tissue metastases CT- or MRI-scans of thorax, abdomen and pelvis were performed. For information on bone metastases bone scans were used. Imaging was repeated during the course of the Abiraterone-therapy whenever clinically indicated but not in routine fashion. PD was defined according to RECIST 1.1 criteria for axial cross sectional imaging [23] and by PCWG2 criteria for bone scans [22].

The current response status, i.e., complete remission (CR), partial remission (PR), stable disease (SD) or PD was assessed after each visit. PD was considered evident upon deterioration of general condition or worsening of pain when unequivocally caused be prostate cancer, when PSA-progression occurred or when radiographic progression was detected. The baseline cohort was subdivided according to NLR (group-1 (favorable): NLR < 5; group-2 (non favorable): NLR > 5).

4.2. Statistical Methods

For descriptive statistics we report medians with 95% CI or IQR for continuous variables and as populations and frequencies for categorical variables. We used the χ^2-test, Fisher's exact-test or Mann-Whitney U-test for determining the significance of differences between categorical and continuous variables. For analysis of survival outcomes were applied Kaplan-Meier-analysis. For uni- and multivariate analysis of significance for survival outcomes we used Cox-regression-models. All reported *p*-values are two-sided and we assumed statistical significance when $p \leq 0.05$. We used SPSS-Statistics V.23 (IBM Inc., Armonk, NY, USA) for statistical assessment.

5. Conclusions

In patients with mCRPC under Abiraterone treatment a pre-treatment NLR < 5 is associated with better survival outcomes. Unexpectedly, after eight weeks of Abiraterone therapy, a change of initially elevated NLR of >5 to <5 was associated with worse survival. The immunomodulating effect of concomitant Prednisone next to Abiraterone combined with insufficient bone marrow reserves in patients with advanced (bony) disease may be the reason for this phenomenon. Thus, a putatively favourable change of NLR to <5 may represent worsening of disease and should be interpreted cautiously. A deeper understanding of the underlying immune mechanisms in this setting is highly warranted. Therefore, larger prospective trials on the matter are needed prior to application of change of NLR in mCRPC.

Author Contributions: Martin Boegemann conceived of the study and participated in its design and coordination, performed part of the statistical analysis and drafted the manuscript. Katrin Schlack helped in acquisition of the data, participated in the study design, performed part of the statistical analysis and participated in critical revision of the manuscript. Stefan Thomes and Julie Steinestel helped in acquisition of the data and participated in critical revision of the manuscript. Kambiz Rahbar participated in critical revision of the manuscript. Axel Semjonow and Andres Jan Schrader participated in critical revision of the manuscript. Martin Aringer helped in drafting and participated in critical revision of the manuscript. Laura-Maria Krabbe helped in acquisition of the data, in statistical analysis and critical revision of the manuscript. All authors read and approved the final manuscript.

Conflicts of Interest: The authors declare no conflict of interest.

Abbreviations

mCRPC	Metastatic castration resistant prostate cancer
OS	Overall survival
PD	Progressive disease
NLR	Neutrophil-to-lymphocyte ratio
PFS	Progression-free survival
PSA	Prostate specific antigen
IQR	Interquartile range
ALP	Alkaline phosphatase
LDH	Lactate dehydrogenase
ULN	Upper normal limit
PCWG2	Prostate cancer working group 2
HR	Hazard ratio
95%CI	95% confidence interval
ECOG	Eastern Collaborative Oncology Group Performance Status
CR	Complete remission
PR	Partial remission
SD	Stable disease

References

1. Ryan, C.J.; Smith, M.R.; Fizazi, K.; Saad, F.; Mulders, P.F.; Sternberg, C.N.; Miller, K.; Logothetis, C.J.; Shore, N.D.; Small, E.J.; et al. Abiraterone acetate plus prednisone versus placebo plus prednisone in chemotherapy-naive men with metastatic castration-resistant prostate cancer (COU-AA-302): Final overall survival analysis of a randomised, double-blind, placebo-controlled phase 3 study. *Lancet Oncol.* **2015**, *16*, 152–160. [CrossRef]

2. Beer, T.M.; Tombal, B. Enzalutamide in metastatic prostate cancer before chemotherapy. *N. Engl. J. Med.* **2014**, *371*, 1755–1756. [CrossRef] [PubMed]

3. Kantoff, P.W.; Higano, C.S.; Shore, N.D.; Berger, E.R.; Small, E.J.; Penson, D.F.; Redfern, C.H.; Ferrari, A.C.; Dreicer, R.; Sims, R.B.; et al. Sipuleucel-T immunotherapy for castration-resistant prostate cancer. *N. Engl. J. Med.* **2010**, *363*, 411–422. [CrossRef] [PubMed]

4. Fizazi, K.; Scher, H.I.; Molina, A.; Logothetis, C.J.; Chi, K.N.; Jones, R.J.; Staffurth, J.N.; North, S.; Vogelzang, N.J.; Saad, F.; et al. Abiraterone acetate for treatment of metastatic castration-resistant prostate cancer: Final overall survival analysis of the COU-AA-301 randomised, double-blind, placebo-controlled phase 3 study. *Lancet Oncol.* **2012**, *13*, 983–992. [CrossRef]

5. Scher, H.I.; Fizazi, K.; Saad, F.; Taplin, M.E.; Sternberg, C.N.; Miller, K.; de Wit, R.; Mulders, P.; Chi, K.N.; Shore, N.D.; et al. Increased survival with enzalutamide in prostate cancer after chemotherapy. *N. Engl. J. Med.* **2012**, *367*, 1187–1197. [PubMed]

6. De Bono, J.S.; Oudard, S.; Ozguroglu, M.; Hansen, S.; Machiels, J.P.; Kocak, I.; Gravis, G.; Bodrogi, I.; Mackenzie, M.J.; Shen, L.; et al. Prednisone plus cabazitaxel or mitoxantrone for metastatic castration-resistant prostate cancer progressing after docetaxel treatment: A randomised open-label trial. *Lancet* **2010**, *376*, 1147–1154. [CrossRef]

7. Parker, C.; Nilsson, S.; Heinrich, D.; Helle, S.I.; O'Sullivan, J.M.; Fossa, S.D.; Chodacki, A.; Wiechno, P.; Logue, J.; Seke, M.; et al. α emitter radium-223 and survival in metastatic prostate cancer. *N. Engl. J. Med.* **2013**, *369*, 213–223. [CrossRef] [PubMed]

8. Ryan, C.J.; Shah, S.; Efstathiou, E.; Smith, M.R.; Taplin, M.E.; Bubley, G.J.; Logothetis, C.J.; Kheoh, T.; Kilian, C.; Haqq, C.M.; et al. Phase II study of abiraterone acetate in chemotherapy-naive metastatic castration-resistant prostate cancer displaying bone flare discordant with serologic response. *Clin. Cancer Res.* **2011**, *17*, 4854–4861. [CrossRef] [PubMed]

9. Hanahan, D.; Weinberg, R.A. Hallmarks of cancer: The next generation. *Cell* **2011**, *144*, 646–674. [CrossRef] [PubMed]

10. Lorente, D.; Mateo, J.; Templeton, A.J.; Zafeiriou, Z.; Bianchini, D.; Ferraldeschi, R.; Bahl, A.; Shen, L.; Su, Z.; Sartor, O.; et al. Baseline neutrophil-lymphocyte ratio (NLR) is associated with survival and response to treatment with second-line chemotherapy for advanced prostate cancer independent of baseline steroid use. *Ann. Oncol.* **2015**, *26*, 750–755. [CrossRef] [PubMed]

11. Luo, Y.; She, D.L.; Xiong, H.; Fu, S.J.; Yang, L. Pretreatment Neutrophil to Lymphocyte Ratio as a Prognostic Predictor of Urologic Tumors: A Systematic Review and Meta-Analysis. *Medicine* **2015**, *94*, e1670. [CrossRef] [PubMed]

12. Xin-Ji, Z.; Yong-Gang, L.; Xiao-Jun, S.; Xiao-Wu, C.; Dong, Z.; Da-Jian, Z. The prognostic role of neutrophils to lymphocytes ratio and platelet count in gastric cancer: A meta-analysis. *Int. J. Surg.* **2015**, *21*, 84–91. [CrossRef] [PubMed]

13. Yodying, H.; Matsuda, A.; Miyashita, M.; Matsumoto, S.; Sakurazawa, N.; Yamada, M.; Uchida, E. Prognostic significance of neutrophil-to-lymphocyte ratio and platelet-to-lymphocyte ratio in oncologic outcomes of esophageal cancer: A systematic review and meta-analysis. *Ann. Surg. Oncol.* **2016**, *23*, 646–654. [CrossRef] [PubMed]

14. Keizman, D.; Gottfried, M.; Ish-Shalom, M.; Maimon, N.; Peer, A.; Neumann, A.; Rosenbaum, E.; Kovel, S.; Pili, R.; Sinibaldi, V.; et al. Pretreatment neutrophil-to-lymphocyte ratio in metastatic castration-resistant prostate cancer patients treated with ketoconazole: Association with outcome and predictive nomogram. *Oncologist* **2012**, *17*, 1508–1514. [CrossRef] [PubMed]

15. Sonpavde, G.; Pond, G.R.; Armstrong, A.J.; Clarke, S.J.; Vardy, J.L.; Templeton, A.J.; Wang, S.L.; Paolini, J.; Chen, I.; Chow-Maneval, E.; et al. Prognostic impact of the neutrophil-to-lymphocyte ratio in men with metastatic castration-resistant prostate cancer. *Clin. Genitourin. Cancer* **2014**, *12*, 317–324. [CrossRef] [PubMed]

16. Van Soest, R.J.; Templeton, A.J.; Vera-Badillo, F.E.; Mercier, F.; Sonpavde, G.; Amir, E.; Tombal, B.; Rosenthal, M.; Eisenberger, M.A.; Tannock, I.F.; et al. Neutrophil-to-lymphocyte ratio as a prognostic biomarker for men with metastatic castration-resistant prostate cancer receiving first-line chemotherapy: Data from two randomized phase III trials. *Ann. Oncol.* **2015**, *26*, 743–749. [CrossRef] [PubMed]

17. Leibowitz-Amit, R.; Templeton, A.J.; Omlin, A.; Pezaro, C.; Atenafu, E.G.; Keizman, D.; Vera-Badillo, F.; Seah, J.A.; Attard, G.; Knox, J.J.; et al. Clinical variables associated with PSA response to abiraterone acetate in patients with metastatic castration-resistant prostate cancer. *Ann. Oncol.* **2014**, *25*, 657–662. [CrossRef] [PubMed]

18. Donskov, F. Immunomonitoring and prognostic relevance of neutrophils in clinical trials. *Semin. Cancer Biol* **2013**, *23*, 200–207. [CrossRef] [PubMed]

19. Guthrie, G.J.; Charles, K.A.; Roxburgh, C.S.; Horgan, P.G.; McMillan, D.C.; Clarke, S.J. The systemic inflammation-based neutrophil-lymphocyte ratio: Experience in patients with cancer. *Crit. Rev. Oncol. Hematol.* **2013**, *88*, 218–230. [CrossRef] [PubMed]

20. Nakagawa, M.; Terashima, T.; D'Yachkova, Y.; Bondy, G.P.; Hogg, J.C.; van Eeden, S.F. Glucocorticoid-induced granulocytosis: Contribution of marrow release and demargination of intravascular granulocytes. *Circulation* **1998**, *98*, 2307–2313. [CrossRef] [PubMed]

21. Facchini, G.; Caffo, O.; Ortega, C.; D'Aniello, C.; Di Napoli, M.; Cecere, S.C.; Della Pepa, C.; Crispo, A.; Maines, F.; Ruatta, F.; et al. Very Early PSA Response to Abiraterone in mCRPC Patients: A Novel Prognostic Factor Predicting Overall Survival. *Front. Pharmacol.* **2016**, *7*, 123. [CrossRef] [PubMed]

22. Scher, H.I.; Halabi, S.; Tannock, I.; Morris, M.; Sternberg, C.N.; Carducci, M.A.; Eisenberger, M.A.; Higano, C.; Bubley, G.J.; Dreicer, R.; et al. Design and end points of clinical trials for patients with progressive prostate cancer and castrate levels of testosterone: Recommendations of the Prostate Cancer Clinical Trials Working Group. *J. Clin. Oncol.* **2008**, *26*, 1148–1159. [CrossRef] [PubMed]
23. Eisenhauer, E.A.; Therasse, P.; Bogaerts, J.; Schwartz, L.H.; Sargent, D.; Ford, R.; Dancey, J.; Arbuck, S.; Gwyther, S.; Mooney, M.; et al. New response evaluation criteria in solid tumours: Revised RECIST guideline (version 1.1). *Eur. J. Cancer* **2009**, *45*, 228–247. [CrossRef] [PubMed]

International Journal of
Molecular Sciences

MDPI

Review

Prostate Cancer Detection and Prognosis: From Prostate Specific Antigen (PSA) to Exosomal Biomarkers

Xavier Filella * and Laura Foj

Department of Biochemistry and Molecular Genetics (CDB), Hospital Clínic, IDIBAPS, C/Villarroel, 170, 08036 Barcelona, Catalonia, Spain; lfoj@clinic.cat
* Correspondence: xfilella@clinic.cat; Tel.: +34-93-227-93-75, Fax: +34-93-227-93-76

Academic Editor: Carsten Stephan
Received: 13 September 2016; Accepted: 14 October 2016; Published: 26 October 2016

Abstract: Prostate specific antigen (PSA) remains the most used biomarker in the management of early prostate cancer (PCa), in spite of the problems related to false positive results and overdiagnosis. New biomarkers have been proposed in recent years with the aim of increasing specificity and distinguishing aggressive from non-aggressive PCa. The emerging role of the prostate health index and the 4Kscore is reviewed in this article. Both are blood-based tests related to the aggressiveness of the tumor, which provide the risk of suffering PCa and avoiding negative biopsies. Furthermore, the use of urine has emerged as a non-invasive way to identify new biomarkers in recent years, including the *PCA3* and *TMPRSS2:ERG* fusion gene. Available results about the PCA3 score showed its usefulness to decide the repetition of biopsy in patients with a previous negative result, although its relationship with the aggressiveness of the tumor is controversial. More recently, aberrant microRNA expression in PCa has been reported by different authors. Preliminary results suggest the utility of circulating and urinary microRNAs in the detection and prognosis of PCa. Although several of these new biomarkers have been recommended by different guidelines, large prospective and comparative studies are necessary to establish their value in PCa detection and prognosis.

Keywords: prostate cancer detection; biomarker; prostate specific antigen (PSA); prostate health index; 4Kscore; PCA3 score; miRNAs; exosomal biomarkers

1. Introduction

Prostate cancer (PCa) remains a medical challenge, since it is one of the most frequently diagnosed tumors and a common cause of cancer death among men in western countries [1]. The introduction of the measurement of prostate specific antigen (PSA) in the mid-eighties of last century represented a major innovation in the management of patients with PCa. PSA, or human kallicrein 3 (hK3), is a glandular kallikrein with abundant expression in the prostate encoded by the *KLK3* gene. The sequential measurement of PSA allows the monitoring of the response to treatment and the assessment of its effectiveness. PSA is also widely used in the detection of PCa, despite its low specificity, with false positive results in patients with benign prostatic hyperplasia (BPH). Biopsy is only positive in around 25% of patients with PSA in the range between 2 and 10 µg/L. Elevated levels of PSA are particularly found in patients with enlarged prostate volume [2]. Age-specific PSA reference ranges and PSA density have been proposed to avoid this effect, increasing the specificity of PSA. Moreover, prostate volume is a key element in multivariate PCa risk calculators [3].

More recently, several studies have shown that single nucleotide polymorphisms (SNPs) in the *KLK3* gene or other related genes can influence PSA serum levels. Gudmundsson et al. [4] detected a significant association between PSA serum levels and SNPs at six loci, finding the strongest association for SNPs located near or in the *KLK3* gene. Furthermore, some of these SNPs could be associated with predisposition to PCa and PCa aggressiveness [5]. Genetic effect on PSA expression advocates a genetic correction of PSA serum levels. Unnecessary biopsies could be avoided in genetically high PSA producers, while necessary biopsies could be performed in genetically low PSA producers. Helfand et al. [6] suggested a genetically personalized interpretation of PSA serum levels to obtain more accurate results. According to these authors, genetic correction of PSA serum levels could decrease the number of potentially unnecessary biopsies by approximately 18% to 22%, improving the use of PSA as a screening tool.

A major harm of PCa screening is overdiagnosis, defined as the diagnosis of patients whose cancer never surfaces clinically during lifetime. Active surveillance is now accepted as a valuable strategy in the management of low-risk PCa, decreasing the negative effects of overdiagnosis and overtreatment. The criteria for the selection of patients for active surveillance include several variables, such as a PSA lower than 10 µg/L, a Gleason score lower than 7 and a low number of cylinders affected by the tumor [7]. However, PCa characterization based on these findings is not ideal [8]. The availability of more accurate inclusion criteria would lead to an improvement in the management of patients with early PCa.

The introduction of a new biomarker in PCa should fulfil the dual purpose of providing increased specificity and—being related to the aggressiveness of the tumor—differentiating aggressive cancers from non-aggressive cancers. Several PSA derivatives have been proposed as PCa biomarkers with this aim. The percentage of free PSA to total PSA (%fPSA) was introduced three decades ago for the detection of PCa, but this test improves clinical information only when levels reach extreme values [9].

More recently, fPSA has been found to include the isoforms BPSA, proPSA and iPSA [10] (Figure 1), with usefulness in the detection of PCa. A commercial immunoassay for the measurement of [−2]proPSA, the most stable form of proPSA, has been developed. The Prostate Health Index has been proposed as a new index for PCa detection, combining serum concentrations of [−2]proPSA, fPSA and total PSA, according to the formula ([−2]proPSA/fPSA)* $\sqrt{}$ total PSA. In addition, an index based on the measurement of total PSA, fPSA, intact PSA (iPSA) and human kallicrein 2 (hK2) in combination with clinical and demographic data, called the 4Kscore, has been proposed for the detection of high-risk PCa.

Furthermore, the development of molecular biology has allowed the study of genes associated with PCa. Numerous studies have evaluated the utility of the *PCA3* gene and more recently the *TMPRSS2:ERG* fusion gene has also been investigated. On the other hand, the study of microRNAs (miRNAs) has opened an emerging field for the research of these biomarkers in the detection and prognosis of PCa.

Several sources for the study of these potential novel biomarkers for PCa are being analyzed, each one with its advantages and disadvantages. The measurement in plasma or serum has the advantage of being a consolidated media, for which commercial automated assays are available. However, the study of the *PCA3* gene in urine obtained after performing a prostate massage has opened a new way for the study of new biomarkers for PCa. Urine, due to the anatomic proximity of the prostate gland to the urethra, may allow the detection of biomarkers for PCa, particularly when the sample is enriched with prostate cells by performing a slight prostate massage. In addition, the protein content of the urine is lower than in serum and plasma, so that some interferences caused by excess of proteins can be reduced, thus facilitating the detection of new biomarkers. Finally, exosomal biomarkers obtained from blood or urine have been suggested as novel diagnostic and prognostic indicators for several cancers, e.g., PCa. Figure 2 shows a timeline for the identification of PCa biomarkers from the discovery of PSA till the current moment.

The aim of our article was to review the usefulness of blood and urine biomarkers in the detection and prognosis of PCa.

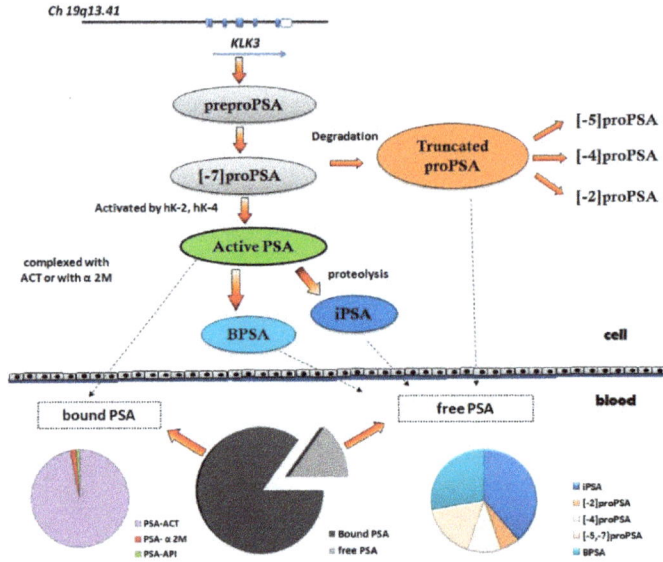

Figure 1. Molecular forms of PSA. The arrows with dashed line mean the forms of PSA that go from the cell to the blood. PSA: prostate specific antigen, BPSA: benign PSA, iPSA: intact PSA, PSA-ACT: Alpha 1-antichymotrypsin-PSA, PSA-API: alpha1-trypsin inhibitor PSA, PSA-A2M: alpha 2 macroglobulin, hK-2: human kallicrein 2, hk-4: human kallicrein 4.

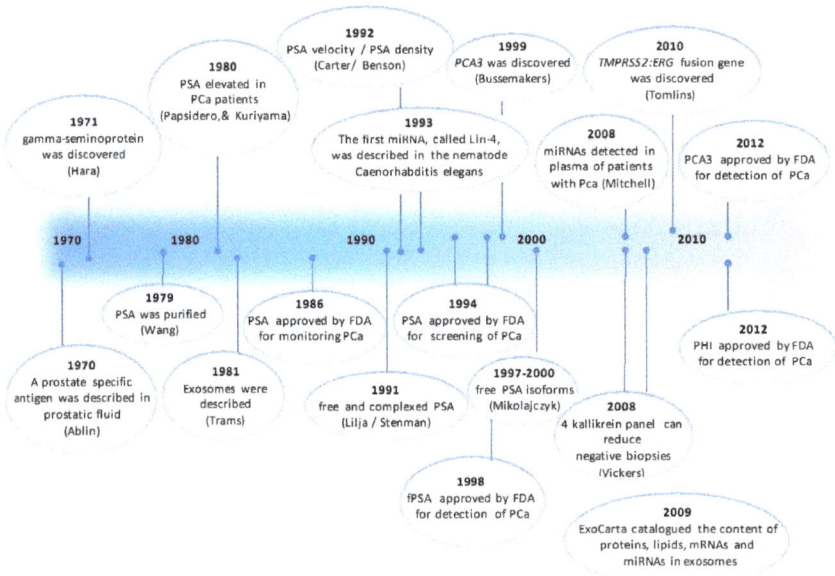

Figure 2. Timeline for the identification of PCa biomarkers. Most important events related to PCa markers from 1970 till the present. PSA: prostate specific antigen, miRNA: microRNA, PCa: prostate cancer, fPSA: free PSA, PHI: prostate health index.

2. Blood-Based PCa Biomarkers

2.1. Prostate Health Index (PHI)

The Prostate Health Index (PHI) was approved by the Food and Drug Administration (FDA) in June 2012 for the detection of PCa in men aged 50 or older, with a PSA between 4 and 10 µg/L and a non-suspicious digital rectal examination (DRE). Also, this test is recommended by the National Comprehensive Cancer Network for patients who have never undergone a biopsy or after a negative biopsy, considering that results higher than 35 are related to a high probability of PCa [11]. Several authors agree with the usefulness of the PHI and percentage of [−2]proPSA to fPSA (%[−2]proPSA) to predict the biopsy outcome. Two multicenter studies, including 1362 and 646 patients, highlight the value of these tests in patients with PSA values between 2 and 10 µg/L. Stephan et al. [12] reported an area under the curve (AUC) of 0.72 and 0.74 for %[−2]proPSA and the PHI, respectively, while Lazzeri et al. [13] reported an AUC of 0.67 for both biomarkers.

Two meta-analyses published in 2013 and 2014 corroborate these results, documenting AUCs from 0.635 to 0.78 for %[−2]proPSA and from 0.69 to 0.781 for the PHI [14,15]. Both meta-analyses underlined that these tests outperform the results obtained with PSA and %fPSA. This point has been confirmed by a subsequent meta-analysis comparing the PHI and %fPSA in patients with total PSA between 2 and 10 µg/L [16], showing an AUCs of 0.74 and 0.63, respectively for the PHI and %fPSA.

The three mentioned meta-analyses also agree regarding the relation of the PHI and %[−2]proPSA with the aggressiveness of the tumor. This fact has been documented by Loeb et al. [17] in a study that evaluated 658 male candidates to biopsy with PSA values between 4 and 10 µg/L. The authors noted that the AUC to identify a clinically significant PCa was 0.698 for the PHI, while it was only 0.654 for %fPSA. Also, Fossati et al. [18] indicated that both the PHI and %[−2]proPSA predicted a pathologic stage T3 and a Gleason score ≥7 in the surgical specimen after evaluating a series of 489 patients treated with radical prostatectomy.

Moreover, several studies have suggested that the addition of the PHI and %[−2]proPSA to multivariate models based on the combination of PSA and various clinical and demographic variables provides better clinical performance for the prediction of PCa. Stephan et al. [19] reported that the AUC increased from 0.69 to 0.75 when [−2]proPSA or the PHI were added to a multivariable model based on patient age, prostate volume, DRE, PSA and %fPSA. Similarly, Guazzoni et al. [20] showed an improvement in the AUC by including %[−2]proPSA (0.82) or the PHI (0.83) in a model based on the patient age, prostate volume, PSA and %fPSA (0.72). Finally, Filella et al. [21] showed that the AUC increased from 0.762 to 0.802 (using a logistic regression analysis) or 0.815 (using an artificial neural network) when the PHI and %[−2]proPSA were included in a multivariate model based on patient age, prostate volume, PSA and %fPSA. Moreover, this study showed a relationship between prostate volume and the PHI values, highlighting that prostate volume is a key factor in the interpretation of PHI results. According to these authors, the performance of the PHI changed in relation to prostate volume, finding AUCs of 0.818, 0.716 and 0.654 for patients with small, medium and large prostate volume, respectively.

Lughezzani et al. [22] developed a PHI based nomogram for predicting PCa evaluating 729 patients undergoing prostate biopsy. The PHI increased significantly the accuracy obtained using a multivariable logistic regression model based on patient age, prostate volume, DRE and biopsy history from 0.73 to 0.80. External validation of this model was provided by a multicenter European study based on 833 patients, showing an accuracy of 0.752 [23]. More recently, Roobol et al. [24] showed analogous results validating this nomogram on 1185 men from four European sites, obtaining an accuracy of 0.75 for all PCa and 0.69 for clinically relevant PCa (clinical stage > T2b and/or a biopsy Gleason score ≥ 7). Similar results were found in this study by adding the PHI to the European Randomized Study of Prostate Cancer (ERSPC) risk calculator, obtaining an accuracy of 0.72 for all PCa and 0.68 for clinically relevant PCa. Despite these positive results, the authors concluded that only limited reductions in the rate of unnecessary biopsies are possible when models are updated adding the PHI. More optimistic are the conclusions of a recent cost-effectiveness study of PCa detection using the PHI [25]. The authors used a micro simulation model based on the results of the ERSPC trial to compare the effects of screening using the PHI and those using only PSA. The model predicted a reduction of 23% in negative biopsies for men with PSA between 3 and 10 µg/L, concluding that PHI testing is 11% more cost-effective than screening only based on PSA.

2.2. Kallikrein Panel

The 4Kscore is a risk calculation for the detection of PCa on the biopsy based on the measurement of a 4-kallikrein panel combined with the patient age, DRE and biopsy history. The 4-kallikrein panel includes the measurement of total PSA, fPSA, iPSA and hK2, a kallikrein with high homology with PSA. The test provides information about the probability of having a high-risk PCa. Therefore, the aim is no longer to indicate a biopsy to detect PCa, but it is intended only to detect aggressive tumors with a Gleason score of 7 or higher. Although the 4Kscore does not have the approval of the FDA, in June 2015 the National Comprehensive Cancer Network recommended the use of this test for the detection of high-risk PCa for patients who have never undergone biopsy or after a negative biopsy [11].

The test was developed after studies performed by the group led by Lilja and Vickers, from Memorial Sloan-Kettering Cancer Center, who conducted an extensive research in collaboration with several European centers. The AUCs for the 4-kallikrein panel reported in all these studies for the detection of high-risk PCa were higher than those for a PSA based model [26–32]. The AUCs obtained for the 4-kallikrein panel ranged between 0.793 and 0.870, while the AUCs for the PSA based model ranged from 0.658 to 0.816. Similar differences were observed when the corresponding clinical models to predict high-risk PCa were compared. The AUCs for the model combining PSA, patient age and DRE ranged from 0.709 to 0.868, while the AUCs ranged from 0.798 to 0.903 when fPSA, iPSA and hK2 were also included.

Prostate volume is a crucial factor influencing PSA serum levels and it is usually included in PCa risk calculators. However, according to results published by Carlsson et al. [33], prostate volume is not affecting the performance of a 4-kallicrein panel in PCa detection. This group studied these biomarkers in two cohorts of 2914 and 740 patients with PSA $\geq$3 µg/L. In the first cohort, the AUC increased from 0.856 to 0.860 when the prostate volume was added to a model based on a 4-kallikrein panel, patient age and DRE, while in the second cohort the AUC was identical including or not prostate volume (0.802).

All results published about the 4Kscore before 2014 were obtained in retrospective European cohorts. In 2015, a prospective multicenter study developed in United States was published [34]. This evaluation, which included 1012 patients scheduled for prostate biopsy, confirmed the results shown in previous studies. The AUC obtained for the 4Kscore test in the detection of high-risk PCa was 0.82. The study documented that 58% of biopsies could be saved using a cut-off value of 15%, although 4.7% of high-risk PCa would not be detected. Using a cut-off value of 6%, the reduction in the number of biopsies was of 30%, but in contrast, only 1.3% of high-risk tumors were not detected.

On the other hand, a study based on a cohort collected in Sweden since 1986, in which samples of included subjects were collected at the age of 40, 50 and 60, showed the predictive value of the 4Kscore [35]. The result of this test assessed at 50 and 60 years old allowed the classification of the patients into two groups according to the probability of developing distant metastasis 20 years later. Patients with PSA $\geq$3 µg/L at 60 years old were classified using a cut-off for the 4Kscore of 7.5 in two groups with a significantly different risk of developing distant metastasis. Similarly, the 4Kscore was also predictive using a cut-off of 5 considering patients with PSA $\geq$2 µg/L at 50 years old.

3. Urine-Based PCa Biomarkers

3.1. PCA3

PCA3 (prostate cancer gene 3, previously referred as *DD3*) is a gene that transcribes a long non-coding mRNA that is overexpressed in PCa tissue (Figure 3). The PCA3 test is the score calculated measuring the concentration of *PCA3* mRNA in relation to PSA mRNA, which is used to normalize *PCA3* signals. Measurements are performed in the urine obtained after performing a prostate massage to enrich prostate cell content. This test, based on quantitative real time polymerase chain reaction (qRT-PCR) technology, obtained the European Conformity in November 2006 and was approved by the FDA in 2012 with the aim of deciding the repetition of a prostate biopsy in men more than 50 years old who have one or more previous negative biopsies and results of PCA3 score higher than 25. The 2015 clinical guide of the National Comprehensive Cancer Network for the early detection of PCa [11] considered PCA3 score as a useful test to decide the repetition of a biopsy in patients with a previous negative biopsy, suggesting the use of 35 as the discriminating value.

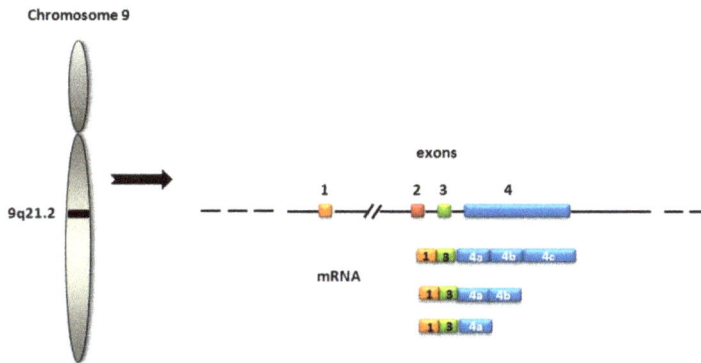

Figure 3. *PCA3* gene structure. The most frequent transcript contains exons 1, 3, 4a, and 4b.

The available results suggest that the PCA3 score can be useful in detecting PCa, especially in patients with a previous negative biopsy [36]. However, the definition of the best discriminating value is controversial. A multicenter study led by Haese [37] indicated that a score of 35 provides the optimal balance between sensitivity (47%) and specificity (72%). Similar results have been recently reported in a meta-analysis published by Hu et al. [38] including 16 studies that used 35 as the cut-off to decide when to perform a biopsy. These authors documented an overall sensitivity of 57% and an overall specificity of 71%, concluding that, although high quality further studies are missing, the PCA3 score is a useful test in the detection of PCa. However, a large group of tumors were not diagnosed when a biopsy was indicated in patients with a PCA3 score higher than 35. In the aforementioned study of Haese et al. [37], 67% of biopsies would have been saved using the discriminating value of 35, but 21% of high-grade PCa would not have been diagnosed. In the same study, using a discriminating value of 20, the saved biopsies were only from 44%, but 9% of high-grade PCa were not detected.

These data have been confirmed by Crawford et al. [39] in a multicenter study involving 1913 patients, 802 of which with PCa. These authors point out that the traditional score of 35 allowed a reduction of 77% in biopsies, but the number of false negatives was large, because PCa was detected in 413 of 1275 patients (32%) with a PCA3 score lower than 35, of which 195 had a Gleason score of 7 or higher. However, by lowering the discriminating value to 10, the number of undiagnosed tumors was reduced to 108, of which 52 had a Gleason score lower than 7. This study also points out that PCa was only detected in 86 of 114 patients with a PCA3 score ≥100. It is somewhat paradoxical because initial studies presented *PCA3* as an overexpressed gene in PCa tissue between 10- and 100-fold relative to non-neoplastic tissue [40]. In this way, Schröder et al. [41] showed a low positive predictive value (38.9%) for a PCA3 score ≥100, even when significant efforts to detect a PCa were performed.

Finally, the relationship between PCA3 score and the aggressiveness of the tumor is also controversial. Merola et al. [42], in a group of 114 patients with PCa, observed a significant association between the PCA3 score and Gleason score (p = 0.02). Moreover, Chevli et al. [43] in a study based on 3073 patients observed that the PCA3 score is related to the Gleason score, with mean values of 47.5 and 58.5 for patients with a Gleason score of 6 and 7 or higher, respectively. However, this study also shows that the AUC to predict high-grade PCa is lower than that obtained with PSA (0.679 vs. 0.682, respectively). Other authors do not find any correlation between the PCA3 score and Gleason score, without explanations for this divergence [44,45].

The accumulated data on the PCA3 score generate an important expectation. Nevertheless, the available evidence about its usefulness is not enough, as noted by a comparative analysis published by Bradley et al. [46] based on 34 observational studies. The majority of studies suggest that the diagnostic accuracy of the PCA3 score is clearly superior to that of PSA. In this regard, however, Roobol et al. [47] highlight the influence of the bias caused by the use of PSA in the selection of the population included in *PCA3* studies. In their report, the authors try to minimize this bias by not only performing the biopsy when PSA was greater than 3 µg/L, but also when the PCA3 score was lower than 10. The obtained data showed less difference in the AUCs than in other studies (0.581 and 0.635 for PSA and PCA3 score, respectively).

3.2. TMPRSS2:ERG Fusion Gene

The gene fusion involving *ERG* (v-ets erythroblastosis virus E26 oncogene homologs) and the androgen regulated gene *TMPRSS2* (transmembrane serine protease isoform 2) (Figure 4) was reported in 2005 by Tomlins et al. [48]. Aberrant fusion of the *TMPRSS2* gene with the *ERG* gene is observed in 15%–59% of PCa [49], but there are not unanimous results regarding its association with disease prognosis. Differences could be explained in relation to the exon involved in the fusion [50] or due to the presence of multiple copies of the fusion gene, showing a reduced prostate cancer-specific survival in those patients whose tumor had multiple copies of the fusion gene [51].

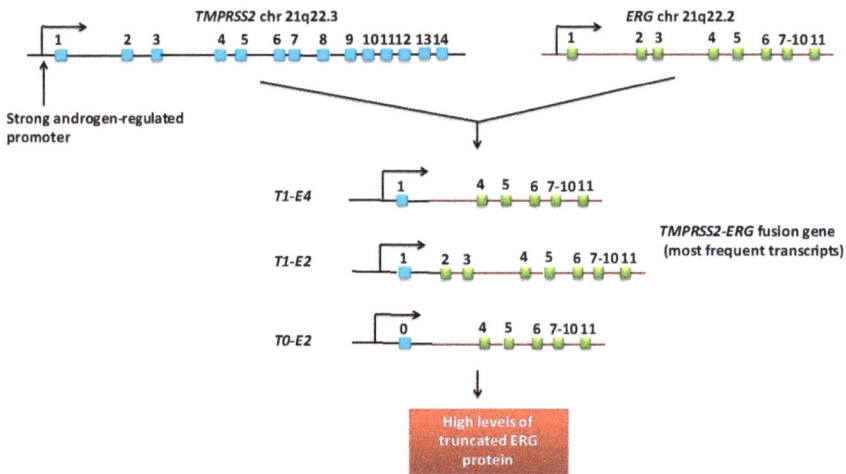

Figure 4. The *TMPRSS2:ERG* fusion gene and its most frequent transcripts. The most frequent transcripts of the *TMPRSS2:ERG* fusion gene consist in the fusion of exon 1 of *TMPRSS2* to either exon 2 or 4 of *ERG* (T1-E4 or T1-E2). These transcripts encode for N-terminal truncated ERG proteins. Exon 1 from *TPRSS2* is non-codifying. Transcription of the gene *TMPRSS2* can start not only from its first exon, exon 1, but also from an alternative first exon (exon 0).

The *TMPRSS2:ERG* gene rearrangements are studied in urine samples obtained after a prostate massage using qRT-PCR. A ratio with PSA mRNA is used to normalize *TMPRSS2:ERG* signals. The combination of the TMPRSS2:ERG score with the PCA3 score has been proposed as a way to improve the prediction of the presence of PCa on the biopsy. A multicenter study published by

Leyten et al. [52] prospectively evaluated the diagnostic utility of PCA3 and TMPRSS2:ERG scores in 443 patients who underwent a biopsy and found that both scores significantly increased the predictive value obtained with the risk calculator ERSPC, which includes PSA and several clinical variables. The AUC increased from 0.799 to 0.833 for the risk calculator ERSPC when PCA3 score was added, and to 0.842 when PCA3 and TMPRSS2:ERG scores were added. In addition, TMPRSS2-ERG, but not PCA3, was associated with the Gleason score and the tumor clinical stage.

More recently, Tomlins et al. [53] also proposed the combination of PCA3 and TMPRSS2:ERG scores with PSA serum levels as a useful tool for detecting PCa. The study, prospectively conducted in three centers and based on 1244 patients who underwent biopsy, showed the value of PCA3 and TMPRSS2:ERG scores when they were added to the PCPT (Prostate Cancer Prevention Trial) risk calculator, also based on PSA and several clinical variables. The AUC increased from 0.639 for the PCPT risk calculator to 0.739 and to 0.762 by adding the TMPRSS2:ERG score alone or the TMPRSS2:ERG score plus the PCA3 score, respectively. Both biomarkers were also evaluated to predict high-risk PCa, being the respective AUCs of 0.707 (for the PCPT risk calculator), 0.752 (plus *TMPRSS2:ERG*) and 0.779 (plus *TMPRSS2:ERG* and PCA3). The model that brings together the PCPT risk calculator, PCA3 and TMPRSS2:ERG scores, called My Prostate score (MiPS), would allow the avoidance of 36% of biopsies using a discriminating value of 15%, although it would cease to diagnose 1.6% of high-risk tumors.

The value of MiPS has been questioned by Stephan et al. [54] comparing PCA3, *TMPRSS2:ERG*, PSA and PHI in a series of 246 patients, including 110 patients with PCa. The authors obtained an AUC for MiPS of 0.748. The highest accuracy was found for a model including PCA3 and the PHI, with AUCs of 0.757 and 0.752, when an artificial neural network and the logistic regression analysis were used, respectively. The authors argued that the clinical potential of the TMPRSS2:ERG score is limited because of its low prevalence in PCa patients.

4. miRNAs

miRNAs are small (17–22 nucleotides) non-coding RNAs, which negatively regulate the gene expression at the posttranscriptional level by base-pairing to the complementary sites in their target mRNAs, resulting in a repression or degradation of the target. The number of identified miRNAs has progressively increased since 1993, when the first miRNA was described in the nematode *Caenorhabditis elegans*. Currently, the number of miRNAs described according to the database mirbase.org is 28,645 miRNAs (miRBase, release 21 June 2014) [55], 2588 of which are found in humans, targeting the vast majority of mRNAs.

Several studies have shown their participation in the development and progression of cancer. Calin et al. [56] showed in 2002 that some miRNAs were suppressed or downregulated in chronic lymphocytic leukemia. The number of studies that evaluate their usefulness as biomarkers has been increasing since 2008, when Lawrie et al. [57] proposed a profile of circulating miRNAs as a diagnostic tool in patients with B-cell lymphoma. Many research findings prove the usefulness of circulating miRNAs in the detection and prognosis of various tumors and in the prediction of the response to treatment, revealing robust signatures of miRNAs that distinguish healthy controls from patients with cancer. However, these results should be contextualized because of differences in the methodology used in the available studies. Witwer [58] underlines that limited overlap has been observed between the findings of similar studies in the same disease, probably due to methodological reasons. Indeed, although miRNAs are stable enough in plasma and serum, optimization and standardization of the methodology used for the miRNAs measurement is required to obtain high quality results. So far, there are still differences due to the type of sample (with differences between serum and plasma, and also related to plasma, among the different anticoagulants); the specific reagent used in the isolation of miRNA; the performance of a preamplification to obtain more Cdna; and the platform used for the qRT-PCR (AbiPrism, LightCycle, etc.), the use of endogenous (none of which behaves as ideal control) or exogenous controls (e.g., cel-miR-39) to normalize the results, and the calculation method used, of which probably $\Delta\Delta C_t$ is the most frequent (Figure 5).

The development of digital PCR could improve the performance of qRT-PCR and would remove the dependence for a reference miRNA for normalization. Besides, the introduction of digital count technologies, such as next-generation sequencing (NGS) and the NanoString nCounter System has supplied new tools for miRNA profiling [59]. As opposed to microarrays and qRT-PCR, NGS enables the discovery of new miRNAs in addition to the confirmation of known miRNAs. It beats the limitations of microarrays that have background signal and cross-hybridization problems. Another recent novelty for miRNA profiling is the NanoString nCounter system, a hybridization-based technology, which consists on direct digital detection of RNA molecules of interest using target-specific, colour-coded probe pairs without the requirement of reverse transcription or cDNA amplification.

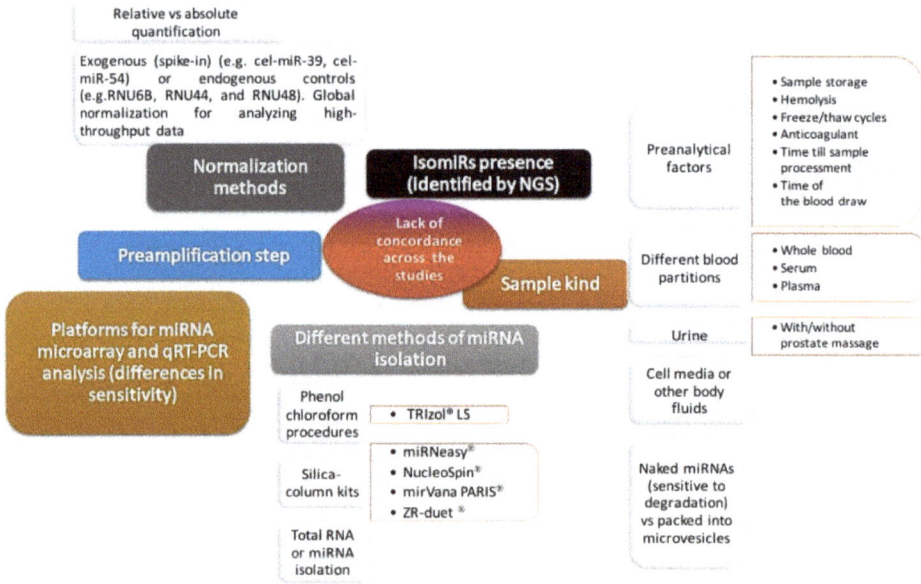

Figure 5. Sources of lack of concordance across the different studies. miRNA: microRNA, NGS: next generation sequencing.

The expression of aberrant miRNAs has been demonstrated in PCa, playing a critical role in tumor initiation, development and progression [60]. Mitchell et al. [61] were the first to show that miRNAs are present in the plasma of patients with PCa in a remarkably stable manner, noting that miR-141 is significantly elevated in patients with advanced PCa compared to healthy controls. Subsequently, several studies based on microarrays have identified several miRNAs signatures with utility in the diagnosis and prognosis of PCa. However, the differences among these panels are substantial and only miR-141, miR-375 and miR-21 repeatedly appear in various studies [62]. Prominently, Mihelich et al. [63] retrospectively studied the serum levels of 21 miRNAs in 100 PCa patients in stages T1–T2. The authors found a signature composed by 14 miRNAs (let-7a, miR-24, miR-26b, miR-30c, miR-93, miR-100, miR-103, miR-106a, miR-107, miR-130b, miR-146a, miR-223, miR-451, and miR-874) to distinguish accurately (negative predictive value of 0.939) high-grade PCa from low-grade PCa. Seven of these miRNAs (miR-451, miR-106a, miR-223, miR-107, miR-130b, let-7a and miR-26b) were also significantly lower in PCa patients with biochemical recurrence after radical prostatectomy compared with those without biochemical recurrence. Similarly, Chen et al. [64] identified, in plasma, a panel of five miRNAs (miR-622, miR-1285, let-7e, let-7c, and miR-30c) to discriminate CaP from BPH and healthy controls with high sensitivity and specificity. On the other hand, Moltzahn et al. [65] found AUCs from 0.812 to 0.928 for miR-106a, miR-1274, miR-93, miR-223, miR-874, miR-1207 and miR-24, showing a trend to correlate with the CAPRA score, a PCa risk assessment based on patient age, PSA serum levels, clinical tumor stage, the Gleason score and the percentage of positive biopsy cores. The authors reported the up-regulation of miR-30c, miR-93, miR-106a, miR-223 and miR-451 in patients with high-risk PCa compared with low-risk PCa. It should be noted that opposite results have been indicated by Mihelich et al. [63], showing that these miRNAs were highly expressed in BPH and low-risk PCa compared with high-risk PCa. These differences could be attributed to deficiencies in the standardization of the collection and measurement of miRNAs.

Although the clinical usefulness of miRNAs in urine has been investigated by several authors, no comprehensive evaluation has been reported till the present moment. Studies have been done in several urine fractions, including whole urine, urinary pellet and the cell-free urinary fraction. The enrichment with prostatic cells after a prostate massage has been also assayed. Preliminary results suggest that the measurement of urinary miRNAs may be valuable as biomarkers in the management of patients with early PCa, although conclusions of the available studies are based on a short series of patients [62]. Recently, Salido-Guadarrama et al. [66] identified a miR-100/200b signature in the urine obtained after prostate massage comparing 73 patients with high-risk PCa and 70 patients with BPH. The AUC for this signature (0.738) was higher than the obtained AUCs for total PSA (0.681) and %fPSA (0.710). Adding the miR-100/200b signature to a multivariate model based on age, DRE, total PSA and %fPSA the AUC increased from 0.816 to 0.876.

Contributions regarding the clinical usefulness of miRNAs in blood and urine are summarized in Table 1.

5. Exosomal Biomarkers

Exosomes are small (30–150 nm) double lipid membrane vesicles of endocytic origin secreted by most cell types, that contain proteins, lipids and nucleic acids. Information about the molecules identified in exosomes from multiple organisms is provided in the ExoCarta database (http://exocarta. org/). Currently, ExoCarta contains information on 41,860 protein entries, 4946 mRNA entries and 2838 miRNA entries that have been identified from 286 exosomal studies. Exosomes were initially described by Trams et al. [67] as exfoliated membrane vesicles, obtained from neoplasic cell cultures. Their presence has later been confirmed in serum and plasma as well as in other biological fluids, including urine, saliva, ascites, breast milk and amniotic fluid. The study of new biomarkers in exosomes is a promising field because they are remarkably stable in body fluids and their content is protected from enzymatic degradation by the exosomal lipid bilayer [68,69].

Cancer cells have been shown to release high levels of exosomes, which may influence tumor initiation, growth and progression as well as drug resistance. Actually, a significantly higher amount of exosomes has been found in serum from ovarian cancer patients in comparison with healthy subjects or patients with benign ovarian diseases. Also, the amount of exosomes has been correlated with the clinical stage in ovarian cancer, showing high levels in patients at an advanced stage [70]. Duijvesz et al. [71] showed that CD9 and CD63 exosomal biomarkers in urine collected after DRE were significantly higher in men with PCa. Analogous results have been found in other tumors, including lung cancer, colorectal cancer and chronic lymphocytic leukemia [72–74].

The emerging involvement of exosomes in intercellular communication focuses on their role in carcinogenesis. Tumor derived exosomes exchange information with other cancer cells and also with other cell types, including stroma and extracellular matrix cells, establishing favorable conditions for tumor growth and invasion. Exosomes can stimulate target cells through different ways, including the interaction with specific membrane receptors, endocitosis and the horizontal transfer of proteins and RNA species, such as mRNA, large non-coding RNA molecules and miRNAs [75]. Available evidence showed that exosomes derived from cancer cells actively contribute to the progression of disease [76]. Franzen et al. [77] reported that epithelial-to-mesenchymal transition in urothelial cells is induced by muscle invasive bladder cancer exosomes. Similar data have been shown in PCa. Abd Elmageed et al. [78] demonstrated that exosomes from PCa cells can induce the

Int. J. Mol. Sci. **2016**, *17*, 1784

neoplastic transformation of adipose derived stem cells through the horizontal transfer of several oncogenic factors, including the oncomiRs miR-125b, miR-130b and miR-155. Besides, according to Hosseini-Beheshti et al. [79], exosomes derived from PCa cells could contribute to cancer progression, reducing apoptosis, increasing cancer cell proliferation, and inducing cell migration in LNCaP and RWPE-1 cells.

In spite of the growing interest in the role of exosomes in several diseases and their use for functional studies and biomarker discovery, no standard method is available to acquire highly pure and well-characterized exosomes. The most common method for exosomes isolation is differential ultracentrifugation, which consists of multiple centrifugation steps with increasing centrifugal strength to sequentially pellet cells (300 g), microvesicles (10,000 g) and exosomes (100,000 g) [80]. Many variations to these speeds are implemented in practice. Serial filtration through 0.45 and 0.2 mm filters is used optionally before exosome pelleting. Furthermore, density gradient based isolation, using sucrose or iodixanol (OptiPrep™, Sigma-Aldrich, St. Louis, MO, USA), can be used to obtain more pure exosome preparations. Commercial easy-to-use reagents have recently been developed, including precipitation solutions, such as ExoQuick™ or Total Exosome Isolation™ kits, and column-based assays, such as exoRNeasy Serum/Plasma. Transmission electron microscopy or immuno electron microscopy can be performed to check the presence of exosomes. Furthermore, Western blot and ELISA assays can be used to detect several exosomal markers (such as CD9, CD63 and CD81) in order to identify the isolated vesicles as exosomes (Figure 6).

Few studies have evaluated exosomal biomarkers in PCa detection and prognosis. Proteome of urinary exosomes has been analyzed using mass spectrometry to identify proteins by Øverbye et al. [81] in 15 healthy subjects and 16 PCa patients. This study showed that 246 proteins were significantly altered in urinary exosomes of PCa patients. The authors found that 17 proteins presented sensitivities above 60% at 100% specificity. The highest sensitivities were observed for transmembrane protein 256 (TM256) (94%), LAMTOR1 (81%) and ADIRF (81%). Furthermore, the authors reported an AUC of 0.87 for TM256, which increased to 0.94 by combining with LAMTOR1.

Comparison of *PCA3* and *TMPRSS2:ERG* between urinary sediment and exosomes has been reported by Dijkstra et al. [82], concluding that exosomes seem to be a more robust source of biomarkers, although a significant proportion of samples are not assessable because they do not reach the analytical detection limit. Exosomal *PCA3* and *TMPRSS2:ERG* levels were significantly higher when urine was collected after a prostate massage. More recently, published results by this group based on 29 men undergoing prostate biopsies showed that *PCA3* and *ERG* perform better in whole urine than in urinary sediment or exosomes [83]. No significant differences were found for both biomarkers in exosomes comparing PCa and non-PCa patients.

Opposite results have been reported by Donovan et al. [84] evaluating the combination of exosomal *PCA3* and *ERG* mRNA in the detection of high-grade PCa. The authors studied both biomarkers in first-catch urine samples obtained without prostate massage from 195 patients submitted for initial biopsy because of PSA serum levels within the gray zone. They showed that the EXO106 score (the sum of normalized exosomal *PCA3* and *ERG* mRNA levels) is related to the Gleason score, showing higher results in patients with a Gleason score $\geq$7. The authors proposed to combine the EXO106 score with PSA, age, race and PCa family history for the detection of high PCa, showing negative and positive predictive values of 97.5% and 34.5%, respectively (AUC: 0.803).

Several authors suggested that exosomes obtained from blood and urine are a consistent source of miRNA for disease biomarker detection [85–88], although some doubts have been presented by other researchers underlying that exosomes in standard preparations do not carry a biologically significant amount of miRNAs [89]. Moreover, according to Arroyo et al. [90], vesicle associated miRNAs only represent a minority, while around 90% of miRNAs in the circulation are present in a non-membrane-bound form. Instead, Gallo et al. [91] showed that the majority of miRNAs detectable in serum and saliva are concentrated in exosomes. Moreover, Cheng et al. [85] showed that in urine the highest proportion of miRNA was extracted from exosomes.

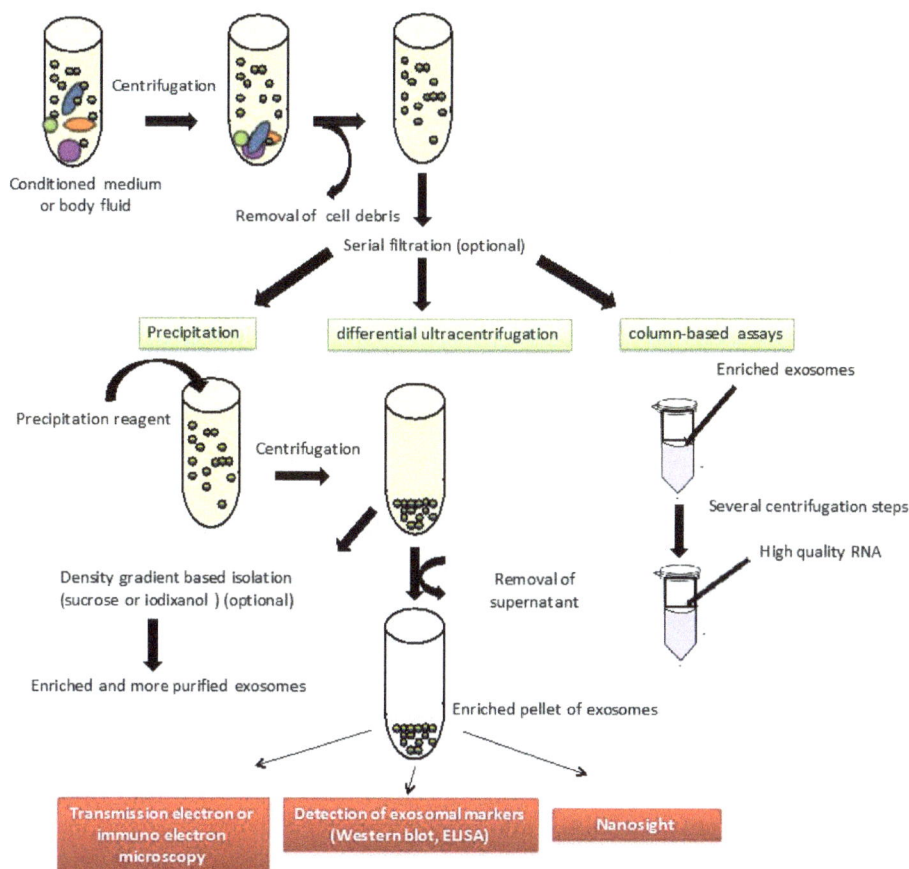

Figure 6. Exosome isolation procedures. The different size particles contained in body fluids or conditioned media are drawn as different color circles.

Actually, some studies have evaluated the usefulness of exosomes' miRNAs in PCa management (Table 1). Li et al. [92] showed that the level of the miR-141 was significantly higher in exosomes compared with the whole serum. Also, these authors reported that the level of serum exosomal miR-141 was significantly higher in PCa patients compared with BPH patients and healthy controls, finding the most elevated levels in patients with metastatic PCa. Moreover, Huang et al. [93] found that the levels of plasma exosomal miR-1290 and miR-375 were significantly associated with poor overall survival. The addition of these new biomarkers into a clinical prognostic model improved predictive performance with a time-dependent AUC increase from 0.66 to 0.73. On the other hand, Bryzgunova et al. [94] demonstrated promising results in the detection of PCa, analyzing miR-19b in urinary exosomes isolated by differential centrifugation. Finally, Samsonov et al. [95] indicated that miR-21, miR-141 and miR-574 were upregulated in PCa patients compared with healthy controls in urinary exosomes isolated by a lectin-based exosomes agglutination method. Nevertheless, only miR-141 was found to be significantly upregulated when urinary exosomes were isolated by differential centrifugation.

Table 1. microRNAs studies in PCa patients.

Reference	Body Fluid	miRNAs Analyzed	Methodology	Patients	Clinical Results
Mitchell et al. 2008 [61]	Serum	miR-100, -125b, -141, -143, -205, and -296	qRT-PCR	25 metastatic PCa and 25 matched healthy controls	AUC of 0.907 for miR-141 comparing PCa and healthy
Mihelich et al. 2015 [63]	Serum	21 miRNAs	qRT-PCR	100 no treated PCa (50 low-grade, 50 high-grade) and 50 BPH	A panel combining let-7a, miR-103, -451, -24, -26b, -30c, -93, -106a, -223, -874, -146a, -125b, -100, -107 and -130b distinguish high-grade PCa from low-grade PCa and BPH
Chen et al. 2012 [64]	Plasma	1146 miRNAs, 8 selected miRNAs for validation study	Illumina's Human miRNA microarray, qRT-PCR	Screening set: 17 BPH and 25 CaP. Validation set: 44 BPH, 54 healthy controls and 80 CaP	A panel combining miR-622, -1285, -30c, let-7e and let-7c discriminate CaP from BPH (AUC: 0.924) or healthy controls (AUC: 0.860)
Moltzahn et al. 2011 [65]	Serum	384 miRNAs, 12 miRs selected for validation study	multiplex qRT-PCR	12 low-risk PCa, 12 intermediate-risk PCa, 12 high-risk PCa and 12 healthy controls	AUCs: miR-106a, 0.928; miR-1274, 0.928; miR-93, 0.907; miR-223, 0.876; miR-874, 0.845; miR-1207, 0.812; miR-24: 0.778. miR-93, -106a and -24 differentiate healthy and metastatic groups
Salido-Guadarrama et al. 2016 [66]	urine obtained after prostate massage	364 miRNAs	MicroRNA TaqMan Low Density Array, qRT-PCR	73 patients with high-risk PCa and 70 patients with BPH	AUC for miR-100/200b signature was 0.738. Adding the miR-100/200b signature to a multivariate model based on age, DRE, total PSA and %fPSA the AUC increased from 0.316 to 0.876
Li et al. 2015 [92]	serum exosomes	miR-141	qRT-PCR	Serum vs. exosomes cohort: 20 PCa, 20 BPH, 20 healthy controls	Serum exosomal miR-141 was significantly higher in PCa patients compared with BPH patients and healthy controls
Huang et al. 2015 [93]	Plasma exosomes	let-7c, miR-30a/e, -99a, -1246, -1290, -16, -125a, and -375	Illumina HiSeq2000 platform, qRT-PCR	Screening cohort: 23 CRPC patients. Follow-up cohort: 100 CRPC patients	Plasma exosomal miR-1290 and miR-375 were significantly associated with poor overall survival
Bryzgunova et al. 2016 [94]	total extracellular vesicles and exosome-enriched fractions	miR-19b, -25, -125b, and -205	qRT-PCR	20 healthy controls and 14 untreated PCa patients	Detection of miR-19b versus miR-16 in total vesicles and exosome-enriched fractions achieved 100%/93% and 95%/79% specificity/sensitivity in distinguishing cancer patients from healthy individuals, respectively, demonstrating the diagnostic value of urine extracellular vesicles. miR-19b in total extracellular vesicles distinguishes cancer patients from healthy individuals with a sensitivity of 93% and a specificity of 100%
Samsonov et al. 2016 [95]	Urinary exosomes	miR-21, -107, -141, -221, -298, -326, -375, -432, -574, -2110, -625, -301a, -191	qRT-PCR	35 PCa patients and 35 healthy controls	miR-21, -141 and -574 were upregulated in PCa patients compared with healthy controls in urinary exosomes

AUC: area under the curve, BPH: benign prostatic hyperplasia, CRPC: castration resistant prostate cancer, PCa: prostate cancer.

More recently, NGS technologies revealed the presence of sequence variants of miRNAs, called isomiRs, showing their utility as biomarkers. They are generated from a single miRNA locus through the miRNA processing and maturation process. Koppers-Lalic et al. [96] identified that isomiRs of miR-21, miR-375 and miR-204 measured in urinary extracellular vesicles could distinguish control men from PCa patients. The authors found an AUC of 0.821 for these set of isomiRs, meanwhile the AUC corresponding to the mature miRNA was 0.661. The existence of isomiRs could explain disagreements about the usefulness of miRNAs in PCa detection.

Preliminary results about exosomal biomarkers are promising, although contradictory results have been published. New advances in standardization of isolation and characterization procedures together with larger clinical studies are required to assess the clinical usefulness of exosomes.

6. Conclusions

PCa is a very heterogeneous tumor, including patients with a low-risk of progression, in which cancer-specific survival rates exceeded 99% at a 15-year follow-up. The percentage of low-risk PCa has been estimated by Klotz [97] to be between 50% and 60% of new diagnosed cases. However, the survival rate decreases considerably for men with aggressive PCa. Specific gene-expression patterns have been identified for PCa subclasses. Recently, Rubin et al. [98] have reported specific genomic profiles related to the Gleason score, showing that few driver mutations and no polyploidy are associated with tumors with a Gleason score lower than 7. In this regard, to label those patients with a Gleason 6 or lower as having cancer has been put into question [99,100].

New biomarkers have been recently introduced for the management of early PCa, offering improvement in detection and some are useful in differentiating between aggressive and non-aggressive PCa. In this article, we have underlined recent advances in the discovery of PCa biomarkers related to aggressiveness. The 4kscore is defined to identify with high accuracy an individual patient's risk for aggressive PCa. Also, high values of PHI are associated with tumor aggressiveness. Besides, both tests outperform the specificity of tPSA and %fPSA. Furthermore, although the most appropriate cut-off for the PCA3 score still has to be established, it outperforms better than tPSA and %fPSA, although its relationship with the aggressiveness of the tumor is controversial. More results are necessary to identify more accurately the usefulness of emerging biomarkers based on molecular techniques, including the *TMPRSS2:ERG* fusion gene and exosomal and non-exosomal miRNAs. Furthermore, new efforts in the standardization of these methods are necessary to use these novel biomarkers routinely.

Available studies comparing PCa biomarkers showed non-conclusive results, although several studies have evaluated these biomarkers. Scattoni et al. [101] evaluated the PHI and PCA3 score in 211 patients undergoing initial (116) or repeat (95) prostate biopsy, finding that the PHI was significantly more accurate than the PCA3 score for predicting PCa (AUC 0.70 vs. 0.59). However, Stephan et al. [102] evaluated 246 patients, showing no significant differences in accuracy between the PCA3 score and the PHI (AUC 0.74 vs. 0.68), whereas the urinary *TMPRSS2:ERG* fusion gene failed to significantly improve the ability to detect PCa. On the other hand, only PSA and the PHI correlated with the Gleason score, whereas *PCA3*, %fPSA, and the *TMPRSS2:ERG* fusion gene did not. Instead, Tallon et al. [103] concluded that the PHI and urinary *PCA3* and *TMPRSS2:ERG* fusion gene are complementary predictors of cancer aggressiveness at radical prostatectomy. According to this group, the PCA3 score was related to tumor volume $\geq$0.5 mL and multifocality, while the PHI was related to tumor volume $\geq$0.5 mL, the Gleason score $\geq$7 and extracapsular extension. The *TMPRSS2:ERG* fusion gene was only related to the pathological T stage.

A comparison between the 4Kscore and the PHI was provided by Nordström et al. [104] evaluating the performance of these biomarkers in a series of 531 men with PSA levels between 3 and 15 µg/L. No significant differences were found between both tests in the detection of any-grade PCa as well as in the detection of high-grade PCa (AUCs: 0.69 and 0.718, respectively, for the 4Kscore; and 0.704 and 0.711, respectively, for the PHI). On the other hand, Vedder et al. [105] compared PCA3 and the 4Kscore in 708 patients in which biopsy was done when PSA was $\geq$3 µg/L or PCA3 score was $\geq$10.

Int. J. Mol. Sci. **2016**, *17*, 1784

The 4Kscore outperforms the PCA3 score (AUC 0.78 vs. 0.62) in men with elevated PSA, although the accuracy of the PCA3 score was higher in the global population (AUC 0.63 for PCA3 vs. 0.56 for the 4Kscore). Additionally, the authors showed that both tests increased the value of a base multivariate model (from 0.70 to 0.73 adding the PCA3 score and 0.71 adding the 4kscore), but significant differences were only found adding the PCA3 score ($p = 0.02$).

To summarize, in recent years, several new promising PCa biomarkers have been identified and found to be associated with tumor aggressiveness (Box 1). Multicenter prospective studies showed the utility of the PHI, the 4Kscore and the PCA3 score to reduce the number of unnecessary prostate biopsies in PSA tested men. Actually, these biomarkers have been recommended for different guidelines [11,106,107]. However, large prospective studies, avoiding bias due to preselection of patients according to PSA serum levels, are necessary to compare the value of these biomarkers. Also, new efforts are necessary to standardize the methodology for the measurement of exosomal and non-exosomal miRNAs to analyze accurately their usefulness in the management of patients with early PCa. Finally, the combined role of these biomarkers together with magnetic resonance imaging data would be elucidated [108–110]. Additionally, results obtained using blood and urinary biomarkers must be compared with promising data obtained with Prostate-specific membrane antigen (PSMA). PSMA is a transmembrane glycoprotein overexpressed in PCa cells. Results of PSMA serum levels in the detection of the PCa are inconclusive, but its expression is correlated with a higher Gleason score, PSA at diagnosis and advanced clinical stage. PSMA features enable this biomarker as an optimal target for developing imaging strategies for PCa. Recent results show that the use of PET probes targeting PSMA for imaging prostate cancer might improve detection of cancerous foci within the prostate, especially in patients with a previous negative prostate biopsy [111]. Further studies are necessary to better characterize new PCa biomarkers, but available publications show promising results to detect PCa and to distinguish patients with aggressive and non-aggressive PCa.

Box 1. Biomarkers in PCa detection and prognosis.

Prostate health index
Biomarkers measured: PSA, fPSA, [−2]proPSA
Sample: serum
Approved by the Food and Drug Administration (FDA)
Recommended by the National Comprehensive Cancer Network
Related to PCa aggressiveness

4Kscore
Biomarkers measured: PSA, fPSA, iPSA, hK2
Sample: serum
The test provides information about the probability of having a high-risk PCa
Recommended by the National Comprehensive Cancer Network
Related to PCa aggressiveness

PCA3 score
Biomarkers measured: mRNA *PCA3* in relation to mRNA *PSA*
Sample: urine obtained after prostate massage
Approved by the Food and Drug Administration (FDA)
Recommended by the National Comprehensive Cancer Network
Inconclusive results about its relationship with PCa aggressiveness

***TMPRSS2:ERG* fusion gene**
Biomarkers measured: mRNA *TMPRSS2:ERG* in relation to mRNA *PSA*
Sample: urine obtained after prostate massage
Preliminary results

miRNAs and other exosomal biomarkers
Sample: blood and urine
Directly related to development and progression of cancer
No standardized methodology
Preliminary results

Conflicts of Interest: The authors declare no conflict of interest.

References

1. Zhou, C.K.; Check, D.P.; Lortet-Tieulent, J.; Laversanne, M.; Jemal, A.; Ferlay, J.; Bray, F.; Cook, M.B.; Devesa, S.S. Prostate cancer incidence in 43 populations worldwide: An analysis of time trends overall and by age group. *Int. J. Cancer* **2016**, *138*, 1388–1400. [CrossRef] [PubMed]
2. Bohnen, A.M.; Groeneveld, F.P.; Bosch, J.L. Serum prostate-specific antigen as a predictor of prostate volume in the community: The Krimpen study. *Eur. Urol.* **2007**, *51*, 1645–1652. [CrossRef] [PubMed]
3. Roobol, M.J.; van Vugt, H.A.; Loeb, S.; Zhu, X.; Bul, M.; Bangma, C.H.; van Leenders, A.G.; Steyerberg, E.W.; Schröder, F.H. Prediction of prostate cancer risk: The role of prostate volume and digital rectal examination in the ERSPC risk calculators. *Eur. Urol.* **2012**, *61*, 577–583. [CrossRef] [PubMed]
4. Gudmundsson, J.; Besenbacher, S.; Sulem, P.; Gudbjartsson, D.F.; Olafsson, I.; Arinbjarnarson, S.; Agnarsson, B.A.; Benediktsdottir, K.R.; Isaksson, H.J.; Kostic, J.P.; et al. Genetic correction of PSA values using sequence variants associated with PSA levels. *Sci. Transl. Med.* **2010**, *2*, 62ra92. [CrossRef] [PubMed]
5. Helfand, B.; Roehl, K.A.; Cooper, P.R.; McGuire, B.B.; Fitzgerald, L.M.; Cancel-Tassin, G.; Cornu, J.N.; Bauer, S.; van Blarigan, E.L.; Chen, X.; et al. Associations of prostate cancer risk variants with disease aggressiveness: Results of the NCI-SPORE Genetics Working Group analysis of 18,343 cases. *Hum. Genet.* **2015**, *134*, 439–450. [CrossRef] [PubMed]
6. Helfand, B.T.; Loeb, S.; Hu, Q.; Cooper, P.R.; Roehl, K.A.; McGuire, B.B.; Baumann, N.A.; Catalona, W.J. Personalized prostate specific antigen testing using genetic variants may reduce unnecessary prostate biopsies. *J. Urol.* **2013**, *189*, 1697–1701. [CrossRef] [PubMed]
7. Heidenreich, A.; Bastian, P.J.; Bellmunt, J.; Bolla, M.; Joniau, S.; van der Kwast, T.; Mason, M.; Matveev, V.; Wiegel, T.; Zattoni, F.; et al. EAU guidelines on prostate cancer. Part 1: Screening; diagnosis; and local treatment with curative intent-update 2013. *Eur. Urol.* **2014**, *65*, 124–137. [CrossRef] [PubMed]
8. Palisaar, J.R.; Noldus, J.; Löppenberg, B.; von Bodman, C.; Sommerer, F.; Eggert, T. Comprehensive report on prostate cancer misclassification by 16 currently used low-risk and active surveillance criteria. *BJU Int.* **2012**, *110*, E172–E181. [CrossRef] [PubMed]
9. Lee, R.; Localio, A.R.; Armstrong, K.; Malkowicz, S.B.; Schwartz, J.S. A meta-analysis of the performance characteristics of the free prostate-specific antigen test. *Urology* **2006**, *67*, 762–768. [CrossRef] [PubMed]
10. Mikolajczyk, S.D.; Marks, L.S.; Partin, A.W.; Rittenhouse, H.G. Free prostate-specific antigen in serum is becoming more complex. *Urology* **2002**, *59*, 797–802. [CrossRef]
11. Prostate Cancer Early Detection. National Cancer Comprehensive Network Clinical Practice Guidelines in Oncology. Version 2. 2016. Available online: https://www.nccn.org/professionals/physician_gls/pdf/prostate_detection.pdf (accessed on 9 September 2016).
12. Stephan, C.; Vincendeau, S.; Houlgatte, A.; Cammann, H.; Jung, K.; Semjonow, A. Multicenter evaluation of [−2]proprostate-specific antigen and the prostate health index for detecting prostate cancer. *Clin. Chem.* **2013**, *59*, 306–314. [CrossRef] [PubMed]
13. Lazzeri, M.; Haese, A.; de la Taille, A.; Palou Redorta, J.; McNicholas, T.; Lughezzani, G.; Scattoni, V.; Bini, V.; Freschi, M.; Sussman, A.; et al. Serum isoform [−2]proPSA derivatives significantly improve prediction of prostate cancer at initial biopsy in a total PSA range of 2–10 ng/mL: A multicentric European study. *Eur. Urol.* **2013**, *63*, 986–994. [CrossRef] [PubMed]
14. Filella, X.; Giménez, N. Evaluation of [−2]proPSA and Prostate Health Index (phi) for the detection of prostate cancer: A systematic review and meta-analysis. *Clin. Chem. Lab. Med.* **2012**, *15*, 1–11. [CrossRef] [PubMed]
15. Wang, W.; Wang, M.; Wang, L.; Adams, T.S.; Tian, Y.; Xu, J. Diagnostic ability of %p2PSA and prostate health index for aggressive prostate cancer: A meta-analysis. *Sci. Rep.* **2014**, *4*, 5012. [CrossRef] [PubMed]
16. Bruzzese, D.; Mazzarella, C.; Ferro, M.; Perdonà, S.; Chiodini, P.; Perruolo, G.; Terracciano, D. Prostate health index vs. percent free prostate-specific antigen for prostate cancer detection in men with "gray" prostate-specific antigen levels at first biopsy: Systematic review and meta-analysis. *Transl. Res.* **2014**, *164*, 444–451. [PubMed]

17. Loeb, S.; Sanda, M.G.; Broyles, D.L.; Shin, S.S.; Bangma, C.H.; Wei, J.T.; Partin, A.W.; Klee, G.G.; Slawin, K.M.; Marks, L.S.; et al. The prostate health index selectively identifies clinically significant prostate cancer. *J. Urol.* **2015**, *193*, 1163–1169. [CrossRef] [PubMed]

18. Fossati, N.; Buffi, N.M.; Haese, A.; Stephan, C.; Larcher, A.; McNicholas, T.; de la Taille, A.; Freschi, M.; Lughezzani, G.; Abrate, A.; et al. Preoperative Prostate-specific Antigen isoform p2PSA and its derivatives; %p2PSA and Prostate Health Index; predict pathologic outcomes in patients undergoing radical prostatectomy for prostate cancer: Results from a multicentric European prospective study. *Eur. Urol.* **2015**, *68*, 132–138. [CrossRef] [PubMed]

19. Stephan, C.; Kahrs, A.M.; Cammann, H.; Lein, M.; Schrader, M.; Deger, S.; Miller, K.; Jung, K. A [−2]proPSA-based artificial neural network significantly improves differentiation between prostate cancer and benign prostatic diseases. *Prostate* **2009**, *69*, 198–207. [CrossRef] [PubMed]

20. Guazzoni, G.; Nava, L.; Lazzeri, M.; Scattoni, V.; Lughezzani, G.; Maccagnano, C.; Dorigatti, F.; Ceriotti, F.; Pontillo, M.; Bini, V.; et al. Prostate-specific antigen (PSA) isoform p2PSA significantly improves the prediction of prostate cancer at initial extended prostate biopsies in patients with total PSA between 2.0 and 10 ng/mL: Results of a prospective study in a clinical setting. *Eur. Urol.* **2011**, *60*, 214–222.

21. Filella, X.; Foj, L.; Alcover, J.; Augé, J.M.; Molina, R.; Jiménez, W. The influence of prostate volume in prostate health index performance in patients with total PSA lower than 10 μg/L. *Clin. Chim. Acta* **2014**, *436*, 303–307. [CrossRef] [PubMed]

22. Lughezzani, G.; Lazzeri, M.; Larcher, A.; Lista, G.; Scattoni, V.; Cestari, A.; Buffi, N.M.; Bini, V.; Guazzoni, G. Development and internal validation of a Prostate Health Index based nomogram for predicting prostate cancer at extended biopsy. *J. Urol.* **2012**, *188*, 1144–1150. [CrossRef] [PubMed]

23. Lughezzani, G.; Lazzeri, M.; Haese, A.; McNicholas, T.; de la Taille, A.; Buffi, N.M.; Fossati, N.; Lista, G.; Larcher, A.; Abrate, A.; et al. Multicenter European external validation of a prostate health index-based nomogram for predicting prostate cancer at extended biopsy. *Eur. Urol.* **2014**, *66*, 906–912. [CrossRef] [PubMed]

24. Roobol, M.J.; Moniek, M.; Vedder, M.M.; Nieboer, D.; Houlgatte, A.; Vincendeau, S.; Lazzeri, M.; Guazzoni, G.; Stephan, C.; Semjonow, A.; et al. Comparison of two prostate cancer risk calculators that include the Prostate Health Index. *Eur. Urol. Foucs* **2015**, *1*, 185–190.

25. Heijnsdijk, E.A.; Denham, D.; de Koning, H.J. The cost-efectiveness of prostate cancer detection with the use of Prostate Health Index. *Value Health* **2016**, *19*, 153–157. [CrossRef] [PubMed]

26. Vickers, A.J.; Cronin, A.M.; Aus, G.; Pihl, C.G.; Becker, C.; Pettersson, K.; Scardino, P.T.; Hugosson, J.; Lilja, H. A panel of kallikrein markers can reduce unnecessary biopsy for prostate cancer: Data from the European Randomized Study of Prostate Cancer Screening in Göteborg, Sweden. *BMC Med.* **2008**, *6*, 19. [CrossRef] [PubMed]

27. Vickers, A.; Cronin, A.; Roobol, M.; Savage, C.; Peltola, M.; Pettersson, K.; Scardino, P.T.; Schröder, F.; Lilja, H. Reducing unnecessary biopsy during prostate cancer screening using a four-kallikrein panel: An independent replication. *J. Clin. Oncol.* **2010**, *28*, 2493–2498. [CrossRef] [PubMed]

28. Vickers, A.J.; Cronin, A.M.; Roobol, M.J.; Savage, C.J.; Peltola, M.; Pettersson, K.; Scardino, P.T.; Schröder, F.H.; Lilja, H. A four-kallikrein panel predicts prostate cancer in men with recent screening: Data from the European Randomized Study of Screening for Prostate Cancer, Rotterdam. *Clin. Cancer Res.* **2010**, *16*, 3232–3239. [CrossRef] [PubMed]

29. Vickers, A.J.; Cronin, A.M.; Aus, G.; Pihl, C.G.; Becker, C.; Pettersson, K.; Scardino, P.T.; Hugosson, J.; Lilja, H. Impact of recent screening on predicting the outcome of prostate cancer biopsy in men with elevated prostate-specific antigen: Data from the European Randomized Study of Prostate Cancer Screening in Gothenburg, Sweden. *Cancer* **2010**, *116*, 2612–2620. [CrossRef] [PubMed]

30. Gupta, A.; Roobol, M.J.; Savage, C.J.; Peltola, M.; Pettersson, K.; Scardino, P.T.; Vickers, A.J.; Schröder, F.H.; Lilja, H. A four-kallikrein panel for the prediction of repeat prostate biopsy: Data from the European Randomized Study of Prostate Cancer screening in Rotterdam, Netherlands. *Br. J. Cancer* **2010**, *103*, 708–714. [CrossRef] [PubMed]

31. Benchikh, A.; Savage, C.; Cronin, A.; Salama, G.; Villers, A.; Lilja, H.; Vickers, A. A panel of kallikrein markers can predict outcome of prostate biopsy following clinical work-up: An independent validation study from the European Randomized Study of Prostate Cancer screening, France. *BMC Cancer* **2010**, *10*, 635. [CrossRef] [PubMed]

32. Vickers, A.J.; Gupta, A.; Savage, C.J.; Pettersson, K.; Dahlin, A.; Bjartell, A.; Manjer, J.; Scardino, P.T.; Ulmert, D.; Lilja, H. A panel of kallikrein marker predicts prostate cancer in a large; population-based cohort followed for 15 years without screening. *Cancer Epidemiol. Biomark. Prev.* **2011**, *20*, 255–261. [CrossRef] [PubMed]

33. Carlsson, S.V.; Peltola, M.T.; Sjoberg, D.; Schröder, F.H.; Hugosson, J.; Pettersson, K.; Scardino, P.T.; Vickers, A.J.; Lilja, H.; Roobol, M.J. Can one blood draw replace transrectal ultrasonography-estimated prostate volume to predict prostate cancer risk? *BJU Int.* **2013**, *112*, 602–609. [CrossRef] [PubMed]

34. Parekh, D.J.; Punnen, S.; Sjoberg, D.D.; Asroff, S.W.; Bailen, J.L.; Cochran, J.S.; Concepcion, R.; David, R.D.; Deck, K.B.; Dumbadze, I.; et al. A Multi-institutional prospective trial in the USA confirms that the 4K score accurately identifies men with high-grade prostate cancer. *Eur. Urol.* **2015**, *68*, 464–470. [CrossRef] [PubMed]

35. Stattin, P.; Vickers, A.J.; Sjoberg, D.D.; Johansson, R.; Granfors, T.; Johansson, M.; Pettersson, K.; Scardino, P.T.; Hallmans, G.; Lilja, H. Improving the specificity of screening for lethal prostate cancer using Prostate-specific Antigen and a panel of kallikrein markers: A nested case-control study. *Eur. Urol.* **2015**, *68*, 207–213. [CrossRef] [PubMed]

36. Filella, X.; Foj, L.; Milà, M.; Augé, J.M.; Molina, R.; Jiménez, W. PCA3 in the detection and management of early prostate cancer. *Tumour Biol.* **2013**, *34*, 1337–1347. [CrossRef] [PubMed]

37. Haese, A.; de la Taille, A.; van Poppel, H.; Marberger, M.; Stenzl, A.; Mulders, P.F.; Huland, H.; Abbou, C.C.; Remzi, M.; Tinzl, M. Clinical utility of the PCA3 urine assay in European men scheduled for repeat biopsy. *Eur. Urol.* **2008**, *54*, 1081–1088. [CrossRef] [PubMed]

38. Hu, B.; Yang, H.; Yang, H. Diagnostic value of urine prostate cancer antigen 3 test using a cutoff value of 35 µg/L in patients with prostate cancer. *Tumour Biol.* **2014**, *35*, 8573–8580. [CrossRef] [PubMed]

39. Crawford, E.D.; Rove, K.O.; Trabulsi, E.J.; Qian, J.; Drewnowska, K.P.; Kaminetsky, J.C.; Huisman, T.K.; Bilowus, M.L.; Freedman, S.J.; Glover, W.L.; et al. Diagnostic performance of PCA3 to detect prostate cancer in men with increased prostate specific antigen: A prospective study of 1962 cases. *J. Urol.* **2012**, *188*, 1726–1731. [CrossRef] [PubMed]

40. Bussemakers, M.J.; van Bokhoven, A.; Verhaegh, G.W.; Smit, F.P.; Karthaus, H.F.; Schalken, J.A.; Debruyne, F.M.; Ru, N.; Isaacs, W.B. DD3: A new prostate-specific gene, highly overexpressed in prostate cancer. *Cancer Res.* **1999**, *59*, 5975–5979. [PubMed]

41. Schröder, F.H.; Venderbos, L.D.; van den Bergh, R.C.; Hessels, D.; van Leenders, G.J.; van Leeuwen, P.J.; Wolters, T.; Barentsz, J.; Roobol, M.J. Prostate cancer antigen 3: Diagnostic outcomes in men presenting with urinary prostate cancer antigen 3 scores ≥100. *Urology* **2014**, *83*, 613–616. [CrossRef] [PubMed]

42. Merola, R.; Tomao, L.; Antenucci, A.; Sperduti, I.; Sentinelli, S.; Masi, S.; Mandoj, C.; Orlandi, G.; Papalia, R.; Guaglianone, S.; et al. PCA3 in prostate cancer and tumor aggressiveness detection on 407 high-risk patients: A National Cancer Institute experience. *J. Exp. Clin. Cancer Res.* **2015**, *34*, 15. [CrossRef] [PubMed]

43. Chevli, K.K.; Duff, M.; Walter, P.; Yu, C.; Capuder, B.; Elshafei, A.; Malczewski, S.; Kattan, M.W.; Jones, J.S. Urinary PCA3 as a predictor for prostate cancer in a cohort of 3073 men undergoing initial prostate biopsy. *J. Urol.* **2014**, *191*, 1743–1748. [CrossRef] [PubMed]

44. Hessels, D.; van Gils, M.P.; van Hooij, O.; Jannink, S.A.; Witjes, J.A.; Verhaegh, G.W.; Schalken, J.A. Predictive value of PCA3 in urinary sediments in determining clinico-pathological characteristics of prostate cancer. *Prostate* **2010**, *70*, 10–16. [CrossRef] [PubMed]

45. Foj, L.; Milà, M.; Mengual, L.; Luque, P.; Alcaraz, A.; Jiménez, W.; Filella, X. Real-time PCR PCA3 assay is a useful test measured in urine to improve prostate cancer detection. *Clin. Chim. Acta* **2014**, *435*, 53–58. [CrossRef] [PubMed]

46. Bradley, L.A.; Palomaki, G.E.; Gutman, S.; Samson, D.; Aronson, N. Comparative effectiveness review: Prostate cancer antigen 3 testing for the diagnosis and management of prostate cancer. *J. Urol.* **2013**, *190*, 389–398. [CrossRef] [PubMed]

47. Roobol, M.J.; Schröder, F.H.; van Leeuwen, P.; Wolters, T.; van den Bergh, R.C.; van Leenders, G.J.; Hessels, D. Performance of the prostate cancer antigen 3 (*PCA3*) gene and prostate-specific antigen in prescreened men: Exploring the value of *PCA3* for a first-line diagnostic test. *Eur. Urol.* **2010**, *58*, 475–481. [CrossRef] [PubMed]

48. Tomlins, S.A.; Rhodes, D.R.; Perner, S.; Dhanasekaran, S.M.; Mehra, R.; Sun, X.W.; Varambally, S.; Cao, X.; Tchinda, J.; Kuefer, R.; et al. Recurrent fusion of TMPRSS2 and ETS transcription factor genes in prostate cancer. *Science* **2005**, *310*, 644–648. [CrossRef] [PubMed]

49. Boström, P.J.; Bjartell, A.S.; Catto, J.W.; Eggener, S.E.; Lilja, H.; Loeb, S.; Schalken, J.; Schlomm, T.; Cooperberg, M.R. Genomic predictors of outcome in prostate cancer. *Eur. Urol.* **2015**, *68*, 1033–1044. [CrossRef] [PubMed]

50. Boormans, J.L.; Porkka, K.; Visakorpi, T.; Trapman, J. Confirmation of the association of *TMPRSS2 (exon 0):ERG* expression and a favorable prognosis of primary prostate cancer. *Eur. Urol.* **2011**, *60*, 183–184. [CrossRef] [PubMed]

51. FitzGerald, L.M.; Agalliu, I.; Johnson, K.; Miller, M.A.; Kwon, E.M.; Hurtado-Coll, A.; Fazli, L.; Rajput, A.B.; Gleave, M.E.; Cox, M.E.; et al. Association of *TMPRSS2-ERG* gene fusion with clinical characteristics and outcomes: Results from a population-based study of prostate cancer. *BMC Cancer* **2008**, *8*, 230. [CrossRef] [PubMed]

52. Leyten, G.H.; Hessels, D.; Jannink, S.A.; Smit, F.P.; de Jong, H.; Cornel, E.B.; de Reijke, T.M.; Vergunst, H.; Kil, P.; Knipscheer, B.C.; et al. Prospective multicentre evaluation of *PCA3* and *TMPRSS2-ERG* gene fusions as diagnostic and prognostic urinary biomarkers for prostate cancer. *Eur. Urol.* **2014**, *65*, 534–542. [CrossRef] [PubMed]

53. Tomlins, S.A.; Day, J.R.; Lonigro, R.J.; Hovelson, D.H.; Siddiqui, J.; Kunju, L.P.; Dunn, R.L.; Meyer, S.; Hodge, P.; Groskopf, J.; et al. Urine *TMPRSS2:ERG* Plus PCA3 for individualized prostate cancer risk assessment. *Eur. Urol.* **2016**, *70*, 45–53. [CrossRef] [PubMed]

54. Stephan, C.; Cammann, H.; Jung, K. Urine *TMPRSS2:ERG* Plus PCA3 for individualized prostate cancer risk assessment. *Eur. Urol.* **2015**, *68*, e106–e107. [CrossRef] [PubMed]

55. Griffiths-Jones, S. miRBase: The microRNA sequence database. *Methods Mol. Biol. Clifton NJ* **2006**, *342*, 129–138.

56. Calin, G.A.; Dumitru, C.D.; Shimizu, M.; Bichi, R.; Zupo, S.; Noch, E.; Aldler, H.; Rattan, S.; Keating, M.; Rai, K.; et al. Frequent deletions and down-regulation of micro-RNA genes *miR15* and *miR16* at 13q14 in chronic lymphocytic leukemia. *Proc. Natl. Acad. Sci. USA* **2002**, *99*, 15524–15529. [CrossRef] [PubMed]

57. Lawrie, C.H.; Gal, S.; Dunlop, H.M.; Pushkaran, B.; Liggins, A.P.; Pulford, K.; Banham, A.H.; Pezzella, F.; Boultwood, J.; Wainscoat, J.S.; et al. Detection of elevated levels of tumour-associated microRNAs in serum of patients with diffuse large B-cell lymphoma. *Br. J. Haematol.* **2008**, *141*, 672–675. [CrossRef] [PubMed]

58. Witwer, K.W. Circulating microRNA biomarker studies: Pitfalls and potential solutions. *Clin. Chem.* **2015**, *61*, 56–63. [CrossRef] [PubMed]

59. Tam, S.; de Borja, R.; Tsao, M.S.; McPherson, J.D. Robust global microRNA expression profiling using next-generation sequencing technologies. *Lab. Investig.* **2014**, *94*, 350–358. [CrossRef] [PubMed]

60. ChunJiao, S.; Huan, C.; ChaoYang, X.; GuoMei, R. Uncovering the roles of miRNAs and their relationship with androgen receptor in prostate cancer. *IUBMB Life* **2014**, *66*, 379–386. [CrossRef] [PubMed]

61. Mitchell, P.S.; Parkin, R.K.; Kroh, E.M.; Fritz, B.R.; Wyman, S.K.; Pogosova-Agadjanyan, E.L.; Peterson, A.; Noteboom, J.; O'Briant, K.C.; Allen, A.; et al. Circulating microRNAs as stable blood-based markers for cancer detection. *Proc. Natl. Acad. Sci. USA* **2008**, *105*, 10513–10518. [CrossRef] [PubMed]

62. Filella, X.; Foj, L. miRNAs as novel biomarkers in the management of prostate cancer. *Clin. Chem. Lab. Med.* **2016**. [CrossRef] [PubMed]

63. Mihelich, B.L.; Maranville, J.C.; Nolley, R.; Peehl, D.M.; Nonn, L. Elevated serum microRNA levels associate with absence of high-grade prostate cancer in a retrospective cohort. *PLoS ONE* **2015**, *10*, e0124245. [CrossRef] [PubMed]

64. Chen, Z.H.; Zhang, G.L.; Li, H.R.; Luo, J.D.; Li, Z.X.; Chen, G.M.; Yang, J. A panel of five circulating microRNAs as potential biomarkers for prostate cancer. *Prostate* **2012**, *72*, 1443–1452. [CrossRef] [PubMed]

65. Moltzahn, F.; Olshen, A.B.; Baehner, L.; Peek, A.; Fong, L.; Stöppler, H.; Simko, J.; Hilton, J.F.; Carroll, P.; Blelloch, R.; et al. Microfluidic-based multiplex qRT-PCR identifies diagnostic and prognostic microRNA signatures in the sera of prostate cancer patients. *Cancer Res.* **2011**, *71*, 550–560. [CrossRef] [PubMed]

66. Salido-Guadarrama, A.I.; Morales-Montor, J.G.; Rangel-Escareño, C.; Langley, E.; Peralta-Zaragoza, O.; Cruz-Colín, J.L.; Rodriguez-Dorantes, M. Urinary microRNA-based signature improves accuracy of detection of clinically relevant prostate cancer within the prostate-specific antigen grey zone. *Mol. Med. Rep.* **2016**, *13*, 4549–4560. [CrossRef] [PubMed]

67. Trams, E.G.; Lauter, C.J.; Salem, N., Jr.; Heine, U. Exfoliation of membrane ecto-enzymes in the form of micro-vesicles. *Biochim. Biophys. Acta* **1981**, *645*, 63–70. [CrossRef]

68. Zhou, H.; Yuen, P.S.; Pisitkun, T.; Gonzales, P.A.; Yasuda, H.; Dear, J.W.; Gross, P.; Knepper, M.A.; Star, R.A. Collection; storage; preservation; and normalization of human urinary exosomes for biomarker discovery. *Kidney Int.* **2006**, *69*, 1471–1476. [CrossRef] [PubMed]

69. Ge, Q.; Zhou, Y.; Lu, J.; Bai, Y.; Xie, X.; Lu, Z. miRNA in plasma exosome is stable under different storage conditions. *Molecules* **2014**, *19*, 1568–1575. [CrossRef] [PubMed]

70. Taylor, D.D.; Gercel-Taylor, C. MicroRNA signatures of tumor-derived exosomes as diagnostic biomarkers of ovarian cancer. *Gynecol. Oncol.* **2008**, *110*, 13–21. [CrossRef] [PubMed]

71. Duijvesz, D.; Versluis, C.Y.; van der Fels, C.A.; Vredenbregt-van den Berg, M.S.; Leivo, J.; Peltola, M.T.; Bangma, C.H.; Pettersson, K.S.; Jenster, G. Immuno-based detection of extracellular vesicles in urine as diagnostic marker for prostate cancer. *Int. J. Cancer* **2015**, *137*, 2869–2878. [CrossRef] [PubMed]

72. Rabinowits, G.; Gercel-Taylor, C.; Day, J.M.; Taylor, D.D.; Kloecker, G.H. Exosomal microRNA: A diagnostic marker for lung cancer. *Clin. Lung Cancer* **2009**, *10*, 42–46. [CrossRef] [PubMed]

73. Silva, J.; Garcia, V.; Rodriguez, M.; Compte, M.; Cisneros, E.; Veguillas, P.; Garcia, J.M.; Dominguez, G.; Campos-Martin, Y.; Cuevas, J.; et al. Analysis of exosome release and its prognostic value in human colorectal cancer. *Genes Chromosomes Cancer* **2012**, *51*, 409–418. [CrossRef] [PubMed]

74. Yeh, Y.Y.; Ozer, H.G.; Lehman, A.M.; Maddocks, K.; Yu, L.; Johnson, A.J.; Byrd, J.C. Characterization of CLL exosomes reveals a distinct microRNA signature and enhanced secretion by activation of BCR signaling. *Blood* **2015**, *125*, 3297–3305. [CrossRef] [PubMed]

75. Ratajczak, M.Z.; Ratajczak, J. Horizontal transfer of RNA and proteins between cells by extracellular microvesicles: 14 years later. *Clin. Transl. Med.* **2016**, *5*, 7. [CrossRef] [PubMed]

76. O'Driscoll, L. Expanding on exosomes and ectosomes in cancer. *N. Engl. J. Med.* **2015**, *372*, 2359–2362. [CrossRef] [PubMed]

77. Franzen, C.A.; Blackwell, R.H.; Todorovic, V.; Greco, K.A.; Foreman, K.E.; Flanigan, R.C.; Kuo, P.C.; Gupta, G.N. Urothelial cells undergo epithelial-to-mesenchymal transition after exposure to muscle invasive bladder cancer exosomes. *Oncogenesis* **2015**, *4*, e163. [CrossRef] [PubMed]

78. Abd Elmageed, Z.Y.; Yang, Y.; Thomas, R.; Ranjan, M.; Mondal, D.; Moroz, K.; Fang, Z.; Rezk, B.M.; Moparty, K.; Sikka, S.C.; et al. Neoplastic reprogramming of patient-derived adipose stem cells by prostate cancer cell-associated exosomes. *Stem Cells* **2014**, *32*, 983–997. [CrossRef] [PubMed]

79. Hosseini-Beheshti, E.; Choi, W.; Weiswald, L.B.; Kharmate, G.; Ghaffari, M.; Roshan-Moniri, M.; Hassona, M.D.; Chan, L.; Chin, M.Y.; Tai, I.T.; et al. Exosomes confer pro-survival signals to alter the phenotype of prostate cells in their surrounding environment. *Oncotarget* **2016**. [CrossRef]

80. Thery, C.; Amigorena, S.; Raposo, G.; Clayton, A. Isolation and characterization of exosomes from cell culture supernatants and biological fluids. *Curr. Protoc. Cell Biol.* **2006**. [CrossRef]

81. Øverbye, A.; Skotland, T.; Koehler, C.J.; Thiede, B.; Seierstad, T.; Berge, V.; Sandvig, K.; Llorente, A. Identification of prostate cancer biomarkers in urinary exosomes. *Oncotarget* **2015**, *6*, 30357–30376. [PubMed]

82. Dijkstra, S.; Birker, I.L.; Smit, F.P.; Leyten, G.H.; de Reijke, T.M.; van Oort, I.M.; Mulders, P.F.; Jannink, S.A.; Schalken, J.A. Prostate cancer biomarker profiles in urinary sediments and exosomes. *J. Urol.* **2014**, *191*, 1132–1138. [CrossRef] [PubMed]

83. Hendriks, R.J.; Dijkstra, S.; Jannink, S.A.; Steffens, M.G.; van Oort, I.M.; Mulders, P.F.; Schalken, J.A. Comparative analysis of prostate cancer specific biomarkers PCA3 and ERG in whole urine; urinary sediments and exosomes. *Clin. Chem. Lab. Med.* **2016**, *54*, 483–492. [CrossRef] [PubMed]

84. Donovan, M.J.; Noerholm, M.; Bentink, S.; Belzer, S.; Skog, J.; O'Neill, V.; Cochran, J.S.; Brown, G.A. A molecular signature of PCA3 and ERG exosomal RNA from non-DRE urine is predictive of initial prostate biopsy result. *Prostate Cancer Prostatic Dis.* **2015**, *18*, 370–375. [CrossRef] [PubMed]

85. Cheng, L.; Sun, X.; Scicluna, B.J.; Coleman, B.M.; Hill, A.F. Characterization and deep sequencing analysis of exosomal and non-exosomal miRNA in human urine. *Kidney Int.* **2014**, *86*, 433–444. [CrossRef] [PubMed]

86. Cheng, L.; Sharples, R.A.; Scicluna, B.J.; Hill, A.F. Exosomes provide a protective and enriched source of miRNA for biomarker profiling compared to intracellular and cell-free blood. *J. Extracell. Vesicles* **2014**, *3*. [CrossRef] [PubMed]

87. Mall, C.; Rocke, D.M.; Durbin-Johnson, B.; Weiss, R.H. Stability of miRNA in human urine supports its biomarker potential. *Biomark. Med.* **2013**, *7*, 623–631. [CrossRef] [PubMed]

88. Hessvik, N.P.; Sandvigm, K.; Llorente, A. Exosomal miRNAs as Biomarkers for Prostate Cancer. *Front. Genet.* **2013**, *4*, 36. [CrossRef] [PubMed]

89. Chevillet, J.R.; Kang, Q.; Ruf, I.K.; Briggs, H.A.; Vojtech, L.N.; Hughes, S.M.; Cheng, H.H.; Arroyo, J.D.; Meredith, E.K.; Gallichotte, E.N.; et al. Quantitative and stoichiometric analysis of the microRNA content of exosomes. *Proc. Natl. Acad. Sci. USA* **2014**, *111*, 14888–14893. [CrossRef] [PubMed]

90. Arroyo, J.D.; Chevillet, J.R.; Kroh, E.M.; Ruf, I.K.; Pritchard, C.C.; Gibson, D.F.; Mitchell, P.S.; Bennett, C.F.; Pogosova-Agadjanyan, E.L.; Stirewalt, D.L.; et al. Argonaute2 complexes carry a population of circulating microRNAs independent of vesicles in human plasma. *Proc. Natl. Acad. Sci. USA* **2011**, *108*, 5003–5008. [CrossRef] [PubMed]

91. Gallo, A.; Tandon, M.; Alevizos, I.; Illei, G.G. The majority of microRNAs detectable in serum and saliva is concentrated in exosomes. *PLoS ONE* **2012**, *7*, e30679. [CrossRef] [PubMed]

92. Li, Z.; Ma, Y.Y.; Wang, J.; Zeng, X.F.; Li, R.; Kang, W.; Hao, X.K. Exosomal microRNA-141 is upregulated in the serum of prostate cancer patients. *Onco. Targets Ther.* **2015**, *9*, 139–148. [PubMed]

93. Huang, X.; Yuan, T.; Liang, M.; Du, M.; Xia, S.; Dittmar, R.; Wang, D.; See, W.; Costello, B.A.; Quevedo, F.; et al. Exosomal miR-1290 and miR-375 as prognostic markers in castration-resistant prostate cancer. *Eur. Urol.* **2015**, *67*, 33–41. [CrossRef] [PubMed]

94. Bryzgunova, O.E.; Zaripov, M.M.; Skvortsova, T.E.; Lekchnov, E.A.; Grigor'eva, A.E.; Zaporozhchenko, I.A.; Morozkin, E.S.; Ryabchikova, E.I.; Yurchenko, Y.B.; Voitsitskiy, V.E.; et al. Comparative Study of Extracellular Vesicles from the Urine of Healthy Individuals and Prostate Cancer Patients. *PLoS ONE* **2016**, *11*, e0157566. [CrossRef] [PubMed]

95. Samsonov, R.; Shtam, T.; Burdakov, V.; Glotov, A.; Tsyrlina, E.; Berstein, L.; Nosov, A.; Evtushenko, V.; Filatov, M.; Malek, A. Lectin-induced agglutination method of urinary exosomes isolation followed by mi-RNA analysis: Application for prostate cancer diagnostic. *Prostate* **2016**, *76*, 68–79. [CrossRef] [PubMed]

96. Koppers-Lalic, D.; Hackenberg, M.; Menezes, R.; Misovic, B.; Wachalska, M.; Geldof, A.; Zini, N.; Reijke, T.; Wurdinger, T.; Vis, A.; et al. Non-invasive prostate cancer detection by measuring miRNA variants (isomiRs) in urine extracellular vesicles. *Oncotarget* **2016**. [CrossRef]

97. Klotz, L. Low-risk prostate cancer can and should often be managed with active surveillance and selective delayed intervention. *Nat. Clin. Pract. Urol.* **2008**, *5*, 2–3. [CrossRef] [PubMed]

98. Rubin, M.A.; Girelli, G.; Demichelis, F. Genomic Correlates to the Newly Proposed Grading Prognostic Groups for Prostate Cancer. *Eur. Urol.* **2016**, *69*, 557–560. [CrossRef] [PubMed]

99. Carter, H.B.; Partin, A.W.; Walsh, P.C.; Trock, B.J.; Veltri, R.W.; Nelson, W.G.; Coffey, D.S.; Singer, E.A.; Epstein, J.I. Gleason score 6 adenocarcinoma: Should it be labeled as cancer? *J. Clin. Oncol.* **2012**, *30*, 4294–4296. [CrossRef] [PubMed]

100. Kulac, I.; Haffner, M.C.; Yegnasubramanian, S.; Epstein, J.I.; de Marzo, A.M. Should Gleason 6 be labeled as cancer? *Curr. Opin. Urol.* **2015**, *25*, 238–245. [CrossRef] [PubMed]

101. Scattoni, V.; Lazzeri, M.; Lughezzani, G.; de Luca, S.; Passera, R.; Bollito, E.; Randone, D.; Abdollah, F.; Capitanio, U.; Larcher, A.; et al. Head-to-head comparison of prostate health index and urinary PCA3 for predicting cancer at initial or repeat biopsy. *J. Urol.* **2013**, *190*, 496–501. [CrossRef] [PubMed]

102. Stephan, C.; Jung, K.; Semjonow, A.; Schulze-Forster, K.; Cammann, H.; Hu, X.; Meyer, H.A.; Bögemann, M.; Miller, K.; Friedersdorff, F. Comparative assessment of urinary prostate cancer antigen 3 and *TMPRSS2:ERG* gene fusion with the serum [−2]proprostate-specific antigen-based prostate health index for detection of prostate cancer. *Clin. Chem.* **2013**, *59*, 280–288. [CrossRef] [PubMed]

103. Tallon, L.; Luangphakdy, D.; Ruffion, A.; Colombel, M.; Devonec, M.; Champetier, D.; Paparel, P.; Decaussin-Petrucci, M.; Perrin, P.; Vlaeminck-Guillem, V. Comparative Evaluation of Urinary PCA3 and TMPRSS2:ERG Scores and Serum PHI in Predicting Prostate Cancer Aggressiveness. *Int. J. Mol. Sci.* **2014**, *15*, 13299–13316. [CrossRef] [PubMed]

104. Nordström, T.; Vickers, A.; Assel, M.; Lilja, H.; Grönberg, H.; Eklund, M. Comparison between the four-kallikrein panel and Prostate Health Index for predicting Prostate Cancer. *Eur. Urol.* **2015**, *68*, 139–146. [CrossRef] [PubMed]

105. Vedder, M.M.; de Bekker-Grob, E.W.; Lilja, H.G.; Vickers, A.J.; van Leenders, G.J.; Steyerberg, E.W.; Roobol, M.J. The added value of percentage of free to total Prostate-specific Antigen, PCA3, and a kallikrein panel to the ERSPC Risk Calculator for Prostate Cancer in prescreened Men. *Eur. Urol.* **2014**, *66*, 1109–1115. [CrossRef] [PubMed]

106. Mottet, N.; Bellmunt, J.; Briers, E.; Bolla, M.; Cornford, P.; de Santis, M.; Henry, A.; Joniau, S.; Lam, T.; Mason, M.D.; et al. EAU Guideliness Prostate Cancer. Available online: http://uroweb.org/guideline/prostate-cancer/ (accessed on 9 September 2016).

107. Vickers, A.J.; Eastham, J.A.; Scardino, P.T.; Lilja, H. The Memorial Sloan Kettering Cancer Center Recommendations for Prostate Cancer Screening. *Urology* **2016**, *91*, 12–18. [CrossRef] [PubMed]

108. De Visschere, P.J.; Briganti, A.; Fütterer, J.J.; Ghadjar, P.; Isbarn, H.; Massard, C.; Ost, P.; Sooriakumaran, P.; Surcel, C.I.; Valerio, M.; et al. Role of multiparametric magnetic resonance imaging in early detection of prostate cancer. *Insights Imaging* **2016**, *7*, 205–214. [CrossRef] [PubMed]

109. Porpiglia, F.; Cantiello, F.; de Luca, S.; Manfredi, M.; Veltri, A.; Russo, F.; Sottile, A.; Damiano, R. In-parallel comparative evaluation between multiparametric magnetic resonance imaging; prostate cancer antigen 3 and the prostate health index in predicting pathologically confirmed significant prostate cancer in men eligible for active surveillance. *BJU Int.* **2015**. [CrossRef] [PubMed]

110. Washino, S.; Okochi, T.; Saito, K.; Konishi, T.; Hirai, M.; Kobayashi, Y.; Miyagawa, T. Combination of PI-RADS score and PSA density predicts biopsy outcome in biopsy naïve patients. *BJU Int.* **2016**. [CrossRef] [PubMed]

111. Maurer, T.; Eiber, M.; Schwaiger, M.; Gschwend, J.E. Current use of PSMA-PET in prostate cancer management. *Nat. Rev. Urol.* **2016**, *13*, 226–235. [CrossRef] [PubMed]

International Journal of
Molecular Sciences

MDPI

Review

Epigenetic Signature: A New Player as Predictor of Clinically Significant Prostate Cancer (PCa) in Patients on Active Surveillance (AS)

Matteo Ferro [1], Paola Ungaro [2,*], Amelia Cimmino [3,*], Giuseppe Lucarelli [4], Gian Maria Busetto [5], Francesco Cantiello [6], Rocco Damiano [6] and Daniela Terracciano [7,*]

[1] Urologic Surgery Unit, European Institute of Oncology, 20141 Milan, Italy; matteo.ferro@ieo.it
[2] Institute of Experimental Endocrinology and Oncology (IEOS-CNR) "G. Salvatore", Via Sergio Pansini, 5, 80131 Naples, Italy
[3] Institute of Genetics and Biophysics "A. Buzzati Traverso", National Research Council (CNR), Via Pietro Castellino 111, 80131 Naples, Italy
[4] Department of Emergency and Organ Transplantation-Urology, Andrology and Kidney Transplantation Unit, University of Bari, 70124 Bari, Italy; giuseppe.lucarelli@inwind.it
[5] Department of Gynecological-Obstetrics Sciences and Urological Sciences, Sapienza Rome University Policlinico Umberto I, 00161 Rome, Italy; gianmaria.busetto@uniroma1.it
[6] Department of Urology, Magna Graecia University of Catanzaro, 88100 Catanzaro, Italy; cantiello@unicz.it (F.C.); damiano@unicz.it (R.D.)
[7] Department of Translational Medical Sciences, University of Naples Federico II, Via Sergio Pansini, 5, 80131 Naples, Italy
* Correspondence: pungaro@ieos.cnr.it (P.U.); amelia.cimmino@igb.cnr.it (A.C.); daniela.terracciano@unina.it (D.T.); Tel./Fax: +39-81-746-3617 (P.U. & A.C. & D.T.)

Academic Editor: Carsten Stephan
Received: 11 April 2017; Accepted: 22 May 2017; Published: 27 May 2017

Abstract: Widespread prostate-specific antigen (PSA) testing notably increased the number of prostate cancer (PCa) diagnoses. However, about 30% of these patients have low-risk tumors that are not lethal and remain asymptomatic during their lifetime. Overtreatment of such patients may reduce quality of life and increase healthcare costs. Active surveillance (AS) has become an accepted alternative to immediate treatment in selected men with low-risk PCa. Despite much progress in recent years toward identifying the best candidates for AS in recent years, the greatest risk remains the possibility of misclassification of the cancer or missing a high-risk cancer. This is particularly worrisome in men with a life expectancy of greater than 10–15 years. The Prostate Cancer Research International Active Surveillance (PRIAS) study showed that, in addition to age and PSA at diagnosis, both PSA density (PSA-D) and the number of positive cores at diagnosis (two compared with one) are the strongest predictors for reclassification biopsy or switching to deferred treatment. However, there is still no consensus upon guidelines for placing patients on AS. Each institution has its own protocol for AS that is based on PRIAS criteria. Many different variables have been proposed as tools to enrol patients in AS: PSA-D, the percentage of freePSA, and the extent of cancer on biopsy (number of positive cores or percentage of core involvement). More recently, the Prostate Health Index (PHI), the 4 Kallikrein (4K) score, and other patient factors, such as age, race, and family history, have been investigated as tools able to predict clinically significant PCa. Recently, some reports suggested that epigenetic mapping differs significantly between cancer patients and healthy subjects. These findings indicated as future prospect the use of epigenetic markers to identify PCa patients with low-grade disease, who are likely candidates for AS. This review explores literature data about the potential of epigenetic markers as predictors of clinically significant disease.

Keywords: active surveillance; prostate cancer; epigenetic biomarkers

1. Introduction

AS has recently become an accepted alternative for patients with low-risk prostate cancer (PCa)-related mortality, allowing for delayed curative intervention if there is reclassification of cancer risk or evidence of disease progression [1]. It has been widely accepted that pre-treatment prostate-specific antigen (PSA) below 10 ng/mL, a biopsy Gleason score of 6 or less, and clinical stage T1c or T2a identify low-risk PCa [2]. Moreover, Klotz et al. [3] recently suggested that patients with a Gleason score of 3 + 4 PCa may be selected for active surveillance (AS), since most of these patients are at a low risk of progression. Risk factors for reclassification and progression have still not been adequately characterized. Low-risk PCa patients were monitored through PSA levels, clinical examination, and repeated prostate biopsies. Changes in biopsy results suggest a need for intervention. Widespread AS use has been prevented because under-sampling prostate biopsies may result in occult high-grade cancer. Moreover, biopsies are invasive tests, not free from side effects such as bleeding and infection risks, suggesting a need to avoid unnecessary repeated prostate biopsies in AS regimens. Several approaches to selecting patients for AS have been proposed. One of these includes the addition of biomarkers to currently used clinical and demographical variables [4]. Biomarker levels can ideally be obtained non-invasively, allowing for controlled follow-up of patients in AS in order to avoid overtreatment and the consequent impairment of quality of life.

In this review, we focused our attention on the potential use of epigenetic mapping as prognostic factors in the clinical management of PCa patients.

1.1. Biomarkers and Active Surveillance: Current Status

PCa patients are classified at diagnostic biopsy as very low-risk when they meet the following criteria: clinical stage T1c disease, PSA-D less than 0.15 ng/mL, a Gleason score ≤6, two or fewer biopsy cores with cancer, and a maximum of 50% involvement of any core with cancer [5]. Thaxton et al. showed that the number of patients eligible for AS, who have fatal disease at radical prostatectomy (RP), depends upon the criteria used for AS selection [6]. However, the optimal patient selection and follow-up protocol is still a matter of debate. Prostate biopsy risks, such as bleeding and infections, highlight the need for non-invasive tools for the selection and follow-up of AS patients [7]. The Prostate Health Index (PHI) may play a role in monitoring men under active surveillance (AS) [8,9]. Tosoian et al. [10] found an association between baseline and longitudinal PHI, but not PCA3 [11] values and reclassification during active surveillance. Sottile et al. [12] demonstrated significantly higher p2PSA and PHI levels in men with metastatic disease as compared to those without clinical metastasis. Collectively, PHI and/or PCA3 improve the selection of eligible patients for AS and decision curve analysis demonstrated that PHI outperforms PCA3 [13]. A Canadian report indicated that neutrophil-to-lymphocyte ratio (NLR) is a less expensive and more easily accessible test able to predict Gleason score upgrading and biochemical recurrence in patients with low-risk PCa eligible for AS [14]. Recently, Ferro et al. [15] found a significant association between low serum testosterone levels and upgrading, upstaging, and unfavorable disease, suggesting a new cheap parameter useful to identify low-risk patients eligible for AS. A recently published preliminary study [16] showed that the presence of primary circulating prostate cells is associated with aggressive disease, suggesting that these patients are not eligible for active surveillance. Several authors have shown that, besides the PHI index, the PCA3 score can improve the Epstein and PRIAS protocol's ability to predict insignificant PCa in subjects eligible for AS [17,18]. Lin et al. found that [19] the combination of urine TMPRSS2-ERG and PCA3 were associated with aggressive cancer in men with low-risk PCa on AS. Berg et al. found a significant association between tissue ERG expression and progression during AS [20]. Other authors have recently shown that Ki67 and DLX2, two cancer cell proliferation markers, are predictive of increased metastasis risk and may aid patient selection for AS [21]. It has been repeatedly demonstrated [22–24] that PTEN loss is uncommon in clinically localized PCa, suggesting the potential use of this histopathological biomarker as a predictor of unfavorable prognosis in patients on AS.

Tissue-based prognostic panels as OncotypeDX1 and Prolaris1 have been validated for routine clinical use on men with low-risk PCa. The first is a quantitative RT-PCR assay performed using biopsy samples. This test measures the expression of several genes involved in four different pathways to calculate the Genomic Prostate Score (GPS), which is predictive of aggressive disease in low- and intermediate-risk PCa patients [25]. The Prolaris 1 test provides a proliferative index, called CCP (cell cycle progression) score, on the basis of the expression of 31 cell cycle progression and 15 housekeeping genes [26]. In a multicentric study, the biopsy CCP score from low-risk patients was associated with aggressiveness at radical prostatectomy [27].

1.2. Epigenetic Biomarkers and PCa Prognosis: Future Challenge

The term "epigenetics" defined heritable changes in gene expression that are independent from those occurring in the genome. The epigenetic mechanisms contribute to gene regulation throughout the whole life course of an organism, by changing chromatin architecture and/or access by transcription factors. They include different processes, such as DNA methylation, histone modifications, and post-transcriptional gene regulation by non-coding RNAs. Together, they regulate gene expression by changing chromatin organization and DNA accessibility.

The most well studied epigenetic modification in human diseases is DNA methylation. It involves an enzymatic process mediated by DNA methyltransferases (DNMTs) that catalyze the addition of a methyl group, using S-adenosyl methionine (SAM) as the methyl supplier, to the 5-carbon of the cytosine within CpG dinucleotides to form 5-methylcytosine. CpG are normally methylated when dispersed in the genome or in DNA repetitive region, but remain unmethylated to enable gene expression when they are clustered as a CpG island at the 5′ ends of many genes [28]. The conventional view is that DNMT1 is responsible for the maintenance of tissue-specific methylation patterns during cell replication, while DNMT3A and DNMT3B catalyze the addition of methyl groups de novo during embryogenesis [29,30]. In tumorigenesis, DNA methylation and demethylation are associated with silencing tumor suppressor genes and activating oncogenes, respectively [31].

Histone post-translational modifications (PTMs) include acetylation, biotinylation, methylation, phosphorylation, ubiquitination, SUMOylation, ADP (adenosine diphosphate) ribosylation, proline isomerization, citrulination, butyrylation, propionylation, and glycosylation, which are known as "the histone code" and strongly contribute to the control of gene expression [32].

Such modifications alter the affinity of the histone tails to the DNA and change the conformation of chromatin structure, resulting in transcriptional genes activation or repression. For instance, di- and trimethylation and poor acetylation of lysine 9 residue on histone H3 are associated with the silencing of gene expression. By contrast, the acetylation of histones H3 and H4, together with the methylation of lysine 4 residue on histone H3, results in gene expression. In general, acetylation promotes transcriptional activity and is catalyzed by histone acetyltransferases (HAT). Conversely, histone deacetylases (HDACs) remove acetyl groups leading to a silent chromatin state. Depending on the specific amino acid residues modified and the number of methyl groups added, histone methylation may be associated with the activation or repression of transcription. In general, histone methyltransferases (HMTs) and histone demethylases (HDMs) catalyze the addition or the removal of methyl groups from histone proteins, respectively [33].

There is good evidence that another epigenetic modification, known as noncoding RNAs (ncRNAs), can influence gene expression [34].

The best-characterized class of non-coding RNAs is represented by microRNA (miRNAs), which are single-stranded RNAs, about 19–24 nucleotides in length. They regulate gene expression through the binding to mRNAs, resulting in degradation or translational inhibition [35]. It was estimated that at least 30% of human genes are regulated by miRNA. In metazoans, each mRNA can be combined with multiple miRNAs and each miRNA regulates multiple mRNAs. It is noteworthy that miRNAs regulate a large spectrum of biological processes and play an important role in tumorigenesis, activating oncogenes or restraining tumor suppressor genes [36,37].

Development and progression of PCa are usually associated with global DNA hypomethylation with a lower overall content of 5-methylcytosine (m5C) found in metastatic tissue [38]. The global DNA hypomethylation in PCa causes a loss of IGF2 imprinting (with expression of both parental alleles) both in cancerous and in distant areas within the peripheral zone, indicating that the epigenetic defect in histologically normal tissue might be employed to identify PCa in patients [39] (Figure 1). Conversely, promoter hypermethylation is widespread during neoplastic transformation of prostate cells; indeed, this is one of the first aberrations, seen early in pre-invasive lesions, and appears to be clonally maintained during metastatic progression of PCa [40]. Thus far, several genes, including tumor suppressor genes, have been described as de novo methylated in morphologically normal prostate tissue and in pre-invasive lesions, such as PIN (Prostatic Intraepithelial Neoplasia), and persisting during prostate carcinogenesis [40]. As an example, the relative frequency of methylation of Ras association domain family protein 1, isoform A (*RASSF1A*) promoter was higher in more aggressive tumors compared to less malignant tumors [41,42]. Aberrant promoter methylation was also found in different genes involved in important molecular pathways of carcinogenesis, such as DNA repair/protection, cell cycle regulation, and signal transduction. As an example, the frequency of methylation at the promoter region of Glutathion S-transferase Pi 1 (*GSTP1*), a gene involved in DNA repair, was found elevated not only in more than 90% of PCa cases, but also in over 50% of PCa precursor lesions, confirming that this is an early event in prostate carcinogenesis [42–44]. Further supporting the relevance of DNA methylation in PCa progression, different authors have related CpG methylation patterns to clinical outcomes and revealed that methylation of certain loci (e.g., *AOX1* and *RARB*) predicted disease progression [45,46].

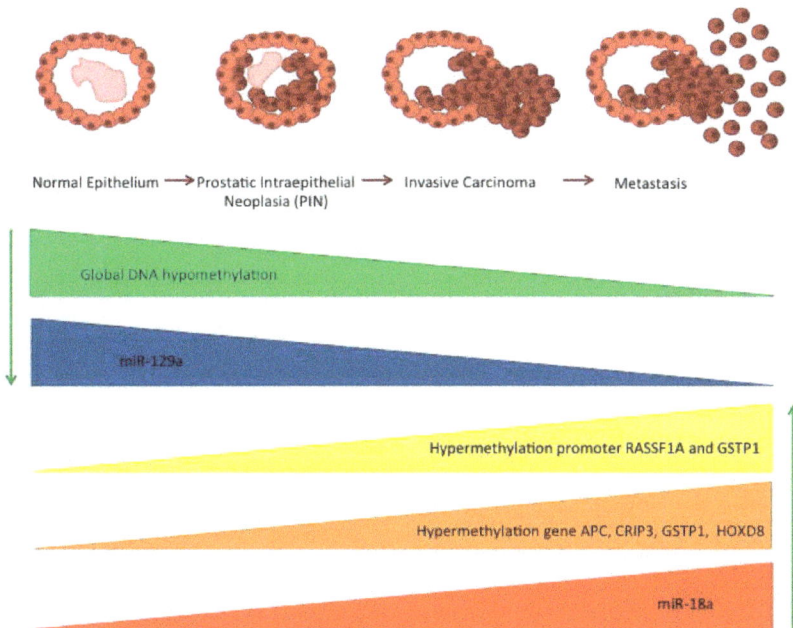

Figure 1. Potential epigenetic biomarkers in patients on AS. The figure shows the epigenetic modifications that have been tested as biomarkers. Green arrows represent down- and up-regulation of epigenetic biomarkers. The aberrant epigenetic changes have been described and cited in the text.

Histone modifications were also found to relate to the pathogenesis of prostate cancer and regulation of cancer cell proliferation. Ellinger and co-authors analyzed H3K4 methylation in patients with advanced PCa and found that this epigenetic modification was a significant predictor of PSA recurrence following radical prostatectomy. Moreover, they also found that H3K4me1, H3K4me2, and H3K4me3 levels were significantly increased in hormone-refractory prostate cancer (HPRC) [47]. One of the most studied epigenetic enzymes in PCa is the histone methyltransferase EZH2 responsible for H3K27 trimethylation. Its overexpression, particularly found in mCRPC [48], correlates with promoter hypermethylation and repression of some tumor suppressor genes [48,49]. Other epigenetic enzymes, such as SET9, SMYD3, JHDM2A, JMJD2C, and LSD1, have been demonstrated to play a role in prostate carcinogenesis [50–54]. LSD1, whose activity is associated to both transcriptional activation or repression, was associated with aggressive CRPC and a high risk of disease relapse [55,56]. Strong expression of all HDAcs was accompanied by enhanced tumor cell proliferation. High rates of HDAC1 and HDAC2 expression were significantly associated with tumor dedifferentiation, and HDAC2 expression is associated with shorter PSA relapse time after radical prostatectomy [57,58]. Moreover, AR transcriptional activity is regulated by HAT or HDAC activities, i.e., acetylation facilitates its binding to target DNA sequences, while HDAC1 and HDAC2 abrogate its activity [59]. These data confirm the role of HATs and HDACs in influencing the acetylation status of non-histone proteins. Additionally, the NAD-dependent deacetylase sirtuin-1 (SIRT1) is involved in PCa where its downregulation leads to the upregulation of different oncogenes as a consequence of H2A.Z overexpression [60].

Important biomarkers for PCa have been identified in microRNAs. Recently, Al-Kafaji et al. demonstrated that miR-18a expression increased in peripheral blood of patients with prostate cancer, indicating miR-18a as a potential noninvasive biomarker for prostate cancer tissue [61] (Figure 1). On the contrary, miRNA-129 was found downregulated in prostate cancer. Thus, its overexpression could prevent prostate cancer growth by developing tumor suppressive functions [62].

Zhao et al. [63] recently reported that a 4-gene methylation classifier panel (*APC, CRIP3, GSTP1,* and *HOXD8*) was able to predict patient reclassification on AS.

Epigenetic alterations are commonly found in PCa and play a role in carcinogenesis and tumor spreading. Moreover, new technologies, such as next-generation sequencing (NGS), have been implemented, allowing us both to expand our knowledge on prostate tumorigenesis and to obtain new epigenetic biomarkers useful to PCa patient clinical management.

Collectively, no single epigenetic biomarker has been identified as a marker of aggressive phenotypes.

However, several studies have identified potentially useful epigenetic biomarkers in PCa (Table 1) mainly in tissue samples.

Table 1. Overview of prostate cancer epigenetic biomarkers.

Biomarker	Type of Epigenetic Modification	Sample	References
IGF2	DNA hypo- and hyper-methylation	Tissue	[38–40]
RSSF1A	DNA hypermethylation	Tissue	[41,42]
GSTP1	DNA hypermethylation	Tissue	[42–44]
AOX1	DNA hypermethylation	Tissue	[45,46]
RARB	DNA hypermethylation	Tissue	[45,46]
EZH2	Increased H3K27 trimethylation	Tissue	[48]
SET9, SMYD3, JHDM2A, JMJD2C, LSD1	Histone modifications	Tissue	[50–54]
HAT	Variation in histone acetylation	Tissue	[59]
HDAC1, HDAC2	Histone deacetylation	Tissue	[59]
SIRT1	Downregulation	Tissue	[60]
miR-18a	Overexpression	Peripheral blood	[62]
miRNA-129	Downregulation	Peripheral blood	[62]
APC, CRIP3, GSTP1, HOXD8	DNA hypermethylation	Urinary	[63]

The most studied epigenetic alteration is DNA methylation. Consequently, it is conceivable that methylation markers will be the first that will be translated into clinical practice for the management of PCa patients.

DNA methylation alterations, measured in cell-free circulating and urinary tumor DNA, can potentially be used as PCa biomarkers. A large number of specific DNA methylation alterations are cancer-specific and not detectable in unaffected subjects. Examples of such alterations may be represented by CpG island methylation in the regulatory regions of *GSTP1*, *APC*, *PTGS2*, *RASSF1A*, and *RARB* [64].

GSTP1, *RARB*, and *RASSF1* DNA promoter methylation has been widely investigated in body fluids as a non-invasive biomarker for the early diagnosis of PCa [65,66]. Other authors [67–69] have suggested that urine cell-free DNA could represent a non-invasive and inexpensive biomarker for assessing specific promoter region methylation.

Chromatin remodeling and non-coding RNA regulation represent an expanding research field.

Collectively, based on literature data and on the improvement of new technologies such as next-generation sequencing (NGS), epigenetic signatures seem to be promising tools for stratifying PCa patients for progression risk. Nevertheless, some obstacles may contribute to the lack of translating such biomarkers for PCa in clinical practice. In particular, an important role was played by the limitation of available methods for analysis, the consistency of experimental design to validate the biomarkers, and the relevance of the epigenetic alteration in prostate carcinogenesis. Moreover, it should be taken into account that epigenetic modifications are affected by aging and prostate cancer is an age-related disease.

Further studies on larger population will define the clinical benefit of epigenetic markers in body fluids.

2. Conclusions

Follow-up of patients on AS can prevent overtreatment and the related impairment of quality of life. Diagnosis of true low-risk PCa is essential to address this therapeutic strategy. As recently reported by Klotz [70], AS is harmless in the medium to long term with a very low cancer-specific mortality of 10–15 years. Many studies have been focused on tools able to provide further improvement of the safety of this conservative therapeutic option.

Several serum and urine biomarkers, including the PHI, the 4K score, and urinary TMPRSS2-ERG or PCA3 mRNA, have been evaluated in men on AS. However, the association with tumor aggressiveness and thus prognostic value remains controversial [71]. Consequently, new players have to be considered to predict cancer progression in an AS regimen.

In this scenario, it could be advantageous to implement epigenetic signature identification of clinically significant PCa in the setting of active surveillance. Such studies represent an urgent need to identify indolent cancer and avoid overtreatment. DNA methylation, histone modifications, and noncoding RNA could potentially provide new tools for prognosis of prostate cancer, affecting clinical management of patients. In particular, since these biomarkers lack specificity, further studies are needed to ascertain if a panel of multiple epigenetic targets may be helpful in planning AS strategies.

Acknowledgments: The authors thank Monica Autiero for critically reading the manuscript. The Italian Association for Cancer Research (MFAG number 11510 to A.C.) supported this work.

Author Contributions: Matteo Ferro and Daniela Terracciano conceived and designed the review; Matteo Ferro, Paola Ungaro, Amelia Cimmino, Daniela Terracciano, Giuseppe Lucarelli, Gian Maria Busetto, Francesco Cantiello and Rocco Damiano performed literature search; Paola Ungaro and Daniela Terracciano wrote the paper.

Conflicts of Interest: The authors declare no conflict of interest.

References

1. Klotz, L. Active surveillance for prostate cancer: Patient selection and management. *Curr. Oncol.* **2010**, *17*, S11–S17. [CrossRef] [PubMed]
2. D'Amico, A.V.; Whittington, R.; Malkowicz, S.B.; Schultz, D.; Blank, K.; Broderick, G.A.; Tomaszewski, J.E.; Renshaw, A.A.; Kaplan, I.; Beard, C.J.; et al. Biochemical outcome after radical prostatectomy, external beam radiation therapy, or interstitial radiation therapy for clinically localized prostate cancer. *JAMA* **1998**, *280*, 969–974. [CrossRef] [PubMed]
3. Klotz, L.; Zhang, L.; Lam, A.; Nam, R.; Mamedov, A.; Loblaw, A. Clinical results of long-term follow-up of a large, active surveillance cohort with localized prostate cancer. *J. Clin. Oncol.* **2010**, *28*, 126–131. [CrossRef] [PubMed]
4. Thomsen, F.B.; Berg, K.D.; Roder, M.A.; Iversen, P.; Brasso, K. Active surveillance for localized prostate cancer: An analysis of patient contacts and utilization of healthcare resources. *Scand. J. Urol.* **2015**, *49*, 43–50. [CrossRef] [PubMed]
5. Weinreb, J.C.; Barentsz, J.O.; Choyke, P.L.; Cornud, F.; Haider, M.A.; Macura, K.J.; Margolis, D.; Schnall, M.D.; Shtern, F.; Tempany, C.M.; et al. Pi-rads prostate imaging—Reporting and data system: 2015, Version 2. *Eur. Urol.* **2016**, *69*, 16–40. [CrossRef] [PubMed]
6. Thaxton, C.S.; Loeb, S.; Roehl, K.A.; Kan, D.; Catalona, W.J. Treatment outcomes of radical prostatectomy in potential candidates for 3 published active surveillance protocols. *Urology* **2010**, *75*, 414–418.
7. Loeb, S.; Carter, H.B.; Berndt, S.I.; Ricker, W.; Schaeffer, E.M. Is repeat prostate biopsy associated with a greater risk of hospitalization? Data from seer-medicare. *J. Urol.* **2013**, *189*, 867–870. [CrossRef] [PubMed]
8. Loeb, S.; Catalona, W.J. The prostate health index: A new test for the detection of prostate cancer. *Ther. Adv. Urol.* **2014**, *6*, 74–77. [CrossRef] [PubMed]
9. Heidegger, I.; Klocker, H.; Pichler, R.; Pircher, A.; Prokop, W.; Steiner, E.; Ladurner, C.; Comploj, E.; Lunacek, A.; Djordjevic, D.; et al. Propsa and the prostate health index as predictive markers for aggressiveness in low-risk prostate cancer-results from an international multicenter study. *Prostate Cancer Prostatic Dis.* **2017**. [CrossRef] [PubMed]
10. Tosoian, J.J.; Loeb, S.; Feng, Z.; Isharwal, S.; Landis, P.; Elliot, D.J.; Veltri, R.; Epstein, J.I.; Partin, A.W.; Carter, H.B.; et al. Association of [−2]proPSA with biopsy reclassification during active surveillance for prostate cancer. *J. Urol.* **2012**, *188*, 1131–1136. [CrossRef] [PubMed]
11. Tosoian, J.J.; Loeb, S.; Kettermann, A.; Landis, P.; Elliot, D.J.; Epstein, J.I.; Partin, A.W.; Carter, H.B.; Sokoll, L.J. Accuracy of PCa3 measurement in predicting short-term biopsy progression in an active surveillance program. *J. Urol.* **2010**, *183*, 534–538. [CrossRef] [PubMed]
12. Sottile, A.; Ortega, C.; Berruti, A.; Mangioni, M.; Saponaro, S.; Polo, A.; Prati, V.; Muto, G.; Aglietta, M.; Montemurro, F. A pilot study evaluating serum pro-prostate-specific antigen in patients with rising PSA following radical prostatectomy. *Oncol. Lett.* **2012**, *3*, 819–824. [PubMed]
13. Hendriks, R.J.; van Oort, I.M.; Schalken, J.A. Blood-based and urinary prostate cancer biomarkers: A review and comparison of novel biomarkers for detection and treatment decisions. *Prostate Cancer Prostatic Dis.* **2017**, *20*, 12–19. [CrossRef] [PubMed]
14. Gokce, M.I.; Tangal, S.; Hamidi, N.; Suer, E.; Ibis, M.A.; Beduk, Y. Role of neutrophil-to-lymphocyte ratio in prediction of gleason score upgrading and disease upstaging in low-risk prostate cancer patients eligible for active surveillance. *Can. Urol. Assoc. J.* **2016**, *10*, E383–E387. [CrossRef] [PubMed]
15. Ferro, M.; Lucarelli, G.; Bruzzese, D.; di Lorenzo, G.; Perdona, S.; Autorino, R.; Cantiello, F.; La Rocca, R.; Busetto, G.M.; Cimmino, A.; et al. Low serum total testosterone level as a predictor of upstaging and upgrading in low-risk prostate cancer patients meeting the inclusion criteria for active surveillance. *Oncotarget* **2016**, *8*, 18424–18434. [CrossRef] [PubMed]
16. Murray, N.P.; Reyes, E.; Fuentealba, C.; Aedo, S.; Jacob, O. The presence of primary circulating prostate cells is associated with upgrading and upstaging in patients eligible for active surveillance. *Ecancermedicalscience* **2017**, *11*, 711. [CrossRef] [PubMed]
17. Cantiello, F.; Russo, G.I.; Ferro, M.; Cicione, A.; Cimino, S.; Favilla, V.; Perdona, S.; Bottero, D.; Terracciano, D.; de Cobelli, O.; et al. Prognostic accuracy of prostate health index and urinary prostate cancer antigen 3 in predicting pathologic features after radical prostatectomy. *Urol. Oncol.* **2015**, *33*, 163.e15–163.e23. [CrossRef] [PubMed]

18. Porpiglia, F.; Cantiello, F.; de Luca, S.; Manfredi, M.; Veltri, A.; Russo, F.; Sottile, A.; Damiano, R. In-parallel comparative evaluation between multiparametric magnetic resonance imaging, prostate cancer antigen 3 and the prostate health index in predicting pathologically confirmed significant prostate cancer in men eligible for active surveillance. *BJU Int.* **2016**, *118*, 527–534. [CrossRef] [PubMed]
19. Lin, D.W.; Newcomb, L.F.; Brown, E.C.; Brooks, J.D.; Carroll, P.R.; Feng, Z.; Gleave, M.E.; Lance, R.S.; Sanda, M.G.; Thompson, I.M.; et al. Urinary tmprss2:ERG and PCa3 in an active surveillance cohort: Results from a baseline analysis in the canary prostate active surveillance study. *Clin. Cancer Res.* **2013**, *19*, 2442–2450. [CrossRef] [PubMed]
20. Berg, K.D.; Vainer, B.; Thomsen, F.B.; Roder, M.A.; Gerds, T.A.; Toft, B.G.; Brasso, K.; Iversen, P. ERG protein expression in diagnostic specimens is associated with increased risk of progression during active surveillance for prostate cancer. *Eur. Urol.* **2014**, *66*, 851–860. [CrossRef] [PubMed]
21. Green, W.J.; Ball, G.; Hulman, G.; Johnson, C.; Van Schalwyk, G.; Ratan, H.L.; Soria, D.; Garibaldi, J.M.; Parkinson, R.; Hulman, J.; et al. KI67 and DLX2 predict increased risk of metastasis formation in prostate cancer—A targeted molecular approach. *Br. J. Cancer* **2016**, *115*, 236–242. [CrossRef] [PubMed]
22. Murphy, S.J.; Karnes, R.J.; Kosari, F.; Castellar, B.E.; Kipp, B.R.; Johnson, S.H.; Terra, S.; Harris, F.R.; Halling, G.C.; Klein, J.L.; et al. Integrated analysis of the genomic instability of pten in clinically insignificant and significant prostate cancer. *Mod. Pathol.* **2016**, *29*, 143–156. [CrossRef] [PubMed]
23. Lotan, T.L.; Carvalho, F.L.; Peskoe, S.B.; Hicks, J.L.; Good, J.; Fedor, H.L.; Humphreys, E.; Han, M.; Platz, E.A.; Squire, J.A.; et al. Pten loss is associated with upgrading of prostate cancer from biopsy to radical prostatectomy. *Mod. Pathol.* **2015**, *28*, 128–137. [CrossRef] [PubMed]
24. Mithal, P.; Allott, E.; Gerber, L.; Reid, J.; Welbourn, W.; Tikishvili, E.; Park, J.; Younus, A.; Sangale, Z.; Lanchbury, J.S.; et al. Pten loss in biopsy tissue predicts poor clinical outcomes in prostate cancer. *Int. J. Urol.* **2014**, *21*, 1209–1214. [CrossRef] [PubMed]
25. Knezevic, D.; Goddard, A.D.; Natraj, N.; Cherbavaz, D.B.; Clark-Langone, K.M.; Snable, J.; Watson, D.; Falzarano, S.M.; Magi-Galluzzi, C.; Klein, E.A.; et al. Analytical validation of the oncotype dx prostate cancer assay—A clinical RT-PCR assay optimized for prostate needle biopsies. *BMC Genom.* **2013**, *14*, 690. [CrossRef] [PubMed]
26. Cuzick, J.; Swanson, G.P.; Fisher, G.; Brothman, A.R.; Berney, D.M.; Reid, J.E.; Mesher, D.; Speights, V.O.; Stankiewicz, E.; Foster, C.S.; et al. Prognostic value of an RNA expression signature derived from cell cycle proliferation genes in patients with prostate cancer: A retrospective study. *Lancet Oncol.* **2011**, *12*, 245–255. [CrossRef]
27. Bishoff, J.T.; Freedland, S.J.; Gerber, L.; Tennstedt, P.; Reid, J.; Welbourn, W.; Graefen, M.; Sangale, Z.; Tikishvili, E.; Park, J.; et al. Prognostic utility of the cell cycle progression score generated from biopsy in men treated with prostatectomy. *J. Urol.* **2014**, *192*, 409–414. [CrossRef] [PubMed]
28. Suzuki, M.; Bird, A. DNA methylation landscapes: Provocative insights from epigenomics. *Nat. Rev. Genet.* **2008**, *9*, 465–476. [CrossRef] [PubMed]
29. Jurkowska, R.Z.; Jukowski, T.P.; Jeltsch, A. Structure and function of mammalian DNA methyltransferases. *ChemBioChem* **2011**, *12*, 206–222. [CrossRef] [PubMed]
30. Okano, M.; Xie, S.; Li, E. Cloning and characterization of a family of novel mammalian DNA (cytosine-5) methyltransferases. *Nat. Genet.* **1998**, *19*, 219–220. [PubMed]
31. Waddington, C.H. "The epigenotype 1942". *Int. J. Epidemiol.* **2012**, *41*, 10–13. [CrossRef] [PubMed]
32. Kouzarides, T. Chromatin modifications and their function. *Cell* **2007**, *128*, 693–705. [CrossRef] [PubMed]
33. Haberland, M.; Montgomery, R.L.; Olson, E.N. The many roles of histone deacetylases in development and physiology: Implications for disease and therapy. *Nat. Rev. Genet.* **2009**, *10*, 32–42. [CrossRef] [PubMed]
34. Rouhi, A.; Mager, D.L.; Humphries, R.K.; Kuchenbauer, F. MiRNAs, epigenetics, and cancer. *Mamm. Genome* **2008**, *19*, 517–525. [CrossRef] [PubMed]
35. He, L.; Hannon, G.J. MicroRNAs: Small RNAs with a big role in gene regulation. *Nat. Rev. Genet.* **2004**, *5*, 522–531. [CrossRef] [PubMed]
36. Farazi, T.A.; Hoell, J.I.; Morozov, P.; Tuschl, T. MicroRNAs in human cancer. *Adv. Exp. Med. Biol.* **2013**, *774*, 1–20.
37. Yoo, C.B.; Jones, P.A. Epigenetic therapy of cancer: Past, present and future. *Nat. Rev. Drug Discov.* **2006**, *5*, 37–50. [CrossRef] [PubMed]

38. Yegnasubramanian, S.; Haffner, M.C.; Zhang, Y.; Gurel, B.; Cornish, T.C.; Wu, Z.; Irizarry, R.A.; Morgan, J.; Hicks, J.; DeWeese, T.L.; et al. DNA hypomethylation arises later in prostate cancer progression than CpG island hypermethylation and contributes to metastatic tumor heterogeneity. *Cancer Res.* **2008**, *68*, 8954–8967. [CrossRef] [PubMed]

39. Jarrad, D.F.; Bussemakers, M.J.; Bova, G.S.; Isaacs, W.B. Regional loss of imprinting of the insulin-like growth factor II gene occurs in human prostate tissues. *Clin. Cancer Res.* **1995**, *12*, 1471–1478.

40. Perry, A.S.; Watson, R.W.; Lawler, M.; Hollywood, D. The epigenome as a therapeutic target in prostate cancer. *Nat. Rev. Urol.* **2010**, *7*, 668–680. [CrossRef] [PubMed]

41. Liu, L.; Yoon, J.H.; Dammann, R.; Pfeifer, G.P. Frequent hypermethylation of the *RASSF1A* gene in prostate cancer. *Oncogene* **2002**, *21*, 6835–6840. [CrossRef] [PubMed]

42. Henrique, R.; Jeronimo, C. Molecular detection of prostate cancer: A role for gstp1 hypermethylation. *Eur. Urol.* **2004**, *46*, 660–669. [CrossRef] [PubMed]

43. Jerónimo, C.; Usadel, H.; Henrique, R.; Oliveira, J.; Lopes, C.; Nelson, W.G.; Sidransky, D. Quantitation of gstp1 methylation in non-neoplastic prostatic tissue and organ-confined prostate adenocarcinoma. *J. Natl. Cancer Inst.* **2001**, *93*, 1747–1752. [CrossRef] [PubMed]

44. Millar, D.S.; Ow, K.K.; Paul, C.L.; Russell, P.J.; Molloy, P.L.; Clark, S.J. Detailed methylation analysis of the glutathione S-transferase-π gene (*GSTP1*) in prostate cancer. *Oncogene* **1999**, *18*, 1313–1324. [CrossRef] [PubMed]

45. Litovkin, K.; van Eynde, A.; Joniau, S.; Lerut, E.; Laenen, A.; Gevaert, T.; Gevaert, O.; Spahn, M.; Kneitz, B.; Gramme, P.; et al. DNA methylation-guided prediction of clinical failure in high-risk prostate cancer. *PLoS ONE* **2015**, *10*, e01300651. [CrossRef] [PubMed]

46. Haldrup, C.; Mundbjerg, K.; Vestergaard, E.M.; Lamy, P.; Wild, P.; Schulz, W.A.; Arsov, C.; Visakorpi, T.; Borre, M.; Høyer, S.; et al. DNA methylation signatures for prediction of biochemical recurrence after radical prostatectomy of clinically localized prostate cancer. *J. Clin. Oncol.* **2013**, *31*, 3250–3258. [CrossRef] [PubMed]

47. Ellinger, J.; Kahl, P.; von der Gathen, J.; Rogenhofer, S.; Heukamp, L.C.; Gutgemann, I.; Walter, B.; Hofstadter, F.; Buttner, R.; Muller, S.C.; et al. Global levels of histone modifications predict prostate cancer recurrence. *Prostate* **2010**, *70*, 61–69. [CrossRef] [PubMed]

48. Varambally, S.; Dhanasekaran, S.M.; Zhou, M.; Barrette, T.R.; Kumar-Sinha, C.; Sanda, M.G.; Ghosh, D.; Pienta, K.J.; Sewalt, R.G.; Otte, A.P.; et al. The polycomb group protein EZH2 is involved in progression of prostate cancer. *Nature* **2002**, *419*, 624–629. [CrossRef] [PubMed]

49. Yu, J.; Rhodes, D.R.; Tomlins, S.A.; Cao, X.; Chen, G.; Mehra, R.; Wang, X.; Ghosh, D.; Shah, R.B.; Varambally, S.; et al. A polycomb repression signature in metastatic prostate cancer predicts cancer outcome. *Cancer Res.* **2007**, *67*, 10657–10663. [CrossRef] [PubMed]

50. Wissmann, M.; Yin, N.; Müller, J.M.; Greschik, H.; Fodor, B.D.; Jenuwein, T.; Vogler, C.; Schneider, R.; Günther, T.; Buettner, R.; et al. Cooperative demethylation by *JMJD2C* and *LSD1* promotes androgen receptor-dependent gene expression. *Nat. Cell Biol.* **2007**, *9*, 347–353. [CrossRef] [PubMed]

51. Gaughan, L.; Stockley, J.; Wang, N.; McCracken, S.R.; Treumann, A.; Armstrong, K.; Shaheen, F.; Watt, K.; McEwan, I.J.; Wang, C.; et al. Regulation of the androgen receptor by set9-mediated methylation. *Nucleic Acids Res.* **2011**, *39*, 1266–1279. [CrossRef] [PubMed]

52. Suikki, H.E.; Kujala, P.M.; Tammela, T.L.; van Weerden, W.M.; Vessella, R.L.; Visakorpi, T. Genetic alterations and changes in expression of histone demethylases in prostate cancer. *Prostate* **2010**, *70*, 889–898. [CrossRef] [PubMed]

53. Vieira, F.Q.; Costa-Pinheiro, P.; Almeida-Rios, D.; Graça, I.; Monteiro-Reis, S.; Simões-Sousa, S.; Carneiro, I.; Sousa, E.J.; Godinho, M.I.; Baltazar, F.; et al. Smyd3 contributes to a more aggressive phenotype of prostate cancer and targets cyclin D2 through H4K20ME3. *Oncotarget* **2015**, *6*, 13644–13657. [CrossRef] [PubMed]

54. Vieira, F.Q.; Costa-Pinheiro, P.; Ramalho-Carvalho, J.; Pereira, A.; Menezes, F.D.; Antunes, L.; Carneiro, I.; Oliveira, J.; Henrique, R.; Jerónimo, C. Deregulated expression of selected histone methylases and demethylases in prostate carcinoma. *Endocr. Relat. Cancer* **2013**, *21*, 51–61. [CrossRef] [PubMed]

55. Kahl, P.; Gullotti, L.; Heukamp, L.C.; Wolf, S.; Friedrichs, N.; Vorreuther, R.; Solleder, G.; Bastian, P.J.; Ellinger, J.; Metzger, E.; et al. Androgen receptor coactivators lysine-specific histone demethylase 1 and four and a half lim domain protein 2 predict risk of prostate cancer recurrence. *Cancer Res.* **2006**, *66*, 11341–11347. [CrossRef] [PubMed]

56. Metzger, E.W.; Wissmann, M.; Yin, N.; Müller, J.M.; Schneider, R.; Peters, A.H.; Günther, T.; Buettner, R.; Schüle, R. LSD1 demethylates repressive histone marks to promote androgen-receptor-dependent transcription. *Nature* **2005**, *437*, 436–439. [CrossRef] [PubMed]

57. Haikidou, K.; Gaughan, L.; Cook, S.; Leung, H.Y.; Neal, D.E.; Robson, C.N. Upregulation and nuclear recruitment of HDAC1 in hormone refractory prostate cancer. *Prostate* **2004**, *59*, 177–189. [CrossRef] [PubMed]

58. Weichert, W.; Röske, A.; Gekeler, V.; Beckers, T.; Stephan, C.; Jung, K.F.; Fritzsche, F.R.N.; Niesporek, S.; Denkert, C.; Dietel, M.; et al. Histone deacetylases 1, 2 and 3 are highly expressed in prostate cancer and hdac2 expression is associated with shorter PSA relapse time after radical prostatectomy. *Br. J. Cancer* **2008**, *98*, 604–610. [CrossRef] [PubMed]

59. Korkmaz, C.G.; Frønsdal, K.; Zhang, Y.; Lorenzo, P.I.; Saatcioglu, F. Potentiation of androgen receptor transcriptional activity by inhibition of histone deacetylation—Rescue of transcriptionally compromised mutants. *J. Endocrinol.* **2004**, *182*, 377–389. [CrossRef] [PubMed]

60. Baptista, T.; Graça, I.; Sousa, E.J.; Oliveira, A.I.; Costa, N.R.; Costa-Pinheiro, P.; Amado, F.; Henrique, R.; Jerónimo, C. Regulation of histone H2A.Z expression is mediated by sirtuin 1 in prostate cancer. *Oncotarget* **2013**, *4*, 1673–1685. [CrossRef] [PubMed]

61. Al-Kafaji, G.; Al-Naieb, Z.T.; Bakhiet, M. Increased oncogenic microRNA-18a expression in the peripheral blood of patients with prostate cancer: A potential novel non-invasive biomarker. *Oncol. Lett.* **2016**, *11*, 1201–1206. [CrossRef] [PubMed]

62. Xu, S.; Yi, X.M.; Zhou, W.Q.; Cheng, W.; Ge, J.P.; Zhang, Z.Y. Downregulation of miR-129 in peripheral blood mononuclear cells is a diagnostic and prognostic biomarker in prostate cancer. *Int. J. Clin. Exp. Pathol.* **2015**, *8*, 14335–14344. [PubMed]

63. Zhao, F.; Olkhov-Mitsel, E.; van der Kwast, T.; Sykes, J.; Zdravic, D.; Venkateswaran, V.; Zlotta, A.R.; Loblaw, A.; Fleshner, N.E.; Klotz, L.; et al. Urinary DNA methylation biomarkers for noninvasive prediction of aggressive disease in patients with prostate cancer on active surveillance. *J. Urol.* **2017**, *197*, 335–341. [CrossRef] [PubMed]

64. Yegnasubramanian, S. Prostate cancer epigenetics and its clinical implications. *Asian J. Androl.* **2016**, *18*, 549–558. [CrossRef] [PubMed]

65. Sunami, E.; Shinozaki, M.; Higano, C.S.; Wollman, R.; Dorff, T.B.; Tucker, S.J.; Martinez, S.R.; Mizuno, R.; Singer, F.R.; Hoon, D.S. Multimarker circulating DNA assay for assessing blood of prostate cancer patients. *Clin. Chem.* **2009**, *55*, 559–567. [CrossRef] [PubMed]

66. Ahmed, H. Promoter methylation in prostate cancer and its application for the early detection of prostate cancer using serum and urine samples. *Biomark. Cancer* **2010**, *2*, 17–33. [CrossRef] [PubMed]

67. Partin, A.W.; Van Neste, L.; Klein, E.A.; Marks, L.S.; Gee, J.R.; Troyer, D.A.; Rieger-Christ, K.; Jones, J.S.; Magi-Galluzzi, C.; Mangold, L.A.; et al. Clinical validation of an epigenetic assay to predict negative histopathological results in repeat prostate biopsies. *J. Urol.* **2014**, *192*, 1081–1087. [CrossRef] [PubMed]

68. Ellinger, J.; Muller, S.C.; Stadler, T.C.; Jung, A.; von Ruecker, A.; Bastian, P.J. The role of cell-free circulating DNA in the diagnosis and prognosis of prostate cancer. *Urol. Oncol.* **2011**, *29*, 124–129. [CrossRef] [PubMed]

69. Roupret, M.; Hupertan, V.; Yates, D.R.; Catto, J.W.; Rehman, I.; Meuth, M.; Ricci, S.; Lacave, R.; Cancel-Tassin, G.; de la Taille, A.; et al. Molecular detection of localized prostate cancer using quantitative methylation-specific PCR on urinary cells obtained following prostate massage. *Clin. Cancer Res.* **2007**, *13*, 1720–1725. [CrossRef] [PubMed]

70. Klotz, L. Active surveillance for low-risk prostate cancer. *Curr. Opin. Urol.* **2017**, *27*, 225–230. [CrossRef] [PubMed]

71. Loeb, S.; Bruinsma, S.M.; Nicholson, J.; Briganti, A.; Pickles, T.; Kakehi, Y.; Carlsson, S.V.; Roobol, M.J. Active surveillance for prostate cancer: A systematic review of clinicopathologic variables and biomarkers for risk stratification. *Eur. Urol.* **2015**, *67*, 619–626. [CrossRef] [PubMed]

International Journal of
Molecular Sciences

MDPI

Review

Nanoparticles as Theranostic Vehicles in Experimental and Clinical Applications—Focus on Prostate and Breast Cancer

Jörgen Elgqvist [1,2]

[1] Department of Medical Physics and Biomedical Engineering, Sahlgrenska University Hospital,
     413 45 Gothenburg, Sweden; jorgen.elgqvist@vgregion.se or jorgen.elgqvist@gu.se
[2] Department of Physics, University of Gothenburg, 412 96 Gothenburg, Sweden

Academic Editor: Carsten Stephan
Received: 10 April 2017; Accepted: 15 May 2017; Published: 20 May 2017

Abstract: Prostate and breast cancer are the second most and most commonly diagnosed cancer in men and women worldwide, respectively. The American Cancer Society estimates that during 2016 in the USA around 430,000 individuals were diagnosed with one of these two types of cancers, and approximately 15% of them will die from the disease. In Europe, the rate of incidences and deaths are similar to those in the USA. Several different more or less successful diagnostic and therapeutic approaches have been developed and evaluated in order to tackle this issue and thereby decrease the death rates. By using nanoparticles as vehicles carrying both diagnostic and therapeutic molecular entities, individualized targeted theranostic nanomedicine has emerged as a promising option to increase the sensitivity and the specificity during diagnosis, as well as the likelihood of survival or prolonged survival after therapy. This article presents and discusses important and promising different kinds of nanoparticles, as well as imaging and therapy options, suitable for theranostic applications. The presentation of different nanoparticles and theranostic applications is quite general, but there is a special focus on prostate cancer. Some references and aspects regarding breast cancer are however also presented and discussed. Finally, the prostate cancer case is presented in more detail regarding diagnosis, staging, recurrence, metastases, and treatment options available today, followed by possible ways to move forward applying theranostics for both prostate and breast cancer based on promising experiments performed until today.

Keywords: nanoparticles; theranostics; nanomedicine; prostate cancer; breast cancer

1. Introduction

In order to be able to combine both a therapeutic and a diagnostic function into a single molecular entity, the research on and development of theranostic nanoparticles (TNPs) have increased continuously during the last couple of years (the MeSH (Medical Subject Headings) term "theranostics" gave 0 hits for 2004 but 801 for 2016 on Pubmed) [1–8]. These type of nanosized drug vehicles have been developed and tested in different settings such as, for example, iron oxide, gadolinium, gold, manganese, or polymeric nanoparticles (NPs), quantum dots, and liposomes for the diagnosis and treatment of various diseases [1–18]. The dimensions of these different molecular nanostructures is generally less than 100 nm, and two of the common goals for all different settings is to maximize the drug loading capacity, and increase the specificity of the TNPs towards cancer cells [1]. The key benefit of TNPs is that they have the possibility to increase the therapeutic efficacy, partly due to ameliorated drug circulation times increasing the tumor uptake, but also to lessen the risk of unwanted toxic effects on healthy tissue [16–20]. Another important feature of the TNPs is that the diagnostic and therapeutic functionality could be localized to the exact same position due to the attachment of both these agents on the same molecular drug vehicle. Compared to other larger molecular vehicles these small NPs has a larger surface-area-to-volume ratio. This feature allows the TNPs to reach the capillary bed while at the same time carry a large variety of therapeutic and diagnostic agents themselves.

When considering the use of NPs for cancer treatment or diagnosis, the enhanced permeability and retention (EPR) effect plays a central role, although this effect seems to vary among individuals in the human population [2,19,21]. The EPR effect is also believed not to be present or even similar for all types of tumors, as well as not being the only parameter responsible for the efficacy of a certain NP application [22]. Parameters such as the level of unwanted drug release into the systemic circulation and the intra-tumoral allocation as well as the amount and kinetics of the intra-tumoral drug release will also determine the final efficacy [22].

However, the EPR effect can make TNPs accumulate, and be retained, to a higher degree in tumor tissue as compared to normal tissue, due to leakiness of the increased vasculature in the tumor tissue [19,20,23]. Magnetic resonance imaging (MRI) is a valuable tool for image and evaluate the degree of accumulation due to its high spatial resolution in both tumors and healthy tissue [24]. During MRI, Gd(III) (Gd^{3+}) based contrast agents are very effective and highly paramagnetic substances, enhancing the contrast by increasing the T1 relaxation rate R1 (=1/T1), due to its rare electronic configuration of seven unpaired electron spins. By combining Gd(III) with different kind of NPs it would be possible to increase the accumulation and retention time in tumors, and therefore even more increase the MRI contrast [25–34]. TNPs could then be developed with the Gd(III)-based MRI contrast agent combined with, for example, a therapeutic drug for cancer such as gemcitabine [35,36].

Another strong argument for developing a NP based anti-cancer technology is that it could enable earlier detection of the disease. Early detection is most often absolutely crucial and decides whether a cancer patient will have the possibility to be cured, or just receive a treatment extending survival before finally succumbing to the disease. It is a well-known fact that the major cause of mortality in cancer is due to tumor metastasis [37–39]. In most cases, the dissemination of the cancerous disease is caused by tumor cells that have shed from the primary tumor and enters into the systemic circulation, i.e., circulating tumor cells CTCs. The idea of CTCs was first noticed by Dr. Tomas Ashworth already in 1869 [40]. However, while CTCs can only be confirmed and monitored in patients having a more or less advanced cancerous disease [39,41–54], the goal for a nanomedical approach would be to detect the CTCs at a much earlier time point compared to what is possible today [55,56].

The significance of the TNP technology regarding cancer in general is that it potentially could diagnose better, using multimodal imaging, and at the same time more effectively treat these diseases, especially in the disseminated cases. To achieve this, different types of isotopes could be attached to the TNPs. By attaching, for example, specific monoclonal antibodies (mAbs), fraction of mAbs, or peptides to these multifunctional TNPs highly specific and therefore targeted imaging and radiation therapies could be achieved. Personalized medicine might be possible using TNPs, since the imaging of drug accumulation in individual patients would be possible. Such imaging would refer to both tumor as well as healthy tissue and would then make it possible to estimate and predict, to a better degree than what is possible today, the therapeutic efficacy, and also make possible adjustments of ongoing regimens [2,19–21,23].

This article presents different type of NPs, imaging, and therapy options, as well as promising theranostic applications utilizing those techniques, with a focus on prostate cancer (PCa) and breast cancer (BC). Relevant PCa and BC references are presented and discussed in every section, but towards the end of the article the PCa case is given extra focus regarding diagnosis, staging, recurrence, metastases, and treatment options available today. Finally, possible ways to move forward applying theranostics for both PCa and BC are suggested and discussed based on promising experiments performed until today.

2. Nanoparticles for Prostate and Breast Cancer

The development of different and innovative NPs for various medical applications has increased tremendously during the past couple of years. Despite some problems with relatively low tumor uptake in some cancer applications, there is today a number of promising alternatives that have been tested or are under the development [57–90]. Below follows a presentation of important and interesting NP alternatives suitable for PCa and BC, some of which already have been tested for these type of diseases while some are still to be explored more thoroughly. For example, since the Nobel Prize in 2010 to Dr. Geim and Dr. Novoselov for their pioneering work on the two-dimensional material graphene, several promising NPs have since been suggested, developed, and tested based on that material. A prerequisite although, before any translation into clinical use of any NPs, is a meticulous survey of their pharmacological and toxicological properties. The presentation below summarizes the most important general characteristics of the NP options, and to what extent they have been used in the PCa and BC context so far. When possible, theranostic applications are referenced to and discussed shortly for each type of NP. In addition to theranostic applications, some studies in which only an imaging or therapeutic approach has been used are also referenced to for each type of NP.

2.1. Iron Oxide Nanoparticles

Due to their magnetic properties, with a diameter ranging from a few nanometers up to approximately 100 nm, iron oxide particles have been evaluated and used in several magnetic resonance technology-based biomedical applications such as multifunctional theranostic complexes combining tumor targeting, imaging as well as cancer nanotherapy in personalized cancer treatment [91–100]. The superparamagnetic iron oxide NPs (SPIONs) are most often used, and the subpopulation of ultrasmall SPIONs denoted as USPIOs defined as having a diameter of less than ~20 nm. A schematic representation of an iron oxide NP is shown in Figure 1. Three main variants of this NP are magnetite (Fe_3O_4), hematite (α-Fe_2O_3), and maghemite (γ-Fe_2O_3). The two latter being differently structured (rhombohedral and cubic, respectively) allotropic oxidized forms of magnetite. γ-Fe_2O_3 and Fe_3O_4 are preferred in medical applications due to their lack of toxicity and good biocompatibility in humans. Since SPIONSs tend to aggregate due to magnetic, van der Waals, and/or hydrophilic/hydrophobic forces it is important to minimize this effect in biomedical applications by different kind of surface modifications, for example, by PEGylation. PEGylation of a NP means attaching polyethylene glycol (PEG) molecules to its surface, thereby not only hindering aggregation but also masking the NP from the immune system. Other type of coatings can also be necessary for certain applications. For example, if SPIONs is used as a contrast agent during photoacoustic imaging, using near-infrared (IR) light, the NPs could be coated with silica (SiO_2), enhancing the light absorption compared to the bare iron oxide NP [101–103]. Regarding diagnosis iron oxide NPs have been used for atherosclerotic evaluation, gene expression analysis, inflammation, angiogenesis, stem-cell tracking and also for cancer diagnosis.

Specifically, for PCa and BC there is ongoing research on different applications using these kind of NPs [93,104–115]. For example, regarding PCa, Zhu et al. have performed synthesis, characterization, an in vitro binding assay, and an in vivo magnetic resonance imaging (MRI) evaluation of prostate specific membrane antigen (PSMA) targeting SPIONs. They showed specific uptake of their polypeptide-based SPIONs by the PSMA expressing cells, and that the MRI signal could be specifically enhanced. They concluded that PSMA-targeting SPIONs might provide a new strategy for imaging PCa [107]. As an example regarding BC, Pasha Shaik et al. recently performed experiments on blocking the IL4-α receptor (IL4Rα) using PEGylated SPIONs to inhibit BC cell proliferation [111]. They found that for 4T1 cells, blocking of this receptor caused a significant decrease in cell viability and induced apoptosis. They also concluded that a combined treatment using SPION-IL4Rα-doxorubicin caused significant increases in cell death, apoptosis, and oxidative stress, compared to either SPION-IL4Rα or doxorubicin alone.

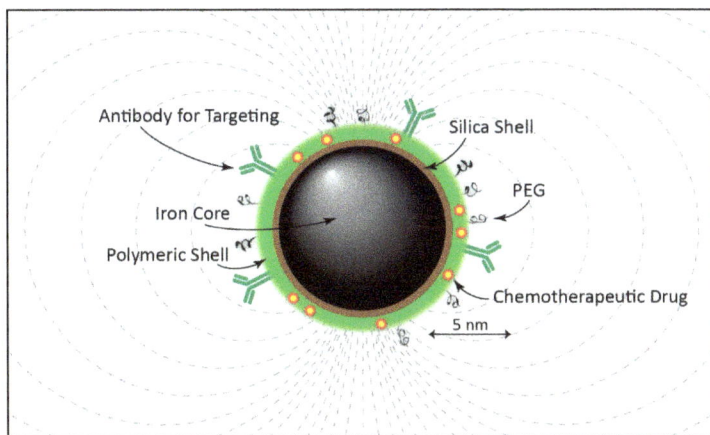

Figure 1. Functionalized iron-oxide nanoparticle. Functional biocompatible polymers are grafted onto an inorganic core of magnetite (Fe_3O_4) or maghemite (Fe_2O_3) through an anchoring group, such as an amine, carboxylic acid or phosphonic acid group. The polymeric shell improves the stability of the iron oxide nanoparticles (NPs) in solution, and also allows the encapsulation of, for example, therapeutic agents. An alternative coating in the form of a fluorescent silica (SiO_2) produces a type of iron oxide NPs often referred to as SCIONs, i.e., silica coated iron oxide NPs. It should be noted that the polymeric shell has to be very opaque in order not to block the fluorescent silica-based core, if used simultaneous. Most commonly, experiments with only a polymeric or a silica core have been evaluated. The custom size of an iron oxide NP is in the range of 10–20 nm, as exemplified by a 10 nm in diameter iron core in the figure. Indicated in the figure are also the magnetic field lines created by this type of NP.

2.2. Gadolinium, Manganese, Gold, Silver, and Platinum Nanoparticles

For many years, gadolinium (Gd^{3+}, Gd(III)) has been used in different contrast media for magnetic resonance imaging (MRI) during, for example, angiography or brain tumor imaging due to its paramagnetic characteristics, ability to shorten the T1 relaxation time, and to cross a degraded blood-brain barrier [116]. The research on contrast media based on the paramagnetic element manganese (Mn^{2+}, Mn(II)) intended for MRI and NPs has just started to accelerate in recent years. For example, carbon and Mn^{2+} based NPs have been evaluated as contrast agents for MRI, or as manganese enhance MRI (MEMRI) during in vivo studies or for functional brain imaging [117–122]. Different applications such as Mn(II)–Au NPs as MRI contrast agents in stem cell labeling or Mn(II) based prussian blue($Fe_7(CN)_{18}$)-based NPs as a theranostic agent having ultrahigh pH-responsive longitudinal relaxivity have also been investigated [123,124].

Gold NPs (AuNPs) in the form of nano-cages, -spheres, -beacons, -stars, -shells, -seeds, -sheets, or nanorods is being evaluated in many different settings, for example, for both PCa and BC (Figure 2) [8,125–141]. By changing the AuNPs' shape, size, or surface characteristics it is possible to fine-tune their properties in order to maximize their applicability as a tool for cancer diagnosis, photo-dynamic/thermal therapy, therapy-drug carrier, radiotherapy drug enhancer, targeted gene therapy, or as a combined theranostic nanovehicle [142–155]. Regarding the gene therapy approach, in which for example, small-interference RNA (siRNA) could be utilized in order to knock out specific gene expressions in cancer cells, it has received increased attention recent years. For example, Guo et al. have recently shown interesting results on the PCa cells PC-3 and LNCaP indicating that two of their investigated formulations with transferrin and folate-receptor targeting ligands respectively (AuNPs-PEG-Tf and AuNPs-PEI-FA) show potential as non-viral gene delivery vectors in the treatment of PCa [141]. The photo-dynamic/thermal technique has also shown some progress recent years for both PCa and BC [156–165]. For example, Oh et al. have shown promising PCa cell killing efficacy by using a 55 nm small icosahedral phage that was engineered to display a gold-binding peptide as well as a PCa cell-binding peptide and applying a 60 mW/cm^2 light irradiation [156]. Mkandawire et al. have recently investigated an alternative way to treat BC cells inducing apoptosis by targeting their mitochondria using AuNPs during photothermal treatment [164]. Regarding tumor detection, surface enhanced Raman spectroscopy (SERS) can be used for in vivo applications, and some studies have been performed recent years investigating its applicability for PCa and BC [129,161,166–170]. Ramaya et al. investigated the overexpression of prostate specific antigen (PSA) in LNCaP cells by using tetraphenylethylene (TPE) appended organic fluorogens adsorbed on AuNPs. Indoline-based TPE-AuNPs were efficient recognizing PSA overexpressing LNCaP cells using SERS mapping. For BC for example, Butler et al. investigated MCF-7 (Michigan Cancer Foundation-7) cells incubated with 150 nm AuNPs and concluded that they were a good starting point for near-infrared (NIR) or infrared (IR) SERS analysis [168].

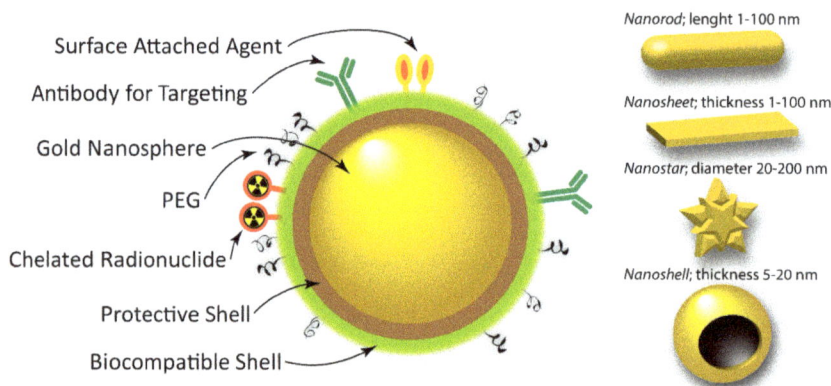

Figure 2. Image showing an example of a nanosphere of gold (Au). Gold nanoparticles (NPs) can also be based on, for example, nano-rods, -sheets, -stars, -or nanoshells, as indicated to the right with typical dimensions. The Au-nanospheres are most often produced in the size interval of 5 to 400 nm in diameter, approximately. The characteristics of AuNPs, and therefore their application possibilities, strongly depend on size, shape, and surface functionalities. Indicated in the image are also a protective and a biocompatible shell encapsulating the Au nanosphere, enabling the attachment of different kind of targeting vectors, imaging agents, therapeutic drugs, as well as polyethylene glycol (PEG) molecules. The latter an important parameter in order to protect the NPs from the immune system, to avoid reticuloendothelial system (RES) uptake, and to minimize nonspecific binding.

The use of silver NPs (AgNPs) for medical applications has not been as intense as that of AuNPs. Regarding PCa and BC, however, there have been some studies published for different applications during recent years [171–182]. For example, Wang et al. developed a Ag-hybridized-silica-NP-based electrochemical immunosensor for the sensing of PSA in human serum with promising results [172], and Swanner et al. investigated the radiosensitizing and cytotoxic effect on triple-negative BC using AgNPs with good results at doses that have little effect on nontumorigenic BC cells [178].

Platinum-based NPs (PtNPs) is also a technique with a relatively low number of publications for PCa and BC applications, and then most often in the context of AuNPs, FeONPs, or for immunosensor applications [183–192]. Cui et al. developed an immunoassay based on mesoporous PtNPs, evaluated its efficacy against the BC tumor markers CA125, CA153, and CEA, and concluded its high potential clinic value [185]. Spain et al. recently proposed an electrochemical immunosensor based on PtNPs conjugated to a recombinant scFv antibody for the detection of PSA during PCa diagnosis, and showed that picomolar PSA concentrations could be detected without the need for further PCR or nucleic-acid-sequence-based amplification (NASBA) techniques [188].

2.3. Quantum and Cornell Dots

The semi-conductor metal based NPs called quantum dots (QDs) have, due to their adjustable properties, been used and tested in many electronic applications such as LCD displays and solar cells [193,194]. However, lately they are also evaluated in cancer research and medical-imaging applications and are usually fabricated in sizes of approximately 2–7 nm in diameter [195–197]. Examples of such, from blood rapidly excreted NPs based on combinations of different semi-conductor and heavy metals, are CdSe, CdS, PbS ZnS, InP, and CdTe, having fluorescence emission spectra in the region of 450–950 nm depending mainly on size and which type of coating the QDs have for a certain application [198,199]. A common coating is polyethylene glycol (PEG), which has the effect of increasing the blood circulation time by decreasing the kidney clearance.

Due to concerns regarding the heavy metal involvement in the QDs alternatives have been developed. One such option is the silica based Cornell dots (C dots) [200], first invented by Uli Wiesner (Wiesner Group, Cornell University, USA). The spherically shaped C dots are constructed with a silica based core, in which fluorescent molecules (fluorophores) are embedded, surrounded by a PEGylated silica shell [201]. In order to turn these C dots into targeted cancer probes one possibility is then to label chains in the PEG molecule with peptides or (fraction of) monoclonal antibodies (mAbs) that are site specific for certain cancer cell receptors. If the cancer cell bound C dots are illuminated using a near-infrared light source they fluoresce and can then serve as, for example, optical guidance for surgeons. Besides this diagnostic and surgical tool the C dots can of course also be labeled with suitable anticancer drugs or radioactive isotopes, and therefore also serve as a nanovehicle for targeted cancer therapy. It should be noted that PEGylation of C dots causes them to be rapidly excreted via the kidneys, as opposed to the QDs, decreasing blood circulation times [202–205]. A first clinical trial performed at the Memorial Sloan Kettering Cancer Center (MSKCC) in New York, USA, investigated radioactive iodine-labeled 7 nm C dots on five metastatic melanoma patients regarding positron emission tomography (PET) traceability and toxicity. The results showed that, under the U.S. Food and Drug Administration's (FDA's) Investigational New Drug (IND) guidelines, these type of NPs were safe for the use in humans [204].

Regarding applications for PCa and BC there are some publications using QDs or C dots [86,206–209]. For example, Zhao et al. recently evaluated QDs in a Cerenkov-imaging PCa model. [210]. They developed three different near-infrared QDs and ^{89}Zr dual-labeled NPs and demonstrated the applicability of such self-illuminating NPs for imaging of lymph nodes and PCa tumors. For BC, applications of QDs is exemplified with a paper by Bwatanglang et al., in which they present results after investigating folic-acid functionalized chitosan-encapsulated QDs [208]. They found both enhanced binding affinity and internalization of their NP platform for folate receptor-overexpressing MCF-7 and MDA-MB-231 cells, and therefore concluded it to be a promising candidate for theranostic applications.

2.4. Carbon Based Nanoparticles

The research on NPs based on carbon and allotropes of carbon such as fullerenes and graphene (e.g., nanohorns or nanotubes) have increased in recent years (Figure 3) [57,58,211–215]. These kind of NPs have received increased attention due to their chemical stability, favorable surface chemistry, high drug loading capacity, as well as high degree of variability. When it comes to medical applications attention has been particularly directed towards applications such as drug delivery, photothermal therapy, and imaging. Examples of this kind of NPs are fullerenes (spherical (i.e., buckyballs), ellipsoidal, or tube-shaped), carbon nanotubes (CNTs) such as single-walled CNTs (SWCNTs), double-walled CNTs (DWCNTs), or multi-walled CNTs (MWCNTs), carbon quantum dots (CQDs), graphene quantum dots (GQDs), and graphene oxide (GO). Regarding fullerenes they have been evaluated and utilized as for both X-ray and MRI imaging contrast agents, but also in applications for bringing a therapeutic substance to its target, such as gene delivery [216]. Different forms of CNTs can all be produced and chemically modified enabling labeling with, for example, radioactive isotopes [217,218]. Although promising applications of CNTs have been shown the question regarding toxicity of this nanovehicle is still under debate [219]. For example, it has been demonstrated that under certain circumstances the CNTs are able to cross the cell membranes of healthy tissue and induce harmful inflammatory and fibrotic responses, as well as cell death [220–224]. It should be noted though, that elevated risks are especially connected to chronic exposure to CNTs, which is not the case for medical applications for which the administration is performed under a limited period of time.

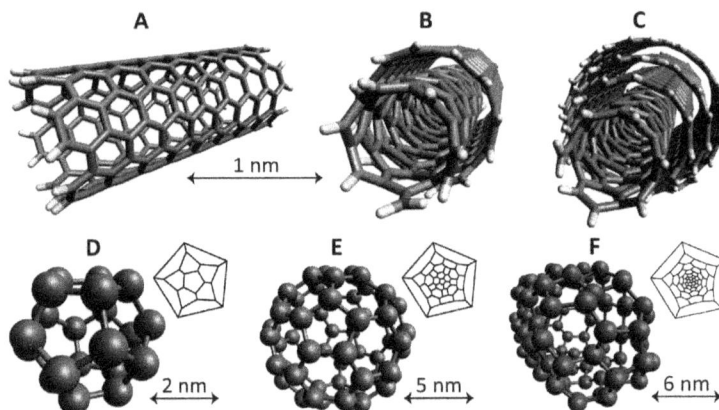

Figure 3. Examples of carbon-based nanotubes and fullerenes. (**A**) Single-walled carbon nanotube (SWCNT); (**B**) Double-walled carbon nanotube (DWCNT); (**C**) Multi-walled carbon nanotube (MWCNT); (**D**) Fullerene based on 20 carbon atoms (20-fullerene); (**E**) Fullerene based on 60 carbon atoms (60-fullerene); and (**F**) Fullerene based on 100 carbon atoms (100-fullerene). The top size indicator applies for panels **A**–**C**. The 3D-structures were created using Avogadro molecule editor. For panels **A**–**C** a rod-based representation, and for panels **D**–**F** a ball-based representation, was chosen for best clarity of the 3D-distribution of carbon atoms and the covalent bindings between them. For panels **D**–**F** is also shown 2D-representations of the fullerene structures, as well as individual size indicators. Note, the 100-fullerene has an obloid-like structure for its global energy minima, compared to the spherical 20- and 60-fullerenes, as discussed, for example, by Yoshida et al. [225].

Regarding PCa and BC applications utilizing carbon-based NPs there is an increasing number of publications during the last couple of years, both for imaging and therapy but also for different kind of electrochemical biosensor systems [226–234]. For instance, regarding PCa, Heydari-Bafrooei et al. and Pan et al. have both developed different kind of CNT-based biosensor systems able to detect prostate specific antigen (PSA) in serum and vascular endothelial growth factor (VEGF) and PSA in serum, respectively, for early diagnosis of PCa [227,229]. For example, regarding BC, Misra et al. developed a carbon NP-DNA complex (CNPLex) used to transfect green fluorescent protein (GFP) reporter gene containing plasmid DNA (pDNA) pEGFP-N1 targeting BC cells MCF-7 and MDA-MB231 with promising results [234].

2.5. Liposomes

The phospholipid, principally phosphatidylcholine, based bilayer structure, constituting the body of the spherical vesicle called liposome, can be arranged in such a way as to produce a small unilamellar liposome vesicle (SUV) (Ø < 100 nm), large unilamellar vesicle (LUV) (Ø = 100–1000 nm), giant unilamellar vesicle (GUV) (Ø > 1000 nm), multilamellar vesicle (MLV), or a cylindrically shaped nanocochleate vesicle (NCV). An example of a hypothetical spherical unilamellar liposome is shown in Figure 4. The MLV's are constructed by one or more unilamellar liposomes being encapsulated within a larger one. By disrupting the bilayer structure by ultra-sonication the liposomes can be prepared and loaded with, for example, different pharmaceutical drugs, either hydrophobic or hydrophilic. In such a way, the liposomes can act as vehicles for drugs directed against different diseases. The persistent and important work over several decades by the biophysicist Alec D Bengham on liposomes paved the way for its application as NPs in current biomedical research [235]. Depending on which type of liposome under consideration, the size ranges from around 20–100 nm (SUV), 100–1000 nm (LUV), to over 1000 nm for the MLV, and up to 200 μm for the GUV. The negative charges of the hydrophilic phospholipid heads on the surface of the liposome could be utilized for binding positively charged molecules and/or radioisotopes by electrostatic interaction [236]. Also, by specifically blocking these negatively charged heads using PEGylation, it is possible to increase the blood circulation time by minimizing the kidney clearance rate [237].

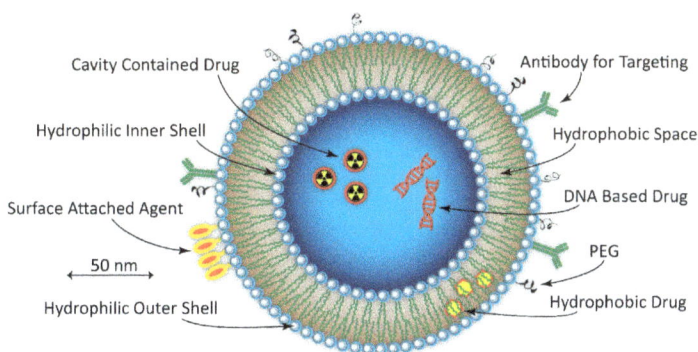

Figure 4. Schematic presentation of a spherical PEGylated liposome. Indicated are the different available spaces and surfaces that could carry different kind of targeting molecules as well as therapeutic drugs and imaging agents. In this hypothetical example, an antibody chelated to the outer surface is used for targeting. On that surface is also attached an imaging agent. Some kind of therapeutic drug is carried by the hydrophobic space between outer and inner shell. And in the core cavity, two hydrophilic drugs are situated, here exemplified by a radionuclide and a DNA-based drug such as strings of DNA, RNA, or small-interference RNA (siRNA). Unilamellar liposomes lies mostly in the size range of >150 nm, exemplified by a 200 nm liposome in the figure.

Studies utilizing liposomal-based NPs for PCa and BC theranostic techniques are few [238–247], although applications for therapy or imaging alone are significantly more frequently published [248–261]. For example, Yeh et al. recently published results from experiments investigating peptide-conjugated liposomal NPs in a theranostic approach for PCa. They found that the administration of liposomal doxorubicin and vinorelbine conjugated with targeting peptides increased the inhibition human PCa growth, and concluded that the targeting peptide SP204 has significant potential for both targeted imaging and therapy of PCa [238]. A liposomal-based theranostic approach against BC could be exemplified by the work by Rizzitelli et al. in which they evaluated the release of doxorubicin from liposomes monitored by MRI and triggered by ultrasound stimuli. The treatment led to a complete tumor regression in their BC mouse model [241].

2.6. Polymer Based Nanoparticles

These type of biodegradable block-copolymer based NPs can encapsulate and carry relatively large pharmaceutical molecules such as proteins, individual genes, or pieces of DNA [262]. Examples of polymers used for these types of NPs are polycyanoacrylate (PCA), poly D,L-glycolide (PLG), polylactic acid (PLA), polylactide-*co*-glycolide (PLGA), poly(isohexyl cyanoacrylate) (PIHCA), or polybutyl cyanoacrylate (PBCA) [262]. Among these polymers, PBCA and PIHCA have shown to be the fastest regarding biodegradability. For example, 24 h post an intravenous injection of PBCA it showed a level of reduction in the order of 80% [263]. PIHCA is currently undergoing clinical trials in phase III for hepatocellular carcinomas using the drug doxorubicin (Livatag® (Doxorubicin Transdrug™), Paris, France). Three main NP structures, or micelles, can be achieved using polymers, namely nanocapsules, nanospheres, and nanoparticles. In the first case the pharmaceutical is encapsulated and completely surrounded by a spherical or rod-shaped shell of block-copolymers. In the second case, the pharmaceutical is embedded in a polymeric spherically shaped matrix. In the last case, the pharmaceutical is attached on the surface of the polymer based nanostructure [262]. The encapsulated, embedded, or attached pharmaceutical can of course be a radioactive agent of some sort [264]. Covalent pegylation of polymeric NPs can increase blood circulation times as well as facilitate uptake of the drug at targets aimed for [265]. Dendrons and symmetrical dendrimers (Figure 5) are a type of branched polymer based macromolecule that could be used as NPs [266,267]. Dendrimers have attracted much attention regarding theranostic applications, and the subject could easily fill a review on its own [268–270]. For example, hyperbranched PAMAM (polyamidoamine) dendrimer, based on hydrophilic ethylene diamine and investigated for medical purposes, can be labelled with monoclonal antibodies due to suitable amino groups in its structure [271]. The PAMAMs can be produced with multifunctional terminal surfaces and a narrow molecular weight distribution [272]. Finally, polymersomes is a type of polymeric NP for which amphiphilic synthetic block-copolymers are utilized to construct the membrane of the vesicle. Polymersomes have many similarities with liposomes, built by natural lipids (see above), but exhibit decreased permeability and increased stability compared to liposomes. The polymersomes have a size span of approximately 50–5000 nm.

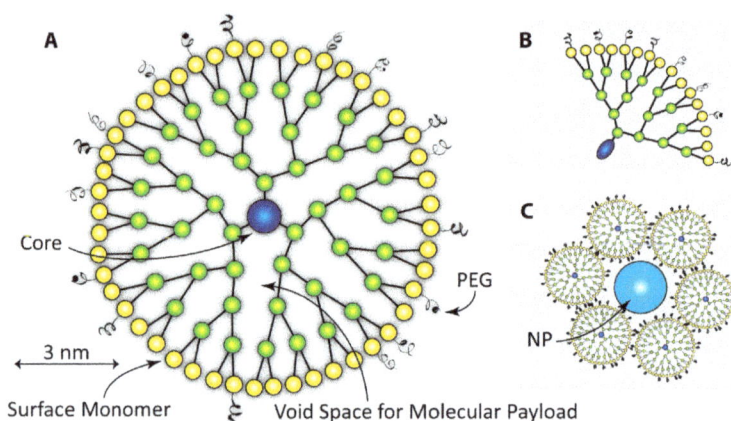

Figure 5. Schematic presentation of one type of spherical PEGylated dendrimer based on a central core (**A**), which in itself can be a dendrimer or some other type of NP (i.e., if so, a NP-cored dendrimer (NPCD)). A standard size of a dendrimer shown in A is 5–10 nm, exemplified by a 10 nm dendrimer in the figure. The surface monomers enable attachment of imaging and/or drug payloads, also able to be entrapped in the void space. The network of covalently bound interior monomers connects the surface monomers to the core. The number of radially emerging branch points defines the generation of the dendrimer; in this case four, denoted G-4. If instead, NPs are situated inside the network of monomers the term used is dendrimer-encapsulated nanoparticles (DENPs). Hyper-branched polymers are a variant similar to dendrimers, except that the branches emanating from the core differ from each other with regard to number of branch points. To the right is shown an example of a dendron (**B**), i.e., a small fragment of a dendrimer that also could be used as a NP. An example of a dendrimer-stabilized nanoparticle (DSNP) is shown in the lower right corner (**C**). Depending on if the functionalization is located on the surface, in the void space, or at the core of the dendrimer it is usually denoted as surface, interior, or core functionalized, or as a combination of all three possibilities.

There are very few theranostic polymer-based NP studies reported for both PCa and BC. In one such study Ling et al. evaluated multifunctional dual docetaxel/superparamagnetic iron oxide (SPIONs) loaded polymer vesicles (147 nm in diameter) for both imaging and therapy of PCa [273]. Enhanced cellular uptake and anti-proliferative effect for the PC3 cell line was observed which, in conjunction with the SPION-based MRI possibility, made the authors conclude that these polymer-based NP vesicles were promising for simultaneous imaging, drug delivery, and real-time monitoring of the therapeutic effect. For BC, a theranostic polymer based technique has been published by Abbasi et al. in which they experimented with Mn-oxide and docetaxel co-loaded fluorescent polymer-based NPs for dual-modal imaging and therapy of BC [274]. The authors concluded this type of polymer-based NP as a good candidate for cancer theranostic applications. Other interesting, however not yet fully theranostic, applications of polymer-based NP for PCa and BC have been published [275–279].

2.7. Solid Lipid Nanoparticles

The methodology of solid lipid NPs (SLNPs, or SLNs) is a promising emerging research field of lipid nanotechnology [280–285], which offers good possibilities to incorporate drugs into nanosized targeted vehicles having great biotolerability and low biotoxicity due to their constitution of physiological lipids [286]. Examples of such lipids are mono-, di-, and tri-glycerides, steroids, and fatty acids. Among the advantages of SLNPs could be mentioned the possibility of incorporating both hydro- and lipophilic drugs, good drug stability, and the lack of organic solvents in its composition [286]. The size of the SLNPs vary between 10 and 1000 nm, and the most common geometrical shape is spherical (Figure 6). The core is composed of solid lipids which is stabilized by emulsifiers, which also has the effect of decreasing the risk of NP agglomeration [286,287].

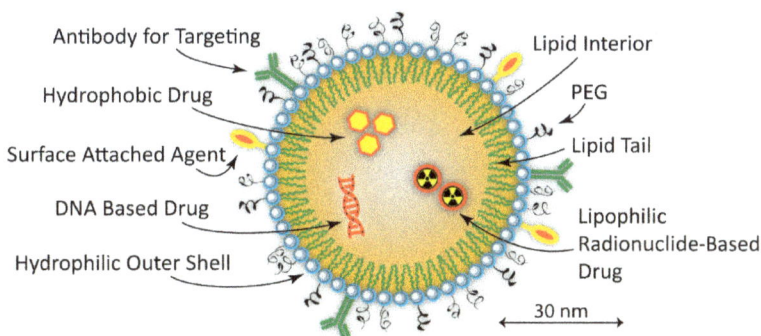

Figure 6. Solid lipid NP (SLNP). Although liquid lipid NPs are possible to produce, the most common form of lipid-based NPs investigated for medical purposes are SLNPs. The lipids most often used are fatty acids or different forms of glycerides. The smallest forms of SLNPs are in the shape of micelles, in which the fully dehydrated tails of the phosphatidylcholine molecules meet in the center, producing SLNPs in the size of 10 nm in diameter. Typical sizes for SLNPs are, however, most commonly in the interval of 5 to 500 nm in radii, exemplified by a hypothetical SLNP with a radius of approximately 30 nm in the figure.

Until today, no fully theranostic application has been reported for neither PCa nor BC utilizing SLNPs. However, some studies have been published investigating the possibility using this NP system for either imaging or therapy alone [288–298]. For example, Radaic et al. investigated the possibility of gene therapy using SLNPs and tested the capacity of their NPs to accommodate DNA (and withstand DNase degradation), their colloidal stability and in vitro cytotoxicity, as well as the transfection efficiency in PCa cells [288]. For BC, Jain et al. recently published results investigating the anticancer efficacy of lycopene loaded SLNPs [294]. They found that the concentration and time dependent cell survival of MCF-7 BC cells was significantly reduced by LYC-SLNPs, as compared to their free lycopene counterparts.

3. Multimodal Imaging Options for Prostate and Breast Cancer

The use of various biomedical imaging techniques in preclinical and clinical applications, during both diagnosis and treatment has increased tremendously during the past decades, and is now considered a central part in many of such applications. Computed tomography, positron emission tomography, single photon emission computed tomography, magnetic resonance imaging, ultrasound imaging, Cherenkov luminescence imaging, photoacoustic imaging, and optical imaging are examples of images techniques used, and still under development. The images these different technologies produce enable early detection, screening, image-guided treatment, as well as the possibility of estimating the level of progression or retrogression of the disease investigated [299]. By using a NP-based targeted vehicle it is possible to use combinations of these imaging techniques simultaneously in order to increase the level of accuracy, possibly also at a cellular or even a molecular level. This NP-based multifunctionality in biomedical imaging could be referred to as multimodal imaging (MMI). Most often, only two imaging techniques is utilized simultaneously, which then is referred to as bimodal imaging (BMI). Below follows a condensed presentation of each of these different biomedical imaging techniques and their basic characteristics. When applicable, examples on how they have been used in NP-settings for PCa and BC applications so far, e.g., in MMI or BMI contexts, are shortly mentioned and referenced to.

3.1. Computed Tomography and Tomosynthesis

Regarding computed tomography (CT), contrast enhancers such as barium or iodine compounds (e.g., Gastrografin) could be used to increase the absorption of X-rays and thereby enhancing the contrast of tissues in an image, taking up these contrast agents. A targeted NP platform carrying such contrast enhancers could thus improve the anatomical visualization of organs and other structures, compared to non-targeted regions. But also some NPs themselves can improve the contrast, exemplified by AuNPs increasing the contrast approximately three times as compared to the same amount of iodine [300]. Differently sized AuNPs have been evaluated for micro CT for example, and some are under the consideration for approval for use in the clinic [301–303]. A low-density lipoprotein (LDL)-based iodinated nanoparticle targeting the LDL receptor (LDLR), expressed in PCa, for example, has also been evaluated for CT imaging [304–306], as well as polyvinyl pyrrolidone coated bismuth sulfide (Bi_2S_3) based nanocrystals [307]. For tomosynthesis, the same general principle applies as described above for CT, namely that the contrast in an image could be increased approximately three times using AuNPs as compared to using the same amount of iodine-based contrast agent. For contrast-enhanced breast tomosynthesis (CE DBT), temporal or dual-energy subtraction techniques are of course still possible to use regardless of which type of contrast agent is utilized [308].

Most applications of CT in NP contexts are based on positron emission and single photon emission computed tomography (see below). However, one example in which CT has been used is described by Kim et al., in which they investigated a multifunctional gold-based NP system for both contrast-enhanced imaging and therapy of PCa [309]. RNA-aptamer functionalized gold NPs targeting PSMA enabled specific imaging of PCa. When also loaded with doxorubicin their theranostic NPs showed good therapeutic efficacy against LNCaP cells. For BC, a study performed by Naha et al. evaluated gold silver alloy NPs as an imaging probe for BC screening using conventional CT as well as dual-energy mammography (DEM) [310]. In vivo experiments in mice exhibited good tumor accumulation of the NPs and produced high contrast DEM and CT images, enabling the authors to conclude that their NP system has potential for both blood pool imaging and BC screening.

3.2. Positron Emission and Single Photon Emission Computed Tomography

Radio imaging using positron emission tomography (PET) and/or single photon emission computed tomography (SPECT) they have been evaluated for several NP applications using both organic and inorganic NPs, for example, in PET/CT or SPECT/CT settings [311–314]. Using a NP platform makes it possible to increase the contrast in an image due to the possibility to label each NP with a large number of radionuclides. For PET, examples of radionuclides used in such applications are ^{18}F, ^{124}I and ^{64}Cu, and for SPECT examples are ^{125}I and ^{125}Cd.

Regarding PCa, PET (or PET/CT) and SPECT (or SPECT/CT) has been used in several NP applications [131,315–321]. For example, Pressley et al. evaluated an amphiphilic NP for natriuretic peptide clearance receptor (NPRC) targeting and DOTA (1,4,7,10-tetraazacyclododecane-1,4,7,10-tetraacetic acid) chelator for high specific activity ^{64}Cu PET-radiolabeling. PET/CT images revealed high blood pool retention, low renal clearance, enhanced tumor uptake, and decreased hepatic burden relative to a nontargeted NP version [317], indicating the possibility of a new nanoagent for PCa PET imaging, according to the authors. In NP-based BC applications, PET and SPECT have been utilized in several studies [322–326]. For example, Lee et al. recently published a study in which they assessed the EPR effect in nineteen patients with HER2-positive metastatic BC using a ^{64}Cu-labeled NP (^{64}Cu-labeled HER2-targeted PEGylated liposomal doxorubicin) using PET/CT [322]. The authors concluded that the results provide evidence and quantification of the EPR effect in human metastatic BC tumors, as well as support NP-deposition imaging as a potential technique for identifying patients well-suited for NP-based therapeutics.

3.3. Magnetic Resonance Imaging

The magnetic resonance imaging (MRI), or magnetic resonance tomography (MRT), technology is today widely used for imaging physiological processes as well as the anatomy during both preclinical research and in the clinic [327]. Varieties of this technique includes functional MRI (fMRI), measuring levels of cerebral blood flow, as well as techniques to increase the contrast in MRI images such as dynamic contrast-enhanced MRI (DCE-MRI), utilizing a contrast agent such as gadolinium, and diffusion-weighted MRI (DW-MRI), utilizing the Brownian motion of water molecules. The MRI tool plays an important role in the staging of PCa (see below in section Prostate Cancer) and will most likely be applicable and show an increased importance also for other types of cancers, such as BC, for both diagnosis and staging, especially if used in a NP setting, discussed in this paper.

The ^{19}F-based MRI [328–332] offers many advantages, despite the difficulty of providing suitable non-toxic ^{19}F-based compounds in sufficient amounts for in vivo imaging, compared to common proton-based ^{1}H-MRI such as decreased background levels, quantitative determination of pharmacokinetics, and estimation of tissue oxygenation [300,333,334]. The use of ^{19}F-MRI in NP applications is a developing research field investigating, for example, the applicability compared to SPIONs, the efficacy of fluorinated dendrimers and multifunctional micelle-based core–shell NPs, and the detection of folate-receptor positive tumors in a ^{19}F/florescence-based bimodal imaging setting [331,335–337]. Just recently, a ^{19}F based nanoemulsion has been FDA approved for noninvasive clinical cell-tracking imaging [338].

Regarding MRI/MRT used in PCa related NP settings the number of publications is constantly increasing [104,339–344]. Jin et al., for example, evaluated MRI-guided focal NP-based (porphysomes) photothermal therapy (see below) in an orthotropic PCa mouse model, and concluded that it might be an effective and safe technique to treat PCa, with a low risk of progression of disease [343]. For BC, there is also a large number of publications [91,345–349]. In a study by Turino et al. L-ferritine-coated paclitaxel- and Gd-loaded NPs were evaluated for simultaneous delivery of a therapeutic drug and a MRI-contrast agent in a MCF7 BC model [345]. According to the authors, the theranostic potential of this NP system was demonstrated by, for example, evaluating signal-intensity enhancements in T1-weighted MRI images.

As for CT, by combining the MR-imaging mode with PET enables localization and biodistribution at the same time. Very few studies are however reported until today using this technique in NP settings for PCa and BC [350–352].

3.4. Ultrasound Imaging

During ultrasound imaging (USI), utilizing contrast agents to improve the contrast, particle sizes used often exceed 250 nm. Although the most commonly used size definition for NPs is 1–100 nm, the USI contrast agents is still appropriate to mention here since USI can play an important role in MMI. Two examples of commercially available USI contrast agents are Definity and Optison based on microbubbles. Both type of bubbles contain octafluoropropane gas, while Definity uses a phospholipid spherical shell and Optison an albumin based shell to enclose the gas. Experimentally, SPIONs, liposomes, AuNPs, and nanodroplets have also been evaluated as contrast agents and drug carriers during NP-based US applications [353–358]. All types of microbubbles can serve as platforms for not only imaging, but also for the distribution of drugs for therapy by encapsulating the drug into the bubbles [359,360]. Targeted microbubbles for use during both imaging and therapy, or at the same time as targeted theranostic vehicles, can be achieved by attaching ligands on the surface of the bubbles which specifically bind to receptors on, for example, tumor cells.

In the study by Tong et al. mentioned above for PCa, a protamine cationic microbubble ($\varnothing \approx$ 500 nm) was constructed for simultaneous gene therapy (androgen-receptor siRNA) and ultrasound imaging [360]. The authors concluded that the gene-transfection efficiency was better than that of a liposomes-based comparable system, and that their microbubble system could be used as a gene-loading and ultrasound-imaging technique of tumors. For BC, Zhao et al. recently investigated a near-infrared (808 nm) photothermal responsive dual aptamers-targeted docetaxel-containing NPs for both cancer therapy and USI [353]. The dual-ligand functionalization increased uptake in MCF-7 cells and made USI possible at tumor site. The authors therefore concluded their system to be a promising theranostic option involving light-thermal response, dual ligand targeted triplex therapy (chemotherapy, photothermal therapy, and biological therapy), and USI.

3.5. Cherenkov Luminescence Imaging

The Cherenkov luminescence imaging (CLI) technology, named after Pavel Alekseyevich Cherenkov who shared the Nobel Prize in physics in 1958 for the discovery of the now called Cherenkov effect, has during recent years emerged as an alternative imaging technique for several different applications [361–367]. The electromagnetic Cherenkov radiation, emitted when charged particles passes through a medium at a speed greater than light propagates through the same medium, could be used to image, for example, the uptake of a charged-particle emitter in tumors during radioimmunotherapy. There are very few publications on PCa and BC using CLI [368–371], among which very few have a NP approach. Lohrman et al. found a positive correlation between the radioactivity uptake and CLI signal from the ^{90}Y labeled gastrin-releasing peptide-receptor (GRPR) antagonist DOTA-AR in xenografted PCa tumors [370]. Regarding using CLI in NP applications besides PCa and BC there are some publications [372–375]. For example, Madru et al. investigated the usability ^{68}Ga-labeled SPIONs for multi-modality PET/MR/CLI imaging of sentinel lymph nodes (SLNs). Based on promising biodistribution experiments, the authors concluded that ^{68}Ga-SPIONs can enhance the identification of SLNs by combining PET and MR imaging, and potential also enable Cherenkov luminescent-guided resection of SLNs [372].

3.6. Photoacoustic Imaging

A relatively new emerging imaging modality is photoacoustic imaging (PAI), although already with some PCa and BC orientated studies published the past decade [376–383]. PAI could be achieved by irradiating a biological site with a pulsed laser-beam in the megahertz range, which energy is then absorbed by, for example, targeted NPs. The absorbed energy creates acoustic pressure waves, caused by thermoelastic expansion, in the irradiated site. These waves could then be detected by using an ultrasonic transducer, enabling an image to be constructed. If instead of using a laser, pulses in radio-frequency range are used, the technique is called thermoacoustic imaging (TAI). Su et al. investigated recently mesoporous-silica coated and PEG modified multifunctional doxorubicin-loaded prussian-blue nanocubes (PB@mSiO$_2$-PEG/DOX). Good both MRI and PAI ability, as well as a synergistic photothermal and chemical therapeutic efficacy, for BC was found [377]. For PCa, Levi et al. evaluated a PAI agent (AA3G-740) targeting the gastrin-releasing peptide receptor (GRPR), highly overexpressed in PCa [380]. The study showed that, even for poorly vascularized tumors, AA3G-740 was able to bind to GRPR and led to a significantly higher photoacoustic signal relative to a control agent.

Regarding NP applications using PAI there are some studies published for both PCa and BC [230,382,384–391]. By using PCa cells, Tian et al., for example, constructed PEGylated and RGD (Arginine-Glycine-Aspartic)-peptide functionalized AuNPs and evaluated them as a contrast agent for PAI of single PCa cells. The authors concluded that these NPs provide a platform for detection and imaging of individual cancer cells, with a potential impact on clinical diagnostic [384]. Pham et al. used mouse models based on orthotopic primary triple-negative BC xenografts (including patient-derived xenografts) to evaluate the efficacy of bevacizumab (a VEGF pathway-targeting antiangiogenic drug) in combination with CRLX101, an NP-drug conjugate containing camptothecin (a cytotoxic quinoline alkaloid inhibiting DNA topoisomerase). PAI was used in the study to show that the use of CRLX101 led to an improved tumor perfusion as well as reduced hypoxia. The authors concluded that pairing antiangiogenic therapy with a cytotoxic NP construct may be a promising way to treat metastatic BC [391].

3.7. Optical Imaging

The Optical imaging (OI) technology (sometimes referred to as biophotonics), the inclusive term often used for different types of infrared, ultraviolet, and visible light, as well as sometimes also photoacoustic (see above) applications in biomedical imaging, have a number of interesting publications regarding NP applications for both PCa and BC [319,392–400]. Generally, a NP-based OI approach could enable or optimize an optical excitation energy in, for example, tumor tissue, enable multispectral imaging by combining spectroscopy and OI, or make possible multiplex imaging by using different color emitters for different targets simultaneously [300]. For cancer applications, Ahir et al. recently developed a copper oxide-nanowire NP decorated with folic acid and studied its effect on triple negative BC (TNBC). They found that their NPs induced apoptosis and retarded migration of the TNBC cells, and used optical fluorescence imaging to monitor its distribution in tumors and different organs [392]. Regarding PCa, Behnam et al. constructed and investigated a PSMA-targeted bionized nanoferrite (BNF) NP in an experimental PCa model [319]. The study used near-infrared fluorescence microscopy, SPECT, Prussian blue staining, immunohistochemistry, and biodistribution to show an enhanced NP uptake in PSMA-positive tumors, with a maximum uptake 48 h post injection.

3.8. Electron Microscopy

Regarding electron microscopy (EM) in general, but especially transmission electron microscopy (TEM), it has an important role to play for the in vitro and ex vivo analyses due to its often sub-nanometer spatial resolution. Nanoparticles based on heavy elements such as gold could therefore be used for TEM applications in order to retrieve information on, for example, NP distribution on the organelle level. So far though, this technique has only been used occasionally in PCa and BC NP applications. Since EM is not considered to be an imaging technique possible to use in theranostic contexts, its value lies instead in in vitro and ex vivo analyses, as mentioned above, or during the production process of the NPs in order to be able to characterize them properly [401–408].

4. Multimodal Therapy Options for Prostate and Breast Cancer

Due to differences in metabolic and chemical stability, level of solubility in blood serum and interstitial fluid, degree of toxicity, and most important level of specificity for a certain tumor type as well as potency once properly targeted, several different drugs have been evaluated and some approved for targeted therapies against cancer [409,410]. For a detailed compilation of the NP-based technology and therapy of cancer in general the reader is referred to Professor P.N. Prasad's fine textbook on the subject [300]. A NP-based therapeutic, in some cases also potentially synergistic, multifunctionality can be referred to as multimodal therapy (MMT), or if only two therapies are used simultaneously, as bimodal therapy (BMT). Below follows a presentation of different therapy options that all could be implemented in various NP settings. The basic characteristics and principles of each modality are presented briefly. References are also listed and some specific examples on how some of the available therapy modalities have been utilized in NP-settings for PCa and BC are discussed shortly, e.g., in MMT or BMT contexts.

4.1. Chemotherapy

Treatments using chemotherapy (CTH), since many years successfully applied and still under development for a wide category of cancers including PCa and BC, has limitations due its relatively high degree of non-specificity, inducing toxicity [411–415]. A targeted NP-based approach has shown to be beneficial and has been evaluated by many research teams. Some such NP-based CTH drugs have been FDA approved; the albumin-paclitaxel-based Abraxane® and the PEG-doxorubicin-based Doxil® for metastasized BC, the latter being the first FDA approved nanodrug [416,417]. But also other formulations are being evaluated in the clinic, or have already been approved or are being marketed in, for example, Europe. Such examples for BC is the non-pegylated liposomal-doxorubicin-based Myocet® or the polymeric micelle-paclitaxel-based Genexol-PM® [418,419]. An update from 2014 of FDA approved NP-based cancer drugs, and also others at various stages of development, has been published [420]. Regarding targeting of NP-based CTH drugs research are ongoing in order to, instead of relying on the passive targeting caused by the EPR effect, develop strategies for active targeting using, for example, mAbs (or fraction of mAbs) directed against the PSA receptor in the PCa case [421].

However, a large number of studies, using different techniques, have been published with a NP and CTH-based approach for both PCa and BC, of which only a tiny fraction are listed here [105,253,315,353,422–431]. For example, Belz et al. recently designed ultra-small silica NPs containing the radiosensitizing drug docetaxel for combined chemoradiation therapy, with potential benefit for patients with PCa [425]. For BC, Li et al. found synergistic inhibition of both migration and invasion of 4T1 BC cells by doubly loaded NPs (docetaxel + the Akt inhibitor quercetin), via the Akt/matrix metallopeptidase 9 (MMP-9) pathway [431].

4.2. Gene Therapy

The gene therapy (GTH) alternative has during the last two decades evolved as a promising tool for the treatment of cancer, either as a stand-alone therapy or in conjunction with chemotherapy, surgery, and/or radiation therapy [432,433]. The development of GTH towards treatment based on an individual's specific genome, immune status, and tumor characteristics, together with new vectors for transferring the genetic material such as synthetic viruses as well as non-viral methods will further refine this still experimental, treatment option [434]. By adopting an NP-based GTH approach, the treatment is believed to be improved even further, especially when implemented as TNPs enabling imaging simultaneous with therapy. For the two main groups of genes associated with cancer, i.e., tumor-suppressor genes and oncogenes, examples of nucleic-acid based therapeutic molecules that are being evaluated for GTH are cytotoxic or corrective genes, small interfering RNA (siRNA) or short hairpin RNA (shRNA) [435].

Regarding NP-based GTH techniques for BC there are very few, however an increasing number, publications the last decade [234,436–444]. Su et al., for example, investigated recently the efficacy of a combinational technique including photothermal therapy (see below), CTH, and GTH for triple negative BC [442]. Indocyanine green, paclitaxel, and survivin siRNA was integrated into a NP and was found to exhibit very good tumor efficacy with low toxicity. The protein survivin, encoded by the *BIRC5* oncogene in the human genome, and which inhibits the caspase activation and therefore downregulates the apoptotic pathway, has received much attention lately. Several attempts, also including NPs, have been made to distribute anti-survivin siRNA in order to silence the BIRC5 gene [445–447]. For NP-based GTH applications for PCa, there are also quite few publications the last decade [141,448–452]. Guo et al. investigated, for example, gene silencing using siRNA-based AuNPs for LNCaP cells, overexpressing PSMA. With AuNPs conjugated with folate-receptor targeting ligands it was found that siRNA was specifically delivered into the LNCaP cells, and produced enhanced endogenous gene silencing [141].

4.3. Photon Activation Therapy

The photon activation therapy (PAT), sometimes also referred to as photon activated therapy, involving Auger electrons and mentioned for the first time for medical applications over three decades ago, shows a limited amount of publications but is an interesting option for TNPs and therefore discussed here [453–461]. The PAT technique is based on the principle of specific tumor localization of a high-Z compound such as platinum (Pt), incorporated in, for example, a CTH drug, after which synchrotron radiation or X-rays directed against the tumor site is used to, via the photoelectric effect, trigger a cascade release of high linear energy transfer (high-LET) Auger electrons. Except Pt, other nuclides investigated for PAT have been Au, Tl, Gd, I, and Fe. In the Pt case, the photon energy suitable for triggering this effect should be just above the binding energy of the K-shell electrons, i.e., 78.4 keV. As for α-particles, the mean LET value for Auger electrons is considerable higher than that of, for example, beta-particles; ~100, ~15, and ~0.2 keV/μm, respectively. This means that, as for α-particles, the Auger electrons will create densely ionization tracks causing damages in the cells, such as double strand breaks (DSB), which are very difficult to repair. An additional advantage with Auger electrons, compared to α-particles, is that their range in tissue is on the nanometer scale, compared to 50–100 μm for α-particles. So, provided that the Auger-electron emitting nuclide is being properly targeted in close proximity to, or incorporated into, the DNA of the tumor cells, a highly targeted high-LET irradiation will be achieved.

Regarding PCa and BC utilizing a NP-based PAT approach there is only one study published, having only a tentative BC relevance [462]. In that experiment Choi et al. investigated the therapeutic efficacy on colon cancer tumor-bearing mice injected with FeO NPs and irradiated using 7.1 keV synchrotron X-rays, an energy near the Fe K-shell binding energy. For example, one group that received FeO NPs and an absorbed tumor dose of 10 Gy showed 80% complete tumor regression after 15–35 days. As noted by the authors however, the use of 7.1 keV X-rays, having a high tissue attenuation, makes the treatment only suitable for superficial skin malignancies, and possibly also for superficial chest wall recurrence of BC.

4.4. Photodynamic Therapy

Photodynamic therapy (PDT), also referred to as photochemotherapy, utilizes a photosensitizing chemical substance called a photosensitizer (PS) that is irradiated with light at certain wavelengths to induce the production of molecular oxygen in the form of reactive oxygen species (ROS). The ROS, e.g., superoxide, peroxide, singlet oxygen, or hydroxyl radicals, have the capacity to induce cell death at the site of production and can therefore, if targeted properly, be used as a therapeutic option against several diseases [463]. Acne and psoriasis, or to some extent even herpes experimentally, are treated using PDT. But also different type of cancers, in particular different types of skin cancer, are being treated with PDT techniques [464]. Both wavelength and fluence of the light are important parameters

Int. J. Mol. Sci. **2017**, *18*, 1102

to monitor in order to target and trigger the PS properly, using a laser-equipped endoscope as a special case for reaching, for example, intestinal cancers [465].

An interesting version of PDT is the two-photon excitation (TPE) based PDT for the treatment of cancer. This technique combine the advantages of TPE near-infrared (NIR) photosensitizers and nanotechnology and has been reviewed by Shen et al. [466]. The absorption of two relatively low-energy NIR photons will enable the emission of high-energy photons in the visible spectrum, which in its turn will sensitize oxygen producing singlet oxygen and reactive oxygen species (ROS) able to kill targeted cancer cells due to its cytotoxic effect. Compared to single-photon based PDT the possibility of reaching further into tissues, due to the relatively long wavelength of the light used in TPE PDT, has great advantages enabling to reach tumors more deep seated. There are some publications using the TPE technique, both for imaging and therapy, in different NP and theranostic settings [467–471]. The paper by Gary-Bobo et al. [471] was the first two-photon based PDT experiment in vivo using NPs.

Regarding PCa, PDT approaches have been evaluated both preclinically and for patients [472,473]. Especially, studies using PDT as a theranostic approach for PCa has lately also been published. For example, Chen et al. recently investigated a low-molecular-weight theranostic photosensitizer denoted YC-9 for PSMA-targeted optical imaging and PDT [474]. The study indicates that YC-9 is a promising therapeutic agent for targeted PDT of PSMA-expressing tissues, such as PCa. Similarly, Wang et al. synthesized two PSMA-targeting PDT conjugates (PSMA-1-Pc413 and PSMA-1-IR700), both having the potential to aid in the detection and resection of PCa [475]. Lin et al. developed a novel nano-platform for targeted delivery of heat, ROS, as well as the heat shock-protein 90 (Hsp90) inhibitor for the treatment of PCa [476]. Vaillant et al. investigated targeting a membrane lectin using a mannose-6-phosphate analogue grafted onto the surface of functionalized mesoporous silica NPOs [477].

For BC, several approaches have been evaluated. For example, Feng et al. investigated a multimodality theranostic agent based on mesoporous copper sulfide NPs encapsulating doxorubicin, enabling both PAI as well as chemo- and ROS generating phototherapy of BC [400]. Against TNBC, Choi et al. developed photosensitizer-conjugated and camptothecin-encapsulated hyaluronic acid NPs as enzyme-activatable theranostic NPs for near-infrared fluorescence imaging and photodynamic/chemo dual therapy [478]. Both in vitro and in vivo, Wang et al. performed experiments evaluating the effects of sinoporphyrin sodium-mediated PDT on tumor cell proliferation and metastasis for the highly metastatic 4T1 BC cells and a mouse xenograft model [479]. Targeting the TrkC (tropomyosin receptor kinase C) receptor, which tends to be overexpressed in metastatic BC, with a ROS photosensitizer-labeled small molecule enabled Kue et al. to investigate the therapeutic efficacy of PDT in nude mice [480]. Finally, Shemesh et al. used a liposomal-based theranostic delivery system, with indocyanine green as a photosensitizer, for investigating real-time biodistribution monitoring as well as the efficacy of PDT against TNBC [244].

4.5. Photothermal Therapy

The photothermal therapy (PTT) approach builds on the PDT principle (see above) in that it via passive (e.g., via the EPR effect) or active (e.g., via mAbs) tumor accumulation of nanoheaters/photosensitizers enables a localized temperature increase. This could cause the destruction of DNA/RNA molecules as well as proteins, leading to cell death by membrane rupture or necrosis [300,481]. The difference of PTT compared to PDT is that the former does not need oxygen present in order to induce cell killing. Especially one version of PTT has attracted increased attention, namely plasmonic PTT (PPTT) [482–484]. The PPTT technology is based on the principle that when AuNPs are irradiated using infrared or near-infrared light coherent excitation of its conduction electrons at the surface will take place, due to the surface plasmon resonance (SPR) effect. When these electrons deexcite, they will produce localized heat waves causing wanted cell destruction.

Regarding PCa and BC there are several publications investigating PTT in different NP settings [91,156,228,230,247,343,422,485,486]. For example, Hosoya et al. evaluated a theranostic hydrogel-based NP platform combining both targeting of the tumor cells, photon-to-heat conversion, as well as triggered drug delivery enabling controlled release of the anticancer drug and multimodal imaging [247]. Their results showed the possibility of simultaneous targeted delivery of an anticancer drug and noninvasive imaging for both PCa and BC in a mouse model. Also, Cantu et al. investigated polymeric NPs (<100 nm in diameter) in a photothermal ablation setting. When experimenting on MDA-MB-231 BC cells they were able to show complete cancer cell ablation in vitro using an 808-nm laser, indicating the potential benefit of their NP platform utilizing the PTT technology.

4.6. Radioimmunotherapy

The cancer treatment modality termed radioimmunotherapy (RIT) is since many years a well-established technique to specifically irradiate targeted tumor cells using monoclonal antibodies (mAbs), or fraction of mAbs, labeled with suitable radioactive isotopes such as α-, β-, or Auger-electron emitters. Review papers regarding the current RIT status for PCa and BC is referred to for further reading [487–492]. Regarding NP-based platforms for cancer utilizing RIT, sometimes referred to as radioimmunonanoparticles (RINPs), there is a small but increasing number of publications [493–496]. For PCa and BC, there have been only a few papers presenting some promising results [497–501]. Natarajan et al., for example, published a paper in 2008 presenting a potential theranostic approach in a PCa and BC experiment [501]. They developed a novel ^{111}In-radioconjugate NP based on anti-MUC-1-scFv antibody fragments and functionalized NPs.

4.7. Neutron Capture Therapy

The radiation-based technique called neutron capture therapy (NCT) is based on a neutron source in order to generate a targeted internal radiation therapy at the specific tumor site, and has been described in several publications [300,502–504]. The technique is still a highly active research area, and applications in which it is evaluated now also includes theranostic NP-based settings [505]. Most applications so far have been exploiting the nuclear reaction ^{10}B(n,α)^{7}Li, i.e., bombarding boron atoms with thermal neutrons to produce internally emitted α-particles. This technique is called boron neutron capture therapy (BNCT). The isotope ^{157}Gd has also been evaluated for NCT, although it has been questioned due to toxicity concerns regarding Gd^{3+} ions. However, chelation using DTPA has been promising and capable of producing stable Gd-DTPA complexes, and therefore nontoxic. The isotope ^{157}Gd has some advantages over ^{10}B, including, for example, a 67 times higher cross section for thermal neutrons as well as Gd^{3+} ions being paramagnetic and thereby able to function as contrast enhancers during MRI [300]. Regarding NP-based applications of NCT, it could help to increase the accumulation of ^{10}B or ^{157}Gd in the targeted tumor tissue. Liposome-based NP techniques is a possible approach and some experiments have shown promising results [504,506–508].

Regarding PCa and BC there are very few studies published using NP-based NCT techniques. Only two BC-related publications are to be found on PubMed, investigating dendrimer- and lipid-based gadolinium NPs [509,510].

4.8. Magnetic Therapy

In addition, for magnetic NPs being able to serve as contrast enhancers during MRI (see above), these type of NPs could also be used for therapy, i.e., magnetic therapy (MTH), and thus used as a theranostic platform. Both alternating-current (AC) and direct-current (DC) based magnetic fields could be utilized for this type of technique, although the most commonly used option called magnetic hyperthermia uses AC-based magnetic fields [300]. The Brownian and Neel relaxation processes are the two sources of heat generation during the AC-based magnetic hyperthermia option [511]. The smaller the NP used the more the Neel relaxation process will dominate over the Brownian in contributing to the heat generation in targeted tissue. If instead using a DC-based magnetic field is used the process of magnetocytolysis is utilized in order to induce cellular disruption.

Regarding PCa and BC related applications using a NP-based MTH technique there are only a few publications available [493,512–516]. For example, Han et al. recently evaluated a theranostic strategy based on Fe_3O_4/Au NPs used for prostate-specific antigen detection, MRI, as well as magnetic hyperthermia [512]. For BC, Yao et al. recently investigated a multifunction therapy platform based on silica NP and quantum dots for controlled and targeted drug (doxorubicin) release, NIR-based PTT, and AC-based magnetic hyperthermia in a 4T1 BC model, indicating a significant synergistic therapeutic effect [515].

5. The Prostate Cancer Case

TNPs might play an important role in the future for the detection, diagnosis, and staging, as well as for the therapy of cancerous diseases at different stages. In order to specifically up-date the reader on the current situation regarding the statistics, diagnosis options, staging, recurrence, metastases, as well as some available treatment options for PCa, a short presentation of this is given below.

5.1. Background Statistics

Worldwide, PCa is the second most frequently diagnosed cancer in men, and the fourth most common in both sexes combined. Approximately 1.1 million men were diagnosed with PCa in the world during 2012, which is approximately 15% of all cancers diagnosed in men [517]. Prostate cancer is also the fifth leading cause of death related to cancer in men, with 307,000 deaths worldwide during 2012 [517]. In the USA, PCa is the second leading cause of cancer-related deaths and the second most frequently diagnosed cancer, while in Europe it is number one. The American Cancer Society estimated that 180,890 men would be diagnosed with PCa during 2016 in USA, and about 26,120 would die from the diseases [518]. The International Agency for Research on Cancer (IARC) concluded that during 2012 in Europe close to 345,000 men were diagnosed with PCa. Although more effort has been directed towards early detection through screening, 72,000 men died of PCa in Europe in 2012 [519–526]. Notably, there is less variation in mortality worldwide than is observed for the incidence. This is explained by the PSA testing having greater effect on incidence than on the mortality [517]. The development of improved therapy modalities should therefore be prioritized and targeted therapies based on TNPs are promising candidates to increase the therapeutic efficacy and chance for survival of this category of patients. Several studies of the therapeutic efficacy and toxicity of RIT against PCa have been performed [492,527–544].

5.2. Diagnosis, Staging, Recurrence, and Metastases

A transrectal ultrasonography—guided pathologic examinational procedure is applied during tumor diagnosis of PCa. The extent of the localized PCa tumor is also estimated by digital rectal examination and PSA testing, sometimes supplemented using CT, bone scanning, or multiparametric MRI [545]. The staging procedure for malignant PCa tumors as outlined in the National Comprehensive Cancer Network (NCCN) guidelines [546], should follow the TNM (Tumor—regional lymph Nodes—Metastasis) classification developed by the Union for International Cancer Control (UICC) and published by the American Joint Committee on Cancer (AJCC), as well as the International Federation of Gynecology and Obstetrics (FIGO), staging manuals [547].

Positron emission tomography in combination with CT is used for the staging of lymph-node metastasis involvement. Depending on stage, ^{18}F-FDG, ^{18}F-choline, or ^{11}C PET could be used [548–556]. The use of PET in combination with MRI may also help detect PCa as well as improve accuracy of staging [557–560]. Regarding the estimation if the PCa under investigation is clinically insignificant/indolent (CIPC) or significant/aggressive the Epstein criteria could be used, taken into account its limitations and many variations [561–566]. Better criteria deciding CIPC or not CIPC could help minimize the amount of under- and overtreated men having PCa [521]. Regardless whether CIPC or not CIPC, monitoring the disease is usually performed using PSA testing, complemented with MRI and/or PET/CT if PSA level is rising [567].

Regarding the treatment of localized PCa radiation therapy (RT) and radical prostectomy (RP) are established protocols, although resulting in up to 50% of PSA recurrence often referred to as biochemical recurrence (BCR) [568]. The PSA doubling time, the Gleason score, and the pathologic T-stage determines the time between BCR and when metastases are confirmed [488,569]. In order to be able to determine if the recurrence is local or in the form of metastases, ^{11}C-choline based PET/CT and/or MRI are often used [488,545,551,570–573]. Regarding metastases, the skeleton and regional lymph nodes are the most common sites, with >80% of the men succumbing to PCa having metastases in the skeleton [574]. Bone scintigraphy, as SPECT in conjunction with CT (SPECT/CT), using Technetium-99 (^{99m}Tc)-methylene di-phosphonate is often used to estimate the degree of metastases in the skeleton [567,575]. The use of ^{18}F-fluoride PET/CT might also be an option to be used for detecting and classifying metastases in the skeleton [545].

6. Theranostic Nanoparticles for Prostate and Breast Cancer

For cancer applications in general there is an increasing number of publications regarding multifunctional TNPs, exemplified by a limited selection of references [1,76,576–591]. For PCa and BC there is only a limited number of publications, some of which already have been mentioned above in conjunction with the presentation of the different type of NPs as well as different imaging and therapy options available for TNPs [1,131,230,345,358,476–478,592,593]. As TNPs combine into one nanovehicle both imaging and therapy, the presentation of selected representative examples below illustrate some of these combinations evaluated so far for PCa and BC.

Lin et al. developed and evaluated a novel multifunctional NP-based platform for simultaneous imaging and therapy of PCa using LNCaP and PC3 cells in a mouse model [476]. The imaging was achieved by using NIR activatable fluorescence NPs enabling optical imaging and therefore real-time monitoring of the drug delivery. The therapy was achieved by simultaneous targeted delivery of heat, ROS, and heat-chock protein 90 (Hsp90) inhibitor. Their porphyrine-based system was able to generate enough heat and ROS simultaneous with light activation in order to achieve a dual PTT/PDT therapy. The developed formulations of Hsp90 inhibitors also enabled a decrease of the level of pro-survival and angiogenic signaling induced by the PTT and PDT treatment, which sensitizes the tumor cells to the phototherapy. The authors concluded that by using their PCa-specific and image-guided minimally invasive NP-based PTT/PDT drug delivery system, in conjunction with the Hsp90 inhibitors, could enhance the therapeutic efficacy for PCa.

In a paper by Vaillant et al. it was investigated the possibility of developing and using a targeting molecule against the cation-independent mannose 6-phosphate receptor (M6PR), over-expressed in especially the LNCaP cell line [477]. The targeting molecule was a mannose 6-phosphate analogue, synthesized in six steps, which was grafted onto functionalized silica NPs. Experiments were performed both in vitro and in vivo using PDT and showed promising results regarding both targeting, imaging, and therapy of PCa. Regarding the developed biomarker and the M6PR investigated, the authors especially emphasize that the target fulfill important characteristics, namely (i) over-expression in 84% of PCa tissues; (ii) no expression in normal tissues or non-cancerous hypertrophy of prostate; and (iii) over-expression in low-grade cancers. Therefore, M6PR is according to the authors a promising target for non-invasive personalized therapy of PCa, with the possibility of future theranostic applications.

In a paper by Agemy et al. the targeting of the tumor vasculature was investigated for both therapy and imaging of PCa [358]. By screening phage-displayed peptide libraries they identified specific targets in the vessels of PCa tumors. One such peptide, the penta-peptide Cys-Arg-Glu-Lys-Ala, recognizes a fibrin-fibronectin complex located in tumor vasculature. By using SPIONs coated with this peptide in 22Rv1-and LAPC9-PCa cells xenograft mice models an accumulation in tumor vessels was achieved after intravenous injection, which in its turn caused additional clotting and thereby additional sites for the TNPs to bind to in the tumors. No clotting was seen in other parts of the body. Imaging was performed by MRI. The addition of an anti-cancer drug, to these tumor vasculature-blocking TNPs, is hypothesized by the authors to increase the therapeutic efficacy even further.

In a study by Li et al. a BC xenograft-mice model was used to evaluate the imaging and therapeutic efficacy of self-assembled gemcitabine-Gd(III)-based pegylated 50 nm TNPs [1]. The anti-cancer drug gemcitabine combined with the MRI contrast agent Gd(III) used in this setting for the BC cell line MDA-MB-231 showed a high in vivo antitumor efficacy compared to saline control; median tumor volume equal to 188 and 695 mm^3 28 days post injection, respectively. The level of toxicity was indistinguishable compared to controls, drug loading capacity of the TNPs higher than compared to other systems [35,36], and the in vivo MRI-signaling efficacy comparable with other similar NPs [30,594].

In vitro experiments were performed by Choi et al. in which they evaluated enzyme-activatable TNPs for NIR-fluorescence imaging and a combination of PDT and CTH of TNBC [478]. The photosensitizer chlorin e6 (Ce6) conjugated to hyaluronic acid (HA) formed Ce6-HA NPs by self-assembly. Then, the anticancer topoisomerase-1 inhibitor camptothecin (CPT) was encapsulated inside these NPs forming the final TNPs. Treatment using the enzyme hyaluronidase induced activation of singlet oxygen generation and NIR fluorescence, as well as the release of CPT from the TNPs. The light irradiation of treated TNBC cells further enhanced the therapeutic efficacy significantly. An up-dated and well written review on the subject of targeted NPs for image-guided treatment of TNBC has been written by Miller-Kleinhenz et al. and discusses, for example, subtypes, biomarkers, and potential surface targets for TNBC [242].

Ansari et al. demonstrated in a study the feasibility of a TNP incorporating both tumor specificity, enzyme-activated prodrugs, and in vivo imaging possibilities by conjugating the FDA-approved magnetic iron-oxide NP ferumoxytol to a matrix metalloproteinase-activatable peptide conjugate of the colchicine analogue azademethyl-colchicine [592]. Intravenous injections of the TNPs into MMTV-PyMT (mouse mammary tumor virus-polyoma middle-T-antigen) BC tumor-bearing mice resulted in a significant anti-tumor efficacy compared to controls, with no detectable normal-tissue toxicity, explained by a significant tumor accumulation of the TNPs shown by MRI. The results are important since the MMTV-PyMT cells are considered to be a good model for BC metastasis [595]. It should be noted that by March 30, 2015, FDA changed its prescription instructions for the ferumoxytol-based anemia drug Feraheme® due to risk of serious allergic reactions [596].

In conclusion, theranostic NPs applied in an individualized targeted nanomedicine setting have a high potential to become one of the most valuable technologies for the detection, diagnosis, and therapy of PCa and BC. The tumor cell specific multifunctionality of such nanovehicles could enable earlier detection of the diseases, as well as increased sensitivity and specificity during diagnosis. The TNPs also have the potential to increase the likelihood of survival as well as decreasing systemic toxicity for treated patients, compared to the options available today. There are many combinatorial possibilities when constructing TNPs, and all of them have pros and cons as illustrated in this paper. Clinical trials need to be performed in order to give the U.S. Food and Drug Administration, its European Union equivalence European Medicines Agency, and other national medicine-regulatory authorities, the possibility to evaluate relevant TNP options further. This will hopefully add to the list of NP-based drugs under clinical evaluation or already clinically approved, and listed in Table 1 below, also theranostic applications for both PCa and BC.

Table 1. Nanoparticle-based drugs for PCa and BC, approved or under clinical evaluation. Listed are also examples of drugs for solid cancers in general, since they also might be applicable to PCa and BC in the future.

Cancer	Specific Indication	Nanoparticle	Drug	Product	Phase	Company
PCa	US enhancement imaging	Phospholipid microbubbles	-	SonoVue®	Phase III	Bracco Diagnostics Inc.
	Metastatic CRPC	Polymeric	Docetaxel	BIND-014 (Accurin™)	Phase II	Pfizer Inc/BIND Therapeutics Inc.
	Hormone refractive PCa	Albumin-based NP	Paclitaxel	Abraxane®	Phase II	Celgene Corporation
	Androgen independant PCa	Liposome	Doxorubicin	Doxil®	Phase II	Janssen Pharmaceuticals Inc.
		Iron NP	Iron NP	Magnablate	Phase I/0	University College London Hospitals
BC	Metastatic BC	Liposome	Paclitaxel	Myocet™	Approved	Teva UK Ltd.
	Metastatic BC	Albumin-based NP	Paclitaxel	Abraxane™	Approved	Celgene Corporation
		Micelle (polymeric)	-	Genexol-PM™	Approved	Samyang Pharmaceuticals Co.
	US enhancement imaging	Lipid microspheres	-	Definity®	Approved	Lantheus Medical Imaging Inc.
	Metastatic BC	Liposome	Paclitaxel	LIPUSU®	Phase IV	Nanjing Luye Sike Pharmaceutical Co., Ltd.
	Metastatic BC	Polymeric conjugate	Irinotecan	NKTR-102	Phase III	Nektar Therapeutics
	Refractory chest wall BC	Liposome	Doxorubicin	ThermoDox™	Phase II/I	Celsion Co.
	Advanced/metastatic BC	Micelle (polymeric)	Paclitaxel	NK105	Phase III	NanoCarrier Co., Ltd.
	Tripple-negative metastatic BC	HER2-liposome	Doxorubicin	-MM-302	Phase III/II/I	Merrimack Pharmaceuticals Inc.
	Tripple-negative metastatic BC	Liposome	Doxorubicin	Doxil®	Phase II	Janssen Pharmaceuticals Inc.
	Metastatic	Liposome	Doxorubicin	Caelyx®	Phase II	Janssen-Cilag Ltd.
		Liposome	Paclitaxel	EndoTAG-1	Phase II	MediGene AG
	Advanced recurrent/metastatic BC	Liposome	Paclitaxel	LEP-ETU	Phase II	Insys Therapeutics Inc.
		Liposome	Mitoxantrone	Mitoxantrone HCL Liposome	Phase I	CSPC ZhongQi Pharmaceutical Technology
	Metastatic BC	Micelle (polymeric)	Paclitaxel	Nanoxel™	Phase I	Samyang Pharmaceuticals Co.
	US enhancement imaging	Phospholipid microbubbles	-	SonoVue®	Pilot	Bracco Imaging Inc.
	Metastatic/locally recurrent	Micelle (polymeric)	Paclitaxel	Cynviloq	Not provided	Sorrento Therapeutics Inc.
Solid cancers	Advanced tumors	Liposome	Curcumin	Lipocurc	Phase II/I	SignPath Pharma Inc.
	Advanced tumors	Micelle	Gemcitabine/Cisplatin	NC-6004 Nanoplatin	Phase II/I	NanoCarrier Co., Ltd.
	Advanced tumors	Cyclodextrin-based NP	Docetaxel	CRLX301	Phase II/I	Cerulean Pharma Inc.
		Micelle (polymeric)	Docetaxel/Taxotere	Docetaxel-PM DOPNP201	Phase I	Samyang Pharmaceuticals Co.
		Micelle	Docetaxel	CriPec	Phase I	Cristal Therapeutics
	Advanced tumors	Micelle (polymeric)	Cisplatin/paclitaxel	NC-4016 DACH-Platin micelle	Phase I	NanoCarrier Co Ltd/MD Anderson Cancer Center
		Liposome	Eribulin mesylate	Halaven E7389-LF	Phase I	Eisai Co., Ltd.
		Liposome	Mitomycin-C	Promitil®	Phase I	LipoMedix Pharmaceuticals Inc.
	Refractory/recurrent tumors	Liposome	Topotecan, docetaxel, CP	SGT-53	Phase I	SynerGene Therapeutics Inc.
		Liposome	R894 plasmid DNA	SGT-94	Phase I	SynerGene Therapeutics Inc.
	Advanced tumors	Liposome	188Re-BMEDA	188Re-BMEDA	Phase I	INER, Taiwan
	Advanced tumors	Albumin-based NP	Thiocolchicine	ABI-011	Phase I	NantBioScience Inc.
		Lipid	DsiRNA	DCR-MYC	Phase I	Dicerna Pharmaceuticals Inc.
	Advanced recurrent tumors	Liposome	siRNA	siRNA-EphA2-DOPC	Phase I	MD Anderson Cancer Center
	Advanced/refractory tumors	Liposome	Cisplatin	LiPlaCis	Phase I	Oncology Venture/LiPlasome Pharma A/S
	Advanced solid tumors	Polymeric	AZD2811, Irinotecan	AZD2811 (Accurin™)	Phase I	AztraZeneca/BIND Therapeutics Inc.

US = Ultra sound; CRPC = Castration resistant prostate cancer; PSMA = Prostate-specific membrane antigen; AZD2811—Aurora B kinase inhibitor; DsiRNA = Double stranded small interfering RNA; siRNA = small interfering RNA; CP = Cyclophosphamide; NP = Nanoparticle; EndoTag; EndoTag = Endothelial targeting agent; LEP-ETU = liposome entrapped paclitaxel easy-to-us; LIPUSU = Paclitaxel liposome for injection; BMEDA = (2-mercaptoethyl)-N',N'-diethylethylenediamine.

Conflicts of Interest: The author declares no conflict of interest.

1.   Li, L.; Tong, R.; Li, M.; Kohane, D.S. Self-assembled gemcitabine-gadolinium nanoparticles for magnetic resonance imaging and cancer therapy. *Acta Biomater.* **2016**, *33*, 34–39. [CrossRef] [PubMed]
2.   Rizzo, L.Y.; Theek, B.; Storm, G.; Kiessling, F.; Lammers, T. Recent progress in nanomedicine: Therapeutic, diagnostic and theranostic applications. *Curr. Opin. Biotechnol.* **2013**, *24*, 1159–1166. [CrossRef] [PubMed]
3.   Lee, G.Y.; Qian, W.P.; Wang, L.; Wang, Y.A.; Staley, C.A.; Satpathy, M.; Nie, S.; Mao, H.; Yang, L. Theranostic nanoparticles with controlled release of gemcitabine for targeted therapy and MRI of pancreatic cancer. *ACS Nano* **2013**, *7*, 2078–2089. [CrossRef] [PubMed]
4.   Tian, Q.; Hu, J.; Zhu, Y.; Zou, R.; Chen, Z.; Yang, S.; Li, R.; Su, Q.; Han, Y.; Liu, X. Sub-10 nm $Fe_3O_4@Cu_2-xS$ core-shell nanoparticles for dual-modal imaging and photothermal therapy. *J. Am. Chem. Soc.* **2013**, *135*, 8571–8577. [CrossRef] [PubMed]
5.   Bardhan, R.; Lal, S.; Joshi, A.; Halas, N.J. Theranostic nanoshells: From probe design to imaging and treatment of cancer. *Acc. Chem. Res.* **2011**, *44*, 936–946. [CrossRef] [PubMed]
6.   Huang, P.; Rong, P.; Lin, J.; Li, W.; Yan, X.; Zhang, M.G.; Nie, L.; Niu, G.; Lu, J.; Wang, W.; et al. Triphase interface synthesis of plasmonic gold bellflowers as near-infrared light mediated acoustic and thermal theranostics. *J. Am. Chem. Soc.* **2014**, *136*, 8307–8313. [CrossRef] [PubMed]
7.   Kim, J.; Piao, Y.; Hyeon, T. Multifunctional nanostructured materials for multimodal imaging, and simultaneous imaging and therapy. *Chem. Soc. Rev.* **2009**, *38*, 372–390. [CrossRef] [PubMed]
8.   Giljohann, D.A.; Seferos, D.S.; Daniel, W.L.; Massich, M.D.; Patel, P.C.; Mirkin, C.A. Gold nanoparticles for biology and medicine. *Angew. Chem. Int. Ed. Engl.* **2010**, *49*, 3280–3294. [CrossRef] [PubMed]
9.   Liu, G.; Zhang, G.; Hu, J.; Wang, X.; Zhu, M.; Liu, S. Hyperbranched self-immolative polymers (HSIPS) for programmed payload delivery and ultrasensitive detection. *J. Am. Chem. Soc.* **2015**, *137*, 11645–11655. [CrossRef] [PubMed]
10.   Sanna, V.; Pala, N.; Sechi, M. Targeted therapy using nanotechnology: Focus on cancer. *Int. J. Nanomed.* **2014**, *9*, 467–483.
11.   Doane, T.L.; Burda, C. The unique role of nanoparticles in nanomedicine: Imaging, drug delivery and therapy. *Chem. Soc. Rev.* **2012**, *41*, 2885–2911. [CrossRef] [PubMed]
12.   Kim, B.Y.; Rutka, J.T.; Chan, W.C. Nanomedicine. *N. Engl. J. Med.* **2010**, *363*, 2434–2443. [CrossRef] [PubMed]
13.   Riehemann, K.; Schneider, S.W.; Luger, T.A.; Godin, B.; Ferrari, M.; Fuchs, H. Nanomedicine—Challenge and perspectives. *Angew. Chem. Int. Ed. Engl.* **2009**, *48*, 872–897. [CrossRef] [PubMed]
14.   Zhang, L.; Gu, F.X.; Chan, J.M.; Wang, A.Z.; Langer, R.S.; Farokhzad, O.C. Nanoparticles in medicine: Therapeutic applications and developments. *Clin. Pharmacol. Ther.* **2008**, *83*, 761–769. [CrossRef] [PubMed]
15.   Peer, D.; Karp, J.M.; Hong, S.; Farokhzad, O.C.; Margalit, R.; Langer, R. Nanocarriers as an emerging platform for cancer therapy. *Nat. Nanotechnol.* **2007**, *2*, 751–760. [CrossRef] [PubMed]
16.   Langer, R. Polymer-controlled drug delivery systems. *Acc. Chem. Res.* **1993**, *26*, 537–542. [CrossRef]
17.   Tong, R.; Chiang, H.H.; Kohane, D.S. Photoswitchable nanoparticles for in vivo cancer chemotherapy. *Proc. Natl. Acad. Sci. USA* **2013**, *110*, 19048–19053. [CrossRef] [PubMed]
18.   Cheng, L.; Wang, C.; Feng, L.; Yang, K.; Liu, Z. Functional nanomaterials for phototherapies of cancer. *Chem. Rev.* **2014**, *114*, 10869–10939. [CrossRef] [PubMed]
19.   Carmeliet, P.; Jain, R.K. Principles and mechanisms of vessel normalization for cancer and other angiogenic diseases. *Nat. Rev. Drug Discov.* **2011**, *10*, 417–427. [CrossRef] [PubMed]
20.   Perrault, S.D.; Walkey, C.; Jennings, T.; Fischer, H.C.; Chan, W.C. Mediating tumor targeting efficiency of nanoparticles through design. *Nano Lett.* **2009**, *9*, 1909–1915. [CrossRef] [PubMed]
21.   Miller, M.A.; Gadde, S.; Pfirschke, C.; Engblom, C.; Sprachman, M.M.; Kohler, R.H.; Yang, K.S.; Laughney, A.M.; Wojtkiewicz, G.; Kamaly, N.; et al. Predicting therapeutic nanomedicine efficacy using a companion magnetic resonance imaging nanoparticle. *Sci. Transl. Med.* **2015**, *7*, 314ra183. [CrossRef] [PubMed]
22.   Hare, J.I.; Lammers, T.; Ashford, M.B.; Puri, S.; Storm, G.; Barry, S.T. Challenges and strategies in anti-cancer nanomedicine development: An industry perspective. *Adv. Drug Deliv. Rev.* **2017**, *108*, 25–38. [CrossRef] [PubMed]

23. Matsumura, Y.; Maeda, H. A new concept for macromolecular therapeutics in cancer chemotherapy: Mechanism of tumoritropic accumulation of proteins and the antitumor agent smancs. *Cancer Res.* **1986**, *46*, 6387–6392. [PubMed]

24. Caravan, P.; Ellison, J.J.; McMurry, T.J.; Lauffer, R.B. Gadolinium(III) chelates as MRI contrast agents: Structure, dynamics, and applications. *Chem. Rev.* **1999**, *99*, 2293–2352. [CrossRef] [PubMed]

25. Walker, E.A.; Fenton, M.E.; Salesky, J.S.; Murphey, M.D. Magnetic resonance imaging of benign soft tissue neoplasms in adults. *Radiol. Clin. N. Am.* **2011**, *49*, 1197–1217. [CrossRef] [PubMed]

26. Liu, T.; Li, X.; Qian, Y.; Hu, X.; Liu, S. Multifunctional pH-disintegrable micellar nanoparticles of asymmetrically functionalized β-cyclodextrin-based star copolymer covalently conjugated with doxorubicin and DOTA-Gd moieties. *Biomaterials* **2012**, *33*, 2521–2531. [CrossRef] [PubMed]

27. Kircher, M.F.; de la Zerda, A.; Jokerst, J.V.; Zavaleta, C.L.; Kempen, P.J.; Mittra, E.; Pitter, K.; Huang, R.; Campos, C.; Habte, F.; et al. A brain tumor molecular imaging strategy using a new triple-modality MRI-photoacoustic-raman nanoparticle. *Nat. Med.* **2012**, *18*, 829–834. [CrossRef] [PubMed]

28. Kielar, F.; Tei, L.; Terreno, E.; Botta, M. Large relaxivity enhancement of paramagnetic lipid nanoparticles by restricting the local motions of the Gd(III) chelates. *J. Am. Chem. Soc.* **2010**, *132*, 7836–7837. [CrossRef] [PubMed]

29. Mi, P.; Kokuryo, D.; Cabral, H.; Kumagai, M.; Nomoto, T.; Aoki, I.; Terada, Y.; Kishimura, A.; Nishiyama, N.; Kataoka, K. Hydrothermally synthesized pegylated calcium phosphate nanoparticles incorporating Gd-DTPA for contrast enhanced MRI diagnosis of solid tumors. *J. Control. Release* **2014**, *174*, 63–71. [CrossRef] [PubMed]

30. Mi, P.; Cabral, H.; Kokuryo, D.; Rafi, M.; Terada, Y.; Aoki, I.; Saga, T.; Takehiko, I.; Nishiyama, N.; Kataoka, K. Gd-DTPA-loaded polymer-metal complex micelles with high relaxivity for MR cancer imaging. *Biomaterials* **2013**, *34*, 492–500. [CrossRef] [PubMed]

31. Frias, J.C.; Williams, K.J.; Fisher, E.A.; Fayad, Z.A. Recombinant hdl-like nanoparticles: A specific contrast agent for MRI of atherosclerotic plaques. *J. Am. Chem. Soc.* **2004**, *126*, 16316–16317. [CrossRef] [PubMed]

32. Li, X.; Qian, Y.; Liu, T.; Hu, X.; Zhang, G.; You, Y.; Liu, S. Amphiphilic multiarm star block copolymer-based multifunctional unimolecular micelles for cancer targeted drug delivery and mr imaging. *Biomaterials* **2011**, *32*, 6595–6605. [CrossRef] [PubMed]

33. Fossheim, S.L.; Fahlvik, A.K.; Klaveness, J.; Muller, R.N. Paramagnetic liposomes as MRI contrast agents: Influence of liposomal physicochemical properties on the in vitro relaxivity. *Magn. Reson. Imaging* **1999**, *17*, 83–89. [CrossRef]

34. Perrier, M.; Gallud, A.; Ayadi, A.; Kennouche, S.; Porredon, C.; Gary-Bobo, M.; Larionova, J.; Goze-Bac, C.; Zanca, M.; Garcia, M.; et al. Investigation of cyano-bridged coordination nanoparticles Gd(3+)/[Fe(Cn)6](3-)/D-mannitol as T1-weighted MRI contrast agents. *Nanoscale* **2015**, *7*, 11899–11903. [CrossRef] [PubMed]

35. Hu, X.; Liu, G.; Li, Y.; Wang, X.; Liu, S. Cell-penetrating hyperbranched polyprodrug amphiphiles for synergistic reductive milieu-triggered drug release and enhanced magnetic resonance signals. *J. Am. Chem. Soc.* **2015**, *137*, 362–368. [CrossRef] [PubMed]

36. Lee, S.M.; Song, Y.; Hong, B.J.; MacRenaris, K.W.; Mastarone, D.J.; O'Halloran, T.V.; Meade, T.J.; Nguyen, S.T. Modular polymer-caged nanobins as a theranostic platform with enhanced magnetic resonance relaxivity and pH-responsive drug release. *Angew. Chem. Int. Ed. Engl.* **2010**, *49*, 9960–9964. [CrossRef] [PubMed]

37. Budd, G.T. Let me do more than count the ways: What circulating tumor cells can tell us about the biology of cancer. *Mol. Pharm.* **2009**, *6*, 1307–1310. [CrossRef] [PubMed]

38. Bray, F.; Moller, B. Predicting the future burden of cancer. *Nat. Rev. Cancer* **2006**, *6*, 63–74. [CrossRef] [PubMed]

39. Danila, D.C.; Fleisher, M.; Scher, H.I. Circulating tumor cells as biomarkers in prostate cancer. *Clin. Cancer Res.* **2011**, *17*, 3903–3912. [CrossRef] [PubMed]

40. Ashworth, T.R. A case of cancer in which cells similar to those in the tumours were seen in the blood after death. *Med. J. Aust.* **1869**, *14*, 146–147.

41. Pantel, K.; Alix-Panabieres, C.; Riethdorf, S. Cancer micrometastases. *Nat. Rev. Clin. Oncol.* **2009**, *6*, 339–351. [CrossRef] [PubMed]

42. Gupta, G.P.; Massague, J. Cancer metastasis: Building a framework. *Cell* **2006**, *127*, 679–695. [CrossRef] [PubMed]

43. Fehm, T.; Sagalowsky, A.; Clifford, E.; Beitsch, P.; Saboorian, H.; Euhus, D.; Meng, S.; Morrison, L.; Tucker, T.; Lane, N.; et al. Cytogenetic evidence that circulating epithelial cells in patients with carcinoma are malignant. *Clin. Cancer Res.* **2002**, *8*, 2073–2084. [PubMed]

44. Sleijfer, S.; Gratama, J.W.; Sieuwerts, A.M.; Kraan, J.; Martens, J.W.; Foekens, J.A. Circulating tumour cell detection on its way to routine diagnostic implementation? *Eur. J. Cancer* **2007**, *43*, 2645–2650. [CrossRef] [PubMed]

45. Fidler, I.J. The pathogenesis of cancer metastasis: The 'seed and soil' hypothesis revisited. *Nat. Rev. Cancer* **2003**, *3*, 453–458. [CrossRef] [PubMed]

46. Hayes, D.F.; Smerage, J. Is there a role for circulating tumor cells in the management of breast cancer? *Clin. Cancer Res.* **2008**, *14*, 3646–3650. [CrossRef] [PubMed]

47. Pantel, K.; Riethdorf, S. Pathology: Are circulating tumor cells predictive of overall survival? *Nat. Rev. Clin. Oncol.* **2009**, *6*, 190–191. [CrossRef] [PubMed]

48. Aceto, N.; Bardia, A.; Miyamoto, D.T.; Donaldson, M.C.; Wittner, B.S.; Spencer, J.A.; Yu, M.; Pely, A.; Engstrom, A.; Zhu, H.; et al. Circulating tumor cell clusters are oligoclonal precursors of breast cancer metastasis. *Cell* **2014**, *158*, 1110–1122. [CrossRef] [PubMed]

49. Miller, M.C.; Doyle, G.V.; Terstappen, L.W. Significance of circulating tumor cells detected by the cellsearch system in patients with metastatic breast colorectal and prostate cancer. *J. Oncol.* **2010**, *2010*, 617421. [CrossRef] [PubMed]

50. Swaby, R.F.; Cristofanilli, M. Circulating tumor cells in breast cancer: A tool whose time has come of age. *BMC Med.* **2011**, *9*, 43. [CrossRef] [PubMed]

51. Hekimian, K.; Meisezahl, S.; Trompelt, K.; Rabenstein, C.; Pachmann, K. Epithelial cell dissemination and readhesion: Analysis of factors contributing to metastasis formation in breast cancer. *ISRN Oncol.* **2012**, *2012*, 601810. [CrossRef] [PubMed]

52. O'Hara, S.M.; Moreno, J.G.; Zweitzig, D.R.; Gross, S.; Gomella, L.G.; Terstappen, L.W. Multigene reverse transcription-PCR profiling of circulating tumor cells in hormone-refractory prostate cancer. *Clin. Chem.* **2004**, *50*, 826–835. [CrossRef] [PubMed]

53. Attard, G.; Swennenhuis, J.F.; Olmos, D.; Reid, A.H.; Vickers, E.; A'Hern, R.; Levink, R.; Coumans, F.; Moreira, J.; Riisnaes, R.; et al. Characterization of *ERG, AR* and *PTEN* gene status in circulating tumor cells from patients with castration-resistant prostate cancer. *Cancer Res.* **2009**, *69*, 2912–2918. [CrossRef] [PubMed]

54. Coumans, F.A.; Doggen, C.J.; Attard, G.; de Bono, J.S.; Terstappen, L.W. All circulating EpCam+CK+CD45-objects predict overall survival in castration-resistant prostate cancer. *Ann. Oncol.* **2010**, *21*, 1851–1857. [CrossRef] [PubMed]

55. Chen, F.; Hong, H.; Zhang, Y.; Valdovinos, H.F.; Shi, S.; Kwon, G.S.; Theuer, C.P.; Barnhart, T.E.; Cai, W. In vivo tumor targeting and image-guided drug delivery with antibody-conjugated, radiolabeled mesoporous silica nanoparticles. *ACS Nano* **2013**, *7*, 9027–9039. [CrossRef] [PubMed]

56. Gupta, P.B.; Onder, T.T.; Jiang, G.; Tao, K.; Kuperwasser, C.; Weinberg, R.A.; Lander, E.S. Identification of selective inhibitors of cancer stem cells by high-throughput screening. *Cell* **2009**, *138*, 645–659. [CrossRef] [PubMed]

57. Zhu, S.; Xu, G. Single-walled carbon nanohorns and their applications. *Nanoscale* **2010**, *2*, 2538–2549. [CrossRef] [PubMed]

58. Kaur, R.; Badea, I. Nanodiamonds as novel nanomaterials for biomedical applications: Drug delivery and imaging systems. *Int. J. Nanomed.* **2013**, *8*, 203–220.

59. Clift, M.J.; Stone, V. Quantum dots: An insight and perspective of their biological interaction and how this relates to their relevance for clinical use. *Theranostics* **2012**, *2*, 668–680. [CrossRef] [PubMed]

60. Taylor, A.; Wilson, K.M.; Murray, P.; Fernig, D.G.; Levy, R. Long-term tracking of cells using inorganic nanoparticles as contrast agents: Are we there yet? *Chem. Soc. Rev.* **2012**, *41*, 2707–2717. [CrossRef] [PubMed]

61. Bae, K.H.; Chung, H.J.; Park, T.G. Nanomaterials for cancer therapy and imaging. *Mol. Cells* **2011**, *31*, 295–302. [CrossRef] [PubMed]

62. Wilhelm, S.; Tavares, A.J.; Dai, Q.; Ohta, S.; Audet, J.; Dvorak, H.F.; Chan, W.C.W. Analysis of nanoparticle delivery to tumours. *Nat. Rev. Mater.* **2016**, *1*, 1–12. [CrossRef]

63. Cheng, C.J.; Tietjen, G.T.; Saucier-Sawyer, J.K.; Saltzman, W.M. A holistic approach to targeting disease with polymeric nanoparticles. *Nat. Rev. Drug Discov.* **2015**, *14*, 239–247. [CrossRef] [PubMed]

64. Ding, Y.; Jiang, Z.; Saha, K.; Kim, C.S.; Kim, S.T.; Landis, R.F.; Rotello, V.M. Gold nanoparticles for nucleic acid delivery. *Mol. Ther.* **2014**, *22*, 1075–1083. [CrossRef] [PubMed]

65. Kasprzak, B.; Miskiel, S.; Markowska, J. Nanooncology in ovarian cancer treatment. *Eur. J. Gynaecol. Oncol.* **2016**, *37*, 161–163. [PubMed]

66. Hu, J.J.; Xiao, D.; Zhang, X.Z. Advances in peptide functionalization on mesoporous silica nanoparticles for controlled drug release. *Small* **2016**, *12*, 3344–3359. [CrossRef] [PubMed]

67. Mocan, L.; Matea, C.T.; Bartos, D.; Mosteanu, O.; Pop, T.; Mocan, T.; Iancu, C. Advances in cancer research using gold nanoparticles mediated photothermal ablation. *Clujul Med.* **2016**, *89*, 199–202. [CrossRef] [PubMed]

68. Lu, B.; Huang, X.; Mo, J.; Zhao, W. Drug delivery using nanoparticles for cancer stem-like cell targeting. *Front. Pharmacol.* **2016**, *7*, 84. [CrossRef] [PubMed]

69. Genchi, G.G.; Marino, A.; Rocca, A.; Mattoli, V.; Ciofani, G. Barium titanate nanoparticles: Promising multitasking vectors in nanomedicine. *Nanotechnology* **2016**, *27*, 232001. [CrossRef] [PubMed]

70. Dolati, S.; Sadreddini, S.; Rostamzadeh, D.; Ahmadi, M.; Jadidi-Niaragh, F.; Yousefi, M. Utilization of nanoparticle technology in rheumatoid arthritis treatment. *Biomed. Pharmacother.* **2016**, *80*, 30–41. [CrossRef] [PubMed]

71. Santoso, M.R.; Yang, P.C. Magnetic nanoparticles for targeting and imaging of stem cells in myocardial infarction. *Stem Cells Int.* **2016**, *2016*. [CrossRef] [PubMed]

72. Li, X.; Tsibouklis, J.; Weng, T.; Zhang, B.; Yin, G.; Feng, G.; Cui, Y.; Savina, I.N.; Mikhalovska, L.I.; Sandeman, S.R.; et al. Nano carriers for drug transport across the blood brain barrier. *J. Drug Target.* **2017**, *25*, 17–28. [CrossRef] [PubMed]

73. Beloqui, A.; des Rieux, A.; Preat, V. Mechanisms of transport of polymeric and lipidic nanoparticles across the intestinal barrier. *Adv. Drug Deliv. Rev.* **2016**, *106*, 242–255. [CrossRef] [PubMed]

74. Nishiyama, N.; Matsumura, Y.; Kataoka, K. Development of polymeric micelles for targeting intractable cancers. *Cancer Sci.* **2016**, *107*, 867–874. [CrossRef] [PubMed]

75. Ulbrich, K.; Hola, K.; Subr, V.; Bakandritsos, A.; Tucek, J.; Zboril, R. Targeted drug delivery with polymers and magnetic nanoparticles: Covalent and noncovalent approaches, release control, and clinical studies. *Chem. Rev.* **2016**, *116*, 5338–5431. [CrossRef] [PubMed]

76. Shahbazi, R.; Ozpolat, B.; Ulubayram, K. Oligonucleotide-based theranostic nanoparticles in cancer therapy. *Nanomedicine (lond.)* **2016**, *11*, 1287–1308. [CrossRef] [PubMed]

77. Yuan, Y.; Cai, T.; Xia, X.; Zhang, R.; Cai, Y.; Chiba, P. Nanoparticle delivery of anticancer drugs overcomes multidrug resistance in breast cancer. *Drug Deliv.* **2016**, *23*, 3350–3357. [CrossRef] [PubMed]

78. Zhou, M.; Tian, M.; Li, C. Copper-based nanomaterials for cancer imaging and therapy. *Bioconjug. Chem.* **2016**, *27*, 1188–1199. [CrossRef] [PubMed]

79. Rajabi, M.; Mousa, S.A. Lipid nanoparticles and their application in nanomedicine. *Curr. Pharm. Biotechnol.* **2016**, *17*, 662–672. [CrossRef] [PubMed]

80. Rao, P.V.; Nallappan, D.; Madhavi, K.; Rahman, S.; Jun Wei, L.; Gan, S.H. Phytochemicals and biogenic metallic nanoparticles as anticancer agents. *Oxid. Med. Cell. Longev.* **2016**, *2016*. [CrossRef] [PubMed]

81. Ma, D.D.; Yang, W.X. Engineered nanoparticles induce cell apoptosis: Potential for cancer therapy. *Oncotarget* **2016**, *7*, 40882–40903. [CrossRef] [PubMed]

82. Shabestari Khiabani, S.; Farshbaf, M.; Akbarzadeh, A.; Davaran, S. Magnetic nanoparticles: Preparation methods, applications in cancer diagnosis and cancer therapy. *Artif. Cells Nanomed. Biotechnol.* **2016**, *45*, 6–17. [CrossRef] [PubMed]

83. Lemaster, J.E.; Jokerst, J.V. What is new in nanoparticle-based photoacoustic imaging? *Wiley Interdiscip. Rev. Nanomed. Nanobiotechnol.* **2016**, *9*, e1404. [CrossRef] [PubMed]

84. Liu, H.; Zhang, J.; Chen, X.; Du, X.S.; Zhang, J.L.; Liu, G.; Zhang, W.G. Application of iron oxide nanoparticles in glioma imaging and therapy: From bench to bedside. *Nanoscale* **2016**, *8*, 7808–7826. [CrossRef] [PubMed]

85. Fathi Karkan, S.; Mohammadhosseini, M.; Panahi, Y.; Milani, M.; Zarghami, N.; Akbarzadeh, A.; Abasi, E.; Hosseini, A.; Davaran, S. Magnetic nanoparticles in cancer diagnosis and treatment: A review. *Artif. Cells Nanomed. Biotechnol.* **2016**, *45*, 1–5. [CrossRef] [PubMed]

86. Radenkovic, D.; Kobayashi, H.; Remsey-Semmelweis, E.; Seifalian, A.M. Quantum dot nanoparticle for optimization of breast cancer diagnostics and therapy in a clinical setting. *Nanomedicine* **2016**, *12*, 1581–1592. [CrossRef] [PubMed]

87. Pratt, E.C.; Shaffer, T.M.; Grimm, J. Nanoparticles and radiotracers: Advances toward radionanomedicine. *Wiley Interdiscip. Rev. Nanomed. Nanobiotechnol.* **2016**, *8*, 872–890. [CrossRef] [PubMed]

88. Pasqua, L.; Leggio, A.; Sisci, D.; Ando, S.; Morelli, C. Mesoporous silica nanoparticles in cancer therapy: Relevance of the targeting function. *Mini Rev. Med. Chem.* **2016**, *16*, 743–753. [CrossRef] [PubMed]

89. Rancoule, C.; Magne, N.; Vallard, A.; Guy, J.B.; Rodriguez-Lafrasse, C.; Deutsch, E.; Chargari, C. Nanoparticles in radiation oncology: From bench-side to bedside. *Cancer Lett.* **2016**, *375*, 256–262. [CrossRef] [PubMed]

90. Alam, F.; Naim, M.; Aziz, M.; Yadav, N. Unique roles of nanotechnology in medicine and cancer-II. *Indian J. Cancer* **2015**, *52*, 1–9. [CrossRef] [PubMed]

91. Yang, R.M.; Fu, C.P.; Fang, J.Z.; Xu, X.D.; Wei, X.H.; Tang, W.J.; Jiang, X.Q.; Zhang, L.M. Hyaluronan-modified superparamagnetic iron oxide nanoparticles for bimodal breast cancer imaging and photothermal therapy. *Int. J. Nanomed.* **2017**, *12*, 197–206. [CrossRef] [PubMed]

92. Oddo, L.; Cerroni, B.; Domenici, F.; Bedini, A.; Bordi, F.; Chiessi, E.; Gerbes, S.; Paradossi, G. Next generation ultrasound platforms for theranostics. *J. Colloid Interface Sci.* **2017**, *491*, 151–160. [CrossRef] [PubMed]

93. Dadras, P.; Atyabi, F.; Irani, S.; Ma'mani, L.; Foroumadi, A.; Mirzaie, Z.H.; Ebrahimi, M.; Dinarvand, R. Formulation and evaluation of targeted nanoparticles for breast cancer theranostic system. *Eur. J. Pharm. Sci.* **2017**, *97*, 47–54. [CrossRef] [PubMed]

94. Huang, Y.; Mao, K.; Zhang, B.; Zhao, Y. Superparamagnetic iron oxide nanoparticles conjugated with folic acid for dual target-specific drug delivery and MRI in cancer theranostics. *Mater. Sci. Eng. C Mater. Biol. Appl.* **2017**, *70*, 763–771. [CrossRef] [PubMed]

95. Sun, L.; Joh, D.Y.; Al-Zaki, A.; Stangl, M.; Murty, S.; Davis, J.J.; Baumann, B.C.; Alonso-Basanta, M.; Kaol, G.D.; Tsourkas, A.; et al. Theranostic application of mixed gold and superparamagnetic iron oxide nanoparticle micelles in glioblastoma multiforme. *J. Biomed. Nanotechnol.* **2016**, *12*, 347–356. [CrossRef] [PubMed]

96. Shevtsov, M.; Multhoff, G. Recent developments of magnetic nanoparticles for theranostics of brain tumor. *Curr. Drug Metab.* **2016**, *17*, 737–744. [CrossRef] [PubMed]

97. Zarrin, A.; Sadighian, S.; Rostamizadeh, K.; Firuzi, O.; Hamidi, M.; Mohammadi-Samani, S.; Miri, R. Design, preparation, and in vitro characterization of a trimodally-targeted nanomagnetic onco-theranostic system for cancer diagnosis and therapy. *Int. J. Pharm.* **2016**, *500*, 62–76. [CrossRef] [PubMed]

98. Bakhtiary, Z.; Saei, A.A.; Hajipour, M.J.; Raoufi, M.; Vermesh, O.; Mahmoudi, M. Targeted superparamagnetic iron oxide nanoparticles for early detection of cancer: Possibilities and challenges. *Nanomedicine* **2016**, *12*, 287–307. [CrossRef] [PubMed]

99. Kandasamy, G.; Maity, D. Recent advances in superparamagnetic iron oxide nanoparticles (SPIONs) for in vitro and in vivo cancer nanotheranostics. *Int. J. Pharm.* **2015**, *496*, 191–218. [CrossRef] [PubMed]

100. Bulte, J.W.; Kraitchman, D.L. Iron oxide mr contrast agents for molecular and cellular imaging. *NMR Biomed.* **2004**, *17*, 484–499. [CrossRef] [PubMed]

101. Alwi, R.; Telenkov, S.; Mandelis, A.; Leshuk, T.; Gu, F.; Oladepo, S.; Michaelian, K. Silica-coated super paramagnetic iron oxide nanoparticles (SPION) as biocompatible contrast agent in biomedical photoacoustics. *Biomed. Opt. Express* **2012**, *3*, 2500–2509. [CrossRef] [PubMed]

102. Bohmer, N.; Jordan, A. Caveolin-1 and CDC42 mediated endocytosis of silica-coated iron oxide nanoparticles in HeLa cells. *Beilstein J. Nanotechnol.* **2015**, *6*, 167–176. [CrossRef] [PubMed]

103. Nyalosaso, J.L.; Rascol, E.; Pisani, C.; Dorandeu, C.; Dumail, X.; Maynadier, M.; Gary-Bobo, M.; Kee Him, J.L.; Bron, P.; Garcia, M.; et al. Synthesis, decoration, and cellular effects of magnetic mesoporous silica nanoparticles. *RSC Adv.* **2016**, *6*, 7275–7283. [CrossRef]

104. Winter, A.; Engels, S.; Kowald, T.; Paulo, T.S.; Gerullis, H.; Chavan, A.; Wawroschek, F. Magnetic sentinel lymph node detection in prostate cancer after intraprostatic injection of superparamagnetic iron oxide nanoparticles. *Aktuelle Urol.* **2017**. [CrossRef]

105. Sabnis, S.; Sabnis, N.A.; Raut, S.; Lacko, A.G. Superparamagnetic reconstituted high-density lipoprotein nanocarriers for magnetically guided drug delivery. *Int. J. Nanomed.* **2017**, *12*, 1453–1464. [CrossRef] [PubMed]

106. Nagesh, P.K.; Johnson, N.R.; Boya, V.K.; Chowdhury, P.; Othman, S.F.; Khalilzad-Sharghi, V.; Hafeez, B.B.; Ganju, A.; Khan, S.; Behrman, S.W.; et al. PSMA targeted docetaxel-loaded superparamagnetic iron oxide nanoparticles for prostate cancer. *Colloids Surf. B Biointerfaces* **2016**, *144*, 8–20. [CrossRef] [PubMed]

107. Zhu, Y.; Sun, Y.; Chen, Y.; Liu, W.; Jiang, J.; Guan, W.; Zhang, Z.; Duan, Y. In vivo molecular MRI imaging of prostate cancer by targeting psma with polypeptide-labeled superparamagnetic iron oxide nanoparticles. *Int. J. Mol. Sci.* **2015**, *16*, 9573–9587. [CrossRef] [PubMed]

108. Yu, M.K.; Kim, D.; Lee, I.H.; So, J.S.; Jeong, Y.Y.; Jon, S. Image-guided prostate cancer therapy using aptamer-functionalized thermally cross-linked superparamagnetic iron oxide nanoparticles. *Small* **2011**, *7*, 2241–2249. [CrossRef] [PubMed]

109. Min, K.; Jo, H.; Song, K.; Cho, M.; Chun, Y.S.; Jon, S.; Kim, W.J.; Ban, C. Dual-aptamer-based delivery vehicle of doxorubicin to both PSMA (+) and PSMA (−) prostate cancers. *Biomaterials* **2011**, *32*, 2124–2132. [CrossRef] [PubMed]

110. Prabhu, S.; Ananthanarayanan, P.; Aziz, S.K.; Rai, S.; Mutalik, S.; Sadashiva, S.R. Enhanced effect of geldanamycin nanocomposite against breast cancer cells growing in vitro and as xenograft with vanquished normal cell toxicity. *Toxicol. Appl. Pharmacol.* **2017**, *320*, 60–72. [CrossRef] [PubMed]

111. Shaik, A.P.; Shaik, A.S.; Majwal, A.A.; Faraj, A.A. Blocking IL4-α receptor using polyethylene glycol functionalized superparamagnetic iron oxide nanocarriers to inhibit breast cancer cell proliferation. *Cancer Res. Treat.* **2016**, *49*, 322–329. [CrossRef] [PubMed]

112. Chiappi, M.; Conesa, J.J.; Pereiro, E.; Sorzano, C.O.; Rodriguez, M.J.; Henzler, K.; Schneider, G.; Chichon, F.J.; Carrascosa, J.L. Cryo-soft X-ray tomography as a quantitative three-dimensional tool to model nanoparticle:Cell interaction. *J. Nanobiotechnol.* **2016**, *14*, 15. [CrossRef] [PubMed]

113. Stapf, M.; Pompner, N.; Teichgraber, U.; Hilger, I. Heterogeneous response of different tumor cell lines to methotrexate-coupled nanoparticles in presence of hyperthermia. *Int. J. Nanomed.* **2016**, *11*, 485–500. [CrossRef] [PubMed]

114. Almaki, J.H.; Nasiri, R.; Idris, A.; Majid, F.A.; Salouti, M.; Wong, T.S.; Dabagh, S.; Marvibaigi, M.; Amini, N. Synthesis, characterization and in vitro evaluation of exquisite targeting SPIONs-PEG-HER in HER2+ human breast cancer cells. *Nanotechnology* **2016**, *27*, 105601. [CrossRef] [PubMed]

115. Kievit, F.M.; Stephen, Z.R.; Veiseh, O.; Arami, H.; Wang, T.; Lai, V.P.; Park, J.O.; Ellenbogen, R.G.; Disis, M.L.; Zhang, M. Targeting of primary breast cancers and metastases in a transgenic mouse model using rationally designed multifunctional spions. *ACS Nano* **2012**, *6*, 2591–2601. [CrossRef] [PubMed]

116. Lentschig, M.G.; Reimer, P.; Rausch-Lentschig, U.L.; Allkemper, T.; Oelerich, M.; Laub, G. Breath-hold gadolinium-enhanced MR angiography of the major vessels at 1.0 t: Dose-response findings and angiographic correlation. *Radiology* **1998**, *208*, 353–357. [CrossRef] [PubMed]

117. Lin, Y.J.; Koretsky, A.P. Manganese ion enhances t1-weighted MRI during brain activation: An approach to direct imaging of brain function. *Magn. Reson. Med.* **1997**, *38*, 378–388. [CrossRef] [PubMed]

118. Zhen, Z.; Xie, J. Development of manganese-based nanoparticles as contrast probes for magnetic resonance imaging. *Theranostics* **2012**, *2*, 45–54. [CrossRef] [PubMed]

119. Silva, A.C.; Lee, J.H.; Aoki, I.; Koretsky, A.P. Manganese-enhanced magnetic resonance imaging (MEMRI): Methodological and practical considerations. *NMR Biomed.* **2004**, *17*, 532–543. [CrossRef] [PubMed]

120. Koretsky, A.P.; Silva, A.C. Manganese-enhanced magnetic resonance imaging (MEMRI). *NMR Biomed.* **2004**, *17*, 527–531. [CrossRef] [PubMed]

121. Paratala, B.S.; Jacobson, B.D.; Kanakia, S.; Francis, L.D.; Sitharaman, B. Physicochemical characterization, and relaxometry studies of micro-graphite oxide, graphene nanoplatelets, and nanoribbons. *PLoS ONE* **2012**, *7*, e38185. [CrossRef] [PubMed]

122. Harisinghani, M.G.; Jhaveri, K.S.; Weissleder, R.; Schima, W.; Saini, S.; Hahn, P.F.; Mueller, P.R. MRI contrast agents for evaluating focal hepatic lesions. *Clin. Radiol.* **2001**, *56*, 714–725. [CrossRef] [PubMed]

123. Hunyadi Murph, S.; Jacobs, S.; Liu, J.; Hu, T.; Siegfried, M.; Serkiz, S.; Hudson, J. Manganese–gold nanoparticles as an MRI positive contrast agent in mesenchymal stem cell labeling. *J. Nanopart. Res.* **2012**, *14*, 658. [CrossRef]

124. Cai, X.; Gao, W.; Ma, M.; Wu, M.; Zhang, L.; Zheng, Y.; Chen, H.; Shi, J. Nanoparticles: A prussian blue-based core-shell hollow-structured mesoporous nanoparticle as a smart theranostic agent with ultrahigh pH-responsive longitudinal relaxivity (adv. Mater. 41/2015). *Adv. Mater.* **2015**, *27*, 6382–6389. [CrossRef] [PubMed]

125. Peng, G.; Tisch, U.; Adams, O.; Hakim, M.; Shehada, N.; Broza, Y.Y.; Billan, S.; Abdah-Bortnyak, R.; Kuten, A.; Haick, H. Diagnosing lung cancer in exhaled breath using gold nanoparticles. *Nat. Nanotechnol.* **2009**, *4*, 669–673. [CrossRef] [PubMed]

126. Samadian, H.; Hosseini-Nami, S.; Kamrava, S.K.; Ghaznavi, H.; Shakeri-Zadeh, A. Folate-conjugated gold nanoparticle as a new nanoplatform for targeted cancer therapy. *J. Cancer Res. Clin. Oncol.* **2016**, *142*, 2217–2229. [CrossRef] [PubMed]

127. Gossai, N.P.; Naumann, J.A.; Li, N.S.; Zamora, E.A.; Gordon, D.J.; Piccirilli, J.A.; Gordon, P.M. Drug conjugated nanoparticles activated by cancer cell specific mRNA. *Oncotarget* **2016**, *7*, 38243–38256. [CrossRef] [PubMed]

128. Gupta, A.; Moyano, D.F.; Parnsubsakul, A.; Papadopoulos, A.; Wang, L.S.; Landis, R.F.; Das, R.; Rotello, V.M. Ultra-stable biofunctionalizable gold nanoparticles. *ACS Appl. Mater. Interfaces* **2016**, *8*, 14096–14101. [CrossRef] [PubMed]

129. Ramya, A.N.; Joseph, M.M.; Nair, J.B.; Karunakaran, V.; Narayanan, N.; Maiti, K.K. New insight of tetraphenylethylene-based raman signatures for targeted SERS nanoprobe construction toward prostate cancer cell detection. *ACS Appl. Mater. Interfaces* **2016**, *8*, 10220–10225. [CrossRef] [PubMed]

130. Spaliviero, M.; Harmsen, S.; Huang, R.; Wall, M.A.; Andreou, C.; Eastham, J.A.; Touijer, K.A.; Scardino, P.T.; Kircher, M.F. Detection of lymph node metastases with SERRS nanoparticles. *Mol. Imaging Biol.* **2016**, *18*, 677–685. [CrossRef] [PubMed]

131. Moeendarbari, S.; Tekade, R.; Mulgaonkar, A.; Christensen, P.; Ramezani, S.; Hassan, G.; Jiang, R.; Oz, O.K.; Hao, Y.; Sun, X. Theranostic nanoseeds for efficacious internal radiation therapy of unresectable solid tumors. *Sci. Rep.* **2016**, *6*. [CrossRef] [PubMed]

132. Tsai, L.C.; Hsieh, H.Y.; Lu, K.Y.; Wang, S.Y.; Mi, F.L. EGCG/gelatin-doxorubicin gold nanoparticles enhance therapeutic efficacy of doxorubicin for prostate cancer treatment. *Nanomedicine (Lond.)* **2016**, *11*, 9–30. [CrossRef] [PubMed]

133. Morshed, R.A.; Muroski, M.E.; Dai, Q.; Wegscheid, M.L.; Auffinger, B.; Yu, D.; Han, Y.; Zhang, L.; Wu, M.; Cheng, Y.; et al. Cell penetrating peptide-modified gold nanoparticles for the delivery of doxorubicin to brain metastatic breast cancer. *Mol. Pharm.* **2016**, *13*, 1843–1854. [CrossRef] [PubMed]

134. Her, S.; Cui, L.; Bristow, R.G.; Allen, C. Dual action enhancement of gold nanoparticle radiosensitization by pentamidine in triple negative breast cancer. *Radiat. Res.* **2016**, *185*, 549–562. [CrossRef] [PubMed]

135. Rizk, N.; Christoforou, N.; Lee, S. Optimization of anti-cancer drugs and a targeting molecule on multifunctional gold nanoparticles. *Nanotechnology* **2016**, *27*, 185704. [CrossRef] [PubMed]

136. Zhou, F.; Feng, B.; Yu, H.; Wang, D.; Wang, T.; Liu, J.; Meng, Q.; Wang, S.; Zhang, P.; Zhang, Z.; et al. Cisplatin prodrug-conjugated gold nanocluster for fluorescence imaging and targeted therapy of the breast cancer. *Theranostics* **2016**, *6*, 679–687. [CrossRef] [PubMed]

137. Yook, S.; Lu, Y.; Jeong, J.J.; Cai, Z.; Tong, L.; Alwarda, R.; Pignol, J.P.; Winnik, M.A.; Reilly, R.M. Stability and biodistribution of thiol-functionalized and ^{177}Lu-labeled metal chelating polymers bound to gold nanoparticles. *Biomacromolecules* **2016**, *17*, 1292–1302. [CrossRef] [PubMed]

138. Huang, X.; Jain, P.K.; El-Sayed, I.H.; El-Sayed, M.A. Determination of the minimum temperature required for selective photothermal destruction of cancer cells with the use of immunotargeted gold nanoparticles. *Photochem. Photobiol.* **2006**, *82*, 412–417. [CrossRef] [PubMed]

139. Huang, X.; El-Sayed, I.H.; Qian, W.; El-Sayed, M.A. Cancer cell imaging and photothermal therapy in the near-infrared region by using gold nanorods. *J. Am. Chem. Soc.* **2006**, *128*, 2115–2120. [CrossRef] [PubMed]

140. Chen, W.; Zhang, S.; Yu, Y.; Zhang, H.; He, Q. Structural-engineering rationales of gold nanoparticles for cancer theranostics. *Adv. Mater.* **2016**, *28*, 8567–8585. [CrossRef] [PubMed]

141. Guo, J.; O'Driscoll, C.M.; Holmes, J.D.; Rahme, K. Bioconjugated gold nanoparticles enhance cellular uptake: A proof of concept study for siRNA delivery in prostate cancer cells. *Int. J. Pharm.* **2016**, *509*, 16–27. [CrossRef] [PubMed]

142. Stuchinskaya, T.; Moreno, M.; Cook, M.J.; Edwards, D.R.; Russell, D.A. Targeted photodynamic therapy of breast cancer cells using antibody-phthalocyanine-gold nanoparticle conjugates. *Photochem. Photobiol. Sci.* **2011**, *10*, 822–831. [CrossRef] [PubMed]

143. Brown, S.D.; Nativo, P.; Smith, J.A.; Stirling, D.; Edwards, P.R.; Venugopal, B.; Flint, D.J.; Plumb, J.A.; Graham, D.; Wheate, N.J. Gold nanoparticles for the improved anticancer drug delivery of the active component of oxaliplatin. *J. Am. Chem. Soc.* **2010**, *132*, 4678–4684. [CrossRef] [PubMed]

144. Chen, Y.W.; Liu, T.Y.; Chen, P.J.; Chang, P.H.; Chen, S.Y. A high-sensitivity and low-power theranostic nanosystem for cell sers imaging and selectively photothermal therapy using anti-EGFR-conjugated reduced graphene oxide/mesoporous silica/aunps nanosheets. *Small* **2016**, *12*, 1458–1468. [CrossRef] [PubMed]

145. Ashraf, S.; Pelaz, B.; del Pino, P.; Carril, M.; Escudero, A.; Parak, W.J.; Soliman, M.G.; Zhang, Q.; Carrillo-Carrion, C. Gold-based nanomaterials for applications in nanomedicine. *Top. Curr. Chem.* **2016**, *370*, 169–202. [PubMed]

146. Conde, J.; de la Fuente, J.M.; Baptista, P.V. Nanomaterials for reversion of multidrug resistance in cancer: A new hope for an old idea? *Front. Pharmacol.* **2013**, *4*, 134. [CrossRef] [PubMed]

147. Han, G.; Ghosh, P.; Rotello, V.M. Multi-functional gold nanoparticles for drug delivery. *Adv. Exp. Med. Biol.* **2007**, *620*, 48–56. [PubMed]

148. Conde, J.; Tian, F.; Hernandez, Y.; Bao, C.; Cui, D.; Janssen, K.P.; Ibarra, M.R.; Baptista, P.V.; Stoeger, T.; de la Fuente, J.M. In vivo tumor targeting via nanoparticle-mediated therapeutic siRNA coupled to inflammatory response in lung cancer mouse models. *Biomaterials* **2013**, *34*, 7744–7753. [CrossRef] [PubMed]

149. McMahon, S.J.; Hyland, W.B.; Muir, M.F.; Coulter, J.A.; Jain, S.; Butterworth, K.T.; Schettino, G.; Dickson, G.R.; Hounsell, A.R.; O'Sullivan, J.M.; et al. Biological consequences of nanoscale energy deposition near irradiated heavy atom nanoparticles. *Sci. Rep.* **2011**, *1*. [CrossRef] [PubMed]

150. Conde, J.; Larguinho, M.; Cordeiro, A.; Raposo, L.R.; Costa, P.M.; Santos, S.; Diniz, M.S.; Fernandes, A.R.; Baptista, P.V. Gold-nanobeacons for gene therapy: Evaluation of genotoxicity, cell toxicity and proteome profiling analysis. *Nanotoxicology* **2014**, *8*, 521–532. [CrossRef] [PubMed]

151. Conde, J.; Rosa, J.; de la Fuente, J.M.; Baptista, P.V. Gold-nanobeacons for simultaneous gene specific silencing and intracellular tracking of the silencing events. *Biomaterials* **2013**, *34*, 2516–2523. [CrossRef] [PubMed]

152. Cabral, R.M.; Baptista, P.V. Anti-cancer precision theranostics: A focus on multifunctional gold nanoparticles. *Expert Rev. Mol. Diagn.* **2014**, *14*, 1041–1052. [CrossRef] [PubMed]

153. Song, J.; Wang, F.; Yang, X.; Ning, B.; Harp, M.G.; Culp, S.H.; Hu, S.; Huang, P.; Nie, L.; Chen, J.; et al. Gold nanoparticle coated carbon nanotube ring with enhanced raman scattering and photothermal conversion property for theranostic applications. *J. Am. Chem. Soc.* **2016**, *138*, 7005–7015. [CrossRef] [PubMed]

154. Croissant, J.G.; Qi, C.; Maynadier, M.; Cattoen, X.; Wong Chi Man, M.; Raehm, L.; Mongin, O.; Blanchard-Desce, M.; Garcia, M.; Gary-Bobo, M.; et al. Multifunctional gold-mesoporous silica nanocomposites for enhanced two-photon imaging and therapy of cancer cells. *Front. Mol. Biosci.* **2016**, *3*, 1. [CrossRef] [PubMed]

155. Croissant, J.; Maynadier, M.; Mongin, O.; Hugues, V.; Blanchard-Desce, M.; Chaix, A.; Cattoen, X.; Wong Chi Man, M.; Gallud, A.; Gary-Bobo, M.; et al. Enhanced two-photon fluorescence imaging and therapy of cancer cells via gold@bridged silsesquioxane nanoparticles. *Small* **2015**, *11*, 295–299. [CrossRef] [PubMed]

156. Oh, M.H.; Yu, J.H.; Kim, I.; Nam, Y.S. Genetically programmed clusters of gold nanoparticles for cancer cell-targeted photothermal therapy. *ACS Appl. Mater. Interfaces* **2015**, *7*, 22578–22586. [CrossRef] [PubMed]

157. Jimenez-Mancilla, N.; Ferro-Flores, G.; Santos-Cuevas, C.; Ocampo-Garcia, B.; Luna-Gutierrez, M.; Azorin-Vega, E.; Isaac-Olive, K.; Camacho-Lopez, M.; Torres-Garcia, E. Multifunctional targeted therapy system based on (99M)TC/(177) Lu-labeled gold nanoparticles-TAT(49–57)-lys(3)-bombesin internalized in nuclei of prostate cancer cells. *J. Labelled Comp. Radiopharm.* **2013**, *56*, 663–671. [CrossRef] [PubMed]

158. Szlachcic, A.; Pala, K.; Zakrzewska, M.; Jakimowicz, P.; Wiedlocha, A.; Otlewski, J. FGF1-gold nanoparticle conjugates targeting FGFR efficiently decrease cell viability upon NIR irradiation. *Int. J. Nanomed.* **2012**, *7*, 5915–5927.

159. Van de Broek, B.; Devoogdt, N.; D'Hollander, A.; Gijs, H.L.; Jans, K.; Lagae, L.; Muyldermans, S.; Maes, G.; Borghs, G. Specific cell targeting with nanobody conjugated branched gold nanoparticles for photothermal therapy. *ACS Nano* **2011**, *5*, 4319–4328. [CrossRef] [PubMed]

160. Lukianova-Hleb, E.Y.; Oginsky, A.O.; Samaniego, A.P.; Shenefelt, D.L.; Wagner, D.S.; Hafner, J.H.; Farach-Carson, M.C.; Lapotko, D.O. Tunable plasmonic nanoprobes for theranostics of prostate cancer. *Theranostics* **2011**, *1*, 3–17. [CrossRef] [PubMed]

161. Lu, W.; Singh, A.K.; Khan, S.A.; Senapati, D.; Yu, H.; Ray, P.C. Gold nano-popcorn-based targeted diagnosis, nanotherapy treatment, and in situ monitoring of photothermal therapy response of prostate cancer cells using surface-enhanced raman spectroscopy. *J. Am. Chem. Soc.* **2010**, *132*, 18103–18114. [CrossRef] [PubMed]

162. Kang, J.H.; Ko, Y.T. Lipid-coated gold nanocomposites for enhanced cancer therapy. *Int. J. Nanomed.* **2015**, *10*, 33–45.

163. Banu, H.; Sethi, D.K.; Edgar, A.; Sheriff, A.; Rayees, N.; Renuka, N.; Faheem, S.M.; Premkumar, K.; Vasanthakumar, G. Doxorubicin loaded polymeric gold nanoparticles targeted to human folate receptor upon laser photothermal therapy potentiates chemotherapy in breast cancer cell lines. *J. Photochem. Photobiol. B* **2015**, *149*, 116–128. [CrossRef] [PubMed]

164. Mkandawire, M.M.; Lakatos, M.; Springer, A.; Clemens, A.; Appelhans, D.; Krause-Buchholz, U.; Pompe, W.; Rodel, G.; Mkandawire, M. Induction of apoptosis in human cancer cells by targeting mitochondria with gold nanoparticles. *Nanoscale* **2015**, *7*, 10634–10640. [CrossRef] [PubMed]

165. Yang, L.; Tseng, Y.T.; Suo, G.; Chen, L.; Yu, J.; Chiu, W.J.; Huang, C.C.; Lin, C.H. Photothermal therapeutic response of cancer cells to aptamer-gold nanoparticle-hybridized graphene oxide under nir illumination. *ACS Appl. Mater. Interfaces* **2015**, *7*, 5097–5106. [CrossRef] [PubMed]

166. Lechtman, E.; Mashouf, S.; Chattopadhyay, N.; Keller, B.M.; Lai, P.; Cai, Z.; Reilly, R.M.; Pignol, J.P. A Monte Carlo-based model of gold nanoparticle radiosensitization accounting for increased radiobiological effectiveness. *Phys. Med. Biol.* **2013**, *58*, 3075–3087. [CrossRef] [PubMed]

167. Jain, S.; Coulter, J.A.; Hounsell, A.R.; Butterworth, K.T.; McMahon, S.J.; Hyland, W.B.; Muir, M.F.; Dickson, G.R.; Prise, K.M.; Currell, F.J.; et al. Cell-specific radiosensitization by gold nanoparticles at megavoltage radiation energies. *Int. J. Radiat. Oncol. Biol. Phys.* **2011**, *79*, 531–539. [CrossRef] [PubMed]

168. Butler, H.J.; Fogarty, S.W.; Kerns, J.G.; Martin-Hirsch, P.L.; Fullwood, N.J.; Martin, F.L. Gold nanoparticles as a substrate in bio-analytical near-infrared surface-enhanced Raman spectroscopy. *Analyst* **2015**, *140*, 3090–3097. [CrossRef] [PubMed]

169. Kalmodia, S.; Harjwani, J.; Rajeswari, R.; Yang, W.; Barrow, C.J.; Ramaprabhu, S.; Krishnakumar, S.; Elchuri, S.V. Synthesis and characterization of surface-enhanced Raman-scattered gold nanoparticles. *Int. J. Nanomed.* **2013**, *8*, 4327–4338. [CrossRef] [PubMed]

170. Zhu, J.; Zhou, J.; Guo, J.; Cai, W.; Liu, B.; Wang, Z.; Sun, Z. Surface-enhanced Raman spectroscopy investigation on human breast cancer cells. *Chem. Cent. J.* **2013**, *7*, 37. [CrossRef] [PubMed]

171. Firdhouse, M.J.; Lalitha, P. Biosynthesis of silver nanoparticles using the extract of -antiproliferative effect against prostate cancer cells. *Cancer Nanotechnol.* **2013**, *4*, 137–143. [CrossRef] [PubMed]

172. Wang, H.; Zhang, Y.; Yu, H.; Wu, D.; Ma, H.; Li, H.; Du, B.; Wei, Q. Label-free electrochemical immunosensor for prostate-specific antigen based on silver hybridized mesoporous silica nanoparticles. *Anal. Biochem.* **2013**, *434*, 123–127. [CrossRef] [PubMed]

173. Nayak, D.; Minz, A.P.; Ashe, S.; Rauta, P.R.; Kumari, M.; Chopra, P.; Nayak, B. Synergistic combination of antioxidants, silver nanoparticles and chitosan in a nanoparticle based formulation: Characterization and cytotoxic effect on MCF-7 breast cancer cell lines. *J. Colloid Interface Sci.* **2016**, *470*, 142–152. [CrossRef] [PubMed]

174. Karunamuni, R.; Naha, P.C.; Lau, K.C.; Al-Zaki, A.; Popov, A.V.; Delikatny, E.J.; Tsourkas, A.; Cormode, D.P.; Maidment, A.D. Development of silica-encapsulated silver nanoparticles as contrast agents intended for dual-energy mammography. *Eur. Radiol.* **2016**, *26*, 3301–3309. [CrossRef] [PubMed]

175. Farah, M.A.; Ali, M.A.; Chen, S.M.; Li, Y.; Al-Hemaid, F.M.; Abou-Tarboush, F.M.; Al-Anazi, K.M.; Lee, J. Silver nanoparticles synthesized from adenium obesum leaf extract induced DNA damage, apoptosis and autophagy via generation of reactive oxygen species. *Colloids Surf. B Biointerfaces* **2016**, *141*, 158–169. [CrossRef] [PubMed]

176. Jannathul Firdhouse, M.; Lalitha, P. Apoptotic efficacy of biogenic silver nanoparticles on human breast cancer MCF-7 cell lines. *Prog. Biomater.* **2015**, *4*, 113–121.

177. Casanas Pimentel, R.G.; Robles Botero, V.; San Martin Martinez, E.; Gomez Garcia, C.; Hinestroza, J.P. Soybean agglutinin-conjugated silver nanoparticles nanocarriers in the treatment of breast cancer cells. *J. Biomater. Sci. Polym. Ed.* **2016**, *27*, 218–234. [CrossRef] [PubMed]

178. Swanner, J.; Mims, J.; Carroll, D.L.; Akman, S.A.; Furdui, C.M.; Torti, S.V.; Singh, R.N. Differential cytotoxic and radiosensitizing effects of silver nanoparticles on triple-negative breast cancer and non-triple-negative breast cells. *Int. J. Nanomed.* **2015**, *10*, 3937–3953.

179. Gurunathan, S.; Park, J.H.; Han, J.W.; Kim, J.H. Comparative assessment of the apoptotic potential of silver nanoparticles synthesized by bacillus tequilensis and calocybe indica in MDA-MB-231 human breast cancer cells: Targeting p53 for anticancer therapy. *Int. J. Nanomed.* **2015**, *10*, 4203–4222. [CrossRef] [PubMed]

180. Wei, L.; Lu, J.; Xu, H.; Patel, A.; Chen, Z.S.; Chen, G. Silver nanoparticles: Synthesis, properties, and therapeutic applications. *Drug Discov. Today* **2015**, *20*, 595–601. [CrossRef] [PubMed]

181. Thompson, E.A.; Graham, E.; MacNeill, C.M.; Young, M.; Donati, G.; Wailes, E.M.; Jones, B.T.; Levi-Polyachenko, N.H. Differential response of MCF-7, MDA-MB-231, and MCF-10a cells to hyperthermia, silver nanoparticles and silver nanoparticle-induced photothermal therapy. *Int. J. Hyperth.* **2014**, *30*, 312–323. [CrossRef] [PubMed]

182. Chung, I.M.; Park, I.; Seung-Hyun, K.; Thiruvengadam, M.; Rajakumar, G. Plant-mediated synthesis of silver nanoparticles: Their characteristic properties and therapeutic applications. *Nanoscale Res. Lett.* **2016**, *11*, 40. [CrossRef] [PubMed]

183. Ghoneum, A.; Zhu, H.; Woo, J.; Zabinyakov, N.; Sharma, S.; Gimzewski, J.K. Biophysical and morphological effects of nanodiamond/nanoplatinum solution (DPV576) on metastatic murine breast cancer cells in vitro. *Nanotechnology* **2014**, *25*, 465101. [CrossRef] [PubMed]

184. Xiao, C.; Liu, Y.L.; Xu, J.Q.; Lv, S.W.; Guo, S.; Huang, W.H. Real-time monitoring of H_2O_2 release from single cells using nanoporous gold microelectrodes decorated with platinum nanoparticles. *Analyst* **2015**, *140*, 3753–3758. [CrossRef] [PubMed]

185. Cui, Z.; Wu, D.; Zhang, Y.; Ma, H.; Li, H.; Du, B.; Wei, Q.; Ju, H. Ultrasensitive electrochemical immunosensors for multiplexed determination using mesoporous platinum nanoparticles as nonenzymatic labels. *Anal. Chim. Acta* **2014**, *807*, 44–50. [CrossRef] [PubMed]

186. Sengupta, P.; Basu, S.; Soni, S.; Pandey, A.; Roy, B.; Oh, M.S.; Chin, K.T.; Paraskar, A.S.; Sarangi, S.; Connor, Y.; et al. Cholesterol-tethered platinum II-based supramolecular nanoparticle increases antitumor efficacy and reduces nephrotoxicity. *Proc. Natl. Acad. Sci. USA* **2012**, *109*, 11294–11299. [CrossRef] [PubMed]

187. Teow, Y.; Valiyaveettil, S. Active targeting of cancer cells using folic acid-conjugated platinum nanoparticles. *Nanoscale* **2010**, *2*, 2607–2613. [CrossRef] [PubMed]

188. Spain, E.; Gilgunn, S.; Sharma, S.; Adamson, K.; Carthy, E.; O'Kennedy, R.; Forster, R.J. Detection of prostate specific antigen based on electrocatalytic platinum nanoparticles conjugated to a recombinant SCFV antibody. *Biosens. Bioelectron.* **2016**, *77*, 759–766. [CrossRef] [PubMed]

189. Zhang, B.; Liu, B.; Chen, G.; Tang, D. Redox and catalysis 'all-in-one' infinite coordination polymer for electrochemical immunosensor of tumor markers. *Biosens. Bioelectron.* **2015**, *64*, 6–12. [CrossRef] [PubMed]

190. Kumar, A.; Huo, S.; Zhang, X.; Liu, J.; Tan, A.; Li, S.; Jin, S.; Xue, X.; Zhao, Y.; Ji, T.; et al. Neuropilin-1-targeted gold nanoparticles enhance therapeutic efficacy of platinum(IV) drug for prostate cancer treatment. *ACS Nano* **2014**, *8*, 4205–4220. [CrossRef] [PubMed]

191. Taylor, R.M.; Sillerud, L.O. Paclitaxel-loaded iron platinum stealth immunomicelles are potent MRI imaging agents that prevent prostate cancer growth in a psma-dependent manner. *Int. J. Nanomed.* **2012**, *7*, 4341–4352. [CrossRef] [PubMed]

192. Taylor, R.M.; Huber, D.L.; Monson, T.C.; Ali, A.M.; Bisoffi, M.; Sillerud, L.O. Multifunctional iron platinum stealth immunomicelles: Targeted detection of human prostate cancer cells using both fluorescence and magnetic resonance imaging. *J. Nanopart. Res.* **2011**, *13*, 4717–4729. [CrossRef] [PubMed]

193. Chuang, C.H.; Brown, P.R.; Bulovic, V.; Bawendi, M.G. Improved performance and stability in quantum dot solar cells through band alignment engineering. *Nat. Mater.* **2014**, *13*, 796–801. [CrossRef] [PubMed]

194. Sun, Q.; Wang, A.; Li, L.; Wang, D.; Zhu, T.; Xu, J.; Yang, C.; Li, Y. Bright, multicoloured light-emitting diodes based on quantum dots. *Nat. Photonics* **2007**, *1*, 717–722. [CrossRef]

195. Fang, M.; Peng, C.W.; Pang, D.W.; Li, Y. Quantum dots for cancer research: Current status, remaining issues, and future perspectives. *Cancer Biol. Med.* **2012**, *9*, 151–163. [PubMed]

196. Jamieson, T.; Bakhshi, R.; Petrova, D.; Pocock, R.; Imani, M.; Seifalian, A.M. Biological applications of quantum dots. *Biomaterials* **2007**, *28*, 4717–4732. [CrossRef] [PubMed]

197. Hotz, C.Z. Applications of quantum dots in biology: An overview. *Methods Mol. Biol.* **2005**, *303*, 1–17. [PubMed]

198. Barar, J.; Omidi, Y. Surface modified multifunctional nanomedicines for simultaneous imaging and therapy of cancer. *Bioimpacts* **2014**, *4*, 3–14. [PubMed]

199. Choi, H.S.; Frangioni, J.V. Nanoparticles for biomedical imaging: Fundamentals of clinical translation. *Mol. Imaging* **2010**, *9*, 291–310. [PubMed]

200. Ow, H.; Larson, D.R.; Srivastava, M.; Baird, B.A.; Webb, W.W.; Wiesner, U. Bright and stable core-shell fluorescent silica nanoparticles. *Nano Lett.* **2005**, *5*, 113–117. [CrossRef] [PubMed]

201. Larson, D.; Ow, H.; Vishwasrao, H.; Heikal, A.; Wiesner, U.; Webb, W. Silica nanoparticle architecture determines radiative properties of encapsulated fluorophores. *Chem. Mater.* **2008**, *20*, 2677–2684. [CrossRef]

202. Burns, A.A.; Vider, J.; Ow, H.; Herz, E.; Penate-Medina, O.; Baumgart, M.; Larson, S.M.; Wiesner, U.; Bradbury, M. Fluorescent silica nanoparticles with efficient urinary excretion for nanomedicine. *Nano Lett.* **2009**, *9*, 442–448. [CrossRef] [PubMed]

203. Schipper, M.L.; Cheng, Z.; Lee, S.W.; Bentolila, L.A.; Iyer, G.; Rao, J.; Chen, X.; Wu, A.M.; Weiss, S.; Gambhir, S.S. Micropet-based biodistribution of quantum dots in living mice. *J. Nucl. Med.* **2007**, *48*, 1511–1518. [CrossRef] [PubMed]

204. Phillips, E.; Penate-Medina, O.; Zanzonico, P.B.; Carvajal, R.D.; Mohan, P.; Ye, Y.; Humm, J.; Gonen, M.; Kalaigian, H.; Schoder, H.; et al. Clinical translation of an ultrasmall inorganic optical-PET imaging nanoparticle probe. *Sci. Transl. Med.* **2014**, *6*, 260ra149. [CrossRef] [PubMed]

205. Choi, H.S.; Liu, W.; Misra, P.; Tanaka, E.; Zimmer, J.P.; Itty Ipe, B.; Bawendi, M.G.; Frangioni, J.V. Renal clearance of quantum dots. *Nat. Biotechnol.* **2007**, *25*, 1165–1170. [CrossRef] [PubMed]

206. Thakur, M.; Mewada, A.; Pandey, S.; Bhori, M.; Singh, K.; Sharon, M.; Sharon, M. Milk-derived multi-fluorescent graphene quantum dot-based cancer theranostic system. *Mater. Sci. Eng. C Mater. Biol. Appl.* **2016**, *67*, 468–477. [CrossRef] [PubMed]

207. Wang, J.; Wang, F.; Li, F.; Zhang, W.; Shen, Y.; Zhou, D.; Guo, S. A multifunctional poly(curcumin) nanomedicine for dual-modal targeted delivery, intracellular responsive release, dual-drug treatment and imaging of multidrug resistant cancer cells. *J. Mater. Chem. B Mater. Biol. Med.* **2016**, *4*, 2954–2962. [CrossRef] [PubMed]

208. Bwatanglang, I.B.; Mohammad, F.; Yusof, N.A.; Abdullah, J.; Hussein, M.Z.; Alitheen, N.B.; Abu, N. Folic acid targeted MN:Zns quantum dots for theranostic applications of cancer cell imaging and therapy. *Int. J. Nanomed.* **2016**, *11*, 413–428.

209. Lin, Z.; Ma, Q.; Fei, X.; Zhang, H.; Su, X. A novel aptamer functionalized cuins2 quantum dots probe for daunorubicin sensing and near infrared imaging of prostate cancer cells. *Anal. Chim. Acta* **2014**, *818*, 54–60. [CrossRef] [PubMed]

210. Zhao, Y.; Shaffer, T.M.; Das, S.; Perez-Medina, C.; Mulder, W.J.; Grimm, J. Near-infrared quantum dot and ^{89}Zr dual-labeled nanoparticles for in vivo Cerenkov imaging. *Bioconjug. Chem.* **2017**, *28*, 600–608. [CrossRef] [PubMed]

211. Klumpp, C.; Kostarelos, K.; Prato, M.; Bianco, A. Functionalized carbon nanotubes as emerging nanovectors for the delivery of therapeutics. *Biochim. Biophys. Acta* **2006**, *1758*, 404–412. [CrossRef] [PubMed]

212. Partha, R.; Conyers, J.L. Biomedical applications of functionalized fullerene-based nanomaterials. *Int. J. Nanomed.* **2009**, *4*, 261–275.

213. Karousis, N.; Suarez-Martinez, I.; Ewels, C.P.; Tagmatarchis, N. Structure, properties, functionalization, and applications of carbon nanohorns. *Chem. Rev.* **2016**, *116*, 4850–4883. [CrossRef] [PubMed]

214. Serpell, C.J.; Kostarelos, K.; Davis, B.G. Can carbon nanotubes deliver on their promise in biology? Harnessing unique properties for unparalleled applications. *ACS Cent. Sci.* **2016**, *2*, 190–200. [CrossRef] [PubMed]

215. Bhattacharya, K.; Mukherjee, S.P.; Gallud, A.; Burkert, S.C.; Bistarelli, S.; Bellucci, S.; Bottini, M.; Star, A.; Fadeel, B. Biological interactions of carbon-based nanomaterials: From coronation to degradation. *Nanomedicine* **2016**, *12*, 333–351. [CrossRef] [PubMed]

216. Lalwani, G.; Sitharaman, B. Multifunctional fullerene- and metallofullerene-based nanobiomaterials. *Nano Life* **2013**, *3*, 1342003. [CrossRef]

217. Ruggiero, A.; Villa, C.H.; Bander, E.; Rey, D.A.; Bergkvist, M.; Batt, C.A.; Manova-Todorova, K.; Deen, W.M.; Scheinberg, D.A.; McDevitt, M.R. Paradoxical glomerular filtration of carbon nanotubes. *Proc. Natl. Acad. Sci. USA* **2010**, *107*, 12369–12374. [CrossRef] [PubMed]

218. Kostarelos, K.; Bianco, A.; Prato, M. Promises, facts and challenges for carbon nanotubes in imaging and therapeutics. *Nat. Nanotechnol.* **2009**, *4*, 627–633. [CrossRef] [PubMed]

219. Zhang, Y.; Petibone, D.; Xu, Y.; Mahmood, M.; Karmakar, A.; Casciano, D.; Ali, S.; Biris, A.S. Toxicity and efficacy of carbon nanotubes and graphene: The utility of carbon-based nanoparticles in nanomedicine. *Drug Metab. Rev.* **2014**, *46*, 232–246. [CrossRef] [PubMed]

220. Porter, A.E.; Gass, M.; Muller, K.; Skepper, J.N.; Midgley, P.A.; Welland, M. Direct imaging of single-walled carbon nanotubes in cells. *Nat. Nanotechnol.* **2007**, *2*, 713–717. [CrossRef] [PubMed]

221. Kolosnjaj, J.; Szwarc, H.; Moussa, F. Toxicity studies of carbon nanotubes. *Adv. Exp. Med. Biol.* **2007**, *620*, 181–204. [PubMed]

222. Poland, C.A.; Duffin, R.; Kinloch, I.; Maynard, A.; Wallace, W.A.; Seaton, A.; Stone, V.; Brown, S.; Macnee, W.; Donaldson, K. Carbon nanotubes introduced into the abdominal cavity of mice show asbestos-like pathogenicity in a pilot study. *Nat. Nanotechnol.* **2008**, *3*, 423–428. [CrossRef] [PubMed]

223. Lam, C.W.; James, J.T.; McCluskey, R.; Arepalli, S.; Hunter, R.L. A review of carbon nanotube toxicity and assessment of potential occupational and environmental health risks. *Crit. Rev. Toxicol.* **2006**, *36*, 189–217. [CrossRef] [PubMed]

224. Corredor, C.; Hou, W.; Klein, S.; Moghadam, B.; Goryll, M.; Doudrick, K.; Westerhoff, P.; Posner, J. Disruption of model cell membranes by carbon nanotubes. *Carbon* **2013**, *60*, 67–75. [CrossRef]

225. Yoshida, M.; Goto, H.; Hirose, Y.; Zhao, X.; Osawa, E. Prediction of favorable isomeric structures for the c100 to c120 giant fullerenes. An application of the phason line criteria. *Electron. J. Theor. Chem.* **1996**, *1*, 163–171. [CrossRef]

226. Castro Nava, A.; Cojoc, M.; Peitzsch, C.; Cirillo, G.; Kurth, I.; Fuessel, S.; Erdmann, K.; Kunhardt, D.; Vittorio, O.; Hampel, S.; et al. Development of novel radiochemotherapy approaches targeting prostate tumor progenitor cells using nanohybrids. *Int. J. Cancer* **2015**, *137*, 2492–2503. [CrossRef] [PubMed]

227. Heydari-Bafrooei, E.; Shamszadeh, N.S. Electrochemical bioassay development for ultrasensitive aptasensing of prostate specific antigen. *Biosens. Bioelectron.* **2017**, *91*, 284–292. [CrossRef] [PubMed]

228. Thapa, R.K.; Youn, Y.S.; Jeong, J.H.; Choi, H.G.; Yong, C.S.; Kim, J.O. Graphene oxide-wrapped pegylated liquid crystalline nanoparticles for effective chemo-photothermal therapy of metastatic prostate cancer cells. *Colloids Surf. B Biointerfaces* **2016**, *143*, 271–277. [CrossRef] [PubMed]

229. Pan, L.H.; Kuo, S.H.; Lin, T.Y.; Lin, C.W.; Fang, P.Y.; Yang, H.W. An electrochemical biosensor to simultaneously detect vegf and psa for early prostate cancer diagnosis based on graphene oxide/ssdna/plla nanoparticles. *Biosens. Bioelectron.* **2017**, *89*, 598–605. [CrossRef] [PubMed]

230. Yang, L.; Cheng, J.; Chen, Y.; Yu, S.; Liu, F.; Sun, Y.; Chen, Y.; Ran, H. Phase-transition nanodroplets for real-time photoacoustic/ultrasound dual-modality imaging and photothermal therapy of sentinel lymph node in breast cancer. *Sci. Rep.* **2017**, *7*, 45213. [CrossRef] [PubMed]

231. Misra, S.K.; Srivastava, I.; Tripathi, I.; Daza, E.; Ostadhossein, F.; Pan, D. Macromolecularly "caged" carbon nanoparticles for intracellular trafficking via switchable photoluminescence. *J. Am. Chem. Soc.* **2017**, *139*, 1746–1749. [CrossRef] [PubMed]

232. Du, J.; Zhang, Y.; Ming, J.; Liu, J.; Zhong, L.; Liang, Q.; Fan, L.; Jiang, J. Evaluation of the tracing effect of carbon nanoparticle and carbon nanoparticle-epirubicin suspension in axillary lymph node dissection for breast cancer treatment. *World J. Surg. Oncol.* **2016**, *14*, 164. [CrossRef] [PubMed]

233. Wu, X.; Lin, Q.; Chen, G.; Lu, J.; Zeng, Y.; Chen, X.; Yan, J. Sentinel lymph node detection using carbon nanoparticles in patients with early breast cancer. *PLoS ONE* **2015**, *10*, e0135714. [CrossRef] [PubMed]

234. Misra, S.K.; Ohoka, A.; Kolmodin, N.J.; Pan, D. Next generation carbon nanoparticles for efficient gene therapy. *Mol. Pharm.* **2015**, *12*, 375–385. [CrossRef] [PubMed]

235. Bangham, A.D. Lipid bilayers and biomembranes. *Annu. Rev. Biochem.* **1972**, *41*, 753–776. [CrossRef] [PubMed]

236. Abou, D.S.; Thorek, D.L.; Ramos, N.N.; Pinkse, M.W.; Wolterbeek, H.T.; Carlin, S.D.; Beattie, B.J.; Lewis, J.S. [89]Zr-labeled paramagnetic octreotide-liposomes for PET-MR imaging of cancer. *Pharm. Res.* **2013**, *30*, 878–888. [CrossRef] [PubMed]

237. Laverman, P.; Brouwers, A.H.; Dams, E.T.; Oyen, W.J.; Storm, G.; van Rooijen, N.; Corstens, F.H.; Boerman, O.C. Preclinical and clinical evidence for disappearance of long-circulating characteristics of polyethylene glycol liposomes at low lipid dose. *J. Pharmacol. Exp. Ther.* **2000**, *293*, 996–1001. [PubMed]

238. Yeh, C.Y.; Hsiao, J.K.; Wang, Y.P.; Lan, C.H.; Wu, H.C. Peptide-conjugated nanoparticles for targeted imaging and therapy of prostate cancer. *Biomaterials* **2016**, *99*, 1–15. [CrossRef] [PubMed]

239. Lin, Q.; Jin, C.S.; Huang, H.; Ding, L.; Zhang, Z.; Chen, J.; Zheng, G. Nanoparticle-enabled, image-guided treatment planning of target specific rnai therapeutics in an orthotopic prostate cancer model. *Small* **2014**, *10*, 3072–3082. [CrossRef] [PubMed]

240. Yaari, Z.; da Silva, D.; Zinger, A.; Goldman, E.; Kajal, A.; Tshuva, R.; Barak, E.; Dahan, N.; Hershkovitz, D.; Goldfeder, M.; et al. Theranostic barcoded nanoparticles for personalized cancer medicine. *Nat. Commun.* **2016**, *7*, 13325. [CrossRef] [PubMed]

241. Rizzitelli, S.; Giustetto, P.; Faletto, D.; Delli Castelli, D.; Aime, S.; Terreno, E. The release of doxorubicin from liposomes monitored by MRI and triggered by a combination of us stimuli led to a complete tumor regression in a breast cancer mouse model. *J. Control. Release* **2016**, *230*, 57–63. [CrossRef] [PubMed]

242. Miller-Kleinhenz, J.M.; Bozeman, E.N.; Yang, L. Targeted nanoparticles for image-guided treatment of triple-negative breast cancer: Clinical significance and technological advances. *Wiley Interdiscip. Rev. Nanomed. Nanobiotechnol.* **2015**, *7*, 797–816. [CrossRef] [PubMed]

243. Rizzitelli, S.; Giustetto, P.; Cutrin, J.C.; Delli Castelli, D.; Boffa, C.; Ruzza, M.; Menchise, V.; Molinari, F.; Aime, S.; Terreno, E. Sonosensitive theranostic liposomes for preclinical in vivo MRI-guided visualization of doxorubicin release stimulated by pulsed low intensity non-focused ultrasound. *J. Control. Release* **2015**, *202*, 21–30. [CrossRef] [PubMed]

244. Shemesh, C.S.; Moshkelani, D.; Zhang, H. Thermosensitive liposome formulated indocyanine green for near-infrared triggered photodynamic therapy: In vivo evaluation for triple-negative breast cancer. *Pharm. Res.* **2015**, *32*, 1604–1614. [CrossRef] [PubMed]

245. He, Y.; Zhang, L.; Zhu, D.; Song, C. Design of multifunctional magnetic iron oxide nanoparticles/mitoxantrone-loaded liposomes for both magnetic resonance imaging and targeted cancer therapy. *Int. J. Nanomed.* **2014**, *9*, 4055–4066. [CrossRef] [PubMed]

246. Muthu, M.S.; Kulkarni, S.A.; Raju, A.; Feng, S.S. Theranostic liposomes of TPGS coating for targeted co-delivery of docetaxel and quantum dots. *Biomaterials* **2012**, *33*, 3494–3501. [CrossRef] [PubMed]

247. Hosoya, H.; Dobroff, A.S.; Driessen, W.H.; Cristini, V.; Brinker, L.M.; Staquicini, F.I.; Cardo-Vila, M.; D'Angelo, S.; Ferrara, F.; Proneth, B.; et al. Integrated nanotechnology platform for tumor-targeted multimodal imaging and therapeutic cargo release. *Proc. Natl. Acad. Sci. USA* **2016**, *113*, 1877–1882. [CrossRef] [PubMed]

248. Lee, J.B.; Zhang, K.; Tam, Y.Y.; Quick, J.; Tam, Y.K.; Lin, P.J.; Chen, S.; Liu, Y.; Nair, J.K.; Zlatev, I.; et al. A glu-urea-lys ligand-conjugated lipid nanoparticle/siRNA system inhibits androgen receptor expression in vivo. *Mol. Ther. Nucleic Acids* **2016**, *5*, e348. [CrossRef] [PubMed]

249. Sharkey, C.C.; Li, J.; Roy, S.; Wu, Q.; King, M.R. Two-stage nanoparticle delivery of piperlongumine and tumor necrosis factor-related apoptosis-inducing ligand (Trail) anti-cancer therapy. *Technology* **2016**, *4*, 60–69. [CrossRef] [PubMed]

250. Bhosale, R.R.; Gangadharappa, H.V.; Hani, U.; Osmani, R.A.; Vaghela, R.; Kulkarni, P.K.; Venkata, K.S. Current perspectives on novel drug delivery systems and therapies for management of prostate cancer: An inclusive review. *Curr. Drug Targets* **2016**. [CrossRef]

251. Majzoub, R.N.; Wonder, E.; Ewert, K.K.; Kotamraju, V.R.; Teesalu, T.; Safinya, C.R. Rab11 and lysotracker markers reveal correlation between endosomal pathways and transfection efficiency of surface-functionalized cationic liposome-DNA nanoparticles. *J. Phys. Chem. B* **2016**, *120*, 6439–6453. [CrossRef] [PubMed]

252. Wang, F.; Chen, L.; Zhang, R.; Chen, Z.; Zhu, L. RGD peptide conjugated liposomal drug delivery system for enhance therapeutic efficacy in treating bone metastasis from prostate cancer. *J. Control. Release* **2014**, *196*, 222–233. [CrossRef] [PubMed]

253. Nguyen, V.D.; Zheng, S.; Han, J.; Le, V.H.; Park, J.O.; Park, S. Nanohybrid magnetic liposome functionalized with hyaluronic acid for enhanced cellular uptake and near-infrared-triggered drug release. *Colloids Surf. B Biointerfaces* **2017**, *154*, 104–114. [CrossRef] [PubMed]

254. Sneider, A.; Jadia, R.; Piel, B.; VanDyke, D.; Tsiros, C.; Rai, P. Engineering remotely triggered liposomes to target triple negative breast cancer. *Oncomedicine* **2017**, *2*, 1–13. [CrossRef] [PubMed]

255. Bayraktar, R.; Pichler, M.; Kanlikilicer, P.; Ivan, C.; Bayraktar, E.; Kahraman, N.; Aslan, B.; Oguztuzun, S.; Ulasli, M.; Arslan, A.; et al. Microrna 603 acts as a tumor suppressor and inhibits triple-negative breast cancer tumorigenesis by targeting elongation factor 2 kinase. *Oncotarget* **2016**, *8*, 11641–11658. [CrossRef] [PubMed]

256. Fernandes, R.S.; Silva, J.O.; Monteiro, L.O.; Leite, E.A.; Cassali, G.D.; Rubello, D.; Cardoso, V.N.; Ferreira, L.A.; Oliveira, M.C.; de Barros, A.L. Doxorubicin-loaded nanocarriers: A comparative study of liposome and nanostructured lipid carrier as alternatives for cancer therapy. *Biomed. Pharmacother.* **2016**, *84*, 252–257. [CrossRef] [PubMed]

257. Alaarg, A.; Jordan, N.Y.; Verhoef, J.J.; Metselaar, J.M.; Storm, G.; Kok, R.J. Docosahexaenoic acid liposomes for targeting chronic inflammatory diseases and cancer: An in vitro assessment. *Int. J. Nanomed.* **2016**, *11*, 5027–5040. [CrossRef] [PubMed]

258. Amiri, B.; Ebrahimi-Far, M.; Saffari, Z.; Akbarzadeh, A.; Soleimani, E.; Chiani, M. Preparation, characterization and cytotoxicity of silibinin-containing nanoniosomes in T47D human breast carcinoma cells. *Asian Pac. J. Cancer Prev.* **2016**, *17*, 3835–3838. [PubMed]

259. Qian, R.C.; Cao, Y.; Long, Y.T. Binary system for microrna-targeted imaging in single cells and photothermal cancer therapy. *Anal. Chem.* **2016**, *88*, 8640–8647. [CrossRef] [PubMed]

260. Cao, H.; Dan, Z.; He, X.; Zhang, Z.; Yu, H.; Yin, Q.; Li, Y. Liposomes coated with isolated macrophage membrane can target lung metastasis of breast cancer. *ACS Nano* **2016**, *10*, 7738–7748. [CrossRef] [PubMed]

261. Jiang, L.; He, B.; Pan, D.; Luo, K.; Yi, Q.; Gu, Z. Anti-cancer efficacy of paclitaxel loaded in PH triggered liposomes. *J. Biomed. Nanotechnol.* **2016**, *12*, 79–90. [CrossRef] [PubMed]

262. Soppimath, K.S.; Aminabhavi, T.M.; Kulkarni, A.R.; Rudzinski, W.E. Biodegradable polymeric nanoparticles as drug delivery devices. *J. Control. Release* **2001**, *70*, 1–20. [CrossRef]

263. Kreuter, J. Drug delivery to the central nervous system by polymeric nanoparticles: What do we know? *Adv. Drug Deliv. Rev.* **2014**, *71*, 2–14. [CrossRef] [PubMed]

264. Rossin, R.; Pan, D.; Qi, K.; Turner, J.L.; Sun, X.; Wooley, K.L.; Welch, M.J. ^{64}Cu-labeled folate-conjugated shell cross-linked nanoparticles for tumor imaging and radiotherapy: Synthesis, radiolabeling, and biologic evaluation. *J. Nucl. Med.* **2005**, *46*, 1210–1218. [PubMed]

265. Salmaso, S.; Caliceti, P. Stealth properties to improve therapeutic efficacy of drug nanocarriers. *J. Drug Deliv.* **2013**, *2013*. [CrossRef] [PubMed]

266. Roberts, J.C.; Adams, Y.E.; Tomalia, D.; Mercer-Smith, J.A.; Lavallee, D.K. Using starburst dendrimers as linker molecules to radiolabel antibodies. *Bioconjug. Chem.* **1990**, *1*, 305–308. [CrossRef] [PubMed]

267. Tomalia, D.; Naylor, A.; Goddard, W. Starburst dendrimers: Molecular-level control of size, shape, surface-chemistry, topology, and flexibility from atoms to macroscopic matter. *Angew. Chem. Int. Ed. Engl.* **1990**, *29*, 138–175. [CrossRef]

268. Sk, U.H.; Kojima, C. Dendrimers for theranostic applications. *Biomol. Concepts* **2015**, *6*, 205–217. [CrossRef] [PubMed]

269. Sharma, A.; Mejia, D.; Maysinger, D.; Kakkar, A. Design and synthesis of multifunctional traceable dendrimers for visualizing drug delivery. *RSC Adv.* **2014**, *4*, 19242–19245. [CrossRef]

270. Sharma, A.; Khatchadourian, A.; Khanna, K.; Sharma, R.; Kakkar, A.; Maysinger, D. Multivalent niacin nanoconjugates for delivery to cytoplasmic lipid droplets. *Biomaterials* **2011**, *32*, 1419–1429. [CrossRef] [PubMed]

271. Wu, C.; Brechbiel, M.; Kozak, R.; Gansow, O. Metal-chelate-dendrimer-antibody constructs for use in radioimmunotherapy and imaging. *Bioorg. Med. Chem. Lett.* **1994**, *4*, 449–454. [CrossRef]

272. Wangler, C.; Moldenhauer, G.; Saffrich, R.; Knapp, E.M.; Beijer, B.; Schnolzer, M.; Wangler, B.; Eisenhut, M.; Haberkorn, U.; Mier, W. Pamam structure-based multifunctional fluorescent conjugates for improved fluorescent labelling of biomacromolecules. *Chemistry* **2008**, *14*, 8116–8130. [CrossRef]

273. Ling, Y.; Wei, K.; Luo, Y.; Gao, X.; Zhong, S. Dual docetaxel/superparamagnetic iron oxide loaded nanoparticles for both targeting magnetic resonance imaging and cancer therapy. *Biomaterials* **2011**, *32*, 7139–7150. [CrossRef]

274. Abbasi, A.Z.; Prasad, P.; Cai, P.; He, C.; Foltz, W.D.; Amini, M.A.; Gordijo, C.R.; Rauth, A.M.; Wu, X.Y. Manganese oxide and docetaxel co-loaded fluorescent polymer nanoparticles for dual modal imaging and chemotherapy of breast cancer. *J. Control. Release* **2015**, *209*, 186–196. [CrossRef] [PubMed]

275. Farokhzad, O.C.; Jon, S.; Khademhosseini, A.; Tran, T.N.; Lavan, D.A.; Langer, R. Nanoparticle-aptamer bioconjugates: A new approach for targeting prostate cancer cells. *Cancer Res.* **2004**, *64*, 7668–7672. [CrossRef] [PubMed]

276. He, C.; Cai, P.; Li, J.; Zhang, T.; Lin, L.; Abbasi, A.Z.; Henderson, J.T.; Rauth, A.M.; Wu, X.Y. Blood-brain barrier-penetrating amphiphilic polymer nanoparticles deliver docetaxel for the treatment of brain metastases of triple negative breast cancer. *J. Control. Release* **2017**, *246*, 98–109. [CrossRef]

277. Pramanik, A.; Laha, D.; Dash, S.K.; Chattopadhyay, S.; Roy, S.; Das, D.K.; Pramanik, P.; Karmakar, P. An in vivo study for targeted delivery of copper-organic complex to breast cancer using chitosan polymer nanoparticles. *Mater. Sci. Eng. C Mater. Biol. Appl.* **2016**, *68*, 327–337. [CrossRef]

278. Zhou, Z.; Munyaradzi, O.; Xia, X.; Green, D.; Bong, D. High-capacity drug carriers from common polymer amphiphiles. *Biomacromolecules* **2016**, *17*, 3060–3066. [CrossRef] [PubMed]

279. Danafar, H.; Sharafi, A.; Kheiri Manjili, H.; Andalib, S. Sulforaphane delivery using MPEG-PCL co-polymer nanoparticles to breast cancer cells. *Pharm. Dev. Technol.* **2016**, 1–10. [CrossRef] [PubMed]
280. Rostami, E.; Kashanian, S.; Azandaryani, A.H.; Faramarzi, H.; Dolatabadi, J.E.; Omidfar, K. Drug targeting using solid lipid nanoparticles. *Chem. Phys. Lipids* **2014**, *181*, 56–61. [CrossRef] [PubMed]
281. Rostami, E.; Kashanian, S.; Azandaryani, A.H. Preparation of solid lipid nanoparticles as drug carriers for levothyroxine sodium with in vitro drug delivery kinetic characterization. *Mol. Biol. Rep.* **2014**, *41*, 3521–3527. [CrossRef] [PubMed]
282. Mashaghi, S.; Jadidi, T.; Koenderink, G.; Mashaghi, A. Lipid nanotechnology. *Int. J. Mol. Sci.* **2013**, *14*, 4242–4282. [CrossRef] [PubMed]
283. Uner, M.; Yener, G. Importance of solid lipid nanoparticles (SLN) in various administration routes and future perspectives. *Int. J. Nanomed.* **2007**, *2*, 289–300.
284. Muller, R.H.; Mader, K.; Gohla, S. Solid lipid nanoparticles (SLN) for controlled drug delivery—A review of the state of the art. *Eur. J. Pharm. Biopharm.* **2000**, *50*, 161–177. [CrossRef]
285. Zur Muhlen, A.; Schwarz, C.; Mehnert, W. Solid lipid nanoparticles (SLN) for controlled drug delivery—Drug release and release mechanism. *Eur. J. Pharm. Biopharm.* **1998**, *45*, 149–155. [CrossRef]
286. Mehnert, W.; Mader, K. Solid lipid nanoparticles: Production, characterization and applications. *Adv. Drug Deliv. Rev.* **2001**, *47*, 165–196. [CrossRef]
287. Jenning, V.; Thunemann, A.F.; Gohla, S.H. Characterisation of a novel solid lipid nanoparticle carrier system based on binary mixtures of liquid and solid lipids. *Int. J. Pharm.* **2000**, *199*, 167–177. [CrossRef]
288. Radaic, A.; de Paula, E.; de Jesus, M.B. Factorial design and development of solid lipid nanoparticles (SLN) for gene delivery. *J. Nanosci. Nanotechnol.* **2015**, *15*, 1793–1800. [CrossRef] [PubMed]
289. Akanda, M.H.; Rai, R.; Slipper, I.J.; Chowdhry, B.Z.; Lamprou, D.; Getti, G.; Douroumis, D. Delivery of retinoic acid to LNCaP human prostate cancer cells using solid lipid nanoparticles. *Int. J. Pharm.* **2015**, *493*, 161–171. [CrossRef] [PubMed]
290. Swami, R.; Singh, I.; Jeengar, M.K.; Naidu, V.G.; Khan, W.; Sistla, R. Adenosine conjugated lipidic nanoparticles for enhanced tumor targeting. *Int. J. Pharm.* **2015**, *486*, 287–296. [CrossRef] [PubMed]
291. de Jesus, M.B.; Radaic, A.; Hinrichs, W.L.; Ferreira, C.V.; de Paula, E.; Hoekstra, D.; Zuhorn, I.S. Inclusion of the helper lipid dioleoyl-phosphatidylethanolamine in solid lipid nanoparticles inhibits their transfection efficiency. *J. Biomed. Nanotechnol.* **2014**, *10*, 355–365. [CrossRef] [PubMed]
292. Carbone, C.; Tomasello, B.; Ruozi, B.; Renis, M.; Puglisi, G. Preparation and optimization of PIT solid lipid nanoparticles via statistical factorial design. *Eur. J. Med. Chem.* **2012**, *49*, 110–117. [CrossRef] [PubMed]
293. De Jesus, M.B.; Ferreira, C.V.; de Paula, E.; Hoekstra, D.; Zuhorn, I.S. Design of solid lipid nanoparticles for gene delivery into prostate cancer. *J. Control. Release* **2010**, *148*, e89–e90. [CrossRef] [PubMed]
294. Jain, A.; Sharma, G.; Kushwah, V.; Thakur, K.; Ghoshal, G.; Singh, B.; Jain, S.; Shivhare, U.S.; Katare, O.P. Fabrication and functional attributes of lipidic nanoconstructs of lycopene: An innovative endeavour for enhanced cytotoxicity in MCF-7 breast cancer cells. *Colloids Surf. B Biointerfaces* **2017**, *152*, 482–491. [CrossRef] [PubMed]
295. Campos, J.; Varas-Godoy, M.; Haidar, Z.S. Physicochemical characterization of chitosan-hyaluronan-coated solid lipid nanoparticles for the targeted delivery of paclitaxel: A proof-of-concept study in breast cancer cells. *Nanomedicine (Lond.)* **2017**, *12*, 473–490. [CrossRef] [PubMed]
296. Wang, F.; Li, L.; Liu, B.; Chen, Z.; Li, C. Hyaluronic acid decorated pluronic p85 solid lipid nanoparticles as a potential carrier to overcome multidrug resistance in cervical and breast cancer. *Biomed. Pharmacother.* **2017**, *86*, 595–604. [CrossRef] [PubMed]
297. Liu, J.; Meng, T.; Yuan, M.; Wen, L.; Cheng, B.; Liu, N.; Huang, X.; Hong, Y.; Yuan, H.; Hu, F. MicroRNA-200c delivered by solid lipid nanoparticles enhances the effect of paclitaxel on breast cancer stem cell. *Int. J. Nanomed.* **2016**, *11*, 6713–6725. [CrossRef] [PubMed]
298. Cavaco, M.C.; Pereira, C.; Kreutzer, B.; Gouveia, L.F.; Silva-Lima, B.; Brito, A.M.; Videira, M. Evading p-glycoprotein mediated-efflux chemoresistance using solid lipid nanoparticles. *Eur. J. Pharm. Biopharm.* **2017**, *110*, 76–84. [CrossRef] [PubMed]
299. Hou, A.H.; Swanson, D.; Barqawi, A.B. Modalities for imaging of prostate cancer. *Adv. Urol.* **2009**. [CrossRef] [PubMed]
300. Prasad, P. *Introduction to Nanomedicine and Nanobioengineering*; John Wiley & Sons, Inc.: Hoboken, NJ, USA, 2012; p. 590.

301. Hainfeld, J.F.; Slatkin, D.N.; Focella, T.M.; Smilowitz, H.M. Gold nanoparticles: A new X-ray contrast agent. *Br. J. Radiol.* **2006**, *79*, 248–253. [CrossRef] [PubMed]

302. Nebuloni, L.; Kuhn, G.A.; Muller, R. A comparative analysis of water-soluble and blood-pool contrast agents for in vivo vascular imaging with micro-ct. *Acad. Radiol* **2013**, *20*, 1247–1255. [CrossRef] [PubMed]

303. Clark, D.P.; Ghaghada, K.; Moding, E.J.; Kirsch, D.G.; Badea, C.T. In vivo characterization of tumor vasculature using iodine and gold nanoparticles and dual energy micro-CT. *Phys. Med. Biol.* **2013**, *58*, 1683–1704. [CrossRef] [PubMed]

304. Wan, F.; Qin, X.; Zhang, G.; Lu, X.; Zhu, Y.; Zhang, H.; Dai, B.; Shi, G.; Ye, D. Oxidized low-density lipoprotein is associated with advanced-stage prostate cancer. *Tumour Biol.* **2015**, *36*, 3573–3582. [CrossRef] [PubMed]

305. Hill, M.L.; Corbin, I.R.; Levitin, R.B.; Cao, W.; Mainprize, J.G.; Yaffe, M.J.; Zheng, G. In vitro assessment of poly-iodinated triglyceride reconstituted low-density lipoprotein: Initial steps toward ct molecular imaging. *Acad. Radiol.* **2010**, *17*, 1359–1365. [CrossRef] [PubMed]

306. Furuya, Y.; Sekine, Y.; Kato, H.; Miyazawa, Y.; Koike, H.; Suzuki, K. Low-density lipoprotein receptors play an important role in the inhibition of prostate cancer cell proliferation by statins. *Prostate Int.* **2016**, *4*, 56–60. [CrossRef] [PubMed]

307. Ai, K.; Liu, Y.; Liu, J.; Yuan, Q.; He, Y.; Lu, L. Large-scale synthesis of Bi_2S_3 nanodots as a contrast agent for in vivo X-ray computed tomography imaging. *Adv. Mater.* **2011**, *23*, 4886–4891. [CrossRef] [PubMed]

308. Carton, A.K.; Gavenonis, S.C.; Currivan, J.A.; Conant, E.F.; Schnall, M.D.; Maidment, A.D. Dual-energy contrast-enhanced digital breast tomosynthesis—A feasibility study. *Br. J. Radiol.* **2010**, *83*, 344–350. [CrossRef] [PubMed]

309. Kim, D.; Jeong, Y.Y.; Jon, S. A drug-loaded aptamer-gold nanoparticle bioconjugate for combined ct imaging and therapy of prostate cancer. *ACS Nano* **2010**, *4*, 3689–3696. [CrossRef] [PubMed]

310. Naha, P.C.; Lau, K.C.; Hsu, J.C.; Hajfathalian, M.; Mian, S.; Chhour, P.; Uppuluri, L.; McDonald, E.S.; Maidment, A.D.; Cormode, D.P. Gold silver alloy nanoparticles (GSAN): An imaging probe for breast cancer screening with dual-energy mammography or computed tomography. *Nanoscale* **2016**, *8*, 13740–13754. [CrossRef] [PubMed]

311. Devaraj, N.K.; Keliher, E.J.; Thurber, G.M.; Nahrendorf, M.; Weissleder, R. [18]F labeled nanoparticles for in vivo PET-CT imaging. *Bioconjug. Chem.* **2009**, *20*, 397–401. [CrossRef] [PubMed]

312. Welch, M.J.; Hawker, C.J.; Wooley, K.L. The advantages of nanoparticles for pet. *J. Nucl. Med.* **2009**, *50*, 1743–1746. [CrossRef] [PubMed]

313. Chrastina, A.; Schnitzer, J.E. Iodine-125 radiolabeling of silver nanoparticles for in vivo SPECT imaging. *Int. J. Nanomed.* **2010**, *5*, 653–659.

314. Woodward, J.; Kennel, S.; Mirzadeh, S.; Dai, S.; Wall, J.; Richey, T.; Avanell, J.; Rondinone, A. In vivo SPECT/CT imaging and biodistribution using radioactive 125-CDTE/ZnS nanoparticles. *Nanotechnology* **2007**, *18*, 175103. [CrossRef]

315. Wong, P.; Li, L.; Chea, J.; Delgado, M.K.; Crow, D.; Poku, E.; Szpikowska, B.; Bowles, N.; Channappa, D.; Colcher, D.; et al. Pet imaging of ^{64}Cu-DOTA-SCFV-anti-psma lipid nanoparticles (LNPS): Enhanced tumor targeting over anti-PSMA SCFV or untargeted LNPS. *Nucl. Med. Biol.* **2017**, *47*, 62–68. [CrossRef] [PubMed]

316. Zhao, Y.; Sultan, D.; Detering, L.; Luehmann, H.; Liu, Y. Facile synthesis, pharmacokinetic and systemic clearance evaluation, and positron emission tomography cancer imaging of (6)(4)Cu-Au alloy nanoclusters. *Nanoscale* **2014**, *6*, 13501–13509. [CrossRef] [PubMed]

317. Pressly, E.D.; Pierce, R.A.; Connal, L.A.; Hawker, C.J.; Liu, Y. Nanoparticle PET/CT imaging of natriuretic peptide clearance receptor in prostate cancer. *Bioconjug. Chem.* **2013**, *24*, 196–204. [CrossRef] [PubMed]

318. Shen, Y.; Ma, Z.; Chen, F.; Dong, Q.; Hu, Q.; Bai, L.; Chen, J. Effective photothermal chemotherapy with docetaxel-loaded gold nanospheres in advanced prostate cancer. *J. Drug Target.* **2015**, *23*, 568–576. [CrossRef] [PubMed]

319. Behnam Azad, B.; Banerjee, S.R.; Pullambhatla, M.; Lacerda, S.; Foss, C.A.; Wang, Y.; Ivkov, R.; Pomper, M.G. Evaluation of a PSMA-targeted BNF nanoparticle construct. *Nanoscale* **2015**, *7*, 4432–4442. [CrossRef] [PubMed]

320. Lee, C.; Lo, S.T.; Lim, J.; da Costa, V.C.; Ramezani, S.; Oz, O.K.; Pavan, G.M.; Annunziata, O.; Sun, X.; Simanek, E.E. Design, synthesis and biological assessment of a triazine dendrimer with approximately 16 paclitaxel groups and 8 PEG groups. *Mol. Pharm.* **2013**, *10*, 4452–4461. [CrossRef] [PubMed]

321. Mendoza-Sanchez, A.N.; Ferro-Flores, G.; Ocampo-Garcia, B.E.; Morales-Avila, E.; de, M.R.F.; De Leon-Rodriguez, L.M.; Santos-Cuevas, C.L.; Medina, L.A.; Rojas-Calderon, E.L.; Camacho-Lopez, M.A. Lys3-bombesin conjugated to 99mTc-labelled gold nanoparticles for in vivo gastrin releasing peptide-receptor imaging. *J. Biomed. Nanotechnol.* **2010**, *6*, 375–384. [CrossRef] [PubMed]

322. Lee, H.; Shields, A.F.; Siegel, B.A.; Miller, K.D.; Krop, I.; Ma, C.X.; LoRusso, P.M.; Munster, P.N.; Campbell, K.; Gaddy, D.F.; et al. ^{64}Cu-MM-302 positron emission tomography quantifies variability of enhanced permeability and retention of nanoparticles in relation to treatment response in patients with metastatic breast cancer. *Clin. Cancer Res.* **2017**. [CrossRef] [PubMed]

323. Aanei, I.L.; ElSohly, A.M.; Farkas, M.E.; Netirojjanakul, C.; Regan, M.; Taylor Murphy, S.; O'Neil, J.P.; Seo, Y.; Francis, M.B. Biodistribution of antibody-MS2 viral capsid conjugates in breast cancer models. *Mol. Pharm.* **2016**, *13*, 3764–3772. [CrossRef] [PubMed]

324. Perez-Medina, C.; Abdel-Atti, D.; Zhang, Y.; Longo, V.A.; Irwin, C.P.; Binderup, T.; Ruiz-Cabello, J.; Fayad, Z.A.; Lewis, J.S.; Mulder, W.J.; et al. A modular labeling strategy for in vivo pet and near-infrared fluorescence imaging of nanoparticle tumor targeting. *J. Nucl. Med.* **2014**, *55*, 1706–1711. [CrossRef] [PubMed]

325. Wang, Y.; Black, K.C.; Luehmann, H.; Li, W.; Zhang, Y.; Cai, X.; Wan, D.; Liu, S.Y.; Li, M.; Kim, P.; et al. Comparison study of gold nanohexapods, nanorods, and nanocages for photothermal cancer treatment. *ACS Nano* **2013**, *7*, 2068–2077. [CrossRef] [PubMed]

326. Tseng, Y.C.; Xu, Z.; Guley, K.; Yuan, H.; Huang, L. Lipid-calcium phosphate nanoparticles for delivery to the lymphatic system and SPECT/CT imaging of lymph node metastases. *Biomaterials* **2014**, *35*, 4688–4698. [CrossRef] [PubMed]

327. Kilcoyne, A.; Price, M.C.; McDermott, S.; Harisinghani, M.G. Imaging on nodal staging of prostate cancer. *Future Oncol.* **2017**, *13*, 551–565. [CrossRef] [PubMed]

328. Ruiz-Cabello, J.; Barnett, B.P.; Bottomley, P.A.; Bulte, J.W. Fluorine (^{19}F) MRS and MRI in biomedicine. *NMR Biomed.* **2011**, *24*, 114–129. [CrossRef] [PubMed]

329. Schmieder, A.H.; Caruthers, S.D.; Keupp, J.; Wickline, S.A.; Lanza, G.M. Recent advances in fluorine magnetic resonance imaging with perfluorocarbon emulsions. *Engineering (Beijing)* **2015**, *1*, 475–489. [PubMed]

330. Chen, H.; Song, M.; Tang, J.; Hu, G.; Xu, S.; Guo, Z.; Li, N.; Cui, J.; Zhang, X.; Chen, X.; et al. Ultrahigh (19)f loaded cu1.75s nanoprobes for simultaneous ^{19}F magnetic resonance imaging and photothermal therapy. *ACS Nano* **2016**, *10*, 1355–1362. [CrossRef] [PubMed]

331. Muhammad, G.; Jablonska, A.; Rose, L.; Walczak, P.; Janowski, M. Effect of MRI tags: Spio nanoparticles and ^{19}F nanoemulsion on various populations of mouse mesenchymal stem cells. *Acta Neurobiol. Exp. (Wars)* **2015**, *75*, 144–159. [PubMed]

332. Amiri, H.; Srinivas, M.; Veltien, A.; van Uden, M.J.; de Vries, I.J.; Heerschap, A. Cell tracking using ^{19}F magnetic resonance imaging: Technical aspects and challenges towards clinical applications. *Eur. Radiol.* **2015**, *25*, 726–735. [CrossRef] [PubMed]

333. Maxwell, R.J.; Frenkiel, T.A.; Newell, D.R.; Bauer, C.; Griffiths, J.R. [19]F nuclear magnetic resonance imaging of drug distribution in vivo: The disposition of an antifolate anticancer drug in mice. *Magn. Reson. Med.* **1991**, *17*, 189–196. [CrossRef] [PubMed]

334. Kim, J.G.; Zhao, D.; Song, Y.; Constantinescu, A.; Mason, R.P.; Liu, H. Interplay of tumor vascular oxygenation and tumor PO2 observed using near-infrared spectroscopy, an oxygen needle electrode, and 19f mr po2 mapping. *J. Biomed. Opt.* **2003**, *8*, 53–62. [CrossRef] [PubMed]

335. Yu, W.; Yang, Y.; Bo, S.; Li, Y.; Chen, S.; Yang, Z.; Zheng, X.; Jiang, Z.X.; Zhou, X. Design and synthesis of fluorinated dendrimers for sensitive ^{19}F MRI. *J. Org. Chem.* **2015**, *80*, 4443–4449. [CrossRef] [PubMed]

336. Matsushita, H.; Mizukami, S.; Sugihara, F.; Nakanishi, Y.; Yoshioka, Y.; Kikuchi, K. Multifunctional core-shell silica nanoparticles for highly sensitive ^{19}F magnetic resonance imaging. *Angew. Chem. Int. Ed. Engl.* **2014**, *53*, 1008–1011. [CrossRef] [PubMed]

337. Bae, P.K.; Jung, J.; Lim, S.J.; Kim, D.; Kim, S.K.; Chung, B.H. Bimodal perfluorocarbon nanoemulsions for nasopharyngeal carcinoma targeting. *Mol. Imaging Biol.* **2013**, *15*, 401–410. [CrossRef] [PubMed]

338. Ahrens, E.T.; Helfer, B.M.; O'Hanlon, C.F.; Schirda, C. Clinical cell therapy imaging using a perfluorocarbon tracer and fluorine-19 MRI. *Magn. Reson. Med.* **2014**, *72*, 1696–1701. [CrossRef] [PubMed]

339. Cho, S.; Park, W.; Kim, D.H. Silica-coated metal chelating-melanin nanoparticles as a dual-modal contrast enhancement imaging and therapeutic agent. *ACS Appl. Mater. Interfaces* **2017**, *9*, 101–111. [CrossRef] [PubMed]

340. Moghanaki, D.; Turkbey, B.; Vapiwala, N.; Ehdaie, B.; Frank, S.J.; McLaughlin, P.W.; Harisinghani, M. Advances in prostate cancer magnetic resonance imaging and positron emission tomography-computed tomography for staging and radiotherapy treatment planning. *Semin. Radiat. Oncol.* **2017**, *27*, 21–33. [CrossRef] [PubMed]

341. Jayapaul, J.; Arns, S.; Bunker, M.; Weiler, M.; Rutherford, S.; Comba, P.; Kiessling, F. In vivo evaluation of riboflavin receptor targeted fluorescent USPIO in mice with prostate cancer xenografts. *Nano Res.* **2016**, *9*, 1319–1333. [CrossRef] [PubMed]

342. Kilcoyne, A.; Harisinghani, M.G.; Mahmood, U. Prostate cancer imaging and therapy: Potential role of nanoparticles. *J. Nucl. Med.* **2016**, *57*, 105S–110S. [CrossRef] [PubMed]

343. Jin, C.S.; Overchuk, M.; Cui, L.; Wilson, B.C.; Bristow, R.G.; Chen, J.; Zheng, G. Nanoparticle-enabled selective destruction of prostate tumor using MRI-guided focal photothermal therapy. *Prostate* **2016**, *76*, 1169–1181. [CrossRef] [PubMed]

344. Hurley, K.R.; Ring, H.L.; Etheridge, M.; Zhang, J.; Gao, Z.; Shao, Q.; Klein, N.D.; Szlag, V.M.; Chung, C.; Reineke, T.M.; et al. Predictable heating and positive MRI contrast from a mesoporous silica-coated iron oxide nanoparticle. *Mol. Pharm.* **2016**, *13*, 2172–2183. [CrossRef] [PubMed]

345. Turino, L.N.; Ruggiero, M.R.; Stefania, R.; Cutrin, J.C.; Aime, S.; Geninatti Crich, S. Ferritin decorated PLGA/paclitaxel loaded nanoparticles endowed with an enhanced toxicity toward MCF-7 breast tumor cells. *Bioconjug. Chem.* **2017**, *28*, 1283–1290. [CrossRef] [PubMed]

346. Shamsi, M.; Pirayesh Islamian, J. Breast cancer: Early diagnosis and effective treatment by drug delivery tracing. *Nucl. Med. Rev. Cent. East. Eur.* **2017**, *20*, 45–48. [CrossRef] [PubMed]

347. Shan, X.H.; Wang, P.; Xiong, F.; Lu, H.Y.; Hu, H. Detection of human breast cancer cells using a 2-deoxy-D-glucose-functionalized superparamagnetic iron oxide nanoparticles. *Cancer Biomark.* **2017**. [CrossRef] [PubMed]

348. Keshtkar, M.; Shahbazi-Gahrouei, D.; Khoshfetrat, S.M.; Mehrgardi, M.A.; Aghaei, M. Aptamer-conjugated magnetic nanoparticles as targeted magnetic resonance imaging contrast agent for breast cancer. *J. Med. Signals Sens.* **2016**, *6*, 243–247. [PubMed]

349. Zhang, L.; Varma, N.R.; Gang, Z.Z.; Ewing, J.R.; Arbab, A.S.; Ali, M.M. Targeting triple negative breast cancer with a small-sized paramagnetic nanoparticle. *J. Nanomed. Nanotechnol.* **2016**, *7*, 404. [CrossRef] [PubMed]

350. Moon, S.H.; Yang, B.Y.; Kim, Y.J.; Hong, M.K.; Lee, Y.S.; Lee, D.S.; Chung, J.K.; Jeong, J.M. Development of a complementary PET/MR dual-modal imaging probe for targeting prostate-specific membrane antigen (PSMA). *Nanomedicine* **2016**, *12*, 871–879. [CrossRef] [PubMed]

351. Hu, H.; Li, D.; Liu, S.; Wang, M.; Moats, R.; Conti, P.S.; Li, Z. Integrin α2β1 targeted GdVO4:Eu ultrathin nanosheet for multimodal PET/MR imaging. *Biomaterials* **2014**, *35*, 8649–8658. [CrossRef] [PubMed]

352. Aryal, S.; Key, J.; Stigliano, C.; Landis, M.D.; Lee, D.Y.; Decuzzi, P. Positron emitting magnetic nanoconstructs for PET/MR imaging. *Small* **2014**, *10*, 2688–2696. [CrossRef] [PubMed]

353. Zhao, F.; Zhou, J.; Su, X.; Wang, Y.; Yan, X.; Jia, S.; Du, B. A smart responsive dual aptamers-targeted bubble-generating nanosystem for cancer triplex therapy and ultrasound imaging. *Small* **2017**. [CrossRef] [PubMed]

354. Xu, L.; Wan, C.; Du, J.; Li, H.; Liu, X.; Yang, H.; Li, F. Synthesis, characterization, and in vitro evaluation of targeted gold nanoshelled poly(D,L-lactide-*co*-glycolide) nanoparticles carrying anti p53 antibody as a theranostic agent for ultrasound contrast imaging and photothermal therapy. *J. Biomater. Sci. Polym. Ed.* **2017**, *28*, 415–430. [CrossRef] [PubMed]

355. Baghbani, F.; Moztarzadeh, F.; Mohandesi, J.A.; Yazdian, F.; Mokhtari-Dizaji, M. Novel alginate-stabilized doxorubicin-loaded nanodroplets for ultrasounic theranosis of breast cancer. *Int. J. Biol. Macromol.* **2016**, *93*, 512–519. [CrossRef] [PubMed]

356. Lee, J.Y.; Carugo, D.; Crake, C.; Owen, J.; de Saint Victor, M.; Seth, A.; Coussios, C.; Stride, E. Nanoparticle-loaded protein-polymer nanodroplets for improved stability and conversion efficiency in ultrasound imaging and drug delivery. *Adv. Mater.* **2015**, *27*, 5484–5492. [CrossRef] [PubMed]

357. Wang, S.; Dai, Z.; Ke, H.; Qu, E.; Qi, X.; Zhang, K.; Wang, J. Contrast ultrasound-guided photothermal therapy using gold nanoshelled microcapsules in breast cancer. *Eur. J. Radiol.* **2014**, *83*, 117–122. [CrossRef] [PubMed]

358. Agemy, L.; Sugahara, K.N.; Kotamraju, V.R.; Gujraty, K.; Girard, O.M.; Kono, Y.; Mattrey, R.F.; Park, J.H.; Sailor, M.J.; Jimenez, A.I.; et al. Nanoparticle-induced vascular blockade in human prostate cancer. *Blood* **2010**, *116*, 2847–2856. [CrossRef] [PubMed]

359. Yang, H.; Deng, L.; Li, T.; Shen, X.; Yan, J.; Zuo, L.; Wu, C.; Liu, Y. Multifunctional PLGA nanobubbles as theranostic agents: Combining doxorubicin and P-GP siRNA co-delivery into human breast cancer cells and ultrasound cellular imaging. *J. Biomed. Nanotechnol.* **2015**, *11*, 2124–2136. [CrossRef] [PubMed]

360. Tong, H.P.; Wang, L.F.; Guo, Y.L.; Li, L.; Fan, X.Z.; Ding, J.; Huang, H.Y. Preparation of protamine cationic nanobubbles and experimental study of their physical properties and in vivo contrast enhancement. *Ultrasound Med. Biol.* **2013**, *39*, 2147–2157. [CrossRef] [PubMed]

361. Thorek, D.; Robertson, R.; Bacchus, W.A.; Hahn, J.; Rothberg, J.; Beattie, B.J.; Grimm, J. Cerenkov imaging —A new modality for molecular imaging. *Am. J. Nucl. Med. Mol. Imaging* **2012**, *2*, 163–173. [PubMed]

362. Black, K.C.; Ibricevic, A.; Gunsten, S.P.; Flores, J.A.; Gustafson, T.P.; Raymond, J.E.; Samarajeewa, S.; Shrestha, R.; Felder, S.E.; Cai, T.; et al. In vivo fate tracking of degradable nanoparticles for lung gene transfer using pet and cerenkov imaging. *Biomaterials* **2016**, *98*, 53–63. [CrossRef] [PubMed]

363. Spinelli, A.E.; Schiariti, M.P.; Grana, C.M.; Ferrari, M.; Cremonesi, M.; Boschi, F. Cerenkov and radioluminescence imaging of brain tumor specimens during neurosurgery. *J. Biomed. Opt.* **2016**, *21*, 50502. [CrossRef] [PubMed]

364. Schwenck, J.; Fuchs, K.; Eilenberger, S.H.; Rolle, A.M.; Castaneda Vega, S.; Thaiss, W.M.; Maier, F.C. Fluorescence and Cerenkov luminescence imaging. Applications in small animal research. *Nuklearmedizin* **2016**, *55*, 63–70. [PubMed]

365. Pandya, D.N.; Hantgan, R.; Budzevich, M.M.; Kock, N.D.; Morse, D.L.; Batista, I.; Mintz, A.; Li, K.C.; Wadas, T.J. Preliminary therapy evaluation of ^{225}Ac-DOTA-c(RGDyK) demonstrates that cerenkov radiation derived from ^{225}Ac daughter decay can be detected by optical imaging for in vivo tumor visualization. *Theranostics* **2016**, *6*, 698–709. [CrossRef] [PubMed]

366. Shimamoto, M.; Gotoh, K.; Hasegawa, K.; Kojima, A. Hybrid light imaging using cerenkov luminescence and liquid scintillation for preclinical optical imaging in vivo. *Mol. Imaging Biol.* **2016**, *18*, 500–509. [CrossRef] [PubMed]

367. Andreozzi, J.M.; Zhang, R.; Gladstone, D.J.; Williams, B.B.; Glaser, A.K.; Pogue, B.W.; Jarvis, L.A. Cherenkov imaging method for rapid optimization of clinical treatment geometry in total skin electron beam therapy. *Med. Phys.* **2016**, *43*, 993–1003. [CrossRef] [PubMed]

368. Wibmer, A.G.; Burger, I.A.; Sala, E.; Hricak, H.; Weber, W.A.; Vargas, H.A. Molecular imaging of prostate cancer. *Radiographics* **2016**, *36*, 142–159. [CrossRef] [PubMed]

369. Vargas, H.A.; Grimm, J.; O, F.D.; Sala, E.; Hricak, H. Molecular imaging of prostate cancer: Translating molecular biology approaches into the clinical realm. *Eur. Radiol.* **2015**, *25*, 1294–1302. [CrossRef] [PubMed]

370. Lohrmann, C.; Zhang, H.; Thorek, D.L.; Desai, P.; Zanzonico, P.B.; O'Donoghue, J.; Irwin, C.P.; Reiner, T.; Grimm, J.; Weber, W.A. Cerenkov luminescence imaging for radiation dose calculation of a (9)(0)Y-labeled gastrin-releasing peptide receptor antagonist. *J. Nucl. Med.* **2015**, *56*, 805–811. [CrossRef] [PubMed]

371. Hu, Z.; Chi, C.; Liu, M.; Guo, H.; Zhang, Z.; Zeng, C.; Ye, J.; Wang, J.; Tian, J.; Yang, W.; et al. Nanoparticle-mediated radiopharmaceutical-excited fluorescence molecular imaging allows precise image-guided tumor-removal surgery. *Nanomedicine* **2017**, *13*, 1323–1331. [CrossRef] [PubMed]

372. Madru, R.; Tran, T.A.; Axelsson, J.; Ingvar, C.; Bibic, A.; Stahlberg, F.; Knutsson, L.; Strand, S.E. ^{68}Ga-labeled superparamagnetic iron oxide nanoparticles (SPIONs) for multi-modality PET/MR/Cherenkov luminescence imaging of sentinel lymph nodes. *Am. J. Nucl. Med. Mol. Imaging* **2013**, *4*, 60–69. [PubMed]

373. Hu, Z.; Zhao, M.; Qu, Y.; Zhang, X.; Zhang, M.; Liu, M.; Guo, H.; Zhang, Z.; Wang, J.; Yang, W.; et al. In vivo 3-dimensional radiopharmaceutical-excited fluorescence tomography. *J. Nucl. Med.* **2017**, *58*, 169–174. [CrossRef] [PubMed]

374. Lee, S.B.; Yoon, G.; Lee, S.W.; Jeong, S.Y.; Ahn, B.C.; Lim, D.K.; Lee, J.; Jeon, Y.H. Combined positron emission tomography and Cerenkov luminescence imaging of sentinel lymph nodes using pegylated radionuclide-embedded gold nanoparticles. *Small* **2016**, *12*, 4894–4901. [CrossRef] [PubMed]

375. Tanha, K.; Pashazadeh, A.M.; Pogue, B.W. Review of biomedical Cerenkov luminescence imaging applications. *Biomed. Opt. Express* **2015**, *6*, 3053–3065. [CrossRef] [PubMed]

376. Zhang, M.; Kim, H.S.; Jin, T.; Yi, A.; Moon, W.K. Ultrasound-guided photoacoustic imaging for the selective detection of EGFR-expressing breast cancer and lymph node metastases. *Biomed. Opt. Express* **2016**, *7*, 1920–1931. [CrossRef] [PubMed]

377. Su, Y.; Teng, Z.; Yao, H.; Wang, S.; Tian, Y.; Zhang, Y.; Liu, W.; Tian, W.; Zheng, L.; Lu, N.; et al. A multifunctional PB@mSiO$_2$-PEG/DOX nanoplatform for combined photothermal-chemotherapy of tumor. *ACS Appl. Mater. Interfaces* **2016**, *8*, 17038–17046. [CrossRef] [PubMed]

378. Feng, H.; Xia, X.; Li, C.; Song, Y.; Qin, C.; Zhang, Y.; Lan, X. Tyr as a multifunctional reporter gene regulated by the tet-on system for multimodality imaging: An in vitro study. *Sci. Rep.* **2015**, *5*, 15502. [CrossRef] [PubMed]

379. Zhang, T.; Cui, H.; Fang, C.Y.; Cheng, K.; Yang, X.; Chang, H.C.; Forrest, M.L. Targeted nanodiamonds as phenotype-specific photoacoustic contrast agents for breast cancer. *Nanomedicine (Lond.)* **2015**, *10*, 573–587. [CrossRef] [PubMed]

380. Levi, J.; Sathirachinda, A.; Gambhir, S.S. A high-affinity, high-stability photoacoustic agent for imaging gastrin-releasing peptide receptor in prostate cancer. *Clin. Cancer Res.* **2014**, *20*, 3721–3729. [CrossRef] [PubMed]

381. Dogra, V.S.; Chinni, B.K.; Valluru, K.S.; Joseph, J.V.; Ghazi, A.; Yao, J.L.; Evans, K.; Messing, E.M.; Rao, N.A. Multispectral photoacoustic imaging of prostate cancer: Preliminary ex vivo results. *J. Clin. Imaging Sci.* **2013**, *3*, 41. [CrossRef] [PubMed]

382. Kim, G.; Huang, S.W.; Day, K.C.; O'Donnell, M.; Agayan, R.R.; Day, M.A.; Kopelman, R.; Ashkenazi, S. Indocyanine-green-embedded pebbles as a contrast agent for photoacoustic imaging. *J. Biomed. Opt.* **2007**, *12*, 044020. [CrossRef] [PubMed]

383. Kim, C.; Favazza, C.; Wang, L.V. In vivo photoacoustic tomography of chemicals: High-resolution functional and molecular optical imaging at new depths. *Chem. Rev.* **2010**, *110*, 2756–2782. [CrossRef] [PubMed]

384. Tian, C.; Qian, W.; Shao, X.; Xie, Z.; Cheng, X.; Liu, S.; Cheng, Q.; Liu, B.; Wang, X. Plasmonic nanoparticles with quantitatively controlled bioconjugation for photoacoustic imaging of live cancer cells. *Adv. Sci. (Weinh)* **2016**, *3*, 1600237. [CrossRef] [PubMed]

385. Olafsson, R.; Bauer, D.R.; Montilla, L.G.; Witte, R.S. Real-time, contrast enhanced photoacoustic imaging of cancer in a mouse window chamber. *Opt. Express* **2010**, *18*, 18625–18632. [CrossRef] [PubMed]

386. Zhang, H. Cyclic arg-gly-asp-polyethyleneglycol-single-walled carbon nanotubes. In *Molecular Imaging and Contrast Agent Database (Micad)*; Bethesda: Washington DC MD, USA, 2004.

387. Xia, J.; Feng, G.; Xia, X.; Hao, L.; Wang, Z. Nh4hco3 gas-generating liposomal nanoparticle for photoacoustic imaging in breast cancer. *Int. J. Nanomed.* **2017**, *12*, 1803–1813. [CrossRef] [PubMed]

388. Biffi, S.; Petrizza, L.; Garrovo, C.; Rampazzo, E.; Andolfi, L.; Giustetto, P.; Nikolov, I.; Kurdi, G.; Danailov, M.B.; Zauli, G.; et al. Multimodal near-infrared-emitting plus silica nanoparticles with fluorescent, photoacoustic, and photothermal capabilities. *Int. J. Nanomed.* **2016**, *11*, 4865–4874.

389. Hu, D.; Liu, C.; Song, L.; Cui, H.; Gao, G.; Liu, P.; Sheng, Z.; Cai, L. Indocyanine green-loaded polydopamine-iron ions coordination nanoparticles for photoacoustic/magnetic resonance dual-modal imaging-guided cancer photothermal therapy. *Nanoscale* **2016**, *8*, 17150–17158. [CrossRef] [PubMed]

390. Cai, X.; Liu, X.; Liao, L.D.; Bandla, A.; Ling, J.M.; Liu, Y.H.; Thakor, N.; Bazan, G.C.; Liu, B. Encapsulated conjugated oligomer nanoparticles for real-time photoacoustic sentinel lymph node imaging and targeted photothermal therapy. *Small* **2016**, *12*, 4873–4880. [CrossRef] [PubMed]

391. Pham, E.; Yin, M.; Peters, C.G.; Lee, C.R.; Brown, D.; Xu, P.; Man, S.; Jayaraman, L.; Rohde, E.; Chow, A.; et al. Preclinical efficacy of bevacizumab with crlx101, an investigational nanoparticle-drug conjugate, in treatment of metastatic triple-negative breast cancer. *Cancer Res.* **2016**, *76*, 4493–4503. [CrossRef] [PubMed]

392. Feng, Q.; Zhang, Y.; Zhang, W.; Shan, X.; Yuan, Y.; Zhang, H.; Hou, L.; Zhang, Z. Tumor-targeted and multi-stimuli responsive drug delivery system for near-infrared light induced chemo-phototherapy and photoacoustic tomography. *Acta Biomater.* **2016**, *38*, 129–142. [CrossRef] [PubMed]

393. Ahir, M.; Bhattacharya, S.; Karmakar, S.; Mukhopadhyay, A.; Mukherjee, S.; Ghosh, S.; Chattopadhyay, S.; Patra, P.; Adhikary, A. Tailored-CuO-nanowire decorated with folic acid mediated coupling of the mitochondrial-ROS generation and MIR425-PTEN axis in furnishing potent anti-cancer activity in human triple negative breast carcinoma cells. *Biomaterials* **2016**, *76*, 115–132. [CrossRef] [PubMed]

394. Zevon, M.; Ganapathy, V.; Kantamneni, H.; Mingozzi, M.; Kim, P.; Adler, D.; Sheng, Y.; Tan, M.C.; Pierce, M.; Riman, R.E.; et al. CXCR-4 targeted, short wave infrared (SWIR) emitting nanoprobes for enhanced deep tissue imaging and micrometastatic cancer lesion detection. *Small* **2015**, *11*, 6347–6357. [CrossRef] [PubMed]

395. Ozel, T.; White, S.; Nguyen, E.; Moy, A.; Brenes, N.; Choi, B.; Betancourt, T. Enzymatically activated near infrared nanoprobes based on amphiphilic block copolymers for optical detection of cancer. *Lasers Surg. Med.* **2015**. [CrossRef]

396. Yuan, J.P.; Wang, L.W.; Qu, A.P.; Chen, J.M.; Xiang, Q.M.; Chen, C.; Sun, S.R.; Pang, D.W.; Liu, J.; Li, Y. Quantum dots-based quantitative and in situ multiple imaging on ki67 and cytokeratin to improve ki67 assessment in breast cancer. *PLoS ONE* **2015**, *10*, e0122734. [CrossRef] [PubMed]

397. D'Angelis do, E.S.B.C.; Correa, J.R.; Medeiros, G.A.; Barreto, G.; Magalhaes, K.G.; de Oliveira, A.L.; Spencer, J.; Rodrigues, M.O.; Neto, B.A. Carbon dots (C-dots) from cow manure with impressive subcellular selectivity tuned by simple chemical modification. *Chemistry* **2015**, *21*, 5055–5060. [CrossRef] [PubMed]

398. Li, J.; Jiang, X.; Guo, Y.; An, S.; Kuang, Y.; Ma, H.; He, X.; Jiang, C. Linear-dendritic copolymer composed of polyethylene glycol and all-trans-retinoic acid as drug delivery platform for paclitaxel against breast cancer. *Bioconjug. Chem.* **2015**, *26*, 418–426. [CrossRef] [PubMed]

399. Montecinos, V.P.; Morales, C.H.; Fischer, T.H.; Burns, S.; San Francisco, I.F.; Godoy, A.S.; Smith, G.J. Selective targeting of bioengineered platelets to prostate cancer vasculature: New paradigm for therapeutic modalities. *J. Cell Mol. Med.* **2015**, *19*, 1530–1537. [CrossRef] [PubMed]

400. Yu, Y.; Huang, T.; Wu, Y.; Ma, X.; Yu, G.; Qi, J. In Vitro and in vivo imaging of prostate tumor using NAYF4: Yb, Er up-converting nanoparticles. *Pathol. Oncol. Res.* **2014**, *20*, 335–341. [CrossRef] [PubMed]

401. Laprise-Pelletier, M.; Lagueux, J.; Cote, M.F.; LaGrange, T.; Fortin, M.A. Low-dose prostate cancer brachytherapy with radioactive palladium-gold nanoparticles. *Adv. Healthc. Mater.* **2017**, *6*. [CrossRef] [PubMed]

402. Li, W.; Yalcin, M.; Lin, Q.; Ardawi, M.M.; Mousa, S.A. Self-assembly of green tea catechin derivatives in nanoparticles for oral lycopene delivery. *J. Control. Release* **2017**, *248*, 117–124. [CrossRef] [PubMed]

403. Netala, V.R.; Bethu, M.S.; Pushpalatha, B.; Baki, V.B.; Aishwarya, S.; Rao, J.V.; Tartte, V. Biogenesis of silver nanoparticles using endophytic fungus pestalotiopsis microspora and evaluation of their antioxidant and anticancer activities. *Int. J. Nanomed.* **2016**, *11*, 5683–5696. [CrossRef] [PubMed]

404. Jazayeri, M.H.; Amani, H.; Pourfatollah, A.A.; Avan, A.; Ferns, G.A.; Pazoki-Toroudi, H. Enhanced detection sensitivity of prostate-specific antigen via PSA-conjugated gold nanoparticles based on localized surface plasmon resonance: GNP-coated anti-PSA/LSPR as a novel approach for the identification of prostate anomalies. *Cancer Gene. Ther.* **2016**, *23*, 365–369. [CrossRef] [PubMed]

405. Ray, S.; Ghosh Ray, S.; Mandal, S. Development of bicalutamide-loaded PLGA nanoparticles: Preparation, characterization and in vitro evaluation for the treatment of prostate cancer. *Artif. Cells Nanomed. Biotechnol.* **2016**, 1–11. [CrossRef] [PubMed]

406. Huo, Y.; Singh, P.; Kim, Y.J.; Soshnikova, V.; Kang, J.; Markus, J.; Ahn, S.; Castro-Aceituno, V.; Mathiyalagan, R.; Chokkalingam, M.; et al. Biological synthesis of gold and silver chloride nanoparticles by glycyrrhiza uralensis and in vitro applications. *Artif. Cells Nanomed. Biotechnol.* **2017**, 1–13. [CrossRef] [PubMed]

407. Mokhtari, M.J.; Koohpeima, F.; Mohammadi, H. A comparison inhibitory effects of cisplatin and MNPS-PEG-cisplatin on the adhesion capacity of bone metastatic breast cancer. *Chem. Biol. Drug Des.* **2017**. [CrossRef] [PubMed]

408. Bhuvaneswari, R.; Xavier, R.J.; Arumugam, M. Facile synthesis of multifunctional silver nanoparticles using mangrove plant *Excoecaria agallocha* L. For its antibacterial, antioxidant and cytotoxic effects. *J. Parasit. Dis.* **2017**, *41*, 180–187. [CrossRef] [PubMed]

409. Lacroix, M. *Targeted Therapies in Cancer*; Nova Sciences Publishers: Hauppauge, NY, USA, 2014.

410. Abramson, R. Overview of Targeted Therapies for Cancer. Available online: https://www.mycancergenome.org/content/molecular-medicine/overview-of-targeted-therapies-for-cancer/ (accessed on 13 June 2016).

411. Collignon, J.; Lousberg, L.; Schroeder, H.; Jerusalem, G. Triple-negative breast cancer: Treatment challenges and solutions. *Breast Cancer (Dove Med Press)* **2016**, *8*, 93–107. [PubMed]

412. Kontani, K.; Hashimoto, S.I.; Murazawa, C.; Norimura, S.; Tanaka, H.; Ohtani, M.; Fujiwara-Honjo, N.; Date, M.; Teramoto, K.; Houchi, H.; et al. Indication of metronomic chemotherapy for metastatic breast cancer: Clinical outcomes and responsive subtypes. *Mol. Clin. Oncol.* **2016**, *4*, 947–953. [CrossRef] [PubMed]

413. Bateman, J.C.; Carlton, H.N. The role of chemotherapy in the treatment of breast cancer. *Surgery* **1960**, *47*, 895–907. [PubMed]

414. Nakano, M.; Shoji, S.; Higure, T.; Kawakami, M.; Tomonaga, T.; Terachi, T.; Uchida, T. Low-dose docetaxel, estramustine and prednisolone combination chemotherapy for castration-resistant prostate cancer. *Mol. Clin. Oncol.* **2016**, *4*, 942–946. [CrossRef] [PubMed]

415. Herbst, W.P. The present picture in chemotherapy in prostatic carcinoma. *J. Urol.* **1947**, *57*, 296–299. [PubMed]

416. Albumin-bound paclitaxel (abraxane) for advanced breast cancer. *Med. Lett. Drugs Ther.* **2005**, *47*, 39–40.

417. Barenholz, Y. Doxil—The first FDA-approved nano-drug: Lessons learned. *J. Control. Release* **2012**, *160*, 117–134. [CrossRef] [PubMed]

418. Chan, S.; Davidson, N.; Juozaityte, E.; Erdkamp, F.; Pluzanska, A.; Azarnia, N.; Lee, L.W. Phase III trial of liposomal doxorubicin and cyclophosphamide compared with epirubicin and cyclophosphamide as first-line therapy for metastatic breast cancer. *Ann. Oncol.* **2004**, *15*, 1527–1534. [CrossRef] [PubMed]

419. Lee, K.S.; Chung, H.C.; Im, S.A.; Park, Y.H.; Kim, C.S.; Kim, S.B.; Rha, S.Y.; Lee, M.Y.; Ro, J. Multicenter phase II trial of genexol-pm, a cremophor-free, polymeric micelle formulation of paclitaxel, in patients with metastatic breast cancer. *Breast Cancer Res. Treat.* **2008**, *108*, 241–250. [CrossRef] [PubMed]

420. Pillai, G. Nanomedicines for cancer therapy: An update of FDA approved and those under various stages of development. *SOJ Pharm. Pharm. Sci.* **2014**, *1*, 13.

421. Wang, M.; Thanou, M. Targeting nanoparticles to cancer. *Pharmacol. Res.* **2010**, *62*, 90–99. [CrossRef] [PubMed]

422. Zhang, C.; Zhao, X.; Guo, S.; Lin, T.; Guo, H. Highly effective photothermal chemotherapy with pH-responsive polymer-coated drug-loaded melanin-like nanoparticles. *Int. J. Nanomed.* **2017**, *12*, 1827–1840. [CrossRef] [PubMed]

423. Bakht, M.K.; Oh, S.W.; Hwang, D.W.; Lee, Y.S.; Youn, H.; Porter, L.A.; Cheon, G.J.; Kwak, C.; Lee, D.S.; Kang, K.W. The potential roles of radionanomedicine and radioexosomic in prostate cancer research and treatment. *Curr. Pharm. Des.* **2017**. [CrossRef] [PubMed]

424. Lopes, A.M.; Chen, K.Y.; Kamei, D.T. A transferrin variant as the targeting ligand for polymeric nanoparticles incorporated in 3-D PLGA porous scaffolds. *Mater. Sci. Eng. C Mater. Biol. Appl.* **2017**, *73*, 373–380. [CrossRef] [PubMed]

425. Belz, J.; Castilla-Ojo, N.; Sridhar, S.; Kumar, R. Radiosensitizing silica nanoparticles encapsulating docetaxel for treatment of prostate cancer. *Methods Mol. Biol.* **2017**, *1530*, 403–409. [PubMed]

426. Yan, J.; Wang, Y.; Jia, Y.; Liu, S.; Tian, C.; Pan, W.; Liu, X.; Wang, H. Co-delivery of docetaxel and curcumin prodrug via dual-targeted nanoparticles with synergistic antitumor activity against prostate cancer. *Biomed. Pharmacother.* **2017**, *88*, 374–383. [CrossRef] [PubMed]

427. Huang, W.Y.; Lin, J.N.; Hsieh, J.T.; Chou, S.C.; Lai, C.H.; Yun, E.J.; Lo, U.G.; Pong, R.C.; Lin, J.H.; Lin, Y.H. Nanoparticle targeting CD44-positive cancer cells for site-specific drug delivery in prostate cancer therapy. *ACS Appl. Mater. Interfaces* **2016**, *8*, 30722–30734. [CrossRef] [PubMed]

428. Bharali, D.J.; Sudha, T.; Cui, H.; Mian, B.M.; Mousa, S.A. Anti-CD24 nano-targeted delivery of docetaxel for the treatment of prostate cancer. *Nanomedicine* **2017**, *13*, 263–273. [CrossRef] [PubMed]

429. Qu, N.; Lee, R.J.; Sun, Y.; Cai, G.; Wang, J.; Wang, M.; Lu, J.; Meng, Q.; Teng, L.; Wang, D.; et al. Cabazitaxel-loaded human serum albumin nanoparticles as a therapeutic agent against prostate cancer. *Int. J. Nanomed.* **2016**, *11*, 3451–3459.

430. Pirayesh Islamian, J.; Hatamian, M.; Aval, N.A.; Rashidi, M.R.; Mesbahi, A.; Mohammadzadeh, M.; Asghari Jafarabadi, M. Targeted superparamagnetic nanoparticles coated with 2-deoxy-d-gloucose and doxorubicin more sensitize breast cancer cells to ionizing radiation. *Breast* **2017**, *33*, 97–103. [CrossRef] [PubMed]

431. Li, J.; Zhang, J.; Wang, Y.; Liang, X.; Wusiman, Z.; Yin, Y.; Shen, Q. Synergistic inhibition of migration and invasion of breast cancer cells by dual docetaxel/quercetin-loaded nanoparticles via Akt/MMP-9 pathway. *Int. J. Pharm.* **2017**, *523*, 300–309. [CrossRef] [PubMed]

432. Stratton, M.R. Exploring the genomes of cancer cells: Progress and promise. *Science* **2011**, *331*, 1553–1558. [CrossRef] [PubMed]

433. Merz, B. Gene therapy may have future role in cancer treatment. *JAMA* **1987**, *257*, 150–151. [CrossRef] [PubMed]

434. Kozielski, K.L.; Rui, Y.; Green, J.J. Non-viral nucleic acid containing nanoparticles as cancer therapeutics. *Expert. Opin. Drug Deliv.* **2016**, *13*, 1475–1487. [CrossRef] [PubMed]

435. Gogtay, N.J.; Sridharan, K. Therapeutic nucleic acids: Current clinical status. *Br. J. Clin. Pharmacol.* **2016**, *82*, 659–672.

436. Yu, Y.; Yao, Y.; Yan, H.; Wang, R.; Zhang, Z.; Sun, X.; Zhao, L.; Ao, X.; Xie, Z.; Wu, Q. A tumor-specific microRNA recognition system facilitates the accurate targeting to tumor cells by magnetic nanoparticles. *Mol. Ther. Nucleic Acids* **2016**, *5*, e318. [CrossRef] [PubMed]

437. Rejeeth, C.; Vivek, R. Comparison of two silica based nonviral gene therapy vectors for breast carcinoma: Evaluation of the p53 delivery system in balb/C mice. *Artif. Cells Nanomed. Biotechnol.* **2017**, *45*, 489–494. [CrossRef] [PubMed]

438. Rejeeth, C.; Salem, A. Novel luminescent silica nanoparticles (LSN): p53 Gene delivery system in breast cancer in vitro and in vivo. *J. Pharm. Pharmacol.* **2016**, *68*, 305–315. [CrossRef] [PubMed]

439. Li, T.; Shen, X.; Chen, Y.; Zhang, C.; Yan, J.; Yang, H.; Wu, C.; Zeng, H.; Liu, Y. Polyetherimide-grafted $Fe_3O_4@SiO_2$ nanoparticles as theranostic agents for simultaneous VEGF siRNA delivery and magnetic resonance cell imaging. *Int. J. Nanomed.* **2015**, *10*, 4279–4291. [CrossRef] [PubMed]

440. Zhou, H.; Wei, J.; Dai, Q.; Wang, L.; Luo, J.; Cheang, T.; Wang, S. $CaCO_3/CaIP6$ composite nanoparticles effectively deliver AKT1 small interfering rna to inhibit human breast cancer growth. *Int. J. Nanomed.* **2015**, *10*, 4255–4266.

441. Nourbakhsh, M.; Jaafari, M.R.; Lage, H.; Abnous, K.; Mosaffa, F.; Badiee, A.; Behravan, J. Nanolipoparticles-mediated MDR1 siRNA delivery reduces doxorubicin resistance in breast cancer cells and silences MDR1 expression in xenograft model of human breast cancer. *Iran J. Basic Med. Sci.* **2015**, *18*, 385–392. [PubMed]

442. Su, S.; Tian, Y.; Li, Y.; Ding, Y.; Ji, T.; Wu, M.; Wu, Y.; Nie, G. "Triple-punch" strategy for triple negative breast cancer therapy with minimized drug dosage and improved antitumor efficacy. *ACS Nano* **2015**, *9*, 1367–1378. [CrossRef] [PubMed]

443. Dong, D.; Gao, W.; Liu, Y.; Qi, X.R. Therapeutic potential of targeted multifunctional nanocomplex co-delivery of siRNA and low-dose doxorubicin in breast cancer. *Cancer Lett.* **2015**, *359*, 178–186. [CrossRef] [PubMed]

444. Huang, Y.P.; Hung, C.M.; Hsu, Y.C.; Zhong, C.Y.; Wang, W.R.; Chang, C.C.; Lee, M.J. Suppression of breast cancer cell migration by small interfering RNA delivered by polyethylenimine-functionalized graphene oxide. *Nanoscale Res. Lett.* **2016**, *11*, 247. [CrossRef] [PubMed]

445. Arami, S.; Rashidi, M.R.; Mahdavi, M.; Fathi, M.; Entezami, A.A. Synthesis and characterization of Fe_3O_4-PEG-LAC-chitosan-PEI nanoparticle as a survivin siRNA delivery system. *Hum. Exp. Toxicol.* **2017**, *36*, 227–237. [CrossRef] [PubMed]

446. Unsoy, G.; Gunduz, U. Targeted silencing of survivin in cancer cells by siRNA loaded chitosan magnetic nanoparticles. *Expert Rev. Anticancer Ther.* **2016**, *16*, 789–797. [CrossRef] [PubMed]

447. Park, D.H.; Cho, J.; Kwon, O.J.; Yun, C.O.; Choy, J.H. Biodegradable inorganic nanovector: Passive versus active tumor targeting in siRNA transportation. *Angew. Chem. Int. Ed. Engl.* **2016**, *55*, 4582–4586. [CrossRef] [PubMed]

448. McBride, J.W.; Massey, A.S.; McCaffrey, J.; McCrudden, C.M.; Coulter, J.A.; Dunne, N.J.; Robson, T.; McCarthy, H.O. Development of TMTP-1 targeted designer biopolymers for gene delivery to prostate cancer. *Int. J. Pharm.* **2016**, *500*, 144–153. [CrossRef] [PubMed]

449. Xing, Z.; Gao, S.; Duan, Y.; Han, H.; Li, L.; Yang, Y.; Li, Q. Delivery of dnazyme targeting aurora kinase a to inhibit the proliferation and migration of human prostate cancer. *Int. J. Nanomed.* **2015**, *10*, 5715–5727.

450. Fitzgerald, K.A.; Guo, J.; Tierney, E.G.; Curtin, C.M.; Malhotra, M.; Darcy, R.; O'Brien, F.J.; O'Driscoll, C.M. The use of collagen-based scaffolds to simulate prostate cancer bone metastases with potential for evaluating delivery of nanoparticulate gene therapeutics. *Biomaterials* **2015**, *66*, 53–66. [CrossRef] [PubMed]

451. Zhang, T.; Xue, X.; He, D.; Hsieh, J.T. A prostate cancer-targeted polyarginine-disulfide linked pei nanocarrier for delivery of microrna. *Cancer Lett.* **2015**, *365*, 156–165. [CrossRef] [PubMed]

452. Huang, R.Y.; Chiang, P.H.; Hsiao, W.C.; Chuang, C.C.; Chang, C.W. Redox-sensitive polymer/SPIO nanocomplexes for efficient magnetofection and mr imaging of human cancer cells. *Langmuir* **2015**, *31*, 6523–6531. [CrossRef] [PubMed]

453. Ceresa, C.; Nicolini, G.; Semperboni, S.; Requardt, H.; Le Duc, G.; Santini, C.; Pellei, M.; Bentivegna, A.; Dalpra, L.; Cavaletti, G.; et al. Synchrotron-based photon activation therapy effect on cisplatin pre-treated human glioma stem cells. *Anticancer Res.* **2014**, *34*, 5351–5355. [PubMed]

454. Bakhshabadi, M.; Ghorbani, M.; Meigooni, A.S. Photon activation therapy: A Monte Carlo study on dose enhancement by various sources and activation media. *Australas. Phys. Eng. Sci. Med.* **2013**, *36*, 301–311. [CrossRef] [PubMed]

455. Ceresa, C.; Nicolini, G.; Requardt, H.; Le Duc, G.; Cavaletti, G.; Bravin, A. The effect of photon activation therapy on cisplatin pre-treated human tumour cell lines: Comparison with conventional X-ray irradiation. *J. Biol. Regul. Homeost. Agents* **2013**, *27*, 477–485. [PubMed]

456. Ceberg, C.; Jonsson, B.A.; Prezado, Y.; Pommer, T.; Nittby, H.; Englund, E.; Grafstrom, G.; Edvardsson, A.; Stenvall, A.; Stromblad, S.; et al. Photon activation therapy of RG2 glioma carrying fischer rats using stable thallium and monochromatic synchrotron radiation. *Phys. Med. Biol.* **2012**, *57*, 8377–8391. [CrossRef] [PubMed]

457. Laster, B.H.; Dixon, D.W.; Novick, S.; Feldman, J.P.; Seror, V.; Goldbart, Z.I.; Kalef-Ezra, J.A. Photon activation therapy and brachytherapy. *Brachytherapy* **2009**, *8*, 324–330. [CrossRef] [PubMed]

458. Suortti, P.; Thomlinson, W. Medical applications of synchrotron radiation. *Phys Med Biol* **2003**, *48*, R1–R35. [CrossRef] [PubMed]

459. Laster, B.H.; Thomlinson, W.C.; Fairchild, R.G. Photon activation of iododeoxyuridine: Biological efficacy of auger electrons. *Radiat. Res.* **1993**, *133*, 219–224. [CrossRef] [PubMed]

460. Miller, R.W.; DeGraff, W.; Kinsella, T.J.; Mitchell, J.B. Evaluation of incorporated iododeoxyuridine cellular radiosensitization by photon activation therapy. *Int. J. Radiat. Oncol. Biol. Phys.* **1987**, *13*, 1193–1197. [CrossRef]

461. Fairchild, R.G.; Bond, V.P. Photon activation therapy. *Strahlentherapie* **1984**, *160*, 758–763. [PubMed]

462. Choi, G.H.; Seo, S.J.; Kim, K.H.; Kim, H.T.; Park, S.H.; Lim, J.H.; Kim, J.K. Photon activated therapy (PAT) using monochromatic synchrotron X-rays and iron oxide nanoparticles in a mouse tumor model: Feasibility study of PAT for the treatment of superficial malignancy. *Radiat. Oncol.* **2012**, *7*, 184. [CrossRef] [PubMed]

463. Dolmans, D.E.; Fukumura, D.; Jain, R.K. Photodynamic therapy for cancer. *Nat. Rev. Cancer* **2003**, *3*, 380–387. [CrossRef] [PubMed]

464. Chen, J.; Keltner, L.; Christophersen, J.; Zheng, F.; Krouse, M.; Singhal, A.; Wang, S.S. New technology for deep light distribution in tissue for phototherapy. *Cancer J.* **2002**, *8*, 154–163. [CrossRef] [PubMed]

465. Sneider, A.; VanDyke, D.; Paliwal, S.; Rai, P. Remotely triggered nano-theranostics for cancer applications. *Nanotheranostics (Syd.)* **2017**, *1*, 1–22. [CrossRef] [PubMed]

466. Shen, Y.; Shuhendler, A.J.; Ye, D.; Xu, J.J.; Chen, H.Y. Two-photon excitation nanoparticles for photodynamic therapy. *Chem. Soc. Rev.* **2016**, *45*, 6725–6741. [CrossRef] [PubMed]

467. Chen, R.; Zhang, J.; Chelora, J.; Xiong, Y.; Kershaw, S.V.; Li, K.F.; Lo, P.K.; Cheah, K.W.; Rogach, A.L.; Zapien, J.A.; et al. Ruthenium(ii) complex incorporated uio-67 metal-organic framework nanoparticles for enhanced two-photon fluorescence imaging and photodynamic cancer therapy. *ACS Appl. Mater. Interfaces* **2017**, *9*, 5699–5708. [CrossRef] [PubMed]

468. Shen, X.; Li, S.; Li, L.; Yao, S.Q.; Xu, Q.H. Highly efficient, conjugated-polymer-based nano-photosensitizers for selectively targeted two-photon photodynamic therapy and imaging of cancer cells. *Chemistry* **2015**, *21*, 2214–2221. [CrossRef] [PubMed]

469. Secret, E.; Maynadier, M.; Gallud, A.; Chaix, A.; Bouffard, E.; Gary-Bobo, M.; Marcotte, N.; Mongin, O.; El Cheikh, K.; Hugues, V.; et al. Two-photon excitation of porphyrin-functionalized porous silicon nanoparticles for photodynamic therapy. *Adv. Mater.* **2014**, *26*, 7643–7648. [CrossRef] [PubMed]

470. Zhao, T.; Yu, K.; Li, L.; Zhang, T.; Guan, Z.; Gao, N.; Yuan, P.; Li, S.; Yao, S.Q.; Xu, Q.H.; et al. Gold nanorod enhanced two-photon excitation fluorescence of photosensitizers for two-photon imaging and photodynamic therapy. *ACS Appl. Mater. Interfaces* **2014**, *6*, 2700–2708. [CrossRef] [PubMed]

471. Gary-Bobo, M.; Mir, Y.; Rouxel, C.; Brevet, D.; Basile, I.; Maynadier, M.; Vaillant, O.; Mongin, O.; Blanchard-Desce, M.; Morere, A.; et al. Mannose-functionalized mesoporous silica nanoparticles for efficient two-photon photodynamic therapy of solid tumors. *Angew. Chem. Int. Ed. Engl.* **2011**, *50*, 11425–11429. [CrossRef] [PubMed]

472. Swartling, J.; Axelsson, J.; Ahlgren, G.; Kalkner, K.M.; Nilsson, S.; Svanberg, S.; Svanberg, K.; Andersson-Engels, S. System for interstitial photodynamic therapy with online dosimetry: First clinical experiences of prostate cancer. *J. Biomed. Opt.* **2010**, *15*, 058003. [CrossRef] [PubMed]

473. Swartling, J.; Hoglund, O.V.; Hansson, K.; Sodersten, F.; Axelsson, J.; Lagerstedt, A.S. Online dosimetry for temoporfin-mediated interstitial photodynamic therapy using the canine prostate as model. *J. Biomed. Opt.* **2016**, *21*, 28002. [CrossRef] [PubMed]

474. Chen, Y.; Chatterjee, S.; Lisok, A.; Minn, I.; Pullambhatla, M.; Wharram, B.; Wang, Y.; Jin, J.; Bhujwalla, Z.M.; Nimmagadda, S.; et al. A psma-targeted theranostic agent for photodynamic therapy. *J. Photochem. Photobiol. B* **2017**, *167*, 111–116. [CrossRef] [PubMed]

475. Wang, X.; Tsui, B.; Ramamurthy, G.; Zhang, P.; Meyers, J.; Kenney, M.E.; Kiechle, J.; Ponsky, L.; Basilion, J.P. Theranostic agents for photodynamic therapy of prostate cancer by targeting prostate-specific membrane antigen. *Mol. Cancer Ther.* **2016**, *15*, 1834–1844. [CrossRef] [PubMed]

476. Lin, T.Y.; Guo, W.; Long, Q.; Ma, A.; Liu, Q.; Zhang, H.; Huang, Y.; Chandrasekaran, S.; Pan, C.; Lam, K.S.; et al. Hsp90 inhibitor encapsulated photo-theranostic nanoparticles for synergistic combination cancer therapy. *Theranostics* **2016**, *6*, 1324–1335. [CrossRef] [PubMed]

477. Vaillant, O.; El Cheikh, K.; Warther, D.; Brevet, D.; Maynadier, M.; Bouffard, E.; Salgues, F.; Jeanjean, A.; Puche, P.; Mazerolles, C.; et al. Mannose-6-phosphate receptor: A target for theranostics of prostate cancer. *Angew. Chem. Int. Ed. Engl.* **2015**, *54*, 5952–5956. [CrossRef] [PubMed]

478. Choi, J.; Kim, H.; Choi, Y. Theranostic nanoparticles for enzyme-activatable fluorescence imaging and photodynamic/chemo dual therapy of triple-negative breast cancer. *Quant. Imaging Med. Surg.* **2015**, *5*, 656–664. [PubMed]

479. Wang, X.; Hu, J.; Wang, P.; Zhang, S.; Liu, Y.; Xiong, W.; Liu, Q. Analysis of the in vivo and in vitro effects of photodynamic therapy on breast cancer by using a sensitizer, sinoporphyrin sodium. *Theranostics* **2015**, *5*, 772–786. [CrossRef] [PubMed]

480. Kue, C.S.; Kamkaew, A.; Lee, H.B.; Chung, L.Y.; Kiew, L.V.; Burgess, K. Targeted PDT agent eradicates trkc expressing tumors via photodynamic therapy (PDT). *Mol. Pharm.* **2015**, *12*, 212–222. [CrossRef] [PubMed]

481. El-Sayed, I.H. Nanotechnology in head and neck cancer: The race is on. *Curr Oncol Rep* **2010**, *12*, 121–128. [CrossRef] [PubMed]

482. Huang, X.; El-Sayed, M.A. Plasmonic photo-thermal therapy (PPTT). *Alex. J. Med.* **2011**, *47*, 1–9. [CrossRef]

483. Huang, X.; Jain, P.K.; El-Sayed, I.H.; El-Sayed, M.A. Plasmonic photothermal therapy (PPTT) using gold nanoparticles. *Lasers Med. Sci.* **2008**, *23*, 217–228. [CrossRef] [PubMed]

484. Turcheniuk, K.; Dumych, T.; Bilyy, R.; Turcheniuk, V.; Bouckaert, J.; Vovk, V.; Chopyak, V.; Zaitsev, V.; Mariot, P.; Prevarskaya, N.; et al. Plasmonic photothermal cancer therapy with gold nanorods/reduced graphene oxide core/shell nanocomposites. *RSC Adv.* **2016**, *6*, 1600–1610. [CrossRef]

485. Peng, J.; Dong, M.; Ran, B.; Li, W.; Hao, Y.; Yang, Q.; Tan, L.; Shi, K.; Qian, Z. "One-for-all" type, biodegradable prussian blue/manganese dioxide hybrid nanocrystal for tri-modal imaging guided photothermal therapy and oxygen regulation of breast cancer. *ACS Appl. Mater. Interfaces* **2017**, *9*, 13875–13886. [CrossRef] [PubMed]

486. Cantu, T.; Walsh, K.; Pattani, V.P.; Moy, A.J.; Tunnell, J.W.; Irvin, J.A.; Betancourt, T. Conductive polymer-based nanoparticles for laser-mediated photothermal ablation of cancer: Synthesis, characterization, and in vitro evaluation. *Int. J. Nanomed.* **2017**, *12*, 615–632. [CrossRef] [PubMed]

487. Haberkorn, U.; Eder, M.; Kopka, K.; Babich, J.W.; Eisenhut, M. New strategies in prostate cancer: Prostate-specific membrane antigen (PSMA) ligands for diagnosis and therapy. *Clin. Cancer Res.* **2016**, *22*, 9–15. [CrossRef] [PubMed]

488. Bouchelouche, K.; Tagawa, S.T.; Goldsmith, S.J.; Turkbey, B.; Capala, J.; Choyke, P. PET/CT imaging and radioimmunotherapy of prostate cancer. *Semin. Nucl. Med.* **2011**, *41*, 29–44. [CrossRef] [PubMed]

489. Awada, G.; Gombos, A.; Aftimos, P.; Awada, A. Emerging drugs targeting human epidermal growth factor receptor 2 (Her2) in the treatment of breast cancer. *Expert Opin. Emerg. Drugs* **2016**, *21*, 91–101. [CrossRef] [PubMed]

490. Lluch, A.; Eroles, P.; Perez-Fidalgo, J.A. Emerging EGFR antagonists for breast cancer. *Expert Opin. Emerg. Drugs* **2014**, *19*, 165–181. [CrossRef] [PubMed]

491. Doddamane, I.; Butler, R.; Jhaveri, A.; Chung, G.G.; Cheng, D. Where does radioimmunotherapy fit in the management of breast cancer? *Immunotherapy* **2013**, *5*, 895–904. [CrossRef] [PubMed]

492. Evans-Axelsson, S.; Timmermand, O.V.; Bjartell, A.; Strand, S.E.; Elgqvist, J. Radioimmunotherapy for prostate cancer—Current status and future possibilities. *Semin. Nucl. Med.* **2016**, *46*, 165–179. [CrossRef] [PubMed]

493. Zolata, H.; Afarideh, H.; Davani, F.A. Triple therapy of HER2+ cancer using radiolabeled multifunctional iron oxide nanoparticles and alternating magnetic field. *Cancer Biother. Radiopharm.* **2016**, *31*, 324–329. [CrossRef] [PubMed]

494. Wang, X.; Sun, Q.; Shen, S.; Xu, Y.; Huang, L. Nanotrastuzumab in combination with radioimmunotherapy: Can it be a viable treatment option for patients with HER2-positive breast cancer with brain metastasis? *Med. Hypotheses* **2016**, *88*, 79–81. [CrossRef] [PubMed]

495. Parakh, S.; Parslow, A.C.; Gan, H.K.; Scott, A.M. Antibody-mediated delivery of therapeutics for cancer therapy. *Expert Opin. Drug Deliv.* **2016**, *13*, 401–419. [CrossRef] [PubMed]

496. Yang, Q.; Parker, C.L.; McCallen, J.D.; Lai, S.K. Addressing challenges of heterogeneous tumor treatment through bispecific protein-mediated pretargeted drug delivery. *J. Control. Release* **2015**, *220*, 715–726. [CrossRef] [PubMed]

497. Yook, S.; Cai, Z.; Lu, Y.; Winnik, M.A.; Pignol, J.P.; Reilly, R.M. Radiation nanomedicine for egfr-positive breast cancer: Panitumumab-modified gold nanoparticles complexed to the beta-particle-emitter, ^{177}Lu. *Mol. Pharm.* **2015**, *12*, 3963–3972. [CrossRef] [PubMed]

498. Rasaneh, S.; Rajabi, H.; Johari Daha, F. Activity estimation in radioimmunotherapy using magnetic nanoparticles. *Chin. J. Cancer Res.* **2015**, *27*, 203–208. [PubMed]

499. Bushman, J.; Vaughan, A.; Sheihet, L.; Zhang, Z.; Costache, M.; Kohn, J. Functionalized nanospheres for targeted delivery of paclitaxel. *J. Control. Release* **2013**, *171*, 315–321. [CrossRef] [PubMed]

500. D'Huyvetter, M.; Aerts, A.; Xavier, C.; Vaneycken, I.; Devoogdt, N.; Gijs, M.; Impens, N.; Baatout, S.; Ponsard, B.; Muyldermans, S.; et al. Development of ^{177}Lu-nanobodies for radioimmunotherapy of her2-positive breast cancer: Evaluation of different bifunctional chelators. *Contrast Media Mol. Imaging* **2012**, *7*, 254–264. [CrossRef] [PubMed]

501. Natarajan, A.; Xiong, C.Y.; Gruettner, C.; DeNardo, G.L.; DeNardo, S.J. Development of multivalent radioimmunonanoparticles for cancer imaging and therapy. *Cancer Biother. Radiopharm.* **2008**, *23*, 82–91. [CrossRef] [PubMed]

502. Nedunchezhian, K.; Aswath, N.; Thiruppathy, M.; Thirugnanamurthy, S. Boron neutron capture therapy—A literature review. *J. Clin. Diagn. Res.* **2016**, *10*, ZE01–ZE04. [CrossRef] [PubMed]

503. Mirzaei, H.R.; Sahebkar, A.; Salehi, R.; Nahand, J.S.; Karimi, E.; Jaafari, M.R.; Mirzaei, H. Boron neutron capture therapy: Moving toward targeted cancer therapy. *J. Cancer Res. Ther.* **2016**, *12*, 520–525. [CrossRef] [PubMed]

504. Luderer, M.J.; de la Puente, P.; Azab, A.K. Advancements in tumor targeting strategies for boron neutron capture therapy. *Pharm. Res.* **2015**, *32*, 2824–2836. [CrossRef] [PubMed]

505. Alberti, D.; Protti, N.; Franck, M.; Stefania, R.; Bortolussi, S.; Altieri, S.; Deagostino, A.; Aime, S.; Geninatti Crich, S. Theranostic nanoparticles loaded with imaging probes and rubrocurcumin for combined cancer therapy by folate receptor targeting. *Chem. Med. Chem.* **2017**, *12*, 502–509. [CrossRef] [PubMed]

506. Peters, T.; Grunewald, C.; Blaickner, M.; Ziegner, M.; Schutz, C.; Iffland, D.; Hampel, G.; Nawroth, T.; Langguth, P. Cellular uptake and in vitro antitumor efficacy of composite liposomes for neutron capture therapy. *Radiat. Oncol.* **2015**, *10*, 52. [CrossRef] [PubMed]

507. Heber, E.M.; Hawthorne, M.F.; Kueffer, P.J.; Garabalino, M.A.; Thorp, S.I.; Pozzi, E.C.; Monti Hughes, A.; Maitz, C.A.; Jalisatgi, S.S.; Nigg, D.W.; et al. Therapeutic efficacy of boron neutron capture therapy mediated by boron-rich liposomes for oral cancer in the hamster cheek pouch model. *Proc. Natl. Acad. Sci. USA* **2014**, *111*, 16077–16081. [CrossRef] [PubMed]

508. Tachikawa, S.; Miyoshi, T.; Koganei, H.; El-Zaria, M.E.; Vinas, C.; Suzuki, M.; Ono, K.; Nakamura, H. Spermidinium closo-dodecaborate-encapsulating liposomes as efficient boron delivery vehicles for neutron capture therapy. *Chem. Commun. (Camb.)* **2014**, *50*, 12325–12328. [CrossRef] [PubMed]

509. Kobayashi, H.; Kawamoto, S.; Bernardo, M.; Brechbiel, M.W.; Knopp, M.V.; Choyke, P.L. Delivery of gadolinium-labeled nanoparticles to the sentinel lymph node: Comparison of the sentinel node visualization and estimations of intra-nodal gadolinium concentration by the magnetic resonance imaging. *J. Control. Release* **2006**, *111*, 343–351. [CrossRef] [PubMed]

510. Oyewumi, M.O.; Liu, S.; Moscow, J.A.; Mumper, R.J. Specific association of thiamine-coated gadolinium nanoparticles with human breast cancer cells expressing thiamine transporters. *Bioconjug. Chem.* **2003**, *14*, 404–411. [CrossRef] [PubMed]

511. Fortin, J.P.; Gazeau, F.; Wilhelm, C. Intracellular heating of living cells through neel relaxation of magnetic nanoparticles. *Eur. Biophys. J.* **2008**, *37*, 223–228. [CrossRef] [PubMed]

512. Han, Y.; Lei, S.L.; Lu, J.H.; He, Y.; Chen, Z.W.; Ren, L.; Zhou, X. Potential use of sers-assisted theranostic strategy based on Fe_3O_4/Au cluster/shell nanocomposites for bio-detection, MRI, and magnetic hyperthermia. *Mater. Sci. Eng. C Mater. Biol. Appl.* **2016**, *64*, 199–207. [CrossRef] [PubMed]

513. Stocke, N.A.; Sethi, P.; Jyoti, A.; Chan, R.; Arnold, S.M.; Hilt, J.Z.; Upreti, M. Toxicity evaluation of magnetic hyperthermia induced by remote actuation of magnetic nanoparticles in 3d micrometastasic tumor tissue analogs for triple negative breast cancer. *Biomaterials* **2017**, *120*, 115–125. [CrossRef] [PubMed]

514. Oh, Y.; Moorthy, M.S.; Manivasagan, P.; Bharathiraja, S.; Oh, J. Magnetic hyperthermia and pH-responsive effective drug delivery to the sub-cellular level of human breast cancer cells by modified $coFe_2O_4$ nanoparticles. *Biochimie* **2017**, *133*, 7–19. [CrossRef] [PubMed]

515. Yao, X.; Niu, X.; Ma, K.; Huang, P.; Grothe, J.; Kaskel, S.; Zhu, Y. Graphene quantum dots-capped magnetic mesoporous silica nanoparticles as a multifunctional platform for controlled drug delivery, magnetic hyperthermia, and photothermal therapy. *Small* **2017**, *13*, 1602225. [CrossRef] [PubMed]

516. Kossatz, S.; Grandke, J.; Couleaud, P.; Latorre, A.; Aires, A.; Crosbie-Staunton, K.; Ludwig, R.; Dahring, H.; Ettelt, V.; Lazaro-Carrillo, A.; et al. Efficient treatment of breast cancer xenografts with multifunctionalized iron oxide nanoparticles combining magnetic hyperthermia and anti-cancer drug delivery. *Breast Cancer Res.* **2015**, *17*, 66. [CrossRef] [PubMed]

517. Ferlay, J.; Soerjomataram, I.; Ervik, M.; Dikshit, R.; Eser, S.; Mathers, C.; Rebelo, M.; Parkin, D.; Forman, D.; Bray, F. *Globocan 2012 v1.0, Cancer Incidence and Mortality Worldwide: Iarc Cancerbase No. 11 [Internet]*; International Agency for Research on Cancer: Lyon, France, 2013.

518. Siegel, R.L.; Miller, K.D.; Jemal, A. Cancer statistics, 2016. *CA Cancer J. Clin.* **2016**, *66*, 7–30. [CrossRef] [PubMed]

519. Hoffman, R.M.; Couper, M.P.; Zikmund-Fisher, B.J.; Levin, C.A.; McNaughton-Collins, M.; Helitzer, D.L.; VanHoewyk, J.; Barry, M.J. Prostate cancer screening decisions: Results from the national survey of medical decisions (decisions study). *Arch. Intern. Med.* **2009**, *169*, 1611–1618. [CrossRef] [PubMed]

520. Bill-Axelson, A.; Bratt, O. Words of wisdom. Re: Screening and prostate cancer mortality: Results of the european randomised study of screening for prostate cancer (ERSPC) at 13 years of follow-up. *Eur. Urol.* **2015**, *67*, 175. [CrossRef] [PubMed]

521. Heidenreich, A.; Bastian, P.J.; Bellmunt, J.; Bolla, M.; Joniau, S.; van der Kwast, T.; Mason, M.; Matveev, V.; Wiegel, T.; Zattoni, F.; et al. Eau guidelines on prostate cancer. Part 1: Screening, diagnosis, and local treatment with curative intent-update 2013. *Eur. Urol.* **2014**, *65*, 124–137. [CrossRef] [PubMed]

522. Kim, S.P.; Karnes, R.J.; Nguyen, P.L.; Ziegenfuss, J.Y.; Thompson, R.H.; Han, L.C.; Shah, N.D.; Smaldone, M.C.; Gross, C.P.; Frank, I.; et al. A national survey of radiation oncologists and urologists on recommendations of prostate-specific antigen screening for prostate cancer. *BJU Int.* **2014**, *113*, E106–E111. [CrossRef] [PubMed]

523. Loeb, S. Prostate cancer screening: Highlights from the 29th european association of urology congress Stockholm, Sweden, april 11–15, 2014. *Rev. Urol.* **2014**, *16*, 90–91. [PubMed]

524. Schroder, F.H. Screening for prostate cancer: Current status of ERSPC and screening-related issues. *Recent Results Cancer Res.* **2014**, *202*, 47–51. [PubMed]

525. Schroder, F.H.; Hugosson, J.; Roobol, M.J.; Tammela, T.L.; Zappa, M.; Nelen, V.; Kwiatkowski, M.; Lujan, M.; Maattanen, L.; Lilja, H.; et al. Screening and prostate cancer mortality: Results of the european randomised study of screening for prostate cancer (ERSPC) at 13 years of follow-up. *Lancet* **2014**, *384*, 2027–2035. [CrossRef]

526. Schroder, F.H. Erspc, plco studies and critique of cochrane review 2013. *Recent Results Cancer Res.* **2014**, *202*, 59–63. [PubMed]

527. Vallabhajosula, S.; Nikolopoulou, A.; Jhanwar, Y.S.; Kaur, G.; Tagawa, S.T.; Nanus, D.M.; Bander, N.H.; Goldsmith, S.J. Radioimmunotherapy of metastatic prostate cancer with ^{177}Lu-DOTA-HUJ591 anti prostate specific membrane antigen specific monoclonal antibody. *Curr. Radiopharm.* **2016**, *9*, 44–53. [CrossRef] [PubMed]

528. Van Rij, C.M.; Frielink, C.; Goldenberg, D.M.; Sharkey, R.M.; Lutje, S.; McBride, W.J.; Oyen, W.J.; Boerman, O.C. Pretargeted radioimmunotherapy of prostate cancer with an anti-TROP-2×-anti-HSG bispecific antibody and a ^{177}Lu-labeled peptide. *Cancer Biother. Radiopharm.* **2014**, *29*, 323–329. [CrossRef] [PubMed]

529. Wang, H.Y.; Lin, W.Y.; Chen, M.C.; Lin, T.; Chao, C.H.; Hsu, F.N.; Lin, E.; Huang, C.Y.; Luo, T.Y.; Lin, H. Inhibitory effects of rhenium-188-labeled herceptin on prostate cancer cell growth: A possible radioimmunotherapy to prostate carcinoma. *Int. J. Radiat. Biol.* **2013**, *89*, 346–355. [CrossRef] [PubMed]

530. Van Rij, C.M.; Lutje, S.; Frielink, C.; Sharkey, R.M.; Goldenberg, D.M.; Franssen, G.M.; McBride, W.J.; Rossi, E.A.; Oyen, W.J.; Boerman, O.C. Pretargeted immuno-pet and radioimmunotherapy of prostate cancer with an anti-TROP-2× anti-HSG bispecific antibody. *Eur. J. Nucl. Med. Mol. Imaging* **2013**, *40*, 1377–1383. [CrossRef] [PubMed]

531. Pan, M.H.; Gao, D.W.; Feng, J.; He, J.; Seo, Y.; Tedesco, J.; Wolodzko, J.G.; Hasegawa, B.H.; Franc, B.L. Biodistributions of ^{177}Lu- and ^{111}In-labeled 7e11 antibodies to prostate-specific membrane antigen in xenograft model of prostate cancer and potential use of ^{111}In-7e11 as a pre-therapeutic agent for ^{177}Lu-7e11 radioimmunotherapy. *Mol. Imaging Biol.* **2009**, *11*, 159–166. [CrossRef] [PubMed]

532. Kelly, M.P.; Lee, S.T.; Lee, F.T.; Smyth, F.E.; Davis, I.D.; Brechbiel, M.W.; Scott, A.M. Therapeutic efficacy of ^{177}Lu-CHX-A″-DTPA-hu3S193 radioimmunotherapy in prostate cancer is enhanced by EGFR inhibition or docetaxel chemotherapy. *Prostate* **2009**, *69*, 92–104. [CrossRef] [PubMed]

533. Pandit-Taskar, N.; O'Donoghue, J.A.; Morris, M.J.; Wills, E.A.; Schwartz, L.H.; Gonen, M.; Scher, H.I.; Larson, S.M.; Divgi, C.R. Antibody mass escalation study in patients with castration-resistant prostate cancer using ^{111}In-j591: Lesion detectability and dosimetric projections for 90y radioimmunotherapy. *J. Nucl. Med.* **2008**, *49*, 1066–1074. [CrossRef] [PubMed]

534. Kimura, Y.; Inoue, K.; Abe, M.; Nearman, J.; Baranowska-Kortylewicz, J. Pdgfrbeta and HIF-1α inhibition with imatinib and radioimmunotherapy of experimental prostate cancer. *Cancer Biol. Ther.* **2007**, *6*, 1763–1772. [CrossRef] [PubMed]

535. Zhao, X.Y.; Schneider, D.; Biroc, S.L.; Parry, R.; Alicke, B.; Toy, P.; Xuan, J.A.; Sakamoto, C.; Wada, K.; Schulze, M.; et al. Targeting tomoregulin for radioimmunotherapy of prostate cancer. *Cancer Res.* **2005**, *65*, 2846–2853. [CrossRef] [PubMed]

536. Vallabhajosula, S.; Goldsmith, S.J.; Kostakoglu, L.; Milowsky, M.I.; Nanus, D.M.; Bander, N.H. Radioimmunotherapy of prostate cancer using 90y- and ^{177}Lu-labeled j591 monoclonal antibodies: Effect of multiple treatments on myelotoxicity. *Clin. Cancer Res.* **2005**, *11*, 7195s–7200s. [CrossRef] [PubMed]

537. Richman, C.M.; Denardo, S.J.; O'Donnell, R.T.; Yuan, A.; Shen, S.; Goldstein, D.S.; Tuscano, J.M.; Wun, T.; Chew, H.K.; Lara, P.N.; et al. High-dose radioimmunotherapy combined with fixed, low-dose paclitaxel in metastatic prostate and breast cancer by using a muc-1 monoclonal antibody, m170, linked to indium-111/yttrium-90 via a cathepsin cleavable linker with cyclosporine to prevent human anti-mouse antibody. *Clin. Cancer Res.* **2005**, *11*, 5920–5927. [PubMed]

538. DeNardo, S.J.; Richman, C.M.; Albrecht, H.; Burke, P.A.; Natarajan, A.; Yuan, A.; Gregg, J.P.; O'Donnell, R.T.; DeNardo, G.L. Enhancement of the therapeutic index: From nonmyeloablative and myeloablative toward pretargeted radioimmunotherapy for metastatic prostate cancer. *Clin. Cancer Res.* **2005**, *11*, 7187s–7194s. [CrossRef] [PubMed]

539. Vallabhajosula, S.; Smith-Jones, P.M.; Navarro, V.; Goldsmith, S.J.; Bander, N.H. Radioimmunotherapy of prostate cancer in human xenografts using monoclonal antibodies specific to prostate specific membrane antigen (PSMA): Studies in nude mice. *Prostate* **2004**, *58*, 145–155. [CrossRef] [PubMed]

540. DeNardo, S.J.; DeNardo, G.L.; Yuan, A.; Richman, C.M.; O'Donnell, R.T.; Lara, P.N.; Kukis, D.L.; Natarajan, A.; Lamborn, K.R.; Jacobs, F.; et al. Enhanced therapeutic index of radioimmunotherapy (RIT) in prostate cancer patients: Comparison of radiation dosimetry for 1,4,7,10-tetraazacyclododecane-*n*,*n*′,*n*″,*n*‴-tetraacetic acid (DOTA)-peptide versus 2IT-DOTA monoclonal antibody linkage for rit. *Clin. Cancer Res.* **2003**, *9*, 3938S–3944S. [PubMed]

541. O'Donnell, R.T.; DeNardo, S.J.; Miers, L.A.; Lamborn, K.R.; Kukis, D.L.; DeNardo, G.L.; Meyers, F.J. Combined modality radioimmunotherapy for human prostate cancer xenografts with taxanes and 90yttrium-DOTA-peptide-CHL6. *Prostate* **2002**, *50*, 27–37. [CrossRef] [PubMed]

542. O'Donnell, R.T.; DeNardo, S.J.; Yuan, A.; Shen, S.; Richman, C.M.; Lara, P.N.; Griffith, I.J.; Goldstein, D.S.; Kukis, D.L.; Martinez, G.S.; et al. Radioimmunotherapy with ^{111}In/^{90}Y-2IT-BAD-m170 for metastatic prostate cancer. *Clin. Cancer Res.* **2001**, *7*, 1561–1568. [PubMed]

543. McDevitt, M.R.; Barendswaard, E.; Ma, D.; Lai, L.; Curcio, M.J.; Sgouros, G.; Ballangrud, A.M.; Yang, W.H.; Finn, R.D.; Pellegrini, V.; et al. An alpha-particle emitting antibody ([213bi]j591) for radioimmunotherapy of prostate cancer. *Cancer Res.* **2000**, *60*, 6095–6100. [PubMed]

544. Rydh, A.; Riklund-Ahlstrom, K.; Widmark, A.; Bergh, A.; Johansson, L.; Tavelin, B.; Nilsson, S.; Stigbrand, T.; Damber, J.E.; Hietala, S.O. Radioimmunotherapy of du-145 tumours in nude mice—A pilot study with e4, a novel monoclonal antibody against prostate cancer. *Acta Oncol.* **1999**, *38*, 1075–1079. [CrossRef] [PubMed]

545. Mottet, N.; Bellmunt, J.; Briers, E.; van den Bergh, R.; Bolla, M.; van Casteren, N.; Cornford, P.; Culine, S.; Joniau, S.; Lam, T.; et al. *Guidelines on Prostate Cancer*; European Association of Urology: Arnhem, the Netherlands, 2015; pp. 1–137.

546. NCCN. NCCN guidelines on Prostate Cancer (version 1.2015). Available online: http://www.nccn.org (accessed on 9 March 2016).

547. Edge, S.; Byrd, D.; Compton, C. *AJCC Cancer Staging*; Springer: New York, NY, USA, 2010.

548. Powles, T.; Murray, I.; Brock, C.; Oliver, T.; Avril, N. Molecular positron emission tomography and PET/CT imaging in urological malignancies. *Eur. Urol.* **2007**, *51*, 1511–1520, discussion 1520–1511. [CrossRef] [PubMed]

549. Turkbey, B.; Pinto, P.A.; Choyke, P.L. Imaging techniques for prostate cancer: Implications for focal therapy. *Nat. Rev. Urol.* **2009**, *6*, 191–203. [CrossRef] [PubMed]

550. Jana, S.; Blaufox, M.D. Nuclear medicine studies of the prostate, testes, and bladder. *Semin. Nucl. Med.* **2006**, *36*, 51–72. [CrossRef] [PubMed]

551. DeGrado, T.R.; Coleman, R.E.; Wang, S.; Baldwin, S.W.; Orr, M.D.; Robertson, C.N.; Polascik, T.J.; Price, D.T. Synthesis and evaluation of ^{18}F-labeled choline as an oncologic tracer for positron emission tomography: Initial findings in prostate cancer. *Cancer Res.* **2001**, *61*, 110–117. [PubMed]

552. Watanabe, H.; Kanematsu, M.; Kondo, H.; Kako, N.; Yamamoto, N.; Yamada, T.; Goshima, S.; Hoshi, H.; Bae, K.T. Preoperative detection of prostate cancer: A comparison with ^{11}C-choline pet, ^{18}F-fluorodeoxyglucose pet and MR imaging. *J. Magn. Reson. Imaging* **2010**, *31*, 1151–1156. [CrossRef] [PubMed]

553. Morris, M.J.; Akhurst, T.; Osman, I.; Nunez, R.; Macapinlac, H.; Siedlecki, K.; Verbel, D.; Schwartz, L.; Larson, S.M.; Scher, H.I. Fluorinated deoxyglucose positron emission tomography imaging in progressive metastatic prostate cancer. *Urology* **2002**, *59*, 913–918. [CrossRef]

554. Schoder, H.; Herrmann, K.; Gonen, M.; Hricak, H.; Eberhard, S.; Scardino, P.; Scher, H.I.; Larson, S.M. 2-[18]Ffluoro-2-deoxyglucose positron emission tomography for the detection of disease in patients with prostate-specific antigen relapse after radical prostatectomy. *Clin. Cancer Res.* **2005**, *11*, 4761–4769. [CrossRef] [PubMed]

555. Poulsen, M.H.; Bouchelouche, K.; Hoilund-Carlsen, P.F.; Petersen, H.; Gerke, O.; Steffansen, S.I.; Marcussen, N.; Svolgaard, N.; Vach, W.; Geertsen, U.; et al. [18]F fluoromethylcholine (FCH) positron emission tomography/computed tomography (PET/CT) for lymph node staging of prostate cancer: A prospective study of 210 patients. *BJU Int.* **2012**, *110*, 1666–1671. [CrossRef] [PubMed]

556. Brogsitter, C.; Zophel, K.; Kotzerke, J. ^{18}F-choline, ^{11}C-choline and ^{11}C-acetate PET/CT: Comparative analysis for imaging prostate cancer patients. *Eur. J. Nucl. Med. Mol. Imaging* **2013**, *40*, S18–S27. [CrossRef] [PubMed]

557. Murphy, G.; Haider, M.; Ghai, S.; Sreeharsha, B. The expanding role of MRI in prostate cancer. *AJR Am. J. Roentgenol.* **2013**, *201*, 1229–1238. [CrossRef] [PubMed]

558. Souvatzoglou, M.; Eiber, M.; Takei, T.; Furst, S.; Maurer, T.; Gaertner, F.; Geinitz, H.; Drzezga, A.; Ziegler, S.; Nekolla, S.G.; et al. Comparison of integrated whole-body[11]C-choline pet/mr with PET/CT in patients with prostate cancer. *Eur. J. Nucl. Med. Mol. Imaging* **2013**, *40*, 1486–1499. [CrossRef] [PubMed]

559. Wetter, A.; Lipponer, C.; Nensa, F.; Heusch, P.; Rubben, H.; Altenbernd, J.C.; Schlosser, T.; Bockisch, A.; Poppel, T.; Lauenstein, T.; et al. Evaluation of the pet component of simultaneous ^{18}F-choline PET/MRI in prostate cancer: Comparison with[18]F-choline PET/CT. *Eur. J. Nucl. Med. Mol. Imaging* **2014**, *41*, 79–88. [CrossRef] [PubMed]

560. Eiber, M.; Takei, T.; Souvatzoglou, M.; Mayerhoefer, M.E.; Furst, S.; Gaertner, F.C.; Loeffelbein, D.J.; Rummeny, E.J.; Ziegler, S.I.; Schwaiger, M.; et al. Performance of whole-body integrated [18]F -FDG PET/MR in comparison to PET/CT for evaluation of malignant bone lesions. *J. Nucl. Med.* **2014**, *55*, 191–197. [CrossRef] [PubMed]

561. Van der Kwast, T.H.; Roobol, M.J. Defining the threshold for significant versus insignificant prostate cancer. *Nat. Rev. Urol.* **2013**, *10*, 473–482. [CrossRef] [PubMed]

562. Steyerberg, E.W.; Roobol, M.J.; Kattan, M.W.; van der Kwast, T.H.; de Koning, H.J.; Schroder, F.H. Prediction of indolent prostate cancer: Validation and updating of a prognostic nomogram. *J. Urol.* **2007**, *177*, 107–112, discussion 112. [CrossRef] [PubMed]

563. Epstein, J.I.; Walsh, P.C.; Carmichael, M.; Brendler, C.B. Pathologic and clinical findings to predict tumor extent of nonpalpable (stage T1C) prostate cancer. *JAMA* **1994**, *271*, 368–374. [CrossRef] [PubMed]

564. Epstein, J.I.; Chan, D.W.; Sokoll, L.J.; Walsh, P.C.; Cox, J.L.; Rittenhouse, H.; Wolfert, R.; Carter, H.B. Nonpalpable stageT1C prostate cancer: Prediction of insignificant disease using free/total prostate specific antigen levels and needle biopsy findings. *J. Urol.* **1998**, *160*, 2407–2411. [CrossRef]

565. Bastian, P.J.; Carter, B.H.; Bjartell, A.; Seitz, M.; Stanislaus, P.; Montorsi, F.; Stief, C.G.; Schroder, F. Insignificant prostate cancer and active surveillance: From definition to clinical implications. *Eur. Urol.* **2009**, *55*, 1321–1330. [CrossRef] [PubMed]

566. Oon, S.F.; Watson, R.W.; O'Leary, J.J.; Fitzpatrick, J.M. Epstein criteria for insignificant prostate cancer. *BJU Int.* **2011**, *108*, 518–525. [CrossRef] [PubMed]

567. Heidenreich, A.; Bastian, P.J.; Bellmunt, J.; Bolla, M.; Joniau, S.; van der Kwast, T.; Mason, M.; Matveev, V.; Wiegel, T.; Zattoni, F.; et al. Eau guidelines on prostate cancer. Part II: Treatment of advanced, relapsing, and castration-resistant prostate cancer. *Eur. Urol.* **2014**, *65*, 467–479. [CrossRef] [PubMed]

568. Mottet, N.; Briers, J.; van den Berg, D. Guidelines on prostate cancer. *Eur. Urol.* **2015**.

569. Pepe, P.; Fraggetta, F.; Galia, A.; Panella, P.; Pennisi, M.; Colecchia, M.; Aragona, F. Preoperative findings, pathological stage PSA recurrence in men with prostate cancer incidentally detected at radical cystectomy: Our experience in 242 cases. *Int. Urol. Nephrol.* **2014**, *46*, 1325–1328. [CrossRef] [PubMed]

570. Picchio, M.; Messa, C.; Landoni, C.; Gianolli, L.; Sironi, S.; Brioschi, M.; Matarrese, M.; Matei, D.V.; De Cobelli, F.; Del Maschio, A.; et al. Value of ^{11}C-choline-positron emission tomography for re-staging prostate cancer: A comparison with ^{18}F-fluorodeoxyglucose-positron emission tomography. *J. Urol.* **2003**, *169*, 1337–1340. [CrossRef] [PubMed]

571. Winter, A.; Uphoff, J.; Henke, R.P.; Wawroschek, F. First results of ^{11}C-choline PET/CT-guided secondary lymph node surgery in patients with psa failure and single lymph node recurrence after radical retropubic prostatectomy. *Urol. Int.* **2010**, *84*, 418–423. [CrossRef] [PubMed]

572. Fuccio, C.; Castellucci, P.; Schiavina, R.; Santi, I.; Allegri, V.; Pettinato, V.; Boschi, S.; Martorana, G.; Al-Nahhas, A.; Rubello, D.; et al. Role of ^{11}C-choline PET/CT in the restaging of prostate cancer patients showing a single lesion on bone scintigraphy. *Ann. Nucl. Med.* **2010**, *24*, 485–492. [CrossRef] [PubMed]

573. Cimitan, M.; Bortolus, R.; Morassut, S.; Canzonieri, V.; Garbeglio, A.; Baresic, T.; Borsatti, E.; Drigo, A.; Trovo, M.G. ^{18}F-fluorocholine PET/CT imaging for the detection of recurrent prostate cancer at psa relapse: Experience in 100 consecutive patients. *Eur. J. Nucl. Med. Mol. Imaging* **2006**, *33*, 1387–1398. [CrossRef] [PubMed]

574. Ibrahim, T.; Flamini, E.; Mercatali, L.; Sacanna, E.; Serra, P.; Amadori, D. Pathogenesis of osteoblastic bone metastases from prostate cancer. *Cancer* **2010**, *116*, 1406–1418. [CrossRef] [PubMed]

575. Helyar, V.; Mohan, H.K.; Barwick, T.; Livieratos, L.; Gnanasegaran, G.; Clarke, S.E.; Fogelman, I. The added value of multislice SPECT/CT in patients with equivocal bony metastasis from carcinoma of the prostate. *Eur. J. Nucl. Med. Mol. Imaging* **2010**, *37*, 706–713. [CrossRef] [PubMed]

576. Herranz-Blanco, B.; Shahbazi, M.A.; Correia, A.R.; Balasubramanian, V.; Kohout, T.; Hirvonen, J.; Santos, H.A. PH-switch nanoprecipitation of polymeric nanoparticles for multimodal cancer targeting and intracellular triggered delivery of doxorubicin. *Adv. Healthc. Mater.* **2016**, *5*, 1904–1916. [CrossRef] [PubMed]

577. Vu-Quang, H.; Vinding, M.S.; Nielsen, T.; Ullisch, M.G.; Nielsen, N.C.; Kjems, J. Theranostic tumor targeted nanoparticles combining drug delivery with dual near infrared and f magnetic resonance imaging modalities. *Nanomedicine* **2016**, *12*, 1873–1884. [CrossRef] [PubMed]

578. Hu, W.; Ma, H.; Hou, B.; Zhao, H.; Ji, Y.; Jiang, R.; Hu, X.; Lu, X.; Zhang, L.; Tang, Y.; et al. Engineering lysosome-targeting bodipy nanoparticles for photoacoustic imaging and photodynamic therapy under near-infrared light. *ACS Appl. Mater. Interfaces* **2016**, *8*, 12039–12047. [CrossRef] [PubMed]

579. Detappe, A.; Lux, F.; Tillement, O. Pushing radiation therapy limitations with theranostic nanoparticles. *Nanomedicine (Lond.)* **2016**, *11*, 997–999. [CrossRef] [PubMed]

580. Hung, C.C.; Huang, W.C.; Lin, Y.W.; Yu, T.W.; Chen, H.H.; Lin, S.C.; Chiang, W.H.; Chiu, H.C. Active tumor permeation and uptake of surface charge-switchable theranostic nanoparticles for imaging-guided photothermal/chemo combinatorial therapy. *Theranostics* **2016**, *6*, 302–317. [CrossRef] [PubMed]

581. Wang, G.; Zhang, F.; Tian, R.; Zhang, L.; Fu, G.; Yang, L.; Zhu, L. Nanotubes-embedded indocyanine green-hyaluronic acid nanoparticles for photoacoustic-imaging-guided phototherapy. *ACS Appl. Mater. Interfaces* **2016**, *8*, 5608–5617. [CrossRef] [PubMed]

582. Ghaemi, B.; Mashinchian, O.; Mousavi, T.; Karimi, R.; Kharrazi, S.; Amani, A. Harnessing the cancer radiation therapy by lanthanide-doped zinc oxide based theranostic nanoparticles. *ACS Appl. Mater. Interfaces* **2016**, *8*, 3123–3134. [CrossRef] [PubMed]

583. Gurka, M.K.; Pender, D.; Chuong, P.; Fouts, B.L.; Sobelov, A.; McNally, M.W.; Mezera, M.; Woo, S.Y.; McNally, L.R. Identification of pancreatic tumors in vivo with ligand-targeted, pH responsive mesoporous silica nanoparticles by multispectral optoacoustic tomography. *J. Control. Release* **2016**, *231*, 60–67. [CrossRef] [PubMed]

584. Andreou, C.; Pal, S.; Rotter, L.; Yang, J.; Kircher, M.F. Molecular imaging in nanotechnology and theranostics. *Mol. Imaging Biol.* **2017**, *19*, 363–372. [CrossRef] [PubMed]

585. Li, J.; Wang, S.; Shi, X.; Shen, M. Aqueous-phase synthesis of iron oxide nanoparticles and composites for cancer diagnosis and therapy. *Adv. Colloid Interface Sci.* **2017**. [CrossRef] [PubMed]

586. Lin, L.S.; Yang, X.; Zhou, Z.; Yang, Z.; Jacobson, O.; Liu, Y.; Yang, A.; Niu, G.; Song, J.; Yang, H.H.; et al. Yolk-shell nanostructure: An ideal architecture to achieve harmonious integration of magnetic-plasmonic hybrid theranostic platform. *Adv. Mater.* **2017**. [CrossRef] [PubMed]

587. Tang, L.; Zhang, F.; Yu, F.; Sun, W.; Song, M.; Chen, X.; Zhang, X.; Sun, X. Croconaine nanoparticles with enhanced tumor accumulation for multimodality cancer theranostics. *Biomaterials* **2017**, *129*, 28–36. [CrossRef] [PubMed]

588. Kuang, Y.; Zhang, K.; Cao, Y.; Chen, X.; Wang, K.; Liu, M.; Pei, R. Hydrophobic IR-780 dye encapsulated in CRGD-conjugated solid lipid nanoparticles for NIR imaging-guided photothermal therapy. *ACS Appl. Mater. Interfaces* **2017**, *9*, 12217–12226. [CrossRef] [PubMed]

589. Owen, J.; Crake, C.; Lee, J.Y.; Carugo, D.; Beguin, E.; Khrapitchev, A.A.; Browning, R.J.; Sibson, N.; Stride, E. A versatile method for the preparation of particle-loaded microbubbles for multimodality imaging and targeted drug delivery. *Drug Deliv. Transl. Res.* **2017**. [CrossRef] [PubMed]

590. Rajasekharreddy, P.; Huang, C.; Busi, S.; Rajkumari, J.; Tai, M.H.; Liu, G. Green synthesized nanomaterials as theranostic platforms for cancer treatment: Principles, challenges and the road ahead. *Curr. Med. Chem.* **2017**. [CrossRef] [PubMed]

591. Wang, Z.; Qiao, R.; Tang, N.; Lu, Z.; Wang, H.; Zhang, Z.; Xue, X.; Huang, Z.; Zhang, S.; Zhang, G.; et al. Active targeting theranostic iron oxide nanoparticles for MRI and magnetic resonance-guided focused ultrasound ablation of lung cancer. *Biomaterials* **2017**, *127*, 25–35. [CrossRef] [PubMed]

592. Ansari, C.; Tikhomirov, G.A.; Hong, S.H.; Falconer, R.A.; Loadman, P.M.; Gill, J.H.; Castaneda, R.; Hazard, F.K.; Tong, L.; Lenkov, O.D.; et al. Development of novel tumor-targeted theranostic nanoparticles activated by membrane-type matrix metalloproteinases for combined cancer magnetic resonance imaging and therapy. *Small* **2014**, *10*, 417, 566–575. [CrossRef]

593. Kiess, A.P.; Banerjee, S.R.; Mease, R.C.; Rowe, S.P.; Rao, A.; Foss, C.A.; Chen, Y.; Yang, X.; Cho, S.Y.; Nimmagadda, S.; et al. Prostate-specific membrane antigen as a target for cancer imaging and therapy. *Q. J. Nucl. Med. Mol. Imaging* **2015**, *59*, 241–268. [PubMed]

594. Hou, Y.; Qiao, R.; Fang, F.; Wang, X.; Dong, C.; Liu, K.; Liu, C.; Liu, Z.; Lei, H.; Wang, F.; et al. Nagdf4 nanoparticle-based molecular probes for magnetic resonance imaging of intraperitoneal tumor xenografts in vivo. *ACS Nano* **2013**, *7*, 330–338. [CrossRef] [PubMed]

595. Franci, C.; Zhou, J.; Jiang, Z.; Modrusan, Z.; Good, Z.; Jackson, E.; Kouros-Mehr, H. Biomarkers of residual disease, disseminated tumor cells, and metastases in the MMTV-PYMT breast cancer model. *PLoS ONE* **2013**, *8*, e58183. [CrossRef] [PubMed]

596. FDA. Fda Drug Safety Communication: Fda Strengthens Warnings and Changes Prescribing Instructions to Decrease the Risk of Serious Allergic Reactions with Anemia Drug Feraheme (Ferumoxytol). Available online: http://www.fda.gov/Drugs/DrugSafety/ucm440138.htm (accessed on 17 June 2016).

International Journal of
Molecular Sciences

MDPI

Review

Current Stem Cell Biomarkers and Their Functional Mechanisms in Prostate Cancer

Kaile Zhang [1,2,†], Shukui Zhou [1,†], Leilei Wang [3], Jianlong Wang [4], Qingsong Zou [1], Weixin Zhao [2], Qiang Fu [1,*] and Xiaolan Fang [5,*,‡]

[1] The Department of Urology, Affiliated Sixth People's Hospital, Shanghai JiaoTong University, Shanghai 200233, China; great_z0313@126.com (K.Z.); 2005507098@163.com (S.Z.); zou_qingsong@126.com (Q.Z.)
[2] Wake Forest Institute for Regenerative Medicine, Winston-Salem, NC 27101, USA; wezhao@wakehealth.edu
[3] VIP Department of Beijing Hospital, Beijing 100730, China; luckyleileiaaa@sina.com
[4] Urology Department of Beijing Hospital, Beijing 100730, China; wjlspplaaa@sina.com
[5] Department of Cancer Biology, Wake Forest University School of Medicine, Wake Forest Institute for Regenerative Medicine, Winston-Salem, NC 27101, USA
* Correspondence: jamesqfu@aliyun.com (Q.F.); afang@nygenome.org (X.F.); Tel.: +86-21-2405-8372 (Q.F.); +1-434-409-8076 (X.F.)
† These authors contributed equally to this work.
‡ Current address: New York Genome Center, 101 Avenue of the Americas, New York, NY 10013, USA.

Academic Editor: Carsten Stephan
Received: 30 May 2016; Accepted: 9 July 2016; Published: 19 July 2016

Abstract: Currently there is little effective treatment available for castration resistant prostate cancer, which is responsible for the majority of prostate cancer related deaths. Emerging evidence suggested that cancer stem cells might play an important role in resistance to traditional cancer therapies, and the studies of cancer stem cells (including specific isolation and targeting on those cells) might benefit the discovery of novel treatment of prostate cancer, especially castration resistant disease. In this review, we summarized major biomarkers for prostate cancer stem cells, as well as their functional mechanisms and potential application in clinical diagnosis and treatment of patients.

Keywords: prostate cancer; cancer stem cell; stem cell biomarker

1. Introduction

Prostate cancer (PCa) is the most common non-skin cancer in American men [1,2]. Standard PCa treatment includes radical prostatectomy, radiotherapy, chemotherapy and castration (either by drug or by surgery, mainly for androgen sensitive PCa), as well as immunotherapy and palliative therapy (mainly for castration resistant PCa (CRPC)). CRPC is responsible for majority of the PCa-related deaths [3], and currently there are two major hypotheses of CRPC carcinogenesis, the adaptive mechanism and the selective mechanism [4]. The adaptive mechanism suggests gene mutations in PCa cells (e.g., mutations of androgen receptor (AR)), dysregulated expression of genes, etc., contribute to CRPC development [5]. The selective mechanism, which is emerged in the last few decades, suggests that pre-existing castration-resistant subclones in primary PCa tissues and cancer stem cell selection dominates CRPC development (Figure 1) [6–8]. Recently, it has been suggested that stem-cell directed differentiation therapy could promote differentiation of cancer stem cells and sensitize them to anticancer drugs (such as synergistic androgen signaling blocking agents) [9].

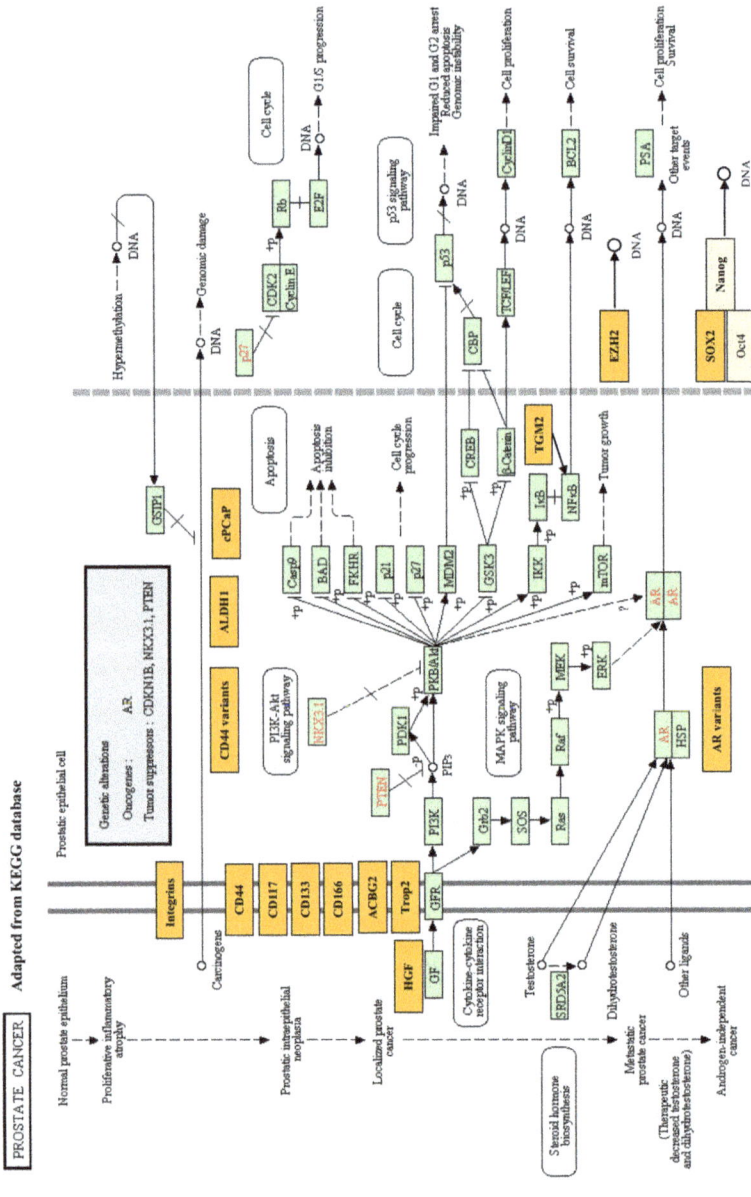

Figure 1. The mechanism and pathway map of the prostate cancer (modified based on KEGG database). Solid line between genes/molecules indicates direct regulation, while dashed lines indicates possible indirect regulation. Circle indicates a group of similar molecules (instead of a specific one). Biomarkers discussed in this review are highlighted in orange and in bold font, related molecules that are newly discovered are in yellow. Classic biomarkers included in KEGG prostate cancer pathway are highlighted in green. Key regulators in classical pathways involved in PCa are displayed in red (e.g., NKX3.1, PTEN, AR).

Cancer stem cells (CSCs) were defined as cells with capacity of self-renewal and proliferation in cancer tissue [8]. Over years, scientists have been arguing about the origin of cancer stem cells. CSCs were suggested to originate from mutated normal stem cells, from mutated progenitor cells in the process of differentiation which re-gains the characteristics of stem cells, or from mature cells that re-acquired self-renewal ability [10]. Various cell surface markers were used to isolate CSCs, whose proliferative potential was verified by in vitro andin vivo assays (Tables 1 and 2). This review summarizes recent research progress of current stem cell markers in PCa.

Table 1. Summary of prostate cancer stem cell biomarkers based on location and function.

Biomarker	Transmembrane Protein	Glycoprotein	Enzyme	Transcription Factor	Extracellular Protein	mRNA
Integrins	Yes	-	-	-	-	-
CD44	Yes	-	-	-	-	-
CD133	Yes	Yes	-	-	-	-
CD166	Yes	-	-	-	-	-
Trop2	Yes	Yes	-	-	-	-
CD117	Yes	-	Yes	-	-	-
ALDH1	-	-	-	Yes	-	-
ABCG2	Yes	-	-	-	-	-
SOX2	-	-	-	Yes	-	-
EZH2	-	-	Yes	-	-	-
cPAcP	-	-	Yes	-	-	-
AR splice variants	-	-	-	-	-	Yes
HGF	-	-	-	-	Yes	-
TGM2	-	-	Yes	-	-	-

Trop2, tumor-associated calcium signal transducer 2; ALDH1, aldehyde dehydrogenase 1; ABCG2, ATP binding membrane transporters; cPAcP, cellular prostatic acid phosphatase; HGF, hepatocyte growth factor; TGM2, transglutaminase II; SOX2, SRY-box 2; EZH2, enhancer of zeste 2 polycomb repressive complex 2 subunit.

Table 2. Summary of verifying studies and possible pathways of prostate cancer stem cell biomarkers.

Markers	PCa Cell Lines	Primary PCa Tissues	Mouse Models	Possible Involved Pathway in PCa
Integrins	Yes	Yes	-	-
CD44	Yes	Yes	-	-
CD133	Yes	Yes	-	-
CD166	-	-	Yes	-
Trop2	-	-	Yes	-
CD117	-	-	Yes	-
ALDH1	Yes	-	-	-
ABCG2	Yes	Yes	-	-
SOX2	-	Yes	-	-
EZH2	-	Yes	-	-
cPAcP	Yes	-	-	-
AR splice variants	Yes	-	-	AR
HGF	Yes	-	-	AR
TGM2	Yes	-	-	NF-κB

PCa, Prostate cancer.

2. Integrins

Integrins are a family of transmembrane receptors known to participate in cell-cell adhesion and cell-surface mediated signaling, serving as bridges for cell-cell and cell-extracellular matrix (ECM) interactions [11]. Integrin could interact with specific ligands to transfer signals through cell-cell or cell-ECM interactions and stimulate expression of downstream target genes. Integrins were generally overexpressed in PCa [12,13]. In PCa, expression of α2-integrin and EZH2 is observed in a small fraction of cancer cells, which is supportive for their role as stem cell marker [14]. α2β1 integrin plays an important role in epithelia-stroma interaction, which is suggested to contribute to selective bone metastasis [12]. In the meantime, it could be a new marker to screen for prostate stem cells. Collins et al. [15] has discovered that the prostate stem cells expressing α2β1 integrin locate at basal epithelial layer. Approximately 1% of basal cells examined by confocal microscopy were integrin positive, and these cells could be isolated directly from the tissue on the basis of rapid adhesion to type I collagen. This isolated cell population displays basal cell phenotype, marked by expression of CK5 and CK14 and lack of expression of differentiation-specific markers (such as prostate specific antigen (PSA) and prostatic acid phosphatase (PAP)). These prostate stem cells could be cultivated in vitro and display much greater capability to form colonies in vitro (comparing to total basal cell population). When α2β1 overexpressing cells and stromal cells were transplanted subcutaneously into nude mouse, they could form structure of normal prostate gland prostate-specific differentiation [15].

Microarray experiments performed by several independent groups found that Integrin-α6 (also known as CD49f) is consistently overexpressed in hematopoietic, neural, and embryonic stem cells, and it is suggested as an effective cell stemness marker [16]. It has been used for characterization of prostatic progenitor cells [17,18], and was suggested as an emerging biomarker for PCa evaluation [14,19,20].

3. CD44

CD44 is a single-pass type I transmembrane protein and an important cellular adhesion molecule related to signaling to extracellular matrix. CD44 was considered as a marker of cancer stem cells from many organs including prostate [21–23]. It was located extensively on cell membrane and is important for cell adhesion and signal transduction. It was reported that CD44 positive cells from primary prostatic tumor tissues possess cell stemness [24]. Molecular studies demonstrated that CD44[+] PCa cells retain certain intrinsic properties of progenitor cells [25]. CD44[+] cells express high levels of stemness genes including *Oct-3/4*, *Bmi*, β-catenin and Smoothened (*SMO*) [2,26,27]. Kasper et al. discovered that in PCa cells (such as LNCaP, DU145 and PC3), CD44 positive cells had much greater proliferative capability than CD44 negative cells [28]. Van et al. isolated the DU145 cells from CD44[+] and CD44[−] Cells and tested the gene expression of stem cells by RT-PCR. Low expression of luminal cell markers (e.g., CK18) and AR were observed in CD44[+] cells, whereas the genes highly related to stem cell proliferation and differentiation were overexpressed [29]. Recently, CD44 expression level was reported to be correlated with PCa grade in prostate biopsy samples [30], and proteomics analysis showed that CD44[+] cells had positive correlation with genes related to cancer proliferation and metastasis [31]. However, Ugolkov discovered that expression of CD44 and Oct4 were observed in large populations of benign and malignant cells in the prostate, which is somewhat contradictory to the definition of stem cells as a small fraction of the total cell population [32]. Their results suggested that combined expression of embryonic stem cell markers EZH2 and SOX2 might be used to identify potential cancer stem cells as a minor (<10%) subgroup in CD44[+] prostatic adenocarcinoma cells [32].

Recently, quite a few thorough analyses have been done on CD44 isoforms that are generated through alternative splicing of CD44 precursor mRNA. Those CD44 variants function distinctly in PCa and might serve as independent markers comparing to total CD44 expression level. For example, CD44v2 correlated with a better recurrence-free survival rate in PCa patients and is underexpressed in metastatic PCa cell lines [33]. Another well-studied isoform is CD44v6, which is associated with PCa proliferation, invasion, adhesion, metastasis, chemo-/radioresistance, and the induction of epithelial–mesenchymal transition (EMT) as well as the activation PI3K/Akt/mTOR and Wnt signaling pathways, and CD44v6 expression was closely associated with conventional prognostic factors and is identified as significant predictor for biochemical recurrence in PCa [34,35]. CD44v7–10 were overexpressed in PCa, and knock-down of CD44v7–10 by RNAi would significantly decrease invasion and migration in PCa cells [36].

Taken together, those results demonstrated CD44 RNA isoforms, but not total CD44 protein, might serve as specific marker for prostate cancer stem cells, though total CD44 protein level might still serve as a stem cell marker for other types of cancers [29].

4. CD133

CD133 is a glycoprotein with five transmembrane domains, generally expressed in various stem cells and endothelial progenitor cells but not in mature endothelial cells [37]. CD133 has been widely used, usually in combination with other stem cell markers such as CD44 and $\alpha 2\beta 1$ integrin, to isolate cancer stem cells from prostate tumors with different Gleason grade, including cells from both primary and metastatic lesions [38–41]. Approximately 0.1% of cells in any prostate tumor displayed this phenotype, though there was no correlation between the number of CD44$^+$/$\alpha 2\beta 1$hi/CD133$^+$ cells and tumor grade [23]. In normal prostate tissues, CD133 expression was observed in both basal and luminal cells [42]. Although its expression in normal prostate tissue is pretty low, CD133 is usually overexpressed in inflammation cell population [43].

Prostatic basal cells could be enriched based on $\alpha 2\beta 1$ integrin (hi) expression and further enriched for stem cells using CD133 in non-tumorigenic BPH-1 cells [44]. It is demonstrated that the tumorigenic potential did not reside in the CD133$^+$ stem cells but was consistently observed in the CD133$^-$ population [45]. These data confirmed that benign basal cells include cells of origin of prostate cancer and suggested that proliferative CD133$^-$ basal cells are more susceptible to tumorigenesis compared to the CD133$^+$-enriched stem cells. These findings challenged the current dogma that normal stem cells and cells of origin of cancer are the same cell type(s) [45]. Intensive studies need to be done to learn more about the role of CD133 in PCa origination.

5. ALDH1

ALDH1 was suggested as a stem cell marker for both normal and tumor tissues [46]. As a cytoplasmic enzyme, ALDH1 has multiple intracellular aldehydes which can be converted into carboxylic acids, and could be involved in intracellular degradation of cell toxic substances [47]. ALDH1 expression was reported to be correlated with tumor grade and prognosis in PCa patients [48]. Burger et al. [49] found that cells with high ALDH enzymatic activity have greater in vitro proliferative potential than cells with low ALDH activity. Similar results were observed in an in vivo prostate reconstitution assay [49]. Thus, ALDH enzymatic activity might be used as a functional marker of prostate stem/progenitor cells and allow for simple, efficient isolation of cells with primitive features. ALDH $\alpha 2+/\alpha 6+/\alpha V$ + CD44$^+$ cells also displayed high colonization in vitro and highly invasive tumorigenesis and aggressive metastasis characteristics in vivo [50]. p63 cytoplasmic aberrance is associated with high ALDH1A1 expression, and it was found that cytoplasmic p63 levels were significantly associated with the frequency of proliferating cells and cells undergoing apoptosis in prostate cancers [51]. These components are suggested to have an important role in prostate cancer progression and may be used as a panel of molecular markers [52].

The aldehyde dehydrogenase enzymes are likely to protect stem cells by detoxification of cell toxic compounds, which indicates that ALDH1 might prevent prostate cancer stem cells from conventional chemotherapy attack, while effective inhibition of ALDH1 could enhance the chemotherapy efficiency. Thus, ALDH1 could not only be used as a prostate cancer stem cell marker for prognosis, but also as a potential drug target in cancer treatment.

6. ATP Binding Membrane Transporters (ABCG2, Also Known as Breast Cancer Resistant Protein or BCRP)

Studies have shown that prostate cancer contains side population cells (SP cells), which could be isolated by flow cytometry techniques based on behavioral characteristics of stem cells. SP cells have stem cell properties that are exclusively mediated by ABCG2. As a result, ABCG2 is considered as a marker of SP cells, as well as a cancer stem cell marker.

ABCG2 is ATP binding membrane transporters, and is related to prostate cancer multi-drug resistance [10]. After castration, ABCG2+/AR− prostate cancer stem cells could be isolated from prostate cancer tissues, and it is suggested that ABCG2 expression might protect prostate cancer stem cells from castration, chemotherapy and hypoxic environment. ABCG2 has been suggested as a biomarker for treatments targeting on prostate cancer stem cells [53]. Interestingly, Patrswala et al. [54] found that ABCG2(+) cells could produce ABCG2(−) cells, and both types of cells have similar tumorigenicity and colony formation ability. 30% of human cancer cell lines (and more in the bone marrow) and xenografts contain 0.04% to 0.20% of SP cells (low but detectable), yet most of the primary tumor cells have only a very small portion of the SP cells, almost impossible to detect [55,56]. Giving the evidence that non-recurrent PCa samples presented relatively lower level of ABCG2, compared to both normal tissue and recurrent samples, it might be associated with chemo-sensitivity [57]. Whether ABCG2 could be used as a specific biomarker in PCa diagnosis and prognosis is still unclear and requires further research.

7. SOX2 and EZH2

SOX2 and EZH2 are essential for the development of human embryonic stem cells. SOX2 is a transcription factor and plays a key role in maintaining undifferentiated status and keeping self-renewal ability of embryonic stem cells [58]. EZH2 is critical for embryonic stem cells rebuilding and embryonic development. Studies show that they play a key role in prostate cancer stem cells [32]. Recently, SOX2 and EZH2 are also suggested as markers in malignant glioma patients [59]. Ugolkov et al. [32] analyzed expression of CD44, CD133, Oct4, SOX2 and EZH2 in benign prostate tissues, high grade prostatic intraepithelial neoplasia (HGPIN) and PCa tissues, and found that EZH2 and SOX2 were expressed in <10% of benign prostate tissue, HGPINs and prostate cancer. In addition, 82% (27/33) of SOX2+ prostate cancer cases were EZH2+ type, and 100% (33/33) of cases were CD44+. On the other hand, CD44 was found in 97% of benign prostate and HGPIN cases, and in 72% of prostate cancer cases. CD133 was found in only a small portion of PCa tissues (6%, 4/67). Oct4 expression was found to be closely correlated with benign and HGPIN, but not with PCa. It is believed that CD44 and Oct4 were expressed in most of benign and malignant prostate cells, which is not likely to be representative for a very small proportion of cancer cells (such as cancer stem cells).

8. CD166

CD166 is a newly discovered molecular surface marker of prostate cancer stem cells [60]. CD166 belongs to the Ig family of type I transmembrane proteins, which mediate cell-cell interactions, and have been used as prognostic markers for a variety of cancers [1]. CD166 was reported to enrich sphere-forming activity of WT LSC (hi) and Pten null LSC (hi), and enhance the sphere-forming ability of benign primary human prostate cells in vitro and induce the formation of tubule-like structures in vivo [60]. CD166 could be used to identify and isolate human, murine prostate cancer stem cells and hormone refractory prostate cancer [61]. CD166 protein level is upregulated in human PCa, especially in CRPC patients. Although genetic deletion of murine CD166 in the Pten null PCa model does not interfere with sphere formation or block prostate cancer progression and CRPC development, the presence of CD166 on prostate stem/progenitors and castration resistant sub-population of cells suggest that it could be a surface marker of cell stemness. It could be a potential therapeutic target for prostate cancer therapies, as reduced expression of CD166 might be able to interfere or reverse prostate cancer metastasis.

9. cPAcP

cPAcP is a prostate specific differentiation antigen. In PCa cells, decreased cPAcP expression is associated with androgen-independent cell proliferation and tumorigenicity as seen in advanced hormone-refractory prostate carcinomas [62]. It was demonstrated that HDAC inhibitor treatment could result in increased cPAcP protein level in cPAcP positive cells, increase androgen responsiveness, and exhibit higher inhibitory activities on AR/cPAcP-positive PCa cells than on AR/cPAcP-negative PCa cells. These data indicate that cPAcP has potential clinical importance serving as a useful biomarker in the identification of PCa patient sub-population suitable for HDAC inhibitor treatment [63,64].

10. Hepatocyte Growth Factor

It was found that prostate cancer stem-like cells (CSCs)/cancer initiating cells (CICs) express hepatocyte growth factor (HGF) and that the HGF/c-MET proto-oncogene product (c-MET) signal has a role in the maintenance of prostate CSCs/CICs in an autocrine fashion. Immunohistochemical staining of HGF was compared to biochemical recurrence after radical prostatectomy, and patients with PCa tumors exhibiting HGF positivity of 5% or more had a significantly shorter biochemical recurrence-free period than that of patients whose tumor HGF positivity was less than 5% ($p = 0.001$). In multivariate Cox regression, preoperative PSA and HGF positivity had the potential to be independent predictors of biochemical recurrence following prostatectomy [65].

11. Tumor-Associated Calcium Signal Transducer 2

Tumor-associated calcium signal transducer 2 (also known as Trop2) is a type I membrane glycoprotein which transduces intracellular calcium signal and acts as a cell surface receptor [66,67]. Trop2 is highly expressed in epithelial related cancers, and its protein level often correlates with poor prognosis [68–73]. Trop2 positive cells could be identified as a subpopulation of prostate basal cells with stem cell characteristics, and it has been used as an effective marker for isolation of basal prostate progenitor cells [74–76]. In prostate cancer, scientists discovered that Trop2 regulate cancer cell proliferation, self-renewal, cell-cell adhesion and metastasis through β-catenin and β1-integrin signaling pathways [77–79]. Interestingly, Trop2 expression in prostate cancer cells was regulated by energy restriction, glucose deprivation and methylation [80–82], making it a potential drug target in cancer treatment. Moreover, anti-Trop2 bispecific antibody was approved to effectively lead pre-targeted immunoPET and radioimmunotherapy of PCa in preclinical models, which significantly increased PCa related survival [83,84].

12. CD117

CD117 (also known as c-Kit) is a receptor tyrosine kinase protein, and has been used as an important cell surface marker to identify hematopoietic progenitors in bone marrow [85–87]. CD117 overexpression was observed in several types of solid tumors including prostate [88,89], and is correlated with the capacity of cell self-renewal and cancer progression [90,91]. Circulating CD117 positive cell percentage is correlated with cancer progression and PSA values in advanced PCa [92]. CD117 could be activated by its ligand, Stem Cell Factor (SCF), to promote bone marrow cell migration, tumor dissemination and potential bone metastasis [91–94].

13. AR Splice Variants

AR splice variants were found to promote EMT as well as induce the expression of stem cell signature genes [95]. Over 10 different AR splice variants were discovered in PCa cell lines, PCa xenografts and human patient samples, and a few of them were dissected to understand their functions in cancer progression [96–103]. More importantly, AR splice variants, such as AR-V7, were suggested to contribute to the drug resistance after suppression of AR signaling, especially in CRPCs [104,105]. High level of AR-V7 was observed in CRPC specimen, but rarely in hormone-naïve specimen [102]. It was suggested that transition from negative to positive status of AR-V7 might reflect the selective pressures on tumor, which makes it a dynamic marker for PCa diagnosis based on liquid biopsy samples, such as circulating tumor cells (CTC) [106].

14. TGM2

Transglutaminases are enzymes that catalyze the crosslinking of proteins by epsilon-γ glutamyl lysine isopeptide bonds. While the primary structure of transglutaminases is not conserved, they all have the same amino acid sequence at their active sites and their activity is calcium-dependent. The protein encoded by this gene acts as a monomer, is induced by retinoic acid, and appears to be involved in apoptosis. TGM2 expression is shown to negatively regulate AR expression and to attenuate androgen sensitivity of prostate cancer cells [107]. TGM2 activation of NF-κB expression induces NF-κB binding to DNA elements in the AR gene to reduce AR gene expression, and triggers epithelial–mesenchymal transition [107]. This suggests that TGM2-regulated inflammatory signaling may contribute to the androgen dependence of prostate cancer cells [107]. Thus, TGM2 is concluded as a cancer stem cell survival factor in various types of cancers, including prostate cancer [108].

15. Conclusions

Studies of prostate cancer stem cells have gained much progress in the past few years and numerous potential approaches were discussed for novel PCa treatment [109,110]. This review summarizes the major intracellular PCa stem cell biomarkers, including a few novel markers discovered recently. The normal or pathological process and potential drug response reflected by those biomarkers were discussed, which might help with early diagnosis, prevention, drug target identification, drug response evaluation and so on. With the progress in study of circulating biomarkers, we expect that more candidates would be identified to facilitate PCa biopsies, especially those soluble markers (circulating tumor cells (CTCs), circulating tumor nucleic acid (ctNAs), miRNA, lncRNA, exosomes, etc.) for liquid biopsies.

Acknowledgments: This work is supported by NIH grant CA079448 to Xiaolan Fang.

Conflicts of Interest: The authors declare no conflict of interest.

References

1. Siegel, R.; Ma, J.; Zou, Z.; Jemal, A. Cancer statistics, 2014. *CA Cancer J. Clin.* **2014**, *64*, 9–29. [CrossRef] [PubMed]
2. Monsef, N.; Soller, M.; Isaksson, M.; Abrahamsson, P.A.; Panagopoulos, I. The expression of pluripotency marker Oct 3/4 in prostate cancer and benign prostate hyperplasia. *Prostate* **2009**, *69*, 909–916. [CrossRef] [PubMed]
3. Attard, G.; Parker, C.; Eeles, R.A.; Schroder, F.; Tomlins, S.A.; Tannock, I.; Drake, C.G.; de Bono, J.S. Prostate Cancer. *Lancet* **2016**, *387*, 70–82. [CrossRef]
4. Zong, Y.; Goldstein, A.S. Adaptation or selection—Mechanisms of castration-resistant prostate cancer. *Nat. Rev. Urol.* **2013**, *10*, 90–98. [CrossRef] [PubMed]
5. Isaacs, J.T.; Coffey, D.S. Adaptation versus selection as the mechanism responsible for the relapse of prostatic cancer to androgen ablation therapy as studied in the Dunning R-3327-H adenocarcinoma. *Cancer Res.* **1981**, *41*, 5070–5075. [PubMed]
6. Blum, R.; Gupta, R.; Burger, P.E.; Ontiveros, C.S.; Salm, S.N.; Xiong, X.; Kamb, A.; Wesche, H.; Marshall, L.; Cutler, G.; et al. Molecular signatures of prostate stem cells reveal novel signaling pathways and provide insights into prostate cancer. *PLoS ONE* **2009**, *4*, e5722. [CrossRef] [PubMed]
7. Lang, S.H.; Frame, F.M.; Collins, A.T. Prostate cancer stem cells. *J. Pathol.* **2009**, *217*, 299–306. [CrossRef] [PubMed]
8. Chen, X.; Rycaj, K.; Liu, X.; Tang, D.G. New insights into prostate cancer stem cells. *Cell Cycle* **2013**, *12*, 579–586. [CrossRef] [PubMed]
9. Rane, J.K.; Pellacani, D.; Maitland, N.J. Advanced prostate cancer—A case for adjuvant differentiation therapy. *Nat. Rev. Urol.* **2012**, *9*, 595–602. [CrossRef] [PubMed]

10. Castillo, V.; Valenzuela, R.; Huidobro, C.; Contreras, H.R.; Castellon, E.A. Functional characteristics of cancer stem cells and their role in drug resistance of prostate cancer. *Int. J. Oncol.* **2014**, *45*, 985–994. [CrossRef] [PubMed]

11. McMillen, P.; Holley, S.A. Integration of cell–cell and cell–ECM adhesion in vertebrate morphogenesis. *Curr. Opin. Cell Biol.* **2015**, *36*, 48–53. [CrossRef] [PubMed]

12. Van Slambrouck, S.; Groux-Degroote, S.; Krzewinski-Recchi, M.A.; Cazet, A.; Delannoy, P.; Steelant, W.F. Carbohydrate-to-carbohydrate interactions between α2,3-linked sialic acids on α2 integrin subunits and asialo-GM1 underlie the bone metastatic behaviour of LNCAP-derivative C4-2B prostate cancer cells. *Biosci. Rep.* **2014**, *34*. [CrossRef] [PubMed]

13. Dedhar, S.; Saulnier, R.; Nagle, R.; Overall, C.M. Specific alterations in the expression of α3β1 and α6β4 integrins in highly invasive and metastatic variants of human prostate carcinoma cells selected by in vitro invasion through reconstituted basement membrane. *Clin. Exp. Metastasis* **1993**, *11*, 391–400. [CrossRef] [PubMed]

14. Hoogland, A.M.; Verhoef, E.I.; Roobol, M.J.; Schroder, F.H.; Wildhagen, M.F.; van der Kwast, T.H.; Jenster, G.; van Leenders, G.J. Validation of stem cell markers in clinical prostate cancer: α 6-integrin is predictive for non-aggressive disease. *Prostate* **2014**, *74*, 488–496. [CrossRef] [PubMed]

15. Collins, A.T.; Habib, F.K.; Maitland, N.J.; Neal, D.E. Identification and isolation of human prostate epithelial stem cells based on α2β1-integrin expression. *J. Cell Sci.* **2001**, *114*, 3865–3872. [PubMed]

16. Fortunel, N.O.; Otu, H.H.; Ng, H.-H.; Chen, J.; Mu, X.; Chevassut, T.; Li, X.; Joseph, M.; Bailey, C.; Hatzfeld, J.A.; et al. Comment on "'Stemness': Transcriptional Profiling of Embryonic and Adult Stem Cells" and "A Stem Cell Molecular Signature" (I). *Science* **2003**, *302*, 393. [CrossRef] [PubMed]

17. Barclay, W.W.; Axanova, L.S.; Chen, W.; Romero, L.; Maund, S.L.; Soker, S.; Lees, C.J.; Cramer, S.D. Characterization of Adult Prostatic Progenitor/Stem Cells Exhibiting Self-Renewal and Multilineage Differentiation. *Stem Cells* **2008**, *26*, 600–610. [CrossRef] [PubMed]

18. Lawson, D.A.; Xin, L.; Lukacs, R.U.; Cheng, D.; Witte, O.N. Isolation and functional characterization of murine prostate stem cells. *Proc. Natl. Acad. Sci. USA* **2007**, *104*, 181–186. [CrossRef] [PubMed]

19. Marthick, J.R.; Dickinson, J.L. Emerging Putative Biomarkers: The Role of α 2 and 6 Integrins in Susceptibility, Treatment, and Prognosis. *Prostate Cancer* **2012**, *2012*, 298732. [CrossRef] [PubMed]

20. Finetti, F.; Terzuoli, E.; Giachetti, A.; Santi, R.; Villari, D.; Hanaka, H.; Radmark, O.; Ziche, M.; Donnini, S. mPGES-1 in prostate cancer controls stemness and amplifies epidermal growth factor receptor-driven oncogenicity. *Endocr. Relat. Cancer* **2015**, *22*, 665–678. [CrossRef] [PubMed]

21. Lokeshwar, B.L.; Lokeshwar, V.B.; Block, N.L. Expression of CD44 in prostate cancer cells: Association with cell proliferation and invasive potential. *Anticancer Res.* **1995**, *15*, 1191–1198. [PubMed]

22. Liu, A.Y. Expression of CD44 in prostate cancer cells. *Cancer Lett.* **1994**, *76*, 63–69. [CrossRef]

23. Collins, A.T.; Berry, P.A.; Hyde, C.; Stower, M.J.; Maitland, N.J. Prospective identification of tumorigenic prostate cancer stem cells. *Cancer Res.* **2005**, *65*, 10946–10951. [CrossRef] [PubMed]

24. Ajani, J.A.; Song, S.; Hochster, H.S.; Steinberg, I.B. Cancer stem cells: The promise and the potential. *Semin. Oncol.* **2015**, *42* (Suppl. S1), S3–S17. [CrossRef] [PubMed]

25. Wang, L.; Huang, X.; Zheng, X.; Wang, X.; Li, S.; Zhang, L.; Yang, Z.; Xia, Z. Enrichment of prostate cancer stem-like cells from human prostate cancer cell lines by culture in serum-free medium and chemoradiotherapy. *Int. J. Biol. Sci.* **2013**, *9*, 472–479. [CrossRef] [PubMed]

26. Yu, J.; Lu, Y.; Cui, D.; Li, E.; Zhu, Y.; Zhao, Y.; Zhao, F.; Xia, S. miR-200b suppresses cell proliferation, migration and enhances chemosensitivity in prostate cancer by regulating Bmi-1. *Oncol. Rep.* **2014**, *31*, 910–918. [PubMed]

27. Ibuki, N.; Ghaffari, M.; Pandey, M.; Iu, I.; Fazli, L.; Kashiwagi, M.; Tojo, H.; Nakanishi, O.; Gleave, M.E.; Cox, M.E. TAK-441, a novel investigational smoothened antagonist, delays castration-resistant progression in prostate cancer by disrupting paracrine hedgehog signaling. *Int. J. Cancer* **2013**, *133*, 1955–1966. [CrossRef] [PubMed]

28. Kasper, S. Identification, characterization, and biological relevance of prostate cancer stem cells from clinical specimens. *Urol. Oncol.* **2009**, *27*, 301–303. [CrossRef] [PubMed]

29. Van Leenders, G.J.; Schalken, J.A. Stem cell differentiation within the human prostate epithelium: Implications for prostate carcinogenesis. *BJU Int.* **2001**, *88* (Suppl. S2), 35–42. [CrossRef] [PubMed]

30. Korski, K.; Malicka-Durczak, A.; Breborowicz, J. Expression of stem cell marker CD44 in prostate cancer biopsies predicts cancer grade in radical prostatectomy specimens. *Pol. J. Pathol.* **2014**, *65*, 291–295. [CrossRef] [PubMed]

31. Liu, C.; Kelnar, K.; Liu, B.; Chen, X.; Calhoun-Davis, T.; Li, H.; Patrawala, L.; Yan, H.; Jeter, C.; Honorio, S.; et al. The microRNA miR-34a inhibits prostate cancer stem cells and metastasis by directly repressing CD44. *Nat. Med.* **2011**, *17*, 211–215. [CrossRef] [PubMed]

32. Ugolkov, A.V.; Eisengart, L.J.; Luan, C.; Yang, X.J. Expression analysis of putative stem cell markers in human benign and malignant prostate. *Prostate* **2011**, *71*, 18–25. [CrossRef] [PubMed]

33. Moura, C.M.; Pontes, J.; Reis, S.T.; Viana, N.I.; Morais, D.R.; Dip, N.; Katz, B.; Srougi, M.; Leite, K.R.M. Expression profile of standard and variants forms of CD44 related to prostate cancer behavior. *Int. J. Biol. Markers* **2015**, *30*, e49–e55. [CrossRef] [PubMed]

34. Tei, H.; Miyake, H.; Harada, K.-I.; Fujisawa, M. Expression profile of CD44s, CD44v6, and CD44v10 in localized prostate cancer: Effect on prognostic outcomes following radical prostatectomy. *Urol. Oncol. Semin. Orig. Investig.* **2014**, *32*, 694–700. [CrossRef] [PubMed]

35. Ni, J.; Cozzi, P.J.; Hao, J.L.; Beretov, J.; Chang, L.; Duan, W.; Shigdar, S.; Delprado, W.J.; Graham, P.H.; Bucci, J.; et al. CD44 variant 6 is associated with prostate cancer metastasis and chemo-/radioresistance. *Prostate* **2014**, *74*, 602–617. [CrossRef] [PubMed]

36. Yang, K.; Tang, Y.; Habermehl, G.K.; Iczkowski, K.A. Stable alterations of CD44 isoform expression in prostate cancer cells decrease invasion and growth and alter ligand binding and chemosensitivity. *BMC Cancer* **2010**, *10*, 16. [CrossRef] [PubMed]

37. Choy, W.; Nagasawa, D.T.; Trang, A.; Thill, K.; Spasic, M.; Yang, I. CD133 as a marker for regulation and potential for targeted therapies in glioblastoma multiforme. *Neurosurg. Clin. N. Am.* **2012**, *23*, 391–405. [CrossRef] [PubMed]

38. Islam, F.; Gopalan, V.; Wahab, R.; Smith, R.A.; Lam, A.K. Cancer stem cells in oesophageal squamous cell carcinoma: Identification, prognostic and treatment perspectives. *Crit. Rev. Oncol./Hematol.* **2015**, *96*, 9–19. [CrossRef] [PubMed]

39. Zhou, Q.; Chen, A.; Song, H.; Tao, J.; Yang, H.; Zuo, M. Prognostic value of cancer stem cell marker CD133 in ovarian cancer: A meta-analysis. *Int. J. Clin. Exp. Med.* **2015**, *8*, 3080–3088. [PubMed]

40. Han, G.W.; Yi, S.H. Prostate stem cells: An update. *Zhonghua Nan Ke Xue* **2014**, *20*, 460–463. [PubMed]

41. Zenzmaier, C.; Untergasser, G.; Berger, P. Aging of the prostate epithelial stem/progenitor cell. *Exp. Gerontol.* **2008**, *43*, 981–985. [CrossRef] [PubMed]

42. Missol-Kolka, E.; Karbanova, J.; Janich, P.; Haase, M.; Fargeas, C.A.; Huttner, W.B.; Corbeil, D. Prominin-1 (CD133) is not restricted to stem cells located in the basal compartment of murine and human prostate. *Prostate* **2011**, *71*, 254–267. [CrossRef] [PubMed]

43. Hsu, W.T.; Jui, H.Y.; Huang, Y.H.; Su, M.Y.; Wu, Y.W.; Tseng, W.Y.; Hsu, M.C.; Chiang, B.L.; Wu, K.K.; Lee, C.M. CXCR4 Antagonist TG-0054 Mobilizes Mesenchymal Stem Cells, Attenuates Inflammation, and Preserves Cardiac Systolic Function in a Porcine Model of Myocardial Infarction. *Cell Transplant.* **2015**, *24*, 1313–1328. [CrossRef] [PubMed]

44. Trerotola, M.; Rathore, S.; Goel, H.L.; Li, J.; Alberti, S.; Piantelli, M.; Adams, D.; Jiang, Z.; Languino, L.R. CD133, Trop-2 and α2β1 integrin surface receptors as markers of putative human prostate cancer stem cells. *Am. J. Transl. Res.* **2010**, *2*, 135–144. [PubMed]

45. Taylor, R.A.; Toivanen, R.; Frydenberg, M.; Pedersen, J.; Harewood, L.; Australian Prostate Cancer Bioresource; Collins, A.T.; Maitland, N.J.; Risbridger, G.P. Human epithelial basal cells are cells of origin of prostate cancer, independent of CD133 status. *Stem Cells* **2012**, *30*, 1087–1096. [CrossRef] [PubMed]

46. Liu, J.F.; Xia, P.; Hu, W.Q.; Wang, D.; Xu, X.Y. Aldehyde dehydrogenase 1 expression correlates with clinicopathologic features of patients with breast cancer: A meta-analysis. *Int. J. Clinl. Exp. Med.* **2015**, *8*, 8425–8432.

47. Schnier, J.B.; Kaur, G.; Kaiser, A.; Stinson, S.F.; Sausville, E.A.; Gardner, J.; Nishi, K.; Bradbury, E.M.; Senderowicz, A.M. Identification of cytosolic aldehyde dehydrogenase 1 from non-small cell lung carcinomas as a flavopiridol-binding protein. *FEBS Lett.* **1999**, *454*, 100–104. [CrossRef]

48. Li, T.; Su, Y.; Mei, Y.; Leng, Q.; Leng, B.; Liu, Z.; Stass, S.A.; Jiang, F. ALDH1A1 is a marker for malignant prostate stem cells and predictor of prostate cancer patients' outcome. *Lab. Investig. J. Tech. Methods Pathol.* **2010**, *90*, 234–244. [CrossRef] [PubMed]

49. Burger, P.E.; Gupta, R.; Xiong, X.; Ontiveros, C.S.; Salm, S.N.; Moscatelli, D.; Wilson, E.L. High aldehyde dehydrogenase activity: A novel functional marker of murine prostate stem/progenitor cells. *Stem Cells* **2009**, *27*, 2220–2228. [CrossRef] [PubMed]

50. Wang, G.; Wang, Z.; Sarkar, F.H.; Wei, W. Targeting prostate cancer stem cells for cancer therapy. *Discov. Med.* **2012**, *13*, 135–142. [PubMed]

51. Gangavarapu, K.J.; Azabdaftari, G.; Morrison, C.D.; Miller, A.; Foster, B.A.; Huss, W.J. Aldehyde dehydrogenase and ATP binding cassette transporter G2 (ABCG2) functional assays isolate different populations of prostate stem cells where ABCG2 function selects for cells with increased stem cell activity. *Stem Cell Res. Ther.* **2013**, *4*, 132. [CrossRef] [PubMed]

52. Ferronika, P.; Triningsih, F.X.; Ghozali, A.; Moeljono, A.; Rahmayanti, S.; Shadrina, A.N.; Naim, A.E.; Wudexi, I.; Arnurisa, A.M.; Nanwani, S.T.; et al. p63 cytoplasmic aberrance is associated with high prostate cancer stem cell expression. *Asian Pac. J. Cancer Prev. (APJCP)* **2012**, *13*, 1943–1948. [CrossRef] [PubMed]

53. An, Y.; Ongkeko, W.M. ABCG2: The key to chemoresistance in cancer stem cells? *Expert Opin. Drug Metab. Toxicol.* **2009**, *5*, 1529–1542. [CrossRef] [PubMed]

54. Patrawala, L.; Calhoun, T.; Schneider-Broussard, R.; Zhou, J.; Claypool, K.; Tang, D.G. Side population is enriched in tumorigenic, stem-like cancer cells, whereas ABCG2$^+$ and ABCG2$^-$ cancer cells are similarly tumorigenic. *Cancer Res.* **2005**, *65*, 6207–6219. [CrossRef] [PubMed]

55. Kim, H.A.; Kim, M.C.; Kim, N.Y.; Kim, Y. Inhibition of hedgehog signaling reduces the side population in human malignant mesothelioma cell lines. *Cancer Gene Ther.* **2015**, *22*, 387–395. [CrossRef] [PubMed]

56. Gao, G.; Sun, Z.; Wenyong, L.; Dongxia, Y.; Zhao, R.; Zhang, X. A preliminary study of side population cells in human gastric cancer cell line HGC-27. *Ann. Transplant. Q. Pol. Transplant. Soc.* **2015**, *20*, 147–153.

57. Guzel, E.; Karatas, O.F.; Duz, M.B.; Solak, M.; Ittmann, M.; Ozen, M. Differential expression of stem cell markers and ABCG2 in recurrent prostate cancer. *Prostate* **2014**, *74*, 1498–1505. [CrossRef] [PubMed]

58. Fu, T.Y.; Hsieh, I.C.; Cheng, J.T.; Tsai, M.H.; Hou, Y.Y.; Lee, J.H.; Liou, H.H.; Huang, S.F.; Chen, H.C.; Yen, L.M.; et al. Association of OCT4, SOX2, and NANOG expression with oral squamous cell carcinoma progression. *J. Oral Pathol. Med.* **2015**, *45*, 89–95. [CrossRef] [PubMed]

59. Li, A.M.; Dunham, C.; Tabori, U.; Carret, A.S.; McNeely, P.D.; Johnston, D.; Lafay-Cousin, L.; Wilson, B.; Eisenstat, D.D.; Jabado, N.; et al. EZH2 expression is a prognostic factor in childhood intracranial ependymoma: A Canadian Pediatric Brain Tumor Consortium study. *Cancer* **2015**, *121*, 1499–1507. [CrossRef] [PubMed]

60. Jiao, J.; Hindoyan, A.; Wang, S.; Tran, L.M.; Goldstein, A.S.; Lawson, D.; Chen, D.; Li, Y.; Guo, C.; Zhang, B.; et al. Identification of CD166 as a surface marker for enriching prostate stem/progenitor and cancer initiating cells. *PLoS ONE* **2012**, *7*, e42564. [CrossRef] [PubMed]

61. Rowehl, R.A.; Crawford, H.; Dufour, A.; Ju, J.; Botchkina, G.I. Genomic analysis of prostate cancer stem cells isolated from a highly metastatic cell line. *Cancer Genom. Proteom.* **2008**, *5*, 301–310.

62. Chuang, T.-D.; Chen, S.-J.; Lin, F.-F.; Veeramani, S.; Kumar, S.; Batra, S.K.; Tu, Y.; Lin, M.-F. Human Prostatic Acid Phosphatase, an Authentic Tyrosine Phosphatase, Dephosphorylates ErbB-2 and Regulates Prostate Cancer Cell Growth. *J. Biol. Chem.* **2010**, *285*, 23598–23606. [CrossRef] [PubMed]

63. Chou, Y.W.; Lin, F.F.; Muniyan, S.; Lin, F.C.; Chen, C.S.; Wang, J.; Huang, C.C.; Lin, M.F. Cellular prostatic acid phosphatase (cPAcP) serves as a useful biomarker of histone deacetylase (HDAC) inhibitors in prostate cancer cell growth suppression. *Cell Biosci.* **2015**, *5*, 38. [CrossRef] [PubMed]

64. Chou, Y.W.; Chaturvedi, N.K.; Ouyang, S.; Lin, F.F.; Kaushik, D.; Wang, J.; Kim, I.; Lin, M.F. Histone deacetylase inhibitor valproic acid suppresses the growth and increases the androgen responsiveness of prostate cancer cells. *Cancer Lett.* **2011**, *311*, 177–186. [CrossRef] [PubMed]

65. Nishida, S.; Hirohashi, Y.; Torigoe, T.; Nojima, M.; Inoue, R.; Kitamura, H.; Tanaka, T.; Asanuma, H.; Sato, N.; Masumori, N. Expression of hepatocyte growth factor in prostate cancer may indicate a biochemical recurrence after radical prostatectomy. *Anticancer Res.* **2015**, *35*, 413–418. [CrossRef]

66. Fornaro, M.; Dell'Arciprete, R.; Stella, M.; Bucci, C.; Nutini, M.; Capri, M.G.; Alberti, S. Cloning of the gene encoding Trop-2, a cell-surface glycoprotein expressed by human carcinomas. *Int. J. Cancer* **1995**, *62*, 610–618. [CrossRef] [PubMed]

67. Ripani, E.; Sacchetti, A.; Corda, D.; Alberti, S. Human Trop-2 is a tumor-associated calcium signal transducer. *Int. J. Cancer* **1998**, *76*, 671–676. [CrossRef]

68. Mühlmann, G.; Spizzo, G.; Gostner, J.; Zitt, M.; Maier, H.; Moser, P.; Gastl, G.; Müller, H.M.; Margreiter, R.; Öfner, D.; et al. TROP2 expression as prognostic marker for gastric carcinoma. *J. Clin. Pathol.* **2009**, *62*, 152–158. [CrossRef] [PubMed]

69. Fang, Y.J.; Lu, Z.H.; Wang, G.Q.; Pan, Z.Z.; Zhou, Z.W.; Yun, J.P.; Zhang, M.F.; Wan, D.S. Elevated expressions of MMP7, TROP2, and survivin are associated with survival, disease recurrence, and liver metastasis of colon cancer. *Int. J. Colorectal. Dis.* **2009**, *24*, 875–884. [CrossRef] [PubMed]

70. Nakashima, K.; Shimada, H.; Ochiai, T.; Kuboshima, M.; Kuroiwa, N.; Okazumi, S.; Matsubara, H.; Nomura, F.; Takiguchi, M.; Hiwasa, T. Serological identification of TROP2 by recombinant cDNA expression cloning using sera of patients with esophageal squamous cell carcinoma. *Int. J. Cancer* **2004**, *112*, 1029–1035. [CrossRef] [PubMed]

71. Ohmachi, T.; Tanaka, F.; Mimori, K.; Inoue, H.; Yanaga, K.; Mori, M. Clinical Significance of TROP2 Expression in Colorectal Cancer. *Clin. Cancer Res.* **2006**, *12*, 3057–3063. [CrossRef] [PubMed]

72. Fong, D.; Moser, P.; Krammel, C.; Gostner, J.M.; Margreiter, R.; Mitterer, M.; Gastl, G.; Spizzo, G. High expression of TROP2 correlates with poor prognosis in pancreatic cancer. *Br. J. Cancer* **2008**, *99*, 1290–1295. [CrossRef] [PubMed]

73. Köbel, M.; Kalloger, S.E.; Boyd, N.; McKinney, S.; Mehl, E.; Palmer, C.; Leung, S.; Bowen, N.J.; Ionescu, D.N.; Rajput, A.; et al. Ovarian carcinoma subtypes are different diseases: Implications for biomarker studies. *PLoS Med.* **2008**, *5*, e232. [CrossRef] [PubMed]

74. Höfner, T.; Eisen, C.; Klein, C.; Rigo-Watermeier, T.; Goeppinger, S.M.; Jauch, A.; Schoell, B.; Vogel, V.; Noll, E.; Weichert, W.; et al. Defined Conditions for the Isolation and Expansion of Basal Prostate Progenitor Cells of Mouse and Human Origin. *Stem Cell Rep.* **2015**, *4*, 503–518. [CrossRef] [PubMed]

75. Goldstein, A.S.; Lawson, D.A.; Cheng, D.; Sun, W.; Garraway, I.P.; Witte, O.N. Trop2 identifies a subpopulation of murine and human prostate basal cells with stem cell characteristics. *Proc. Natl. Acad. Sci. USA* **2008**, *105*, 20882–20887. [CrossRef] [PubMed]

76. Fedr, R.; Pernicová, Z.; Slabáková, E.; Straková, N.; Bouchal, J.; Grepl, M.; Kozubík, A.; Souček, K. Automatic cell cloning assay for determining the clonogenic capacity of cancer and cancer stem-like cells. *Cytom. A* **2013**, *83*, 472–482. [CrossRef] [PubMed]

77. Stoyanova, T.; Goldstein, A.S.; Cai, H.; Drake, J.M.; Huang, J.; Witte, O.N. Regulated proteolysis of Trop2 drives epithelial hyperplasia and stem cell self-renewal via β-catenin signaling. *Genes Dev.* **2012**, *26*, 2271–2285. [CrossRef] [PubMed]

78. Trerotola, M.; Jernigan, D.L.; Liu, Q.; Siddiqui, J.; Fatatis, A., Languino, L.R. Trop-2 promotes prostate cancer metastasis by modulating β$_1$ integrin functions. *Cancer Res.* **2013**, *73*, 3155–3167. [CrossRef] [PubMed]

79. Trerotola, M.; Li, J.; Alberti, S.; Languino, L.R. Trop-2 inhibits prostate cancer cell adhesion to fibronectin through the β1 integrin-RACK1 axis. *J. Cell. Physiol.* **2012**, *227*, 3670–3677. [CrossRef] [PubMed]

80. Ibragimova, I.; de Cáceres, I.I.; Hoffman, A.M.; Potapova, A.; Dulaimi, E.; Al-Saleem, T.; Hudes, G.R.; Ochs, M.F.; Cairns, P. Global Reactivation of Epigenetically Silenced Genes in Prostate Cancer. *Cancer Prev. Res.* **2010**, *3*, 1084–1092. [CrossRef] [PubMed]

81. Jerónimo, C.; Esteller, M. DNA Methylation Markers for Prostate Cancer with a Stem Cell Twist. *Cancer Prev. Res.* **2010**, *3*, 1053–1055. [CrossRef] [PubMed]

82. Lin, H.-Y.; Kuo, Y.-C.; Weng, Y.-I.; Lai, I.L.; Huang, T.H.M.; Lin, S.-P.; Niu, D.-M.; Chen, C.-S. Activation of Silenced Tumor Suppressor Genes in Prostate Cancer Cells by a Novel Energy Restriction-Mimetic Agent. *Prostate* **2012**, *72*. [CrossRef] [PubMed]

83. Van Rij, C.M.; Frielink, C.; Goldenberg, D.M.; Sharkey, R.M.; Lutje, S.; McBride, W.J.; Oyen, W.J.; Boerman, O.C. Pretargeted Radioimmunotherapy of Prostate Cancer with an Anti-TROP-2 × Anti-HSG Bispecific Antibody and a ^{177}Lu-Labeled Peptide. *Cancer Biother. Radiopharm.* **2014**, *29*, 323–329. [CrossRef] [PubMed]

84. Van Rij, C.M.; Lutje, S.; Frielink, C.; Sharkey, R.M.; Goldenberg, D.M.; Franssen, G.M.; McBride, W.J.; Rossi, E.A.; Oyen, W.J.; Boerman, O.C. Pretargeted immuno-PET and radioimmunotherapy of prostate cancer with an anti-TROP-2 × anti-HSG bispecific antibody. *Eur. J. Nucl. Med. Mol. Imaging* **2013**, *40*, 1377–1383. [CrossRef] [PubMed]

85. Yarden, Y.; Kuang, W.J.; Yang-Feng, T.; Coussens, L.; Munemitsu, S.; Dull, T.J.; Chen, E.; Schlessinger, J.; Francke, U.; Ullrich, A. Human proto-oncogene c-kit: A new cell surface receptor tyrosine kinase for an unidentified ligand. *EMBO J.* **1987**, *6*, 3341–3351. [PubMed]

86. Matthews, W.; Jordan, C.T.; Wiegand, G.W.; Pardoll, D.; Lemischka, I.R. A receptor tyrosine kinase specific to hematopoietic stem and progenitor cell-enriched populations. *Cell* **1991**, *65*, 1143–1152. [CrossRef]

87. Broxmeyer, H.E.; Maze, R.; Miyazawa, K.; Carow, C.; Hendrie, P.C.; Cooper, S.; Hangoc, G.; Vadhan-Raj, S.; Lu, L. The kit receptor and its ligand, steel factor, as regulators of hemopoiesis. *Cancer Cells* **1991**, *3*, 480–487. [PubMed]

88. Chi, P.; Chen, Y.; Zhang, L.; Guo, X.; Wongvipat, J.; Shamu, T.; Fletcher, J.A.; Dewell, S.; Maki, R.G.; Zheng, D.; et al. ETV1 is a lineage-specific survival factor in GIST and cooperates with KIT in oncogenesis. *Nature* **2010**, *467*, 849–853. [CrossRef] [PubMed]

89. Di Lorenzo, G.; Autorino, R.; D'Armiento, F.P.; Mignogna, C.; de Laurentiis, M.; de Sio, M.; D'Armiento, M.; Damiano, R.; Vecchio, G.; de Placido, S. Expression of proto-oncogene c-kit in high risk prostate cancer. *Eur. J. Surg. Oncol.* **2004**, *30*, 987–992. [CrossRef] [PubMed]

90. Leong, K.G.; Wang, B.E.; Johnson, L.; Gao, W.Q. Generation of a prostate from a single adult stem cell. *Nature* **2008**, *456*, 804–808. [CrossRef] [PubMed]

91. Wiesner, C.; Nabha, S.M.; Dos Santos, E.B.; Yamamoto, H.; Meng, H.; Melchior, S.W.; Bittinger, F.; Thüroff, J.W.; Vessella, R.L.; Cher, M.L.; et al. C-Kit and Its Ligand Stem Cell Factor: Potential Contribution to Prostate Cancer Bone Metastasis. *Neoplasia* **2008**, *10*, 996–1003. [CrossRef] [PubMed]

92. Kerr, B.A.; Miocinovic, R.; Smith, A.K.; West, X.Z.; Watts, K.E.; Alzayed, A.W.; Klink, J.C.; Mir, M.C.; Sturey, T.; Hansel, D.E.; et al. CD117⁺ cells in the circulation are predictive of advanced prostate cancer. *Oncotarget* **2015**, *6*, 1889–1897. [CrossRef] [PubMed]

93. Okumura, N.; Tsuji, K.; Ebihara, Y.; Tanaka, I.; Sawai, N.; Koike, K.; Komiyama, A.; Nakahata, T. Chemotactic and chemokinetic activities of stem cell factor on murine hematopoietic progenitor cells. *Blood* **1996**, *87*, 4100–4108. [PubMed]

94. Blume-Jensen, P.; Claesson-Welsh, L.; Siegbahn, A.; Zsebo, K.M.; Westermark, B.; Heldin, C.H. Activation of the human c-kit product by ligand-induced dimerization mediates circular actin reorganization and chemotaxis. *EMBO J.* **1991**, *10*, 4121–4128. [PubMed]

95. Kong, D.; Sethi, S.; Li, Y.; Chen, W.; Sakr, W.A.; Heath, E.; Sarkar, F.H. Androgen receptor splice variants contribute to prostate cancer aggressiveness through induction of EMT and expression of stem cell marker genes. *Prostate* **2015**, *75*, 161–174. [CrossRef] [PubMed]

96. Guo, Z.; Yang, X.; Sun, F.; Jiang, R.; Linn, D.E.; Chen, H.; Chen, H.; Kong, X.; Melamed, J.; Tepper, C.G.; et al. A novel androgen receptor splice variant is up-regulated during prostate cancer progression and promotes androgen depletion–resistant growth. *Cancer Res.* **2009**, *69*, 2305–2313. [CrossRef] [PubMed]

97. Marcias, G.; Erdmann, E.; Lapouge, G.; Siebert, C.; Barthélémy, P.; Duclos, B.; Bergerat, J.-P.; Céraline, J.; Kurtz, J.-E. Identification of novel truncated androgen receptor (AR) mutants including unreported pre-mRNA splicing variants in the 22Rv1 hormone-refractory prostate cancer (PCa) cell line. *Hum. Mutat.* **2010**, *31*, 74–80. [CrossRef] [PubMed]

98. Sun, S.; Sprenger, C.C.T.; Vessella, R.L.; Haugk, K.; Soriano, K.; Mostaghel, E.A.; Page, S.T.; Coleman, I.M.; Nguyen, H.M.; Sun, H.; et al. Castration resistance in human prostate cancer is conferred by a frequently occurring androgen receptor splice variant. *J. Clin. Investig.* **2010**, *120*, 2715–2730. [CrossRef] [PubMed]

99. Hu, R.; Isaacs, W.B.; Luo, J. A snapshot of the expression signature of androgen receptor splicing variants and their distinctive transcriptional activities. *Prostate* **2011**, *71*, 1656–1667. [CrossRef] [PubMed]

100. Watson, P.A.; Chen, Y.F.; Balbas, M.D.; Wongvipat, J.; Socci, N.D.; Viale, A.; Kim, K.; Sawyers, C.L. Constitutively active androgen receptor splice variants expressed in castration-resistant prostate cancer require full-length androgen receptor. *Proc. Natl. Acad. Sci. USA* **2010**, *107*, 16759–16765. [CrossRef] [PubMed]

101. Dehm, S.M.; Schmidt, L.J.; Heemers, H.V.; Vessella, R.L.; Tindall, D.J. Splicing of a Novel Androgen Receptor Exon Generates a Constitutively Active Androgen Receptor that Mediates Prostate Cancer Therapy Resistance. *Cancer Res.* **2008**, *68*, 5469–5477. [CrossRef] [PubMed]

102. Hu, R.; Dunn, T.A.; Wei, S.; Isharwal, S.; Veltri, R.W.; Humphreys, E.; Han, M.; Partin, A.W.; Vessella, R.L.; Isaacs, W.B.; et al. Ligand-Independent Androgen Receptor Variants Derived from Splicing of Cryptic Exons Signify Hormone-Refractory Prostate Cancer. *Cancer Res.* **2009**, *69*, 16–22. [CrossRef] [PubMed]

103. Dehm, S.M.; Tindall, D.J. Alternatively spliced androgen receptor variants. *Endocr. Relat. Cancer* **2011**, *18*, R183–R196. [CrossRef] [PubMed]

104. Nakazawa, M.; Antonarakis, E.; Luo, J. Androgen Receptor Splice Variants in the Era of Enzalutamide and Abiraterone. *Horm. Cancer* **2014**, *5*, 265–273. [CrossRef] [PubMed]

105. Onstenk, W.; Sieuwerts, A.M.; Kraan, J.; van, M.; Nieuweboer, A.J.M.; Mathijssen, R.H.J.; Hamberg, P.; Meulenbeld, H.J.; de Laere, B.; Dirix, L.Y.; et al. Efficacy of cabazitaxel in castration-resistant prostate cancer is independent of the presence of AR-V7 in circulating tumor cells. *Eur. Urol.* **2015**, *68*, 939–945. [CrossRef] [PubMed]

106. Nakazawa, M.; Lu, C.; Chen, Y.; Paller, C.J.; Carducci, M.A.; Eisenberger, M.A.; Luo, J.; Antonarakis, E.S. Serial blood-based analysis of AR-V7 in men with advanced prostate cancer. *Ann. Oncol.* **2015**. [CrossRef] [PubMed]

107. Han, A.L.; Kumar, S.; Fok, J.Y.; Tyagi, A.K.; Mehta, K. Tissue transglutaminase expression promotes castration-resistant phenotype and transcriptional repression of androgen receptor. *Eur. J. Cancer* **2014**, *50*, 1685–1696. [CrossRef] [PubMed]

108. Eckert, R.L.; Fisher, M.L.; Grun, D.; Adhikary, G.; Xu, W.; Kerr, C. Transglutaminase is a tumor cell and cancer stem cell survival factor. *Mol. Carcinog.* **2015**, *54*, 947–958. [CrossRef] [PubMed]

109. Mayer, M.J.; Klotz, L.H.; Venkateswaran, V. Metformin and prostate cancer stem cells: A novel therapeutic target. *Prostate Cancer Prostatic Dis.* **2015**, *18*, 303–309. [CrossRef] [PubMed]

110. Yu, V.Y.; Nguyen, D.; Pajonk, F.; Kupelian, P.; Kaprealian, T.; Selch, M.; Low, D.A.; Sheng, K. Incorporating cancer stem cells in radiation therapy treatment response modeling and the implication in glioblastoma multiforme treatment resistance. *Int. J. Radiat. Oncol. Biol. Phys.* **2015**, *91*, 866–875. [CrossRef] [PubMed]

International Journal of
Molecular Sciences

MDPI

Article

Akt Activation Correlates with Snail Expression and Potentially Determines the Recurrence of Prostate Cancer in Patients at Stage T2 after a Radical Prostatectomy

Wei-Yu Chen [1,2,3], Kuo-Tai Hua [4], Wei-Jiunn Lee [5,6], Yung-Wei Lin [7], Yen-Nien Liu [8], Chi-Long Chen [1,2,*], Yu-Ching Wen [6,7,*] and Ming-Hsien Chien [1,5,*]

[1] Graduate Institute of Clinical Medicine, College of Medicine, Taipei Medical University, Taipei 110, Taiwan; 1047@tmu.edu.tw
[2] Department of Pathology, School of Medicine, College of Medicine, Taipei Medical University, Taipei 110, Taiwan
[3] Department of Pathology, Wan Fang Hospital, Taipei Medical University, Taipei 116, Taiwan
[4] Graduate Institute of Toxicology, College of Medicine, National Taiwan University, Taipei 100, Taiwan; d94447003@gmail.com
[5] Department of Medical Education and Research, Wan Fang Hospital, Taipei Medical University, Taipei 116, Taiwan; lwj5905@gmail.com
[6] Department of Urology, School of Medicine, College of Medicine, Taipei Medical University, Taipei 110, Taiwan
[7] Department of Urology, Wan Fang Hospital, Taipei Medical University, Taipei 116, Taiwan; highwei168@gmail.com
[8] Graduate Institute of Cancer Biology and Drug Discovery, College of Medical Science and Technology, Taipei Medical University, Taipei 110, Taiwan; liuy@tmu.edu.tw
* Correspondence: chcl0997@yahoo.com.tw (C.-L.C.); s811007@yahoo.com.tw (Y.-C.W.); mhchien1976@gmail.com (M.-H.C.); Tel.: +886-2-2736-1661 (ext. 3139) (C.-L.C.); +886-2-2930-7930 (ext. 1867) (Y.-C.W.); +886-2-2736-1661 (ext. 3237) (M.-H.C.); Fax: +886-2-2377-0054 (C.-L.C.); +886-2-6628-0167 (Y.-C.W.); +886-2-2739-0500 (M.-H.C.)

Academic Editor: Carsten Stephan
Received: 8 June 2016; Accepted: 20 July 2016; Published: 23 July 2016

Abstract: Our previous work demonstrated the epithelial-mesenchymal transition factor, Snail, is a potential marker for predicting the recurrence of localized prostate cancer (PCa). Akt activation is important for Snail stabilization and transcription in PCa. The purpose of this study was to retrospectively investigate the relationship between the phosphorylated level of Akt (p-Akt) in radical prostatectomy (RP) specimens and cancer biochemical recurrence (BCR). Using a tissue microarray and immunohistochemistry, the expression of p-Akt was measured in benign and neoplastic tissues from RP specimens in 53 patients whose cancer was pathologically defined as T2 without positive margins. Herein, we observed that the p-Akt level was higher in PCa than in benign tissues and was significantly associated with the Snail level. A high p-Akt image score ($\geqslant 8$) was significantly correlated with a higher histological Gleason sum, Snail image score, and preoperative prostate-specific antigen (PSA) value. Moreover, the high p-Akt image score and Gleason score sum ($\geqslant 7$) showed similar discriminatory abilities for BCR according to a receiver-operator characteristic curve analysis and were correlated with worse recurrence-free survival according to a log-rank test ($p < 0.05$). To further determine whether a high p-Akt image score could predict the risk of BCR, a Cox proportional hazard model showed that only a high p-Akt image score (hazard ratio (HR): 3.12, $p = 0.05$) and a high Gleason score sum ($\geqslant 7$) (HR: 1.18, $p = 0.05$) but not a high preoperative PSA value (HR: 0.62, $p = 0.57$) were significantly associated with a higher risk of developing BCR. Our data indicate that, for localized PCa patients after an RP, p-Akt can serve as a potential prognostic marker that improves predictions of BCR-free survival.

Keywords: prostate cancer; radical prostatectomy; stage T2; Akt; Snail; biochemical recurrence

---

1. Introduction

Globally, prostate cancer (PCa) accounts for 15% of male cancers and 6.6% of total male cancer mortality [1]. A radical prostatectomy (RP) is recognized as the gold standard for treating patients with localized PCa. The most important advantage of an RP is its potential to cure without damaging adjacent tissues and provide accurate staging because of the total removal of the organ. Although most patients are cured after surgery, around 23%–35% of PCa patients progress to biochemical recurrence (BCR) due to serum prostate-specific antigen (PSA) elevation, indicating that they have an increased risk of developing advanced PCa among 10 years after an RP [2,3]. To now, the challenge of PCa patients after an RP has been to determine which patients harbor high-risk disease requiring aggressive/curative therapy and which patients harbor indolent disease that can be managed with active surveillance.

Clinical prognostic risk factors such as the Gleason score, pathological stage, a positive surgical margin, and preoperative PSA value are used to estimate patient outcomes postoperatively [4,5]. However, the sensitivity of predicting BCR of individual patients using such parameters is insufficient [4,5]. Hence, novel biomarkers are needed to predict BCR in PCa patients after an RP to help provide better patient counseling, to help with more-precise clinical decision-making, and to search for therapeutic targets. Recently, studies have identified several molecular alterations involved in prostate recurrence. For example, we previously identified that the epithelial-mesenchymal transition (EMT) factor, Snail, is upregulated in PCa and is a predictive factor for subsequent localized PCa recurrence after an RP [6]. However, the precise mechanisms underlying Snail expression in this malignancy has not been fully elucidated.

Activation of the serine threonine kinase, Akt (phosphorylated (p)-Akt), was reported to regulate the stability and transcription of Snail in several cancer types, such as colorectal [7], oral [8], and prostate [9] cancers. A previous report indicated that p-Akt was expressed in around 8% of non-neoplastic prostate and 50% of PCa cases, indicative of its overexpression in cancer [10]. Increased Akt phosphorylation was observed in high-Gleason-score PCa and was correlated with proliferation in human PCa as estimated by the expression of the cell proliferation antigen, Ki67 [11,12]. Bedolla et al. recruited 65 PCa patients including T1~T3 stages with positive margins and showed that p-Akt is an important predictor of the risk of BCR [13]. Based on these results, we hypothesized an important role for Akt in PCa recurrence.

To further investigate the role of Akt activation in localized PCa recurrence, this study recruited 53 PCa patients at the T2 stage without positive margins after an RP. We evaluated the p-Akt expression pattern in these PCa patients using immunohistochemistry (IHC), and correlated expression levels with Snail and other clinicopathological parameters. We report for the first time that expression of p-Akt was highly correlated with Snail expression in localized PCa, and the cytoplasmic p-Akt protein level has potential to serve as an independent biomarker to improve estimation of localized PCa prognoses.

2. Results

In this study, we recruited 76 PCa patients who had not received neoadjuvant therapy and had undergone a whole-mount pathological assessment of their tumor after an RP. Next, we further excluded patients with a positive surgical margin and seminal vesicle invasion, and 53 of 76 patients who had organ-confined disease were ultimately recruited. Demographic and clinical characteristics are summarized in Table 1. Among the 53 PCa patients, the age at the time of the RP ranged 48–88 (mean, 70.7 ± 15.2) years. The histologic type of all tumors was an adenocarcinoma. According to the American Joint Committee on Cancer (AJCC) TNM staging system, tumors were classified into T2a ($n = 6$), T2b ($n = 4$), and T2c ($n = 43$). At a mean follow-up time of 71 months, 25 of 53 patients had BCR.

Table 1. Characteristics of prostate cancer (PCa) patients at the pT2 stage who underwent a radical prostatectomy (RP).

Characteristic	Total (%)
Total number of patients	53
Median age at RP (years)	71, mean 70.7 ± 15.2 (48–88 y/o)
Mean preoperative PSA level (ng/mL)	10.31 (1–21.64 ng/mL)
Biochemical failure	25 (47.2)
Pathological stage	
T2a	6 (11.3)
T2b	4 (7.5)
T2c	43 (81.2)
Gleason score	
$\geqslant 7$	34 (64.2)
$\leqslant 6$	19 (35.8)
Snail image score	
$\geqslant 8$	35 (66)
$\leqslant 6$	18 (34)
Phosphorylated-Akt image score	
$\geqslant 8$	32 (60.4)
$\leqslant 6$	21 (39.1)
Median follow-up time (months)	99, mean: 71 ± 49.5 (53–184 m)

RP, radical prostatectomy; PSA, prostate specific antigen; y/o, years old; m, months.

Figure 1A–D shows that p-Akt expression was observed in PCa tissue with a wide distribution of IHC scores. Immunostaining was almost completely restricted to the cytoplasm of epithelial tumor cells, and the pattern of expression was usually homogeneous. The p-Akt score was determined by multiplying the staining intensity (1–3) by the distribution rate (1–4) to represent p-Akt expression in PCa tissues, and representative examples of tumors showing overall low (with an image score of $\leqslant 6$) and high (with an image score of $\geqslant 8$) p-Akt expressions are illustrated in Figure 1A–D. In contrast to PCa, non-tumor adjacent tissues or benign prostatic hyperplasia (BPH) expressed p-Akt very weakly or not at all (Figure 1D,E), indicating that high levels of p-Akt were almost exclusively expressed in cancer tissues.

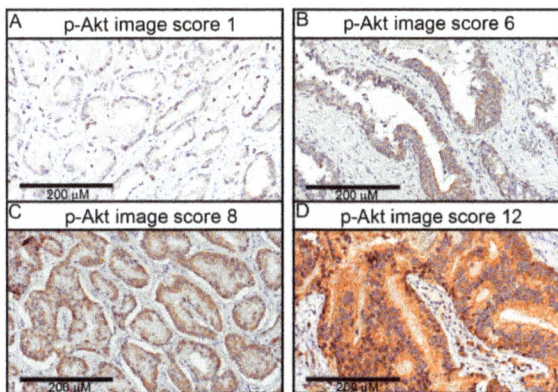

Figure 1. *Cont.*

E

F

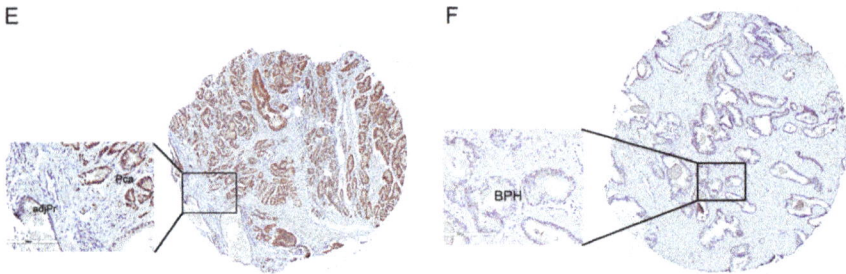

Figure 1. Phosphorylated (p)-Akt expression levels in representative primary prostate cancer (PCa) and non-neoplastic prostate tissues. Tissue microarrays (TMAs) of primary PCa and non-neoplastic prostate (benign prostatic hyperplasia; BPH) tissues were immunohistochemistry (IHC) analyzed for p-Akt. (**A**) Patient with a weak p-Akt expression level (intensity score 1 × extent score 1 = p-Akt image score 1); (**B**) Patient with T2cN0M0 cancer, a Gleason score of 3 + 4 = 7, and a moderate p-Akt expression level (intensity score 2 × extent score 3 = p-Akt image score 6); (**C**) Patient with T2cN0M0 cancer, a Gleason score of 4 + 3 = 7, and marked p-Akt immunostaining in the cytoplasm (intensity score 2 × extent score 4 = Snail image score 8); (**D**) Patient with T2cN0M0 cancer and a Gleason score of 4 + 5 = 9 and who displayed marked p-Akt immunostaining in the cytoplasm and discrete, diffuse staining in the nucleus (intensity score 3 × extent score 4 = image score 12) (200×); (**E,F**) No p-Akt immunostaining signal was detected in non-tumor adjacent tissues (**E**) or BPH (**F**). The high-power fields (200×) are magnified fields in the black boxed area in the right panel.

Previous studies indicated that Akt activation is important for Snail stabilization and transcription in PCa cells [9,14]. To further examine the correlation between expression levels of p-Akt and Snail in PCa, the same PCa TMA cohort was used. Representative IHC staining of p-Akt and Snail with different image scores on serial section of the same patients are shown in Figure 2A. IHC analysis of PCa specimens revealed a significant positive correlation between p-Akt and Snail expressions (Spearman correlation coefficient $r = 0.851$, $p < 0.0001$; Figure 2B).

A

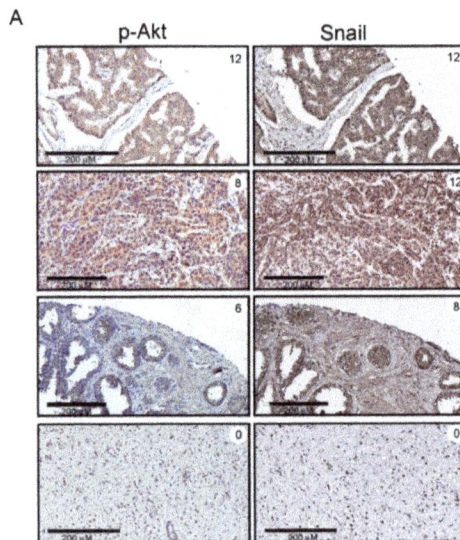

Figure 2. *Cont.*

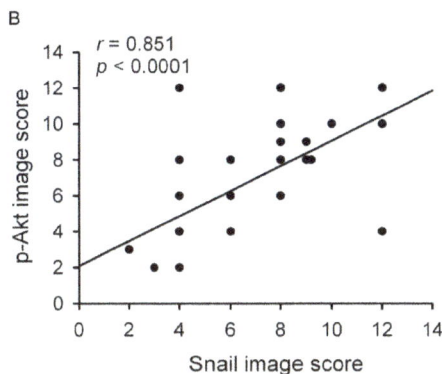

Figure 2. Phosphorylated (p)-Akt expression is positively correlated with Snail protein levels of patients with localized prostate cancer (PCa). (**A**) IHC staining analysis of p-Akt and Snail proteins in serial sections (200× magnification). Note the positive correlation of p-Akt and Snail protein expressions in tumor cells; (**B**) A significant positive correlation was observed between p-Akt expression levels and Snail expression levels (Spearman's correlation coefficients: $r = 0.851$, $p < 0.0001$).

As we showed earlier [6], staining for Snail was significantly correlated with postoperative BCR of PCa. We further investigated relationships between p-Akt expression and selected clinicopathologic factors. Table 2 shows that among the 53 recruited patients, 32 patients (60.4%) were identified as having a high p-Akt image score (of $\geqslant 8$), and the remaining 21 patients had a low p-Akt image score (of $\leqslant 6$). High p-Akt (score of $\geqslant 8$) expression was significantly associated with a higher histological Gleason sum (score of $\geqslant 7$) ($p = 0.024$), Snail image score (score of $\geqslant 8$) ($p = 0.035$), and preoperative PSA value ($p = 0.026$). Moreover, we also observed that high p-Akt expression was significantly correlated with postoperative BCR ($p = 0.012$). Moreover, according to an ROC analysis, the areas under the ROC curve for high p-Akt image score (score of $\geqslant 8$) and Gleason score sum (score of $\geqslant 7$) were similar, indicating that the high p-Akt image score and Gleason score sum showed similar discriminatory capacities for BCR (Figure 3).

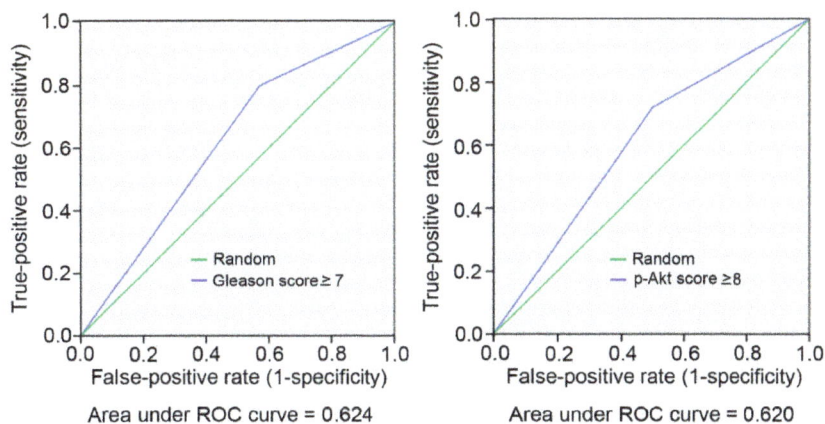

Figure 3. Sensitivity and specificity of gain of a high Gleason score or high p-Akt in specimens with respect to biochemical recurrence (BRC). Areas under the ROC (AUC) for high p-Akt image score ($\geqslant 8$) and high Gleason score ($\geqslant 7$) were 0.62 and 0.624, indicating similar discriminatory abilities for BRC.

Table 2. The association of phosphorylated (p)-Akt staining and clinicopathological features of prostate cancer (PCa) patients.

Characteristic	No. of Patients (%)		*p* Value
	pAkt Score $\geqslant$ 8	pAkt Score $\leqslant$ 6	
Total number of patients	32 (60.4)	21 (39.6)	
Age (years)			
<50	1	0	
50–59	5	3	
60–69	11	6	
$\geqslant$70	15	12	
Pathological stage			
T2a	2 (6.25)	3 (14.3)	0.540
T2b	2 (6.25)	2 (9.5)	
T2c	28 (87.5)	16 (76.2)	
Gleason score			
$\geqslant$7	28 (87.5)	8 (38.1)	0.024
$\leqslant$6	4 (12.5)	13 (61.9)	
Snail image score			
$\geqslant$8	29 (90.6)	6 (28.6)	0.035
$\leqslant$6	3 (9.4)	15 (71.4)	
Recurrence	18 (56.3)	7 (33.3)	0.012
PSA, mean (ng/mL)	29.9	12.1	0.026

PSA: prostate specific antigen.

According to the Kaplan-Meier test, we observed that patients with higher p-Akt expression (with scores of $\geqslant$8) had shorter recurrence-free survival times compared to those with lower expression (with scores of $\leqslant$6) of the protein (Figure 4A). For patients who had higher p-Akt tumor expression, the median recurrence-free survival was 62 months, whereas for those who demonstrated lower p-Akt tumor expression, it was 88 months ($p = 0.03$) (Figure 4A). Moreover, results of the Kaplan-Meier test also showed that patients with a higher Gleason score sum (of $\geqslant$7) or a higher Snail expression (with a score of $\geqslant$8) all had significantly shorter recurrence-free survival times ($p = 0.03$ and 0.05) (Figure 4B,C). These results showed that the *p* value of the Kaplan-Meier test used to compare the higher p-Akt group was the same and smaller than the higher Gleason score group and higher Snail group, respectively.

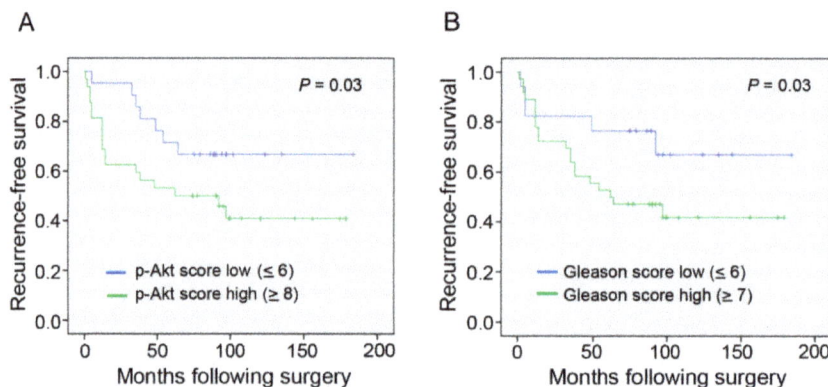

Figure 4. *Cont.*

C

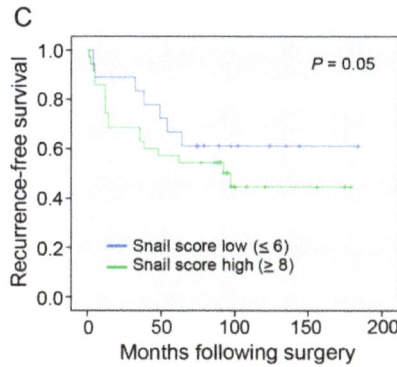

Figure 4. Kaplan-Meier survival curves showing relationships of the phosphorylated (p)-Akt image score (**A**), Gleason score sum (**B**), and Snail image score (**C**) in primary tumors with recurrence-free survival in 53 patients with clinically-localized prostate cancer. The recurrence-free survival of patients with a higher p-Akt, Snail image score ($\geqslant$8) or Gleason score sum ($\geqslant$7) was significantly lower than that of patients with a lower p-Akt, Snail image score ($\leqslant$6) or Gleason score sum ($\leqslant$6) ($p \leqslant 0.05$, log-rank test).

A Cox proportional hazard model was conducted to further explore relationships of p-Akt and Snail expressions with recurrence-free survival of the 53 patients with PCa after an RP. Table 3 summarizes the associations between the recurrence-free survival rate of the 53 patients with PCa and clinicopathologic parameters. In this analysis, we only observed that a high p-Akt image score ($\geqslant$8) (hazard ratio (HR): 3.12, $p = 0.05$) or a high Gleason score sum ($\geqslant$7) (HR: 1.18, $p = 0.05$), but not a high pre-operative PSA value (>10 ng/mL) (HR: 0.62, $p = 0.57$), was significantly correlated with worse recurrence-free survival (Table 3). In conclusion, our results suggest that a high p-Akt image score and a high histological Gleason score sum but not the preoperative PSA value can predict organ-confined PCa recurrence in our study. Moreover, we observed that patients with a high Snail image score ($\geqslant$8) also tended to correlate with BCR (HR: 1.31, $p = 0.06$). Furthermore, our data indicated that patients with a high p-Akt image score (HR: 3.12) showed a higher risk for BRC than patients with a high Gleason score sum (HR: 1.18) or a high Snail image score (HR: 1.31) (Table 3).

Table 3. Survival analyses of biochemical progression predictors in patients with prostate cancer at pT2 who underwent a radical prostatectomy (RP) according to a Cox proportional hazards regression model.

Factor	Hazard Ratio (95% CI)	*p* Value
PSA > 10 ng/mL	0.62 (0.12–3.19)	0.57
Pathological Gleason sum $\geqslant$ 7	1.18 (0.23–1.32)	0.05
Snail image score $\geqslant$ 8	1.31 (0.09–3.03)	0.06
P-Akt image score $\geqslant$ 8	3.12 (0.95–10.27)	0.05

3. Discussion

PCa is the most common cancer and the second leading cause of male cancer deaths in the United States [15]. This underscores the need for a more-thorough molecular understanding of this resilient disease. Generally, patients with clinically-localized PCa will be cured after receiving radical surgery. However, a fraction of patients with localized PCa harbor microscopic localized or metastatic residual disease. The lethal consequences of PCa are related to its metastasis to other organ sites. Although the preoperative PSA value, surgical margin status, and Gleason score sum have been extensively used in assessing biochemical disease recurrence risk after RP, the sensitivities of these approaches are insufficient [16,17]. Therefore, it is of critical significance to discover a new marker for the early prediction of tumor recurrence, and earlier adjuvant therapy is very important for clinicians.

The EMT is a critical cellular mechanism during tumor progression and development of metastasis. It was suggested that the EMT is co-opted by PCa cells during their metastatic dissemination from a primary organ to secondary sites [18]. We previously showed that increased expression of the EMT promoter, Snail, in the prostatic epithelium is a good predictor of BCR following a prostatectomy [6]. The phosphatidylinositol 3′-kinase (PI3K)/Akt pathway is frequently activated in various cancers and plays an important role in promoting the EMT through regulating Snail stability in PCa [9,19]. Our current results indicated that the Akt activation status was significantly correlated with Snail expression levels in tissues from patients with clinically-localized PCa (T2 stage only). Representative IHC staining patterns of p-Akt and Snail from consecutive serial sections were nearly identical in PCa specimens, further implying their highly correlated expressions. Compared to our previous study [6], we extended the postoperative follow-up time (an average of 51 to 71 months) of localized PCa patients and further investigated the correlation between the Akt activation status in PCa specimens and BRC of patients. Our data showed that p-Akt was predominantly expressed in PCa, but not in non-neoplastic tissues. Cox proportional hazard models suggested that the p-Akt index (HR: 3.12, $p = 0.05$) is a better postoperative marker than the preoperative PSA value (HR: 0.62, $p = 0.57$) in localized PCa in our patient cohort. Although the Gleason score sum showed a similar discriminatory capacity with the p-Akt index and was also a useful predictor of BRC in our patient cohort, it was not as good as p-Akt. Compared to a high p-Akt image score, a high Gleason score sum showed a lower HR (HR: 1.18, $p = 0.05$) for BRC. Moreover, our previous study [6] indicated that the Snail index might be a useful predictor of BRC in the same patient cohort. However, after we extended the postoperative follow-up time in this study, the high Snail image score only showed a borderline significant trend ($p = 0.06$) of correlating with BRC, suggesting that p-Akt might show higher sensitivity than Snail for predicting organ-confined PCa recurrence. In addition to Snail, other transcription factors such as Slug, Zeb1, and Zeb2 are also involved in the control of EMT [20]. A previous report indicated that Akt activation can upregulate Snail and Zeb2 and promote EMT in squamous cell carcinoma [21]. However, the roles of Akt and Zeb2 in prostate cancer progression and recurrence are still unclear and worth further investigation in our future work.

High levels of p-Akt are associated with earlier recurrence, clearly indicating that p-Akt is associated with aggressiveness and disease progression in PCa. In addition to the Akt-mediated Snail expression and EMT induction in an androgen-independent PC3 cell line [14], Akt was also shown to be involved in a number of proliferative, metabolic, and antiapoptotic pathways that are dependent on PI3K signaling for activation [22]. Activated Akt was suggested to regulate a number of intracellular targets such as p27^{Kip1}, Bcl-2-associated death promoter (BAD), and caspase-9 which are involved in PCa progression and androgen independence [23–25]. Androgen deprivation therapy on androgen-dependent PCa cells such as LNCaP was reported to stimulate Akt activation, which finally resulted in androgen independence of the cells [26]. The pro-survival role of Akt activities was further shown in several clinical studies. For instance, increased levels of Akt or p-Akt expression were associated with a high Gleason grade and a worse prognosis in PCa [27–29]. Herein, we also showed that high levels of p-Akt were associated with more-aggressive features of the disease, as patients with high levels of p-Akt were identified as having a high PSA level and a high Gleason score.

Although the clinical application of p-Akt in predicting BCR of PCa was previously reported [10,13], the patient cohort recruited in this study was totally different from those of previous studies. Previous studies [10,13] enrolled PCa patients after an RP at different stages (pT1–pT3), and included those with lymph node metastasis, extracapsular extension, positive margins, and seminal vesicle invasion. However, in our study, we excluded PCa patients with positive margins or seminal vesicle involvement and only included true organ-confined PCa patients (pT2a–pT2c) after an RP. To our best knowledge, this is the first report to investigate the relationships of p-Akt and Snail with the prognostic role of p-Akt in patients with clinically-localized PCa. The pathological definition of

T2a–T2c stages is involvement of tumor cells in the prostate: in T2a, the tumor involves ⩽50% of one lobe; in T2b, the tumor involves >50% of one lobe, but not both lobes; and in T2c, the tumor involves both lobes [30]. A previous report indicated that tumor focality can significantly influence the BCR-free survival rate [31]. However, more-recent reports indicated that tumor focality did not predict the risk of BCR after an RP in men with clinically-localized PCa, even if the tumor involved both lobes of the prostate [32], suggesting that tumor focality might not be a suitable predictive marker for BCR in patients with organ-confined PCa. Our study observed that p-Akt expression level should be a better predictive marker of BCR in this PCa population. In addition to Akt, its downstream signaling pathways such as GSK-3β inactivation and NF-κB activation were also involved in Akt-mediated Snail expression in prostate cancer cells as well as in other cancer types [7,14,21]. Our future work will further investigate the correlation between these downstream effectors and BCR-free survival rate in patients with localized PCa.

4. Materials and Methods

4.1. Patient Selection and Specimen Collection

Pathology files of Wan Fang Hospital and Taipei Medical University Hospital were searched, and 76 radical prostatectomy specimens with a pathologic diagnosis of prostatic adenocarcinoma were found from March 1999 to December 2011. The pathologic diagnosis and Gleason scoring were microscopically reconfirmed by pathologists. Each case was pathologically staged using the 2002 American Joint Committee on Cancer TNM staging system. In our recruited patients, 13 patients with advanced stage were excluded. Another 10 patients who had positive surgical margins were also excluded from the study. Ultimately, 53 cases fulfilling the selection criteria were included for further study. Follow-up information was obtained from a cancer registration database. A PSA level of ⩾0.2 ng/mL on at least two occasions over a 2-month period was used to define biochemical failure [33].

4.2. Tissue Microarrays (TMAs)

Two independent PCa TMA sets were used in this study. PCa samples from patients were obtained with informed consent (Taipei Medical University-Wan Fang Hospital Institutional Review Board No. 99049). TMAs were constructed using a manual tissue-arraying instrument (Beecher Instruments, Sun Prairie, WI, USA). Briefly, carcinoma areas were manually punched, and duplicate tissue cores measuring 2.5 mm in diameter were inserted into recipient paraffin blocks. Sections measuring 5 μm in thickness were cut and transferred to glass slides. The presence of tumor tissue was further verified on a hematoxylin and eosin (H and E)-stained section.

4.3. Immunohistochemical (IHC) Staining

In brief, tissue microarray (TMA) sections were deparaffinized and immersed in 10 mM sodium citrate buffer (pH 6.0) in a microwave oven twice for 5 min to enhance antigen retrieval. After washing, slides were incubated with 0.3% H_2O_2 in methanol to quench the endogenous peroxidase activity. Slides were washed with phosphate-buffered saline (PBS) and incubated with anti-p-Akt (rabbit polyclonal antibody, Santa Cruz Biotechnology, Santa Cruz, CA, USA), anti-Snail (monoclonal mouse anti-Snail antibody, Biorbyt, Cambridge, UK), and appropriate negative control antibodies for 2 h at room temperature. After washing in PBS, slides were developed with a VECTASTAIN® ABC (avidin-biotin complex) peroxidase kit (Vector Laboratories, Burlingame, CA, USA) and a 3,3,9-diaminobenzidine (DAB) peroxidase substrate kit (Vector Laboratories) according to the manufacturer's instructions. All specimens were stained with H and E, which was used as a light counterstain.

4.4. Scoring of Immunoexpression

IHC results of p-Akt and Snail were classified into two groups according to the intensity and extent of staining. The intensity was scored semi-quantitatively as 0, negative; 1 point, weakly positive;

2 points, moderately positive; or 3 points, strongly positive. To determine the extent of Snail expression, 1000 consecutive malignant cells were counted in the area of the strongest staining. Numbers of cells with positive cytoplasmic staining for p-Akt and positive cytoplasmic and nuclear staining for Snail were recorded. The extent of p-Akt and Snail staining was semi-quantitatively scored as 0, positive in <1% of cells; 1 point, positive in 1%–25% of cells; 2 points, positive in 25%–50% of cells; 3 points, positive in 50%–75% of cells; or 4 points, positive in 75%–100% of cells. We then developed a p-Akt or Snail image score as previously described [6] by multiplying the intensity score (0–3 points) by the extent score (0–4 points) to represent the expression of p-Akt or Snail in cancer tissues. Low and high expression levels of p-Akt or Snail were respectively defined as 0–6 and 8–12 points. All sections were independently scored by the authors.

4.5. Statistical Analysis

SPSS 17.0 statistical software (SPSS, Chicago, IL, USA) was used for all statistical analyses. Differences in the clinicopathological features and Akt image scores of the tumors were assessed using paired t-tests for continuous and categorical variables. A Cox proportional hazards regression model was used for a univariate analysis when assessing predictors of biochemical progression. The Kaplan-Meier method was used to compare the time to recurrence among the groups. The diagnostic value of potential biomarkers as predictors of biochemical failure was evaluated with receiver-operator characteristic (ROC) curves. The area under the ROC curve (AUC) was determined from the plot of sensitivity versus 1–specificity (true positive rate versus false positive rate) and is a measure of the predictability of a test. Statistical significance was defined at $p < 0.05$.

5. Conclusions

Our data demonstrate, for the first time, that p-Akt expression is highly correlated with Snail expression in a Taiwanese population with primary localized PCa. We also documented that p-Akt exerts its tumor-promoting role because of its associations with various aggressive clinicopathological characteristics and BCR in men with clinically-localized PCa. Our results highlight that, in patients with clinically-localized PCa, and a high p-Akt image score in cancer tissues, adjuvant radiotherapy or hormone therapy might be suggested to prevent early BCR. However, larger prospective cohorts and experimental studies are needed for comprehensive functional validation and better understanding of the clinical significance of p-Akt and Snail expression in PCa.

Acknowledgments: This study was supported by grant numbers DOH-TD-B-111-003 and DOH102-TD-C-111-008 from Taipei Medical University and 105TMU-WFH-04 from Wan Fang Hospital, Taipei Medical University.

Author Contributions: Yu-Ching Wen, Kuo-Tai Hua, and Ming-Hsien Chien conceived and designed the experiments, contributed materials, and analyzed the data; Wei-Yu Chen and Chi-Long Chen performed the IHC experiments; Yen-Nien Liu contributed analytical tools; Ming-Hsien Chien, Yu-Ching Wen, and Wei-Jiunn Lee contributed reagents and wrote the paper.

Conflicts of Interest: The authors state that there are no conflicts of interest.

Abbreviations

BPH	Benign prostatic hyperplasia
BCR	Biochemical recurrence
EMT	Epithelial-mesenchymal transition
PCa	Prostate cancer
PI3K	Phosphatidylinositol 3'-kinase
PSA	Prostate-specific antigen
RP	Radical prostatectomy
TMA	Tissue microarrays

References

1. Ferlay, J.; Soerjomataram, I.; Dikshit, R.; Eser, S.; Mathers, C.; Rebelo, M.; Parkin, D.M.; Forman, D.; Bray, F. Cancer incidence and mortality worldwide: Sources, methods and major patterns in GLOBOCAN 2012. *Int. J. Cancer* **2015**, *136*, E359–E386. [CrossRef] [PubMed]

2. Stephenson, A.J.; Scardino, P.T.; Eastham, J.A.; Bianco, F.J., Jr.; Dotan, Z.A.; Fearn, P.A.; Kattan, M.W. Preoperative nomogram predicting the 10-year probability of prostate cancer recurrence after radical prostatectomy. *J. Natl. Cancer Inst.* **2006**, *98*, 715–717. [CrossRef] [PubMed]

3. Hull, G.W.; Rabbani, F.; Abbas, F.; Wheeler, T.M.; Kattan, M.W.; Scardino, P.T. Cancer control with radical prostatectomy alone in 1000 consecutive patients. *J. Urol.* **2002**, *167*, 528–534. [CrossRef]

4. Stephenson, A.J.; Scardino, P.T.; Eastham, J.A.; Bianco, F.J., Jr.; Dotan, Z.A.; DiBlasio, C.J.; Reuther, A.; Klein, E.A.; Kattan, M.W. Postoperative nomogram predicting the 10-year probability of prostate cancer recurrence after radical prostatectomy. *J. Clin. Oncol.* **2005**, *23*, 7005–7012. [CrossRef] [PubMed]

5. Kotb, A.F.; Elabbady, A.A. Prognostic factors for the development of biochemical recurrence after radical prostatectomy. *Prostate Cancer* **2011**, *2011*, 485189. [CrossRef] [PubMed]

6. Wen, Y.C.; Chen, W.Y.; Lee, W.J.; Yang, S.F.; Lee, L.M.; Chien, M.H. Snail as a potential marker for predicting the recurrence of prostate cancer in patients at stage T2 after radical prostatectomy. *Clin. Chim. Acta* **2014**, *431*, 169–173. [CrossRef] [PubMed]

7. Wang, H.; Wang, H.S.; Zhou, B.H.; Li, C.L.; Zhang, F.; Wang, X.F.; Zhang, G.; Bu, X.Z.; Cai, S.H.; Du, J. Epithelial-mesenchymal transition (EMT) induced by TNF-α requires AKT/GSK-3β-mediated stabilization of snail in colorectal cancer. *PLoS ONE* **2013**, *8*, e56664. [CrossRef] [PubMed]

8. Grille, S.J.; Bellacosa, A.; Upson, J.; Klein-Szanto, A.J.; van Roy, F.; Lee-Kwon, W.; Donowitz, M.; Tsichlis, P.N.; Larue, L. The protein kinase Akt induces epithelial mesenchymal transition and promotes enhanced motility and invasiveness of squamous cell carcinoma lines. *Cancer Res.* **2003**, *63*, 2172–2178. [PubMed]

9. Liu, Z.C.; Wang, H.S.; Zhang, G.; Liu, H.; Chen, X.H.; Zhang, F.; Chen, D.Y.; Cai, S.H.; Du, J. AKT/GSK-3β regulates stability and transcription of snail which is crucial for bFGF-induced epithelial-mesenchymal transition of prostate cancer cells. *Biochim. Biophys. Acta* **2014**, *1840*, 3096–3105. [CrossRef] [PubMed]

10. Ayala, G.; Thompson, T.; Yang, G.; Frolov, A.; Li, R.; Scardino, P.; Ohori, M.; Wheeler, T.; Harper, W. High levels of phosphorylated form of Akt-1 in prostate cancer and non-neoplastic prostate tissues are strong predictors of biochemical recurrence. *Clin. Cancer Res.* **2004**, *10*, 6572–6578. [CrossRef] [PubMed]

11. Malik, S.N.; Brattain, M.; Ghosh, P.M.; Troyer, D.A.; Prihoda, T.; Bedolla, R.; Kreisberg, J.I. Immunohistochemical demonstration of phospho-Akt in high Gleason grade prostate cancer. *Clin. Cancer Res.* **2002**, *8*, 1168–1171. [PubMed]

12. Ghosh, P.M.; Malik, S.N.; Bedolla, R.G.; Wang, Y.; Mikhailova, M.; Prihoda, T.J.; Troyer, D.A.; Kreisberg, J.I. Signal transduction pathways in androgen-dependent and -independent prostate cancer cell proliferation. *Endocr. Relat. Cancer* **2005**, *12*, 119–134. [CrossRef] [PubMed]

13. Bedolla, R.; Prihoda, T.J.; Kreisberg, J.I.; Malik, S.N.; Krishnegowda, N.K.; Troyer, D.A.; Ghosh, P.M. Determining risk of biochemical recurrence in prostate cancer by immunohistochemical detection of PTEN expression and Akt activation. *Clin. Cancer Res.* **2007**, *13*, 3860–3867. [CrossRef] [PubMed]

14. Wang, H.; Fang, R.; Wang, X.F.; Zhang, F.; Chen, D.Y.; Zhou, B.; Wang, H.S.; Cai, S.H.; Du, J. Stabilization of Snail through AKT/GSK-3β signaling pathway is required for TNF-α-induced epithelial-mesenchymal transition in prostate cancer PC3 cells. *Eur. J. Pharmacol.* **2013**, *714*, 48–55. [CrossRef] [PubMed]

15. Siegel, R.L.; Miller, K.D.; Jemal, A. Cancer statistics, 2015. *CA Cancer J. Clin.* **2015**, *65*, 5–29. [CrossRef] [PubMed]

16. Kattan, M.W. Nomograms are superior to staging and risk grouping systems for identifying high-risk patients: preoperative application in prostate cancer. *Curr. Opin. Urol.* **2003**, *13*, 111–116. [CrossRef] [PubMed]

17. Chun, F.K.; Steuber, T.; Erbersdobler, A.; Currlin, E.; Walz, J.; Schlomm, T.; Haese, A.; Heinzer, H.; McCormack, M.; Huland, H.; et al. Development and internal validation of a nomogram predicting the probability of prostate cancer Gleason sum upgrading between biopsy and radical prostatectomy pathology. *Eur. Urol.* **2006**, *49*, 820–826. [CrossRef] [PubMed]

18. Matuszak, E.A.; Kyprianou, N. Androgen regulation of epithelial-mesenchymal transition in prostate tumorigenesis. *Expert Rev. Endocrinol. Metab.* **2011**, *6*, 469–482. [CrossRef] [PubMed]

19. Larue, L.; Bellacosa, A. Epithelial-mesenchymal transition in development and cancer: Role of phosphatidylinositol 3′ kinase/AKT pathways. *Oncogene* **2005**, *24*, 7443–7454. [CrossRef] [PubMed]
20. Peinado, H.; Olmeda, D.; Cano, A. Snail, Zeb and bHLH factors in tumour progression: An alliance against the epithelial phenotype? *Nat. Rev. Cancer* **2007**, *7*, 415–428. [CrossRef] [PubMed]
21. Julien, S.; Puig, I.; Caretti, E.; Bonaventure, J.; Nelles, L.; van Roy, F.; Dargemont, C.; de Herreros, A.G.; Bellacosa, A.; Larue, L. Activation of NF-κB by Akt upregulates Snail expression and induces epithelium mesenchyme transition. *Oncogene* **2007**, *26*, 7445–7456. [CrossRef] [PubMed]
22. Alessi, D.R.; Cohen, P. Mechanism of activation and function of protein kinase B. *Curr. Opin. Genet. Dev.* **1998**, *8*, 55–62. [CrossRef]
23. Graff, J.R.; Konicek, B.W.; McNulty, A.M.; Wang, Z.; Houck, K.; Allen, S.; Paul, J.D.; Hbaiu, A.; Goode, R.G.; Sandusky, G.E.; et al. Increased AKT activity contributes to prostate cancer progression by dramatically accelerating prostate tumor growth and diminishing p27Kip1 expression. *J. Biol. Chem.* **2000**, *275*, 24500–24505. [CrossRef] [PubMed]
24. Graff, J.R. Emerging targets in the AKT pathway for treatment of androgen-independent prostatic adenocarcinoma. *Expert Opin. Ther. Targets* **2002**, *6*, 103–113. [CrossRef] [PubMed]
25. Datta, S.R.; Dudek, H.; Tao, X.; Masters, S.; Fu, H.; Gotoh, Y.; Greenberg, M.E. Akt phosphorylation of BAD couples survival signals to the cell-intrinsic death machinery. *Cell* **1997**, *91*, 231–241. [CrossRef]
26. Murillo, H.; Huang, H.; Schmidt, L.J.; Smith, D.I.; Tindall, D.J. Role of PI3K signaling in survival and progression of LNCaP prostate cancer cells to the androgen refractory state. *Endocrinology* **2001**, *142*, 4795–4805. [CrossRef] [PubMed]
27. Le Page, C.; Koumakpayi, I.H.; Alam-Fahmy, M.; Mes-Masson, A.M.; Saad, F. Expression and localisation of Akt-1, Akt-2 and Akt-3 correlate with clinical outcome of prostate cancer patients. *Br. J. Cancer* **2006**, *94*, 1906–1912. [CrossRef] [PubMed]
28. Liao, Y.; Grobholz, R.; Abel, U.; Trojan, L.; Michel, M.S.; Angel, P.; Mayer, D. Increase of AKT/PKB expression correlates with Gleason pattern in human prostate cancer. *Int. J. Cancer* **2003**, *107*, 676–680. [CrossRef] [PubMed]
29. Kreisberg, J.I.; Malik, S.N.; Prihoda, T.J.; Bedolla, R.G.; Troyer, D.A.; Kreisberg, S.; Ghosh, P.M. Phosphorylation of Akt (Ser473) is an excellent predictor of poor clinical outcome in prostate cancer. *Cancer Res.* **2004**, *64*, 5232–5236. [CrossRef] [PubMed]
30. Koh, H.; Maru, N.; Muramoto, M.; Wheeler, T.M.; Scardino, P.T.; Ohori, M. The 1992 TNM classification of T2 prostate cancer predicts pathologic stage and prognosis better than the revised 1997 classification. *Nihon Hinyokika Gakkai Zasshi* **2002**, *93*, 595–601. [CrossRef] [PubMed]
31. Rice, K.R.; Furusato, B.; Chen, Y.; McLeod, D.G.; Sesterhenn, I.A.; Brassell, S.A. Clinicopathological behavior of single focus prostate adenocarcinoma. *J. Urol.* **2009**, *182*, 2689–2694. [CrossRef] [PubMed]
32. Masterson, T.A.; Cheng, L.; Mehan, R.M.; Koch, M.O. Tumor focality does not predict biochemical recurrence after radical prostatectomy in men with clinically localized prostate cancer. *J. Urol.* **2011**, *186*, 506–510. [CrossRef] [PubMed]
33. Freedland, S.J.; Sutter, M.E.; Dorey, F.; Aronson, W.J. Defining the ideal cutpoint for determining PSA recurrence after radical prostatectomy: Prostate-specific antigen. *Urology* **2003**, *61*, 365–369. [CrossRef]

International Journal of
Molecular Sciences

MDPI

Article

EpCAM Expression in Lymph Node and Bone Metastases of Prostate Carcinoma: A Pilot Study

Anna K. Campos [1], Hilde D. Hoving [2,*], Stefano Rosati [3], Geert J. L. H. van Leenders [4] and Igle J. de Jong [2]

[1] Laboratory of Neuroimmunology, National Institute of Neurology and Neurosurgery "Manuel Velasco Suárez", Avenida Insurgentes Sur 3877, La Fama, Tlalpan, 14269 Mexico City, Mexico; camposakaren@gmail.com
[2] Department of Urology, University Medical Center Groningen, University of Groningen, P.O. Box 30.001, Groningen 9700 RB, The Netherlands; i.j.de.jong@umcg.nl
[3] Department of Pathology, University Medical Center Groningen, University of Groningen, P.O. Box 30.001, Groningen 9700 RB, The Netherlands; s.rosati@umcg.nl
[4] Department of Pathology, Josephine Nefkens Institute, Erasmus MC, P.O. Box 2040, Rotterdam 3000 CA, The Netherlands; g.vanleenders@erasmusmc.nl
* Correspondence: h.d.hoving@umcg.nl; Tel.: +31-50-361-2801

Academic Editor: Carsten Stephan
Received: 12 June 2016; Accepted: 15 September 2016; Published: 29 September 2016

Abstract: There is an urgent need for new imaging modalities in prostate carcinoma staging. A non-invasive modality that can assess lymph node and bone metastases simultaneously is preferred. Epithelial cell adhesion molecule (EpCAM) is a membranous protein of interest as an imaging target since it is overexpressed in prostatic carcinoma compared with benign prostate epithelium and compared with stroma. However, EpCAM expression in lymph node metastases is sparsely available in the literature and EpCAM expression in bone metastases is yet unknown. The current study evaluates the expression of EpCAM in prostate carcinoma lymph nodes, in matched normal lymph nodes, in prostate carcinoma bone metastases, and in normal bone by immunohistochemistry. EpCAM was expressed in 100% of lymph node metastases (21 out of 21), in 0% of normal lymph nodes (0 out of 21), in 95% of bone metastases (19 out of 20), and in 0% of normal bone (0 out of 14). Based on these results, EpCAM may be a feasible imaging target in prostate carcinoma lymph node and bone metastases. Prospective clinical trials are needed to confirm current results. Preoperative visualization of prostate carcinoma metastases will improve disease staging and will prevent unnecessary invasive surgery.

Keywords: molecular imaging; imaging target; prostate carcinoma; metastases; epithelial cell adhesion molecule (EpCAM); immunohistochemistry

1. Introduction

Epithelial cell adhesion molecule (EpCAM), also known as CD326 and 17-1A antigen, is a transmembrane glycoprotein originally identified as a marker for carcinoma [1]. EpCAM functions as a cell adhesion molecule in benign and malignant epithelial cells [2]. However, its role also includes signaling, cell migration, proliferation, and differentiation [1,3,4].

EpCAM has been found expressed in various types of carcinoma, including colon and rectum, gallbladder, liver, esophagus, lung, head and neck, pancreas, ovarian, breast, and prostate carcinoma [3–5].

There are several studies that show a significantly elevated expression of EpCAM in prostatic carcinoma compared with benign prostate epithelium [5–7]. Next, Poczatek et al. and Benko et al.

found that there was no expression of EpCAM in prostate stroma [5,7]. Based on these results, EpCAM would be an attractive specific target for imaging purposes in prostate carcinoma. Several preclinical trials targeted EpCAM for fluorescent and nuclear imaging [8–11]. Results of preclinical fluorescent imaging are promising for the accurate assessment of tumor margins intraoperatively [11,12]. Thus, nuclear imaging of EpCAM would be of special interest for prostate carcinoma staging and for monitoring of treatment response. The importance of correct staging lies in the fact that staging has huge treatment implications. Metastasized prostate carcinoma cannot be treated curatively. However, current imaging modalities in prostate carcinoma staging have limitations.

Prostate carcinoma lymph node staging is limited by poor sensitivity and specificity of anatomical imaging modalities [13–15]. Molecular imaging of lymph node metastases by positron emission tomography (PET) is booming nowadays. ^{11}C- or ^{18}F-choline PET/CT has good specificity for the detection of lymph node metastases. However, the sensitivity ranges from 10% to 73% [16,17]. This might result from the target choline, which is not prostate carcinoma-specific. Pelvic lymph node dissection remains the gold standard for nodal staging. Lymph node dissection has been associated with regional disease control and better outcome, but this can lead to poor patient outcome (complications such as lymph edema and thrombosis) and the overtreatment of patients with a low risk of metastases [18–20]. Therefore, a noninvasive imaging modality to assess lymph node metastases would be preferred.

For the assessment of bone metastases, skeletal scintigraphy is the most sensitive method. However, false positive skeletal scintigraphy occurs, for example, from degenerative disease, inflammation, and trauma. Furthermore, skeletal scintigraphy is hampered by the osteoblastic response that accompanies bone healing, which can also lead to a false positive diagnosis of disease progression [21–23]. Furthermore, skeletal scintigraphy lacks anatomical detail, and treatment response can take about 6 to 8 months before response can be visualized [24,25].

To summarize, there is an urgent need for new imaging modalities in prostate carcinoma staging. A non-invasive modality, that can assess lymph node and bone metastases simultaneously, is preferred. This imaging modality should be based on a prostate carcinoma-specific target such as EpCAM.

However, EpCAM expression in prostate carcinoma metastases has been evaluated in only one study. EpCAM was significantly overexpressed in metastasized prostate carcinoma compared with benign prostate hyperplasia, which served as a normal control [6]. Since specimens were collected at autopsies, results should be interpreted with caution. A decay of antigens can be seen in postmortem specimens.

The current research evaluates the expression of EpCAM in prostate carcinoma lymph nodes, in matched normal lymph nodes, in prostate carcinoma bone metastases, and in normal bone in order to determine the feasibility of EpCAM as a specific marker for prostate carcinoma staging.

EpCAM was expressed in 100% of lymph node metastases (21 out of 21), in 0% of normal lymph nodes (0 out of 21), in 95% of bone metastases (19 out of 20), and in 0% of normal bone (0 out of 14). Based on these results, EpCAM may be a feasible imaging target in prostate carcinoma lymph node and bone metastases. Preoperative visualization of these metastases will improve disease staging and will prevent unnecessary invasive surgery.

2. Results

After immunohistochemistry, one prostate carcinoma lymph node metastasis and four prostate carcinoma bone metastases were not evaluable due to bad morphology. The matched normal lymph node was also excluded from statistical analysis. In the end, 20 bone metastases (of 20 patients), 14 normal bone (of 14 patients) and 21 lymph node metastases, and 21 normal lymph node metastases (of 16 patients) were available for statistical analysis. Clinicopathological parameters of these patients are presented in Tables 1–3.

Table 1. Normal lymph nodes and lymph node metastases.

Patient	Tissue Type	Hormonal Therapy	Radiotherapy	PS	IS	TIS
1	Lymph node normal	No	No	0	0	0
	Lymph node metastasis	No	No	4	2	8
2	Lymph node normal	No	No	0	0	0
	Lymph node normal	No	No	0	0	0
	Lymph node metastasis	No	No	4	2	8
	Lymph node metastasis	No	No	4	2	8
3	Lymph node normal	No	No	0	0	0
	Lymph node metastasis	No	No	4	3	12
4	Lymph node normal	No	No	0	0	0
	Lymph node metastasis	No	No	3	1	3
5	Lymph node normal	No	No	0	0	0
	Lymph node metastasis	No	No	4	3	12
6	Lymph node normal	No	No	0	0	0
	Lymph node metastasis	No	No	3	3	9
7	Lymph node normal	No	No	0	0	0
	Lymph node metastasis	No	No	4	2	8
8	Lymph node normal	No	No	0	0	0
	Lymph node metastasis	No	No	4	3	12
9	Lymph node normal	No	No	0	0	0
	Lymph node normal	No	No	0	0	0
	Lymph node normal	No	No	0	0	0
	Lymph node metastasis	No	No	4	2	8
	Lymph node metastasis	No	No	4	2	8
	Lymph node metastasis	No	No	4	3	12
10	Lymph node normal	No	No	0	0	0
	Lymph node normal	No	No	0	0	0
	Lymph node metastasis	No	No	3	2	6
	Lymph node metastasis	No	No	3	1	3
11	Lymph node normal	No	No	0	0	0
	Lymph node metastasis	No	No	3	1	3
12	Lymph node normal	No	No	0	0	0
	Lymph node metastasis	No	No	4	3	12
13	Lymph node normal	No	No	0	0	0
	Lymph node metastasis	No	No	4	2	8
14	Lymph node normal	No	No	0	0	0
	Lymph node metastasis	No	No	3	3	9
15	Lymph node normal	Yes	Yes	0	0	0
	Lymph node normal	Yes	Yes	0	0	0
	Lymph node metastasis	Yes	Yes	4	3	12
	Lymph node metastasis	Yes	Yes	3	3	9
16	Lymph node normal	No	No	0	0	0
	Lymph node metastasis	No	No	4	2	8

PS: proportion score; IS: intensity score; TIS: total immunostaining score.

Table 2. Bone metastases.

Patient	Tissue Type	Hormonal Therapy	Radiotherapy	PS	IS	TIS
1	Bone metastasis	Unknown	Unknown	4	3	12
2	Bone metastasis	Unknown	Unknown	4	3	12
3	Bone metastasis	Yes	No	4	3	12
4	Bone metastasis	Unknown	Unknown	4	3	12
5	Bone metastasis	Yes	No	4	3	12
6	Bone metastasis	No	No	4	2	8
7	Bone metastasis	Unknown	Unknown	0	0	0
8	Bone metastasis	Yes	Yes	4	2	8
9	Bone metastasis	Unknown	Unknown	4	3	12
10	Bone metastasis	Yes	No	3	2	6
11	Bone metastasis	Unknown	Yes	4	3	12
12	Bone metastasis	Yes	Yes	4	3	12
13	Bone metastasis	Yes	Yes	4	3	12
14	Bone metastasis	No	No	4	3	12
15	Bone metastasis	Yes	No	3	3	9
16	Bone metastasis	Unknown	Yes	4	3	12
17	Bone metastasis	Unknown	Unknown	1	1	1
18	Bone metastasis	No	No	4	3	12
19	Bone metastasis	No	No	4	3	12
20	Bone metastasis	No	No	4	3	12

PS: proportion score; IS: intensity score; TIS: total immunostaining score.

Table 3. Normal bone.

Patient	Tissue Type	Hormonal Therapy	Radiotherapy	PS	IS	TIS
1	Normal bone	Unknown	Unknown	0	0	0
2	Normal bone	Unknown	Unknown	0	0	0
3	Normal bone	Unknown	Unknown	0	0	0
4	Normal bone	Unknown	Unknown	0	0	0
5	Normal bone	Unknown	Unknown	0	0	0
6	Normal bone	Unknown	Unknown	0	0	0
7	Normal bone	Unknown	Unknown	0	0	0
8	Normal bone	Unknown	Unknown	0	0	0
9	Normal bone	Unknown	Unknown	0	0	0
10	Normal bone	Unknown	Unknown	0	0	0
11	Normal bone	Unknown	Unknown	0	0	0
12	Normal bone	Unknown	Unknown	0	0	0
13	Normal bone	Unknown	Unknown	0	0	0
14	Normal bone	Unknown	Unknown	0	0	0

PS: proportion score; IS: intensity score; TIS: total immunostaining score. Epithelial cell adhesion molecule (EpCAM) expression was observed in 21 out of 21 lymph node metastases (100%) and was absent in 21 out of 21 matched normal lymph nodes (100%). Median EpCAM expression (TIS) in lymph node metastases was 8. Even after hormonal therapy and radiotherapy EpCAM expression was high in 2 lymph node metastases of a single patient (patient 15). Heterogeneous EpCAM expression was seen in 3 lymph node metastases of patient 9, in 2 lymph node metastases of patient 10, and in 2 lymph node metastases of patient 15.

EpCAM expression was observed in 19 out of 20 bone metastases (95%) and was absent in 14 out of 14 cases of normal bone (100%). Median EpCAM expression in bone metastases (TIS) was 12. After either hormonal therapy or radiotherapy, TIS was high. Even after both hormonal therapy and radiotherapy, EpCAM expression was high (Table 2).

EpCAM expression was membranous in prostate carcinoma cells of lymph node (Figure 1) and bone metastases (Figure 2).

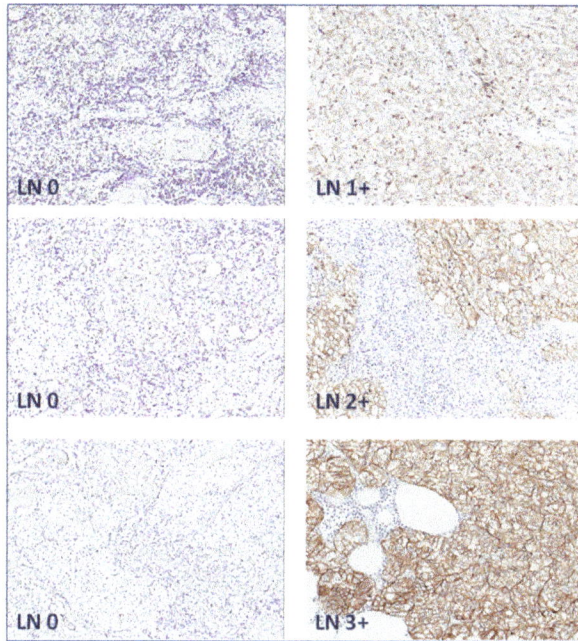

Figure 1. EpCAM expression in lymph node (LN) metastases. Matched normal lymph nodes without EpCAM expression (**left** panel) and lymph node metastases of prostate carcinoma with EpCAM expression of intensity score 1: weak; 2: moderate; and 3: strong (**right** panel). Original magnification: 200×.

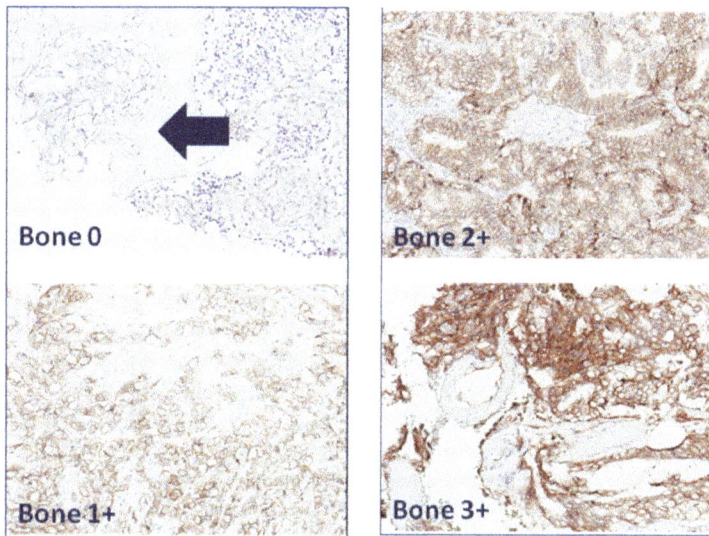

Figure 2. EpCAM expression in prostate carcinoma bone metastases. Bone metastases with EpCAM expression of intensity score 1: weak; 2: moderate; and 3: strong. Arrow points towards an area of bone metastasis in bone without EpCAM expression. Original magnification: 200×.

3. Discussion

In the current study, EpCAM was expressed in 100% of prostate carcinoma lymph node metastases and 95% of prostate carcinoma bone metastases. The median EpCAM expression was high, with a TIS of 8 and 12 for lymph node and bone metastases, respectively. EpCAM was not expressed in matched normal lymph nodes and in non-matched normal bone.

EpCAM expression in prostate carcinoma metastases has been evaluated in only one study previously. However, bone metastases were not included. EpCAM was significantly overexpressed in metastasized prostate carcinoma compared with benign prostate hyperplasia, which served as a normal control [6]. Since specimens were collected at autopsies, results should be interpreted with caution.

The current study is the first to compare EpCAM expression between matched normal lymph nodes and lymph node metastases and to evaluate EpCAM expression in bone metastases and normal bone. Absent EpCAM expression in normal lymph nodes and normal bone supports the use of EpCAM as a target with high specificity for prostate carcinoma.

Went et al. and Benko et al. investigated a possible correlation between EpCAM expression and nodal stage (N0, N1, or N2) in primary prostate carcinoma. EpCAM expression was defined as negative, weak to moderate and strong [1]. No correlation was found, presumably caused by small sample sizes of 10 and 3 primary prostate carcinoma cases, respectively. Moreover, lymph node specimens were not available in both studies [1,7].

In previous literature EpCAM overexpression was described as TIS > 4 [26]. In the current study, EpCAM overexpression was found in both lymph node and bone metastases regardless of previous treatment. This is in agreement with a study from Benko et al. in which a correlation was found between high EpCAM expression and biochemical recurrence of prostate carcinoma [7]. Next, EpCAM expression was seen in 82.3% of salvage prostatectomy specimens taken from patients with locally recurrent prostate carcinoma after external beam radiotherapy or brachytherapy [27].

4. Materials and Methods

Formalin-fixed paraffin-embedded prostate carcinoma lymph node metastases, normal lymph nodes, prostate carcinoma bone metastases, samples of normal bone, and samples of normal colon were retrieved from the archives of the Department of Pathology of University Medical Center Groningen. Lymph node metastases (n = 22) and matched normal lymph nodes (n = 22) were available in 17 patients who underwent a pelvic lymph node dissection because of a suspicion of nodal involvement on computed tomography (CT). Of 12 patients, 1 prostate carcinoma lymph node metastasis and 1 normal lymph node was included; of 3 patients, 2 prostate carcinoma lymph node metastases and 2 normal lymph nodes were included; of 1 patient, 3 prostate carcinoma lymph node metastases and 3 normal lymph nodes were included.

Prostate carcinoma bone metastases (n = 24) were available in 24 patients who underwent surgery to confirm clinical suspicion of prostate carcinoma or to treat skeletal related events. Bone metastases were taken from several surgical procedures (biopsy (n = 6), osteosynthesis (n = 1), laminectomy (n = 6), extramedullary excision (n = 1), corporectomy (n = 2), during procedures of hip prosthesis implantation (n = 1), and other, unknown procedures (n = 7)). Matched normal bone was not available. Fourteen cases of non-matched normal bone were available in 14 patients who underwent hip prosthesis implantation. All tissue specimens were anonymously coded. According to Dutch law, no further Institutional Review Board approval was required (http://www.federa.org/).

Immunohistochemistry was performed in order to determine EpCAM expression. Normal colon was used as a positive control and omission of the primary antibody on positive control specimens served as a negative control. After deparaffinization with decreasing grades of alcohol, antigen retrieval was performed by incubation with 0.1% protease for 30 min at room temperature. Endogeneous peroxidase was blocked with 0.3% hydrogen peroxide in phosphate buffered saline (PBS) for 20 min. Slides were incubated with primary mouse monoclonal antibody anti-EpCAM (clone VU-1D9, Leica Biosystems, Newcastle, UK) diluted at 1:100 in 1% bovine serum albumin and phosphate buffered saline (BSA/PBS) for 1 h at room temperature. In the secondary step, slides were incubated with a rabbit anti-mouse antibody conjugated to polymer-horseradish peroxidase (DAKO, Glostrup, Denmark), diluted at 1:100 in 1% BSA/PBS with 1% AB serum. In the tertiary step, a goat anti-rabbit antibody conjugated to polymer-horseradish peroxidase (DAKO, Glostrup, Denmark) was used, diluted at 1:100 in 1% BSA/PBS with 1% AB serum. Both secondary and tertiary steps required incubation for 30 min at room temperature. Next, slides were immersed for 10 min in a solution of 0.05% 3,3'-diaminobenzidine (Sigma-Aldrich, Steinheim, Germany) and 0.03% hydrogen peroxide in PBS for visualization of the signal as brown staining. After washing with demineralized water, slides were slightly counterstained with hematoxylin, dehydrated, and mounted with Tissue Tec film (Sakura Finetek, Leiden, The Netherlands).

A pathologist (GJLHvL), blinded to clinical data, scored EpCAM immunoreactivity according to a previous established method [28]. This method determines a total immunostaining score (TIS), which is the product of a proportion score (PS) and an intensity score (IS). The PS represents the estimated amount of positively stained cells (0: none; 1: <10%; 2: 10%–50%; 3: 51%–80%; 4: >80%). IS describes the estimated staining intensity (0: no staining; 1: weak; 2: moderate; 3: strong). TIS (TIS = PS × IS) ranges from 0 to 12 with only 9 possible values (0, 1, 2, 3, 4, 6, 8, 9, or 12) [28].

Sensitivity and specificity of EpCAM expression were determined. Specificity of EpCAM expression could be determined for lymph node specimens only. For bone specimens, specificity could not be determined due to the lack of negative control bone specimens. Sensitivity is the percentage of cases with EpCAM expression out of the total amount of either histopathological proven lymph node metastases or histopathological proven bone metastases. Sensitivity is the percentage of cases without EpCAM expression in histopathological proven lymph nodes without prostate carcinoma metastasis.

Descriptive statistics were used to describe the results. For ordinal data, the median is presented.

5. Conclusions

In conclusion, the purpose of the current study was to establish the feasibility of EpCAM as an imaging target for prostate carcinoma lymph node and bone metastases. EpCAM has proved to be a target with high sensitivity (100%) and specificity (100%) for lymph node metastases and high sensitivity (95%) and specificity (100%) for bone metastases. Based on previous literature, EpCAM may be used as imaging target for locally recurrent prostate carcinoma [7,27]. Based on the current study, EpCAM may be additionally used as an imaging target for prostate carcinoma lymph node and bone metastases. Preoperative visualization of metastases will improve disease staging and will prevent unnecessary invasive surgery. Prospective clinical trials are needed to confirm current results.

Author Contributions: Anna K. Campos, Hilde D. Hoving and Igle J. de Jong conceived and designed the experiments. Anna K. Campos and Hilde D. Hoving performed the immunohistochemistry. Stefano Rosati contributed materials. Geert J. L. H. van Leenders scored EpCAM immunoreactivity. Anna K. Campos, Hilde D. Hoving and Igle J. de Jong analyzed data and wrote the manuscript.

Conflicts of Interest: The authors declare no conflict of interest.

References

1. Went, P.; Vasei, M.; Bubendorf, L.; Terracciano, L.; Tornillo, L.; Riede, U.; Kononen, J.; Simon, R.; Sauter, G.; Baeuerle, P.A. Frequent high-level expression of the immunotherapeutic target Ep-CAM in colon, stomach, prostate and lung cancers. *Br. J. Cancer* **2006**, *94*, 128–135. [CrossRef] [PubMed]
2. Litvinov, S.V.; Velders, M.P.; Bakker, H.A.; Fleuren, G.J.; Warnaar, S.O. Ep-CAM: A human epithelial antigen is a homophilic cell–cell adhesion molecule. *J. Cell Biol.* **1994**, *125*, 437–446. [CrossRef] [PubMed]
3. Went, P.T.; Lugli, A.; Meier, S.; Bundi, M.; Mirlacher, M.; Sauter, G.; Dirnhofer, S. Frequent EpCAM protein expression in human carcinomas. *Hum. Pathol.* **2004**, *35*, 122–128. [CrossRef] [PubMed]
4. Trzpis, M.; McLaughlin, P.M.; de Leij, L.M.; Harmsen, M.C. Epithelial cell adhesion molecule: More than a carcinoma marker and adhesion molecule. *Am. J. Pathol.* **2007**, *171*, 386–395. [CrossRef] [PubMed]

5. Poczatek, R.B.; Myers, R.B.; Manne, U.; Oelschlager, D.K.; Weiss, H.L.; Bostwick, D.G.; Grizzle, W.E. Ep-CAM levels in prostatic adenocarcinoma and prostatic intraepithelial neoplasia. *J. Urol.* **1999**, *162*, 1462–1466. [CrossRef]

6. Zellweger, T.; Ninck, C.; Bloch, M.; Mirlacher, M.; Koivisto, P.A.; Helin, J.H.; Mihatsch, M.J.; Gasser, T.C.; Bubendorf, L. Expression patterns of potential therapeutic targets in prostate cancer. *Int. J. Cancer* **2005**, *113*, 619–628. [CrossRef] [PubMed]

7. Benko, G.; Spajic, B.; Kruslin, B.; Tomas, D. Impact of the EpCAM expression on biochemical recurrence-free survival in clinically localized prostate cancer. *Urol. Oncol.* **2013**, *31*, 468–474. [CrossRef] [PubMed]

8. Leung, K. DiD-Labeled anti-EpCAM-directed NK-92-scFv(MOC31) zeta cells. In *Molecular Imaging and Contrast Agent Database (MICAD) Bethesda (MD)*; National Center for Biotechnology Information (US): Bethesda, MD, USA, 2004.

9. Hall, M.A.; Pinkston, K.L.; Wilganowski, N.; Robinson, H.; Ghosh, P.; Azhdarinia, A.; Vazquez-Arreguin, K.; Kolonin, A.M.; Harvey, B.R.; Sevick-Muraca, E.M. Comparison of mAbs targeting epithelial cell adhesion molecule for the detection of prostate cancer lymph node metastases with multimodal contrast agents: Quantitative small-animal PET/CT and NIRF. *J. Nucl. Med.* **2012**, *53*, 1427–1437. [CrossRef] [PubMed]

10. Hall, M.A.; Kwon, S.; Robinson, H.; Lachance, P.-A.; Azhdarinia, A.; Ranganathan, R.; Price, R.E.; Chan, W.; Sevick-Muraca, E.M. Imaging prostate cancer lymph node metastases with a multimodality contrast agent. *Prostate* **2012**, *72*, 129–146. [CrossRef] [PubMed]

11. Zhu, B.; Wu, G.; Robinson, H.; Wilganowski, N.; Hall, M.A.; Ghosh, S.C.; Pinkston, K.L.; Azhdarinia, A.; Harvey, B.R.; Sevick-Muraca, E.M. Tumor margin detection using quantitative NIRF molecular imaging targeting EpCAM validated by far red gene reporter iRFP. *Mol. Imaging Biol.* **2013**, *15*, 560–568. [CrossRef] [PubMed]

12. Tavri, S.; Jha, P.; Meier, R.; Henning, T.D.; Müller, T.; Hostetter, D.; Knopp, C.; Johansson, M.; Reinhart, V.; Boddington, S. Optical imaging of cellular immunotherapy against prostate cancer. *Mol. Imaging* **2009**, *8*, 15–26. [PubMed]

13. Pinto, F.; Totaro, A.; Palermo, G.; Calarco, A.; Sacco, E.; D'Addessi, A.; Racioppi, M.; Valentini, A.L.; Gui, B.; Bassi, P.F. Imaging in prostate cancer staging: Present role and future perspectives. *Urol. Int.* **2012**, *88*, 125–136. [CrossRef] [PubMed]

14. Jung, A.J.; Westphalen, A.C. Imaging prostate cancer. *Radiol. Clin. N. Am.* **2012**, *50*, 1043–1059. [CrossRef] [PubMed]

15. Talab, S.S.; Preston, M.A.; Elmi, A.; Tabatabaei, S. Prostate cancer imaging: What the urologist wants to know. *Radiol. Clin. N. Am.* **2012**, *50*, 1015–1041. [CrossRef] [PubMed]

16. Brogsitter, C.; Zophel, K.; Kotzerke, J. ^{18}F-Choline, ^{11}C-choline and ^{11}C-acetate PET/CT: Comparative analysis for imaging prostate cancer patients. *Eur. J. Nucl. Med. Mol. Imaging* **2013**, *40* (Suppl. 1), 18–27. [CrossRef] [PubMed]

17. Poulsen, M.H.; Bouchelouche, K.; Hoilund-Carlsen, P.F.; Petersen, H.; Gerke, O.; Steffansen, S.I.; Marcussen, N.; Svolgaard, N.; Vach, W.; Geertsen, U.; et al. [18]F]fluoromethylcholine (FCH) positron emission tomography/computed tomography (PET/CT) for lymph node staging of prostate cancer: A prospective study of 210 patients. *BJU Int.* **2012**, *110*, 1666–1671. [CrossRef] [PubMed]

18. Heidenreich, A.; Ohlmann, C.H.; Polyakov, S. Anatomical extent of pelvic lymphadenectomy in patients undergoing radical prostatectomy. *Eur. Urol.* **2007**, *52*, 29–37. [CrossRef] [PubMed]

19. Briganti, A.; Blute, M.L.; Eastham, J.H.; Graefen, M.; Heidenreich, A.; Karnes, J.R.; Montorsi, F.; Studer, U.E. Pelvic lymph node dissection in prostate cancer. *Eur. Urol.* **2009**, *55*, 1251–1265. [CrossRef] [PubMed]

20. Berney, D.M.; Wheeler, T.M.; Grignon, D.J.; Epstein, J.I.; Griffiths, D.F.; Humphrey, P.A.; van der Kwast, T.; Montironi, R.; Delahunt, B.; Egevad, L.; et al. International Society of Urological Pathology (ISUP) Consensus Conference on Handling and Staging of Radical Prostatectomy Specimens. Working group 4: Seminal vesicles and lymph nodes. *Mod. Pathol.* **2011**, *24*, 39–47. [CrossRef] [PubMed]

21. McGregor, B.; Tulloch, A.G.; Quinlan, M.F.; Lovegrove, F. The role of bone scanning in the assessment of prostatic carcinoma. *Br. J. Urol.* **1978**, *50*, 178–181. [CrossRef] [PubMed]

22. Messiou, C.; Cook, G.; deSouza, N.M. Imaging metastatic bone disease from carcinoma of the prostate. *Br. J. Cancer* **2009**, *101*, 1225–1232. [CrossRef] [PubMed]

23. Cook, G.J.; Venkitaraman, R.; Sohaib, A.S.; Lewington, V.J.; Chua, S.C.; Huddart, R.A.; Parker, C.C.; Dearnaley, D.D.; Horwich, A. The diagnostic utility of the flare phenomenon on bone scintigraphy in staging prostate cancer. *Eur. J. Nucl. Med. Mol. Imaging* **2011**, *38*, 7–13. [CrossRef] [PubMed]

24. Smith, P.H.; Bono, A.; Calais da Silva, F.; Debruyne, F.; Denis, L.; Robinson, P.; Sylvester, R.; Armitage, T.G. Some limitations of the radioisotope bone scan in patients with metastatic prostatic cancer. A subanalysis of EORTC trial 30853. The EORTC Urological Group. *Cancer* **1990**, *66* (Suppl. 5), 1009–1016. [PubMed]

25. Scher, H.I. Prostate carcinoma: Defining therapeutic objectives and improving overall outcomes. *Cancer* **2003**, *97* (Suppl. 3), 758–771. [CrossRef] [PubMed]

26. Gastl, G.; Spizzo, G.; Obrist, P.; Dünser, M.; Mikuz, G. Ep-CAM overexpression in breast cancer as a predictor of survival. *Lancet* **2000**, *356*, 1981–1982. [CrossRef]

27. Rybalov, M.; Ananias, H.J.; Hoving, H.D.; van der Poel, H.G.; Rosati, S.; de Jong, I.J. PSMA, EpCAM, VEGF and GRPR as imaging targets in locally recurrent prostate cancer after radiotherapy. *Int. J. Mol. Sci.* **2014**, *15*, 6046–6061. [CrossRef] [PubMed]

28. Spizzo, G.; Fong, D.; Wurm, M.; Ensinger, C.; Obrist, P.; Hofer, C.; Mazzoleni, G.; Gastl, G.; Went, P. EpCAM expression in primary tumour tissues and metastases: An immunohistochemical analysis. *J. Clin. Pathol.* **2011**, *64*, 415–420. [CrossRef] [PubMed]

International Journal of
Molecular Sciences

MDPI

Article

Neuroendocrine Differentiation in Metastatic Conventional Prostate Cancer Is Significantly Increased in Lymph Node Metastases Compared to the Primary Tumors

Vera Genitsch [1], Inti Zlobec [1], Roland Seiler [2], George N. Thalmann [2] and Achim Fleischmann [1,*

[1] Institute of Pathology, University of Bern, Bern 3008, Switzerland; vera.genitsch@pathology.unibe.ch (V.G.); inti.zlobec@pathology.unibe.ch (I.Z.)

[2] Department of Urology, University of Bern, Bern 3010, Switzerland; roland.seiler@insel.ch (R.S.); george.thalmann@insel.ch (G.N.T.)

* Correspondence: achim.fleischmann@stgag.ch; Tel.: +41-(71)-6862285

Received: 27 April 2017; Accepted: 7 July 2017; Published: 28 July 2017

Abstract: Neuroendocrine serum markers released from prostate cancers have been proposed for monitoring disease and predicting survival. However, neuroendocrine differentiation (NED) in various tissue compartments of metastatic prostate cancer is poorly described and its correlation with specific tumor features is unclear. NED was determined by Chromogranin A expression on immunostains from a tissue microarray of 119 nodal positive, hormone treatment-naïve prostate cancer patients who underwent radical prostatectomy and extended lymphadenectomy. NED in the primary cancer and in the metastases was correlated with tumor features and survival. The mean percentage of NED cells increased significantly ($p < 0.001$) from normal prostate glands (0.4%), to primary prostate cancer (1.0%) and nodal metastases (2.6%). In primary tumors and nodal metastases, tumor areas with higher Gleason patterns tended to display a higher NED, although no significance was reached. The same was observed in patients with a larger primary tumor volume and higher total size and number of metastases. NED neither in the primary tumors nor in the metastases predicted outcome significantly. Our data suggest that (a) increasing levels of neuroendocrine serum markers in the course of prostate cancer might primarily derive from a poorly differentiated metastatic tumor component; and (b) NED in conventional hormone-naïve prostate cancers is not significantly linked to adverse tumor features.

Keywords: prostate cancer; lymph node metastases; neuroendocrine; chromogranin A; prognosis

1. Introduction

The current World Health Organization (WHO) classification of prostate neoplasms with neuroendocrine (NE) differentiation (NED) comprises of: (1) adenocarcinomas with NED; (2) well-differentiated NE tumors (carcinoid); (3) small-cell NE carcinomas; and (4) large cell NE carcinomas [1]. While the last three entities are exceedingly rare, the first occurs frequently. In 10–100% of the conventional adenocarcinomas, NED can be demonstrated immunohistochemically in the form of scattered NE cancer cells, depending on the number of slides evaluated and the number of antibodies used [1].

NE cells in prostate cancer most likely emerge from the secretory prostate cancer cells by trans-differentiation [2–4]. Each NE cell may store a single, or a mix of neuropeptides in cytoplasmic granules, including Chromogranin A, the most frequently detected and most intensely studied NE product in prostate tissue [5], serotonin, somatostatin and bombesin [6]. The exact biological function of neuropeptides in prostate cancer is largely unknown; however, data indicate that they may stimulate growth, differentiation and secretory processes [5,7]. While small and large cell NE carcinomas are particularly aggressive [8], the prognostic significance of NE cells in conventional adenocarcinomas of the prostate is still controversial [4,9,10]. Importantly, neuropeptides released from the NE prostate cancer cells may appear in the circulation [6]. These serum markers have recently attracted considerable attention for their ability to monitor disease [6,11] and predict survival [12,13]. NE serum markers have been suggested as beneficial surrogates for tumor burden [6] and mirror prostate cancer progression when raising. In line with this, serum levels of Chromogranin A are significantly higher in metastatic compared to non-metastatic prostate cancers [14]. However, despite this interest in NE serum markers, little is known about the distribution of their source, which are the NE tumor cells, in the various growth patterns and in the metastases of prostate cancer. In this study, we more accurately describe the extent of NED in the different tissue compartments of metastasizing prostate cancer, and determine its correlations with different tumor features and survival.

2. Results

2.1. Patient Characteristics and Expression of Chromogranin A in Benign Prostate, Primary Tumors and Lymph Node Metastases Considering the Gleason Patterns

The patient, prostatectomy and lymphadenectomy characteristics are specified in Table 1. A higher proportion of 92% of primary tumors displayed any positivity for Chromogranin A compared to lymph node metastases with a positive expression in 77%. When the density of NE cancer cells was recorded, a progressive and significant increase in expression from non-neoplastic prostate glands (0.4% mean of Chromogranin A positive cells) to primary tumors (1.0%) and lymph node metastases (2.6%; $p < 0.001$) was noted for Chromogranin A (Figure 1A).

A tendency for higher Chromogranin A expression in less-differentiated tumor areas (reflected by a higher Gleason pattern (GP)) was observed in the primary tumors (GP3: 0.8% mean of Chromogranin A positive cells; GP4: 1.0%; GP5: 1.4%; $p > 0.05$) and in the nodal metastases (GP3: 0.0%; GP4: 1.8%; GP5: 7.8%; P = NE), but no statistical significance was reached (Figure 1B).

Table 1. Characteristics of 119 nodal positive prostate cancer patients.

Patient Data (n = 119)	
Age (median, range) at surgery (years)	65 (45–75)
Follow-up (median, range) (years)	5.9 (0.1–15.2)
Patients with biochemical failure at last follow-up (n)	103
Patients dead of disease at last follow-up (n)	33
Patients dead at last follow-up (n)	40
Prostatectomy Data	
pT2 (n)	14
pT3a (n)	55
pT3b (n)	50
Prostate cancer volume (median, range) (cm^3)	12.6 (0.66–127)
Gleason score 6 (n)	12
Gleason score 7 (n)	63
Gleason score 8 (n)	21
Gleason score 9 (n)	23
Lymphadenectomy Data	
Evaluated nodes per patient (median, range) (n)	22 (9–68)
Positive nodes per patient (median, range) (n)	2 (1–24)

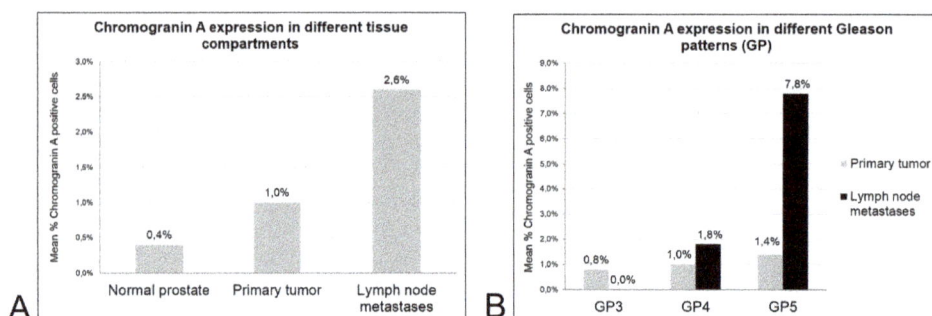

Figure 1. Mean density of Chromogranin A positive cells is significantly different between normal prostate glands, primary prostate cancer and matched lymph node metastases ((**A**) $p < 0.001$). The difference between the Gleason patterns is not significant ((**B**) $p > 0.05$).

2.2. Correlations of Chromogranin A Expression in Primary Tumors and Lymph Node Metastases with Clinico-Pathological Tumor Characteristics and Survival

Primary tumors with Chromogranin A expression were larger (mean 21.5 ± 24.9 cm^3 versus 18.0 ± 15.4 cm^3; $p = 0.821$) and the tumor burden of a Chromogranin A positive metastasizing component was higher for mean total size and number of metastases (36.4 ± 49.4 mm versus 19.4 ± 31.7 mm; $p = 0.458$ and 5.3 ± 6.9 versus 3.3 ± 3.4; $p = 0.279$) (Table 2); however, these differences were not statistically significant. Chromogranin A expression in primary tumors or lymph node metastases was not associated with categorical tumor characteristics as stage of the primary tumor. In univariate analysis, Chromogranin A expression in primary tumors or lymph node metastases did not significantly predict biochemical recurrence-free, cancer-specific, or overall survival (Figure 2). Only the total size of metastases independently predicted all three endpoints in a multivariate analysis (Table 3).

Table 2. Tumor features according to Chromogranin A expression.

CgA Expression	Parameters of the Primary Tumor (Mean $\pm$ SD)				Parameters of Nodal Metastases (Mean $\pm$ SD)			
	Age	p	Tumor volume (cm^3)	p	Total size (mm)	p	Total number	p
Primary Tumor								
CgA negative	64.4 ± 6.1	0.978	18.0 ± 15.4	0.821	19.6 ± 34.8	0.989	3.3 ± 3.8	0.813
CgA positive	64.3 ± 5.8		21.5 ± 24.9		17.2 ± 24.4		3.0 ± 3.3	
Nodal Metastases								
CgA negative	64.3 ± 5.9	0.027	19.1 ± 19.5	0.819	19.4 ± 31.7	0.458	3.3 ± 3.4	0.279
CgA positive	59.3 ± 6.3		18.9 ± 13.7		36.4 ± 49.4		5.3 ± 6.9	

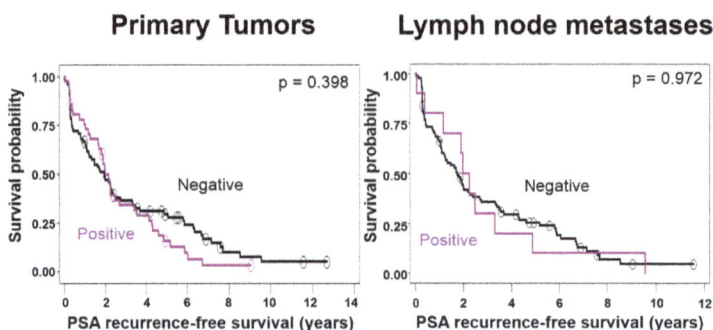

Figure 2. *Cont.*

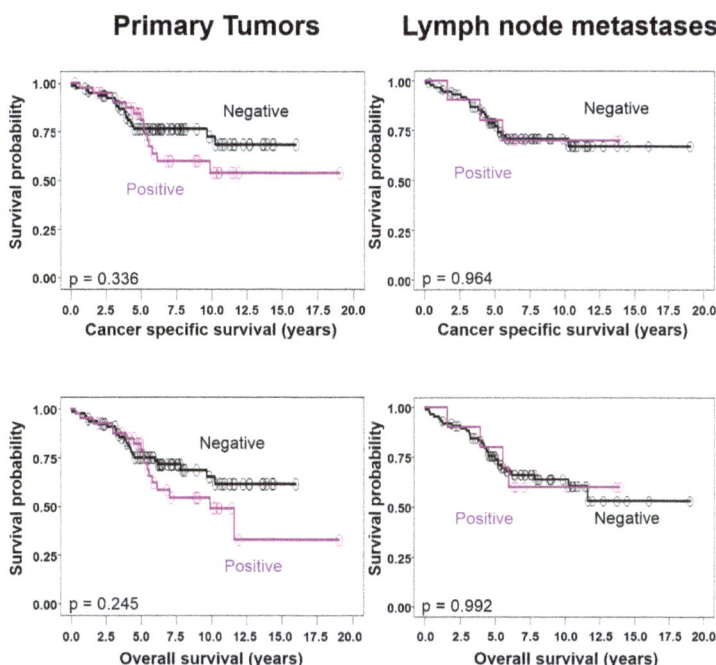

Figure 2. Chromogranin A expression in primary tumors and metastases is not significantly correlated with outcome.

Table 3. Multivariate analyses for the prognostic impact of Chromogranin A (CgA) expression in primary prostate cancer (upper half) and in lymph node metastases (lower half), after adjustment for total size of metastases and Gleason score of primary tumor: Only nodal tumor burden predicts survival independently. HR, hazard ratio; and CI, confidence interval.

Parameter	Cut-Off	Overall Survival		Disease-Specific Survival		Recurrence-Free Survival	
		HR (95% CI)	*p*	HR (95% CI)	*p*	HR (95% CI)	*p*
CgA in Primary Tumor	Positive	1.0	0.132	1.0	0.241	1.0	0.66
	Negative	1.65 (0.9–3.1)		1.54 (0.8–3.1)		1.1 (0.7–1.7)	
Metastases size	<7.5 mm	1.0	**<0.001**	1.0	**0.002**	1.0	**0.036**
	≥7.5 mm	4.34 (2.0–9.6)		4.12 (1.7–10.0)		1.58 (1.1–2.4)	
Gleason score	6 to 8	1.0	0.571	1.0	0.375	1.0	0.074
	9 to 10	1.23 (0.6–2.5)		1.41 (0.7–3.0)		1.57 (1.0–2.6)	
CgA in Nodal Metastases	Positive	1.0	0.571	1.0	0.5	1.0	0.327
	Negative	0.73 (0.2–2.2)		0.65 (0.2–2.8)		0.69 (0.3–1.4)	
Metastases size	<7.5 mm	1.0	**<0.001**	1.0	**0.003**	1.0	0.063
	≥7.5 mm	5.3 (2.0–14.1)		6.44 (1.9–22.1)		1.58 (1.0–2.5)	
Gleason score	6 to 8	1.0	0.365	1.0	1.88	1.0	0.082
	9 to 10	1.43 (0.7–3.1)		1.75 (0.8–4.0)		1.62 (0.9–2.8)	

3. Discussion

Only a few studies on prostate cancers have evaluated NED in metastatic tissues from lymph nodes and various other organs with immunohistochemistry [15–20]. Reported incidences for bone metastases were 19% [18] and 52% [16], those for lymph node metastases 12% [19], 37.5% [17] and 46% [16]. A wide range in the extent of NED in metastases was also noted in an autopsy series by Roudier et al. [20], specifically between patients, and also between different metastases of a single patient. In our series, NED in lymph node metastases was present in 77% of the patients. The metastases had a lower prevalence for NED positivity compared to the primary tumors, which showed NE differentiation in 92%. This decrease was consistent with the only two series on NED in surgically treated nodal positive prostate cancer reported by Bostwick et al. [17] and Quek et al. [19]. However, when considering not only the presence or absence of NED, but also the density of positive cells in primary tumors and metastases, we noticed a significant increase in NED in metastases when compared to primary tumors. Furthermore, NED increased in higher Gleason patterns in the primary tumors, and was even more striking in the metastases where tumor growths of Gleason pattern 5 showed the highest levels of NED among all evaluated cancer components. Our findings were consistent with reports on a positive correlation of the extent of NED and the Gleason score in primary tumors [21,22]. Together with the previously described correlation of Chromogranin A expression by the tumor tissue with its serum level [23], our data might suggest that elevated NE serum markers in metastatic prostate cancer [14] may primarily reflect the metastatic, frequently poorly differentiated tumor burden [24–26].

The presence of NED in our prostate cancer patients showed a tendency for association with adverse tumor characteristics. Patients with detected NED in primary tumors had larger tumors, and those with NED present in metastases had a greater nodal tumor burden, indicated by more metastases and greater total diameter of metastases when compared with patients without NED. Consistent with our data, Quek et al. [19] reported the association of high NED with an advanced tumor stage. Furthermore, NED in the primary tumors of our patients translated into long-term survival. After five years, the curves for disease-specific and overall survival segregate clearly indicated a poorer outcome for patients with NED when compared to those without NED. However, this was not significant, most likely due to the size of our cohort. Contrarily, survival curves based on NED in lymph node metastases intersected repeatedly. Only two other studies have evaluated NED in nodal positive prostate cancer patients treated by radical prostatectomy and bilateral lymphadenectomy. NED detected by Chromogranin A was not a risk factor in the study by Bostwick et al. [17], neither in the primary tumors nor in the metastases, whereas Queck et al. [19] reported significantly poorer median recurrence-free and overall survival for patients with high NED in the primary tumor and metastases, respectively, when compared to patients with low NED. However, survival curves were not presented in the latter study and other outcome measures were not significantly different.

Previous studies on NED in prostate cancer tissues assessed expression on large sections (for comprehensive review of the literature see Table 3 in Bostwick et al. [17]) and cancers were categorized as negative (absence of NE cells), or positive (presence of NE cells). While it was generally noticed that NED in prostate cancer is a very focal, dispersed phenomenon, reported incidences for NED varied between 24 and 98.5% [17]. This wide range was attributed to differences between the cohorts, sample types, types and extent of fixation, the antibodies used in determining the presence of malignant NE cells, variance in interpretation and, most importantly, a sampling error related to the focal and unequal distribution of NE cells in most tumors [27]. It is evident that the amount of tumor tissue evaluated may impact on reported prevalence in cases of only focally expressed biomarkers like NED. We determined NED in primary tumors and metastases by tissue microarray (TMA). This technology has also been considered to be useful for these focally expressed biomarkers in prostate cancer by a study comparing the expression of NE markers on whole tissue sections to a TMA [28]. Investigating these focally expressed biomarkers on large sections may have also been problematic as tissue slides from primary prostate cancer generally contain much greater amounts of tumor tissue than the usually scarce metastatic tissue that makes the comparison of incidences difficult. However, the use of a TMA certainly remains a limiting factor in our study. Finally, for a delicate biomarker like NED in prostate cancer, the size of the cohort may play a major role in detecting significant correlations between tumor features and survival. Our cohort was comparably large for surgically treated nodal positive prostate cancer and therefore allowed detection of a significant increase in NED in nodal metastases and trends between biomarker expression levels, tumor features and survival. However, it may have been too small to demonstrate these trends as significant.

4. Materials and Methods

4.1. Patients

In total, 119 consecutive prostate cancer patients without demonstrable metastases (physical and radiological examination), but with nodal metastases upon histological investigation of the lymphadenectomy specimens were studied. All patients had undergone standardized surgery at the Department of Urology, University of Bern between 1989 and 2006 with bilateral extended pelvic lymphadenectomy and radical prostatectomy as a single procedure. Follow-up was performed prospectively. Neoadjuvant therapy was not implemented and no adjuvant treatment, especially androgen deprivation, was suggested before symptomatic disease progression.

4.2. Surgical Technique of Lymphadenectomy

A bilateral pelvic lymphadenectomy was performed in all patients as previously described [29]. Summarized, lymph nodes were dissected along the external iliac vein down to the deep circumflex iliac vein and femoral canal, up to the bifurcation of the common iliac artery and the obturator fossa. Thereafter, the lymphatic tissue along the medial and lateral aspect of the internal iliac artery and vein was excised. Three tissue samples from each side were submitted separately for pathological examination. Frozen sections were not carried out.

4.3. Pathology

All specimens were processed at the Institute of Pathology, University of Bern [24,30]. The prostatectomies were completely embedded as described in references [24,30]. The following microscopic tumor characteristics were noted: type, Gleason score [31], tertiary Gleason pattern, tumor stage, and the percentage of prostate tissue area on the sections occupied by the tumor. NE tumors/carcinomas of the prostate were excluded. Tumor volume was estimated by multiplying the percentage of the specimen involved by cancer by the prostate volume.

The fatty tissue of lymphadenectomy specimens was dissolved in aceton after formalin fixation and all lymph nodes were entirely embedded. One section per paraffin block was stained with hematoxylin and eosin. The length and width of the metastatic deposits were measured. A Gleason score (primary and secondary pattern) and a tertiary Gleason pattern (if present), were determined based on the entire metastatic tissue.

All Gleason patterns present in the primary tumors and lymph node metastases were accurately marked for subsequent TMA construction. Staging was completed according to the 8th edition of the International Union Against Cancer (UICC) TNM Classification [32].

4.4. Tissue Microarray

For TMA construction, one 0.6 mm tissue core of benign prostatic tissue (peripheral zone) and every Gleason pattern present in primary tumors and matched lymph node metastases was retrieved from the paraffin blocks. The TMA contains overall 403 prostate tissue samples, 119 normal prostate tissues and 284 primary cancers (mean per patient, 3.3; range, 2–4; including 101, 112 and 71 samples from Gleason patterns 3, 4 and 5, respectively) and 167 lymph node metastases (mean per patient, 1.4; range, 1–3; including 35, 103 and 29 samples from Gleason patterns 3, 4 and 5, respectively). In the vast majority of primary tumors, all Gleason patterns sampled were present in the index tumor. Additional tissue from separate tumor foci was included only rarely, when a Gleason pattern not present in the index tumor was detected here. Although sampling from the primary tumor was more extensive, the relative tumor amount in the TMA was larger from the metastases due to their smaller volume.

4.5. Immunohistochemistry

Freshly cut TMA sections were pre-treated by steam with target retrieval solution, pH 9 (Dako, Glostrup, Denmark). For Chromogranin A detection, a monoclonal mouse antibody cocktail (clone LK2H10 + PHE5; Bicarta; Hamburg, Germany) was used at 1:500 antibody dilution. Bound primary antibodies were detected using the Envision Plus system (Dako, Glostrup, Denmark). Chromogranin A was expressed in the cytoplasm of the prostate cancer cells (Figure 3). The percentage of positive neoplastic cells was determined for every tissue sample.

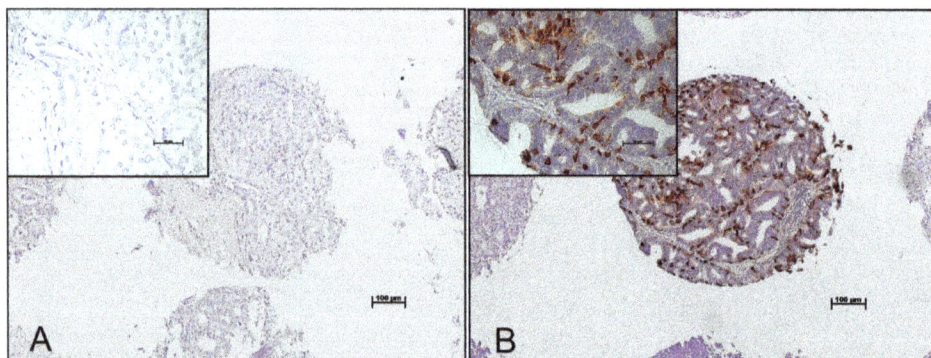

Figure 3. No Chromogranin A expression in (**A**) primary prostate cancer; and (**B**) high Chromogranin A expression in a lymph node metastasis.

4.6. Statistical Analysis

Chromogranin A expression in normal prostate, primary tumors and lymph node metastases was evaluated using the Wilcoxon Signed Rank test and the Friedman test for differences between Gleason pattern 3, 4 and 5 within primary carcinomas and nodal metastases. Chromogranin A expression was compared with normally distributed quantitative and categorical tumor attributes using Wilcoxon Signed Rank test and χ-Square test, respectively. Suitable cut-off values for positive (more than 0 positive cells) and negative (0 positive cells) Chromogranin A expression in primary tumors and lymph node metastases were defined using Receiver-operating characteristic curves [33]. Outcome was analyzed for Prostate-Specific Antigen (PSA) recurrence-free, cancer-specific and overall survival defined as the intervals from surgery to the date of biochemical recurrence (PSA failure defined as values >0.2 ng/mL), death from prostate carcinoma, and death from any cause, respectively. Patients without event for the respective endpoints were censored at the date of last follow-up. The above time-to-events were performed using log-rank test; p values < 0.05 were regarded as significant. The Cox proportional hazards model was used to identify independent prognostic factors for all three endpoints. Statistical analysis was made using SAS 9.2 (The SAS Institute, Cary, NC, USA).

5. Conclusions

Our data suggest that, firstly, increasing serum levels of neuroendocrine serum markers in prostate cancer primarily mirror growth of a poorly differentiated metastatic tumor component and, secondly, NED in early metastasizing, hormone-naïve prostate cancer is only weakly linked to adverse tumor features.

Acknowledgments: Project received funding from the Bernische Krebsliga and the Krebsliga Thurgau (Achim Fleischmann).

Author Contributions: Achim Fleischmann conceived and designed the experiments; Inti Zlobec performed the statistical analysis; Vera Genitsch, Roland Seiler, George N. Thalmann and Achim Fleischmann analyzed the data; Vera Genitsch and Achim Fleischmann wrote the paper; Roland Seiler and George N. Thalmann revised the manuscript for important intellectual content.

Conflicts of Interest: The authors declare no conflict of interest.

Abbreviations

NED	Neuroendocrine Differentiation
NE	Neuroendocrine
GP	Gleason Pattern
CgA	Chromogranin A
HR	Hazard Ratio
CI	Confidence Interval
TMA	Tissue Microarray
PSA	Prostate-Specific Antigen

References

1. Epstein, J.I.; Amin, M.B.; Evans, A.J.; Huang, J.; Rubin, M.A. Neuroendocrine tumours. In *WHO Classification of Tumours of the Urinary System and Male Genital Organs*; Moch, H., Humphrey, P.A., Ulbright, T.M., Reuter, V.E., Eds.; IARC: Lyon, France, 2016.
2. Bonkhoff, H.; Stein, U.; Remberger, K. Multidirectional differentiation in the normal, hyperplastic, and neoplastic human prostate: Simultaneous demonstration of cell-specific epithelial markers. *Hum. Pathol.* **1994**, *25*, 42–46. [CrossRef]
3. Yuan, T.C.; Veeramani, S.; Lin, M.F. Neuroendocrine-like prostate cancer cells: Neuroendocrine transdifferentiation of prostate adenocarcinoma cells. *Endocr. Relat. Cancer* **2007**, *14*, 531–547.
4. Priemer, D.S.; Montironi, R.; Wang, L.; Williamson, S.R.; Lopez-Beltran, A.; Cheng, L. Neuroendocrine tumors of the prostate: Emerging insights from molecular data and updates to the 2016 world health organization classification. *Endocr. Pathol.* **2016**, *27*, 123–135. [CrossRef] [PubMed]
5. Sciarra, A.; Cardi, A.; Dattilo, C.; Mariotti, G.; di Monaco, F.; di Silverio, F. New perspective in the management of neuroendocrine differentiation in prostate adenocarcinoma. *Int. J. Clin. Pract.* **2006**, *60*, 462–470. [CrossRef] [PubMed]
6. Komiya, A.; Suzuki, H.; Imamoto, T.; Kamiya, N.; Nihei, N.; Naya, Y.; Ichikawa, T.; Fuse, H. Neuroendocrine differentiation in the progression of prostate cancer. *Int. J. Urol.* **2008**, *16*, 37–44. [CrossRef] [PubMed]
7. Grigore, A.D.; Ben-Jacob, E.; Farach-Carson, M.C. Prostate cancer and neuroendocrine differentiation: More neuronal, less endocrine? *Front. Oncol.* **2015**, *5*, 37. [CrossRef] [PubMed]
8. Wang, H.T.; Yao, Y.H.; Li, B.G.; Tang, Y.; Chang, J.W.; Zhang, J. Neuroendocrine prostate cancer (NEPC) progressing from conventional prostatic adenocarcinoma: Factors associated with time to development of NEPC and survival from NEPC diagnosis—A systematic review and pooled analysis. *J. Clin. Oncol.* **2014**, *32*, 3383–3390. [PubMed]
9. Epstein, J.I.; Amin, M.B.; Beltran, H.; Lotan, T.L.; Mosquera, J.-M.; Reuter, V.E.; Robinson, B.D.; Troncoso, P.; Rubin, M.A. Proposed morphologic classification of prostate cancer with neuroendocrine differentiation. *Am. J. Surg. Pathol.* **2014**, *38*, 756–767. [CrossRef] [PubMed]
10. Berruti, A.; Vignani, F.; Russo, L.; Bertaglia, V.; Tullio, M.; Tucci, M.; Poggio, M.; Dogliotti, L. Prognostic role of neuroendocrine differentiation in prostate cancer, putting together the pieces of the puzzle. *Open Access J. Urol.* **2010**, *2*, 109–124. [CrossRef] [PubMed]
11. Appetecchia, M.; Meçule, A.; Pasimeni, G.; Iannucci, C.V.; de Carli, P.; Baldelli, R.; Barnabei, A.; Cigliana, G.; Sperduti, I.; Gallucci, M. Incidence of high Chromogranin A serum levels in patients with non metastatic prostate adenocarcinoma. *J. Exp. Clin. Cancer Res.* **2010**, *29*, 166. [CrossRef] [PubMed]
12. Burgio, S.L.; Conteduca, V.; Menna, C.; Carretta, E.; Rossi, L.; Bianchi, E.; Kopf, B.; Fabbri, F.; Amadori, D.; de Giorgi, U. Chromogranin A predicts outcome in prostate cancer patients treated with abiraterone. *Endocr. Relat. Cancer* **2014**, *21*, 487–493. [CrossRef] [PubMed]
13. Conteduca, V.; Burgio, S.L.; Menna, C.; Carretta, E.; Rossi, L.; Bianchi, E.; Masini, C.; Amadori, D.; de Giorgi, U. Chromogranin A is a potential prognostic marker in prostate cancer patients treated with enzalutamide. *Prostate* **2014**, *74*, 1691–1696. [CrossRef] [PubMed]

14. Sciarra, A.; di Silverio, F.; Autran, A.M.; Salciccia, S.; Gentilucci, A.; Alfarone, A.; Gentile, V. Distribution of high Chromogranin A serum levels in patients with nonmetastatic and metastatic prostate adenocarcinoma. *Urol. Int.* **2009**, *82*, 147–151. [CrossRef] [PubMed]

15. Aprikian, A.G.; Cordon-Cardo, C.; Fair, W.R.; Reuter, V.E. Characterization of neuroendocrine differentiation in human benign prostate and prostatic adenocarcinoma. *Cancer* **1993**, *71*, 3952–3965. [CrossRef]

16. Aprikian, A.G.; Cordon-Cardo, C.; Fair, W.R.; Zhang, Z.F.; Bazinet, M.; Hamdy, S.M.; Reuter, V.E. Neuroendocrine differentiation in metastatic prostatic adenocarcinoma. *J. Urol.* **1994**, *151*, 914–919.

17. Bostwick, D.G.; Qian, J.; Pacelli, A.; Zincke, H.; Blute, M.; Bergstralh, E.J.; Slezak, J.M.; Cheng, L. Neuroendocrine expression in node positive prostate cancer: Correlation with systemic progression and patient survival. *J. Urol.* **2002**, *168*, 1204–1211. [CrossRef]

18. Cheville, J.C.; Tindall, D.; Boelter, C.; Jenkins, R.; Lohse, C.M.; Pankratz, V.S.; Sebo, T.J.; Davis, B.; Blute, M.L. Metastatic prostate carcinoma to bone. *Cancer* **2002**, *95*, 1028–1036. [CrossRef] [PubMed]

19. Quek, M.L.; Daneshmand, S.; Rodrigo, S.; Cai, J.; Dorff, T.B.; Groshen, S.; Skinner, D.G.; Lieskovsky, G.; Pinski, J. Prognostic significance of neuroendocrine expression in lymph node-positive prostate cancer. *Urology* **2006**, *67*, 1247–1252. [CrossRef] [PubMed]

20. Roudier, M.P.; True, L.D.; Higano, C.S.; Vesselle, H.; Ellis, W.; Lange, P.; Vessella, R.L. Phenotypic heterogeneity of end-stage prostate carcinoma metastatic to bone. *Hum. Pathol.* **2003**, *34*, 646–653. [CrossRef]

21. McWilliam, L.J.; Manson, C.; George, N.J. Neuroendocrine differentiation and prognosis in prostatic adenocarcinoma. *Br. J. Urol.* **1997**, *80*, 287–290. [CrossRef] [PubMed]

22. Pruneri, G.; Galli, S.; Rossi, R.S.; Roncalli, M.; Coggi, G.; Ferrari, A.; Simonato, A.; Siccardi, A.G.; Carboni, N.; Buffa, R. Chromogranin A and B and secretogranin II in prostatic adenocarcinomas: Neuroendocrine expression in patients untreated and treated with androgen deprivation therapy. *Prostate* **1998**, *34*, 113–120.

23. Angelsen, A.; Syversen, U.; Haugen, O.A.; Stridsberg, M.; Mjølnerød, O.K.; Waldum, H.L. Neuroendocrine differentiation in carcinomas of the prostate: Do neuroendocrine serum markers reflect immunohistochemical findings? *Prostate* **1997**, *30*, 1–6. [CrossRef]

24. Fleischmann, A.; Schobinger, S.; Schumacher, M.; Thalmann, G.N.; Studer, U.E. Survival in surgically treated, nodal positive prostate cancer patients is predicted by histopathological characteristics of the primary tumor and its lymph node metastases. *Prostate* **2009**, *69*, 352–362. [CrossRef] [PubMed]

25. Brawn, P.N.; Speights, V.O. The dedifferentiation of metastatic prostate carcinoma. *Br. J. Cancer* **1989**, *59*, 85–88. [CrossRef] [PubMed]

26. Cheng, L.; Slezak, J.; Bergstralh, E.J.; Cheville, J.C.; Sweat, S.; Zincke, H.; Bostwick, D.G. Dedifferentiation in the metastatic progression of prostate carcinoma. *Cancer* **1999**, *86*, 657–663. [CrossRef]

27. Abrahamsson, P.A. Neuroendocrine differentiation in prostatic carcinoma. *Prostate* **1999**, *39*, 135–148.

28. Mucci, N.R.; Akdas, G.; Manely, S.; Rubin, M.A. Neuroendocrine expression in metastatic prostate cancer: Evaluation of high throughput tissue microarrays to detect heterogeneous protein expression. *Hum. Pathol.* **2000**, *31*, 406–414. [CrossRef] [PubMed]

29. Bader, P.; Burkhard, F.C.; Markwalder, R.; Studer, U.E. Disease progression and survival of patients with positive lymph nodes after radical prostatectomy. Is there a chance of cure? *J. Urol.* **2003**, *169*, 849–854.

30. Fleischmann, A.; Schobinger, S.; Markwalder, R.; Schumacher, M.; Burkhard, F.; Thalmann, G.N.; Studer, U.E. Prognostic factors in lymph node metastases of prostatic cancer patients: The size of the metastases but not extranodal extension independently predicts survival. *Histopathology* **2008**, *53*, 468–475. [CrossRef] [PubMed]

31. Epstein, J.I.; Helpap, B.; Algaba, F.; Humphrey, P.A.; Allsbrook, W.C.; Iczkowski, K.A.; Bastacky, S.; Lopez-Beltran, A.; Boccon-Gibod, L.; Montironi, R.; et al. Acinar adenocarcinoma. In *Pathology and Genetics of Tumours of the Urinary System and Male Genital Organs*; Eble, J.N., Sauter, G., Epstein, J.I., Sesterhenn, I.A., Eds.; IARC Press: Lyon, France, 2004.

32. Green, F.L. *TNM Classification of Malignant Tumors*; Bierley, J., Gospodarowicz, M.K., Wittekind, C., Eds.; Wiley-Blackwell: Oxford, UK, 2017.

33. Søreide, K. Receiver-operating characteristic curve analysis in diagnostic, prognostic and predictive biomarker research. *J. Clin. Pathol.* **2008**, *62*, 1–5. [CrossRef] [PubMed]

International Journal of
Molecular Sciences

MDPI

Article

Sensitivity of HOXB13 as a Diagnostic Immunohistochemical Marker of Prostatic Origin in Prostate Cancer Metastases: Comparison to PSA, Prostein, Androgen Receptor, *ERG*, *NKX3.1*, PSAP, and PSMA

Ilka Kristiansen [1], **Carsten Stephan** [2], **Klaus Jung** [3], **Manfred Dietel** [4], **Anja Rieger** [4], **Yuri Tolkach** [1,†] **and Glen Kristiansen** [1,*,†]

[1] Institute of Pathology, University Hospital Bonn, 53127 Bonn, Germany; ilka.kristiansen@gmx.de (I.K.); Yuri.tolkach@ukbonn.de (Y.T.)
[2] Department of Urology, Charité-Universitätsmedizin Berlin, 10117 Berlin, Germany; carsten.stephan@charite.de
[3] Berlin Institute of Urologic Research, 10117 Berlin, Germany; klaus.jung@charite.de
[4] Institute of Pathology, Charité-Universitätsmedizin Berlin, 10117 Berlin, Germany; Manfred.dietel@charite.de (M.D.); Anja.rieger@charite.de (A.R.)
* Correspondence: glen.kristiansen@ukbonn.de; Tel.: +49-228-287-15375, Fax: +49-228-287-15030
† These authors contributed equally to this work.

Academic Editor: William Chi-shing Cho
Received: 16 April 2017; Accepted: 24 May 2017; Published: 29 May 2017

Abstract: Aims: Determining the origin of metastases is an important task of pathologists to allow for the initiation of a tumor-specific therapy. Recently, homeobox protein Hox-B13 (HOXB13) has been suggested as a new marker for the detection of prostatic origin. The aim of this study was to evaluate the diagnostic sensitivity of HOXB13 in comparison to commonly used immunohistochemical markers for prostate cancer. Materials and methods: Histologically confirmed prostate cancer lymph node metastases from 64 cases were used to test the diagnostic value of immunohistochemical markers: prostate specific antigen (PSA), Prostatic acid phosphatase (PSAP), prostate specific membrane antigen (PSMA), homeobox gene *NKX3.1*, prostein, androgen receptor (AR), HOXB13, and ETS-related gene (ERG). All markers were evaluated semi-quantitatively using Remmele's immune reactive score. Results: The detection rate of prostate origin of metastasis for single markers was 100% for NKX3.1, 98.1% for AR, 84.3% for PSMA, 80.8% for PSA, 66% for PSAP, 60.4% for HOXB13, 59.6% for prostein, and 50.0% for ERG. Conclusions: Our data suggest that HOXB13 on its own lacks sensitivity for the detection of prostatic origin. Therefore, this marker should be only used in conjunction with other markers, preferably the highly specific PSA. The combination of PSA with NKX3.1 shows a higher sensitivity and thus appears preferable in this setting.

Keywords: prostate cancer; metastasis; immunohistochemistry; detection; PSA; PSAP; PSMA; NKX3.1; prostein; HOXB13; ERG; AR

1. Introduction

Determining the origin of the primary tumor in metastases is an important task of pathologists to allow for the initiation of a tumor-specific therapy. Since its description in the 1970s, prostate specific antigen (PSA) has in due course become the dominant prostate marker in serum [1,2]. PSA could be detected by immunohistochemistry in tissue and was quickly adopted by pathologists as a diagnostic marker [3,4]. PSA is still regarded as a highly specific marker of prostatic origin, but due to its decreased or even lost expression in higher grade or metastatic tumors, which was already noted in initial studies, its sensitivity is clearly limited, which necessitates the use of additional markers in PSA negative cases [5]. Prostein, coded by the *SLC45A3* gene and identified by transcript profiling studies of prostate cancer, was found to be highly specific for prostatic origin and displays a characteristic Golgi-type cytoplasmic staining pattern [6,7]. As seen with PSA, prostein expression also shows a slightly diminished expression with tumor progression, but in combination with PSA it proved to be a valuable tool to identify prostatic origin [8,9]. An alternative marker for prostatic origin was the protein coded by the homeobox gene *NKX3.1*, which shows nuclear staining on immunohistochemistry and which was also found to be highly though not exclusively specific for prostate tissues [10,11]. At the 2014 consensus conference of the International Society of Urological Pathology (ISUP), it was recommended to use immunohistochemistry for PSA, NKX3.1 and prostein to ascertain a prostatic origin in doubtful cases [5].

Another diagnostic candidate marker that was not considered at that conference was homeobox protein Hox-B13 (HOXB13), which was characterized as a marker for prostate cancer by Edwards et al. [12]. HOXB13 also has a nuclear staining pattern and has been recommended by several studies as a prostate specific marker [13–16]. To date, the diagnostic value and especially the performance of HOXB13 immunohistochemistry as a marker of prostatic origin in direct comparison to the commonly used prostate markers (PSA, prostein, ERG, the androgen receptor (AR), NKX3.1, Prostatic acid phosphatase (PSAP) and prostate specific membrane antigen (PSMA)) in metastases has not been evaluated, which was therefore the aim of this study.

2. Results

2.1. Immunohistochemical Staining Patterns

All markers showed the expected patterns of immunoreactivity in metastatic prostate cancer. Prostate specific antigen (PSA) and prostatic phosphatase (PSAP) displayed diffuse cytoplasmic staining (Figure 1A,C), where prostein showed the characteristic Golgi-type staining (Figure 1B). Prostate specific membrane antigen (PSMA, Figure 1D) showed a predominantly membranous but also cytoplasmic immunoreactivity. Nuclear staining was seen for the androgen receptor (AR), ERG, NKX3.1, and HOXB13 (Figure 1E–H).

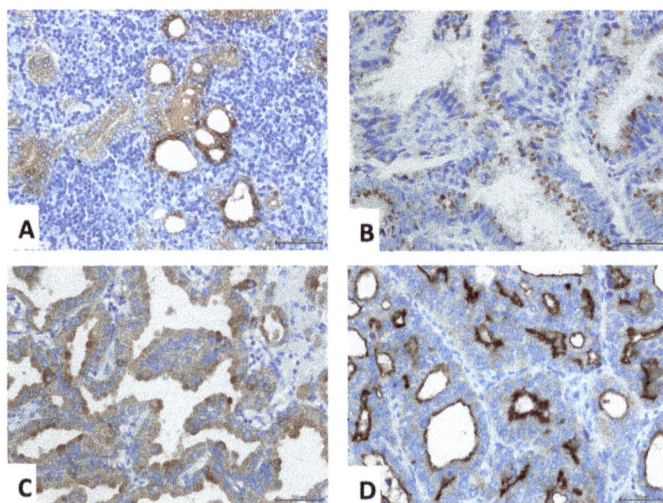

Figure 1. *Cont.*

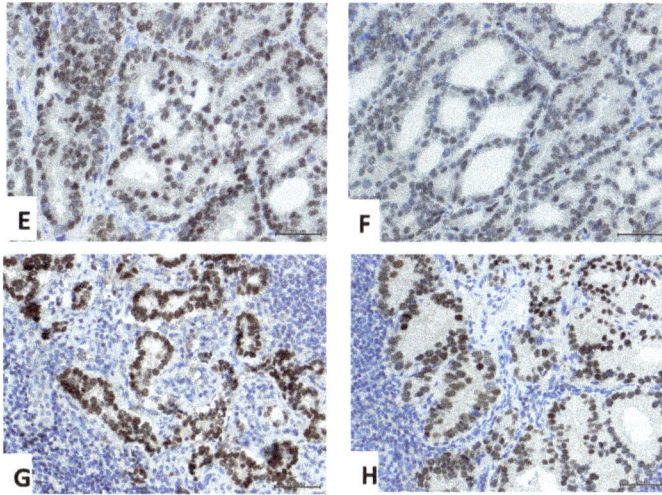

Figure 1. Immunohistochemistry of candidate diagnostic prostate markers. (**A**) Prostate specific antigen (PSA) shows diffuse cytoplasmic staining with a higher degree of heterogeneity in this case. (**B**) Prostein immunoreactivity is restricted to the Golgi apparatus area. (**C**) Prostate specific alkaline phosphatase (PSAP) also displays a diffuse cytoplasmic staining pattern. (**D**) Prostate specific membrane antigen (PSMA) has characteristic cytoplasmic but also membranous staining. (**E**) Androgen receptor shows epithelial nuclear staining. (**F**) ERG is also located in the nucleus and has a higher degree of heterogeneity, as seen here. (**G**) Homeobox gene *NKX3.1* commonly displays strong nuclear staining. (**H**) Homeobox protein HOXB3 is also seen in nuclei, however, the staining intensity is weaker. All images are taken in 400×.

2.2. Statistical Evaluation of Prostate Markers

The highest detection rate was seen with NKX3.1 followed by the androgen receptor and PSMA. PSA, as the most commonly used antibody to specifically recognize prostate cancer, correctly detected nearly 81% of cases. ERG labeled, as expected, only 50% of prostate cancer metastases. Prostein, PSAP, and HOXB13 were only slightly superior, detecting approximately two thirds of cases (Table 1).

Table 1. Detection rates of prostate markers in prostate cancer metastases.

Marker	Detection Rate (%)	Mean IRS	Number of Cases
PSA	80.8	6.3	52
PSAP	66.0	3.5	53
PSMA	84.3	6.0	51
Prostein	59.6	4.2	52
Androgen receptor (AR)	98.1	6.7	53
ERG	50.0	2.6	52
NKX3.1	100.0	8.0	50
HOXB13	60.4	4.7	53

Abbreviations: IRS = immunoreactive Score according to Remmele.

A Spearman rank correlation analysis of IRS values of these markers revealed the following significant associations: AR and HOXB13 (correlation coefficient (CC) = 0.358, p = 0.009), AR and NKX3.1 (CC = 0.505, p = 0.001), AR and PSAP (CC = -0.277, p = 0.047), ERG and HOXB13 (CC = 0.283, p = 0.044), PSA and Prostein (CC = 0.367, p = 0.008), PSA and PSAP (CC = 0.623, p = 0.001), PSA and PSMA (CC = 0.350, p = 0.012), prostein and PSAP (CC = 0.277, p = 0.049), and prostein and HOXB13 (CC = -0.296, p = 0.035).

We then investigated which combination of prostate markers would achieve the highest detection rate by combining the most commonly used marker PSA with the other markers in cross tables. PSAP recognized 10% of PSA negative cases, which did not add significant information. PSMA performed slightly better, labelling 50% of PSA negative cases. HOXB13 correctly recognized 70%, and finally *NKX3.1* correctly detected all PSA negative cases, leading to a detection rate (of the combination of PSA and *NKX3.1*) of 100% in this dataset. Prostein and PSA correctly detected 82% of cases, HOXB13 and PSA labeled 94.2% of cases correctly, AR and PSA detected 100% of cases, and ERG and PSA recognized 88% of prostate cancer metastases.

In a direct analysis of HOXB13 and NKX3.1 expression, both markers together recognized 100%, as NKX3.1 labeled 100% of HOXB13 negative cases. NKX3.1 combined with AR also achieved a detection rate of 100%.

3. Material and Methods

3.1. Case Selection and Construction of Tissue Microarray

First, prostate cancer patients who underwent radical prostatectomy (RP) and lymphadenectomy and who were diagnosed at the Institute of Pathology, Charité–Universitätsmedizin Berlin between 1998 and 2005 were identified. Histologically confirmed metastases from 64 cases were used for tissue microarray construction with one spot per case and a punch size of 1 mm. The tissue microarray comprised of two blocks. The metastases consisted of 57 local lymph node metastases and seven systemic metastases ($4\times$ bone, $1\times$ penis, $1\times$ oral mucosa, $1\times$ cervical lymph node). Lymph nodes were submitted to frozen section analysis prior to RP, which was usually not performed in case of a positive node. Only 13 RPs were performed. Therefore, histopathological RP data on the primary tumor was not available for the majority of cases. For the thirteen RP cases that were completed, the Gleason scores (GS) were: GS7-2, GS8-4, GS9-4, GS10-3. The pT categories (according to Union for International Cancer Control (UICC) TNM classification of malignant tumors, 8th edition) were: pT2-2, pT3a-1, pT3b-10. Positive margins were seen in 11 cases. The bilateral lymphonodectomy specimens ($n = 57$) contained a median of 12 lymph nodes (range 1 to 23). Twenty-nine cases had a single positive lymph node, 15 cases had two metastases, seven cases had three metastases and six cases had four or more positive nodes (maximum 11). The use of this tissue was approved by the Charité University Ethics Committee (EA1/06/2004). No animals were involved in this study.

3.2. Immunohistochemistry

Immunohistochemistry was conducted using semi-automated platforms (Benchmark Ultra, Roche, and Autostainer, Medac) with protocols that are in routine use at the Institute of Pathology, University of Bonn (Table 2).

Table 2. Antibodies used for immunohistochemistry in this study.

Antigen	Clone	Provider	Dilution	Platform	Protocol
PSA	Polyclonal, rabbit	DAKO	1:20,000	Autostainer	No pretreatment
PSAP	PASE/4LJ	Cell Marque	1:6000	Autostainer	HIER (pH 6, 20 s 98 °C)
PSMA	3E6	DAKO	1:500	Benchmark	CC1 (pH 8), ultraview
Prostein	10E3	DAKO	1:100	Benchmark	CC1 (pH 8), ultraview
Androgen Receptor (AR)	AR441	DAKO	1:400	Autostainer	HIER (pH 6, 20 s 98 °C)
ERG	EPR3864	Biologo	1:100	Autostainer	HIER (pH 6, 20 s 98 °C)
NKX3.1	EP356	Cell Marque	1:200	Benchmark	CC1 (pH 8), ultraview
HOXB13	F-9	Santa Cruz	1:50	Benchmark	CC1 (pH 8), optiview

Abbreviations: HIER-Heat Induced Epitope Retrieval.

3.3. Evaluation of Immunohistochemistry

All markers were evaluated semi-quantitatively using the immune reactive score (IRS), which gives the product of categorized percentage of stained cells (0: negative, 1: 1–10%, 2: 11–50%, 3:

51–80%, and 4: >80%) and staining intensity (ranging from negative (0) to strong (3)), and hence has a range from 0 to 12. For statistical evaluation, an IRS below 3 was considered a negative test and higher values were considered as positive, as suggested by Remmele and Stegner in their original proposal of the IRS scoring system [17]. The evaluation was carried out under the immediate supervision of a Genito–Urinary pathologist with broad expertise in immunohistochemistry (Glen Kristiansen).

3.4. Statistics

All data were processed using SPSS (IBM SPSS Statistics for Macintosh, Version 22.0, IBM Corp., Armonk, NY, USA). Spearman Rank correlations were used to analyze the associations among markers.

4. Discussion

This study evaluates the sensitivity of prostate markers to detect lymph node and distant metastases of prostate cancer. It confirms that PSA, as the most commonly used antibody for routine purposes, has a relatively high detection rate of nearly 81%. The value of PSA is its high specificity, as apart from prostate cancer only breast cancer may be PSA positive, but this is usually not in the closer differential diagnosis [18]. The loss of PSA expression with tumor dedifferentiation and progression is long known and constitutes the necessity to use other markers in a combination.

As prostate cancer is an androgen-driven disease, the high rates of androgen receptor expression found in primary and metastatic prostate tumors are not surprising, and again this study confirms that metastases retain their AR expression well. The disadvantage of AR is the relative lack of specificity, as other neoplasms including urothelial carcinoma or salivary gland tumors may also be AR positive [19,20].

The overexpression of ERG, often resulting from the Transmembrane protease, serine 2-erythroblast transformation-specific-related gene TMPRSS2-ERG translocation which is found in nearly half of primary prostate cancer cases, is therefore typical of prostate cancer, but with a detection rate of 50% it clearly lacks sensitivity [21]. It is less well known that even though the genomic translocation is highly specific for prostate cancer, ERG protein expression is seen in a variety of other tumors, too. This includes, apart from vascular tumors, round cell sarcomas and leukemias, the more common differential diagnosis of urothelial carcinoma and, according to the human protein atlas (hppt://www.proteinatlas.org), also melanoma, testicular, and gastrointestinal tumors [22–24].

Prostein was described and characterized by Xu et al. as a prostate specific protein, which was quickly used by surgical pathologists for the differential diagnosis of prostate cancer, especially to rule out urothelial carcinoma, which is almost consistently prostein-negative [6,7,9,25,26]. Prostein also has the advantage of a distinctive Golgi-type staining pattern, which can be reassuring in cases with only weak positivity. In primary prostate cancer, prostein expression is inversely correlated with Gleason scores and is a prognostic marker of disease progression [27]. As our study confirms, the sensitivity of prostein in metastases is fairly limited with a detection rate of 59%, and even in conjunction with PSA only 82% of cases can be confirmed as prostatic in origin, which equals the detection rate of PSA alone in our cohort. As we found that both markers correlated, this redundancy of diagnostic information is not surprising. The same holds true for PSAP, which we found strongly and highly positively correlated to PSA, and this did not add significant information as only 10% of PSA negative cases were picked up by PSAP. A larger cohort may be necessary to demonstrate the additional diagnostic value of prostein or PSAP to PSA.

Prostate specific membrane antigen (PSMA) is overexpressed during tumor progression, and Bostwick et al. found 82% positivity rates in primary prostate cancer [28]. However, despite its name it is not prostate specific at all, but is also seen in a wide variety of tumors including colon, bladder, and renal cancer, so its use as a marker for prostatic differentiation is now discouraged [29].

NKX3.1 is a homeobox gene that shows a prostate and testis specific expression pattern, but may also be found in breast cancer. Its crisp nuclear staining pattern and its high rate of positivity in metastases (98.6%, Gurel et al. [30]), which this study confirms, have made it a popular diagnostic marker that is already endorsed by ISUP in their recommendations on diagnostic markers for genito–urinary pathology [11,30–33]. In particular, its combination with PSA is promising, as the high detection rate of 100% found in this study witnesses.

HOXB13 is another, so far less acknowledged diagnostic marker candidate, which has been recommended lately as a prostate specific marker [12,13,15,16]. Barresi et al. analyzed 15 prostate cancer metastases and found all of them strongly positive (in >75% of tumor cells) which equates a sensitivity of 100% [15]. Minimally lower rates were reported by Varinot et al., who analyzed 74 cases of lymph node metastases and 15 additional bone metastases. They found HOXB13 positivity in 33% of bone metastases and 93% of lymph node metastases. Interestingly, the expression of HOXB13 was also found as an independent prognostic marker in primary prostate cancer and to correlate with AR expression [14]. While our study confirms the significant association with AR, the diagnostic value of HOBX13 appears less convincing, as only 60.4% of prostate cancer metastases were positively stained. Of course, besides aspects of cohort composition that may influence the tumor biology, technical issues of the immunohistochemistry protocol or the tissue micro array construction may also explain this rather significant discrepancy. This detection rate increased markedly to 94.2% if combined with PSA, but it remains inferior to the combination of PSA with NKX3.1. HOXB13 is also expressed in other carcinomas including endometrial cancer, which in itself does not limit its diagnostic value, but also in pancreatic cancer and hepatocellular carcinoma [34–36]. In light of these data, we do not recommend HOXB13 as a sole marker to detect a prostatic origin in a metastasis, but rather prefer PSA in conjunction with NKX3.1. However, prostein or HOXB13 may well be considered as third line markers in doubtful cases, and here combinations are advisable and often necessary [5,37].

This study has several weaknesses. The number of cases is relatively small, which precludes a more detailed analysis of combinational subgroups. Still, the cohort size is large enough to confirm the known expression rates of the markers under question and it is the first study to critically analyze HOXB13 in direct comparison to these other markers in prostate cancer metastases. This study is restricted to the correct detection of prostate markers in known prostate metastases and hence only evaluates the sensitivity of these markers. Also, this study lacks data on the Gleason scores of biopsies of primary tumors, which precludes further correlation analyses of biomarker expression. Finally, we did not aim to verify the specificity of our candidate biomarkers, which would have necessitated an additional large analysis of non-prostatic neoplasms [38].

In summary, our data suggest that the novel marker candidate HOXB13 alone is not a good diagnostic marker for the detection of prostatic origin. Only in combination with PSA does it achieve satisfactory detection rates. Alternatively, the combination of PSA with the already well-established marker NKX3.1 shows an even higher sensitivity and is therefore recommended in this setting.

Acknowledgments: We are greatly indebted to the Sonnenfeldstiftung, Berlin, who funded the tissue microarrayer. We thank Britta Beyer for excellent technical support and Alfred E. Neumann for his enduring sense of humour and constructive discussions.

Author Contributions: Ilka Kristiansen collected data, prepared the statistics and wrote the paper; Carsten Stephan and Klaus Jung provided clinical information, supervised the statistics and revised the paper; Manfred Dietel and Anja Rieger provided tissues and clinico-pathological data, constructed the tissue micro array (TMA) and revised the paper; Yuri Tolkach and Glen Kristiansen conceived the study, conducted the central review of tissues on the TMA, performed the statistics and supervised writing of the paper and its revision.

Conflicts of Interest: The authors declare no conflict of interest.

References

1. Ablin, R.J.; Bronson, P.; Soanes, W.A.; Witebsky, E. Tissue- and species-specific antigens of normal human prostatic tissue. *J. Immunol.* **1970**, *104*, 1329–1339. [PubMed]
2. Frankel, A.E.; Rouse, R.V.; Wang, M.C.; Chu, T.M.; Herzenberg, L.A. Monoclonal antibodies to a human prostate antigen. *Cancer Res.* **1982**, *42*, 3714–3718. [PubMed]
3. Steffens, J.; Friedmann, W.; Lobeck, H. Immunohistochemical diagnosis of the metastasizing prostatic carcinoma. *Eur. Urol.* **1985**, *11*, 91–94. [PubMed]

4. Stein, B.S.; Vangore, S.; Petersen, R.O.; Kendall, A.R. Immunoperoxidase localization of prostate-specific antigen. *Am. J. Surg. Pathol.* **1982**, *6*, 553–557. [CrossRef] [PubMed]
5. Epstein, J.I.; Egevad, L.; Humphrey, P.A.; Montironi, R.; Members of the ISUP Immunohistochemistry in Diagnostic Urologic Pathology Group. Best practices recommendations in the application of immunohistochemistry in the prostate: Report from the international society of urologic pathology consensus conference. *Am. J. Surg. Pathol.* **2014**, *38*, e6–e19. [PubMed]
6. Kalos, M.; Askaa, J.; Hylander, B.L.; Repasky, E.A.; Cai, F.; Vedvick, T.; Reed, S.G.; Wright, G.L., Jr.; Fanger, G.R. Prostein expression is highly restricted to normal and malignant prostate tissues. *Prostate* **2004**, *60*, 246–256. [CrossRef] [PubMed]
7. Xu, J.; Kalos, M.; Stolk, J.A.; Zasloff, E.J.; Zhang, X.; Houghton, R.L.; Filho, A.M.; Nolasco, M.; Badaro, R.; Reed, S.G. Identification and characterization of prostein, a novel prostate-specific protein. *Cancer Res.* **2001**, *61*, 1563–1568. [PubMed]
8. Yin, M.; Dhir, R.; Parwani, A.V. Diagnostic utility of p501s (prostein) in comparison to prostate specific antigen (PSA) for the detection of metastatic prostatic adenocarcinoma. *Diagn. Pathol.* **2007**, *2*, 41. [CrossRef] [PubMed]
9. Sheridan, T.; Herawi, M.; Epstein, J.I.; Illei, P.B. The role of p501s and PSA in the diagnosis of metastatic adenocarcinoma of the prostate. *Am. J. Surg. Pathol.* **2007**, *31*, 1351–1355. [CrossRef] [PubMed]
10. He, W.W.; Sciavolino, P.J.; Wing, J.; Augustus, M.; Hudson, P.; Meissner, P.S.; Curtis, R.T.; Shell, B.K.; Bostwick, D.G.; Tindall, D.J.; et al. A novel human prostate-specific, androgen-regulated homeobox gene (*NKX3.1*) that maps to 8p21, a region frequently deleted in prostate cancer. *Genomics* **1997**, *43*, 69–77. [CrossRef] [PubMed]
11. Bowen, C.; Bubendorf, L.; Voeller, H.J.; Slack, R.; Willi, N.; Sauter, G.; Gasser, T.C.; Koivisto, P.; Lack, E.E.; Kononen, J.; et al. Loss of NKX3.1 expression in human prostate cancers correlates with tumor progression. *Cancer Res.* **2000**, *60*, 6111–6115. [PubMed]
12. Edwards, S.; Campbell, C.; Flohr, P.; Shipley, J.; Giddings, I.; Te-Poele, R.; Dodson, A.; Foster, C.; Clark, J.; Jhavar, S.; et al. Expression analysis onto microarrays of randomly selected cDNA clones highlights HOXB13 as a marker of human prostate cancer. *Br. J. Cancer* **2005**, *92*, 376–381. [CrossRef] [PubMed]
13. Varinot, J.; Cussenot, O.; Roupret, M.; Conort, P.; Bitker, M.O.; Chartier-Kastler, E.; Cheng, L.; Comperat, E. HOXB13 is a sensitive and specific marker of prostate cells, useful in distinguishing between carcinomas of prostatic and urothelial origin. *Virchows Arch.* **2013**, *463*, 803–809. [CrossRef] [PubMed]
14. Zabalza, C.V.; Adam, M.; Burdelski, C.; Wilczak, W.; Wittmer, C.; Kraft, S.; Krech, T.; Steurer, S.; Koop, C.; Hube-Magg, C.; et al. HOXB13 overexpression is an independent predictor of early PSA recurrence in prostate cancer treated by radical prostatectomy. *Oncotarget* **2015**, *6*, 12822–12834. [CrossRef] [PubMed]
15. Barresi, V.; Ieni, A.; Cardia, R.; Licata, L.; Vitarelli, E.; Reggiani Bonetti, L.; Tuccari, G. HOXB13 as an immunohistochemical marker of prostatic origin in metastatic tumors. *APMIS* **2016**, *124*, 188–193. [CrossRef] [PubMed]
16. Varinot, J.; Furudoi, A.; Drouin, S.; Phe, V.; Penna, R.R.; Roupret, M.; Bitker, M.O.; Cussenot, O.; Comperat, E. HOXB13 protein expression in metastatic lesions is a promising marker for prostate origin. *Virchows Arch.* **2016**, *468*, 619–622. [CrossRef] [PubMed]
17. Remmele, W.; Stegner, H.E. Recommendation for uniform definition of an immunoreactive score (IRS) for immunohistochemical estrogen receptor detection (ER-ICA) in breast cancer tissue. *Der. Pathol.* **1987**, *8*, 138–140.
18. Howarth, D.J.; Aronson, I.B.; Diamandis, E.P. Immunohistochemical localization of prostate-specific antigen in benign and malignant breast tissues. *Br. J. Cancer* **1997**, *75*, 1646–1651. [CrossRef] [PubMed]
19. Ide, H.; Inoue, S.; Miyamoto, H. Histopathological and prognostic significance of the expression of sex hormone receptors in bladder cancer: A meta-analysis of immunohistochemical studies. *PLoS ONE* **2017**, *12*, e0174746. [CrossRef] [PubMed]
20. Udager, A.M.; Chiosea, S.I. Salivary duct carcinoma: An update on morphologic mimics and diagnostic use of androgen receptor immunohistochemistry. *Head Neck Pathol.* **2017**. [CrossRef] [PubMed]
21. van Leenders, G.J.; Boormans, J.L.; Vissers, C.J.; Hoogland, A.M.; Bressers, A.A.; Furusato, B.; Trapman, J. Antibody EPR3864 is specific for *ERG* genomic fusions in prostate cancer: Implications for pathological practice. *Mod. Pathol.* **2011**, *24*, 1128–1138. [CrossRef] [PubMed]

22. Miettinen, M.; Wang, Z.F.; Paetau, A.; Tan, S.H.; Dobi, A.; Srivastava, S.; Sesterhenn, I. ERG transcription factor as an immunohistochemical marker for vascular endothelial tumors and prostatic carcinoma. *Am. J. Surg. Pathol.* **2011**, *35*, 432–441. [CrossRef] [PubMed]

23. Yamada, Y.; Kuda, M.; Kohashi, K.; Yamamoto, H.; Takemoto, J.; Ishii, T.; Iura, K.; Maekawa, A.; Bekki, H.; Ito, T.; et al. Histological and immunohistochemical characteristics of undifferentiated small round cell sarcomas associated with *CIC-DUX4* and *BCOR-CCNB3* fusion genes. *Virchows Arch.* **2017**, *470*, 373–380. [CrossRef] [PubMed]

24. Xu, B.; Naughton, D.; Busam, K.; Pulitzer, M. ERG is a useful immunohistochemical marker to distinguish leukemia cutis from nonneoplastic leukocytic infiltrates in the skin. *Am. J. Dermatopathol.* **2016**, *38*, 672–677. [CrossRef] [PubMed]

25. Chuang, A.Y.; DeMarzo, A.M.; Veltri, R.W.; Sharma, R.B.; Bieberich, C.J.; Epstein, J.I. Immunohistochemical differentiation of high-grade prostate carcinoma from urothelial carcinoma. *Am. J. Surg. Pathol.* **2007**, *31*, 1246–1255. [CrossRef] [PubMed]

26. Srinivasan, M.; Parwani, A.V. Diagnostic utility of p63/P501S double sequential immunohistochemical staining in differentiating urothelial carcinoma from prostate carcinoma. *Diagn. Pathol.* **2011**, *6*, 67. [CrossRef] [PubMed]

27. Perner, S.; Rupp, N.J.; Braun, M.; Rubin, M.A.; Moch, H.; Dietel, M.; Wernert, N.; Jung, K.; Stephan, C.; Kristiansen, G. Loss of SLC45A3 protein (prostein) expression in prostate cancer is associated with *SLC45A3-ERG* gene rearrangement and an unfavorable clinical course. *Int. J. Cancer* **2013**, *132*, 807–812. [CrossRef] [PubMed]

28. Bostwick, D.G.; Pacelli, A.; Blute, M.; Roche, P.; Murphy, G.P. Prostate specific membrane antigen expression in prostatic intraepithelial neoplasia and adenocarcinoma: A study of 184 cases. *Cancer* **1998**, *82*, 2256–2261. [CrossRef]

29. Silver, D.A.; Pellicer, I.; Fair, W.R.; Heston, W.D.; Cordon-Cardo, C. Prostate-specific membrane antigen expression in normal and malignant human tissues. *Clin. Cancer Res.* **1997**, *3*, 81–85. [PubMed]

30. Gurel, B.; Ali, T.Z.; Montgomery, E.A.; Begum, S.; Hicks, J.; Goggins, M.; Eberhart, C.G.; Clark, D.P.; Bieberich, C.J.; Epstein, J.I.; et al. NKX3.1 as a marker of prostatic origin in metastatic tumors. *Am. J. Surg. Pathol.* **2010**, *34*, 1097–1105. [CrossRef] [PubMed]

31. Gelmann, E.P.; Bowen, C.; Bubendorf, L. Expression of *NKX3.1* in normal and malignant tissues. *Prostate* **2003**, *55*, 111–117. [CrossRef] [PubMed]

32. Skotheim, R.I.; Korkmaz, K.S.; Klokk, T.I.; Abeler, V.M.; Korkmaz, C.G.; Nesland, J.M.; Fossa, S.D.; Lothe, R.A.; Saatcioglu, F. NKX3.1 expression is lost in testicular germ cell tumors. *Am. J. Pathol.* **2003**, *163*, 2149–2154. [CrossRef]

33. Conner, J.R.; Hornick, J.L. Metastatic carcinoma of unknown primary: Diagnostic approach using immunohistochemistry. *Adv. Anat. Pathol.* **2015**, *22*, 149–167. [CrossRef] [PubMed]

34. Tong, H.; Ke, J.Q.; Jiang, F.Z.; Wang, X.J.; Wang, F.Y.; Li, Y.R.; Lu, W.; Wan, X.P. Tumor-associated macrophage-derived CXCL8 could induce ERα suppression via HOXB13 in endometrial cancer. *Cancer lette.* **2016**, *376*, 127–136. [CrossRef] [PubMed]

35. Zhu, J.Y.; Sun, Q.K.; Wang, W.; Jia, W.D. High-level expression OF HOXB13 Is closely associated with tumor angiogenesis and poor prognosis of hepatocellular carcinoma. *Int. J. Clin. Exp. Pathol.* **2014**, *7*, 2925–2933. [PubMed]

36. Zhai, L.L.; Wu, Y.; Cai, C.Y.; Tang, Z.G. Overexpression of homeobox B-13 correlates with angiogenesis, aberrant expression of emt markers, aggressive characteristics and poor prognosis in pancreatic carcinoma. *Int. J. Clin. Exp. Pathol.* **2015**, *8*, 6919–6927. [PubMed]

37. Queisser, A.; Hagedorn, S.A.; Braun, M.; Vogel, W.; Duensing, S.; Perner, S. Comparison of different prostatic markers in lymph node and distant metastases of prostate cancer. *Mod. Pathol.* **2015**, *28*, 138–145. [CrossRef] [PubMed]

38. Gown, A.M. Diagnostic immunohistochemistry: What can go wrong and how to prevent it. *Arch. Pathol. Lab. Med.* **2016**, *140*, 893–898. [CrossRef] [PubMed]

International Journal of
Molecular Sciences

MDPI

Article

Protease Expression Levels in Prostate Cancer Tissue Can Explain Prostate Cancer-Associated Seminal Biomarkers—An Explorative Concept Study

Jochen Neuhaus [1],[*],[†], Eric Schiffer [2],[†], Ferdinando Mannello [3], Lars-Christian Horn [4], Roman Ganzer [5] and Jens-Uwe Stolzenburg [5]

[1] Department of Urology, Research Laboratory, University of Leipzig, Liebigstraße 19, 04103 Leipzig, Germany
[2] Numares AG, Regensburg, Am BioPark 9, 93053 Regensburg, Germany; eric.schiffer@numares.com
[3] Department of Biomolecular Sciences, University "Carlo Bo", Via O. Ubaldini 7, 61029 Urbino (PU), Italy;
    ferdinando.mannello@uniurb.it
[4] Institute of Pathology, University Hospital Leipzig, Liebigstraße 24, 04103 Leipzig, Germany;
    Lars-Christian.Horn@uniklinik-leipzig.de
[5] Department of Urology, University Hospital Leipzig, Liebigstraße 20, 04103 Leipzig, Germany;
    Roman.Ganzer@uniklinik-leipzig.de (R.G.); Jens-Uwe.Stolzenburg@uniklinik-leipzig.de (J.-U.S.)
[*] Correspondence: jochen.neuhaus@medizin.uni-leipzig.de; Tel.: +49-341-9717688; Fax: +49-341-9717659
[†] These authors contributed equally to this work.

Academic Editor: Carsten Stephan
Received: 24 March 2017; Accepted: 29 April 2017; Published: 4 May 2017

Abstract: Previously, we described prostate cancer (PCa) detection (83% sensitivity; 67% specificity) in seminal plasma by CE-MS/MS. Moreover, advanced disease was distinguished from organ-confined tumors with 80% sensitivity and 82% specificity. The discovered biomarkers were naturally occurring fragments of larger seminal proteins, predominantly semenogelin 1 and 2, representing endpoints of the ejaculate liquefaction. Here we identified proteases putatively involved in PCa specific protein cleavage, and examined gene expression and tissue protein levels, jointly with cell localization in normal prostate (nP), benign prostate hyperplasia (BPH), seminal vesicles and PCa using qPCR, Western blotting and confocal laser scanning microscopy. We found differential gene expression of chymase (CMA1), matrix metalloproteinases (MMP3, MMP7), and upregulation of MMP14 and tissue inhibitors (TIMP1 and TIMP2) in BPH. In contrast tissue protein levels of MMP14 were downregulated in PCa. MMP3/TIMP1 and MMP7/TIMP1 ratios were decreased in BPH. In seminal vesicles, we found low-level expression of most proteases and, interestingly, we also detected TIMP1 and low levels of TIMP2. We conclude that MMP3 and MMP7 activity is different in PCa compared to BPH due to fine regulation by their inhibitor TIMP1. Our findings support the concept of seminal plasma biomarkers as non-invasive tool for PCa detection and risk stratification.

Keywords: seminal plasma biomarkers; matrix metalloproteinase (MMP); tissue inhibitor of MMP (TIMP); qPCR; confocal laser scanning microscopy; Western blotting

1. Introduction

Prostate cancer (PCa) is the main gender-specific malignancy in men and prostate specific antigen (PSA) testing is the gold standard in PCa detection [1]. However, comprehensive PSA screening resulted in significant overdiagnosis and overtreatment [2] and in consequence PSA screening is recommended only for men aged 55 to 69 years by the AUA [3]. New PCa biomarkers are urgently needed to identify patients who may be candidates for curative intervention and guide clinical decisions [4].

Recently, seminal fluid has been acknowledged as a promising source of PCa related biomarkers as it directly reflects the pathological processes within the prostate [5]. The high biological variability of prostate cancer [6] necessitates a distinct and clearly defined set of biomarkers, rather than a single or a combination of few biomarkers for efficient description of the disease on a molecular level [7]. In a previous study, we defined distinct panels of small peptide biomarkers in seminal plasma by capillary electrophoresis mass spectrometry (CE-MS), which could be used for PCa detection and stratification [8]. The procedure distinguished PCa from benign prostate hyperplasia (BPH), chronic prostatitis (CP) and normal prostate (nP) with 83% sensitivity and 67% specificity (AUC 75%, $p < 0.0001$) in a small clinical validation cohort of 125 patients. A further set of biomarkers correctly identified advanced ($\geq$pT3a) and organ-confined (<pT3a) tumors (AUC 83%, $p = 0.0055$) with 80% sensitivity and 82% specificity [8].

The discovered biomarkers were fragments of larger parental seminal proteins, such as *N*-acetyl lactosaminide β-1,3-*N*-acetyl glucosaminyl transferase, prostatic acid phosphatase, stabilin-2 and most dominantly semenogelin-1 and -2. Although the parental proteins were also detected earlier [9–11], the use of these naturally occurring fragments reflecting natural proteolytic liquefaction is a new concept.

There are mutual activation and inhibition mechanisms within the liquefaction cascade, which could lead to different downstream proteolytic cleavage patterns [12]. The investigation of these disease-associated proteinase networks [13] represents a challenging task to close the gap between the classical protein expression based biomarker concept and the hypothesis-free proteomic profiling concepts.

We suggest that seminal polypeptide panels may be the smoking gun that links the complex proteolytic alterations of seminal proteins to PCa providing novel insights into PCa biology with special emphasis on disease-associated proteolytic activities. The aim of this study was to identify potential proteinases and inhibitors involved in the formation of the specific seminal biomarkers.

2. Results

2.1. Sequence Analysis and Putative Proteolytic Cleavage Sites

Of the 141 seminal peptides discovered in our previous study, representing a total of 47 different parental proteins, almost 60% were fragments of semenogelin-1 or -2 (SEMG1 and SEMG2). Therefore, we focused our data base searches on SEMG1 and SEMG2 cleavage patterns. Using MEROPS database (Available online: http://merops.sanger.ac.uk/cgi-bin/specsearch.pl) we could assign SEMG1 (316–344) to kallikrein-3 (KLK3, PSA) cleavage at site 315 (peptide-SSIY//SQTE-peptide). Limiting searches from octa- to hexamere recognition (peptide-IY//SQ-peptide) resulted in chymase (CMA1) and cathepsin G (CTSG) as potential alternative proteinases (Table 1). In contrast, KLK3 activity may not account for biomarker SEMG2 (194–215) cleaved at position 193 (peptide-SQSS//YVLQ-peptide. Searching cleavage site 193 (peptide-SS//YV-peptide) revealed matrix metallopeptidase-3 (MMP3), -7 (MMP7), -13 (MMP13), -14 (MMP14) or -20 (MMP20) as potentially involved proteases (Table 1). KLK3 activity may also promote for cleavage between Q and T at position 197 (peptide-QVLQ//TEEL-peptide), resulting in biomarkers SEMG1 (198–215) and SEMG2 (198–215) (Table 1). Although KLK3 Q//T-cleavage is reported for SMG1 and SMG2 at position 235, position 197 remains inconclusive (see MEROPS database). KLK cleavage of SEMG1 and SEMG2 at position 194 (peptide-QSSY//VLQT-peptide) between Y and V, results in the observed non-marker fragments starting from position 195 (Table 1).

Table 1. Semenogelin derived seminal peptides and their putative proteolytic cleavage sites.

Peptide ID	PCa Biomarker	Peptide Sequence	Peptide Identification	Cleavage Site	Putative Protease
15331	Yes	YVLQTEELVVN KQQRETKNSHQ	SEMG2 (194-215)	SQSS//YVLQ	MMP3, MMP7, MMP13, MMP14, MMP20
1677	No	VLQTEELVA	SEMG1 (195-203)	QSSY//VLQT	KLK3
6925	No	VLQTEELVANKQQ	SEMG1 (195-207)	QSSY//VLQT	KLK3
7186	No	VLQTEELVVNKQQ	SEMG2 (195-207)	QSSY//VLQT	KLK3
10289	No	VLQTEELVANKQQRET	SEMG1 (195-210)	QSSY//VLQT	KLK3
11260	No	VLQTEELVVNKQQRETK	SEMG2 (195-211)	QSSY//VLQT	KLK3
11899	Yes	TEELVANKQQRETKNSHQ	SEMG1 (198-215)	YVLQ//TEEL	KLK3
12083	Yes	TEELVVNKQQRETKNSHQ	SEMG2 (198-215)	YVLQ//TEEL	KLK3
18990	Yes	SQTEEKAQGKSQKQI TIPSQEQEHSQKAN	SEMG1 (316-344)	SSIY//TEEL	KLK3, CMA1, CTSG

Peptide sequences including LC-MS/MS data have been recently published by our group [8]. ID: polypeptide identifier annotated by the structured query language (SQL) database (ID); known cleavage sites are indicated for semenogelin-1 (P04279, Swiss-Prot Acc.No.) and semenogelin-2 (Q02383) with their putative proteases, as supported by literature research of web databases: http://merops.sanger.ac.uk/, http://pmap.burnham.org/proteases and http://www.proteolysis.org/proteases.

2.2. Gene Expression of Proteases in Human Prostate Samples

Gene expression of KLK3, prostatic acid phosphatase (ACPP) and ACPP-V1 (isoform 1) was high, while matrix metalloproteinases MMP3, MMP7, MMP14 showed low abundance expression by qPCR (Figure 1). KLK3 showed a tendency of increased levels in BPH (Figure 1A). Significant higher transcript levels in BPH were found for MMP7, MMP14 and CMA1 compared to PCa (Figure 1C). Some of the low abundance genes (CTSG, MMP3) also showed a tendency for higher expression in BPH (Figure 1B). MMP13 and MMP20 were at the detection limits (data not shown).

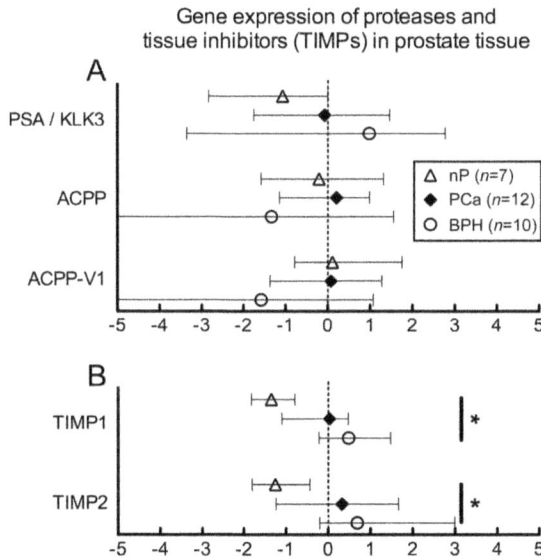

Figure 1. *Cont.*

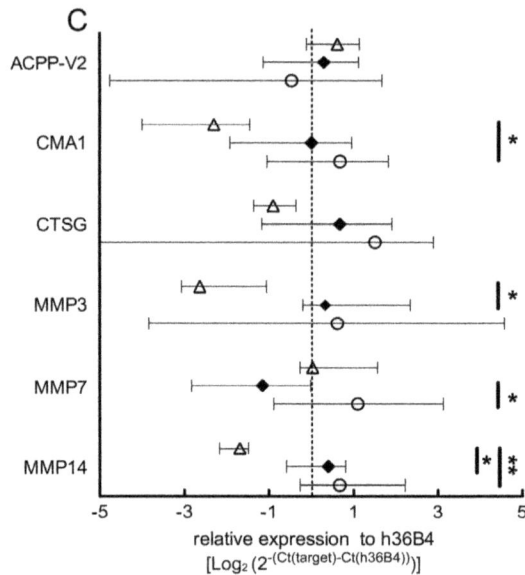

Figure 1. Gene expression analysis. (**A–C**) Expression levels compared to h36B4, global normalization; (**A**) High abundance genes (ratio > 1.0); KLK3 (PSA): kallikrein related peptidase 3; ACPP: total prostatic acid phosphatase; ACPP-V1: splice variant 1 (short secreted isoform of ACPP). The observed differences in mRNA levels between PCa ($n = 12$), benign prostatic hyperplasia (BPH) ($n = 10$) and normal Prostate (nP; $n = 7$) were not significant. However, range of KLK3 expression was much higher in BPH and PCa; (**B**) TIMP1 and TIMP2 expression was significantly different between BPH and nP samples; (**C**) Low abundance genes (ratio < 1.0); ACPP-V2: splice variant 2 (long intracellular isoform); CMA1: chymase 1, CTSG: cathepsin G, and MMP: matrix metalloproteinase. Differential expression levels were detected for CMA1, MMP3, MMP7 and MMP14. Symbols represent medians, whiskers indicate interquartile range. Global normalization; interquartile range: 25%–median–75%; significance: *Kruskal–Wallis* test and *Dunn's* multiple comparison test; * $p < 0.05$, ** $p < 0.01$).

2.3. Protein Levels of Proteinases in Human Prostate and Seminal Vesicles

Highest tissue levels were seen for KLK3 and ACPP-V1 by Western blotting (Figure 2A,B; lanes 1 and 2) and epithelial cells showed highest immunofluorescence in CLSM (Figure 3A,M and Figure 4A,B). Differences in total IF among groups were not significant (Figure 3M), except for TIMP1-IF ($p = 0.0368$) as were the calculated ratios: MMP3/TIMP1 ($p = 0.0119$) and MMP7/TIMP1 ($p = 0.0145$; Figure 3M). TIMP1-IF was especially high in epithelium and in interstitium of the prostate, comprising smooth muscle cells (orange labelling in Figure 3E) and interstitial cells (red labelling) in BPH, while being markedly diminished in epithelium and interstitial cells in PCa (Figure 3H). ACPP-V1 (short, secreted isoform of ACPP) showed significant lower immunofluorescence in grade 4 PCa (Figure 4B). Differences between MMP3-IF, MMP7-IF and TIMP2 were not significant (Figure 4C,D,I), whereas trends of lower TIMP2-IF compared to nP were evident in PCa-gp3 (Figure 4I). Total MMP14-IF was lower in PCa-gp3 ($p < 0.05$) and in PCa-gp4 (trend, not significant), and TIMP1-IF was significantly higher in epithelial cells of BPH compared to PCa-gp3 (Figure 4H).

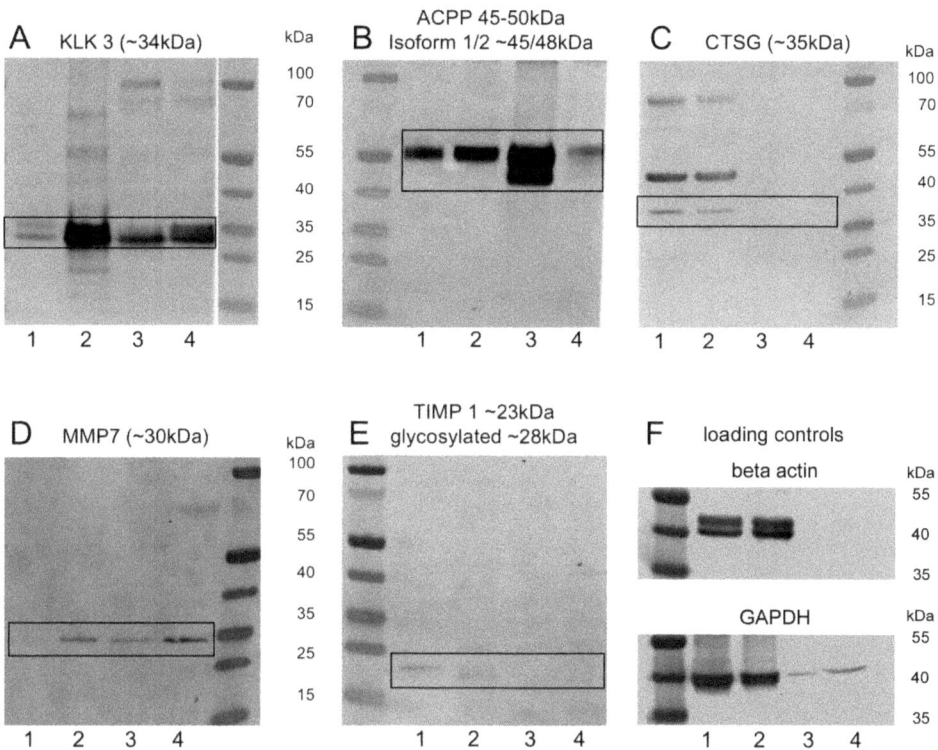

Figure 2. Western blot analysis of prostate and seminal plasma. KLK3 and ACPP were detected in high abundance in prostate tissue (**lanes 1**, **2**) and seminal plasma (**lanes 3**, **4**) at the predicted molecular weight of 34 and 50 kDa, respectively (**A,B**); occasionally multiple banding indicated isoforms or post-translational modification (glycosylation); CTSG was detected exclusively in prostate tissue showing three to four distinct bands at ~35, ~45; ~60 and ~80 kDa (**C**); MMP7 could be detected in prostate tissue and seminal plasma at ~30 kDa (**D**) and we found very low levels of TIMP1 (**E**). MMP14, MMP3, TIMP2 and CMA1 were negative in all examined samples (data not shown; $n = 4$ prostate; $n = 6$ seminal plasma); (**F**) loading controls; note negative staining for β actin in seminal.

CLSM clearly revealed the different contributions of the epithelial cells and smooth muscle cells to the observed alterations in total IF. Interestingly, KLK3-IF (=PSA) was not different between groups (Figure 4A). Also, MMP3-IF and MMP7-IF showed no differences (Figure 4E,F) whereas MMP14-IF was lower in PCa-gp3 than in nP (Figure 4G); and significant differences were detected for TIMP1-IF (Figure 4H). TIMP2-IF was low in PCa-gp3 but did not reach statistical significance (Figure 4I).

We calculated the specific MMP/TIMP ratios as a surrogate of MMP activity. We found that MMPs and TIMP2 were almost balanced in case of MMP3 and MMP7 (values around 1.0; data not shown), while TIMP1 was present in excess indicated by ratios < 1.0 (Figure 5A,B). In contrast, expression of MMP14 well exceeded TIMP2 in nP and PCa as indicated by ratios > 1.0 (Figure 5C), while expression in BPH was balanced.

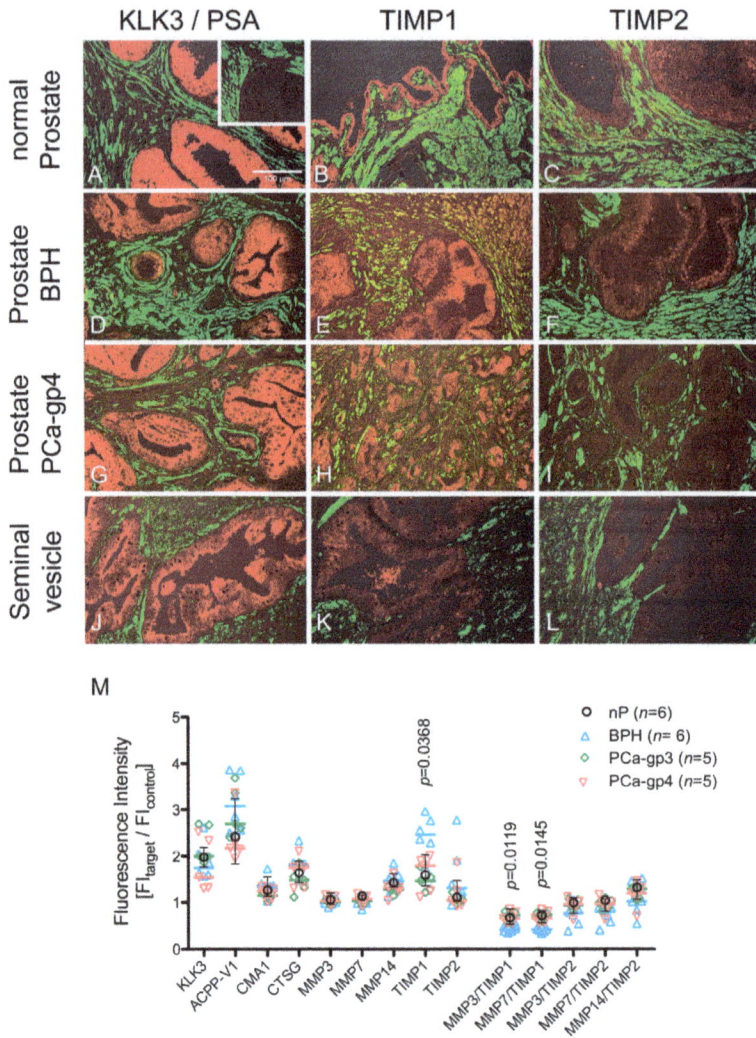

Figure 3. Protein expression analysis using confocal immunofluorescence imaging. Immunohistochemical tissue expression of selected proteases and tissue inhibitors of metalloproteinases. (**A–L**) Confocal images demonstrating tissue distribution of target proteins (red); alpha-smooth muscle cell actin (aSMCA; green). Phase contrast images are merged for tissue structure. Immunofluorescence (IF) of KLK3, TIMP1 and TIMP2 in normal prostate (**A–C**), BPH (**D–F**); PCa Gleason pattern 4 (PCa-gp4; **G–I**) and seminal vesicle (**J–L**). Note the marked upregulation of TIMP1 in BPH especially in SMC (**E**; orange colored cells) and the downregulation of TIMP1-IF in the smooth muscle cells of PCa-gp4 (**H**); TIMP2-IF is almost completely lost in PCa-gp4 tissue (**I**); KLK3-IF in seminal vesicle was low, TIMP1-IF moderate and TIMP2-IF was very low (**J–I**); Scale bar indicating 100 µm in (**A**) applies to all micrographs; staining control (inset in **A**); (**M**) Scatter plot of fluorescence intensities; lines indicate medians. Normal prostate (black symbols) are depicted as median with interquartile range. *p*-Values are based on nonparametric *Kruskal–Wallis* test.

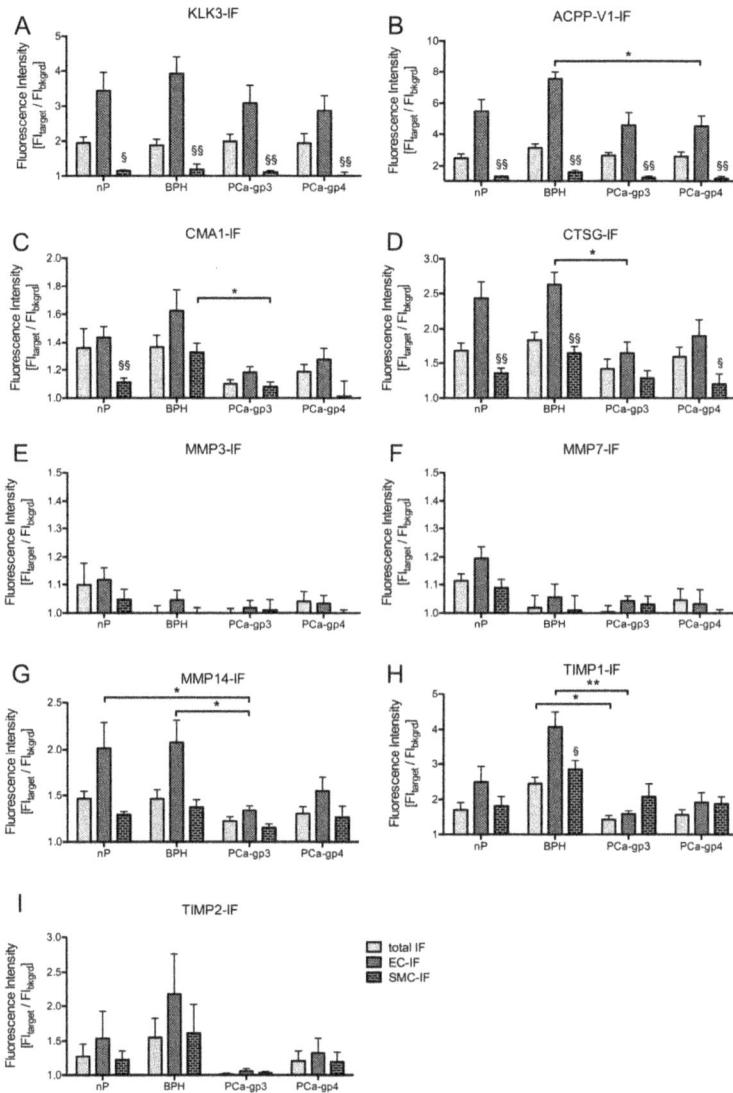

Figure 4. Cellular distribution in prostate tissue. Data (mean (SEM)) are presented as fluorescence intensity [$FI_{target}/FI_{control}$]; background fluorescence (bkgrd) = 1. Total fluorescence included complete tissue area omitting the lumen. Smooth muscle cells (SMC) were identified by anti-smooth muscle cell actin staining (green fluorescence, cf. Figure 2); epithelial cells (EC) area was delineated from DIC images. Note the differences in Y-axis scaling. Differences in fluorescence intensities were not significant for KLK3-IF (**A**), MMP3-IF (**E**), MMP7-IF (**F**) and TIMP2-IF (**I**). Note that the total-IF and SMC-IF is unaltered for ACPP-V1 (**B**), CMA1 (**C**) and CTSG (**D**) whereas epithelial (EC)-IF is significantly lower in PCa compared to BPH, indicating contribution of interstitial cells to total-IF. A comparable trend not reaching significance level ($p < 0.05$) is seen for MMP14 (**G**). In contrast, the significant reduction of TIMP1 total-IF is reflected by significant reduction of EC-IF (**H**). *Kruskal–Wallis* test with post-hoc Dunn's Multiple Comparison test as indicated by bars (* $p < 0.05$; ** $p < 0.01$); § indicates significant difference between EC and SMC; § $p < 0.05$; §§ $p < 0.01$.

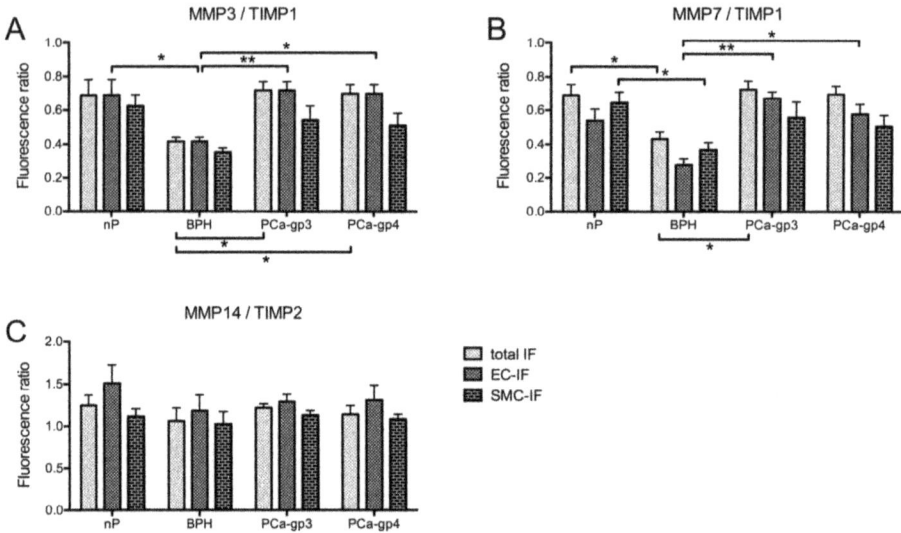

Figure 5. Cellular distribution of MMP/TIMP ratios. Ratios (mean (SEM)) were calculated from means of FI_{target}/FI_{bkgrd} of the respective MMP-TIMP pairs and presented as FI-ratio. Note that MMP14/TIMP2 ratios are >1.0 indicating excess of MMP14 (**C**), while MMP3/TIMP1 and MMP7/TIMP1 ratios are <1.0 indicating excess of TIMP1 (**A**,**B**). *Kruskal–Wallis* test with post-hoc *Dunn's* Multiple Comparison test as indicated by bars (* $p < 0.05$; ** $p < 0.01$); no significant differences were observed between EC and SMC.

To address the contribution of seminal vesicles in protease and inhibitor production, we also investigated seminal vesicles (SV) by CLSM and compared the expression to nP. We found no significant differences (*Mann–Whitney* test) but considerably lower IF for KLK3 and ACPP-V1 especially in epithelial cells (EC IF, Figure 6).

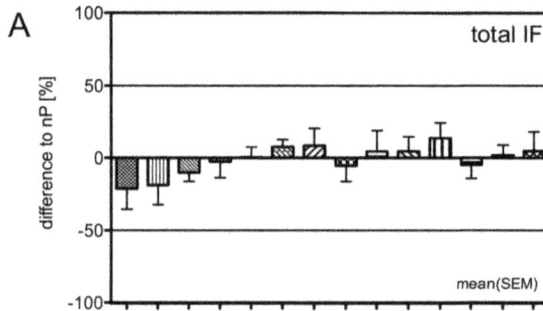

Figure 6. *Cont.*

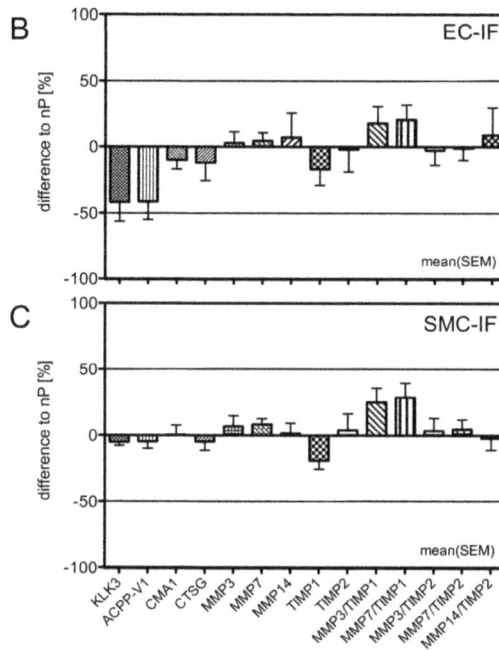

Figure 6. CLSM analysis in seminal vesicles (SV). (**A–C**) Immunofluorescence in SV (*n* = 6) was normalized to the IF in nP (*n* = 6). Parameter-free *Mann–Whitney* test detected no significant differences between expression levels in nP and SV, but considerably lower IF was found for KLK3 and ACPP-V1, especially in epithelial cells (EC, **B**).

3. Discussion

Based on our previous study, which defined small protein fragments within the seminal plasma as biomarker fingerprint of prostate carcinoma [8], we were interested in the molecular background of the production of these small peptides ($\leq$20 kDa). We therefore explored the cellular expression of certain proteases and their specific inhibitors within the prostate and seminal vesicle tissues. Most of the small peptides that make up the biomarker pattern were derived from SEMG1 and SEMG2. Cleavage site analysis revealed, that the observed SEMG1 and SEMG2 derived biomarkers cannot be explained by KLK3 cleavage alone, implicating the presence of additional cleavage events. From the available sequence data and proteolytic cleavage site searches we hypothesized proteolysis schemes for SEMG1 and SEMG2 based on the involvement of KLK3, CMA1, CTSG, MMP3, MMP7, MMP13, MMP14 or MMP20 as well as the pleiotropic functions of TIMP1 and TIMP2 [14].

As major findings of our study we highlighted that biomarkers found in seminal fluids represent the endpoint of a complex liquefaction proteolytic cascade that involves matrix metalloproteinases MMP3, MMP7 and MMP14 particularly when associated with down-regulation of TIMP1. Downregulation of TIMP1 and TIMP2 has been found by in situ hybridization in the stroma of PCa with higher Gleason scores (GS 8–10) compared to tissue of low Gleason scores [15], supporting our immunohistochemical findings. MMPs are generally more active in advanced stages, caused either by upregulation of MMPs and/or downregulation of their specific TIMPs and MMP activity has been linked to metastasis [16]. MMPs can interfere with growth signals, inhibit apoptosis, and induce angiogenesis and lymphangiogenesis all promoting tumor progression and metastasis. However, MMPs may also possess nonproteolytic functions, like triggering cell migration by chemotaxis (MMP14/TIMP2-MMP2 complex) or interfering with the complement proteinase cascade by interaction with C1q [17]. In line with existing literature, we found significant downregulation of TIMP1 in PCa-gp3, while downregulation of TIMP2 did not reach significant levels. TIMP1- and TIMP2-levels were also lower in PCa-gp4 compared to BPH (Figure 4). In most cases, fluorescence intensity was significantly higher in epithelial cells (EC) than in interstitial smooth muscle cells (SMC), supporting the notion, that malignant transformation of epithelial cells is responsible for the altered protease and inhibitor levels. Further studies, using tumor or tumor stem cell specific markers are needed to analyze the subpopulations of epithelial cells and their contribution to the protease pattern [18].

Except for TIMP1 (Figures 3B and 4H) gene expression and protein expression was not correlated. This is in line with the finding that the concordance between transcript levels and protein expression is only 48–64% in prostate specific proteome [19]. We used immunofluorescence techniques to study cellular protein expression levels. Therefore, variations in the protein detection might be due to weak antibody staining in the case of MMPs complex formation with TIMP1 and TIMP2 [20] or due to epitope masking [21].

Our findings that MMP3/TIMP1 and MMP7/TIMP1 ratios were significantly lower in BPH compared to nP and PCa (Figure 5A,B) is in line with the significant higher transcript expression levels in BPH (Figure 1B) and the higher protein expression in confocal analysis (Figure 4H,I). In consequence, activity of those two MMPs would be lower in BPH tissue, resulting in less cleavage activity in seminal plasma as well.

Although western blotting analysis of seminal plasma consistently detected KLK3, ACPP and MMP7 (Figure 2, lanes 3,4), the detection of CTSG, CMA1, MMP3, MMP14, TIMP1 and TIMP2 failed in seminal plasma even though they were detectable in prostate tissue samples. These results could reflect the complexation of MMPs with TIMPs that may mask the antibody recognition site as discussed above. Concerning CTSG and CMA1, it is well known that seminal plasma contains several specific CTSG inhibitors [22,23] and several classes of proteinases (e.g., Metalloproteinases, serine and cysteine proteinases) may be able to activate and degrade CTSG and CMA1 [24,25].

In case of MMP14 discrepancies between gene expression (significantly up regulated in BPH and PCa compared to nP) and protein tissue levels (significantly down regulated in PCa, compared to nP and BPH) cannot be explained by complexation with TIMP2. Our findings indicate that cells of the prostate interstitium (included in "total IF" columns) have increased protein expression and may account for levelling the decrease in protein expression in epithelial cells (Figure 4). Additionally, it should be taken into account that the protein content within cells considerably depends on secretion rates and secretory status. Thus, especially high MMP14 secretion rates from epithelial cells in PCa might account for the low levels of MMP14-IF measured in PCa tissues (Figure 4G).

In addition, we investigated seminal vesicles as a source of proteases and inhibitors. We found low-level expression of most proteases and, interestingly, we also detected TIMP1 and low levels of TIMP2 in seminal vesicles (Figure 3K,L).

So far, TIMP1 has been detected only in bovine tissue [26]. To our best knowledge, this study is the first evidence for TIMP expression in human seminal vesicles. Of noteworthy, TIMPs play an independent role beyond MMP inhibition, e.g., inhibition of tumor growth, invasion and metastasis, growth factor-like activity, inhibition of angiogenesis and suppression of programmed cell death [27,28]. TIMPs have also been detected in seminal plasma and peculiar secretion by human seminal vesicles has been suggested [29,30].

Our findings suggest that the proteinase cocktail provided for semen liquefaction varies considerably between nP (60.71 $\pm$ 6.80 years; mean $\pm$ SD), BPH (70.30 $\pm$ 6.45 years) and PCa (65.17 $\pm$ 6.83 years). Significant alterations in TIMP1 expression strongly suggests that altered proteinase activities could lead to formation of seminal protein fragments and could be a candidate for prostate cancer biomarkers. Although we did not directly measure MMP activity, the observed increases in MMP/TIMP-ratios might indicate disruption/degradation of physiological extracellular matrix associated to tumor growth. Interestingly, MMP3/TIMP1 and MMP7/TIMP1 ratios were not different in nP and PCa, while they were significantly lower in BPH (Figure 5). As the prevalence of

BPH constantly rises with age affecting 50–70% in men in the 5th and 6th decade, respectively [31], it should be differentiated from PCa which occurs in the same age group. This view is supported by our previous study demonstrating the need of a two-step procedure using two different biomarker panels to reach accurate separation of groups: (1) 21PP for separation of PCa + BPH from CP + HC and (2) 5PP for final detection of PCa (PCa vs. BPH) [8].

At the moment the effect of alterations in single proteases cannot be clearly linked to the appearance of a specific cleavage product in seminal plasma due to unidentified co-players (e.g., endopeptidases) in the complex liquefaction cascade. However, elevated serum MMP7 levels were reported to be significantly associated with metastatic and advanced PCa and was therefore considered to be a biomarker candidate to detect metastatic PCa [32]. In addition, TIMP1 was found downregulated in adenocarcinoma compared to BPH [33] and loss of TIMP1 correlated with biochemical recurrence in patients with localized PCa [34]. These results are in line with our findings and support the idea of regulation of MMP activity by TIMPs as crucial event in PCa growth and progression.

The complex network of proteinases found in the seminal plasma during prostate diseases (Figure 7) seems to generate many small proteolytic protein fragments, which represent the prostate cancer microenvironment and may be useful for both understanding prostate biology and as potent novel PCa biomarkers.

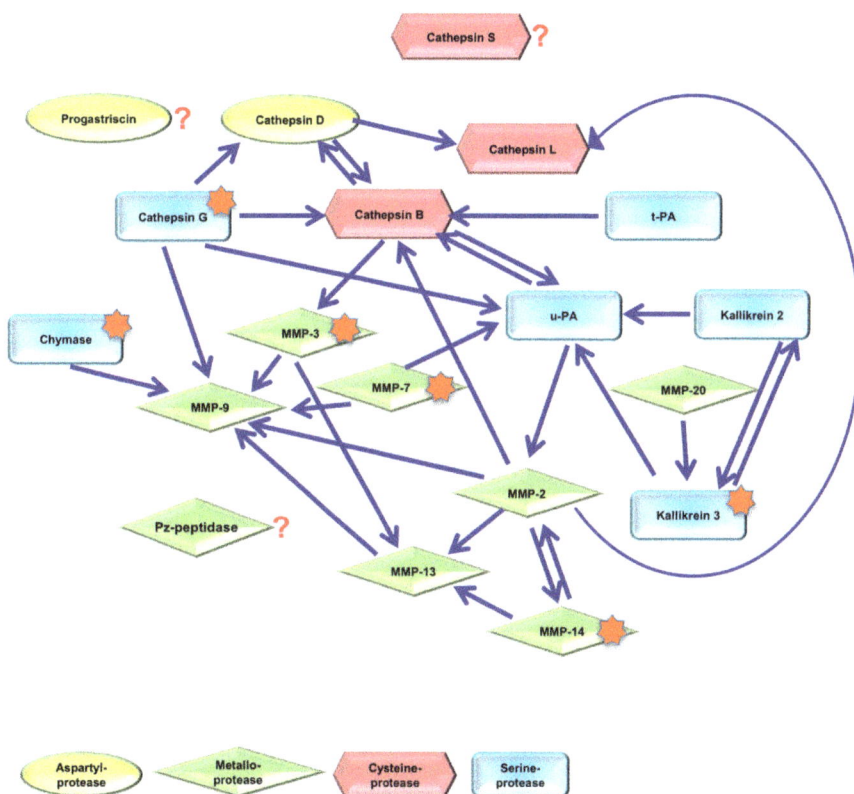

Figure 7. Proteolytic network in human seminal plasma. The role of some enzymes in liquefaction cascade is still obscure (?).

Semen liquefaction is the result of a complex proteolytic cascade involving different and multidirectional protease-protease interactions. Although the interactions involve proteinases from several families, at least three proteinases have been identified in seminal fluid with unknown direct/indirect interactions/functions with other protease classes (adaptation from reference data [13] and web databases: http://merops.sanger.ac.uk/, http://pmap.burnham.org/proteases; asterisks = proteinases with significant changes in MS/MS [8] and qPCR analyses (CMA1, MMP3, MMP7, MMP14, TIMP1, TIMP2; this study).

The major limitation of the present retrospective study was the restricted sample size. However, the aim of this study was to explore the role of MMP/TIMP network in prostate and seminal vesicles in the production of prostate cancer specific peptide patterns found in the seminal plasma. Future extended studies, including metastatic castration-resistant prostate cancer (CRPC) patients, are needed to validate this concept and to encourage the use of urine diagnostic markers as a tool for prostate cancer diagnosis and prognosis.

4. Material and Methods

4.1. Ethics Statement

The study was approved by the Ethics Committee of the University of Leipzig (Reg.No. 084-2009-20042009; 305-12-24092012) and was conducted according to the principles expressed in the Declaration of Helsinki. Written informed consent was obtained from all patients.

4.2. Quantitative Real-Time Polymerase Chain Reaction (qPCR)

We included samples from radical prostatectomies of patients with PCa ($n = 12$), transurethral resection of BPH ($n = 10$) and normal prostate tissue (nP; $n = 7$) after instantaneous section evaluation by our pathologist (LCH; Table 2).

Table 2. Characteristics of patients used for qPCR analysis.

Group	n	Age (Mean ± SD) (95% CI)	PSA [ng/mL] (95% CI)	Gleason Score	Histology
nP	7	60.71 ± 6.80 (54.43–67.00)	n.a.	n.a.	normal prostate
BPH	10	70.30 ± 6.447 (65.69–74.91)	n.a.	n.a.	BPH
PCa	12	65.17 ± 6.834 (60.82–69.51)	6.097 ± 1.868 (4.910–7.284)	≤6 ($n = 6$) 7 ($n = 2$) >7 ($n = 4$)	pT2a ($n = 2$) pT2c ($n = 7$) pT3a ($n = 3$)

Age was not significantly different between groups ($p = 0.0556$; *Kruskal–Wallis* test, *Dunn's* Multiple Comparison Test). PSA values were not available for nP and BPH patients (n.a.).

Total RNA was extracted from deep frozen tissue specimens using peqGOLD TriFast extraction kit (peQLab, Erlangen, Germany) according to the manufacturers protocol and transcribed into cDNA using the Maxima First Strand cDNA Synthesis Kit (Fermentas, St. Leon-Rot, Germany). Quantitative PCR was performed with the real-time PCR-System realplex2 Mastercycler (Eppendorf, Hamburg, Germany) using the SYBR-Green quantitative PCR Mastermix (Fermentas) and custom primers (MWG-Biotech, Ebersberg, Germany; Table 3). Human 36B4 (acidic ribosomal phosphoprotein P0) served as internal standard for normalization using the $2^{-\Delta\Delta Ct}$ method for relative quantification [35].

Table 3. Primer pairs used for qPCR analysis of the liquefaction cascade. KLK3 (PSA): kallikrein related peptidase 3; MMP: matrix metalloproteinase; CMA1: chymase 1; CTSG: cathepsin G; ACPP: prostatic acid phosphatase; h36B4: ribosomal protein P0 (housekeeping gene).

Gene	Acc. No.	Sequence (F) forward, (R) reverse	Binding to Exon
ACPP (both isoforms)	NM_001099.4	(F) 5′-cga agt ccc att gac acc tt-3′ (R) 5′-atc aaa gtc cgg tca acg tc-3′	2 4
ACPP-V1 (transcript variant 1)	NM_001099.4	(F) 5′-tgt gag tgg cct aca gat gg-3′ (R) 5′-tgt act gtc ttc agt acc ttg a-3′	9 10
ACPP-V2 (transcript variant 2)	NM_001134194.1	(F) 5′-gga ctc ctt cct ccc tat gc-3′ (R) 5′-agg caa cag caa aga tga cc-3′	9 11
PSA/KLK3	NM_001030047.1	(F) 5′-cat gct gtg aag gtc atg ga-3′ (R) 5′-agc aca cag cat gaa ctt gg-3′	3 4
MMP3	NM_002422.1	(F) 5′-gca gtt tgc tca gcc tat cc-3′ (R) 5′-gag tgt cgg agt cca gct tc-3′	1 2
MMP7	NM_002423.3	(F) 5′-gag tgc cag atg ttg cag aa-3′ (R) 5′-gcc aat cat gat gtc agc ag-3′	2 3
MMP13	NM_002427.3	(F) 5′-ttg agc tgg act cat tgt cg-3′ (R) 5′-gga gcc tct cag tca tgg ag-3′	1 2
MMP14	NM_004995.2	(F) 5′- caa gca ttg ggt gtt tga tg-3′ (R) 5′-tcc ctt ccc aga ctt tga tg-3′	8 9
MMP20	NM_004771.3	(F) 5′-ctc atc ctt tga cgc tgt ga-3′ (R) 5′-ctt cgt aag ctg cat cca ca-3′	6 7
CMA1	NM_001836.2	(F) 5′-tgc aag agg tga agc tga ga-3′ (R) 5′-gag att cgg gtg aag aca gc-3′	4 5
CTSG	NM_001911.2	(F) 5′-ata atc agc gga cca tcc ag-3′ (R) 5′-tgc cta tcc ctc tgc act ct-3′	3 4
h36B4	NM_002775.1	(F) 5′-ccg act cct ccg act ctt c-3′ (R) 5′-aac atg ctc aac atc tcc cc-3′	6 8

4.3. Protein Analysis by Indirect Immunofluorescence

We performed semi-quantitative analyses of the immunofluorescence (IF) and distribution of target proteins in tissue samples of normal prostate ($n = 6$), BPH ($n = 6$), seminal vesicles ($n = 6$) and PCa ($n = 10$; Table 4) using confocal laser scanning microscopy (CLSM).

Table 4. Characteristics of patients used for immunofluorescence analysis.

Group	n	Age (95% CI)	PSA [ng/mL] (95%CI)	Gleason Score	Histology
nP	6	62.17 ± 4.79 (57.14–67.20)	n.a.	n.a.	normal prostate
BPH	6	67.50 ± 5.93 (61.28–73.72)	n.a.	n.a.	BPH
PCa	10	61.50 ± 9.25 (54.88–68.12)	19.44 ± 12.62 ($n = 8$)	6 ($n = 2$) 7 ($n = 8$)	pT1c ($n = 7$) pT2c ($n = 3$)
SV	6	61.17 ± 8.68 (52.06–70.28)	11.41 ± 4.87 ($n = 5$)	6 ($n = 2$) 7 ($n = 4$)	normal, not infiltrated

Age was not significantly different between groups ($p = 0.3997$; *Kruskal–Wallis* test, *Dunn's* Multiple Comparison Test). PSA values were not available for nP and BPH patients (n.a.).

Seven micron thick paraffin sections of prostate biopsies and material from radical prostatectomies were deparaffinized, rehydrated and processed for antigen retrieval in 10 mM citrate buffer (pH 6.0, 100 °C) in a steamer (Braun, Kronberg, Germany). Slices were then washed in phosphate-buffered saline (pH 7.4) and transferred to Tris-buffered saline (50 mM TBS, pH 7.4). Following treatment with TBS (0.1% Triton X-100) for 10 min and blocking unspecific binding (TBS, 0.1% Triton X-100, 1% bovine serum albumin, 3% fat-free milk powder) for 15 min at room temperature, slices were incubated overnight in a cocktail of anti-alpha smooth muscle cell actin monoclonal mouse IgG2a antibody (1:2000) and target antibody in blocking buffer at 4 °C (Table 5).

Table 5. Antibodies used in indirect immunofluorescence (IF) and Western blotting (WB). Sources: [1] Acris Antibodies GmbH, Herford, Germany; [2] antibodies online, Aachen, Germany; [3] Santa Cruz Biotechnology Inc., Santa Cruz, CA, USA; [4] Proteintech Group Inc., Manchester, UK; [5] Histo-line Laboratories, Milan, Italy; [6] Abcam Inc., Cambridge, MA, USA; [7] QED Bioscience Inc., San Diego, CA, USA; [8] Novus Biologicals Europe, Abingdon, UK; [9] Sigma-Aldrich Chemie GmbH, Steinheim, Germany; [10] LI-COR Biosciences GmbH, Bad Homburg, Germany; Rb (rabbit); Ms (mouse); polyclonal (poly); monoclonal (mono).

Antigen/Primary Antibodies	Source	Type	Order-No.	Dilution
KLK3	1	Rb, poly	AP15748PU-S	1:200 (IF), 1:500 (WB)
CMA1	2	Rb, poly	ABIN679853	1:200 (IF), 1:400 (WB)
CTSG	2	Rb, poly	ABIN731843	1:200 (IF), 1:500 (WB)
MMP3	2	Rb, poly	ABIN668301	1:200 (IF)
MMP3	3	Ms, mono	sc-21732	1:50 (WB)
MMP7	2	Rb, poly	ABIN668451	1:200 (IF)
MMP7	4	Rb, poly	10374-2-AP	1:500 (WB)
MMP14	5	Rb, poly	29025	1:200 (IF)
MMP14	6	Rb, poly	ab3644	1:500 (WB)
ACPP-V1	2	Rb, poly	ABIN966903	1:80 (IF), 1:500 (WB)
TIMP1	7	Rb, poly	29022	1:200 (IF), 1:1000 (WB)
TIMP2	2	Rb, poly	ABIN373976	1:200
β Actin	8	Ms, mono	NB600-501	1:4000 (WB)
GAPDH	3	Ms, mono	sc-47724	1:1000 (WB)
alpha-smooth muscle actin	9	Ms, mono	A2547	1:2000
goat-anti Ms IRDye® 680RD	10	Goat IgG	926-68070	1:8000 (WB)
goat-anti Rb IRDye® 680RD	10	Goat IgG	926-68071	1:8000 (WB)

For visualization, the sections were incubated with Alexa Fluor® 488 goat-anti-mouse IgG2a and Alexa Fluor® 555 goat anti-rabbit antibodies (Invitrogen, Karlsruhe, Germany) at a dilution of 1:500 for 1 h at room temperature. For semi-quantitative analysis scans were acquired at a LSM5 Pascal (Carl Zeiss, Jena, Germany) using a Plan-Neofluar 20x/0.5 Objective (Carl Zeiss, Jena, Germany) at 488 and 543 nm excitation wavelengths. Pinholes were adjusted to give an optical slice of <5.0 μm. Hematoxylin/Eosin stained sections were evaluated by the pathologist and regions of defined histological grading were defined. Five images were taken randomly in corresponding immunofluorescence stained sections and analysed in ImageJ 1.49t (Rasband WS. ImageJ. U S National Institutes of Health, Bethesda, Maryland, USA, available online: http://rsb.info.nih.gov/ij/, 1997–2006) using self-written analysis tools. Data were transferred to Apache OpenOffice™ 3 (Apache Software Foundation, available online: https://www.apache.org) for calculations. Target mean fluorescence intensities were normalized to staining control (omission of primary antibody) fluorescence intensities.

4.4. SDS Page and Western Blotting

We used six samples of normal prostate tissue that was excised following radical resection of the prostate and examined by our pathologist (LCH) to ensure tumor-free samples. Seminal plasma samples were collected and stored in liquid nitrogen at −196 °C until use. For SDS-PAGE (sodium dodecyl sulphate polyacrylamide gel electrophoresis), prostate tissue specimens were thawed on ice and later extracted using 500 μL/100 mg 50 mmol/L Tris-HCL supplemented with 1% sodium dodecyl sulphate (SDS), 5 μL protease inhibitor cocktail (Sigma-Aldrich, Steinheim, Germany) and 5 μL phenylmethylsulfonylfluoride. The tissue was homogenized on ice using an Ultra Turrax T10 (Ika GmbH & Co., Staufen, Germany), further extracted for 1 h at 4°C and thereafter centrifuged at 500× *g* for 2 min at 4 °C. Protein concentration of the supernatant was determined by Pierce™ BCA Protein Assay Kit (Thermo Fischer Scientific, Rockford, IL, USA) using a NanoDrop 1000 spectrophotometer (PEQLAB Biotechnologie GmbH, Erlangen, Germany). Seminal plasma probes were thawed on ice and protein concentration was determined as indicated above. 30 or 50 μg protein was used per lane on NuPage Bis-Tris 4–12% gradient mini gels (Life Technologies, Darmstadt, Germany).

Western blotting was performed onto PVDF membranes according to supplier's protocol in a Bio-Rad Mini Trans-Blot® system (Bio-Rad Laboratories GmbH, München, Germany). Membranes were blocked for 1 h with 2% BSA (Carl Roth GmbH & Co. KG, Karlsruhe, Germany) followed by incubation with primary antibody incubation at 4 °C (Table 5). Anti-beta actin and anti-GAPDH antibodies were used as loading controls. For detection, we used goat-anti mouse IRDye 680 or goat-anti rabbit IRDye 680 secondary antibodies diluted at 1:8000 (LI-COR Biosciences GmbH, Bad Homburg, Germany). Blots were scanned using an Odyssey Infrared Imaging System (LI-COR).

4.5. Statistical Analysis

Statistics and graphing were done with Prism 5 (GraphPad Software Inc., La Jolla, CA, USA). Parameter free tests were used as indicated in the figure legends. All numbers are given as mean ± SEM (standard error of the mean), unless indicated otherwise.

5. Conclusions

We conclude that tissue levels of MMP3 and MMP7 (finely regulated by their inhibitors TIMP1 and TIMP2) are altered in PCa compared to benign prostatic hyperplasia, which is of special relevance in elderly men. These findings link tissue alterations to altered liquefaction cascade, indicating disease-related alterations in prostate tissue and hence in liquefaction cascade. For the first time, we present evidence for a contribution of seminal vesicles to the fine-tuning of the liquefaction protease network. Our findings thus support the concept of seminal plasma biomarkers as non-invasive tool for PCa detection and risk stratification in men aged 60–70 years.

Acknowledgments: The authors gratefully acknowledge the excellent technical support by Annett Weimann, Mandy Berndt-Paetz and Christine Kellner. We thank Vinodh Kumar Adithyaa Arthanareeswaran for language editing and proofreading of the manuscript.

Author Contributions: Conceived and designed the experiments: Jochen Neuhaus, Eric Schiffer. Performed the experiments: Jochen Neuhaus, Eric Schiffer, Ferdinando Mannello. Analyzed the data: Jochen Neuhaus, Eric Schiffer, Lars-Christian Horn. Contributed reagents/materials/analysis tools: Jens-Uwe Stolzenburg, Ferdinando Mannello. Wrote the manuscript: Jochen Neuhaus, Eric Schiffer, Ferdinando Mannello. Edited the manuscript: Jens-Uwe Stolzenburg, Roman Ganzer, Lars-Christian Horn.

Conflicts of Interest: The authors declare no conflict of interest. This study was funded in part by the German Ministry of Economy and Technology BMWi with grant No. KF0362802MD8 to Jens-Uwe Stolzenburg, Jochen Neuhaus and grant No. EP110141 to Eric Schiffer. The founding sponsors had no role in the design of the study; in the collection, analyses, or interpretation of data; in the writing of the manuscript, and in the decision to publish the results.

References

1. Siegel, R.; Naishadham, D.; Jemal, A. Cancer statistics, 2013. *CA Cancer J. Clin.* **2013**, *63*, 11–30. [CrossRef] [PubMed]
2. Hugosson, J.; Carlsson, S.; Aus, G.; Bergdahl, S.; Khatami, A.; Lodding, P.; Pihl, C.G.; Stranne, J.; Holmberg, E.; Lilja, H. Mortality results from the Goteborg randomised population-based prostate-cancer screening trial. *Lancet Oncol.* **2010**, *11*, 725–732. [CrossRef]

3. Carter, H.B. American Urological Association (AUA) guideline on prostate cancer detection: Process and rationale. *BJU Int.* **2013**, *112*, 543–547. [CrossRef] [PubMed]
4. Prensner, J.R.; Rubin, M.A.; Wei, J.T.; Chinnaiyan, A.M. Beyond PSA: The next generation of prostate cancer biomarkers. *Sci. Transl. Med.* **2012**. [CrossRef] [PubMed]
5. Roberts, M.J.; Richards, R.S.; Gardiner, R.A.; Selth, L.A. Seminal fluid: A useful source of prostate cancer biomarkers? *Biomark. Med.* **2015**, *9*, 77–80. [CrossRef] [PubMed]
6. Rajan, P.; Elliott, D.J.; Robson, C.N.; Leung, H.Y. Alternative splicing and biological heterogeneity in prostate cancer. *Nat. Rev. Urol.* **2009**, *6*, 454–460. [CrossRef] [PubMed]
7. Good, D.M.; Thongboonkerd, V.; Novak, J.; Bascands, J.L.; Schanstra, J.P.; Coon, J.J.; Dominiczak, A.; Mischak, H. Body fluid proteomics for biomarker discovery: Lessons from the past hold the key to success in the future. *J. Proteome Res.* **2007**, *6*, 4549–4555. [CrossRef] [PubMed]
8. Neuhaus, J.; Schiffer, E.; von Wilcke, P.; Bauer, H.W.; Leung, H.; Siwy, J.; Ulrici, W.; Paasch, U.; Horn, L.-C.; Stolzenburg, J.U. Seminal Plasma as a Source of Prostate Cancer Peptide Biomarker Candidates for Detection of Indolent and Advanced Disease. *PLoS ONE* **2013**, *8*, e67514. [CrossRef] [PubMed]
9. Batruch, I.; Lecker, I.; Kagedan, D.; Smith, C.R.; Mullen, B.J.; Grober, E.; Lo, K.C.; Diamandis, E.P.; Jarvi, K.A. Proteomic analysis of seminal plasma from normal volunteers and post-vasectomy patients identifies over 2000 proteins and candidate biomarkers of the urogenital system. *J. Proteome Res.* **2011**, *10*, 941–953. [CrossRef] [PubMed]
10. Drake, R.R.; Elschenbroich, S.; Lopez-Perez, O.; Kim, Y.; Ignatchenko, V.; Ignatchenko, A.; Nyalwidhe, J.O.; Basu, G.; Wilkins, C.E.; Gjurich, B.; et al. In-depth proteomic analyses of direct expressed prostatic secretions. *J. Proteome Res.* **2010**, *9*, 2109–2116. [CrossRef] [PubMed]
11. Pilch, B.; Mann, M. Large-scale and high-confidence proteomic analysis of human seminal plasma. *Genome Biol.* **2006**. [CrossRef] [PubMed]
12. Emami, N.; Deperthes, D.; Malm, J.; Diamandis, E.P. Major role of human KLK14 in seminal clot liquefaction. *J. Biol. Chem.* **2008**, *283*, 19561–19569. [CrossRef] [PubMed]
13. Mason, S.D.; Joyce, J.A. Proteolytic networks in cancer. *Trends Cell. Biol.* **2011**, *21*, 228–237. [CrossRef] [PubMed]
14. Murphy, G. Tissue inhibitors of metalloproteinases. *Genome Biol.* **2011**. [CrossRef] [PubMed]
15. Wood, M.; Fudge, K.; Mohler, J.L.; Frost, A.R.; Garcia, F.; Wang, M.; Stearns, M.E. In situ hybridization studies of metalloproteinases 2 and 9 and TIMP-1 and TIMP-2 expression in human prostate cancer. *Clin. Exp. Metastasis* **1997**, *15*, 246–258. [CrossRef] [PubMed]
16. Gong, Y.; Chippada-Venkata, U.D.; Oh, W.K. Roles of matrix metalloproteinases and their natural inhibitors in prostate cancer progression. *Cancers* **2014**, *6*, 1298–1327. [CrossRef] [PubMed]
17. Kessenbrock, K.; Plaks, V.; Werb, Z. Matrix metalloproteinases: Regulators of the tumor microenvironment. *Cell* **2010**, *141*, 52–67. [CrossRef] [PubMed]
18. Taylor, R.A.; Toivanen, R.; Frydenberg, M.; Pedersen, J.; Harewood, L.; Australian, P.C.B.; Collins, A.T.; Maitland, N.J.; Risbridger, G.P. Human epithelial basal cells are cells of origin of prostate cancer, independent of CD133 status. *Stem Cells* **2012**, *30*, 1087–1096. [CrossRef] [PubMed]
19. Varambally, S.; Yu, J.; Laxman, B.; Rhodes, D.R.; Mehra, R.; Tomlins, S.A.; Shah, R.B.; Chandran, U.; Monzon, F.A.; Becich, M.J.; et al. Integrative genomic and proteomic analysis of prostate cancer reveals signatures of metastatic progression. *Cancer Cell* **2005**, *8*, 393–406. [CrossRef] [PubMed]
20. Toth, M.; Chvyrkova, I.; Bernardo, M.M.; Hernandez-Barrantes, S.; Fridman, R. Pro-MMP-9 activation by the MT1-MMP/MMP-2 axis and MMP-3: Role of TIMP-2 and plasma membranes. *Biochem. Biophys. Res. Commun.* **2003**, *308*, 386–395. [CrossRef]
21. Arumugam, S.; Van Doren, S.R. Global orientation of bound MMP-3 and N-TIMP-1 in solution via residual dipolar couplings. *Biochemistry* **2003**, *42*, 7950–7958. [CrossRef] [PubMed]
22. Fritz, H. Human mucus proteinase inhibitor (human MPI). Human seminal inhibitor I (HUSI-I), antileukoprotease (ALP), secretory leukocyte protease inhibitor (SLPI). *Biol. Chem. Hoppe Seyler.* **1988**, *369*, 79–82. [PubMed]
23. Ohlsson, K.; Bjartell, A.; Lilja, H. Secretory leukocyte protease inhibitor in the male genital tract: PSA-induced proteolytic processing in human semen and tissue localization. *J. Androl.* **1995**, *16*, 64–74. [PubMed]

24. Caughey, G.H.; Schaumberg, T.H.; Zerweck, E.H.; Butterfield, J.H.; Hanson, R.D.; Silverman, G.A.; Ley, T.J. The human mast cell chymase gene (*CMA1*): Mapping to the cathepsin G/granzyme gene cluster and lineage-restricted expression. *Genomics* **1993**, *15*, 614–620. [CrossRef] [PubMed]
25. Korkmaz, B.; Horwitz, M.S.; Jenne, D.E.; Gauthier, F. Neutrophil elastase, proteinase 3, and cathepsin G as therapeutic targets in human diseases. *Pharmacol. Rev.* **2010**, *62*, 726–759. [CrossRef] [PubMed]
26. McCauley, T.C.; Zhang, H.M.; Bellin, M.E.; Ax, R.L. Identification of a heparin-binding protein in bovine seminal fluid as tissue inhibitor of metalloproteinases-2. *Mol. Reprod. Dev.* **2001**, *58*, 336–341. [CrossRef]
27. Mannello, F.; Gazzanelli, G. Tissue inhibitors of metalloproteinases and programmed cell death: Conundrums, controversies and potential implications. *Apoptosis* **2001**, *6*, 479–482. [CrossRef] [PubMed]
28. Sinno, M.; Biagioni, S.; Ajmone-Cat, M.A.; Pafumi, I.; Caramanica, P.; Medda, V.; Tonti, G.; Minghetti, L.; Mannello, F.; Cacci, E. The matrix metalloproteinase inhibitor marimastat promotes neural progenitor cell differentiation into neurons by gelatinase-independent TIMP-2-dependent mechanisms. *Stem Cells Dev.* **2013**, *22*, 345–358. [CrossRef] [PubMed]
29. Baumgart, E.; Lenk, S.V.; Loening, S.A.; Jung, K. Tissue inhibitors of metalloproteinases 1 and 2 in human seminal plasma and their association with spermatozoa. *Int. J. Androl.* **2002**, *25*, 369–371. [CrossRef] [PubMed]
30. Shimokawa, K.; Katayama, M.; Matsuda, Y.; Takahashi, H.; Hara, I.; Sato, H. Complexes of gelatinases and tissue inhibitor of metalloproteinases in human seminal plasma. *J. Androl.* **2003**, *24*, 73–77. [PubMed]
31. McVary, K.T. BPH: Epidemiology and comorbidities. *Am. J. Manag. Care* **2006**, *12*, S122–S128. [PubMed]
32. Szarvas, T.; Becker, M.; Vom Dorp, F.; Meschede, J.; Scherag, A.; Bankfalvi, A.; Reis, H.; Schmid, K.W.; Romics, I.; Rubben, H.; et al. Elevated serum matrix metalloproteinase 7 levels predict poor prognosis after radical prostatectomy. *Int. J. Cancer* **2011**, *128*, 1486–1492. [CrossRef] [PubMed]
33. Babichenko, I.I.; Pul'bere, S.A.; Motin, P.I.; Loktev, A.V.; Abud, M. Significance of matrix metalloproteinase-9, tissue inhibitor of metalloproteinase and protein Ki-67 in prostate tumors. *Urologiia.* **2014**, *5*, 82–86.
34. Reis, S.T.; Viana, N.I.; Iscaife, A.; Pontes-Junior, J.; Dip, N.; Antunes, A.A.; Guimarães, V.R.; Santana, I.; Nahas, W.C.; Srougi, M.; et al. Loss of TIMP-1 immune expression and tumor recurrence in localized prostate cancer. *Int. Braz. J. Urol.* **2015**, *41*, 1088–1095. [CrossRef] [PubMed]
35. Livak, K.J.; Schmittgen, T.D. Analysis of relative gene expression data using real-time quantitative PCR and the $2^{-\Delta\Delta Ct}$ Method. *Methods* **2001**, *25*, 402–408. [CrossRef] [PubMed]

International Journal of
Molecular Sciences

MDPI

Article

Fasting Enhances the Contrast of Bone Metastatic Lesions in ^{18}F-Fluciclovine-PET: Preclinical Study Using a Rat Model of Mixed Osteolytic/Osteoblastic Bone Metastases

Shuntaro Oka [1,*], Masaru Kanagawa [1], Yoshihiro Doi [1], David M. Schuster [2], Mark M. Goodman [2] and Hirokatsu Yoshimura [1]

1   Research Center, Nihon Medi-Physics Co., Ltd., 3-1 Kitasode, Sodegaura, Chiba 299-0266, Japan; masaru_kanagawa@nmp.co.jp (M.K.); yoshihiro_doi@nmp.co.jp (Y.D.); hirokatsu_yoshimura@nmp.co.jp (H.Y.)
2   Division of Nuclear Medicine and Molecular Imaging, Department of Radiology and Imaging Sciences, Emory University, Atlanta, GA 30329, USA; dschust@emory.edu (D.M.S.); mgoodma@emory.edu (M.M.G.)
*   Correspondence: shuntaro_oka@nmp.co.jp; Tel.: +81-438-61-1627; Fax: +81-438-62-6969

Academic Editor: Carsten Stephan
Received: 27 March 2017; Accepted: 26 April 2017; Published: 29 April 2017

Abstract: ^{18}F-fluciclovine (*trans*-1-amino-3-^{18}F-fluorocyclobutanecarboxylic acid) is an amino acid positron emission tomography (PET) tracer used for cancer staging (e.g., prostate and breast). Patients scheduled to undergo amino acid-PET are usually required to fast before PET tracer administration. However, there have been no reports addressing whether fasting improves fluciclovine-PET imaging. In this study, the authors investigated the influence of fasting on fluciclovine-PET using triple-tracer autoradiography with ^{14}C-fluciclovine, [5,6-^{3}H]-2-fluoro-2-deoxy-D-glucose (^{3}H-FDG), and ^{99m}Tc-hydroxymethylene diphosphonate (^{99m}Tc-HMDP) in a rat breast cancer model of mixed osteolytic/osteoblastic bone metastases in which the animals fasted overnight. Lesion accumulation of each tracer was evaluated using the target-to-background (muscle) ratio. The mean ratios of ^{14}C-fluciclovine in osteolytic lesions were 4.6 ± 0.8 and 2.8 ± 0.6, respectively, with and without fasting, while those for ^{3}H-FDG were 6.9 ± 2.5 and 5.1 ± 2.0, respectively. In the peri-tumor bone formation regions (osteoblastic), where ^{99m}Tc-HMDP accumulated, the ratios of ^{14}C-fluciclovine were 4.3 ± 1.4 and 2.4 ± 0.7, respectively, and those of ^{3}H-FDG were 6.2 ± 3.8 and 3.3 ± 2.2, respectively, with and without fasting. These results suggest that fasting before ^{18}F-fluciclovine-PET improves the contrast between osteolytic and osteoblastic bone metastatic lesions and background, as well as ^{18}F-FDG-PET.

Keywords: bone metastasis; breast cancer; FACBC; fluciclovine; FDG; positron emission tomography; prostate cancer

1. Introduction

The most used positron emission tomography (PET) tracer for cancer imaging is 2-deoxy-2-^{18}F-fluoro-D-glucose (^{18}F-FDG), which is an analogue of D-glucose. Cancer cells take up this PET tracer because ^{18}F-FDG shares its transport routes (i.e., glucose transporters) with D-glucose. High concentrations of plasma glucose may compete with ^{18}F-FDG at glucose transporters in cancer cells and elevates plasma insulin which unfavorably alters the biodistribution to the radiotracer [1]. Thus, patients scheduled to undergo ^{18}F-FDG-PET are required to fast for 4 h to 6 h to lower plasma glucose concentration before PET imaging.

Several amino acid (AA) PET tracers, including ^{11}C-methionine, have been used for diagnostic purposes in cancer patients [2]. Although studied previously, issues regarding the necessity of fasting in clinical PET with AA tracers [3] have yet to be resolved. Fasting is routinely recommended for patients undergoing AA-PET as well as ^{18}F-FDG-PET. There have been few studies investigating the influence of fasting or food intake before AA-PET. The authors of one study investigating ^{11}C-methionine-PET for head and neck cancer concluded that food ingestion may decrease ^{11}C-methionine uptake in tumors, although the image quality remained satisfactory after food ingestion [4].

[18]F-fluciclovine, also known as *trans*-1-amino-3-^{18}F-fluorocyclobutanecarboxylic acid (FACBC; code name: NMK36, Nihon Medi-Physics; brand name: Axumin, Blue Earth Diagnostics, Burlington, MA, USA), is an AA-PET tracer, and has been approved by the United States Food and Drug Administration for the detection of prostate cancer (PCa) recurrence since 2016. Even in ^{18}F-fluciclovine-PET imaging, patients usually fast for ≥ 4 h before tracer administration without any conclusive evidence supporting the influence of fasting for ^{18}F-fluciclovine uptake in tumors [5]. Although one clinical research study involving non-fasted breast cancer (BCa) patients has been published, PET imaging was performed for the chest only and no comparison was made to control fasted subjects [6]. Thus, it is not known whether fasting influences tumor uptake of ^{18}F-fluciclovine in PCa and BCa patients.

Most patients with advanced PCa and BCa present with mixed osteolytic and osteoblastic bone metastases [7]. To investigate the influence of fasting on ^{18}F-fluciclovine accumulation in bone metastatic lesions, we established a rat bone metastatic model, which forms both osteolytic and osteoblastic lesions in the tibia and/or femur, by intra-arterial injection of MRMT-1 cells, a rat BCa cell line with a 100% success rate for skeletal metastases. Using this animal model, triple-tracer autoradiography was performed using *trans*-1-amino-3-fluoro[1-^{14}C]cyclobutanecarboxylic acid (^{14}C-fluciclovine), [5,6-^{3}H]-2-Fluoro-2-deoxy-d-glucose (^{3}H-FDG), and ^{99m}Tc-hydroxymethylene diphosphonate (^{99m}Tc-HMDP) to compare the tracers' accumulation at identical bony lesions. Our findings suggest that overnight fasting influenced the uptake of ^{18}F-fluciclovine in the bony lesions, and that the tumor-to-muscle uptake ratios in osteolytic and osteoblastic lesions were higher in the fasted condition compared with the fed condition.

2. Results

2.1. In Vitro Experiments

[14]C-fluciclovine is a synthetic AA and is transported by AA transporters (AATs). To investigate which transport system mediates the transport of this tracer in MRMT-1 cells, in vitro uptake inhibition experiments were performed. As shown in Figure 1a, the uptake of ^{14}C-fluciclovine decreased to 38.4% in choline buffer compared with sodium buffer, corresponding to contributions of the Na[+]-dependent and -independent carriers for ^{14}C-fluciclovine transport in MRMT-1 cells of 61.6% and 38.4%, respectively. To narrow the subtypes of AATs involved in ^{14}C-fluciclovine transport in MRMT-1 cells, competitive uptake experiments were performed in the absence or presence of several synthetic and naturally-occurring AAs as inhibitors. As shown in Figure 1b, small neutral AAs, such as glutamine and serine, showed strong inhibitory effects for ^{14}C-fluciclovine uptake (82.4% and 77.0% decreases vs. control, respectively), while phenylalanine, a branched-chain AA, and 2-amino-bicyclo[2,2,1]heptane-2-carboxylic acid (BCH) demonstrated smaller inhibitory effects (33.3% and 37.6% decreases, respectively, vs. control) in the presence of sodium ion. In contrast, phenylalanine and BCH inhibited ^{14}C-fluciclovine transport by more than 90% versus control in the absence of sodium ion. Proline, 2-(methylamino)-isobutyric acid (MeAIB), arginine, and glutamate had no statistically-significant inhibitory effect on ^{14}C-fluciclovine transport into MRMT-1 cells. These results suggest that the strong inhibitory effects of glutamine and serine in the presence of sodium ion mean that the alanine-serine-cysteine (ASC) system, especially the ASC transporter 2 (ASCT2), is the primary carrier in Na[+]-dependent AATs. On the other hand, the intense inhibitory effects of phenylalanine and BCH in the absence of sodium ion suggest that system L, especially the L-type AA

transporter (LAT1), is the primary carrier in Na$^+$-independent AATs for the transport of ^{14}C-fluciclovine in MRMT-1 cells.

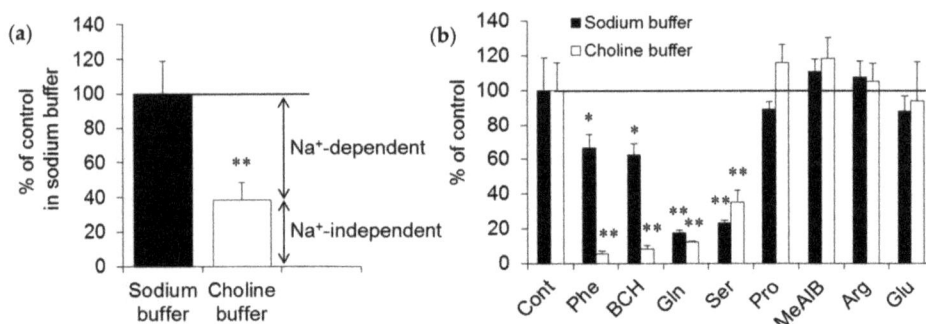

Figure 1. Characteristics of ^{14}C-fluciclovine transport in MRMT-1 cells. (**a**) Contributions of Na$^+$-dependent and -independent carriers on the uptake of ^{14}C-fluciclovine. The transport of ^{14}C-fluciclovine in sodium buffer was normalized to 100%. (**b**) Competitive inhibition of 10 μmol/L ^{14}C-fluciclovine transport by naturally-occurring and synthetic amino acids (2.0 mmol/L). The control transport (absence of inhibitors) of ^{14}C-fluciclovine in sodium and choline buffer was normalized to 100%. Each bar represents the mean ± standard deviation (SD) (*n* = 6). * *p* < 0.05, ** *p* < 0.01. Arg: arginine, BCH: 2-amino-bicyclo[2,2,1]heptane-2-carboxylic acid, Cont: control, Gln: glutamine, Glu: glutamate, MeAIB: 2-(methylamino)-isobutyric acid, Phe: phenylalanine, Pro: proline, Ser: serine.

2.2. In Vivo Experiments

The distribution patterns of ^{14}C-fluciclovine in the tibia and/or femur were basically similar to ^{3}H-FDG in both fasted and fed conditions, as shown in Figure 2. Visualization of the tumor mass in bone marrow cavities was clearer with ^{3}H-FDG because the physiological accumulation of ^{14}C-fluciclovine in bone marrow was higher than ^{3}H-FDG. Although ^{14}C-fluciclovine and ^{3}H-FDG accumulated at the peri-tumor bone formation (pTBF) in osteoblastic lesions, characterized by ^{99m}Tc-HMDP accumulation, the images were obscure compared with the tumor parenchyma (Figure 2).

Figure 3 shows the results of histological examination in osteolytic and osteoblastic bone metastasis lesions in a representative rat that was fed. In the osteolytic lesion (black frame on the toluidine blue (TB) image in Figure 3b), the absorption pits with tartrate-resistant acid phosphatase (TRAP) activity indicating osteoclast infiltration was observed. Hematoxylin and eosin (H&E) staining revealed many osteoclasts, characterized by multiple nuclei on the surface of cortical bone in the pits (Figure 3c). On the other hand, the osteoblastic lesions (the yellow frame on the TB image in Figure 3b) revealed alkaline phosphatase (ALP) activity and pTBF, indicating the appearance of osteoblasts and osteoids, respectively, between tumor mass and the surface of cortical bone (H&E and TB images in Figure 3c). Almost all of the tumor cells in the osteolytic lesion were positive for ASCT2 (upper row in Figure 3c). On the other hand, ASCT2 expression in intra-tumoral cells was scant compared with that in the peripheral cells of the tumor mass and osteoblasts on the bone surface in the osteoblastic lesion (lower row in Figure 3c). LAT1-positive cells were scattered in all of the tumor tissues from both osteolytic and osteoblastic lesions (Figure 3c). No obvious differences were observed in the expression of ASCT2 and LAT1 between the fasted and fed rats (supplementary Figure S1). These results demonstrate that ^{14}C-fluciclovine and ^{3}H-FDG accumulated in histologically confirmed osteolytic and osteoblastic bone metastasis lesions.

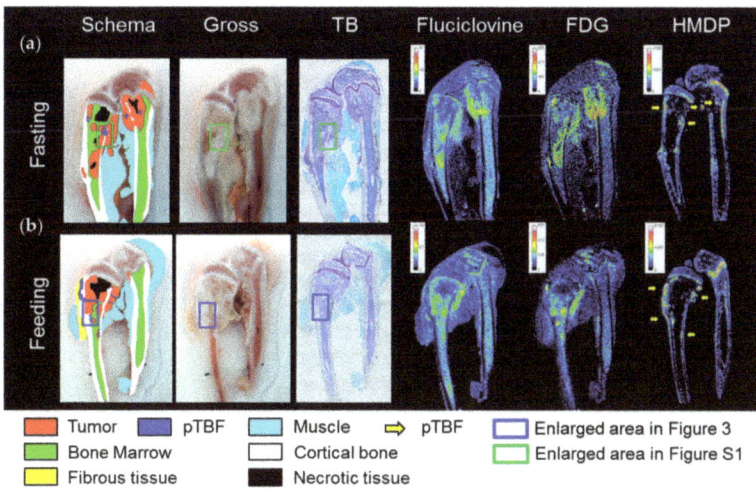

Figure 2. Triple-tracer autoradiography using ^{14}C-fluciclovine, 2-deoxy-2-^{18}F-fluoro-d-glucose (^{3}H-FDG), and ^{99m}Tc-hydroxymethylene diphosphonate (^{99m}Tc-HMDP) in breast cancer bone metastasis model rats. Macroscopic images (schema, gross, toluidine blue (TB) staining) and autoradiograms of each tracer in (**a**) fasting and (**b**) fed rats are shown. Each image was adjusted for optimal contrast and color scale bars on each autoradiogram represent Bq range for each tracer. The high-power microscopic fields of typical osteolytic and osteoblastic lesions corresponding to the blue and green frames on the macroscopic images in Figure 2a,b are shown in Figure 3 and supplementary Figure S1, respectively. The lesions correspond to ^{99m}Tc-HMDP-positive areas, except for physiological accumulation in growth plates, which were considered peri-tumor bone formation (pTBF) in osteoblastic lesions (yellow arrows).

Semi-quantitative analyses were used to evaluate tumor-to-muscle accumulation ratios of ^{14}C-fluciclovine and ^{3}H-FDG are shown in Figure 4 and Table 1. The target$_{mean}$ (metastatic lesion)-to-background$_{mean}$ (muscle) (T/BG) ratio of both tracers were statistically higher in the fasting condition than in the fed condition in osteolytic and pTBF lesions ($p < 0.01$). Comparing tracers, the ratios of ^{3}H-FDG were statistically higher than ^{14}C-fluciclovine in osteolytic lesions ($p < 0.01$), but not in pTBF lesions, under fasting and fed conditions. The distributions of T/BG ratios of ^{3}H-FDG in both lesions were wider than that of ^{14}C-fluciclovine, as shown in Figure 4 and Table 1.

Table 1. Summary of the distributions of T/BG ratios of ^{3}H-FDG and ^{14}C-fluciclovine in the osteolytic lesions and pTBF in osteoblastic lesions of breast cancer bone metastasis model rat with and without overnight fasting.

Lesion	Tracer	Diet	Min.	Max.	Median	1st Qu.	3rd Qu.	Mean $\pm$ S.D.
OL	^{3}H-FDG	Fasting	2.9	13.2	6.2	5.5	7.0	6.9 $\pm$ 2.5
		Feeding	1.6	10.7	5.0	3.7	6.1	5.1 $\pm$ 2.0
	^{14}C-fluciclovine	Fasting	3.1	6.8	4.4	3.9	5.1	4.6 $\pm$ 0.8
		Feeding	1.4	5.9	2.7	2.4	3.1	2.8 $\pm$ 0.6
pTBF	^{3}H-FDG	Fasting	1.6	12.8	5.6	3.7	10.5	6.6 $\pm$ 3.8
		Feeding	0.6	8.2	2.8	1.7	4.3	3.3 $\pm$ 2.2
	^{14}C-fluciclovine	Fasting	2.6	7.6	4.1	3.8	4.9	4.5 $\pm$ 1.4
		Feeding	1.0	3.9	2.5	1.8	2.9	2.4 $\pm$ 0.7

1st Qu.: first quartile, 3rd Qu.: third quartile, ^{3}H-FDG: 2-deoxy-2-^{18}F-fluoro-d-glucose, Max.: maximum value of T/BG ratios, Min.: minimum value of T/BG ratios, pTBF: peri tumor bone formation, OL: osteolytic, S.D.: standard deviation, T/BG: target$_{mean}$ (metastatic lesion)-to-background$_{mean}$ (muscle).

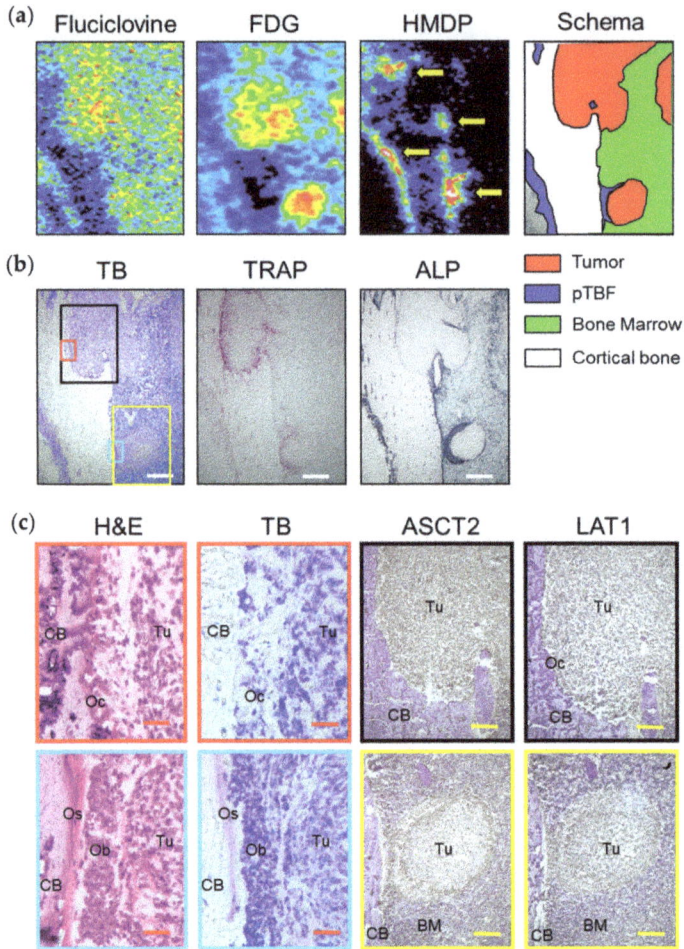

Figure 3. Comparison of the tracer accumulations of ^{14}C-fluciclovine, 2-deoxy-2-^{18}F-fluoro-d-glucose (^{3}H-FDG) and ^{99m}Tc-hydroxymethylene diphosphonate (^{99m}Tc-HMDP), and the histological characteristics of typical osteolytic and osteoblastic lesions in a representative breast cancer bone metastasis model rat that was fed. (**a**) The enlarged autoradiograms and the schema and (**b**) the histological images (toluidine blue (TB), tartrate-resistant acid phosphatase (TRAP), alkaline phosphatase (ALP)) correspond to the blue frame on the schema in Figure 2b are represented. The lesions corresponding to ^{99m}Tc-HMDP-positve were considered peri-tumor bone formation (pTBF) in osteoblastic lesions (yellow arrows). (**c**) The high-power microscopic fields (hematoxylin and eosin (H&E), TB, alanine-serine-cysteine transporter 2 (ASCT2), L-type amino acid transporter 1 (LAT1)) correspond to the black, red, yellow, and cyan frames on the TB image in Figure 3b are shown. The red, yellow, and white scale bars on each panel correspond to 50 μm, 200 μm, and 500 μm, respectively. BM: bone marrow, CB: cortical bone, Ob: osteoblasts, Oc: osteoclasts, Os: osteoids, Tu: tumor.

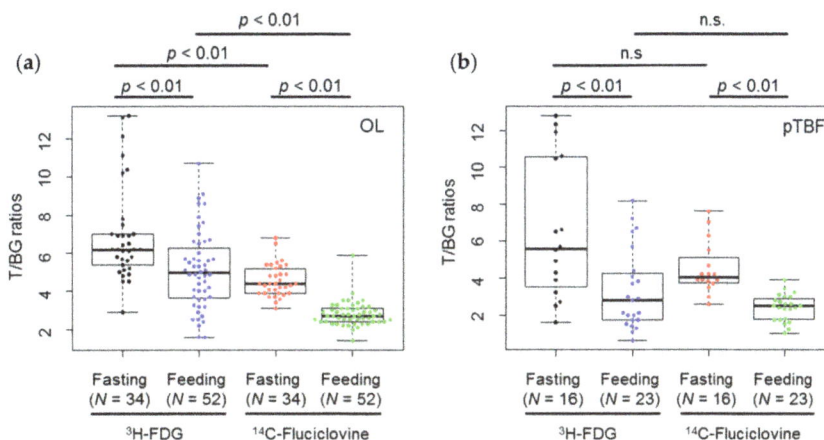

Figure 4. Beeswarm and box plots showing the distributions of target$_{mean}$ (metastatic lesion)-to-background$_{mean}$ (muscle) (T/BG) ratios of 2-deoxy-2-^{18}F-fluoro-D-glucose (^{3}H-FDG) and ^{14}C-fluciclovine in (**a**) osteolytic lesions (OL), and (**b**) peri-tumor bone formation (pTBF) in osteoblastic lesions in breast cancer bone metastasis model rats, with and without overnight fasting. The numbers under each box indicate the number of lesions. n.s.: not significant.

3. Discussion

We aimed to determine whether fasting before ^{18}F-fluciclovine-PET imaging improved visualization of bone metastasis lesions using a rat BCa bone metastasis model exhibiting osteolytic and osteoblastic lesions. Our findings suggest that fasting before fluciclovine-PET imaging improves the background contrast between osteolytic/osteoblastic bony lesions and muscle.

First, we investigated the AATs involved in the uptake of ^{14}C-fluciclovine in MRMT-1 cells, a rat BCa cell line. We confirmed that ASCT2 and LAT1 were the primary AATs for ^{14}C-fluciclovine transport in MRMT-1 cells, as well as prostate and brain cancer cell lines [8–12]. Second, we designed a bone metastasis model and injected MRMT-1 cells into the saphenous artery of the hind legs of rats. In this model, the mixed lesions of osteolytic and osteoblastic metastases, closely mimicking clinical findings [7], were formed in the tibia and/or femur.

Using this animal model, we performed triple-tracer autoradiography with ^{14}C-fluciclovine/^{3}H-FDG/^{99m}Tc-HMDP and evaluated the accumulation of each tracer in identical bony lesions. Comparing the T/BG ratios of ^{14}C-fluciclovine accumulation in rats that were fasted overnight with those in rats fed ad libitum, the ratios were higher in both osteolytic and osteoblastic lesions under the fasted condition than in the fed condition. It has been reported that the sum concentration of all plasma AAs decrease by 10% in rats fasted for one day, although the decreases do not involve drastic changes, unlike blood glucose (−35%) [13]. Among the AAs, proline (−48.9%), alanine (−21.8%), histidine (−18.2%), and glutamine (−13.1%), which are substrates of the ASC or L systems, demonstrated relatively high reduction rates in this experiment [13]. Thus, we expected that the concentration of these neutral AAs in rat plasma would be decreased by fasting before the injection of ^{14}C-fluciclovine and, consequently, that the autoradiography images of tumor lesions would be improved by increased ^{14}C-fluciclovine transport into the cancer cells. The T/BG ratios of ^{14}C-fluciclovine were, in fact, statistically higher under the fasted condition compared with the fed condition. Thus, we believe that it is reasonable to prescribe fasting to PCa patients before fluciclovine-PET, not only to decrease plasma AA concentration, but to force the AA concentration toward favoring fluciclovine transport, thus achieving a more quantitative fluciclovine-PET scan.

There have been some reports describing the influence of meal or plasma AAs on AA-tracer accumulation in tumors. Lindholm et al. [4] performed an intra-individual clinical study to investigate whether the ingestion of a liquid meal influenced the accumulation of ^{11}C-methionine, which is transported by LAT1 [10,11], in patients with head and neck cancer. In that study, initial PET imaging was performed after an overnight fast; the second PET imaging was then performed after ingesting the liquid meal containing L-methionine and branched-chain AAs 6 to 7 days after the first PET imaging. PET imaging was initiated 45 min after ingestion. The standard uptake values of ^{11}C-methionine in the tumor after ingestion decreased to a statistically lower level than that in tumors after fasting, although the tracer entering rate (i.e., K_1) into the tumor was not changed substantially [4]. Additionally, ^{11}C-methionine uptake in gliomas and normal brain tissue, while the patient received an intravenous infusion of branched-chain AAs, decreased [14]. Similar results were observed in glioma patients injected with 3-^{123}I-Iodo-L-α-methyltyrosine (^{123}I-3-IMT), which is known to be transported by LAT1, with intravenous infusion of a mixture of naturally-occurring L-AAs during imaging [15]. The decreasing level of tracer accumulation in tumors observed in these three studies was believed to be caused by competitive inhibition between the exogenous AAs and tracers on LAT1. Thus, it is believed that ingesting food containing abundant AAs before, or intravenous infusion of AAs during, fluciclovine-PET imaging decreases the tumor accumulation of ^{18}F-fluciclovine in PCa patients because ^{18}F-fluciclovine is also transported by LAT1.

ASCT2 and LAT1 are obligatory AA exchangers with a 1:1 stoichiometry. If cancer cells are preloaded with AAs, which can trans-stimulate the influx of extracellular AAs (trans-stimulators), the velocity of AA transport through the AA exchangers is accelerated [8]. Thus, the preinjection of trans-stimulators into cancer patients would be expected to accelerate the uptake of PET tracers into cancer cells. Based on this hypothesis, Lahoutte et al. investigated the effect of intraperitoneal preinjection (30 min before tracer injection) of each naturally-occurring AA (arginine, aspartate, glutamate, phenylalanine, proline, tryptophan) on the uptake of ^{123}I-3-IMT into the subcutaneous tumor of Rhabdomyosarcoma (R1M) tumor-bearing rats fasted for 4 h [16]. They found that the strongest effect was observed with the preinjection of tryptophan, which is a substrate of LAT1, although all AAs used in the study accelerated the uptake of ^{123}I-3-IMT into the tumor. This increase of ^{123}I-3-IMT transport into the subcutaneous tumor is explainable and based on trans-stimulation by preloaded tryptophan via LAT1. Accordingly, it is believed that ^{18}F-fluciclovine transport also would be accelerated by the preinjection of trans-stimulators because this PET tracer is also recognized by ASCT2 and LAT1. On the other hand, Lahoutte's findings are discrepant with findings involving ^{11}C-methionine reported by Lindholm et al. [4]. The reason for the discrepancy may be the result of differences in experimental protocols between the studies: the injected substance (liquid meal vs. AA), the injection route (oral vs. intravenous), the species (rats vs. human), and fasting duration before tracer injection (4 h vs. non-fasting). Although fasting and preloading AAs appear to be conflicting treatments, there is a possibility that both may improve the visualization of PCa on fluciclovine-PET. Sophisticated, non-clinical studies are needed to determine the optimal imaging conditions for fluciclovine-PET, including variables such as fasting duration, type and dose of trans-stimulators, and timing of ^{18}F-fluciclovine administration after fasting, or preinjection of trans-stimulators, because the metabolism of AAs in rats differs from that in humans.

We demonstrated that the T (tumor)/BG (muscle) ratios of ^{14}C-fluciclovine in osteolytic and osteoblastic lesions in fasted rats were higher than in rats that were fed. We believe that the increased rate of ^{14}C-fluciclovine uptake in cancer cells was greater than in muscular cells under experimental starvation. Our hypothesis for these findings is as follows: the most abundant and important neutral AA for mammalian cells is glutamine, which is transported by ASCT2 and sodium-dependent neutral amino acid transporter 2 (SNAT2) [17]. In muscle, SNAT2 is abundantly expressed compared with ASCT2 [18], and regulates intramuscular glutamine concentration [19–21]. A portion of intracellular glutamine is used for transport exchange with extracellular leucine via LAT1, and leucine regulates the production and degradation of protein in muscle as an activator for the mammalian target of rapamycin complex 1 (mTORC1), a key player in cellular functions such as protein and lipid synthesis, glycolysis, autophagy, etc. [22,23]. Decreases in intramuscular leucine represses mTORC1 signaling under fasting conditions, followed by the activation of the autophagic pathway, protein degradation, and the production of AAs [24]. The AAs are released from muscle and delivered to the liver and used for glycogenesis [25]. Thus, the efflux of AAs from muscular cells is facilitated under the starvation condition.

On the other hand, many types of cancers co-express ASCT2 and LAT1 on the cell surface [26], and these AATs function as key players for AA transport including glutamine and leucine in cancer cells [17]. In fact, many studies have reported that the co-expression of ASCT2 and LAT1 correlates with malignancy and prognosis in several types of cancer [27–33]. Although cancer cells can also actuate an autophagic process during starvation, the AAs produced by protein degradation in cancer cells are used for cancer-cell survival [34]. Instead, cancer facilitates the autophagic process in muscle tissue [35]. Thus, it is believed that cancer cells accelerate the influx of AAs rather than efflux, unlike muscle under fasting conditions. Because ASCT2 and LAT1 are the primary AATs mediating ^{14}C-fluciclovine transport, we speculate that the increased rate of ^{14}C-fluciclovine uptake in cancer cells is greater than in muscle cells under conditions of starvation. Therefore, the T/BG ratios of ^{14}C-fluciclovine in the bone metastatic lesions in fasted rats were higher than in rats that were fed.

As mentioned, AA metabolism is a complex process, and different in humans and rats. The duration of fasting for fluciclovine-PET (4 to 6 h) scans in clinical settings and our animal experiment (17 to 18 h) is quite different. Thus, there is a possibility that the influence of fasting on ^{18}F-fluciclovine accumulation in cancer tissue is reduced in clinical practice compared with the influence observed in our animal experiments. Additionally, the imaging time in the current study was 30 min, which falls within the plateau phase of ^{18}F-fluciclovine uptake in orthotopic transplanted PCa of a previously described animal model [36]. On the other hand, imaging results at 30 min correspond to the late phase of a clinical fluciclovine-PET scan (typically 5 to 30 min); by that time, ^{18}F-fluciclovine uptake decreases in prostate, lymph node, and bone lesions [37,38]. Consequently, uptake may have been underestimated in the current animal experiment. Further in vitro and in vivo studies are needed to extrapolate our findings to humans. Furthermore, intra-individual clinical studies are required to elucidate whether fasting and/or preloading trans-stimulators before fluciclovine-PET influences tumor accumulation of ^{18}F-fluciclovine and improves fluciclovine-PET imaging in cancer patients.

4. Materials and Methods

4.1. Reagents

All reagents were purchased from Life Technologies (Carlsbad, CA, USA), Sigma-Aldrich (St. Louis, MO, USA), Nacalai Tesque (Kyoto, Japan), and Wako Pure Chemical Industries (Osaka, Japan), unless otherwise indicated. ^{14}C-labeled fluciclovine (^{14}C-fluciclovine) and ^{3}H-labeled FDG (^{3}H-FDG) were used instead of ^{18}F-labeled tracers because their long half-lives make them more suitable for experiments than tracers labeled with [18]F (t $\frac{1}{2}$ = 110 min). ^{14}C-Fluciclovine (specific activity, 2.08 GBq/mmol; radiochemical purity (RCP) >98%), which has the same chemical structure as ^{18}F-fluciclovine, with the exception of the position of the radioisotope, was commercially synthesized by Nemoto Science (Tokyo, Japan), as described previously [8]. ^{3}H-FDG (specific actvity 2.22 TBq/mmol; RCP, >99%) was purchased from American Radiolabeled Chemicals (St. Louis, MO, USA). The authors' company, Nihon Medi-Physics (Tokyo, Japan), produced ^{99m}Tc-HMDP (740 MBq/vial; RCP, >95%). All AAs used in this study were the L-form.

4.2. Cell Culture

A rat BCa cell line, MRMT-1, was procured from RIKEN BioResource Center (Tsukuba, Japan; obtained in January 2014) and maintained in RPMI 1640 medium supplemented with 10% fetal bovine serum (American Type Culture Collection, Rockville, MD, USA), 100 µg/mL streptomycin, and 100 U/mL penicillin.

4.3. Competitive Inhibition Assay

The contributions of Na^+-dependent and -independent AATs to ^{14}C-fluciclovine transport was estimated using competitive inhibition assays between ^{14}C-fluciclovine and naturally-occurring or synthetic AAs, as described in a previous report [8]. Briefly, MRMT-1 cells were cultured in 24-well, flat-bottom tissue culture plates, and culture medium was replaced with sodium buffer (140 mmol/L NaCl, 5 mmol/L KCl, 5.6 mmol/L glucose, 0.9 mmol/L $CaCl_2$, 1.0 mmol/L $MgCl_2$ and 10 mmol/L HEPES (pH 7.3)) or choline buffer (NaCl of the sodium buffer was preplaced with 140 mmol/L choline chloride) containing 10 µmol/L ^{14}C-fluciclovine in the presence or absence of 2.0 mmol/L inhibitor. Cells were placed in an incubator at 37 °C for 5 min, and radioactivity was measured using a liquid scintillation counter (Tri-Carb 2910TR, Perkin Elmer, Waltham, MA, USA). Protein concentrations of cell lysates were determined using the BCA Protein Assay Kit (Thermo Fisher Scientific, Waltham, MA, USA). Tracer uptake was calculated as pmol/mg protein and the control transport (absence of inhibitors) of ^{14}C-fluciclovine in sodium/choline buffer was normalized to 100%. In this study, the following synthetic and naturally-occurring AAs were used as inhibitors: BCH, MeAIB, phenylalanine, proline, glutamine, serine, arginine, and glutamate; the specificity of each AA to AATs have been described in a previous report [8].

4.4. Animal Handling

All animal handling procedures and experimental protocols were conducted in accordance with Japanese laws as stipulated in the Act on Welfare and Management of Animals, and approved by the Institutional Animal Care and Use Committee of Nihon Medi-Physics (No. 142-019) on 24 October 2014. Male Sprague-Dawley rats (9–12 weeks old; CLEA Japan, Tokyo, Japan) were used for all experiments. Rats were housed in a 12 h light-dark cycle and were maintained on a standard laboratory diet, Labo MR Stock (Nosan Corporation, Kanagawa, Japan) containing approximately 9.2% water, 18.8% crude protein, 3.9% crude fat, 6.6% crude fiber, 6.9% crude ash, 54.7% of nitrogenous compounds, amino acids (1.09% arginine, 0.46% histidine, 0.67% isoleucine, 1.36%, leucine, 0.89%, lysine, 0.26% methionine, 0.84% phenylalanine, 0.64% threonine, 0.21% tryptophan, 0.79% valine, 0.28% cysteine), and drinking water. All animals were anesthetized using 1% isoflurane (Pfizer Japan, Tokyo, Japan). In the surgical procedures, 0.5% meloxicam (Boehringer Ingelheim Vetmedica, Tokyo, Japan) was injected subcutaneously to relieve pain.

4.5. BCa Bone Metastatic Model

A BCa bone metastatic model, described previously [39], was used. Briefly, 100 mL of a cell suspension containing 2.5×10^4 MRMT-1 cells in Hank's Balanced Salt Solution without Ca^{2+} and Mg^{2+} was injected into the saphenous artery of the right hind legs. Development of osteolytic lesions was monitored using a microfocus X-ray imaging system (µFX-1000; FUJIFILM Corporation, Tokyo, Japan) or a preclinical imaging system (FX3000 CT, TriFoil Imaging, Chatsworth, CA, USA) and triple-tracer autoradiography was performed 12 ± 1 days after the cell injection. Osteolytic lesions were induced in the tibia and/or femur of the experimental rats.

4.6. Triple-Tracer Autoradiography

To compare the distribution of each tracer visually in identical lesions, triple-tracer autoradiography was performed using ^{14}C-fluciclovine, ^3H-FDG, and ^{99m}Tc-HMDP in the BCa osteolytic model as described in the supplementary material. Briefly, rats with ($n = 5$) or without ($n = 6$) fasting overnight (17 h to 18 h) were injected (tail vein) with three tracers (^{14}C-fluciclovine: 2.75 MBq/kg, ^3H-FDG: 18.5 MBq/kg, ^{99m}Tc-HMDP: 74 MBq/kg). ^{14}C-Fluciclovine and ^3H-FDG were allowed to remain in circulation for 30 min; ^{99m}Tc-HMDP was allowed to remain for 2 h before euthanizing the animal. The tibiae and femora were removed and frozen in isopentane/dry ice; the frozen bone was then embedded in Super Cryoembedding Medium (SCEM) (Section-Lab, Hiroshima, Japan), followed by freezing again in isopentane/dry ice until the SCEM set. The frozen samples were sectioned using a CM3050S cryostat (Leica Microsystems, Tokyo, Japan) at $-20\ °C$ with an adhesive film (Cryofilm Type 2C(9), Section-Lab) using Kawamoto's film method (5 μm slices and 10 μm slices for the histological and autoradiography specimens, respectively) [40]. The frozen samples were provided for triple-tracer autoradiography and each ^{14}C-, ^{3}H-, and ^{99m}Tc-image was created as described in the supplementary materials. The regions-of-interest (ROIs) were manually drawn around each lesion while referring to the histological images from H&E and TB staining. In the ^{14}C-fluciclovine and ^{3}H-FDG images, tumor lesions with bone absorption were defined as OL (typical osteolytic lesions) and the lesions corresponding to ^{99m}Tc-HMDP accumulation were defined as peri-tumor bone formation (pTBF) in the osteoblastic lesions (typical osteoblastic lesions). Furthermore, three ROIs of random size were manually positioned on the normal regions of muscle surrounding tibiae and/or femora and the average ROI count from the three ROIs was calculated as the background radioactivity. The target$_{mean}$ (metastatic lesion)-to-background$_{mean}$ (muscle) (T/BG) ratios were then calculated and lesion-based analyses were performed.

4.7. Histological Analysis

The following procedures for each histological stain were performed on 5 μm serial sections using general methods: H&E and TB staining for histological changes at the lesion sites; ALP for osteoblast and fibroblast activity; and TRAP staining for osteoclast-activity. Anti-ASCT2 (J-25) polyclonal antibody (1:40; Santa Cruz Biotechnology, Dallas, TX, USA) and anti-LAT1 (H-75) polyclonal antibody (1:200; Santa Cruz Biotechnology) for the amino acid transporters were used in the immunohistochemical assessments with EnVision+ System HR-labeled polymer anti-rabbit (Dako Japan, Tokyo, Japan) and Liquid DAB+ Substrate Chromogen System (Dako Japan). A BZ-9000 HS all-in-one fluorescence microscope (Keyence Corporation, Osaka, Japan) was used for pathological examinations.

4.8. Statistical Analysis

Data are presented as mean $\pm$ the standard deviation. To create beeswarm plots with box plots, R version 3.3.2 (R Foundation, Vienna, Austria) for Windows (Microsoft Corporation, Redmond, WA, USA) was used. In these plots, each dot represents a T/BG ratio in each lesion and the center lines in the boxes represent the medians; box limits indicate the 25th and 75th percentiles; whiskers extend 1.5 times the interquartile range from the 25th and 75th percentiles. All statistical analyses were performed using SAS version 9.4 (SAS Institute, Cary, NC, USA). The two groups were compared using Wilcoxon rank-sum tests for non-normal distribution datasets, or the F-test, followed by two-tailed unpaired Student's *t*-test or Welch's *t*-test for datasets with normal distribution. In all analyses, $p < 0.05$ was considered to be statistically significant.

5. Conclusions

Our findings suggest that fasting influences the uptake of ^{18}F-fluciclovine in osteolytic and osteoblastic bone metastasis lesions, and can facilitate clearer visualization of lesions in fluciclovine-PET imaging. However, because AA metabolism is a complex process, further animal and clinical studies are required to confirm our findings.

Supplementary Materials: Supplementary materials can be found at www.mdpi.com/1422-0067/18/5/934/s1.

Acknowledgments: The authors thank Masahiro Ono for his assistance with the histological examinations; Sachiko Naito for her assistance in the daily maintenance of the cell cultures; and Shiro Yoshida for his assistance in the daily maintenance of the animals. We thank Editage for English language editing.

Author Contributions: Shuntaro Oka conceived and designed the experiments; Shuntaro Oka, Masaru Kanagawa, and Yoshihiro Doi performed the experiments; Shuntaro Oka and Masaru Kanagawa analyzed the data; Yoshihiro Doi contributed analysis tools; All authors wrote, reviewed, and revised the manuscript.

Conflicts of Interest: Shuntaro Oka, Masaru Kanagawa, Yoshihiro Doi, and Hirokatsu Yoshimura are employees of Nihon Medi-Physics Co., Ltd. They are collaborating with Mark M. Goodman and David M. Schuster for nonclinical and clinical studies investigating ^{18}F-fluciclovine. Mark M. Goodman and Emory University (Atlanta, GA, USA) are eligible to receive royalties from Nihon Medi-Physics Co., Ltd. David M. Schuster is receiving research support through Emory University from Nihon Medi-Physics Co., Ltd. and Blue Earth Diagnostics. There is no other potential conflict of interest relevant to this article to declare.

Abbreviations

AA	amino acid
AAT	amino acid transporter
ALP	alkaline phosphatase
ASC	alanine-serine-cysteine
ASCT2	alanine-serine-cysteine transporter 2
BCa	breast cancer
BM	bone marrow
[11]C	carbon-11
[14]C	carbon-14
CB	cortical bone
[18]F	fluorine-18
FDG	2-deoxy-2-fluoro-D-glucose
[3]H	tritium-3
H&E	hematoxylin-eosin
HMDP	hydroxymethylene diphosphonate
LAT1	L-type amino acid transporter 1
mTORC1	mammalian target of rapamycin complex 1
[99m]Tc	technetium-99m
Ob	osteoblasts
Oc	osteoclasts
Os	osteoids
PET	positron emission tomography
PCa	prostate cancer
pTBF	peri-tumor bone formation
RCP	radiochemical purity
ROI	region-of-interest
SCEM	Super Cryoembedding Medium
SNAT2	sodium-dependent neutral amino acid transporter 2
TB	toluidine blue
T/BG	target$_{mean}$ (metastatic leion)-to-background$_{mean}$ (muscle)
TRAP	tartrate-resistant acid phosphatase

References

1. Lindholm, P.; Minn, H.; Leskinen-Kallio, S.; Bergman, J.; Ruotsalainen, U.; Joensuu, H. Influence of the blood glucose concentration on FDG uptake in cancer—A PET study. *J. Nucl. Med.* **1993**, *34*, 1–6. [PubMed]
2. Lewis, D.Y.; Soloviev, D.; Brindle, K.M. Imaging tumor metabolism using positron emission tomography. *Cancer J.* **2015**, *21*, 129–136. [CrossRef] [PubMed]
3. Jager, P.L. Improving amino acid imaging: Hungry or stuffed? *J. Nucl. Med.* **2002**, *43*, 1207–1209. [PubMed]
4. Lindholm, P.; Leskinen-Kallio, S.; Kirvelä, O.; Någren, K.; Lehikoinen, P.; Pulkki, K.; Peltola, O.; Ruotsalainen, U.; Teräs, M.; Joensuu, H. Head and neck cancer: Effect of food ingestion on uptake of C-11 methionine. *Radiology* **1994**, *190*, 863–867. [CrossRef] [PubMed]
5. Schuster, D.M.; Nanni, C.; Fanti, S.; Oka, S.; Okudaira, H.; Inoue, Y.; Sörensen, J.; Owenius, R.; Choyke, P.; Turkbey, B.; et al. Anti-1-amino-3-^{18}F-fluorocyclobutane-1-carboxylic acid: Physiologic uptake patterns, incidental findings, and variants that may simulate disease. *J. Nucl. Med.* **2014**, *55*, 1986–1992. [CrossRef] [PubMed]

6. Ulaner, G.A.; Goldman, D.A.; Gönen, M.; Pham, H.; Castillo, R.; Lyashchenko, S.K.; Lewis, J.S.; Dang, C. Initial Results of a Prospective Clinical Trial of ^{18}F-Fluciclovine PET/CT in Newly Diagnosed Invasive Ductal and Invasive Lobular Breast Cancers. *J. Nucl. Med.* **2016**, *57*, 1350–1356. [CrossRef] [PubMed]

7. Rahim, F.; Hajizamani, S.; Mortaz, E.; Ahmadzadeh, A.; Shahjahani, M.; Shahrabi, S.; Saki, N. Molecular regulation of bone marrow metastasis in prostate and breast cancer. *Bone Marrow Res.* **2014**, *2014*, 405920. [CrossRef] [PubMed]

8. Oka, S.; Okudaira, H.; Yoshida, Y.; Schuster, D.M.; Goodman, M.M.; Shirakami, Y. Transport mechanisms of *trans*-1-amino-3-fluoro[1-(14)C]cyclobutanecarboxylic acid in prostate cancer cells. *Nucl. Med. Biol.* **2012**, *39*, 109–119. [CrossRef] [PubMed]

9. Okudaira, H.; Nakanishi, T.; Oka, S.; Kobayashi, M.; Tamagami, H.; Schuster, D.M.; Goodman, M.M.; Shirakami, Y.; Tamai, I.; Kawai, K. Kinetic analyses of *trans*-1-amino-3-^{18}F]fluorocyclobutanecarboxylic acid transport in *Xenopus laevis* oocytes expressing human ASCT2 and SNAT2. *Nucl. Med. Biol.* **2013**, *40*, 670–675. [CrossRef] [PubMed]

10. Ono, M.; Oka, S.; Okudaira, H.; Schuster, D.M.; Goodman, M.M.; Kawai, K.; Shirakami, Y. Comparative evaluation of transport mechanisms of *trans*-1-amino-3-^{18}F]fluorocyclobutanecarboxylic acid and l-[methyl-^{11}C]methionine in human glioma cell lines. *Brain Res.* **2013**, *1535*, 24–37. [CrossRef] [PubMed]

11. Oka, S.; Okudaira, H.; Ono, M.; Schuster, D.M.; Goodman, M.M.; Kawai, K.; Shirakami, Y. Differences in transport mechanisms of *trans*-1-amino-3-^{18}F]fluorocyclobutanecarboxylic acid in inflammation, prostate cancer, and glioma cells: Comparison with L-[methyl-^{11}C]methionine and 2-deoxy-2-^{18}F]fluoro-D-glucose. *Mol. Imaging Biol.* **2014**, *16*, 322–329. [CrossRef] [PubMed]

12. Liang, Z.; Cho, H.T.; Williams, L.; Zhu, A.; Liang, K.; Huang, K.; Wu, H.; Jiang, C.; Hong, S.; Crowe, R.; et al. Potential Biomarker of L-type Amino Acid Transporter 1 in Breast Cancer Progression. *Nucl. Med. Mol. Imaging* **2011**, *45*, 93–102. [CrossRef] [PubMed]

13. Holecek, M.; Kovarik, M. Alterations in protein metabolism and amino acid concentrations in rats fed by a high-protein (casein-enriched) diet—Effect of starvation. *Food Chem. Toxicol.* **2011**, *49*, 3336–3342. [CrossRef] [PubMed]

14. Bergström, M.; Ericson, K.; Hagenfeldt, L.; Mosskin, M.; von Holst, H.; Norén, G.; Eriksson, L.; Ehrin, E.; Johnström, P. PET study of methionine accumulation in glioma and normal brain tissue: Competition with branched chain amino acids. *J. Comput. Assist. Tomogr.* **1987**, *11*, 208–213. [CrossRef] [PubMed]

15. Langen, K.J.; Roosen, N.; Coenen, H.H.; Kuikka, J.T.; Kuwert, T.; Herzog, H.; Stöcklin, G.; Feinendegen, L.E. Brain and brain tumor uptake of L-3-^{123}I]iodo-α-methyl tyrosine: Competition with natural L-amino acids. *J. Nucl. Med.* **1991**, *32*, 1225–1229. [PubMed]

16. Lahoutte, T.; Caveliers, V.; Franken, P.R.; Bossuyt, A.; Mertens, J.; Everaert, H. Increased tumor uptake of 3-(123)I-Iodo-L-α-methyltyrosine after preloading with amino acids: An in vivo animal imaging study. *J. Nucl. Med.* **2002**, *43*, 1201–1206. [PubMed]

17. Pochini, L.; Scalise, M.; Galluccio, M.; Indiveri, C. Membrane transporters for the special amino acid glutamine: Structure/function relationships and relevance to human health. *Front. Chem.* **2014**, *2*, 61. [CrossRef] [PubMed]

18. Nishimura, M.; Naito, S. Tissue-specific mRNA expression profiles of human ATP-binding cassette and solute carrier transporter superfamilies. *Drug Metab. Pharmacokinet.* **2005**, *20*, 452–477. [CrossRef] [PubMed]

19. Bevington, A.; Brown, J.; Butler, H.; Govindji, S.; M-Khalid, K.; Sheridan, K.; Walls, J. Impaired system A amino acid transport mimics the catabolic effects of acid in L6 cells. *Eur. J. Clin. Investig.* **2002**, *32*, 590–602.

20. Evans, K.; Nasim, Z.; Brown, J.; Butler, H.; Kauser, S.; Varoqui, H.; Erickson, J.D.; Herbert, T.P.; Bevington, A. Acidosis-sensing glutamine pump SNAT2 determines amino acid levels and mammalian target of rapamycin signalling to protein synthesis in L6 muscle cells. *J. Am. Soc. Nephrol.* **2007**, *18*, 1426–1436.

21. Hyde, R.; Hajduch, E.; Powell, D.J.; Taylor, P.M.; Hundal, H.S. Ceramide down-regulates System A amino acid transport and protein synthesis in rat skeletal muscle cells. *FASEB J.* **2005**, *19*, 461–463. [CrossRef] [PubMed]

22. Drummond, M.J.; Glynn, E.L.; Fry, C.S.; Timmerman, K.L.; Volpi, E.; Rasmussen, B.B. An increase in essential amino acid availability upregulates amino acid transporter expression in human skeletal muscle. *Am. J. Physiol. Endocrinol. Metab.* **2010**, *298*, E1011–E1018. [CrossRef] [PubMed]

23. Dodd, K.M.; Tee, A.R. Leucine and mTORC1: A complex relationship. *Am. J. Physiol. Endocrinol. Metab.* **2012**, *302*, E1329–E1342. [CrossRef] [PubMed]

24. Sancak, Y.; Peterson, T.R.; Shaul, Y.D.; Lindquist, R.A.; Thoreen, C.C.; Bar-Peled, L. The Rag GTPases bind raptor and mediate amino acid signaling to mTORC1. *Science* **2008**, *320*, 1496–1501. [CrossRef] [PubMed]

25. Sandri, M. Signaling in muscle atrophy and hypertrophy. *Physiology* **2008**, *23*, 160–170. [CrossRef] [PubMed]

26. Fuchs, B.C.; Bode, B.P. Amino acid transporters ASCT2 and LAT1 in cancer: Partners in crime? *Semin. Cancer Biol.* **2005**, *15*, 254–266. [CrossRef] [PubMed]

27. Wang, Q.; Tiffen, J.; Bailey, C.G.; Lehman, M.L.; Ritchie, W.; Fazli, L. Targeting amino acid transport in metastatic castration-resistant prostate cancer: Effects on cell cycle, cell growth, and tumor development. *J. Natl. Cancer Inst.* **2013**, *105*, 1463–1473. [CrossRef] [PubMed]

28. Toyoda, M.; Kaira, K.; Ohshima, Y.; Ishioka, N.S.; Shino, M.; Sakakura, K.; Takayasu, Y.; Takahashi, K.; Tominaga, H.; Oriuchi, N.; et al. Prognostic significance of amino-acid transporter expression (LAT1, ASCT2, and xCT) in surgically resected tongue cancer. *Br. J. Cancer* **2014**, *110*, 2506–2513. [CrossRef] [PubMed]

29. Namikawa, M.; Kakizaki, S.; Kaira, K.; Tojima, H.; Yamazaki, Y.; Horiguchi, N.; Sato, K.; Oriuchi, N.; Tominaga, H.; Sunose, Y.; et al. Expression of amino acid transporters (LAT1, ASCT2 and xCT) as clinical significance in hepatocellular carcinoma. *Hepatol. Res.* **2014**. [CrossRef] [PubMed]

30. Kaira, K.; Arakawa, K.; Shimizu, K.; Oriuchi, N.; Nagamori, S.; Kanai, Y.; Oyama, T.; Takeyoshi, I. Relationship between CD147 and expression of amino acid transporters (LAT1 and ASCT2) in patients with pancreatic cancer. *Am. J. Transl. Res.* **2015**, *7*, 356–363. [PubMed]

31. Nikkuni, O.; Kaira, K.; Toyoda, M.; Shino, M.; Sakakura, K.; Takahashi, K.; Tominaga, H.; Oriuchi, N.; Suzuki, M.; Iijima, M.; et al. Expression of Amino Acid Transporters (LAT1 and ASCT2) in Patients with Stage III/IV Laryngeal Squamous Cell Carcinoma. *Pathol. Oncol. Res.* **2015**, *21*, 1175–1181.

32. Kaira, K.; Nakamura, K.; Hirakawa, T.; Imai, H.; Tominaga, H.; Oriuchi, N.; Nagamori, S.; Kanai, Y.; Tsukamoto, N.; Oyama, T.; et al. Prognostic significance of L-type amino acid transporter 1 (LAT1) expression in patients with ovarian tumors. *Am. J. Transl. Res.* **2015**, *7*, 1161–1171. [PubMed]

33. Honjo, H.; Kaira, K.; Miyazaki, T.; Yokobori, T.; Kanai, Y.; Nagamori, S.; Oyama, T.; Asao, T.; Kuwano, H. Clinicopathological significance of LAT1 and ASCT2 in patients with surgically resected esophageal squamous cell carcinoma. *J. Surg. Oncol.* **2016**, *113*, 381–389. [CrossRef] [PubMed]

34. Sato, K.; Tsuchihara, K.; Fujii, S.; Sugiyama, M.; Goya, T.; Atomi, Y.; Ueno, T.; Ochiai, A.; Esumi, H. Autophagy is activated in colorectal cancer cells and contributes to the tolerance to nutrient deprivation. *Cancer Res.* **2007**, *67*, 9677–9684. [CrossRef] [PubMed]

35. Luo, Y.; Yoneda, J.; Ohmori, H.; Sasaki, T.; Shimbo, K.; Eto, S.; Kato, Y.; Miyano, H.; Kobayashi, T.; Sasahira, T.; et al. Cancer usurps skeletal muscle as an energy repository. *Cancer Res.* **2014**, *74*, 330–340.

36. Oka, S.; Hattori, R.; Kurosaki, F.; Toyama, M.; Williams, L.A.; Yu, W.; Votaw, J.R.; Yoshida, Y.; Goodman, M.M.; Ito, O. A preliminary study of anti-1-amino-3-^{18}F-fluorocyclobutyl-1-carboxylic acid for the detection of prostate cancer. *J. Nucl. Med.* **2007**, *48*, 46–55. [PubMed]

37. Schuster, D.M.; Votaw, J.R.; Nieh, P.T.; Yu, W.; Nye, J.A.; Master, V.; Bowman, F.D.; Issa, M.M.; Goodman, M.M. Initial experience with the radiotracer anti-1-amino-3-^{18}F-fluorocyclobutane-1-carboxylic acid with PET/CT in prostate carcinoma. *J. Nucl. Med.* **2007**, *48*, 56–63. [PubMed]

38. Inoue, Y.; Asano, Y.; Satoh, T.; Tabata, K.; Kikuchi, K.; Woodhams, R.; Baba, S.; Hayakawa, K. Phase IIa Clinical Trial of Trans-1-Amino-3-^{18}F-Fluoro-Cyclobutane carboxylic acid in metastatic prostate cancer. *Asia Ocean J. Nucl. Med. Biol.* **2014**, *2*, 87–94. [PubMed]

39. Zepp, M.; Bauerle, T.J.; Elazar, V.; Peterschmitt, J.; Lifshitz-Shovali, R.; Adwan, H.; Armbruster, F.P.; Golomb, G.; Berger, M.R. Treatment of breast cancer lytic skeletal metastasis using a model in nude rats. In *Breast Cancer—Current and Alternative Therapeutic Modalities*; Gunduz, E., Ed.; InTech: Rijeka, Croatia, 2011; Available online: http://www.intechopen.com/books/breast-cancer-current-and-alternative-therapeutic-modalities/treatment-of-breast-cancer-lytic-skeletal-metastasis-using-a-model-in-nude-rats (accessed on 28 April 2017).

40. Kawamoto, T. Use of a new adhesive film for the preparation of multi-purpose fresh-frozen sections from hard tissues, whole-animals, insects and plants. *Arch. Histol. Cytol.* **2003**, *66*, 123–413. [CrossRef] [PubMed]

International Journal of
Molecular Sciences

MDPI

Article

CTC-mRNA (AR-V7) Analysis from Blood Samples—Impact of Blood Collection Tube and Storage Time

Alison W. S. Luk [1], Yafeng Ma [1], Pei N. Ding [1,2,3], Francis P. Young [1,4], Wei Chua [2], Bavanthi Balakrishnar [2], Daniel T. Dransfield [5,†], Paul de Souza [1,2,3,4] and Therese M. Becker [1,3,4,*]

[1] Centre for Circulating Tumour Cell Diagnostics and Research, Ingham Institute for Applied Medical Research, 1 Campbell St., Liverpool, NSW 2170, Australia; alisonluk@gmail.com (A.W.S.L.); yafeng.ma@unsw.edu.au (Y.M.); Pei.Ding@sswahs.nsw.gov.au (P.N.D.); francis.young@student.unsw.edu.au (F.P.Y.); P.DeSouza@westernsydney.ed u.au (P.d.S.)

[2] Department of Medical Oncology, Liverpool Hospital, Elizabeth St & Goulburn St, Liverpool, NSW 2170, Australia; Wei.Chua2@sswahs.nsw.gov.au (W.C.); Bavanthi.Balakrishnar@sswahs.nsw.gov.au (B.B.)

[3] Western Sydney University Clinical School, Elizabeth St, Liverpool, NSW 2170, Australia

[4] South Western Clinical School, University of New South Wales, Goulburn St., Liverpool, NSW 2170, Australia

[5] Tokai Pharmaceuticals, Inc., 255 State Street, 6th Floor, Boston, MA 02109, USA; dan@siamab.com

[*] Correspondence: t.becker@unsw.edu.au; Tel.: +61-2-873-89033

[†] Current address: Siamab Therapeutics, 90 Bridge Street, Suite 100, Newton, MA 02458, USA.

Academic Editor: Carsten Stephan
Received: 31 March 2017; Accepted: 8 May 2017; Published: 12 May 2017

Abstract: Circulating tumour cells (CTCs) are an emerging resource for monitoring cancer biomarkers. New technologies for CTC isolation and biomarker detection are increasingly sensitive, however, the ideal blood storage conditions to preserve CTC-specific mRNA biomarkers remains undetermined. Here we tested the preservation of tumour cells and CTC-mRNA over time in common anticoagulant ethylene-diamine-tetra-acetic acid (EDTA) and acid citrate dextrose solution B (Citrate) blood tubes compared to preservative-containing blood tubes. Blood samples spiked with prostate cancer cells were processed after 0, 24, 30, and 48 h storage at room temperature. The tumour cell isolation efficiency and the mRNA levels of the prostate cancer biomarkers androgen receptor variant 7 (AR-V7) and total AR, as well as epithelial cell adhesion molecule (EpCAM) were measured. Spiked cells were recovered across all storage tube types and times. Surprisingly, tumour mRNA biomarkers were readily detectable after 48 h storage in EDTA and Citrate tubes, but not in preservative-containing tubes. Notably, AR-V7 expression was detected in prostate cancer patient blood samples after 48 h storage in EDTA tubes at room temperature. This important finding presents opportunities for measuring AR-V7 expression from clinical trial patient samples processed within 48 h—a much more feasible timeframe compared to previous recommendations.

Keywords: circulating tumour cell; biomarker; androgen receptor; AR-V7; droplet digital polymerase chain reaction (ddPCR); blood storage tube

1. Introduction

Circulating tumour cells (CTCs) are cells shed from tumours into the peripheral blood, and are believed to be the mechanism for metastasis [1,2]. The enumeration of CTCs has great prognostic value, and further molecular profiling of CTCs has great potential in providing insights on cancer progression, identifying CTC specific molecular therapeutic targets, determining prognostic and relapse indicators, and allowing longitudinal monitoring of disease response to treatments [3,4].

Therefore, relatively simple, non-invasive CTC analysis has exciting potential to investigate changes in tumour biomarkers during a cancer patient's disease progression. In prostate cancer, CTC counts have been associated with prognosis and importantly, isolated CTCs can function as a surrogate tumour samples to detect therapy-determining biomarkers, including the mRNA based androgen receptor variant AR-V7 in CTCs [5–8]. However, CTC analysis is still a rapidly-evolving field due to the complexities of isolation and detection of true CTCs. While it is estimated that 10^6 CTCs are shed per 1 g of tumour tissue per day, CTCs are thought to have a short half-life of less than 3 h in the bloodstream [9–12]. Hence, a major technical challenge for CTC analysis has been efficient recovery of rare CTCs from a background of approximately 10^9 erythrocytes and 10^7 leukocytes per mL of blood, and this has spurred the development of improved technologies for CTC isolation [13–15]. Despite these advances, feasible CTC analysis for clinical trials involving multiple sites are particularly challenging, due to strict requirements for pre-analytical conditions of blood samples in transport and storage, and time restrictions to ensure CTC integrity and biomarker detectability is maintained. Parameters, such as blood tube composition, storage or shipping temperature, sample agitation, and delays in sample processing in a large-throughput laboratory need to be considered before such measurements are conducted for clinical trials. Ideally, CTCs should be protected from apoptosis or cell lysis and, importantly, preserve relevant tumour biomarkers, an issue that is considered especially challenging for mRNA-based biomarkers.

Ethylene-diamine-tetra-acetic acid (EDTA) and acid citrate dextrose solution B (Citrate) are commonly used as anticoagulants in blood tubes for pathology blood tests. These tubes are compatible with downstream polymerase chain reaction (PCR) analysis [16,17] and have also been used for CTC isolation. However, as these tubes do not contain fixatives, early studies recommended that blood should be processed within 24 h, and ideally within 5 h, as it was thought that CTCs rapidly enter apoptosis [18]. To target this problem, new blood tubes have been designed with preservatives intended to extend time before sample processing. Cell-free DNA blood collection tubes (DNA BCT) and Cell-free RNA blood collection tubes (RNA BCT) (Streck, Omaha, NE, USA) are proposed to stabilise nucleated blood cells, preventing the release of cellular DNA and RNA into the plasma, and also inhibiting degradation of cell-free DNA and RNA [19,20]. Therefore, these tubes might be advantageous for CTC sample analysis. Additionally, Cyto-Chex blood collection tubes (Cyto-Chex BCT) were originally designed to preserve cell surface antigens for white blood cell immunophenotyping [21] which may allow improved CTC recovery by immunomagnetic isolation.

There have been previous reports suggesting the added preservatives in BCTs indeed aids CTC stability [22,23]. However, the preservative effects on actionable mRNA-based tumour biomarkers, such as AR-V7 has, to our knowledge, not been tested thoroughly in CTCs isolated from any blood collection tube type. We have previously reported a very sensitive method to detect AR-V7, an emerging RNA-based prostate cancer biomarker in prostate cancer patient CTC samples, showing that AR-V7 is not expressed in residual blood cells, and is expressed heterogeneously in CTCs [7]. In this study, we compare the effects of preservative-containing BCTs to commonly used EDTA and Citrate blood tubes on tumour cell isolation and detection of AR-V7 using our sensitive method.

2. Results

2.1. Spiked Cell Recovery

To test CTC preservation and recovery from the five different blood tubes, CTCs were modelled by spiking defined cell numbers of the 22Rv1 prostate cancer cell line into healthy female donor blood. At 0, 24, 30, and 48 h after spiking, tumour cells were enriched and enumerated. Three experiments were performed using 192 mL blood each from three healthy donors. The mean recovery of spiked cells ranged from 13% to 44% across all time points. All blood tubes showed preservation of spiked cells up to 48 h, with mean recovery between 24% and 39% (Figure 1a). The mean recovery of all blood tubes at each time point was compared against recovery at 0 h. Processing samples at 24 and 30 h resulted in significantly lower recovery than when processed at 0 h ($p < 0.05$). Within each time point, the recovery of each blood tube was compared to EDTA. No blood tube had significantly different recovery from EDTA, but the mean recovery of RNA BCT exceeded EDTA at each time point ($p < 0.05$).

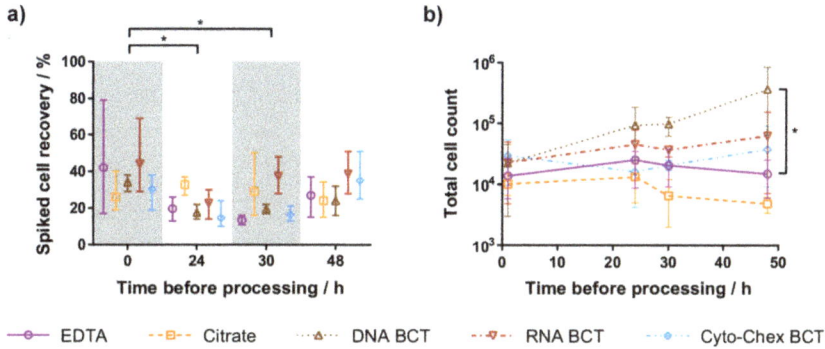

Figure 1. (**a**) Recovery of spiked 22Rv1 cells after tumour cell enrichment. The mean recovery at 24 h and 30 h was significantly different from recovery at 0 h (* $p < 0.05$). (**b**) Total cell count (recovered spiked cells and residual leukocytes after cell enrichment) from samples processed 0 h, 24 h, 30 h, and 48 h after spiking. At 48 h, DNA blood collection tubes (DNA BCT) cell count was significantly different from ethylene-diamine-tetra-acetic acid (EDTA) (* $p < 0.05$). For both (**a,b**), symbols represent the mean from three independent experiments, and whiskers represent the range.

2.2. Leukocyte Contamination

DNA BCT and RNA BCT generally had increased total cell counts (tumour cells plus residual co-purified leukocytes) when samples were processed later (Figure 1b). In comparison, EDTA, Citrate and Cyto-Chex BCT cell counts remained similar across all time points. When compared to EDTA tubes, DNA BCT and RNA BCT had higher mean cell counts at each time point, whereas Citrate tubes gave lower cell counts. A two-way ANOVA of each blood tube compared to EDTA at the same time point indicated that for samples processed at 48 h, DNA BCT total cell counts were significantly higher than for EDTA tubes ($p < 0.05$).

2.3. Cellular RNA Recovery

To evaluate the ability of each blood tube in preserving cellular RNA, tumour cell-specific gene expression (AR-V7, total AR and epithelial cell adhesion molecule (EpCAM)) was measured by droplet digital PCR (ddPCR) for each blood tube at each tumour cell enrichment time point (Figure 2). Cells from EDTA and Citrate tubes generally showed decreased mRNA detection the longer the sample storage time, with mRNA biomarkers still readily detectable even after 48 h. In all BCT samples, gene expression was low when processed immediately and undetectable after any storage duration. Thus, while mRNA detection from Citrate and EDTA blood tube samples was similar, BCT samples showed a striking loss in detectable gene expression compared to EDTA tube samples ($p < 0.01$).

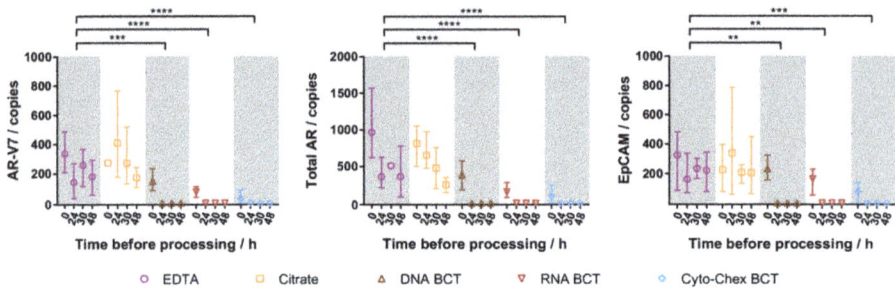

Figure 2. Expression of spiked tumour cell specific genes in samples processed 0 h, 24 h, 30 h, and 48 h after spiking. Symbols represent mean expression from three independent experiments, whiskers represent the range, and are not shown when smaller than the data symbol. The mean gene expression in DNA BCT, RNA BCT, and Cyto-Chex BCT were significantly different from EDTA (** $p < 0.01$, *** $p < 0.001$, **** $p < 0.0001$). AR-V7: androgen receptor variant 7; AR: androgen receptor.

2.4. Increased Proteinase K Treatment

We also investigated the effect of increased proteinase K digestion on the cellular RNA recovery as per manufacturer's suggestions for the BCTs. 22Rv1 cells were spiked into a new set of blood tubes (EDTA, Citrate, DNA BCT, and RNA BCT), followed by enrichment after 48 h of storage. RNA was extracted with and without additional 2 h proteinase K treatment, and gene expression was measured in RNA samples from two independent experiments. In EDTA and Citrate blood tubes, increased proteinase K digestion did not aid RNA recovery but decreased the number of measured copies of all three genes (Figure 3). In DNA BCT and RNA BCT, there was no detectable AR-V7, total AR or EpCAM with standard RNA extraction, and with increased digestion there was a small, but statistically insignificant, increase in the detection of total AR and AR-V7 in DNA BCT and RNA BCT, respectively.

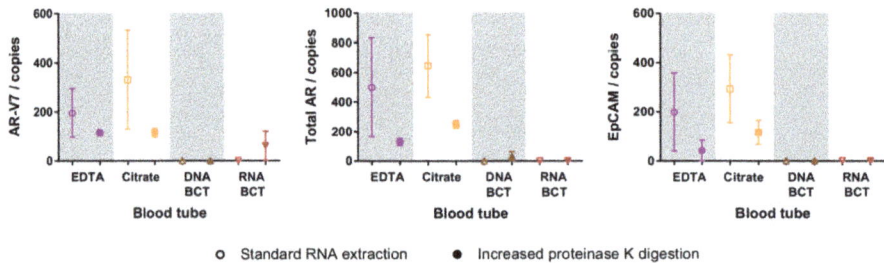

Figure 3. Effect of increased proteinase K treatment on the detection of spiked cell specific genes in samples processed 48 h after spiking. Symbols represent the mean expression from two independent experiments, whiskers represent the range, and are not shown when smaller than the data symbol. EpCAM: epithelial cell adhesion molecule.

2.5. Patient CTC Cellular RNA Detection

Given that spiked cell-specific gene expression was detectable even in samples processed 48 h following spiking of 22Rv1 cells into fresh blood in EDTA and Citrate tubes, we wished to confirm that AR-V7 also remains detectable that long in patient-derived CTCs. Common EDTA tube blood samples from three prostate cancer patients were processed at 4, 24, and 48 h after collection. After CTC enrichment, all patient samples had detectable AR-V7, total AR and EpCAM at all time points (Figure 4). AR-V7 expression varied between the three patients, while being comparable to the expression detected for these patients in a sample collected previously (Table 1). The AR-V7 and total AR detected in patient samples decreased with longer time before processing, but the decrease was not statistically significant.

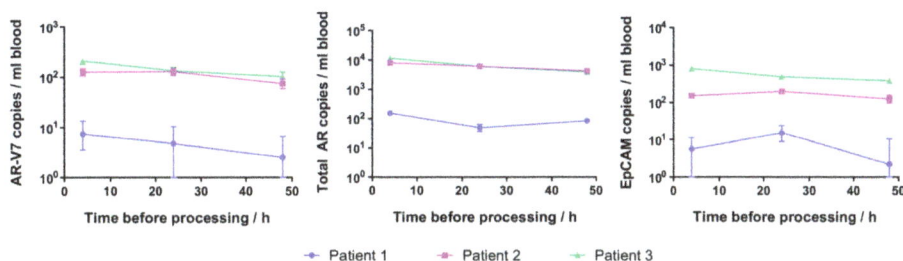

Figure 4. Detection of gene expression in prostate cancer patient circulating tumour cells (CTCs) isolated 4, 24, and 48 h after blood collection in EDTA tubes. Error bars represent 95% confidence intervals of droplet digital polymerase chain reaction (ddPCR) measurements, and error bars smaller than the data symbols are not shown.

Table 1. Androgen receptor variant 7 (AR-V7) and total AR expression in prostate cancer patient blood samples.

Patient	Hormone Sensitivity Status [1]	CTC Count/mL Blood	AR-V7 Copies/mL Blood				Total AR Copies/mL Blood			
		<4 h *	<4 h *	4 h	24 h	48 h	<4 h *	4 h	24 h	48 h
1	CRPC	9 *	2 *	7	5	3	96 *	155	49	84
2	CRPC	2 *	110 *	126	128	74	19,140 *	7963	6083	4191
3	CRPC	6 *	45 *	210	135	102	2610 *	11,392	5994	3848

[1] CRPC = castrate resistant prostate cancer. Patients who had 7–9 month previously high AR-V7 levels detected [7] were chosen for this study (* for comparison previous data are presented; note in the previous study CTC isolation was performed using a different instrument (IsoFlux, Fluxion, San Francisco, CA, USA), CTC counts are normalized per mL blood). New AR-V7 and total AR expression data of the same patients are presented.

3. Discussion

3.1. Tumour Cell Preservation

New blood tubes such as DNA BCT, RNA BCT, and Cyto-Chex BCT contain formaldehyde-free fixatives in addition to traditionally used anticoagulants in blood tubes [19–21]. As these BCTs were previously shown to preserve leukocytes and prevent the release of cellular RNA into plasma, even after over three days of storage at room temperature [20,24,25], our study compared commonly-used EDTA and Citrate tubes with BCTs in terms of the ability to preserve modelled CTCs (spiked cultured tumour cells) and of greater interest, cellular RNA.

From our study, the total cell count by Hoechst staining showed that leukocyte retention after tumour cell enrichment increased over time in DNA BCT compared to other blood tubes. Since BCTs were designed for analysis of plasma cell-free nucleic acids, this has not previously been a concern. However, downstream analysis of CTCs can be adversely affected by contaminating leukocytes, such as for single CTC isolation or CTC nucleic acid analysis. Disregarding background leukocytes, intact tumour cells were visible under bright field microscopy from all blood tubes after 48 h (data not shown), and enumeration of cell tracker-stained cells indicated that the recovery of spiked tumour cells was not significantly different across all blood tube types within each time point. Furthermore, although spiked cell recovery generally decreased with longer storage time before processing, there was no significant difference between recovery at 0 and 48 h, with a mean recovery of 35% and 30%, respectively. Our results are in agreement with a previous study which found that CTC yields from lung cancer patients did not decline significantly when processed after 24, 48, or 72 h storage in EDTA tubes [26]. We suggest that the observed variation in spiked cell recovery in our data can be mostly attributed to the difficulties of spiking exact cell numbers of the 22Rv1 cell line, which shows a very strong tendency towards cell clustering. Nevertheless, this line was chosen due to its high AR-V7 expression, and cell strainers were used to keep the effects of cell aggregates to a minimum. Conversely, our results differ from a recent study which reported that the recovery of 2000 spiked MCF-7 breast cancer cells from DNA BCT after one day and four days was very similar at 60% and 58%, respectively, while the recovery decreased from 32% to 16% for EDTA tubes [23]. The discrepancies between the results may be attributed to the difference in the spiked cell line, spiked cell numbers, method of tumour cell isolation, and detection and blood storage time. Importantly, the observed differences in the limited number of studies in this area highlights the need for more research to further compare and improve methods of CTC isolation after blood storage.

Despite common recommendations to process blood samples as soon as possible to reduce cell lysis, our data indicates that cell fixation is not necessary for the recovery of tumour cells, and common EDTA or Citrate tubes are sufficient for tumour cell detection within 48 h.

3.2. Cellular RNA Preservation

Analysis of gene expression from CTCs can provide valuable information on CTC activity and cancer progression, but is particularly challenging as delays in blood sample processing may cause alterations in CTC gene expression, and existing mRNAs may have a short half-life. DNA BCT and RNA BCT were designed to preserve cellular DNA and RNA, respectively, from being released and, hence, were expected to perform better than commonly-used EDTA and Citrate tubes, which only contain anticoagulants. We used our previously-reported, highly-sensitive and specific ddPCR assay to screen for AR-V7, total AR, and EpCAM; this assay has shown before that AR-V7 and total AR were highly expressed in the 22Rv1 cell line or patient CTCs while having no, or negligible, expression in healthy control peripheral lymphocytes [7]. We were surprised to find AR-V7, total AR, and EpCAM readily detectable after 48 h in common EDTA and Citrate tubes. The data suggest that mRNAs encoding these genes have relatively long half-life and/or are continuously expressed by viable tumour cells during storage in a blood sample. Additionally, the detection of mRNA confirms that cells remain intact in the blood sample as it is well established that any released free RNA will be quickly degraded by RNases in the blood [27]. Since cultured 22Rv1 cells are potentially more robust and survive extended storage in blood samples better than CTCs, we confirmed that AR-V7 detection after 48 h storage was translatable to patient CTC samples by testing three patients previously shown to have high CTC AR-V7 expression [7]. Although a small decrease in AR-V7 levels after 48 h was detected in these patient samples, the decrease was marginal in this time frame and even in patient blood with low AR-V7, CTC derived AR-V7 was still detectable after 48 h blood storage at room temperature. AR-V7 expression was comparable to the previous testing of the same patients [7].

In contrast to our data from blood storage in common blood tubes, the BCT samples processed immediately produced much lower detectable tumour cell specific mRNA, and any storage of spiked blood samples in these tubes prevented detection of gene expression completely. We propose that the preservative in the tubes renders RNA inaccessible, likely due to RNA cross-linking with proteins and DNA, which would be incomplete when samples are processed immediately, explaining the limited gene expression detected for our 0 h samples. This interpretation is supported by a report that spiked breast cancer cells could be retrieved from DNA BCTs after up to 72 h storage and had accessible DNA, however, whole genome amplification (WGA) yielded consistently significantly less DNA than WGA from EDTA tubes [22]. Alternatively, RNA could have degraded in the BCTs in our study, however, mRNA was reported detectable by in situ hybridisation from spiked tumour cells isolated from blood stored for four days at room temperature in DNA BCTs [23], suggesting that RNA remains intact but possibly cross-linked, which would interfere less with in situ hybridisation than RNA extraction. While, according to the manufacturers guidelines, RNA should be extractable from samples stored in BCTs, an extended protein K digest is recommended, which might help to reverse cross-linking effects.

We, therefore, investigated whether increased proteinase K treatment during RNA extraction would allow relevant gene expression detection for samples stored for 48 h. However, increased proteinase K digest failed to produce relevant effects and decreased RNA detectability in EDTA and Citrate tubes.

Previous studies showed BCTs to preserve leukocyte cellular RNA, and prevent cellular RNA contamination of cell-free RNA [20,24]. Additionally, the detection of cellular GAPDH, c-Fos and p53 RNA was reported to show variation in expression in neutrophils from blood samples stored over three days in EDTA tubes, whereas there was no change when stored in RNA BCT [25]. While the reasons for the difference to our data are not completely clear, it stands out that all genes tested in that study are abundantly-expressed genes in common blood cells, which might be easier to detect. We have not extended testing past 48 h, as blood samples from within Australia can be processed in this time frame in our facility; however, it is an area of interest for future studies to determine the maximum storage time for EDTA and Citrate tubes.

From our data, it is evident that CTC-derived AR-V7 can be detected from blood stored in commonly-used EDTA and Citrate tubes for up to 48 h, while blood tubes with preservatives (DNA BCT, RNA BCT, and Cyto-Chex BCT) should not be used for CTC isolation if downstream RNA analysis by PCR is intended.

4. Materials and Methods

4.1. Blood Collection

Prostate cancer patients ($n = 3$) and healthy female blood donors ($n = 4$) provided written informed consent to participate in the study. The study was approved by the South Western Sydney Local Healthy District Ethics Committee, Australia (HREC/13/LPOOL/158; 02/09/2013). Peripheral blood was drawn by venipuncture into blood tubes with the initial 3 mL discarded to prevent keratinocyte contamination and false positive CTCs being present. For the main comparison experiment, a total of 192 mL blood from each healthy donor ($n = 3$) was drawn into a total of 24 blood tubes including: 4×9 mL K_3EDTA tube (Greiner Bio-One, Kremsmünster, Austria), 4×9 mL acid citrate dextrose-B tube (Greiner Bio-One, Kremsmünster, Austria), 4×10 mL Cell-free DNA BCT (Streck, Omaha, NE, USA), 4×10 mL Cell-free RNA BCT (Streck, Omaha, NE, USA), and $4 \times 2 \times 5$ mL Cyto-Chex BCT (Streck, Omaha, NE, USA). In the follow-up experiment to test increased proteinase K treatment, 38 mL blood from two healthy donors was drawn into one set of EDTA, Citrate, and DNA BCT and RNA BCT tubes. In the experiment with prostate cancer patients, blood was drawn into 3×6 mL K_2EDTA tubes (BD, Franklin Lakes, NJ, USA) available in the clinic.

4.2. Cell Spiking

The human prostate cancer cell line 22Rv1 was purchased from the American Type Culture Collection (In Vitro Technologies, Melbourne, Australia). Cells were routinely passaged in Roswell Park Memorial Institute culture medium (RPMI 1640) (Lonza, Basel, Switzerland) supplemented with 10% fetal bovine serum (FBS) (Invitrogen, Carlsbad, CA, USA) in a humidified incubator with 5% CO_2 at 37 °C. For spiking experiments, 22Rv1 cells, cultured for two days after passaging, were gently detached with accutase (Sigma-Aldrich, St. Louis, MO, USA), washed with phosphate buffered saline (PBS), and passed through a 20 μm pre-separation filter "cell strainer" (Miltenyi Biotec, Bergisch Gladbach, Germany) to remove cell aggregates. Cells were then incubated with 15 μM CellTracker Green CMFDA (Life Technologies, Carlsbad, CA, USA) in 100 μL serum-free RPMI media at 37 °C for 1 h. Stained 22Rv1 cells were washed with PBS and a known number (100–200 cells) were spiked into blood tubes. The spiked input cell numbers were verified by aliquoting the same volume onto glass slides in between inoculating blood samples, Hoechst staining, and enumeration by fluorescent microscopy.

4.3. Tumour Cell Enrichment from Whole Blood

Spiked blood samples were enriched for tumour cells after storage (dark, room temperature) for 0, 24, 30, and 48 h. The peripheral blood mononuclear cell (PBMC) layer was extracted using 50 mL SepMate tubes and Lymphoprep according to the manufacturer's instructions (Stemcell Technologies, Vancouver, BC, Canada). PBMCs were washed with separation buffer (PBS with 0.5% FBS and 2 mM EDTA), then incubated at 4 °C for 30 min with 50 μL FcR blocking reagent (Miltenyi Biotec, Bergisch Gladbach, Germany), 50 μL EpCAM conjugated immunomagnetic microbeads (Miltenyi Biotec, Bergisch Gladbach, Germany), 1 μL 50× Hoechst (Fluxion, San Francisco, CA, USA), and made up to a total volume of 500 μL with separation buffer. Cell separation was performed with the Posselds program on the AutoMACS Pro Separator (Miltenyi Biotec, Bergisch Gladbach, Germany). After separation, the positive selected fraction was separated into two aliquots: one was kept on ice until enumeration on the same day, while the other was centrifuged at 400× *g* for 10 min, and the pellet frozen at −80 °C until RNA extraction. For testing of extended proteinase K treatment, samples were enriched after 48 h, with both aliquots frozen at −80 °C until RNA extraction. For prostate cancer patient samples, one blood tube from each patient was processed within 4 h of blood draw, and the other two blood tubes were processed after 24 and 48 h of storage, respectively, before freezing at −80 °C until RNA extraction. Supplementary Figure S1 illustrates the different work flows.

4.4. Cell Enumeration

The enriched enumeration sample was mounted onto slides coated with 2% bovine serum albumine (BSA), then visualised and scanned at 20× magnification with a CellCelector microscope (ALS GmbH, Jena, Thüringen, Germany). The exposure times for the instrument's DAPI and FITC channels were 50 ms and 100 ms, respectively. Scanned images were analysed with ALS CellCelector software v3.0 (ALS GmbH). Nucleated (Hoechst-positive) cells were detected for total cell counts, while cells positive for both Hoechst and CellTracker were considered recovered spiked cells.

To calculate the percentage recovery of spiked cells, the number of cells enumerated after tumour cell enrichment was multiplied by two (to account for CTCs being enumerated in half of the sample while the other half was processed for RNA) and then divided by the original number of spiked cells as determined from input controls.

4.5. Cellular RNA Extraction

Total RNA was extracted from the second enriched sample or from CTC-enriched patient samples with the Total RNA Purification Micro Kit (Norgen Biotek Corp., Thorold, ON, Canada). RNA was double-eluted in 20 μL followed by 10 μL molecular-grade H_2O. For complementary DNA (cDNA) synthesis, 15 μL of eluted RNA was added to form a total volume of 20 μL with the SensiFAST cDNA Synthesis Kit (Bioline, London, UK). For testing increased proteinase K digestion, 0.5 mg of DNAase and RNAase free proteinase K (Bioline) was added with buffer RL at the cell lysate preparation step, and the sample was incubated for 2 h at 60 °C before the addition of ethanol.

4.6. Digital Droplet PCR

Quantification by ddPCR was performed for three tumour cell specific genes. Total androgen receptor (total AR), androgen receptor splice variant 7 (AR-V7), and epithelial cell adhesion molecule (EpCAM) using primers and probes shown in Table 2. In brief, 20 μL ddPCR reactions contained 10 μL ddPCR Supermix for Probes (No dUTP) (Bio-Rad, Hercules, CA, USA), 500 nM of each relevant primer and 250 nM probe Fluorescein (6-FAM) or HEX. Total AR and AR-V7 reactions were multiplexed as previously-described [7]. Droplets were generated with 70 μL oil using a QX200 droplet generator (Bio-Rad). Amplification was performed at 95 °C for 10 min, followed by 40 cycles of 94 °C for 30 s and 55 °C for 1 min using a C1000 Touch Thermo Cycler (Bio-Rad). After amplification, the droplets were read with a QX200 Droplet Reader (Bio-Rad) and analysed with QuantaSoft software v1.7.4.

Int. J. Mol. Sci. **2017**, *18*, 1047

The fluorescence thresholds used were 6-FAM 2000 for AR-V7, HEX 2500 for total AR, and HEX 1500 for EpCAM. Readings with ≥5 droplets were considered positive. The total error calculated by the software was used as the 95% confidence intervals of ddPCR measurements.

Table 2. Primers and probes.

Gene	Primers (5′→3′)	Probes (5′→3′)
Total AR	F: GGA ATT CCT GTG CAT GAA AGC R: CGA TCG AGT TCC TTG ATG TAG TTC	[HEX] CTT CAG CAT TAT TCC AGT G [BHQ1]
AR-V7	F: CGG AAA TGT TAT GAA GCA GGG ATG A R: CTG GTC ATT TTG AGA TGC TTG CAA T	[6FAM] TCT GGG AGA AAA ATT CCG [BHQ1]
EpCAM	F: CGT CAA TGC CAG TGT ACT TCA R: TTT CTG CCT TCA TCA CCA AA	[HEX] TAC TGT CAT TTG CTC AAA GC [BHQ1]

AR: androgen receptor; AR-V7: androgen receptor variant 7; EpCAM: epithelial cell adhesion molecule; 6FAM: Fluorescein; BHQ1:black hole quencher 1. F: Forward, R: Reverse.

4.7. Statistical Analysis

Analysis of cell and RNA recovery was performed by two-way Analysis of Variance (ANOVA), using GraphPad Prism software v6.07 (GraphPad Software Inc., San Diego, CA, USA).

5. Conclusions

While spiked cell recovery was not affected by the blood tube type even after 48 h of storage, tumour cell-specific RNA was undetectable by ddPCR in CTCs from stored blood samples containing preservatives, likely due to crosslinking effects suppressing RNA accessibility. Surprisingly, AR-V7 was readily detectable in patient CTCs enriched from common EDTA blood tubes after up to 48 h. Although BCTs have been thoroughly tested for circulating tumour nucleic acid detection [28–33], and some initial studies of blood storage in BCTs for tumour cell analysis were also encouraging when using image-based cell analysis involving fluorescent probing for proteins or nucleic acids [23], our data suggests that RNA extraction and downstream analysis by PCR-based methods is severely impeded by the preservatives.

Supplementary Materials: Supplementary materials can be found at www.mdpi.com/1422-0067/18/5/1047/s1.

Acknowledgments: This work was supported by the Cancer Institute New South Wales through the Centre for Oncology Education and Research Translation (CONCERT) and the National Breast Cancer Foundation (NBCF). Research was also funded by TOKAI Pharmaceuticals, Boston MA, USA. Francis P. Young is recipient of an Ingham Institute's Honours Scholarship. Human ethics approval, HREC/13/LPOOL/158, was obtained and managed by the CONCERT Biobank.

Author Contributions: Alison W. S. Luk, Therese M. Becker, Daniel T. Dransfield, and Paul de Souza conceived and designed the experiments; Alison W. S. Luk, Yafeng Ma, Pei N. Ding, and Francis P. Young performed the experiments; Alison W. S. Luk analysed the data; Paul de Souza, Wei Chua, Bavanthi Balakrishna, and Daniel T. Dransfield contributed reagents and materials, and coordinated patient recruitment; Alison W. S. Luk and Therese M. Becker wrote the paper and all authors have agreed on the final manuscript version.

Conflicts of Interest: The study was funded by TOKAI Pharmaceuticals to develop optimal blood storage conditions prior to eligibility testing for a clinical trial of one of their drugs. The funding sponsors played a part in the design of the study and in the decision to publish the results, but have no commercial interest in the study outcome.

References

1. Caixeiro, N.J.; Kienzle, N.; Lim, S.H.; Spring, K.J.; Tognela, A.; Scott, K.F.; Souza, P.D.; Becker, T.M. Circulating tumour cells—A bona fide cause of metastatic cancer. *Cancer Metastasis Rev.* **2014**, *33*, 747–756. [CrossRef] [PubMed]
2. Joosse, S.A.; Gorges, T.M.; Pantel, K. Biology, detection, and clinical implications of circulating tumor cells. *EMBO Mol. Med.* **2015**, *7*, 1–11. [CrossRef] [PubMed]

3. Becker, T.M.; Caixeiro, N.J.; Lim, S.H.; Tognela, A.; Kienzle, N.; Scott, K.F.; Spring, K.J.; de Souza, P. New frontiers in circulating tumor cell analysis: A reference guide for biomolecular profiling toward translational clinical use. *Int. J. Cancer* **2014**, *134*, 2523–2533. [CrossRef] [PubMed]

4. Krebs, M.G.; Metcalf, R.L.; Carter, L.; Brady, G.; Blackhall, F.H.; Dive, C. Molecular analysis of circulating tumour cells—Biology and biomarkers. *Nat. Rev. Clin. Oncol.* **2014**, *11*, 129–144. [CrossRef] [PubMed]

5. De Bono, J.S.; Scher, H.I.; Montgomery, R.B.; Parker, C.; Miller, M.C.; Tissing, H.; Doyle, G.V.; Terstappen, L.W.W.M.; Pienta, K.J.; Raghavan, D. Circulating tumor cells predict survival benefit from treatment in metastatic castration-resistant prostate cancer. *Clin. Cancer Res.* **2008**, *14*, 6302–6309. [CrossRef] [PubMed]

6. Antonarakis, E.S.; Lu, C.; Luber, B.; Wang, H.; Chen, Y.; Nakazawa, M.; Nadal, R.; Paller, C.J.; Denmeade, S.R.; Carducci, M.A.; et al. Androgen receptor splice variant 7 and efficacy of taxane chemotherapy in patients with metastatic castration-resistant prostate cancer. *JAMA Oncol.* **2015**, *1*, 582–591. [CrossRef] [PubMed]

7. Ma, Y.; Luk, A.; Young, F.P.; Lynch, D.; Chua, W.; Balakrishnar, B.; de Souza, P.; Becker, T.M. Droplet digital PCR based androgen receptor variant 7 (AR-V7) detection from prostate cancer patient blood biopsies. *Int. J. Mol. Sci.* **2016**, *17*, 1264. [CrossRef] [PubMed]

8. Antonarakis, E.S.; Lu, C.; Luber, B.; Wang, H.; Chen, Y.; Zhu, Y.; Silberstein, J.L.; Taylor, M.N.; Maughan, B.L.; Denmeade, S.R.; et al. Clinical significance of androgen receptor splice variant-7 mRNA detection in circulating tumor cells of men with metastatic castration-resistant prostate cancer treated with first- and second-line abiraterone and enzalutamide. *J. Clin. Oncol.* **2017**. [CrossRef] [PubMed]

9. Liotta, L.A.; Kleinerman, J.; Saidel, G.M. Quantitative relationships of intravascular tumor cells, tumor vessels, and pulmonary metastases following tumor implantation. *Cancer Res.* **1974**, *34*, 997–1004. [PubMed]

10. Butler, T.P.; Gullino, P.M. Quantitation of cell shedding into efferent blood of mammary adenocarcinoma. *Cancer Res.* **1975**, *35*, 512–516. [PubMed]

11. Chang, Y.S.; Tomaso, E.D.; McDonald, D.M.; Jones, R.; Jain, R.K.; Munn, L.L. Mosaic blood vessels in tumors: Frequency of cancer cells in contact with flowing blood. *Proc. Natl. Acad. Sci. USA* **2000**, *97*, 14608–14613. [CrossRef] [PubMed]

12. Meng, S.; Tripathy, D.; Frenkel, E.P.; Shete, S.; Naftalis, E.Z.; Huth, J.F.; Beitsch, P.D.; Leitch, M.; Hoover, S.; Euhus, D.; et al. Circulating tumor cells in patients with breast cancer dormancy. *Clin. Cancer Res.* **2004**, *10*, 8152–8162. [CrossRef] [PubMed]

13. Sun, Y.; Yang, X.; Zhou, J.; Qiu, S.; Fan, J.; Xu, Y. Circulating tumor cells: Advances in detection methods, biological issues, and clinical relevance. *J. Cancer Res. Clin. Oncol.* **2011**, *137*, 1151–1173. [CrossRef] [PubMed]

14. Yu, M.; Stott, S.; Toner, M.; Maheswaran, S.; Haber, D.A. Circulating tumor cells: Approaches to isolation and characterization. *J. Cell Biol.* **2011**, *192*, 373–382. [CrossRef] [PubMed]

15. Alix-Panabières, C.; Pantel, K. Challenges in circulating tumour cell research. *Nat. Rev. Cancer* **2014**, *14*, 623–631. [CrossRef] [PubMed]

16. Lam, N.Y.L.; Rainer, T.H.; Chiu, R.W.K.; Lo, Y.M.D. EDTA is a better anticoagulant than heparin or citrate for delayed blood processing for plasma DNA analysis. *Clin. Chem.* **2004**, *50*, 256–257. [CrossRef] [PubMed]

17. Palmirotta, R.; Ludovici, G.; de Marchis, M.L.; Savonarola, A.; Leone, B.; Spila, A.; de Angelis, F.; Morte, D.D.; Ferroni, P.; Guadagni, F. Preanalytical procedures for DNA studies: The experience of the interinstitutional multidisciplinary BioBank (BioBIM). *Biopreserv. Biobank.* **2011**, *9*, 35–45. [CrossRef] [PubMed]

18. Fehm, T.; Solomayer, E.F.; Meng, S.; Tucker, T.; Lane, N.; Wang, J.; Gebauer, G. Methods for isolating circulating epithelial cells and criteria for their classification as carcinoma cells. *Cytotherapy* **2005**, *7*, 171–185. [CrossRef] [PubMed]

19. Fernando, M.R.; Chen, K.; Norton, S.; Krzyzanowski, G.; Bourne, D.; Hunsley, B.; Ryan, W.L.; Bassett, C. A new methodology to preserve the original proportion and integrity of cell-free fetal DNA in maternal plasma during sample processing and storage. *Prenat. Diagn.* **2010**, *30*, 418–424. [CrossRef] [PubMed]

20. Fernando, M.R.; Norton, S.E.; Luna, K.K.; Lechner, J.M.; Qin, J. Stabilization of cell-free RNA in blood samples using a new collection device. *Clin. Biochem.* **2012**, *45*, 1497–1502. [CrossRef] [PubMed]

21. Warrino, D.E.; DeGennaro, L.J.; Hanson, M.; Swindells, S.; Pirruccello, S.J.; Ryan, W.L. Stabilization of white blood cells and immunologic markers for extended analysis using flow cytometry. *J. Immunol. Methods* **2005**, *305*, 107–119. [CrossRef] [PubMed]

22. Yee, S.S.; Lieberman, D.B.; Blanchard, T.; Rader, J.; Zhao, J.; Troxel, A.B.; DeSloover, D.; Fox, A.J.; Daber, R.D.; Kakrecha, B.; et al. A Novel approach for next-generation sequencing of circulating tumor cells. *Mol. Genet. Genom. Med.* **2016**, *4*, 395–406. [CrossRef] [PubMed]

23. Qin, J.; Alt, J.R.; Hunsley, B.A.; Williams, T.L.; Fernando, M.R. Stabilization of circulating tumor cells in blood using a collection device with a preservative reagent. *Cancer Cell Int.* **2014**, *14*, 23. [CrossRef] [PubMed]

24. Qin, J.; Williams, T.L.; Fernando, M.R. A novel blood collection device stabilizes cell-free RNA in blood during sample shipping and storage. *BMC Res. Notes* **2013**, *6*, 380. [CrossRef] [PubMed]

25. Das, K.; Norton, S.E.; Alt, J.R.; Krzyzanowski, G.D.; Williams, T.L.; Fernando, M.R. Stabilization of cellular RNA in blood during storage at room temperature: A comparison of cell-free RNA BCT with K3EDTA tubes. *Mol. Diagn. Ther.* **2014**, *18*, 647–653. [CrossRef] [PubMed]

26. Flores, L.M.; Kindelberger, D.W.; Ligon, A.H.; Capelletti, M.; Fiorentino, M.; Loda, M.; Cibas, E.S.; Jänne, P.A.; Krop, I.E. Improving the yield of circulating tumour cells facilitates molecular characterisation and recognition of discordant HER2 amplification in breast cancer. *Br. J. Cancer* **2010**, *102*, 1495–1502. [CrossRef] [PubMed]

27. Tsui, N.B.Y.; Ng, E.K.O.; Lo, Y.M.D. Stability of endogenous and added RNA in blood specimens, serum, and plasma. *Clin. Chem.* **2002**, *48*, 1647. [PubMed]

28. Denis, M.G.; Knol, A.; Théoleyre, S.; Vallée, A.; Dréno, B. Efficient detection of BRAF mutation in plasma of patients after long-term storage of blood in cell-free DNA blood collection tubes. *Clin. Chem.* **2015**, *61*, 886–888. [CrossRef] [PubMed]

29. Schiavon, G.; Hrebien, S.; Garcia-Murillas, I.; Cutts, R.J.; Pearson, A.; Tarazona, N.; Fenwick, K.; Kozarewa, I.; Lopez-Knowles, E.; Ribas, R.; et al. Analysis of ESR1 mutation in circulating tumor DNA demonstrates evolution during therapy for metastatic breast cancer. *Sci. Transl. Med.* **2015**, *7*, 313ra182. [CrossRef] [PubMed]

30. Toro, P.V.; Erlanger, B.; Beaver, J.A.; Cochran, R.L.; VanDenBerg, D.A.; Yakim, E.; Cravero, K.; Chu, D.; Zabransky, D.J.; Wong, H.Y.; et al. Comparison of cell stabilizing blood collection tubes for circulating plasma tumor DNA. *Clin. Biochem.* **2015**, *48*, 993–998. [CrossRef] [PubMed]

31. Diaz, I.M.; Nocon, A.; Mehnert, D.H.; Fredebohm, J.; Diehl, F.; Holtrup, F. Performance of streck cfDNA blood collection tubes for liquid biopsy testing. *PLoS ONE* **2016**, *11*, e0166354.

32. Kang, Q.; Henry, N.L.; Paoletti, C.; Jiang, H.; Vats, P.; Chinnaiyan, A.M.; Hayes, D.F.; Merajver, S.D.; Rae, J.M.; Tewari, M. Comparative analysis of circulating tumor DNA stability in K3EDTA, Streck, and CellSave blood collection tubes. *Clin. Biochem.* **2016**, *49*, 1354–1360. [CrossRef] [PubMed]

33. Sherwood, J.L.; Corcoran, C.; Brown, H.; Sharpe, A.D.; Musilova, M.; Kohlmann, A. Optimised pre-analytical methods improve KRAS mutation detection in circulating tumour DNA (ctDNA) from patients with non-small cell lung cancer (NSCLC). *PLoS ONE* **2016**, *11*, e0150197. [CrossRef] [PubMed]

International Journal of
Molecular Sciences

MDPI

Article

Prostate Specific Antigen (PSA) as Predicting Marker for Clinical Outcome and Evaluation of Early Toxicity Rate after High-Dose Rate Brachytherapy (HDR-BT) in Combination with Additional External Beam Radiation Therapy (EBRT) for High Risk Prostate Cancer

Thorsten H. Ecke [1,*], Hui-Juan Huang-Tiel [2], Klaus Golka [3], Silvia Selinski [3], Berit Christine Geis [3], Stephan Koswig [4], Katrin Bathe [4], Steffen Hallmann [1] and Holger Gerullis [5]

1   Department of Urology, HELIOS Hospital, D-15526 Bad Saarow, Germany; steffen.hallmann@helios-kliniken.de
2   Department of Neurology/Emergency Unit, Vivantes Hospital Spandau, D-13585 Berlin, Germany; h.huang-tiel@t-online.de
3   Leibniz Research Centre for Working Environment and Human Factors IfADo, D-44139 Dortmund, Germany; golka@ifado.de (K.G.); selinski@ifado.de (S.S.); berit.geis@tu-dortmund.de (B.C.G.)
4   Department of Radio-Oncology, HELIOS Hospital, D-15525 Bad Saarow, Germany; stephan.koswig@helios-kliniken.de (S.K.); katrin.bathe@helios-kliniken.de (K.B.)
5   School of Medicine and Health Sciences Carl von Ossietzky, University Oldenburg, D-26133 Oldenburg, Germany; holger.gerullis@gmx.net
*   Correspondence: thorsten.ecke@helios-kliniken.de; Tel.: +49-33631-72267; Fax: +49-33631-73136

Academic Editor: Carsten Stephan
Received: 12 September 2016; Accepted: 4 November 2016; Published: 10 November 2016

Abstract: High-dose-rate brachytherapy (HDR-BT) with external beam radiation therapy (EBRT) is a common treatment option for locally advanced prostate cancer (PCa). Seventy-nine male patients (median age 71 years, range 50 to 79) with high-risk PCa underwent HDR-BT following EBRT between December 2009 and January 2016 with a median follow-up of 21 months. HDR-BT was administered in two treatment sessions (one week interval) with 9 Gy per fraction using a planning system and the Ir192 treatment unit GammaMed Plus iX. EBRT was performed with CT-based 3D-conformal treatment planning with a total dose administration of 50.4 Gy with 1.8 Gy per fraction and five fractions per week. Follow-up for all patients was organized one, three, and five years after radiation therapy to evaluate early and late toxicity side effects, metastases, local recurrence, and prostate-specific antigen (PSA) value measured in ng/mL. The evaluated data included age, PSA at time of diagnosis, PSA density, BMI (body mass index), Gleason score, D'Amico risk classification for PCa, digital rectal examination (DRE), PSA value after one/three/five year(s) follow-up (FU), time of follow-up, TNM classification, prostate volume, and early toxicity rates. Early toxicity rates were 8.86% for gastrointestinal, and 6.33% for genitourinary side effects. Of all treated patients, 84.81% had no side effects. All reported complications in early toxicity were grade 1. PSA density at time of diagnosis ($p = 0.009$), PSA on date of first HDR-BT ($p = 0.033$), and PSA on date of first follow-up after one year ($p = 0.025$) have statistical significance on a higher risk to get a local recurrence during follow-up. HDR-BT in combination with additional EBRT in the presented design for high-risk PCa results in high biochemical control rates with minimal side-effects. PSA is a negative predictive biomarker for local recurrence during follow-up. A longer follow-up is needed to assess long-term outcome and toxicities.

Keywords: PSA; toxicity; HDR brachytherapy; prostate cancer

1. Introduction

High-dose-rate brachytherapy (HDR-BT) with additional external-beam radiation therapy (EBRT) is an important therapeutic option for men diagnosed with clinically localized and locally advanced high-risk prostate cancer (PCa) [1–3].

Regarding the actual European guidelines for the treatment of patients with intermediate- and high-risk PCa, a life expectancy of at least 10 years should be mandatory for treatments like radical prostatectomy (RP) or radiation therapy. Nevertheless, until now, there are no randomized clinical trials that compare the oncological outcome of HDR-BT vs. RP [4]. In general, there are not much data available regarding the oncological outcome of HDR-BT [5,6]. Especially, HDR-BT in combination with EBRT seems better than EBRT alone, with respect to biochemical recurrence (BCR)-free survival rates and aspects of quality of life [5,7]. PCa cells seem to have a low α/β ratio. This encourages the use of HDR-BT, where higher doses per fraction can be performed, therefore making it one of the most efficient interventions of hypofractionated radiotherapy. Most reported series combining HDR-BT and EBRT describe impressive results for the treatment of intermediate and high-risk PCa [8–10].

It has been reported by Schiffmann et al. that additional androgen deprivation therapy (ADT) shows higher BCR-free survival rates [11]. In this study, we focused on patients with intermediate- and high-risk PCa that were treated with HDR-BT plus ERBT plus ADT regarding complication rates and oncological outcome. It is difficult to determine an early treatment failure after therapy based on prostate-specific antigen (PSA) fluctuation and a potential benign PSA-rebound phenomenon [12]. In other studies, a benign PSA-rebound rate of up to 30% was described within the first 36 months after treatment [12–14]. The aim of this study was to determine early toxicity rates and the influence of PSA as a predictive marker of clinical outcome.

2. Results

The parameters age, IPSS, PSA (ng/mL) at time of diagnosis, PSA density, BMI, Gleason score, D'Amico risk classification for PCa, PSA value after one year FU, and time of FU are shown in Table 1. In that table, for all main clinical parameters, minimum, median, mean, maximum, standard deviation (SD), and 10%, 25%, 75%, and 90% intervals have been calculated. The median follow up time in our study was 21 months (6–80 months). In total, 8 out of 79 patients (10%) reached a FU time of more than five years. The frequencies of all important clinical parameters—PSA at time of diagnosis, Gleason score, T staging, and D'Amico risk classification for PCa—are detailed in Tables 2–5. Of the study cohort, 64.5% had an initial PSA value of more than 10 ng/mL, the Gleason score of more than 90% of the patients was $\geq$7, more than 80% of the patients had a clinical T staging of 3 (positive digital rectal examination and/or positive for tumor in transrectal ultrasound examination). According to the D'Amico risk classification for PCa, more than 90% are classified to risk group 3. In conclusion, all patients in that study for HDR-Brachytherapy treatment are high-risk PCa patients.

Table 1. Main clinical parameters. BMI: body mass index; FU: follow-up; IPSS: international prostate symptom score; PSA: prostate-specific antigen.

Parameter	Min	10%	25%	Median	Mean	75%	90%	Max	SD
Age	50.000	59.800	66.000	71.000	69.241	74.000	76.000	79.000	6.622
IPSS	0.000	2.000	3.000	5.500	6.271	9.000	11.100	19.000	4.426
PSA Diagnosis	1.360	4.464	7.045	14.550	22.345	24.995	44.736	226.000	29.506
PSA Density	0.053	0.126	0.250	0.463	0.764	0.828	1.760	4.969	0.924
BMI	20.761	23.397	25.282	27.099	27.385	28.572	31.760	44.379	3.713
Gleason Score	6.000	7.000	7.000	7.000	7.354	8.000	9.000	9.000	0.848
D'Amico	2.000	3.000	3.000	3.000	2.962	3.000	3.000	3.000	0.192
PSA FU 1a	0.000	0.010	0.030	0.040	0.167	0.165	0.304	2.300	0.357
time FU	6.000	8.800	11.000	21.000	26.620	35.500	57.400	80.000	18.912

Table 2. Frequency of important clinical parameters for the study cohort. Pre-treatment PSA value.

PSA Diagnosis	N	%
PSA < 10	28	35.44
10 ≤ PSA < 20	22	27.85
PSA ≥ 20	29	36.71
Total	79	100.00

Table 3. Frequency of important clinical parameters for the study cohort. Gleason Score.

Gleason Score	N	%
6	7	8.86
7	49	62.03
8	11	13.92
9	12	15.19
Total	79	100.00

Table 4. Frequency of important clinical parameters for the study cohort. Clinical T Stage.

T Stage	N	%
2a	1	1.27
2b	5	6.33
2c	7	8.86
3	66	83.54
Total	79	100.00

Table 5. Frequency of important clinical parameters for the study cohort. D'Amico risk classification for PCa.

D'Amico	N	%
1	0	0
2	3	3.80
3	76	96.20
Total	79	100.00

During follow-up, one patient (1.27%) died due to progressive disease and bone metastases 63 months after initial diagnosis of PCa. This patient was also the only one with the detection of metastases. In total, a local recurrence was detectable in three patients (3.80%).

Another focus of that study report is the description of side effects regarding early toxicity rates of the demonstrated treatment. Of all treated patients, 84.81% had no side effects. All reported complications in early toxicity were grade 1. The most reported side effects were anal pain (5.06%), symptomatic proctitis (1.27%), and diarrhea (2.53%) for the gastro-intestinal tract; high urinary frequency (3.80%), and urgency (2.53%) were the most complained side effects for the urinary tract. All complications are shown in Table 6.

After descriptive analyses of the documented data, we focused on the influence of PSA value while follow-up for local recurrence, metastases, and/or death. As only one patient died during follow-up, no statistical significance was calculable. However, the presence of local recurrence ($n = 3$) was used for the evaluation of PSA for risk assessment. Table 7 shows the p-value of all important parameters during follow-up regarding the influence on the presence of local recurrence. We could show that PSA density at time of diagnosis ($p = 0.009$), PSA on date of first HDR-BT ($p = 0.033$), and PSA on date of first follow-up after one year ($p = 0.025$) have statistical significance on a higher risk of having a local recurrence during follow-up. We found no statistical significance for Gleason score ($p = 0.463$) or D'Amico risk classification ($p = 0.995$).

Table 6. Early toxicity rates after radiation therapy.

Side Effects	N	%
None	67	84.81
Intestinal		
Pain	4	5.06
Proctitis	1	1.27
Diarrhea	2	2.53
Hemorhage	0	0
Genitourinary		
Frequency	3	3.80
Urgency	2	2.53
Incontinence	0	0
Hematuria	0	0
Renetntion	0	0
Pain	0	0
Total	79	100.00

Table 7. *p*-Value for relevant parameters regarding local recurrence during follow-up.

Variable	OR	2.5%	97.5%	*p*-Value
BMI	1.035	0.778	1.376	0.814
Age	0.876	0.755	1.016	0.080
IPSS	0.401	0.145	1.104	0.077
PSA pre-therapeutic	1.011	0.990	1.033	0.311
PSA density	3.102	1.331	7.228	*0.009*
No. of lymphnodes	0.973	0.795	1.190	0.789
PSA Lymphadenectomy	1.004	0.997	1.032	0.758
PSA HDR1	1.123	1.009	1.250	*0.033*
PSA HDR2	1.095	0.985	1.217	0.093
PSA FU 1a	8.022	1.306	49.287	*0.025*
PSA FU 3a	1.977	0.258	15.134	0.511
PSA FU 5a	4.204	0.422	41.914	0.221

HDR: high dose rate. Statistical significance is written in *italics*.

3. Discussion

HDR brachytherapy is one of the minimally invasive techniques of delivering conformal hypofractionated radiotherapy with steep fall-off of dose beyond the prostate gland. The prostate gland lays very close to critical normal tissues—the anterior rectum wall, urethra, and bladder neck. Because of that biological fact, HDR-BT is ideal for the treatment of PCa [15]. Many groups have shown that HDR-BT boost in combination with EBRT provides better results compared to EBRT alone [2,3]. Moreover, brachytherapy boost has the convenience of decreasing total treatment time, leading to decreased traveling time and expenses.

In the published data describing the experience of HDR-BT boost in combination with EBRT, various fractionation schedules have been used: 15 Gy in three fractions, 11–22 Gy in two fractions, and 12–15 Gy in one fraction. All of them had excellent results, so the Groupe Européen de Curiethérapie—European Society Therapy Radiation Oncology (GECESTRO) and the American Brachytherapy Society (ABS) do not recommend one fractionation schedule over another [16–18]. In this study, all patients were treated with two fractions of 9 Gy each following EBRT as reported above. In our cohort, 84.8% of the treated patients had no side effects. All reported side effects were defined as acute toxicity grade 1; most relevant were pain (5.06%), proctitis (1.27%), and diarrhea (2.53%) as intestinal; and frequency (3.8%) and urgency (2.53%) as genitourinary side effects. None of the patients developed acute GU or GI morbidity higher than Grade 2.

Hoskin et al. [19] reported about early ≥Grade 3 GU and GI morbidity was 3%–7% and 0%, respectively. Late Grade 3 GU toxicity was 3%–16% with no late Grade 3 or 4 GU or GI toxicity. Barkati et al. [20] reported 88% and 85% three-year and five-year biochemical control rates, respectively. They reported all acute GU toxicity as Grade 1. Chronic Grade 3 urinary toxicity was <10% with no Grade 4 toxicity seen.

4. Materials and Methods

4.1. Subjects

In this retrospective study, we report on 79 male patients (median age 71 years, range 50 to 79) who were treated between December 2009 and January 2016 at the Department of Urology and the Department of Radio-Oncology of HELIOS Hospital Bad Saarow, Germany. All patients selected for that treatment have been classified as intermediate and high-risk PCa patients.

4.2. Study Design

All included patients (*n* = 79) underwent HDR-BT after informed patient consent at the time of their treatment. Digital rectal examination, PSA, computerized tomography (CT), and a Technecium-99 bone scan was mandatory. Risk stratification was done as per the National Comprehensive Cancer Network (NCCN), which defines low-risk as PSA ≤ 10 ng/mL, T1c-T2 and a Gleason score (GS) ≤ 6; intermediate risk as PSA 10–20 ng/mL or GS 7; and high risk as a PSA > 20 ng/mL, T3, or GS 8–10. PSA measurements were performed with ElektroChemiLumineszenzImmonoAssay (ELCIA) by Roche Diagnostics GmbH, in accordance with WHO standards. We evaluated and documented the D'Amico risk stratification for PCa for each patient. Exclusion criteria were surgically positive lymph node metastases, distant metastasis, and prior pelvic radiotherapy. Patients with bladder outlet obstruction, patients who already had transurethral operations, and patients with a prostate volume of more than 100 cm^3 were also excluded. All patients had neoadjuvant and adjuvant androgen deprivation therapy (ADT) for at least two years starting after laparoscopic pelvic lymphadenectomy.

HDR-BT was administered before EBRT, based on transrectal ultrasound imaging, using a planning system and the Ir192 treatment unit GammaMed Plus iX (by Varian). HDR-BT was administered in two treatment sessions (one week interval) with 9 Gy per fraction. Overall, 18 Gy was applied to the prostate plus 2 mm margin. The maximal dose for the urethra and rectal wall was 8.0 and 5.0 Gy, respectively.

The HDR-BT procedure was done under general anesthesia. The patient was placed in the lithotomy position, a square lightweight template having a 5 mm grid array was fixed on a stepper stand on which a transrectal ultrasound machine (TRUS) was mounted, and the template was jammed against the perineal skin. There was a grid faceplate fixed onto the template, corresponding to the grid of the TRUS for accurate placement of the ProGuide needles. Under TRUS guidance, metallic trocars were inserted transperineally through the holes in the template to ascertain the position in the prostate as published before by Deger et al. [21]. Seven to twenty needles were inserted into the prostate, then the trocars were removed and replaced by the 6F ProGuide plastic needles in the same position. We always started with the peripheral and anterior needles, and then moved towards the center. As far as possible, the needles were placed at 1 cm intervals. No needles were placed within 7 mm of the urethra, in order to have control over the urethral dose. The needles were pushed beyond the prostate base, and the posterior needles were placed 2–3 mm anterior to the anterior wall of the rectum to avoid overdosing the rectum.

The planning target volume (PTV) was contoured by the radiation oncologist on each ultrasound slice and included the prostate with a 3 mm margin all around, except posteriorly, where no margin was given to avoid overdosing the anterior rectal wall. Superiorly, a margin of 5–7 mm was given to compensate for any post-implant edema and inadvertent caudal movement of the catheters in between the fractions. The PTV constraints were D90 (dose delivered to 90% of PTV) ≥ 97%, V95 ≥ 100%, and V150 ≤ 35%. Isodoses in transrectal ultrasound image are shown in Figure 1. A three-dimensional image with simulation of radiation is shown in Figure 2.

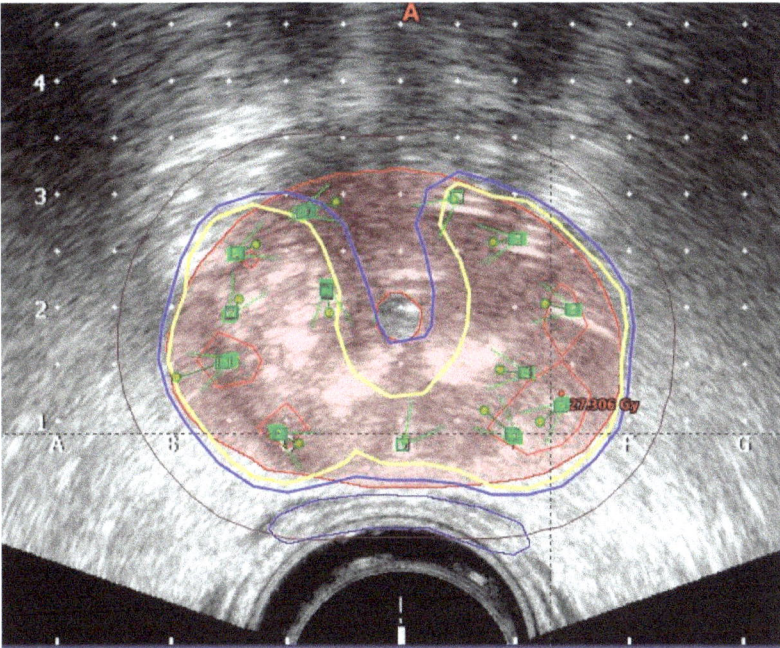

Figure 1. Isodoses in transrectal ultrasound image (red: 15 Gy; yellow: 9 Gy; blue: 8 Gy; brown: 5 Gy).

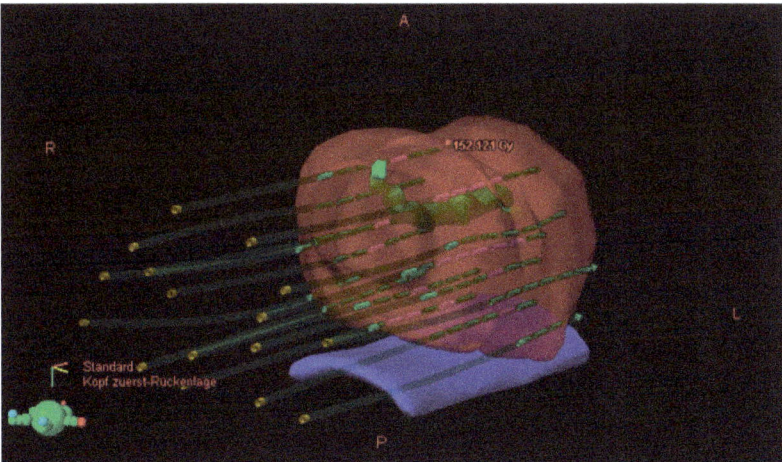

Figure 2. Three-dimensional image with simulation of radiation (red: prostate, green: urethra, blue: rectum, dark green: needle positions).

EBRT started a week after HDR-BT. EBRT was performed according to the standardized protocol with CT-based 3D-conformal treatment planning. The clinical target volume included prostate, the periprostatic region, and the basis of seminal vesicles; the planning target volume included the CTV (clinical target volume) and a margin—margin of 0.6 cm (posterior) and 1.0 cm in all other directions. The reference dose is defined in accordance with the International Commission on Radiation Units and Measurements report 50/63. All patients were irradiated in a supine position with a CT-planned IMRT/VMAT-technique (intensitivity modulated radiotherapy/volume modulated arc therapy) with 6 MV megavoltage photons (Varian, CClinac DHX, PaloAlto, CA, USA). A total dose of 50.4 Gy with 1.8 Gy per fraction and five fractions per week was administered.

Follow-up for all patients was organized one, three, and five years after radiation therapy in the department of Radiooncology to evaluate early and late toxicity side effects, metastases, local recurrence, and PSA value.

4.3. Evaluated Data

The evaluated data included the parameters age, PSA (ng/mL) at time of diagnosis, PSA density, body mass index (BMI), Gleason score, D'Amico risk classification for PCa, digital rectal examination (DRE), PSA value after one/three/five year(s) follow-up (FU), time of follow up, TNM classification, prostate volume, and early toxicity in follow-up. Pretreatment international prostate symptom score (IPSS), uroflow, and rest urine after voiding were also documented. All relevant dates were documented: date of birth, date of death, date of diagnosis, date of lymphadenectomy, date of ADT, date of HDR-BT, and date of follow-up after one, three, and five years.

A radiation oncologist and a urologist performed the follow-up evaluations, including digital rectal examinations and PSA level during follow-up scheme one, three, and five years after initial treatment. PSA failure was defined in terms of the American Society for Therapeutic Radiology and Oncology Consensus Panel recommendations [22]. Acute toxicities were scored according to the Common Terminology Criteria for Adverse Events, Version 4.0 (CTCAE v4.3), by the National Cancer Institute (Common Terminology Criteria for Adverse Events, Version 4.0. Available online: http://evs.nci.nih.gov/ftp1/CTCAE/CTCAE_4.03_2010-06-14_QuickReference_8.5x11.pdf). Acute toxicity was defined as symptoms that were observed during or after treatment and had been completely resolved 6 months after treatment. Following a strict plan for FU, all treated patients were investigated after six months, one, three and five years. Besides clinical investigation including DRE, an interview with a focus on acute toxicities following the Common Terminology Criteria for Adverse Events as written above was included.

4.4. Statistical Analysis

Bravais–Pearson correlation coefficients were estimated for pairs of variables. Odds ratios (OR), 95% confidence intervals (95% CI), and p-values of the Wald test were estimated using unadjusted logistic regression for local recurrence as dependent variable.

The level of significance was $\alpha = 0.05$. All tests and calculations were performed using the software R, version 3.1.2 (R Development Core Team 2014).

In our group, we could show that PSA density at time of diagnosis ($p = 0.009$), PSA on date of first HDR-BT ($p = 0.033$), and PSA on date of first follow-up after one year ($p = 0.025$) have statistical significance with respect to a higher risk of having a local recurrence during follow-up, but not age, Gleason score, or clinical stage. Concerning the number of patients ($n = 79$ in total), we are in average position compared to others and in a good position for a single-center study. Though the series of Yoshiaka et al. performed a monotherapeutic HDR-BT, they found the initial PSA level to be a significant prognostic factor ($p = 0.029$) along with younger age ($p = 0.019$) [23].

In the data published by Hoskin et al., Zwahlen et al., and Kestin et al. [3,5,8,24], they could show a better biological recurrence-free survival after HDR-BT in combination with EBRT, compared to EBRT alone. Combination of both modalities may also improve overall survival (OS) [25]. An explanation could be the high radiation dose that can be prescribed when HDR-BT is combined with EBRT.

Deger et al. [26] presented data of 422 patients with localized PCa treated between 1992 and 2001 with HDR-BT and 3DRT. As also performed in our treatment protocol, all patients underwent laparoscopic pelvic lymph node dissection to have an exact pathological lymph node staging and to be sure to exclude patients with lymphatic involvement. The biological non-evidence of disease (bNED) according to risk group were 100% for low risk, 75% for intermediate risk, and 60% for high risk at 5 years. Five-year bNEDs were 81% in the low risk, 65% in the intermediate risk, and 59% in the high risk group. Five-year OS and bNED were 87% and 94%, respectively. The authors also observed that initial PSA value, risk group, and age were significantly related to bNED. In contrast to our results, we found no statistical significance for D'Amico risk classification ($p = 0.995$) or Gleason score ($p = 0.463$). This could be caused by the fact that the presented cohort consists mainly of patients with high risk PCa and also high Gleason scores.

Most studies of radiotherapy in PCa focus on two points: not only the effectiveness of the treatment, but also its tolerance. However, due to different classifications of radiation reactions, it seems to be difficult to compare the toxicity rates.

There are still different opinions about the use of ADT for patients with intermediate- and high-risk PCa. Martinez et al. [27] published with a large number of patients ($n = 1260$) treated with pelvic RT and HDR-BT. The first group was treated with additional ADT up to six months prior to radiation, and the second group was not. The results for OS, disease free survival (DFS), and bNED have been similar. They observed that additional ADT did not confer a therapeutic advantage, but only side effects and cost. No statistically significant benefit on bNED rates with the use of additional ADT could be shown in any of the groups in that study.

The main limitation of this study is the relatively small number of patients and the short follow-up time regarding the influence on cancer-specific survival, overall survival, and biochemical relapse. Our study adds to the already existing evidence for the effectiveness of HDR-BT combined with EBRT for high-risk PCa.

In conclusion, this study demonstrates that HDR-BT combined with EBRT is effective in the radical radiotherapy of intermediate- and high-risk localized and locally advanced PCa. A longer follow-up is needed to assess long-term outcome and toxicities.

5. Conclusions

HDR-BT in combination with additional EBRT in the presented design for local advanced and high-risk PCa results in high biochemical control rates with minimal side-effects. PSA is a negative predictive biomarker for local recurrence during follow-up.

Author Contributions: Thorsten H. Ecke was mainly collecting data, writing the manuscript and performing the treatment as urologist. Hui-Juan Huang-Tiel was involved in collecting data and writing the manuscript. Klaus Golka, Silvia Selinski and Berit Christine Geis were mainly involved in performance of statistics. Stephan Koswig and Katrin Bathe were performing the treatment as radio-oncologists and collecting data. Steffen Hallmann was involved in writing the manuscript and making treatment decisions. Holger Gerullis was supervisor of writing the manuscript.

Conflicts of Interest: The authors declare no conflict of interest.

References

1. Heidenreich, A.; Bastian, P.J.; Bellmunt, J.; Bolla, M.; Joniau, S.; van der Kwast, T.; Mason, M.; Matveev, V.; Wiegel, T.; Zattoni, F.; et al. Eau guidelines on prostate cancer. Part 1: Screening, diagnosis, and local treatment with curative intent-update 2013. *Eur. Urol.* **2014**, *65*, 124–137. [CrossRef] [PubMed]
2. Sathya, J.R.; Davis, I.R.; Julian, J.A.; Guo, Q.; Daya, D.; Dayes, I.S.; Lukka, H.R.; Levine, M. Randomized trial comparing iridium implant plus external-beam radiation therapy with external-beam radiation therapy alone in node-negative locally advanced cancer of the prostate. *J. Clin. Oncol.* **2005**, *23*, 1192–1199. [CrossRef] [PubMed]

3. Hoskin, P.J.; Rojas, A.M.; Bownes, P.J.; Lowe, G.J.; Ostler, P.J.; Bryant, L. Randomised trial of external beam radiotherapy alone or combined with high-dose-rate brachytherapy boost for localised prostate cancer. *Radiother. Oncol.* **2012**, *103*, 217–222. [CrossRef] [PubMed]

4. Crook, J.M.; Gomez-Iturriaga, A.; Wallace, K.; Ma, C.; Fung, S.; Alibhai, S.; Jewett, M.; Fleshner, N. Comparison of health-related quality of life 5 years after spirit: Surgical prostatectomy versus interstitial radiation intervention trial. *J. Clin. Oncol.* **2011**, *29*, 362–368. [CrossRef] [PubMed]

5. Hoskin, P.J.; Motohashi, K.; Bownes, P.; Bryant, L.; Ostler, P. High dose rate brachytherapy in combination with external beam radiotherapy in the radical treatment of prostate cancer: Initial results of a randomised phase three trial. *Radiother. Oncol.* **2007**, *84*, 114–120. [CrossRef] [PubMed]

6. Galalae, R.M.; Zakikhany, N.H.; Geiger, F.; Siebert, F.A.; Bockelmann, G.; Schultze, J.; Kimmig, B. The 15-year outcomes of high-dose-rate brachytherapy for radical dose escalation in patients with prostate cancer—A benchmark for high-tech external beam radiotherapy alone? *Brachytherapy* **2014**, *13*, 117–122. [CrossRef] [PubMed]

7. Vordermark, D.; Wulf, J.; Markert, K.; Baier, K.; Kolbl, O.; Beckmann, G.; Bratengeier, K.; Noe, M.; Schon, G.; Flentje, M. 3-D conformal treatment of prostate cancer to 74 Gy vs. High-dose-rate brachytherapy boost: A cross-sectional quality-of-life survey. *Acta Oncol.* **2006**, *45*, 708–716. [CrossRef] [PubMed]

8. Zwahlen, D.R.; Andrianopoulos, N.; Matheson, B.; Duchesne, G.M.; Millar, J.L. High-dose-rate brachytherapy in combination with conformal external beam radiotherapy in the treatment of prostate cancer. *Brachytherapy* **2010**, *9*, 27–35. [CrossRef] [PubMed]

9. Kaprealian, T.; Weinberg, V.; Speight, J.L.; Gottschalk, A.R.; Roach, M., 3rd; Shinohara, K.; Hsu, I.C. High-dose-rate brachytherapy boost for prostate cancer: Comparison of two different fractionation schemes. *Int. J. Radiat. Oncol. Biol. Phys.* **2012**, *82*, 222–227. [CrossRef] [PubMed]

10. Wilder, R.B.; Barme, G.A.; Gilbert, R.F.; Holevas, R.E.; Kobashi, L.I.; Reed, R.R.; Solomon, R.S.; Walter, N.L.; Chittenden, L.; Mesa, A.V.; et al. Preliminary results in prostate cancer patients treated with high-dose-rate brachytherapy and intensity modulated radiation therapy (IMRT) vs. IMRT alone. *Brachytherapy* **2010**, *9*, 341–348. [CrossRef] [PubMed]

11. Schiffmann, J.; Lesmana, H.; Tennstedt, P.; Beyer, B.; Boehm, K.; Platz, V.; Tilki, D.; Salomon, G.; Petersen, C.; Krull, A.; et al. Additional androgen deprivation makes the difference: Biochemical recurrence-free survival in prostate cancer patients after hdr brachytherapy and external beam radiotherapy. *Strahlenther. Onkol.* **2015**, *191*, 330–337. [CrossRef] [PubMed]

12. Stephenson, A.J.; Eastham, J.A. Role of salvage radical prostatectomy for recurrent prostate cancer after radiation therapy. *J. Clin. Oncol.* **2005**, *23*, 8198–8203. [CrossRef] [PubMed]

13. Hanlon, A.L.; Pinover, W.H.; Horwitz, E.M.; Hanks, G.E. Patterns and fate of PSA bouncing following 3D-CRT. *Int. J. Radiat. Oncol. Biol. Phys.* **2001**, *50*, 845–849. [CrossRef]

14. Rosser, C.J.; Kuban, D.A.; Levy, L.B.; Chichakli, R.; Pollack, A.; Lee, A.K.; Pisters, L.L. Prostate specific antigen bounce phenomenon after external beam radiation for clinically localized prostate cancer. *J. Urol.* **2002**, *168*, 2001–2005. [CrossRef]

15. Pellizzon, A.C.; Nadalin, W.; Salvajoli, J.V.; Fogaroli, R.C.; Novaes, P.E.; Maia, M.A.; Ferrigno, R. Results of high dose rate afterloading brachytherapy boost to conventional external beam radiation therapy for initial and locally advanced prostate cancer. *Radiother. Oncol.* **2003**, *66*, 167–172. [CrossRef]

16. Roach, M., III; Hanks, G.; Thames, H., Jr.; Schellhammer, P.; Shipley, W.U.; Sokol, G.H.; Sandler, H. Defining biochemical failure following radiotherapy with or without hormonal therapy in men with clinically localized prostate cancer: Recommendations of the RTOG-ASTRO phoenix consensus conference. *Int. J. Radiat. Oncol. Biol. Phys.* **2006**, *65*, 965–974. [CrossRef] [PubMed]

17. Yamada, Y.; Rogers, L.; Demanes, D.J.; Morton, G.; Prestidge, B.R.; Pouliot, J.; Cohen, G.N.; Zaider, M.; Ghilezan, M.; Hsu, I.C. American brachytherapy society consensus guidelines for high-dose-rate prostate brachytherapy. *Brachytherapy* **2012**, *11*, 20–32. [CrossRef] [PubMed]

18. Hoskin, P.J.; Colombo, A.; Henry, A.; Niehoff, P.; Paulsen Hellebust, T.; Siebert, F.A.; Kovacs, G. GEC/ESTRO recommendations on high dose rate afterloading brachytherapy for localised prostate cancer: An update. *Radiother. Oncol.* **2013**, *107*, 325–332. [CrossRef] [PubMed]

19. Hoskin, P.; Rojas, A.; Lowe, G.; Bryant, L.; Ostler, P.; Hughes, R.; Milner, J.; Cladd, H. High-dose-rate brachytherapy alone for localized prostate cancer in patients at moderate or high risk of biochemical recurrence. *Int. J. Radiat. Oncol. Biol. Phys.* **2012**, *82*, 1376–1384. [CrossRef] [PubMed]

20. Barkati, M.; Williams, S.G.; Foroudi, F.; Tai, K.H.; Chander, S.; van Dyk, S.; See, A.; Duchesne, G.M. High-dose-rate brachytherapy as a monotherapy for favorable-risk prostate cancer: A phase II trial. *Int. J. Radiat. Oncol. Biol. Phys.* **2012**, *82*, 1889–1896. [CrossRef] [PubMed]

21. Deger, S.; Dinges, S.; Roigas, J.; Schnorr, D.; Turk, I.; Budach, V.; Hinkelbein, W.; Loening, S.A. High-dose rate iridium192 afterloading therapy in combination with external beam irradiation for localized prostate cancer. *Tech. Urol.* **1997**, *3*, 190–194. [PubMed]

22. Consensus statement: Guidelines for PSA following radiation therapy. American society for therapeutic radiology and oncology consensus panel. *Int. J. Radiat. Oncol. Biol. Phys.* **1997**, *37*, 1035–1041.

23. Yoshioka, Y.; Konishi, K.; Sumida, I.; Takahashi, Y.; Isohashi, F.; Ogata, T.; Koizumi, M.; Yamazaki, H.; Nonomura, N.; Okuyama, A.; et al. Monotherapeutic high-dose-rate brachytherapy for prostate cancer: Five-year results of an extreme hypofractionation regimen with 54 Gy in nine fractions. *Int. J. Radiat. Oncol. Biol. Phys.* **2011**, *80*, 469–475. [CrossRef] [PubMed]

24. Kestin, L.L.; Martinez, A.A.; Stromberg, J.S.; Edmundson, G.K.; Gustafson, G.S.; Brabbins, D.S.; Chen, P.Y.; Vicini, F.A. Matched-pair analysis of conformal high-dose-rate brachytherapy boost versus external-beam radiation therapy alone for locally advanced prostate cancer. *J. Clin. Oncol.* **2000**, *18*, 2869–2880. [PubMed]

25. Pieters, B.R.; de Back, D.Z.; Koning, C.C.; Zwinderman, A.H. Comparison of three radiotherapy modalities on biochemical control and overall survival for the treatment of prostate cancer: A systematic review. *Radiother. Oncol.* **2009**, *93*, 168–173. [CrossRef] [PubMed]

26. Deger, S.; Boehmer, D.; Roigas, J.; Schink, T.; Wernecke, K.D.; Wiegel, T.; Hinkelbein, W.; Budach, V.; Loening, S.A. High dose rate (HDR) brachytherapy with conformal radiation therapy for localized prostate cancer. *Eur. Urol.* **2005**, *47*, 441–448. [CrossRef] [PubMed]

27. Martinez, A.A.; Demanes, D.J.; Galalae, R.; Vargas, C.; Bertermann, H.; Rodriguez, R.; Gustafson, G.; Altieri, G.; Gonzalez, J. Lack of benefit from a short course of androgen deprivation for unfavorable prostate cancer patients treated with an accelerated hypofractionated regime. *Int. J. Radiat. Oncol. Biol. Phys.* **2005**, *62*, 1322–1331. [CrossRef] [PubMed]

International Journal of
Molecular Sciences

MDPI

Article

Perioperative Search for Circulating Tumor Cells in Patients Undergoing Prostate Brachytherapy for Clinically Nonmetastatic Prostate Cancer

Hideyasu Tsumura [1,*], Takefumi Satoh [1], Hiromichi Ishiyama [2], Ken-ichi Tabata [1], Kouji Takenaka [2], Akane Sekiguchi [2], Masaki Nakamura [3], Masashi Kitano [2], Kazushige Hayakawa [2] and Masatsugu Iwamura [1]

[1] Department of Urology, Kitasato University School of Medicine, Sagamihara 252-0374, Japan; tsatoh@kitasato-u.ac.jp (T.S.); ktabata@med.kitasato-u.ac.jp (K.T.); miwamura@med.kitasato-u.ac.jp (M.I.)
[2] Department of Radiology and Radiation Oncology, Kitasato University School of Medicine, Sagamihara 252-0374, Japan; hishiyam@kitasato-u.ac.jp (H.I.); takenaka@kitasato-u.ac.jp (K.T.); akane.o.enaka@gmail.com (A.S.); m-kitano@jcom.home.ne.jp (M.K.); hayakazu@med.kitasato-u.ac.jp (K.H.)
[3] Department of Microbiology, Kitasato University School of Allied Health Sciences, Kanagawa 252-0373, Japan; nakamu7@mac.com
* Correspondence: tsumura@med.kitasato-u.ac.jp; Tel.: +81-42-778-9091; Fax: +81-42-778-9374

Academic Editor: Carsten Stephan
Received: 20 December 2016; Accepted: 4 January 2017; Published: 11 January 2017

Abstract: Despite the absence of local prostate cancer recurrence, some patients develop distant metastases after prostate brachytherapy. We evaluate whether prostate brachytherapy procedures have a potential risk for hematogenous spillage of prostate cancer cells. Fifty-nine patients who were undergoing high-dose-rate (HDR) or low-dose-rate (LDR) brachytherapy participated in this prospective study. Thirty patients with high-risk or locally advanced cancer were treated with HDR brachytherapy after neoadjuvant androgen deprivation therapy (ADT). Twenty-nine patients with clinically localized cancer were treated with LDR brachytherapy without neoadjuvant ADT. Samples of peripheral blood were drawn in the operating room before insertion of needles (preoperative) and again immediately after the surgical manipulation (intraoperative). Blood samples of 7.5 mL were analyzed for circulating tumor cells (CTCs) using the CellSearch System. While no preoperative samples showed CTCs (0%), they were detected in intraoperative samples in 7 of the 59 patients (11.8%; preoperative vs. intraoperative, $p = 0.012$). Positive CTC status did not correlate with perioperative variables, including prostate-specific antigen (PSA) at diagnosis, use of neoadjuvant ADT, type of brachytherapy, Gleason score, and biopsy positive core rate. We detected CTCs from samples immediately after the surgical manipulation. Further study is needed to evaluate whether those CTCs actually can survive and proliferate at distant sites.

Keywords: prostate cancer; brachytherapy; circulating tumor cell

1. Introduction

Brachytherapy approaches have been accepted as a useful method to control localized and locally advanced prostate cancers [1–5]. One of the most appealing reasons for selecting this treatment is favorable long-term outcome with a low degree of toxicity [6]. Low-dose-rate (LDR) brachytherapy provides superior outcomes in patients with low- and intermediate-risk diseases [1,2]. The combination of high-dose-rate (HDR) brachytherapy and external irradiation is an effective treatment for delivering radiation doses more precisely in prostate cancer, even if patients have extracapsular invasion and seminal vesicle invasion [3–5]. Technical modifications for prostate brachytherapy are being developed to obtain the better treatment outcome [2,7–10]. However, approximately 5%–20% of those patients, as it now stands, show recurrence within 5 years after brachytherapy [1,2,5,11,12].

When patients are suspected to have treatment failure, evaluation—including abdominal computed tomography scan, pelvic magnetic resonance imaging, a bone scan, and prostate biopsy—are usually conducted to identify the site of relapse. Some patients develop distant metastases despite the absence of local recurrence. In those cases, micrometastasis that was not detected by radiographic images may have been present at initial diagnosis. Another possibility is that surgical manipulation that involves needles being inserted into prostate tissue may pose a potential risk for hematogenous spillage of prostate cancer cells and play a role in distant metastases in patients undergoing prostate brachytherapy. A no-touch isolation technique, in which vascular control is achieved prior to tumor manipulation, is generally considered to reduce cancer dissemination and subsequently reduce future disease recurrence in various cancer-related surgeries [13–17]. However, this technique is not used during brachytherapy procedures. In addition, needles being inserted into prostate tissue directly penetrate the cancer lesions at a certain rate.

Elucidating the mechanism and causes of relapse is a key challenge for the enhancement of treatment outcome [18,19]. We suspect that iatrogenic circulating tumor cell (CTC) spillage can convert a nonmetastatic cancer to a systemic one. In this study, we evaluated whether brachytherapy procedures can provoke hematogenous spillage of prostate cancer cells. We detected perioperative CTCs using the CellSearch System and compared preoperative CTC counts with intraoperative ones. We analyzed whether intraoperative CTC increases were associated with perioperative clinicopathological features.

2. Results

Characteristics of the 59 patients are shown in Table 1. As shown in Figure 1, no CTCs were detected in preoperative samples. CTCs were detected from samples collected immediately after insertion of needles in 7 of 59 patients (11.8%). Intraoperative CTC detection rates were significantly higher than preoperative ones (11.8% vs. 0%, $p = 0.012$).

Table 1. Patient characteristics ($n = 59$).

Factors	HDR ($n = 30$)		LDR ($n = 29$)		Total ($n = 59$)	
	Median	(Range)	Median	(Range)	Median	(Range)
Age (year)	71.5	(58–82)	70	(51–77)	71	(51–82)
PSA at diagnosis (ng/mL)	26.8	(4.5–396)	6.5	(4.2–14.1)	10.1	(4.2–396)
Prostate volume (cc) *	14.4	(4.6–29.7)	30.7	(20.3–58.4)	22.2	(4.6–58.4)
Number of needles	18	(18–18)	21	(17–29)	–	–
Duration of NHT (months)	16	(7–25)	0	(0)	–	–
	n	(%)	*n*	(%)	*n*	(%)
Gleason Score						
≤6	0	(0)	8	(28)	8	(14)
7	7	(23)	19	(65)	26	(44)
8 to 10	23	(77)	2	(7)	25	(42)
Clinical T Stage						
1c–2a	6	(20)	20	(69)	26	(44)
2b–2c	6	(20)	9	(31)	15	(25)
3a	11	(37)	0	(0)	11	(19)
3b	6	(20)	0	(0)	6	(10)
4	1	(3)	0	(0)	1	(2)

Table 1. Cont.

Factors	HDR (*n* = 30)		LDR (*n* = 29)		Total (*n* = 59)	
	Median	(Range)	Median	(Range)	Median	(Range)
	n	(%)	*n*	(%)	*n*	(%)
Biopsy Positive Core Rate						
<34%	8	(27)	21	(73)	29	(49)
34%–67%	12	(40)	7	(24)	19	(32)
>67%	10	(33)	1	(3)	11	(19)
NCCN Risk Criteria (2015)						
Low	0	(0)	6	(21)	6	(10)
Intermediate	0	(0)	21	(72)	21	(36)
High	20	(67)	2	(7)	22	(37)
Very high	10	(33)	0	(0)	10	(17)

* Prostate volume was measured by transrectal ultrasound sonography immediately before insertion of needles. HDR: high-dose-rate brachytherapy; LDR: low-dose-rate brachytherapy; PSA: prostate-specific antigen; NHT: neoadjuvant hormonal therapy; NCCN: National Comprehensive Cancer Network.

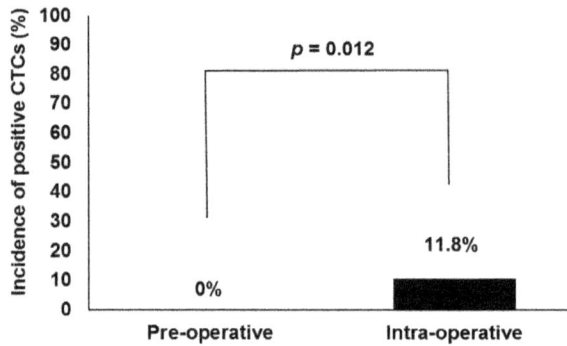

Figure 1. Comparison of circulating tumor cell (CTC) detection rates between pre- and intraoperative blood specimens in all patients undergoing high-dose-rate or low-dose-rate brachytherapy (*n* = 59).

Figure 2A,B showed perioperative CTC detection rates in patients undergoing HDR and LDR brachytherapy, respectively. Intraoperative CTCs were detected in 4 of 30 (13.3%) patients and 3 of 29 (10.5%) patients with HDR and LDR brachytherapy, respectively. While intraoperative CTC detection rates were relatively high when compared with preoperative ones in each group, according to a Fisher's exact test, the differences did not reach statistical significance in the HDR brachytherapy group ($p = 0.112$) or in the LDR brachytherapy group ($p = 0.236$).

Table 2 lists the characteristics of the 7 patients who became positive for CTCs intraoperatively. The intraoperative CTC count was 1 CTC in 3 patients and 2 CTCs in 1 patient treated with HDR brachytherapy, and 1 CTC was detected in 3 patients who underwent LDR brachytherapy.

To investigate the secondary outcome measures, patients were divided into two groups according to positive or negative status for intraoperative CTCs. Positive status did not correlate with clinicopathological and perioperative variables, including use of neoadjuvant hormonal therapy, type of brachytherapy, age, prostate-specific antigen (PSA) at diagnosis, Gleason score, clinical stage, biopsy positive core rates, prostate volume at brachytherapy, or National Comprehensive Cancer Network (NCCN) risk criteria 2015 (Table 3). Neither patients with positive status nor those with negative status for intraoperative CTCs had postoperative clinical progression, with a median follow-up of 18 months (range, 15–24 months).

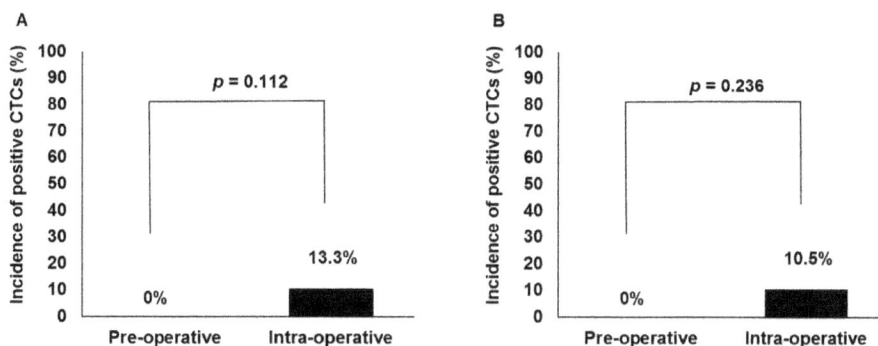

Figure 2. Comparison of circulating tumor cell (CTC) detection rates between pre- and intraoperative blood specimens in patients undergoing high-dose-rate (**A**, *n* = 30) and low-dose-rate (**B**, *n* = 29) brachytherapy.

Table 2. Characteristics of seven patients who changed to positive status for intraoperative circulating tumor cells (CTCs).

Type of Brachytherapy	HDR	HDR	HDR	HDR	LDR	LDR	LDR
Case number	9	26	34	36	8	19	43
Number of CTC counts (/7.5 mL)	2	1	1	1	1	1	1
Age (years)	71	75	65	75	58	65	67
Duration of NHT (months)	17	16	16	17	0	0	0
PSA nadir during NHT (ng/mL)	0.014	<0.008	<0.008	0.14	–	–	–
PSA at diagnosis (ng/mL)	31	13.5	17.6	66.7	8.6	4.6	14.1
Prostate volume (cc) *	7	29.7	21.3	13.9	26.1	38.4	37
Number of needles	18	18	18	18	24	28	18
Gleason score	8	8	7	9	6	7	7
Clinical T stage	1c	3a	3b	2c	2a	2a	2c
Biopsy positive core rate (%)	75	50	25	100	10	16.6	33.3
NCCN risk criteria 2015	H	H	VH	H	L	I	I

* Prostate volume was measured by transrectal ultrasound sonography immediately before insertion of needles; HDR: high-dose-rate brachytherapy; LDR: low-dose-rate brachytherapy; NHT: neoadjuvant hormonal therapy; PSA: prostate-specific antigen; NCCN: National Comprehensive Cancer Network; H: high risk; VH: very high risk; L: low risk; I: intermediate risk.

Table 3. Association of positive status for intraoperative circulating tumor cells (CTCs) with perioperative features (*n* = 59).

Factors	CTC Positive Rates	(*n*)	*p*
Age (>70 vs. ≤70 years)	8.8% vs. 16.0%	(3/34 vs. 4/25)	0.442
Type of brachytherapy (HDR vs. LDR)	13.3% vs. 10.3%	(4/30 vs. 3/29)	>0.999
NHT (yes vs. no)	13.3% vs. 10.3%	(4/30 vs. 3/29)	>0.999
PSA at diagnosis (≥10 vs. <10 ng/mL)	16.1% vs. 7.1%	(5/31 vs. 2/28)	0.424
Prostate volume (cc)	14.8% vs. 9.3%	(4/27 vs. 3/32)	0.691
Prostate volume/number of needle (≥1 vs. <1 cc/needle)	14.2% vs. 8.3%	(5/35 vs. 2/24)	0.689
Gleason score (≥8 vs. <8)	12.0% vs. 11.7%	(3/25 vs. 4/34)	>0.999
Clinical T stage (≥3a vs. ≤2c)	11.7% vs. 11.9%	(2/17 vs. 5/42)	>0.999
Biopsy positive core rate (>34% vs. ≤34%)	10.0% vs. 13.7%	(3/30 vs. 4/29)	0.706
NCCN risk criteria 2015 (H or VH vs. I or L)	12.5% vs. 11.1%	(4/32 vs. 3/27)	>0.999

3. Discussion

In this study of clinically nonmetastatic prostate cancer patients, we detected CTCs from samples immediately after insertion of needles in patients undergoing prostate brachytherapy. Intraoperative

CTC detection rates were significantly higher than preoperative ones. Our results may support a potential risk for hematogenous spread of cancer cells during the procedure.

Historically, transurethral resection of prostate (TURP) was generally performed to relieve the urinary tract obstruction caused by prostate cancer. In 1986, Levine et al. reported the possibility that cancer cells might be disseminated during TURP in patients with clinically evident cancer confined to the prostate [13]. They noted that the 5-year survival rate in those patients undergoing TURP was significantly lower than those not undergoing the procedure ($p = 0.02$). Several investigators then measured perioperative CTCs and reported the possibility of hematogenous spillage of cancer cells during radical prostatectomy in clinically nonmetastatic cancer patients [20–24]. Eschwege et al. investigated the dissemination of malignant prostatic cells during open radical prostatectomy [20], and they confirmed prostate-specific membrane antigen (PSMA) using reverse-transcription nested PCR for CTC detection. The incidence of positive CTC status increased from 21% before the surgery to 86% immediately afterward, supporting the possibility of intraoperative hematogenous dissemination during the open radical prostatectomy. They concluded that surgeons should minimize prostate manipulation to avoid seeding from the gland for the prevention of metastatic disease.

Prostate needle biopsy is one of the most similar procedures to prostate brachytherapy in that needles being inserted into prostate tissue directly penetrate the cancer lesions. Hara et al. examined PSA-mRNA-bearing cells in peripheral blood of the 108 patients before and after prostate biopsy [25]. Of 46 patients who were diagnosed with prostate cancer, the incidence of positive PSA-mRNA-bearing cells increased from 3% before the biopsy to 45% immediately afterward. In addition, the incidence of positive PSA-mRNA status after prostate biopsy in patients diagnosed with prostate cancer were higher than those without prostate cancer (45% vs. 25%, $p < 0.001$). This study supported the possibility of tumor spreading by prostate biopsy.

While we detected intraoperative hematogenous spillage of prostate cancer cells during brachytherapy procedures, it is still controversial whether the CTCs spilled iatrogenically into circulation have the biological capability to implant into distant sites and subsequently develop metastatic foci. Most studies—including the present study that detected the intraoperative CTC increase during radical treatment for primary lesion and prostate biopsy—had a small sample size and lacked a long follow-up period. Thus, the clinical significance of intraoperative CTC increase remains unclear, and we still have the question of whether this kind of iatrogenic CTC is clinically metastable. Eschwege et al. evaluated the cancer-cell seeding impact on recurrence-free survival [23]. Hematogenous spread of prostate cells was assessed by a dual PSA/PSMA PCR assay using very specific PSMA and PSA primers. Ninety-eight patients with negative status for preoperative CTC were divided into two groups according to status for intraoperative CTC: 53 (54%) remained negative and 45 (46%) became positive. Median biological and clinical recurrence-free time did not significantly differ between the two groups (69.6 vs. 65 months). The authors concluded that intraoperative hematogenous spillage of prostate cancer cells does not have a statistically significant adverse effect on recurrence. Their results seem to exclude tumor surgical management as a major cause of metastatic development. Some studies demonstrated that a primary tumor contains subpopulations of metastatic and nonmetastatic cancer cells. Only a restricted fraction of the cells in a primary tumor are considered to be highly metastatic [26,27]. Mareel et al. reported that 0.1% of CTC is responsible for the formation of metastatic foci [28]. When intraoperative hematogenous spillage of prostate cancer cells occurs during surgical procedures, we suspect that it may provoke distant metastases. However, the possibility of metastatic foci formation caused by the iatrogenic CTC spillage from a primary tumor may occur fairly infrequently.

We could not find any association of intraoperative CTC increases with perioperative clinicopathological features in the present study. This may reflect the fact that all patients undergoing brachytherapy have a risk of intraoperative hematogenous spillage of prostate cancer cells, irrespective of use of neoadjuvant hormonal therapy, type of brachytherapy, age, PSA at diagnosis, Gleason score, clinical stage, and biopsy positive core rates. In our regiment of HDR brachytherapy for high-risk and locally advanced cancers, we administered at least 6 months of neoadjuvant androgen deprivation therapy (ADT). Nonetheless, intraoperative CTCs were detected in 13.3% of patients in that group. Although the use of neoadjuvant ADT may reduce the cellular viability of iatrogenic CTCs enough to prevent implantation into distant sites, this could not completely eliminate the hematogenous spillage of CTCs during the procedure.

Several potential limitations of this study must be considered. The sample size for the present study was not calculated to detect a statistical difference in clinical progression between patients with intraoperative CTC increases and those without increases. A large-scale study involving more patients is needed to clarify whether intraoperative CTC increases actually affect the postoperative progression. Second, samples for CTC detection were not drawn before neoadjuvant ADT in patients treated with HDR brachytherapy. These patients were classified with clinically high-risk or locally advanced cancer and were more likely to have occult distant metastasis than lower-risk patients. Some of these patients may have had positive status for CTC before neoadjuvant ADT [29]. In addition, samples for CTC detection were not drawn a few days or months after the brachytherapy procedures in patients with positive status for intraoperative CTCs [20]. Longer detection of CTCs may have a higher risk of later metastases than others. Third, the CellSearch System may lack sensitivity in nonmetastatic cancer patients and consequently underestimates the incidence of perioperative CTCs. In addition, this system only detects the epithelial cancer cells and does not detect the mesenchymal ones. In metastatic formation from primary epithelial cancers, epithelial–mesenchymal transition at primary sites is considered to be important for cancer metastasis [30,31]. This cellular transition allows epithelial cancer cells to acquire the more invasive characteristics. The mesenchymal cancer cells may have a higher metastatic potential than the epithelial ones. In future clinical series, the detection of such highly metastatic potential cells may be helpful in assessing the possibility of metastatic diseases caused by iatrogenic CTCs during prostate brachytherapy procedures.

4. Materials and Methods

4.1. Patient Selection

From October 2014 to July 2015, 59 patients with clinically nonmetastatic prostate cancer who underwent HDR or LDR brachytherapy participated in this prospective study. Thirty patients with high-risk or locally advanced prostate cancer were treated with HDR brachytherapy. Twenty-nine patients with clinically localized prostate cancer were treated with LDR brachytherapy. Samples of peripheral blood were drawn before insertion of needles (preoperative) and again immediately after the surgical manipulation (intraoperative) in each patient. Pretreatment evaluation included clinical history, physical examination, blood laboratory findings, pelvic computed tomography, pelvic magnetic resonance imaging, and a bone scan. Exclusion criteria were a history of cancer, previous surgery for benign prostatic hyperplasia, or concomitant active urinary tract infection. Patients who had a suspicious lesion of cancer other than prostate cancer on the pretreatment evaluation were also excluded. Patients were removed from the study if they wished to discontinue. All biopsy slides were reviewed by our institutional pathologists. Approval was granted by the ethics committee of our institution (B14-20), and all patients signed written informed consent.

4.2. HDR Brachytherapy and Blood Sample Collection

The protocol and procedure for HDR brachytherapy and hormonal therapy in high-risk or locally advanced prostate cancer were reported previously [5,32]. All patients underwent ≥6 months of neoadjuvant ADT, which combined nonsteroidal anti-androgen agents with luteinizing hormone-releasing hormone agonist injections. Either flutamide (375 mg/day) or bicalutamide (80 mg/day) were prescribed as the nonsteroidal anti-androgen agents. Either goserelin (10.8 mg/3 months or 3.6 mg/month) or leuprorelin (11.25 mg/3 months or 3.75 mg/month) were administrated as luteinizing hormone-releasing hormone agonist therapy.

In the operating room, preoperative samples of peripheral blood were drawn in a supine position before epidural anesthesia. Patients were then placed in a lithotomy position. Metallic marker seeds were placed transperineally into the base and apex for the purpose of image-guided external beam radiotherapy following HDR brachytherapy. Treatment was started using placement of a closed transperineal hollow needle under transrectal ultrasound guidance. Multiple 25 cm long, closed-end, 15-G plastic hollow needles were inserted transperineally using a 15-G Prostate Template (Best Medical International Inc., Springfield, VA, USA). Eighteen needles were routinely implanted. Flexible cystoscopy was conducted to check that the urethra had not been penetrated by the implanted needles. The needle tips were left within the urinary bladder, 15 mm above the sonographically or cystoscopically defined base of the prostate. Immediately after all of these procedures had been completed, intraoperative samples of peripheral blood were drawn from each patient again.

4.3. LDR Brachytherapy and Blood Sample Collection

The protocol and procedure for LDR brachytherapy were performed as previously reported [8]. No patients were treated with ADT and external beam radiation therapy before and after LDR brachytherapy.

In the operating room, preoperative samples of peripheral blood were drawn in a supine position before spinal anesthesia. Patients were then placed in a lithotomy position. Results of transrectal ultrasonography in the axial plane were imported into the VariSeed brachytherapy planning system (Varian Medical Systems, Palo Alto, CA, USA). The prostate, urethra, and rectal wall were contoured by radiation oncologists. Seed number and location for both peripheral and centrally located needles were determined manually. The prescribed dose was set at 145 Gy. Dose–volume histograms and isodose lines were evaluated based on predetermined dosimetric parameters. Needle insertion and implantation were done by urologists. As needed, modifications to the plan can be made, and the software recalculates the dose–volume histograms and isodose lines by using a real-time intraoperative dosimetry technique [7]. Patients were assigned to receive loose or intraoperatively built custom-linked (IBCL) seed brachytherapy based on the week of the month, with loose or IBCL seeds used during alternate weeks. In the present study, 14 and 15 patients were implanted with loose and IBCL seeds, respectively. Loose seeds were implanted using a Mick applicator (Mick Radio Nuclear Instruments, Mount Vernon, NY, USA). IBCL seeds were constructed using a Quicklink device (CR Bard, Covington, GA, USA) and implanted. Zauls et al. described the detailed mechanisms of constructing IBCL seeds [33], and we applied the same devices in our study. Immediately after all seeds had been implanted, intraoperative samples of peripheral blood were drawn from each patient.

4.4. CTC Detection

Twenty milliliters of peripheral blood was collected into 10 mL CellSave Preservation Tubes (Immunicon, Hatboro, PA, USA), which contained ethylenediaminetetraacetic acid (EDTA) as an anticoagulant and a cellular preservative. Samples were maintained at room temperature and processed within 72 h of collection. Blood samples of 7.5 mL were analyzed for CTCs.

The CellSearch system was used for the isolation and enumeration of CTCs. This system consists of the CellTracks AutoPrep and the CellTracks Analyzer II unit (Veridex LLC, Raritan, NJ, USA). The AutoPrep is a semiautomated sample preparation for the isolation of CTCs. The procedure enriches the sample for cells expressing epithelial cell adhesion molecule (EpCAM) using antibody-coated magnetic beads. After the magnetic separation, these cells were stained with the fluorescent nucleic acid dye 4′,6-diamidino-2-phenylindole (DAPI) and fluorescently labeled with anticytokeratin 8,18,19-phycoerythrin peridinin and anti-CD45 chlorophyll protein to distinguish epithelial cells from leukocytes. The stained and fluorescently labeled cells were analyzed for the identification of CTCs using the Analyzer II (Veridex LLC). The criteria for CTC included positive staining for DAPI and the cytokeratin and negative staining for the CD45. A CTC must show round or oval morphology.

4.5. Statistical Analysis

The primary outcome measures were changes in CTC detection rates from preoperative to intraoperative blood samples. As secondary outcome measures, incidence of increase in the intraoperative CTC count relative to the preoperative one was tested for association with clinicopathological and perioperative variables. For the purpose of analysis, clinicopathological and perioperative variables including age (>70 vs. ≤70), PSA at diagnosis (≥10 ng/mL vs. <10 ng/mL), Gleason score (≥8 vs. ≤7), clinical T stage (T1c–2c vs. T3a–4), biopsy positive core rates (>34% vs. ≤34%), prostate volume at brachytherapy (>25 cc vs. ≤25 cc), and NCCN risk criteria 2015 (high/very high vs. intermediate/low) were evaluated as dichotomous variables. A Fisher's exact test was used to evaluate the primary and secondary outcome measures.

Sample size calculations determined that 60 patients would be needed to detect a 15% rise from preoperative to intraoperative CTC detection rates with α equal to 0.05 and power equal to 80%. A rise from preoperative to intraoperative CTC detection rates was estimated from the first 30 cases in this study.

Differences were regarded as statistically significant at the $p < 0.05$ level. Analyses were performed using SPSS, version 11.0 for Windows (SPSS, Inc., Chicago, IL, USA) and Microsoft Excel (Microsoft, Redmond, WA, USA).

5. Conclusions

We detected CTCs from samples immediately after insertion of needles in patients undergoing prostate brachytherapy for clinically nonmetastatic prostate cancer. Further research is needed to assess whether those cancer cells actually can survive and proliferate at distant sites. Although the brachytherapy approaches have demonstrated favorable long-term outcomes [6], understanding the mechanism of relapse should lead to the better treatment outcomes in patients undergoing prostate brachytherapy.

Acknowledgments: This work was supported by JSPS KAKENHI Grant Number 26861293 and The Japanese Foundation of Prostate Research.

Author Contributions: Hideyasu Tsumura was mainly collecting data and samples, writing the manuscript and performing the treatment as urologist; Takefumi Satoh was conceived of and designed the experiments; Ken-ichi Tabata was mainly involved in performance of statistics and performing the treatment as urologist; Hiromichi Ishiyama, Kouji Takenaka, Akane Sekiguchi and Masashi Kitano were performing the treatment as radio-oncologists and collecting data; Masaki Nakamura was supervisor of performing the experiments; Kazushige Hayakawa and Masatsugu Iwamura were supervisor of writing the manuscript.

Conflicts of Interest: The authors declare no conflict of interest.

Abbreviations

LDR	low-dose-rate
HDR	high-dose-rate
CTC	circulating tumor cell
PSA	prostate-specific antigen
NCCN	National Comprehensive Cancer Network
TURP	transurethral resection of prostate
PSMA	prostate-specific membrane antigen
ADT	androgen deprivation therapy
IBCL	intraoperatively built custom-linked
EpCAM	expressing epithelial cell adhesion molecule
DAPI	4′,6-diamidino-2-phenylindole

References

1. Zelefsky, M.J.; Kuban, D.A.; Levy, L.B.; Potters, L.; Beyer, D.C.; Blasko, J.C.; Moran, B.J.; Ciezki, J.P.; Zietman, A.L.; Pisansky, T.M.; et al. Multi-institutional analysis of long-term outcome for stages T1-T2 prostate cancer treated with permanent seed implantation. *Int. J. Radiat. Oncol. Biol. Phys.* **2007**, *67*, 327–333. [CrossRef] [PubMed]

2. Stone, N.N.; Stock, R.G.; Cesaretti, J.A.; Unger, P. Local control following permanent prostate brachytherapy: Effect of high biologically effective dose on biopsy results and oncologic outcomes. *Int. J. Radiat. Oncol. Biol. Phys.* **2010**, *76*, 355–360. [CrossRef] [PubMed]

3. Hoskin, P.J.; Rojas, A.M.; Bownes, P.J.; Lowe, G.J.; Ostler, P.J.; Bryant, L. Randomised trial of external beam radiotherapy alone or combined with high-dose-rate brachytherapy boost for localised prostate cancer. *Radiother. Oncol.* **2012**, *103*, 217–222. [CrossRef] [PubMed]

4. Prada, P.J.; Gonzalez, H.; Fernandez, J.; Jimenez, I.; Iglesias, A.; Romo, I. Biochemical outcome after high-dose-rate intensity modulated brachytherapy with external beam radiotherapy: 12 Years of experience. *BJU Int.* **2012**, *109*, 1787–1793. [CrossRef] [PubMed]

5. Ishiyama, H.; Satoh, T.; Kitano, M.; Tabata, K.; Komori, S.; Ikeda, M.; Soda, I.; Kurosaka, S.; Sekiguchi, A.; Kimura, M.; et al. High-dose-rate brachytherapy and hypofractionated external beam radiotherapy combined with long-term hormonal therapy for high-risk and very high-risk prostate cancer: Outcomes after 5-year follow-up. *J. Radiat. Res.* **2014**, *55*, 509–517. [CrossRef] [PubMed]

6. Grimm, P.; Billiet, I.; Bostwick, D.; Dicker, A.P.; Frank, S.; Immerzeel, J.; Keyes, M.; Kupelian, P.; Lee, W.R.; Machtens, S.; et al. Comparative analysis of prostate-specific antigen free survival outcomes for patients with low, intermediate and high risk prostate cancer treatment by radical therapy: Results from the prostate cancer results study group. *BJU Int.* **2012**, *109*, 22–29. [CrossRef] [PubMed]

7. Stock, R.G.; Stone, N.N.; Wesson, M.F.; DeWyngaert, J.K. A modified technique allowing interactive ultrasound-guided three-dimensional transperineal prostate implantation. *Int. J. Radiat. Oncol. Biol. Phys.* **1995**, *32*, 219–225. [CrossRef]

8. Ishiyama, H.; Satoh, T.; Kawakami, S.; Tsumura, H.; Komori, S.; Tabata, K.; Sekiguchi, A.; Takahashi, R.; Soda, I.; Takenaka, K.; et al. A prospective quasi-randomized comparison of intraoperatively built custom-linked seeds versus loose seeds for prostate brachytherapy. *Int. J. Radiat. Oncol. Biol. Phys.* **2014**, *90*, 134–139. [CrossRef] [PubMed]

9. Yoshioka, Y.; Suzuki, O.; Otani, Y.; Yoshida, K.; Nose, T.; Ogawa, K. High-dose-rate brachytherapy as monotherapy for prostate cancer: Technique, rationale and perspective. *J. Contemp. Brachytherapy* **2014**, *6*, 91–98. [CrossRef] [PubMed]

10. Morton, G.C. High-dose-rate brachytherapy boost for prostate cancer: Rationale and technique. *J. Contemp. Brachytherapy* **2014**, *6*, 323–330. [CrossRef] [PubMed]

11. Sekiguchi, A.; Ishiyama, H.; Satoh, T.; Tabata, K.; Komori, S.; Tsumura, H.; Kawakami, S.; Soda, I.; Iwamura, M.; Hayakawa, K. 125Iodine monotherapy for Japanese men with low- and intermediate-risk prostate cancer: Outcomes after 5 years of follow-up. *J. Radiat. Res.* **2014**, *55*, 328–333. [CrossRef] [PubMed]

12. Aoki, M.; Miki, K.; Kido, M.; Sasaki, H.; Nakamura, W.; Kijima, Y.; Kobayashi, M.; Egawa, S.; Kanehira, C. Analysis of prognostic factors in localized high-risk prostate cancer patients treated with HDR brachytherapy, hypofractionated 3D-CRT and neoadjuvant/adjuvant androgen deprivation therapy (trimodality therapy). *J. Radiat. Res.* **2014**, *55*, 527–532. [CrossRef] [PubMed]

13. Levine, E.S.; Cisek, V.J.; Mulvihill, M.N.; Cohen, E.L. Role of transurethral resection in dissemination of cancer of prostate. *Urology* **1986**, *28*, 179–183. [CrossRef]

14. Hayashi, N.; Egami, H.; Kai, M.; Kurusu, Y.; Takano, S.; Ogawa, M. No-touch isolation technique reduces intraoperative shedding of tumor cells into the portal vein during resection of colorectal cancer. *Surgery* **1999**, *125*, 369–374. [CrossRef]

15. Wiggers, T.; Jeekel, J.; Arends, J.W.; Brinkhorst, A.P.; Kluck, H.M.; Luyk, C.I.; Munting, J.D.; Povel, J.A.; Rutten, A.P.; Volovics, A.; et al. No-touch isolation technique in colon cancer: A controlled prospective trial. *Br. J. Surg.* **1988**, *75*, 409–415. [CrossRef] [PubMed]

16. Liu, C.L.; Fan, S.T.; Lo, C.M.; Tung-Ping Poon, R.; Wong, J. Anterior approach for major right hepatic resection for large hepatocellular carcinoma. *Ann. Surg.* **2000**, *232*, 25–31. [CrossRef] [PubMed]

17. Gall, T.M.; Jacob, J.; Frampton, A.E.; Krell, J.; Kyriakides, C.; Castellano, L.; Stebbing, J.; Jiao, L.R. Reduced dissemination of circulating tumor cells with no-touch isolation surgical technique in patients with pancreatic cancer. *JAMA Surg.* **2014**, *149*, 482–485. [CrossRef] [PubMed]

18. Giesing, M.; Suchy, B.; Driesel, G.; Molitor, D. Clinical utility of antioxidant gene expression levels in circulating cancer cell clusters for the detection of prostate cancer in patients with prostate-specific antigen levels of 4–10 ng/mL and disease prognostication after radical prostatectomy. *BJU Int.* **2010**, *105*, 1000–1010. [CrossRef] [PubMed]

19. Forsythe, K.; Burri, R.; Stone, N.; Stock, R.G. Predictors of metastatic disease after prostate brachytherapy. *Int. J. Radiat. Oncol. Biol. Phys.* **2012**, *83*, 645–652. [CrossRef] [PubMed]

20. Eschwege, P.; Dumas, F.; Blanchet, P.; le Maire, V.; Benoit, G.; Jardin, A.; Lacour, B.; Loric, S. Haematogenous dissemination of prostatic epithelial cells during radical prostatectomy. *Lancet* **1995**, *346*, 1528–1530. [CrossRef]

21. Badwe, R.A.; Vaidya, J.S. Haematogenous dissemination of prostate epithelial cells during surgery. *Lancet* **1996**, *347*, 325–326. [PubMed]

22. Davis, J.W.; Nakanishi, H.; Kumar, V.S.; Bhadkamkar, V.A.; McCormack, R.; Fritsche, H.A.; Handy, B.; Gornet, T.; Babaian, R.J. Circulating tumor cells in peripheral blood samples from patients with increased serum prostate specific antigen: Initial results in early prostate cancer. *J. Urol.* **2008**, *179*, 2187–2191. [CrossRef] [PubMed]

23. Eschwege, P.; Moutereau, S.; Droupy, S.; Douard, R.; Gala, J.L.; Benoit, G.; Conti, M.; Manivet, P.; Loric, S. Prognostic value of prostate circulating cells detection in prostate cancer patients: A prospective study. *Br. J. Cancer* **2009**, *100*, 608–610. [CrossRef] [PubMed]

24. Kauffman, E.C.; Lee, M.J.; Alarcon, S.V.; Lee, S.; Hoang, A.N.; Walton Diaz, A.; Chelluri, R.; Vourganti, S.; Trepel, J.B.; Pinto, P.A. Lack of impact of robotic-assisted laparoscopic radical prostatectomy on intraoperative levels of prostate cancer circulating tumor cells. *J. Urol.* **2015**, *195*, 1936–1942. [CrossRef] [PubMed]

25. Hara, N.; Kasahara, T.; Kawasaki, T.; Bilim, V.; Tomita, Y.; Obara, K.; Takahashi, K. Frequency of PSA-mRNA-bearing cells in the peripheral blood of patients after prostate biopsy. *Br. J. Cancer* **2001**, *85*, 557–562. [CrossRef] [PubMed]

26. Fidler, I.J. Selection of successive tumour lines for metastasis. *Nat. New Biol.* **1973**, *242*, 148–149. [CrossRef] [PubMed]

27. Yokota, J. Tumor progression and metastasis. *Carcinogenesis* **2000**, *21*, 497–503. [CrossRef] [PubMed]

28. Mareel, M.M.; van Roy, F.M.; Bracke, M.E. How and when do tumor cells metastasize? *Crit. Rev. Oncog.* **1993**, *4*, 559–594. [PubMed]

29. Meyer, C.P.; Pantel, K.; Tennstedt, P.; Stroelin, P.; Schlomm, T.; Heinzer, H.; Riethdorf, S.; Steuber, T. Limited prognostic value of preoperative circulating tumor cells for early biochemical recurrence in patients with localized prostate cancer. *Urol. Oncol.* **2016**, *34*, 211–236. [CrossRef] [PubMed]

30. Tsai, J.H.; Donaher, J.L.; Murphy, D.A.; Chau, S.; Yang, J. Spatiotemporal regulation of epithelial-mesenchymal transition is essential for squamous cell carcinoma metastasis. *Cancer Cell* **2012**, *22*, 725–736. [CrossRef] [PubMed]

31. Sethi, S.; Macoska, J.; Chen, W.; Sarkar, F.H. Molecular signature of epithelial-mesenchymal transition (EMT) in human prostate cancer bone metastasis. *Am. J. Transl. Res.* **2010**, *3*, 90–99. [PubMed]

32. Tsumura, H.; Satoh, T.; Ishiyama, H.; Tabata, K.; Komori, S.; Sekiguchi, A.; Ikeda, M.; Kurosaka, S.; Fujita, T.; Kitano, M.; et al. Prostate-specific antigen nadir after high-dose-rate brachytherapy predicts long-term survival outcomes in high-risk prostate cancer. *J. Contemp. Brachytherapy* **2016**, *8*, 95–103. [CrossRef] [PubMed]

33. Zauls, A.J.; Ashenafi, M.S.; Onicescu, G.; Clarke, H.S.; Marshall, D.T. Comparison of intraoperatively built custom linked seeds versus loose seed gun applicator technique using real-time intraoperative planning for permanent prostate brachytherapy. *Int. J. Radiat. Oncol. Biol. Phys.* **2011**, *81*, 1010–1016. [CrossRef] [PubMed]